STUDIES OF APPALACHIAN GEOLOGY: NORTHERN AND MARITIME

STUDIES OF APPALACHIAN GEOLOGY:

Northern and Maritime

Edited by

E-AN ZEN, WALTER S. WHITE, AND JARVIS B. HADLEY

U. S. GEOLOGICAL SURVEY, WASHINGTON, D. C.

and

JAMES B. THOMPSON, JR.

DEPARTMENT OF GEOLOGICAL SCIENCES, HARVARD UNIVERSITY,
CAMBRIDGE, MASSACHUSETTS

INTERSCIENCE PUBLISHERS
A DIVISION OF JOHN WILEY & SONS
NEW YORK • LONDON • SYDNEY • TORONTO

MARLAND PRATT BILLINGS

TO

Marland P. Billings

A PIONEER IN THE RENAISSANCE OF
APPALACHIAN GEOLOGY;
A TEACHER AND FRIEND

THIS VOLUME IS AFFECTIONATELY
DEDICATED BY HIS ASSOCIATES AND
FORMER STUDENTS

List of Contributors

ALBEE, ARDEN L., Division of Geological Sciences, California Institute of Technology, Pasadena, California 91109.

BELT, EDWARD S., Department of Geology, Amherst College, Amherst, Massachusetts 01002.

BERRY, W. B. N., Department of Paleontology, University of California, Berkeley, California, 94720.

BIRCH, FRANCIS, Department of Geological Sciences, Harvard University, Cambridge, Massachusetts 02138.

BOUCOT, ARTHUR J., Department of Geology, University of Pennsylvania, Philadelphia, Pennsylvania 19104.

BROMERY, RANDOLPH W., Department of Geology, University of Massachusetts, Amherst, Massachusetts 01002.

CADY, WALLACE M., U. S. Geological Survey, Bldg. 25, Federal Center, Denver, Colorado 80225.

CHAPMAN, CARLETON A., Department of Geology, University of Illinois, Urbana, Illinois 61801.

CHIDESTER, ALFRED H., U. S. Geological Survey, Center for Astrogeology, Box 1906, Flagstaff, Arizona 86001.

CLARK, THOMAS H., Department of Geological Sciences, McGill University, Montreal, Quebec, Canada.

CLIFFORD, TOM N., Department of Earth Sciences, University of Leeds, Leeds, England.

DECKER, EDWARD R., Department of Geological Sciences, Harvard University, Cambridge, Massachusetts 02138.

DIMENT, WILLIAM H., Department of Geological Sciences, University of Rochester, Rochester, New York 14627.

DIXON, H. ROBERTA, U. S. Geological Survey, Bldg. 25, Federal Center, Denver, Colorado 80225.

EAKINS, PETER R., Department of Geological Sciences, McGill University, Montreal, Quebec, Canada.

FAUL, HENRY, Department of Geology, University of Pennsylvania, Philadelphia, Pennsylvania 19104.

GREEN, JOHN C., Department of Geology, University of Minnesota, Duluth, Minnesota 55812.

GRISCOM, ANDREW, U. S. Geological Survey, 345 Middlefield Road, Menlo Park, California 94025.

GUIDOTTI, CHARLES V., Department of Geology, University of California, Davis, California 95616.

HALL, LEO M., Department of Geology, University of Massachusetts, Amherst, Massachusetts 01002.

HATCH, NORMAN L., JR., U. S. Geological Survey, Bldg. 417, Agricultural Research Center, Beltsville, Maryland 20705.

HUSSEY, ARTHUR M., II, Department of Geology, Bowdoin College, Brunswick, Maine 04011.

KANE, MARTIN F., U. S. Geological Survey, Washington, D. C. 20242.

LUNDGREN, LAWRENCE W., JR., Department of Geological Sciences, University of Rochester, Rochester, New York 14627.

LYONS, JOHN B., Department of Geology, Dartmouth College, Hanover, New Hampshire 03755.

MOENCH, ROBERT H., U. S. Geological Survey, Bldg. 25, Federal Center, Denver, Colorado 80225.

MOORE, GEORGE E., JR., Department of Geology, Ohio State University, Columbus, Ohio 43210.

NAYLOR, RICHARD S., Department of Geology and Geophysics, Massachusetts Institute of Technology, Cambridge, Massachusetts 02139.

NEUMAN, ROBERT B., U. S. Geological Survey, Washington, D. C. 20242.

NORTON, STEVEN A., Department of Geological Sciences, Harvard University, Cambridge, Massachusetts 02138.

OSBERG, PHILIP H., Department of Geology, University of Maine, Orono, Maine 04773.

PAGE, LINCOLN R., U. S. Geological Survey, 80 Broad St. Boston, Mass. 02110

PAVLIDES, LOUIS, U. S. Geological Survey, Bldg. 424, Agricultural Research Center, Beltsville, Maryland 20705.

QUINN, ALONZO W., Department of Geological Sciences, Brown University, Providence, Rhode Island 02912.

RANKIN, Douglas W., U. S. Geological
Survey, Washington, D. C. 20242.

ROBINSON, Peter, Department of Geology,
University of Massachusetts, Amherst,
Massachusetts 01002.

RODGERS, John, Department of Geology,
Yale University, New Haven, Connecticut
06520

ROSENFELD, John L., Department of
Geology, University of California, Los
Angeles, California 90024.

ROY, Robert F., Division of Geological
Sciences, California Institute of Technology,
Pasadena, California 91109.

SCHNABEL, Robert W., U. S. Geological
Survey, Bldg. 25, Federal Center, Denver,
Colorado 80225.

SKEHAN, James W., S. J., Department of
Geology, Boston College, Chestnut Hill,
Massachusetts 02167.

SKIDMORE, W. Brian, Quebec Department of
Natural Resources, Quebec, P. Q., Canada.

THEOKRITOFF, George, Department of
Geology, Rutgers University, Newark, New
Jersey 07102.

THOMPSON, James B., Jr., Department of
Geological Sciences, Harvard University,
Cambridge, Massachusetts 02138.

TRASK, Newell J., U. S. Geological Survey,
345 Middlefield Road, Menlo Park, California
94025.

WARNER, Jeffrey, Lunar and Earth Sciences
Division, Manned Spacecraft Center,
NASA (Code TH), Houston, Texas 77058.

WHITTINGTON, Harry B., Sedgwick
Museum, Cambridge University, Cambridge,
England.

WHITE, Walter S., U. S. Geological Survey,
Bldg. 417, Agricultural Research Center,
Beltsville, Maryland 20705.

ZEN, E-an, U. S. Geological Survey,
Washington, D. C. 20242.

Marland Pratt Billings—an Appreciation

We dedicate this book on northern Appalachian geology to Marland Billings, who sought and found the stratigraphic key to New Hampshire geology, and, by leading many others through the door that he unlocked, brought a new era to Appalachian geologic study. As the outgrowth of his labors, a metamorphic terrain that covers many thousands of square miles astride a major orogenic belt is now among the better understood regions of its kind, and provides a useful testing ground for a variety of fundamental studies and speculations.

The situation was very different in 1931 when Billings, still in his twenties, returned to Harvard as an Assistant Professor to teach structural geology. The gross distribution of rock types east of the Green Mountain-Berkshire axis had already been charted, principally by the Hitchcocks and B. K. Emerson, and the petrography of many units had been described in detail. But this earlier work gave only a disjointed two-dimensional view of things; the seeming stratigraphic order was based primarily on tenuous lithologic correlations with rocks at a few well-known localities or on relative metamorphic grade. With rare exceptions, the concept of structural continuity does not appear to have been an important element either in stratigraphic correlation or in the construction of the geologic maps and sections that existed. To make structural and stratigraphic sense of this geologic chaos was a formidable task.

Billings had had brief contact with the metasedimentary gneisses of east-central New Hampshire while studying the petrology of the White Mountain batholith for his doctorate, and the enigma of the age of these gneisses, added to his love of the mountains, presumably led him to look northward for a starting place. The well-known fossil localities near Littleton had made that district "a fossiliferous oasis in a large area of metamorphic rocks," and a logical place to begin. Assisted by two or three graduate students each year, Billings completed geologic mapping of the Littleton-Moosilauke area in three summers. This work conquered the obstacles that wooded terrain, faults, and progressive metamorphism had thrown in the path of previous workers, provided the stratigraphic key to the geology of

western New Hampshire, and laid the foundation upon which a major part of subsequent work in New England has been directly or indirectly built. It provided a convincing demonstration of the rewards to be obtained from careful outcrop mapping in which structural, stratigraphic, and metamorphic problems are solved simultaneously. Billings had a large purpose in mind very early, for after his second summer he wrote his Harvard classmates that he hoped to complete a study of the geology of central New Hampshire in ten years. When the second world war interrupted these studies, he and his students had, in fact, mapped nearly a third of the State and had made a start in eastern Vermont. His work is still continuing, more than a decade, now, after publication of his "Geology of New Hampshire."

Billings' early New Hampshire work had little guidance from others; it was a forward stride of his own making, and an unusually lengthy step it was. He had had good training in conventional field methods, starting with his days as a student of Professor Woodworth. In 1928 he was one of the beneficiaries of Professor Leon Collet's delight in guiding English-speaking colleagues through the structures of the western Alps, which demonstrate so forcefully that structural and stratigraphic solutions are inseparable in areas of great complexity. And finally, some of the individual aspects of his problems in western New Hampshire had been worked out elsewhere, particularly in Europe, where the geologic literature held clues for its conscientious students. In this latter sphere, the scholarship of Professor Daly must surely have been an important influence. Billings' own unique contribution was in successfully combining and applying all these elements to a specific area, wisely chosen for what it might (and did) tell about a much larger region.

Beyond his scientific achievements in New Hampshire and elsewhere, Billings has done much to help geology serve mankind. He began rather casually with some advice on tunnels for the Pennsylvania Turnpike Commission in 1938, an outgrowth of his interest in Appalachian folds. During the war, he participated in the U. S. Geological Survey's search for strategic minerals,

xi

applying his knowledge of New England geology to pegmatite and talc investigations. A stint with the Office of Scientific Research and Development took him to New Caledonia to study the nickel deposits. He has served the State of New Hampshire long and well as a member of its Mineral Resources Advisory Committee. And since 1955, he has closely followed the tunneling operations of the Metropolitan District Commission in the greater Boston area, both as a scientist drawn by the unusual opportunity to study covered parts of the Boston Basin, and as an engineering geologist advising the Commission. He has also given much time to unselfish service to his fellow geologists, serving the Division of Geological Sciences at Harvard as Chairman (1946–51), the AAAS as Vice-President (Chairman of Section E, 1946–47), and the Geological Society of America as President (1959). His contributions have been recognized by honorary doctorates from Washington University (1960) and the University of New Hampshire (1966), and by his election to the National Academy of Sciences.

Impressive as Billings' individual contributions to Appalachian geology may be, for the many who have been closely associated with him, he has an even larger place as a teacher and friend. Being primarily a field geologist, he has designed his courses to give students the background of knowledge and experience necessary for competent field observations: familiarity with the logic of geologic reasoning, alertness to recognize clues to the origin of geologic features, and awareness of the extent to which structural geology draws on evidence from the whole spectrum of geology, particularly stratigraphy. His survey course on the geology of North America, designed for second-year students but commonly attended by graduate students, is based on the primary literature and illustrates how different kinds of evidence have been combined in the solution of specific geologic problems. Those who have learned most from him, however, have learned it during actual field work, or from problems brought in from the field: in his field course, in Easter-vacation field trips to Pennsylvania, in working with him as a field assistant, or in working on a thesis problem under his guidance. Outcrops display the realities with which geologists must deal, and Billings uses them to good effect in helping his students to develop their geologic judgment. The teaching continues as students commit their observations and interpretations to paper. Billings' devotion to clarity and explicitness in thought and expression is evident from the way he strives for precise meaning in geologic description, as in the book, "Structural Geology," his contributions to the AGI Glossary, and other publications. The value of this fundamental training may be measured by the many different kinds of geology, academic and applied, successfully practised by his former graduate students.

Billings is a hard taskmaster and a blunt critic, though this severity tends to be tempered by his warmth and his ready laughter. He presses his students to seek out the weak spots in a geologic map or argument, encourages them over the mental hurdles that interfere with productivity, and does not let them rest with anything less than their best effort. His students and colleagues are most influenced, however, by admiration for his qualities as a man—admiration for his ruggedness and vigor, his dedication, his generosity and unselfishness, and his intellectual honesty. The question, "what holes would Marland pick in this?" is a spur toward excellence when graduate school is many years behind.

WALTER S. WHITE

Bibliography of Marland P. Billings, to 1967

1924, A Geological reconnaissance: Harvard Alumni Bull., v. 26, p. 436–440

1925, On the mechanics of dike intrusion: Jour. Geol., v. 33, p. 140–150

1927, Topaz and phenacite from Baldface Mountain, Chatham, New Hampshire: Am. Mineralogist, v. 12, p. 173–179

1928, The petrology of the North Conway quadrangle in the White Mountains of New Hampshire: Am. Acad. Arts and Sci., Proc., v. 63, p. 67–137

——, (with Collet, L. W., Perret, R., Doggett, R. A.) Sur la presence du cristallin du massif des Aiguilles Rouges dans le Cirque du Fer a Cheval (Hautes Alpes Calcaires de Sixt, Hte. Savoie): Eclogae geologieae Helvetiae, v. 21, p. 349–350

——, The chemistry, optics, and genesis of the hastingsite group of amphiboles: Am. Mineralogist, v. 13, p. 287–296

——, Roof collapse in the late Paleozoic alkaline batholiths (abstract): Geol. Soc. America Bull., v. 39, p. 187–188; Pan-Am. Geologist, v. 49, p. 143

——, Relative abundance of glacial striations, crescentic fractures, and lunoid furrows (abstract): Geol. Soc. America Bull., v. 39, p. 224

1929, Structural geology of the eastern part of the Boston Basin: Am. Jour. Sci., 5th ser., v. 18, p. 97–137; abstracts, Pan-Am. Geologist, v. 51, p. 68; Geol. Soc. America Bull., v. 40, p. 193

——, (with Croneis, C. G.) New areas of alkaline igneous rocks in central Arkansas: Jour. Geol., v. 37, p. 541–561

1930, (with Croneis, C. G.) Igneous rocks in central Arkansas: Arkansas Geol. Survey Bull. 3, p. 149–162

1931, Twenty-seventh annual New England Intercollegiate Geological Excursion, Science, n. s., v. 74, p. 658–659

1932, (with Williams, C. R.) Origin of the Appalachian Highlands: Appalachia, v. 19, p. 1–33

——, Thrusting younger rocks over older: Am. Jour. Sci., 5th ser., v. 25, p. 140–165; abstracts, Pan-Am. Geologist, v. 57, p. 68; Geol. Soc. America Bull., v. 43, p. 141

1933, (with Cleaves, A. B.) Silurian and Devonian strata of the Littleton quadrangle, New Hampshire (abstract): Geol. Soc. America Bull., v. 44, p. 196

——, (with Roy, C. J.) "Weathering of the Medford diabase—pre- or postglacial?" A discussion: Jour. Geology, v. 41, p. 654–661

1934, (with Cleaves, A. B.) Paleontology of the Littleton area, New Hampshire: Am. Jour. Sci., 5th ser., v. 28, p. 412–438

——, Paleozoic age of the rocks of central New Hampshire: Science n. s., v. 79, p. 55–56

——, Review of "Geological structures" by Willis, Bailey, and Willis, Robin: Science, v. 80, p. 562

1935, Geology of the Littleton and Moosilauke quadrangles, New Hampshire: New Hampshire Planning and Development Commission, Concord, New Hampshire, 51 p.

——, (with Williams, C. R.) Geology of the Franconia quadrangle, New Hampshire: New Hampshire Planning and Development Commission, Concord, New Hampshire, 35 p.

——, (with Cleaves, A. B.) Brachiopods from mica schist, Mount Clough, New Hampshire: Am. Jour. Sci., 5th ser., v. 30, p. 530–536

——, Geology and regional metamorphism of the Littleton-Moosilauke area, New Hampshire (abstract): Geol. Soc. America Proc. p. 66

1936, Comments on "Structural and petrologic studies in Dutchess County, New York: Part II, petrology and metamorphism of the Paleozoic rocks," by Tom F. W. Barth: Geol. Soc. America Bull., v. 47, p. 2000–2003

1937, Regional metamorphism of the Littleton-Moosilauke area, New Hampshire: Geol. Soc. America Bull., v. 48, p. 463–566

——, (with Sharp, R. P.) Petrofabric study of a fossiliferous schist, Mount Clough, New Hampshire: Am. Jour. Sci., 5th ser., v. 34, p. 277–292

1938, (with Williams, C. R.) Petrology and structure of the Franconia quadrangle, New Hampshire: Geol. Soc. America Bull., v. 49, p. 1011–1043

——, Physiographic relations of the Lewis overthrust in northern Montana: Am. Jour. Sci., 5th ser., v. 35, p. 260–272

——, Mechanics of batholithic intrusion in New Hampshire (abstract): Geol. Soc. America Bull., v. 49, p. 1867

——, (with Fowler-Billings, Katherine) Progress of studies on the bedrock geology of New Hampshire and eastern Vermont (abstract): Geol. Soc. America Bull., v. 49, p. 1868

——, The geology of western and central New Hampshire (abstract): Washington Acad. Sci., Jour., v. 28, p. 422–423

——, Introduction of potash during regional metamorphism in western New Hampshire: Geol. Soc. America Bull., v. 49, p. 289–302; abstract, Geol. Soc. America Proc. 1937, p. 71

1939, (with Loomis, F. B., Jr., and Stewart, G. W.) Carboniferous topography in the vicinity of

Boston, Masschusetts: Geol. Soc. America Bull., v. 50, p. 1867–1884

——, Chemical changes during regional metamorphism in the Presidential Range of New Hampshire (abstract): Geol. Soc. America Bull., v. 50, p. 1900

1940, Paleozoic igneous activity in New England (abstract): Geol. Soc. America Bull., v. 51, p. 1900

——, (with Spieker, E. N.) Glaciation in the Wasatch Plateau, Utah: Geol. Soc. America Bull., v. 51, p. 1173–1198

1941, Structure and metamorphism in the Mount Washington area, New Hampshire: Geol. Soc. America Bull., v. 52, p. 863–936

——, Pegmatites of Massachusetts: U. S. Geological Survey and Mass. Dept. Public Works, Cooperative Geologic Project, Bull., no. 5, Boston, 22 p.

——, Recent studies on the bed-rock geology of New Hampshire: N. H. Acad. Sci. Proc., v. 1, p. 21–24; abstract, Geol. Soc. America Bull., v. 52, p. 2010–2011

1942, Structural geology: New York: Prentice-Hall, Inc., 473 p.

——, Geology of the central area of the Ossipee Mountains, New Hampshire, earthquakes: Seismol. Soc. America Bull., v. 32, p. 83–92

1943, Ring dikes and their origin: New York Acad. Sci. Trans., ser. II, v. 5, p. 131–144

1944, (with Chapman, C. A., and Chapman, R. W.) Petrology and structure of the Oliverian magma series in the Mount Washington quadrangle, New Hampshire: Geol. Soc. America Bull., v. 55, p. 497–516

——, (with Allen, W. B.) Mica deposits in the Chester-Blandford State Forest: Mass. Dept. Public Works, Inform. Circ., no. 2, 3 p.

——, (with Wolfe, C. W.) Spodumene deposits in the Leominster-Sterling area, Massachusetts: Mass: Dept. Public Works, Inform. Circ., no. 3, 9 p.

1945, New colored geological maps of the Mount Washington, Bellows Falls, and Plymouth quadrangles, New Hampshire (abstract): Geol. Soc. America Bull., v. 56, p. 1148

——, Mechanics of igneous intrusion in New Hampshire: Am. Jour. Sci., v. 243-A (Daly volume), p. 40–68

——, (with Chidester, A. H.) Geologic maps and structure sections of the Johnson Mine, Johnson, Vermont: U. S. Geol. Survey, Strategic Min. Invest., Prelim. Maps.

1946, (with Fowler-Billings, Katherine, Chapman, C. A. Chapman, R. W., and Goldthwait, R. P.) The geology of the Mount Washington quadrangle, New Hampshire: New Hampshire Planning and Development Commission, Concord, New Hampshire, 56 p.

——, (with Chapman, C. A., Chapman, R. W., Fowler-Billings, Katherine, and Loomis, F. B., Jr.) Geology of the Mount Washington quadrangle,

New Hampshire: Geol. Soc. America Bull., v. 57, p. 261–273

——, (with Keevil, N. B.) Petrography and radioactivity of four Paleozoic magma series in New Hampshire: Geol. Soc. America Bull., v. 57, p. 797–828; abstract, v. 56, p. 1148 (1945).

——, Some major problems of New England geology (abstract): Geol. Soc. America Bull., v. 57, p. 1277–1278

——, Areal mapping program in New Hampshire, (abstract): Geol. Soc. America Bull., v. 57, p. 1277

1947, (with Rabbitt, J. C.) Chemical analyses and calculated modes of the Oliverian Magma Series, Mount Washington quadrangle, New Hampshire: (Shaler Memorial Series), Geol. Soc. America Bull., v. 58, p. 573–596

1948, Orogeny in the Appalachian Highlands of New England: Tulsa Geol. Soc. Digest, v. 16, p. 48–53

——, (with Chidester, A. H.) Geologic maps and structure sections of Waterbury Mine, Moretown, Vermont: U. S. Geol. Survey Strategic Min. Invest., Prelim. Maps.

1949, (with Chidester, A. H.) Geologic maps and structure sections of the Carleton talc quarry, Chester, Vermont; Rousseau talc prospect, Cambridge, Vermont; and Mad River talc mine, Fayston, Vermont: U. S. Geol. Survey, Strategic Min. Invest., Perlim. Map H 3-227

1950, Field and laboratory methods in the study of metamorphic rocks: N. Y. Acad. Sci. Trans., ser. II, vol. 13, p. 44–51

——, Stratigraphy and the study of metamorphic rocks: Geol. Soc. America Bull., v. 61, p. 435–448

——, (with White, W. S.) Metamorphosed mafic dikes of the Woodsville quadrangle, Vermont and New Hampshire: Am. Mineralogist, v. 35, p. 629–643

1951, (with White, W. S.) Geology of the Woodsville quadrangle, Vermont-NewHampshire: Geol. Soc. America Bull., v. 62, p. 647–696

——, (with Chidester, A. H., and Cady, W. M.) Talc investigations in Vermont, preliminary report: U. S. Geological Survey, Circ. 95, 33 p.

1952, (with Rodgers, John, and Thompson, J. B., Jr.) Geology of the Appalchian Highlands of east-central New York, southern Vermont, and southern New Hampshire: Geol. Soc. America Guidebook for field trips in New England, p. 1–71

——, Presentation of Penrose Medal to Pentti Eelis Eskola: Geol. Soc. America Proc. for 1951, p. 51–54

1954, Structural geology, 2nd ed.: New York, Prentice-Hall, Inc., 514 p.

1955, Review of "Traité de Tectonique," by Jean Goguel: Jour. Geol., v. 63, p. 397–399

——, (compiler) Geologic map of New Hampshire: U. S. Geol. Survey, scale 1:250,000.

1956, The geology of New Hampshire, Part II, bedrock geology: New Hampshire State Planning and Development Commission, 203 p.

1958, Reginald A. Daly, geologist (1871–1957):
Science, v. 127, p. 19–20

——, Reginald A. Daly, (1871–1957): Geotimes,
v. 2, no. 11, p. 10

——, (with Birch, Francis, Bridgman, P. W., Mather,
K. F.), Article in memory of Reginald Aldworth
Daly: Harvard Univ. Gazette, v. LIII, no. 31,
p. 164–165

1959, Memorial to Reginald Aldworth Daly (1871–
1957): Geol. Soc. America Proc., 1958, p. 115–121

1960, Diastrophism and mountain building: Geol. Soc.
Amer. Bull., v. 71, p. 363–398 (Presidential
Address, Geological Society of America)

——, Diastrofismo y formación de montañas: Notas y
Comuns. Inst. Geol. y Minero de Espana, no.
59, p. 167–252

1961, (with C. R. Williams), Mountain story (revised):
in The world of geology: New York: McGraw-
Hill Book Co., Inc., p. 243–262

——, (with Doll, C. G., Cady, W. M., and Thompson,
J. B., Jr.) Centennial geologic map of Vermont:
Vermont Geol. Survey, Montpelier, scale 1:250,000

1963, Geologia estructural: Eudeba Editorial
Universitaria de Buenos Aires, 564 p.

——, (with Doll, C. G., Cady, W. M., and Thompson,
J. B., Jr.) Reply to Zen's discussion of the
Centennial geologic map of Vermont: Am. Jour.
Sci., v. 261, p. 94–96

1964, (with F. L. Tierney) Geology of the City Tunnel
Extension, greater Boston, Massachusetts: Jour.
Boston Soc. Civil Engineers, v. 51, no. 2, p.
111–154

1965, (with J. R. Wilson) Chemical analyses of rocks
and rock minerals from New Hampshire: New
Hampshire Department of Resources and Economic
Development, Concord, New Hampshire, 104 p.

1965, (As member of North America Geologic Map
Committee, U. S. Geol. Survey, E. H. Goddard,
Chairman), Geologic Map of North America: U. S.
Geol. Survey, scale 1:5,000,000.

1966, (with D. A. Rahm), Geology of the Malden
Tunnel, Massachusetts: Jour. Boston Soc. Civil
Engrs., v. 53, no. 2, p. 116–141.

1967, Introductory remarks to Conference on Economic
Geology in Massachusetts: Economic Geology in
Massachusetts, O. C. Farquhar, ed., Univ.
Mass., p. 5–7.

——, The significance of faults in tunnels: Economic
Geology in Massachusetts, O. C. Farquhar, ed.,
Univ. Mass., p. 267–272.

——, (with R. G. Doyle and others), Preliminary geologic map of Maine: Maine Geol. Survey, scale
1:500,000

——, (with D. C. Noble), Pyroclastic rocks of the White
Mountain magma series: Nature, v. 216, p. 906–907.

(In press) (with F. L. Tierney and M. M. Cassidy),
Geology of the City Tunnel, Greater Boston, Massachusetts: Jour. Boston Soc. Civil Engrs.

(In preparation), Areal distribution of natural radioactivity in the rocks of northern New England.

(In preparation, with Katharine Fowler-Billings),
Geology of the Gorham quadrangle, New
Hampshire.

Contents

Introduction[*]

E-AN ZEN

THE BEDROCK GEOLOGY of the Appalachian area north of the Delaware River has received an enormous amount of intensive field study during the past quarter century. The new data that have accumulated as a result of this concentrated effort permit regional syntheses to be made with an unusual degree of detail and confidence, and make this area one of the better studied and more thoroughly understood orogenic belts of the world. Ths volume brings together representative samples of these efforts, both in the collection of new data and in new interpretations and insights into the factors and processes in the evolution of the northern half of the Appalachian Mountains.

This orogenic belt may be divided into broad zones, forming a series of tectonically distinct geologic units; each of these structural units also tends to have a distinctive stratigraphic sequence. Insofar as feasible, the sequence of presentation of the papers in this book follows the natural sequence of arrangement of these major tectonic units.

1. The foreland includes the Catskill Plateau; the Hudson, Champlain, and St. Lawrence Lowlands; Anticosti Island; and, across the St. Lawrence Gulf, along both shores of the Strait of Belle Isle in Quebec, Labrador, and Newfoundland (Rodgers).[†] Rocks here are chiefly Cambrian and Lower to Middle Ordovician carbonate rock and quartzite overlain by Middle Ordovician shale. Deformation is mild on the west side of this belt, but increases in intensity eastward.

2. A belt of deformed and metamorphosed lower Paleozoic carbonate rocks lies east of the foreland. The two belts are separated by the so-called Logan's Line and its southward continuation approximately along the course of Lake Champlain and the Hudson River. Lithostratigraphically, these rocks correlate with rocks of the foreland, but they have been folded at least twice, once during the Acadian orogeny and once earlier. Recumbent folds locally occur in southwestern New England, but the intensity of folding seems to decrease northward. In western Vermont, these rocks are broken by high-angle faults and thrust faults; in northern Vermont and adjacent Quebec, thrust faults predominate and form a terrane of large-scale imbrication (Clark and Eakins). This zone is not recognized in the area between Quebec City and Newfoundland; it reappears as a zone of folding in western Newfoundland.

2a. Within this broad folded belt is a zone of allochthons, which consist of rocks that have slid into place above the Middle Ordovician shales. These allochthonous rocks are of a facies transitional between the platform sequence of the foreland and the volcanic-bearing eugeosynclinal rocks, and are of Cambrian to Middle Ordovician age. During Middle Ordovician time, they slid as submarine surficial sheets from depositional sites in the next belt east (belt 3) (Zen) into their present locations in the Taconic region, in western Newfoundland, and probably in the Quebec City area and in the Eastern Townships of Quebec. The northern and western limits of the allochthons approximately coincide with Logan's Line.

3. East of zone 2, a zone of Precambrian basement rocks forms a series of discrete "massifs." Within the northern Appalachian region south of the United States–Canadian border, these

---

* Publication authorized by the Director, U. S. Geological Survey.

† Names in parentheses refer to authors of articles in this volume.

are, from south to north, the Hudson massif, the New Milford massif, the Housatonic massif, the Berkshire massif, and the Green Mountain massif. The Precambrian rocks of the northern Long Range and of Indian Head Range in western Newfoundland probably constitute another of these massifs. These massifs are the cores of major anticlinoria which continue beyond the limits of exposure of the basement rocks; for instance, the Sutton Mountain anticlinorium (Clark and Eakins) is a continuation of the Green Mountain anticlinorium. On the east or southeast side of each massif, Paleozoic rocks generally dip conformably away from the older rocks. On the west or northwest side, however, structural complications locally occur that are not yet fully understood. Both high-angle faults and thrust faults doubtless exist here, and the Hudson massif may actually be in part allochthonous. The interaction of these massifs with Paleozoic rocks, at places where the latter rocks undergo facies transition, can render geologic interpretation exceedingly difficult (Hall).

4. East of the massifs of Precambrian basement lies a broad zone of intensely folded, sheared, and variably metamorphosed eugeosynclinal Paleozoic rocks. These rocks constitute the west limb of the Connecticut-Gaspé synclinorium, a major tectonic unit that extends from the shores of Long Island Sound, along the Connecticut valley, through the center of Gaspé peninsula, and probably into central Newfoundland. In New England, the rocks may be assigned to two subzones: a western subzone of homoclinal sequence dipping off the Precambrian masses of zone 3 (Hatch et al.) and an eastern subzone of late domes superimposed upon early isoclinal, and possibly recumbent folds (Rosenfeld). Recent studies in New England and adjacent parts of Quebec have shown that the structure of this zone is exceedingly complex in detail, and some of the new interpretations open up many exciting possibilities; similar complexities may be expected from the Gaspé–Newfoundland area. Syntectonic ultramafic rocks of the northern Appalachian region are concentrated in zone 4 (Chidester).

5. Stratigraphically continuous with zone 4, but in a distinct tectonic band, is a zone of giant penninic nappes. To the east, a second line of gneiss domes is structurally superimposed on the nappes within the Bronson Hill anticlinorium. The nappes involve both lower Paleozoic and possibly Precambrian rocks, and because of the effects of late doming and Triassic faulting, present some of the more intricate geology so far

deciphered in the northern Appalachian region. Large scale overturning of the nappes seems to occur both towards the west (Thompson et al.) and towards the east (Dixon and Lundgren). The gneiss domes might have been the loci of volcanic islands within a broad eugeosynclinal trough during early Paleozoic time (Naylor; Thompson et al.) ; they extend today from northwestern New Hampshire (Naylor; Berry; Thompson et al.) to Long Island Sound (Dixon and Lundgren).

6. East of zone 5, the Merrimack synclinorium extends from northern and western Maine through central New Hampshire into eastern Massachusetts and Connecticut. In a stratigraphic sense, this structure may be regarded as an integral part of zone 4, the Connecticut-Gaspé synclinorium, from which it is separated by the late domes of the Bronson Hill anticlinorium. It is among the less understood of the belts, in part because of the high metamorphic grade, in part because of difficulties of cross-structural stratigraphic correlation, and in part because of large gaps in detailed mapping.

Recent work in eastern Connecticut (Dixon and Lundgren), eastern Massachusetts, and southern New Hampshire (Thompson et al.) brings forth the intriguing possibilty that the synclinorium may be structurally vastly more complex than has hitherto been envisioned; all of it, for example, may eventually prove to be part of the root zone of the penninic nappes of zone 5. Nonetheless, this zone clearly is a key to understanding the structure of the important Acadian orogeny, as well as to correlations with rocks of zone 7. Recent mapping (Osberg et al.) of this zone in Maine indicates some of the complexities of the problems but also shows that rapid and major advances are being made.

7. The coastal belt consists of a heterogeneous and relatively poorly deciphered assemblage of rocks which are partly eugeosynclinal, rich in volcanic rocks, and partly nonmarine (Berry; Boucot; Pavlides et al.). These rocks are probably of Ordovician, Silurian, and early Devonian age. Precambrian rocks crop out in southeastern New England (Quinn and Moore), New Brunswick, Nova Scotia, and Newfoundland; these Precambrian rocks apparently are younger than those of belt 3 (Rodgers; Lyons and Faul). Although fossils are known in the Paleozoic rocks of this belt, they have not yet been effectively used in regional syntheses, and tie-ins with the other belts of the northern Appalachian region depend on difficult and commonly uncertain correlations across the Merrimack synclinorium.

Nevertheless, progress is being made (Hussey), even though the terrane is locally disrupted by transcurrent and high-angle faults of major proportions (Skehan) and by large, crosscutting posttectonic batholiths (Chapman).

8. Superimposed on the above zones are two belts of post-Acadian rocks: (1) the Triassic basins of southeastern New York and northern New Jersey, of central Massachusetts and Connecticut, and of the Bay of Fundy and (2) the series of post-Acadian, largely late Paleozoic basins that extend from the Narragansett basin of Rhode Island (Quinn and Moore), the Boston basin, and the Perry basin on the coast of Maine, to the huge Carboniferous basins of Nova Scotia, New Brunswick, and western Newfoundland (Belt). Although in north-central New Brunswick and in Maine these basins are little affected by deformation and metamorphism, the upper Paleozoic rocks of the Narragansett basin (Quinn and Moore) and adjacent small areas, as well as those exposed locally along the coast of New Brunswick (Belt), are mildly to severely altered and deformed. In contrast, the Triassic rocks are nowhere affected by other than high-angle faulting.

The various geometric combinations of the zones listed have produced relationships of tremendous variety and allow studies of the nature of interactions of tectonic units as well as studies of the original paleogeographic relations. For instance, the absence of zone 2a and of the basement along the United States–Canadian border in western New England brings the sheared and plastically folded eugeosynclinal rocks of zone 4 into juxtaposition with thrust-faulted miogeosynclinal rocks of zone 2, permitting decipherment of the original facies transitions and interdigitations (Clark and Eakins) and providing a "window" through which to discern the proper interfacies correlations (Cady). This particular facies boundary is of especial interest because it might mark the North American continental margin (Rodgers) during early Paleozoic time.

The stratigraphic-structural belts have been variously affected by one or more orogenies since the early Paleozoic. These orogenies provide convenient reference planes to divide the stratigraphic column into gross units, which are shown on the geologic map and which more or less coincide with the major chronostratigraphic series.

In places, one or more orogenies can be recognized within the rocks mapped as Precambrian, separating granulitic rocks from younger, less anhydrous rocks of obviously sedimentary derivation. Thus, although for much of its distance the base of the Paleozoic rocks is sharply defined against the Precambrian, locally this boundary cannot be placed with certainty, and assignment of certain limited rock sequences to the younger Precambrian or Paleozoic age remains a problem (for instance, in certain sequences in Vermont, in some rock units around gneiss domes in Vermont and Massachusetts, in some rocks in western Connecticut, and probably in units in Newfoundland). Undoubtedly, many of these problems stem from our current lack of information and will be solved by additional work; isotopic age determinations can be particularly helpful.

The next major orogenic episode in the northern Appalachian region is the Taconic orogeny, so that the first major subdivision of Paleozoic rocks encompasses those of Cambrian (Theokritoff) and Ordovician (Whittington, Berry, Neuman) ages. The age and nature of the Taconic orogeny has long been a matter of debate in the geologic literature and is discussed in several papers in this volume, especially the summary of Pavlides et al. It appears that the episode started in restricted localities in Middle Ordovician time (Zen) and continued, either in pulses or spatial transgressions, into a wider area; it may include events as late as Late Silurian. The unconformity is locally angular, yet at other places it is absent.

The Silurian and Lower and Middle Devonian rocks (Boucot) span the time interval between the Taconic and the Acadian orogenies. As has been generally recognized (Lyons and Faul), the Acadian orogeny is the most important diastrophic event in the northern Appalachian region in terms of its areal extent, the depth of the crustal units involved, and the intensity of metamorphism. It was the time of at least one major episode of formation of nappes and of evolution of synclinoria and anticlinoria (Cady; Thompson et al.; Dixon and Lundgren). This was also a period of extensive emplacement of igneous rocks, both intrusive and extrusive (Page, Rankin); these igneous activities persisted from preorogenic to postorogenic times (Chapman; Page). These igneous activities may be associated intimately with tectonism (Rankin) although some of them may properly be Taconic rather than Acadian in affinity (Naylor). The igneous rocks of syn- and pretectonic ages have been so altered by repeated deformation and metamorphism, however, that their true nature is locally difficult to decipher and may remain debatable for some time.

Although Paleozoic rocks younger than Middle Devonian are almost entirely restricted to scattered basins along the coast, their ages are well established as ranging from late Middle Devonian to Pennsylvania. The original areal extent of these basins may never be known, but the presence of such isolated patches of rocks as the problematic Pennsylvanian(?) Worcester Formation in eastern Massachusetts suggests that they may have been more extensive than their present distribution would indicate. The distribution of Mississippian and Pennsylvanian rocks in New Brunswick and Nova Scotia, thick in tectonic troughs and thin on shelves of older rocks (Belt), might characterize the relationship that once extended over a much larger area. The Appalachian orogeny severely affected these late Paleozoic rocks in Massachusetts and Rhode Island (Quinn and Moore), but on the whole the intensity of deformation decreases to the northeast: in the Maritime Provinces, the rocks, with some exceptions along the Bay of Fundy, are only mildly folded, though broken by many high-angle, possibly transcurrent faults (Belt). Because of the fragmentary record, the overall effect of the Appalachian orogeny remains largely an enigma in the northern Appalachian region.

The Triassic rocks of the Newark Group occur in discrete, partly fault-bounded basins throughout the area, and are not at all metamorphosed even in Connecticut where they are within 50 mi (80 km) of intensely metamorphosed and deformed Pennsylvanian rocks. These Triassic rocks, then, are presumably truly posttectonic. The later tectonic events are confined to high-angle faulting, block tilting, outpouring of the basalts that mark the Newark Group, emplacement of the postorogenic alkaline rocks of the late Triassic or Jurassic White Mountain Plutonic-Volcanic Series, and the Cretaceous Monteregian series of Quebec.

Calibration of the numerical ages of the diastrophic and metamorphic events is obtained by isotopic measurements on the metamorphic rocks and on igneous rocks showing a known age relationship to the surrounding country rocks (Lyons and Faul). Though it has commonly been assumed that the main event of Paleozoic metamorphism, at least in New England and in the Maritime Provinces, is Acadian, indications of an earlier, possibly Taconic, metamorphic event persist and are accumulating (Albee; Hall; Rosenfeld). At this time it seems probable that the "zone of allochthons," zone 2a, and the bulk of zone 2 itself were unaffected by the earlier metamorphic episode, but farther east, towards the central area of the orogenic belt of higher metamorphic grade (Thompson and Norton), the extent and importance of earlier metamorphism cannot be discounted (Albee).

In addition to the hint of a pre-Acadian metamorphic event, evidence has been accumulating for a major episode of Taconic folding, involving recumbent folds comparable in intensity and style to those of the Acadian event (Zen; Hall; Clark and Eakins). These indications come from many and diverse parts of New England, Quebec, and the Maritime Provinces and can no longer be dismissed; it seems unlikely that they represent merely local deformations around structural obstacles, because the folds follow regionally consistent trends which are nevertheless at variance with the Acadian trend. The deciphering and significance of these early folds are major problems for future workers in the northern Appalachian area.

But however diligently the field geologist maps an area, he is hampered by the two-dimensionality of any land surface, regardless of topography. In rare instances, underground facilities such as tunnels afford continuous sections which can be profitably compared and correlated with results of surface mapping (Skehan). Various types of geophysical surveys afford three-dimensional insights on a regional scale. Knowledge of the northern Appalachian region, especially within the United States, has been significantly advanced by extensive geophysical studies in recent years. Gravity (Diment; Kane and Bromery) and aeromagnetic (Griscom and Bromery) measurements suggest the extent of tectonic and stratigraphic units at depth, and provide clues to the presence of hidden intrusive rocks. Heat flow measurements (Birch et al.) are the basis for fruitful speculation on the sources of the heat that has so profoundly affected this orogenic belt. It may be hoped that thermal measurements will soon be of such detail that correlations with structural zones may become possible.

The physical conditions of deformation and metamorphism and their relative sequence can be specified not only through bedrock mapping and geophysical surveys, but also through recognition of the proper significance of metamorphic mineral assemblages. Such a survey has now become possible in the northern Appalachian region (Thompson and Norton). In one detailed study, however, an attempt to relate specific metamorphic changes to particular orogenies proved inconclusive (Albee). It may be that this sort of approach is inherently limited in its power of resolution, but a final answer must await accumulation of more information.

Some of the papers in this volume propose differing interpretations of certain major geologic features. For instance, the age, mode of emplacement, and geologic significance of the gneisses of the Oliverian Plutonic Series in the Bronson Hill anticlinorium are given conflicting interpretations by Naylor and by Thompson et al. on one hand and by Page on the other. Again, the nature of the junction of the miogeosynclinal and eugeosynclinal belts in northern Vermont, and the implied paleogeography, is portrayed differently by Cady and by Clark and Eakins on one side and by Rodgers and by Zen on the other. The early Paleozoic paleogeographic arrangement in Nova Scotia as given by Pavlides et al. differs from that given by Berry. The location of the root zone for the nappes of zone 5 in the Connecticut valley, as interpreted by Thompson et al., may prove incompatible with the interpretation of Dixon and Lundgren. These are but some principal examples; there are other inconsistencies of a similar sort among both interpretations presented in this book and others presented or to be presented elsewhere. In editing this book, we have attempted to reconcile factual inconsistencies but have not tried to smother conflicting interpretations that, at our present state of understanding, seem equally admissible. Indeed, where different interpretations exist, the authors have been encouraged to state or to refer to all the reasonable hypotheses; naturally, however, the authors alone are responsible for their own favored interpretations. Some of the conflicting hypotheses may be of mere local interest, but others may have far-reaching consequences to future study of northern Appalachian geology. We are hopeful that these conflicting ideas will stimulate further research and lead to fresh new syntheses.

The reader will quickly notice that the distribution of the topics of the articles in terms of geography and geologic units is far from even and far from complete. We have not attempted to produce a regional handbook; rather, we have concentrated on those particular investigations of northern Appalachian geology where significant progress or important new ideas have reached the point for a relatively brief, well-documented summary statement. Some investigations in this category, particularly in Canada, are not represented because their results have already been summarized in other recent publications. We hope that we have had at least a modicum of success in compromising between the demands of a coherent book and the enthusiasms of individual contributors, but assume full responsibility for our editorial shortcomings.

It remains the pleasant duty of the editors to acknowledge the assistance of many people and organizations whose interest, encouragement, and support have aided the creation of this volume and have considerably eased our tasks.

The contributors to this volume have shown toward us a general understanding and courtesy which have been essential to the successful endeavors of four amateurs. While their patience must at times have been worn thin, their enthusiasm rarely waned. For their support, our sincere thanks.

The articles in this volume have been reviewed by many experts in their fields. In addition to numerous contributors to this volume, the following geologists have generously donated their services, for which we acknowledge our debt: P. C. Bateman, J. M. Bird, W. G. Ernst, D. W. Fisher, John Haller, D. S. Harwood, W. B. Joyner, A. H. Lachenbruch, J. C. Maxwell, A. R. Palmer, J. C. Reed Jr., L. V. Rickard, R. L. Smith, O. T. Tobisch, Priestley Toulmin III, and R. E. Zartman.

The wise counsel and continued helpfulness of W. M. Cady, D. W. Rankin, John Rodgers, and Priestley Toulmin in many matters related to this book are remembered with gratitude.

To the U. S. Geological Survey, the editors acknowledge their gratitude for the administrative support which has been absolutely essential for us to carry out the project. It is our special pleasure here to indicate our indebtedness to members of the Geological Survey's editorial staff, who facilitated our work in every way.

Cristina S. Zen volunteered to style-edit manuscripts at a critical moment; her painstaking and voluminous output has greatly reduced our editorial chores. Ruth Abramowitz helped to keep communications among the editors flowing freely and thus played a crucial role in our undertaking. To both of them, our sincere thanks for their help and for their ever-cheerful encouragement.

REGIONAL STRATIGRAPHY
AND PALEOGEOGRAPHY

Chapter 1

Cambrian Biogeography and Biostratigraphy in New England

GEORGE THEOKRITOFF

INTRODUCTION

STRATA OF KNOWN Cambrian age crop out in New England and adjacent areas within two bands. A westerly band lies in the Hudson and Champlain valleys and extends northward into southern Quebec; it has a sand-carbonate facies in its western part grading eastward into a dominantly shale facies, part of which has been thrust westward onto the sand-carbonate facies to form the Taconic klippe. Sporadic outcrops along the coast, in eastern Massachusetts, Maine, New Brunswick, and Nova Scotia, constitute an easterly band dominated by the shale facies. The area in central New England between these two bands may contain some metamorphosed Cambrian strata; much of it is occupied by Precambrian and younger Paleozoic rocks, mainly metamorphosed.

The Cambrian faunas belong to the Acado–Baltic and Pacific provinces[*] as well as to an intermediate zone in which elements of both these provinces occur with endemic elements.

The oldest known faunas in New England are late Early Cambrian. The occurrence of *Fallotaspis* and *Daguinaspis* in eastern California (Cloud and Nelson, 1966) below *Nevadia, Holmia, Nevadella,* and *Paedeumias,* and in Morocco

(Hupé, 1953, p. 44, 46; 1960, p. 76, Tab. 1) below *Callavia, Kjerulfia,* and *Nevadia,* suggests that there may be pre-*Olenellus* Olenellina-bearing strata. By definition, such strata would be Lower Cambrian. The absence of *Fallotaspis* and *Daguinaspis* from New England suggests that the earliest Early Cambrian strata are probably represented there by unknown thicknesses of unfossiliferous strata lying conformably beneath the fossiliferous Lower Cambrian.

Shaw (1961) summarized the extensive literature on the Cambrian of New England. I have not attempted to duplicate or up-date his work; rather, I have concentrated on the paleontological aspects of recent work and have attempted to indicate some of its implications.

ACKNOWLEDGMENTS

I am indebted to more people than could be named in this article; in a very real sense, I am indebted to all who have ever worked in and around New England. Drs. Franco Rasetti and A. B. Shaw freely made unpublished material available to me. Drs. Wallace M. Cady, Franco Rasetti, John Rodgers, H. B. Whittington, and A. R. Palmer critically read the manuscript of this paper; it is a pleasure to acknowledge their many suggestions and discussions.

EASTERN MASSACHUSETTS

Lower and Middle Cambrian strata are known in a number of scattered outcrops in eastern Massachusetts (Fig. 1–1).

Lower Cambrian strata are known at Hoppin Hill in North Attleboro. Here, a quartzite approximately 10–20 feet thick rests unconformably

---

[*] I use the terms *Acado–Baltic* and *Pacific* to denote more or less mutually geographically exclusive associations of organisms having typical occurrences within certain areas; other than this, these terms do not necessarily imply any given paleobiogeographic pattern, whether "bull's eyes" or concentric realms around the craton (Lochman and Wilson, 1958) or "jigsaw pieces" occupying parts of one or more continents as suggested by Whittington (1966) for the Ordovician.

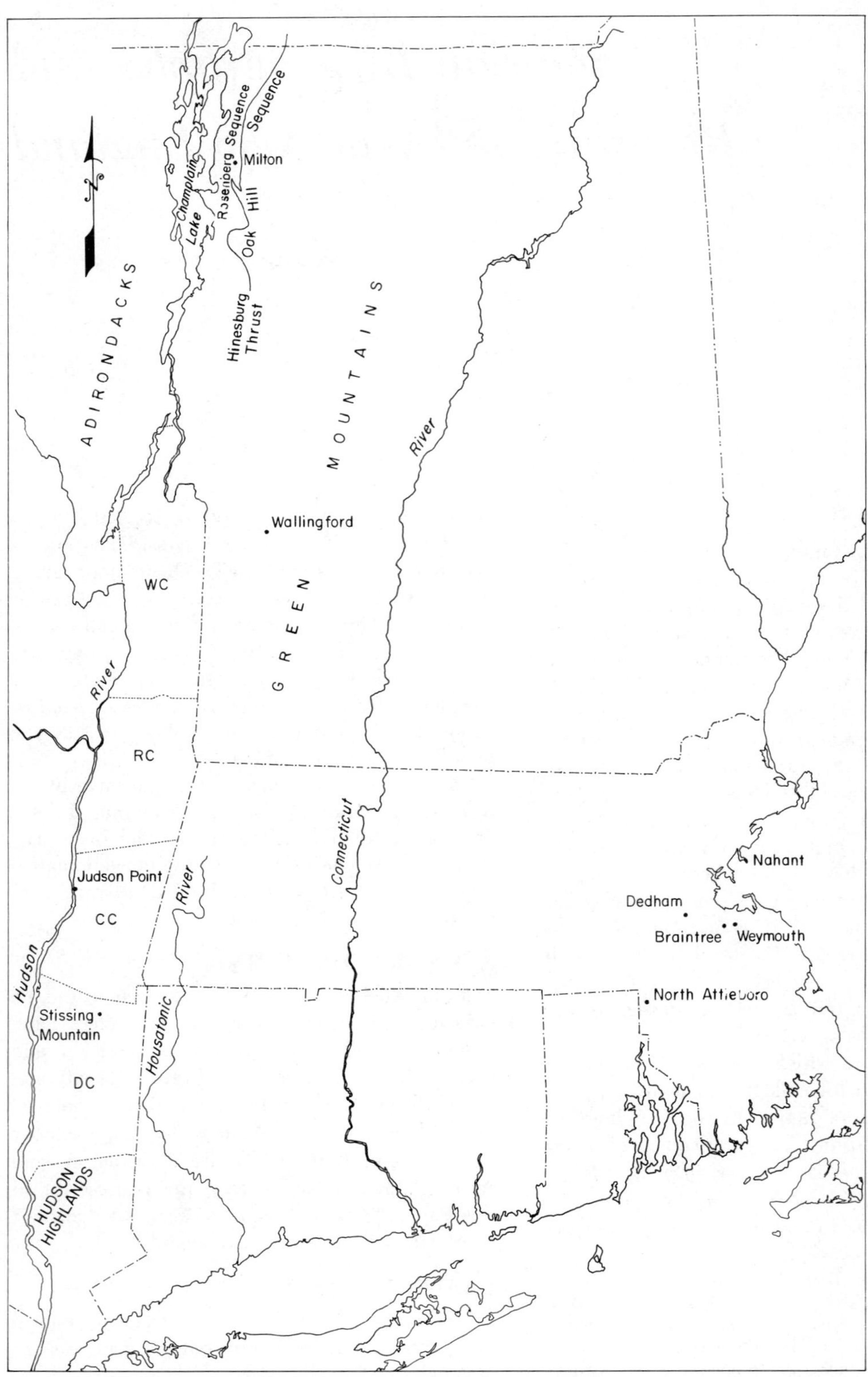

FIGURE 1–1. Index map of western New England and parts of adjoining States and Provinces. WC, Washington County, New York; RC, Rensselaer County, New York; CC, Columbia County, New York; DC, Dutchess County, New York.

on granodiorite (Billings, 1929, p. 103; Dowse, 1950; Shaw, 1950, p. 578). It is, in turn, overlain by the Hoppin Slate, essentially red and gray slates, locally calcareous, with interbedded pink argillaceous fossiliferous limestones showing evidence of water-sorting (Shaw, 1950, p. 578). The fossils have been referred to the Lower Cambrian (Walcott, 1886; Foerste, *in* Shaler, Woodworth, and Foerste, 1899; Grabau, 1900; Shaw, 1950).

Foerste (*in* Shaler, Woodworth, and Foerste 1899) listed the fossils from the Hoppin Slate at North Attleboro and suggested (p. 391) that the fossiliferous strata form two horizons, the lower with conoidal fossils but no trilobites, the upper with trilobites. Grabau (1900, p. 610–611) listed the forms found and stated that there is a decided difference between the faunas. Shaw (1950) restudied the trilobites from the Hoppin Slate; later (1961, p. 436) he changed his earlier stratigraphic assignment of this fauna. The trilobites of the Hoppin Slate belong to a number of families: Eodiscidae represented by *Serrodiscus bellimarginatus* (Shaler and Foerste) (type locality), Pagetiidae by *Hebediscus attleborensis* (Shaler and Foerste) (type locality), and Ellipsocephalidae by *Strenuella strenua* (Billings). Olenellidae are extremely rare; one fragment, now lost, has been reported by Shaler and Foerste (1888, p. 36–37, Pl. 2, Fig. 12) and discussed by Walcott (1890, p. 636–637; 1910, p. 341) and Shaw (1950, p. 581).

Lower Cambrian strata are also known at Weymouth and Nahant, and are referred to the Weymouth Formation. The Weymouth Formation at Weymouth is a greenish, gray, or dark red slate with calcareous lenses, nodules, and argillaceous limestone beds. The calcareous horizons yield Lower Cambrian fossils (Burr, 1900; Grabau, 1900; Shimer, 1907). The Weymouth Formation at Nahant includes a bed of greenish-white limestone exposed at East Point. Fossils from this limestone were first referred to the Lower Cambrian by Foerste (1889) and listed by Grabau (1900, p. 610).

Grabau (1900, p. 612) recognized two fossiliferous horizons in the Weymouth Formation at Weymouth. The lower is a pink argillaceous limestone containing abundant hyolithids and other conoidal fossils, apparently water-sorted. The upper is a medium gray limestone; its fauna includes olenellids referred by Walcott (1910, p. 280–281, 284) to *Callavia burri* Walcott and *C. crosbyi* Walcott, an eodiscid designated the plesiotype of *Weymouthia nobilis* (Ford) (Shaw, 1950, p. 584–585; Rasetti, 1952, p. 446–447),

and an ellipsocephalid identified as *Strenuella strenua* (Billings) by Grabau (1900, p. 672–674), Burr (1900, p. 46–47), and Shimer (1907).

Fossils from the limestone in the Weymouth Formation at Nahant were first referred to the Lower Cambrian by Foerste (1889); Grabau (1900, p. 610) listed a fauna consisting mainly of hyolithids with rare brachiopods and *Stenotheca*.

The sections at Hoppin Hill and Weymouth can be correlated (Shaw, 1961, Fig. 2). The pink argillaceous limestone of the lower fossiliferous horizon at Hoppin Hill is remarkably similar lithologically to that of the lower horizon at Weymouth; both contain abundant conoidal fossils but no trilobites. This fauna may be the correlative of the *Coleoloides* fauna described by Matthew (1899) from southeastern Newfoundland. Because the *Coleoloides* fauna occurs in apparently conformable sections not far below strata with late Early Cambrian faunas, I do not agree that it is the most primitive Cambrian fauna (Hutchinson, 1962, p. 12). I consider the *Coleoloides* fauna to be Early Cambrian, synchronous with olenellid faunas elsewhere; the absence of olenellids from the *Coleoloides* fauna is probably due to ecological factors.

The upper horizon contains *Strenuella strenua* at Hoppin Hill and at Weymouth and the contrast in the number of olenellids, abundant at Weymouth and extremely rare at Hoppin Hill, is probably due to ecological differences. The fauna of the upper horizon includes Eodiscidae, Pagetiidae, Ellipsocephalidae, and Olenellidae, but not the abundance of ptychoparioid and corynexochoid trilobites characteristic of the late Early Cambrian faunas of the carbonate facies of the Appalachian area. Hence, they belong to the Acado–Baltic province. The presence of *Serrodiscus* and *Strenuella* suggests correlation with other faunas that include *Strenuella strenua, Callavia, Serrodiscus,* and *Hebediscus* in southeastern Newfoundland (Hutchinson, 1962, p. 15, 18) and other faunas with *Serrodiscus* in the Acado–Baltic province; such faunas occur in strata not far below strata bearing Protolenidae (see references, *in* Rasetti, 1967, p. 16–17) and are late Early Cambrian.

Although Middle Cambrian fossils (Walcott, 1884; Grabau, 1900; Howell, Shimer, and Lord, 1936) have been obtained in eastern Massachusetts only from the greenish-gray to dark gray non-calcareous Braintree Slate at Braintree, Middle Cambrian strata may have a wider distribution in eastern Massachusetts. The fauna of the Braintree Slate includes *Paradoxides harlani* and ptychoparioid trilobites referred to *Agraulos quadrangularis* and *Ptychoparia rogersi* by Grabau (1900, p. 613). Wheeler (1942) assigned *Ptychoparia rogersi* to a new genus *Braintreëlla*.

Howell, Shimer, and Lord (1936) reported the discovery of a subfauna characterized by *Paradoxides eteminicus* overlying strata with *P. harlani*.

Both Middle Cambrian assemblages include *Paradoxides*; undoubted Pacific province elements are absent. Hence, both assemblages belong to the Acado–Baltic province and have been correlated with the *Paradoxides bennetti* zone of southeastern Newfoundland (Howell, Shimer, and Lord, 1936; Howell et al., 1944; Hutchinson, 1962, p. 21).

Upper Cambrian pebbles have been reported from beaches and conglomerates in eastern Massachusetts (Clark, 1923, p. 481–482; Emerson, 1917, p. 35; Grabau, 1900, p. 613; Shaler, Woodworth, and Foerste, 1899, p. 109–113). Rhodes and Graves (1931) assigned the Green Lodge Formation near Dedham, Massachusetts, to the Upper Cambrian. Shaw (1961, p. 436) considered that there was no reliable evidence for any of these Upper Cambrian assignments.

WESTERN NEW ENGLAND AND ADJACENT AREAS

General Statement

Cambrian strata in western New England and adjacent New York and Quebec belong to two facies. A shelf sequence of sands and carbonates, with subordinate shales, extends from southern Quebec southward along the Champlain, Hudson, and Housatonic valleys into Dutchess County, New York, where it is buried under Ordovician strata. A basin sequence, dominantly of shale, with some volcanics, now metamorphosed into phyllites, schists, greenstones, and locally gneisses, lies immediately east of the sand-carbonate sequence.

The shelf sequence has been driven westward onto the Adirondack foreland by a number of thrusts. In turn, the basin sequence has been thrust westward over the shelf sequence. This thrusting has generally obscured the transition between the two sequences, although this transition may be partly preserved in the Champlain and Rosenberg slices (Rodgers, this volume; Zen, this volume; Cady, 1967, p. 59).

The lithostratigraphic sequence of the several units in the Taconic klippe is closely comparable with that in the basin sequence, especially in the less metamorphosed areas in southern Quebec (Cady, 1960, p. 558; 1967, p. 63; Osberg, 1965). This resemblance is consistent with recent interpretations that the Taconic sequence occupies a klippe; the Taconic sequence is therefore considered to be a remnant of the basin sequence. The relatively abundant paleontological evidence has permitted dating of the several units of the Taconic sequence.

Interpretation of the relationships of the shelf, basin, and Taconic sequences at the time of deposition is presented schematically in Fig. 1–2.

Shelf Sequence

The shelf sequence rests unconformably on the Precambrian basement along the Adirondack border and in the Hudson Highlands; west of the Green Mountains and in the vicinity of the Hinesburg thrust, the shelf sequence rests conformably on slates, graywackes, dolomites, and phyllites.

Cady (1960, p. 536–537) reviewed the shelf sequence; detailed descriptions of parts of the shelf sequence were cited by Doll et al. (1961). Rodgers (this volume) interprets the shelf sequence as a huge carbonate blanket whose eastern margin may have coincided with the continental margin.

In essence, the shelf sequence consists of wedges of sandstone, calcareous sandstone, and sandy carbonate; the carbonates show their maximum development in the eastern part and are replaced westward, toward the craton, by the westward thickening sandstones (Cady, 1945, p. 530–531, 536; 1960, p. 537). The several sandstone wedges probably coalesce at depth and overlap westward onto the basement rocks of the Adirondacks, the basal sandstones getting younger westward (Rodgers, this volume, p. 142); the Potsdam Sandstone, which crops out west of Lake Champlain, yields Dresbachian, Franconian, and Trempealeauan fossils (Fisher, 1956, p. 325, 339) and probably represents the present western edge of such a coalesced sandstone facies.

Although interbedded sandstones and carbonates can be traced southwestward for hundreds of miles along the Appalachians (Rodgers, this volume, p. 142), there are important along-strike variations within the relatively small New England area. The most important, in the present context, is the incursion of basin rocks onto the shelf in northwestern Vermont above the Dunham Dolomite (Fig. 1–2A); these basin rocks are the Parker Slate and its equivalent, the Oak Hill Slate, and the Sweetsburg Slate in the Oak Hill succession, and the Parker Slate, the St. Albans

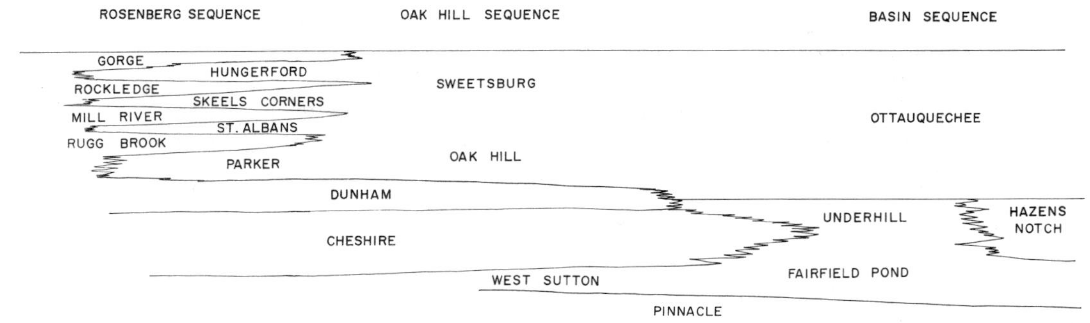

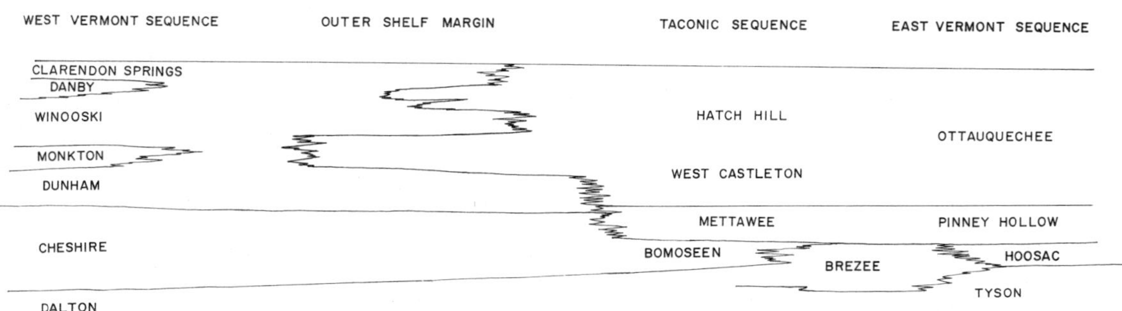

FIGURE 1–2. Schematic west-east section showing inferred relationships of Cambrian shelf and basin sequences in western New England. (For discussions, see Cady, this volume; Doll et al., 1963; Thompson, 1967; Zen, 1967, p. 48–50). Upper section, northern Vermont: lower section, south-central Vermont.

Slate, the Skeels Corners Slate, and the Hungerford Slate in the Rosenberg succession where they are interbedded with limestone conglomerates, tongues, and lenses interpreted by Rodgers (this volume, p. 143) as detritus and slides dropped from a carbonate bank into the deeper water sites of mud deposition.

Basin Sequence

The base of the basin sequence is an unconformity where it rests on the Precambrian on the eastern side of the Green Mountain, Berkshire, Housatonic, and Hudson massifs; elsewhere, it is not fixed stratigraphically.

Detailed descriptions of parts of the basin sequence have been published by Cady, Albee, and Chidester (1963), Cady (*in* Doll, *ed.*, 1961), Dennis (1964), and Stone and Dennis (1964); the earlier references were cited by Doll et al. (1961).

The basin sequence is essentially phyllites, schists, gneisses, graywackes, and in the thicker parts, greenstones. In general, the basin sequence thickens eastward and southeastward. The Taconic sequence, interpreted as a westward thrust remnant of the western part of the basin sequence, is thinner than the shelf sequence it now rests on (Cady, 1967, p. 63), which, in turn, is thinner than the thickest part of the basin sequence (Cady, 1967, p. 59; this volume, p. 155). The Cambrian basin rocks of western New England extend northeastward into Quebec and ultimately Newfoundland; southwestward they are buried under younger strata but appear to extend for hundreds of miles along the Appalachians.

Cambrian fossils are known from the basin sequence where it encroaches onto the shelf in northwestern Vermont. However, Cambrian fossils are not known from the basin sequence east of the Hinesburg thrust. The age assignment of these strata is discussed by Cady (1960, p. 553–554) and Cady, Albee, and Chidester (1963, p. 7).

Paleontology

The Lower Cambrian faunas have been extensively studied by Walcott (1886; 1890; 1910; 1912), Resser (1937), Resser and Howell (1938), and Shaw (1955b; 1957b).

Lower Cambrian fossils occur in the Cheshire Quartzite (Cady, 1945, p. 528; Shaw, 1954, p.

1034; Stone and Dennis, 1964, p. 28) and in its southward extension in Dutchess County, New York, the Poughquag Quartzite (Knopf, 1962, p. 15). Although rare in the Cheshire Quartzite, fossils are abundant locally in the Poughquag Quartzite. Dwight's material from Stissing Mountain, Dutchess County, now in the New York State Museum and at Vassar College, shows abundant disarticulated trilobite fragments, brachiopods, and conoidal fossils preserved as moulds in sandstone. I consider the trilobites to belong to the genus *Olenellus*. The fact that the trilobites are disarticulated, although apparently not badly abraded, suggests sedimentation and burial in gently moving water.

In the summer of 1965, J. B. Thompson, Jr. and I discovered *Skolithos linearis* in the Cheshire Quartzite at Green Hill, near Wallingford, Vt. Here, the Cheshire Quartzite is relatively flat-lying in a gently plunging dome. Erosion has worked along the bedding surfaces, exhuming individual bedding planes over considerable areas. *Skolithos* was noted in large clusters, each separated from adjacent clusters by areas of un-burrowed sandstone. Elsewhere at this locality, ripple-surfaces of 20–30 foot wavelength were noted but *Skolithos* was not noted associated with the rippled surface. The burrows of *Skolithos* may have been destroyed by reworking of the sand; alternatively, *Skolithos* may not have estab-lished itself in areas where the sand was moved by the water but preferred the more stable bottoms.

The general paucity of fossils in the Cheshire Quartzite cannot be entirely due to metamorphism because abundant fossils occur locally in pockets, as at Stissing Mountain. The evidence that depo-sition took place in moving water suggests that constant reworking of the sand may have de-stroyed most of the fossils; preservation took place fortuitously in those few localities where the sand was not reworked significantly. It appears probable that the Cheshire Quartzite represents a strand-line deposit extending per-haps many miles across strike and hundreds of miles along the Appalachians.

The Dunham Dolomite yields a few trilobites, brachiopods, and conoidal fossils (Cady, 1945, p. 530; Shaw, 1954, p. 1035–1037; Stone and Dennis, 1964, p. 35). I have collected hyolithids from limestone in the Dunham Dolomite just be-low the dam on East Creek in the northwestern part of the City of Rutland. Furthermore, Foerste has collected olenellid cephala from the Mallett facies of the Dunham at an unspecified locality in the City of Rutland; this material is in the Museum of Comparative Zoology at Harvard University (MCZ 8820) and I refer the cephala to *Olenellus*.

The Monkton Quartzite yields trilobites, in-articulate and orthoid brachiopods, hyolithids, and a few molluscs from two localities (Kindle and Tasch, 1948; Tasch, 1949; Shaw, 1962b). Although fossils are otherwise absent from the Monkton Quartzite, they are abundant at the two localities. All the fossils are disarticulated, albeit not badly abraded, and are preserved as moulds in the sandstone. It is probable that the circum-stance of their preservation was similar to that of the fossils in the Poughquag Quartzite at Stissing Mountain; the Monkton Quartzite probably rep-resents a strand-line deposit.

The lower Parker Slate yields Lower Cam-brian fossils (Shaw, 1954, p. 1038–1043; 1958, p. 531).

Rasetti (1948, p. 3) suggested that the Lower Cambrian faunules from the conglomerate boulders in Quebec with *Pagetides* are the youngest; in northwestern Vermont, *Pagetides* occurs only in the uppermost *Olenellus*-bearing strata. Hence, the *Olenellus*-bearing fauna of northwestern Vermont may be divided into two: an upper division with *Pagetides*, designated for convenience the *Pagetides-Olenellus* subfauna, and a lower without *Pagetides*, designated the *Olenellus* subfauna.

Pagetides is reported only from the lower Parker Slate (Shaw, 1954, p. 1040–1041; 1955b, p. 795; 1958, p. 531) but not from the Monkton Quartzite, correlated with it (Cady, 1945, p. 531; Kindle and Tasch, 1948, p. 134; Shaw, 1954, p. 1044; 1962b, p. 322–327) on the basis of the common occurrence of trilobites, brachio-pods, and miscellaneous fossils (compare Shaw, 1958, p. 531, Table 2, with Shaw, 1962b, p. 323, Table 1). The absence of *Pagetides* from the Monkton Quartzite may be due to ecological fac-tors. On the other hand, although Shaw (1955b, p. 795) stated that *Pagetides* ranges up from the base of the lower Parker Slate, his fossil lists from localities in the lower Parker Slate (1954, p. 1038–1043) suggest that *Pagetides* may be more common in collections made from 120 to 210 feet above the base of the formation; if so, the faunas of the Monkton Quartzite may be cor-relative of those of the lowest 100 feet or so of the lower Parker Slate.

The *Olenellus*-bearing faunas of the shelf sequence in northwestern Vermont are probably correlative with other late Early Cambrian faunas of the Appalachians; the *Pagetides-*

Olenellus subfauna is probably the latest Early Cambrian assemblage in northwestern Vermont.

Several Middle Cambrian assemblages are known. The first is from the upper Parker Slate (Shaw, 1954, p. 1043–1044; 1957a; 1958, p. 530–532). Although Shaw (1954, p. 1045–1046) could not reconcile the association of all the genera recorded from one of his localities in the upper Parker Slate with their recorded ranges, he gave paramount weight to his identification of *Mexicella stator* in this fauna and hence provisionally correlated it with the *Albertella* zone. Rasetti (1967, oral communication) considers that Shaw's identification of *M. stator* in this fauna is specifically incorrect and the fauna is probably younger than suggested by Shaw.

The second Middle Cambrian assemblage is from the lower part of the St. Albans Slate and includes the Acado–Baltic *Meneviella* (Howell, 1937;* Shaw, 1966b). Shaw (1966b, p. 846) tentatively suggested that this assemblage correlates with the upper part of the *Paradoxides davidis* zone.

The third Middle Cambrian assemblage is from the upper part of the St. Albans Slate and includes the Acado–Baltic paradoxidid *Centropleura* (Howell, 1937; Shaw, 1966b). Shaw (1966b, p. 845–846) correlated this assemblage with the *Paradoxides forchhammeri* zone.

Several assemblages have been correlated with the Dresbachian *Cedaria* zone by Shaw (1952; 1958; 1962a; 1966a). One is from the abundantly fossiliferous limestone pebbles in the Mill River Conglomerate; the second is from the Skeels Corners Slate; two others, differing somewhat from each other, are respectively from the pebbles and matrix of the Rockledge Conglomerate; the last is the sparse fauna from the Hungerford Slate.

Clark and Shaw (1968) decribed a fauna from the lower part of the Gorge Formation that they assigned to the *Dunderbergia* zone.

Faunas correlated with the *Hungaia magnifica* fauna are known from the Gorge Formation (Clark, 1924, p. 11–13; Rasetti, 1944, p. 231; 1946, p. 537; Raymond, 1924, p. 461; Shaw, 1958, p. 549; 1961, p. 456–457; Clark and Shaw, in press) and the Clarendon Springs Dolomite at a locality on Cobble Hill, near Milton, Vt. (Shaw, 1961, p. 456; listed *in* Shaw, 1953, p. 140 and Stone and Dennis, 1964, p. 53). Raymond (1924, p. 459) reported four fossiliferous horizons in the upper part of the Gorge Forma-

tion; because the second and third horizons are separated by only a few feet of strata, he combined them into a "main zone." Later, Raymond (1937, p. 1131–1132) published revised and extended faunal lists based on further collecting. Shaw and Clark (1968) redescribed the type section of the Gorge Formation. Rasetti (1944; 1945; 1946; 1963) and Clark and Shaw (in press) discussed and described some of the trilobites of the *Hungaia magnifica* fauna.

The age of the *Hungaia magnifica* fauna, and hence that of the Gorge Formation, has been the subject of discussion reviewed by Shaw (1961, p. 456–458). Whittington (1966, p. 701) considered the *Hungaia* fauna to be Trempealeauan and also correlative of the Early Tremadoc; he (1966, Fig. 1) showed the geographic distribution of the *Hungaia magnifica* fauna.

Shaw (1951; 1953) described a fauna that he later (1955a; 1958, p. 549, 551) referred to the Highgate Formation and correlated with the Tremadocian (*Bellefontia* zone); he also published a revised and extended faunal list.

TACONIC REGION

Recent studies based on paleontology (Berry, *in* Zen, *ed.*, 1959; 1961; Bird and Rasetti, in press; Rasetti, 1966; 1967; Rasetti and Theokritoff, 1967; Theokritoff, 1959; 1964; Zen, 1961) have clarified the Cambrian stratigraphy.

An extensive graywacke, named the Bomoseen Graywacke, forms the lowest unit along the western margin of the Taconic klippe (Theokritoff, 1964, p. 175; Zen, 1961, p. 302); in Rensselaer County, it interdigitates with a particularly thick section of graywacke named Rensselaer Graywacke (Bird, 1963; Zen and Bird, *in* Bird, *ed.*, 1963, p. 41–42). The graywackes grade up into purple and green argillites named Mettawee Slate. Limestones near the top of the Mettawee Slate contain Lower Cambrian fossils belonging to the *Elliptocephala asaphoides* fauna, the oldest known from the Taconic sequence (Lochman, 1956, p. 1335–1337, 1339–1341; Theokritoff, 1964, p. 175–176). Zen (1961, p. 293, 300) and Theokritoff (1964, p. 175) treated the Bomoseen Graywacke and the Mettawee Slate as members of the Bull Formation. The Bull Formation also includes local lenses of quartzite.

The Bull Formation is overlain by the West Castleton Formation, essentially composed of dark gray shales with limestone conglomerates and bedded limestones (Zen, 1961, p. 304–305). Many of these limestones yield the *E. asaphoides* fauna (Lochman, 1956, p. 1337, 1339–1341; Theokritoff, 1964, p. 177).

---

* Shaw (1966b, p. 843) stated that some of Howell's specimens were from the Skeels Corners Slate.

Rasetti and Theokritoff (1967, p. 191–192) demonstrated that Walcott's (1912), Schuchert's (1937, p. 1027), and Lochman's (1956, p. 1351, 1354–1357) lists of fossils from localities in the Taconic sequence included several younger Early Cambrian faunas with the *Elliptocephala asaphoides* fauna; this failure to differentiate the several Early Cambrian faunas supported the earlier belief that the *E. asaphoides* fauna was the only Lower Cambrian fauna in the Taconics. At present, a comprehensive, even if not complete, list of the *E. asaphoides* fauna can be made from a number of sources: trilobites from Rasetti (1967), brachiopods from Walcott (1912), and miscellaneous fossils from Lochman (1956). Apart from the classic studies of Emmons, Ford, and Hall, the more important descriptions of members of the *E. asaphoides* fauna are those of Lochman (1956), Rasetti (1952), Walcott (1886, 1890; 1910; 1912), and Whittington (1957).

A number of Lower Cambrian assemblages distinctive from the *Elliptocephala asaphoides* fauna are now known from limestones in the West Castleton Formation. Theokritoff (1959; 1964, p. 186) listed the agnostid-bearing *Paedeumias–Bonnia* assemblage. Rasetti (1966; 1967) described the *Acimetopus bilobatus* faunule from Columbia County, New York.

Rasetti and Theokritoff (1967) described the agnostids from the *Paedeumias–Bonnia* assemblage, gave a more extended and corrected list of this assemblage (p. 191), and suggested that this assemblage is essentially contemporaneous with the *Pagetia connexa* faunule of Washington County (Rasetti and Theokritoff, 1967, p. 191) and the *Pagetides elegans* faunule of northern Columbia County (Rasetti, 1967, p. 19–22).

The *Elliptocephala asaphoides* fauna is the oldest so far known from the Taconic sequence; the *Acimetopus bilobatus* is the next younger; the agnostid-bearing *Paedeumias–Bonnia* assemblage and approximately synchronous *Pagetia connexa* and *Pagetides elegans* faunules are the youngest (Rasetti, 1966, p. 4–5; 1967, p. 18–19; Bird and Rasetti, in press; Rasetti and Theokritoff, 1967, p. 192–193; Theokritoff, 1964, p. 183).

Rasetti (1967, p. 22–26) provisionally recognized a number of Middle Cambrian faunules from limestones within the West Castleton Formation of Columbia County (Bird and Rasetti, in press; Rasetti and Bird, *in* Fisher, 1964). He referred the *Ptychagnostus gibbus*, *Pagetia clytioides*, and *Pagetia erratica* faunules to the *Bathyuriscus–Elrathina* zone, and the *Bathyuriscus eboracensis* and the *Centropleura* faunules,

as well as an unnamed faunule, to the *Bolaspidella* zone.

The Hatch Hill Formation is the highest Cambrian unit in the Taconics. It is essentially black, pyritic shales with interbedded dolomitic sandstones (Theokritoff, 1964, p. 178). Dendroid graptolites from near the top of this formation (Berry, *in* Zen, *ed.*, 1959, p. 61; 1961, p. 224; Theokritoff, 1964, p. 179, 189) and from its equivalent in Columbia County, the Germantown Formation (Fisher, 1962, annotation 6) have been compared by Berry (1961, p. 224; *in* Bird, *ed.*, 1963, p. 25) with graptolites from Upper Cambrian strata. The base of the Hatch Hill Formation may be Franconian or Dresbachian (Theokritoff, 1964, p. 179, see also Zen, 1967, p. 54).

Bird and Rasetti (in press) reported a very late Dresbachian faunule, tentatively referred to the *Aphelaspis* zone, and a number of Trempealeauan faunules from limestone conglomerates in undifferentiated West Castleton–Hatch Hill strata in Columbia County.

In the eastern part of the Taconic klippe, units underlying those forming the base of the section in the western part crop out. A chloritoid-bearing purple and green slate, underlying the Bull Formation at the northeastern extremity of the Taconic klippe, has been named the Biddie Knob Formation by Zen (1961, p. 299). Dark gray to black phyllite, geometrically underlying strata referred by Zen to the Biddie Knob Formation, was assigned by Keith (1932, p. 361, 399–400) to the Brezee Phyllite and considered to underlie strata that Zen assigned to the Biddie Knob Formation. Zen (1961), however, assigned the Brezee Phyllite to the West Castleton Formation. On the other hand, Doll et al. (1961; 1963) recognized the Brezee as a valid unit and discussed the alternatives (for further discussion, see Thompson, 1967, p. 84).

Later, Zen (1967, p. 96; see also Cady, this volume, footnote p. 158), in reassigning most of the Brezee to Middle Ordovician autochthonous units, stated that "part of the Brezee Formation can be traced continuously into a belt of the West Castleton Formation which includes limestone outcrops, one of which has yielded Lower Cambrian fossils of the *Elliptocephala asaphoides* fauna which elsewhere has never been found in any rock stratigraphically below the upper part of the Bull Formation."

The Brezee Formation is lithologically like the West Castleton Formation (Doll et al., 1963, p. 95). The assertion that the limestones cited by Zen are in the West Castleton Formation therefore depends on the stated presence of the *E. asaphoides* fauna in one of them. The argument that this fauna "elsewhere has never been found in any rock stratigraphically below the upper part of the Bull Formation" is inconclusive in this context because the area of outcrop of the Brezee is limited. Furthermore, the Hoosac Formation of the Paleozoic sequence on the east side of the Green Mountains is lithologically similar to the Brezee (Doll et al., 1963, p. 95) and also includes limestones. I consider the Brezee Formation probably to be a valid unit, equivalent to the Hoosac Formation. If the fossils cited by Zen have been correctly assigned to the *E. asaphoides* fauna, it is probable that the age span of this fauna includes the Brezee Formation; the Brezee Formation is therefore probably Lower Cambrian.

DISCUSSION

Biogeographic Affinities of the Taconic Faunas

Lochman (1956, p. 1331, 1347, 1350) and Lochman and Wilson (1958, p. 320) characterized the *Elliptocephala asaphoides* fauna as a "mixed" or "transitional" fauna, composed of some Acado–Baltic province, some Pacific province, and some endemic genera. However, the recognition (Rasetti and Theokritoff, 1967, p. 191–192) that earlier lists had confused several distinctive assemblages has shown that almost all the Pacific province genera previously included in the *E. asaphoides* fauna belong to one or more of the younger assemblages. The only known exception is *Kootenia*, represented in the *E. asaphoides* fauna, locally in great abundance, by one authenticated species, *K. troyensis*, which is unlike any species of *Kootenia* known from the Pacific province (Rasetti, 1967, p. 17).

The biogeographic affinities of a biota are determined by the relative abundance and geographic distribution of its component elements. Thus, the trilobites of the *E. asaphoides* fauna are dominated, in point of numbers, by two Acado–Baltic genera, *Calodiscus* and *Serrodiscus*, and two endemic genera, *Elliptocephala* and *Fordaspis*. *Serrodiscus* is represented by *S. speciosus* in the *E. asaphoides* fauna; this species, or very closely related forms, has been widely reported from the Acado–Baltic province (Rasetti, 1967, p. 16–17). *Calodiscus* has also been reported from that province (Rasetti, 1967, p. 16).

Hence, at the specific level, the majority of the trilobites in the *E. asaphoides* fauna are endemic;

only one occurs in the Acado–Baltic province and none in the Pacific. At the generic level, however, some Pacific affinity appears in the occurrence of *Kootenia*. At the familial level, the *E. asaphoides* fauna shares Olenellidae and Pagetiidae with both the Acado–Baltic and Pacific provinces, Dorypygidae and Solenopleuridae with the Pacific province, and Eodiscidae and Conocoryphidae with the Acado–Baltic. On the other hand, Dorypygidae are much less abundant in the *E. asaphoides* fauna than in the Pacific province faunas of northwestern Vermont, whereas the Protolenidae and Ellipsocephalidae, characteristic of the late Early Cambrian faunas of the Acado–Baltic province, are absent from the *E. asaphoides* fauna.

Although the *E. asaphoides* fauna cannot be assigned unequivocally to either the Acado–Baltic or Pacific provinces, it nevertheless does not consist of a mixture of Acado–Baltic and Pacific *species*. Hence it does not seem to have occupied an intermediate or transitional environment between the two provinces, falling within the limits of tolerance of species otherwise characteristic of one or the other. At the specific level, this fauna is not "mixed" or "transitional" in either an environmental or a taxonomic sense; it is "intermediate" only in a geometric sense in that it occupies space between the two major faunal provinces. I propose to characterize such faunas as *intermediate zone faunas*.

The apparent absence of any species in common with the Pacific fauna suggests that the barriers between the *E. asaphoides* and the Pacific faunas were effective at the specific level; the presence of *Serrodiscus speciosus* in the *E. asaphoides* and the Acado–Baltic faunas suggests that at the species level, the barrier between the two was a selective filter. Hence, it is probable that there was a stronger environmental affinity with the Acado–Baltic province than with the Pacific. However, the barrier also excluded many species and higher taxa, and the faunal differences do not seem to justify treating the *E. asaphoides* fauna as a subfacies of the Acado–Baltic province.

Rasetti (1966, p. 6) pointed out that the *Acimetopus bilobatus* faunule resembles the *Elliptocephala asaphoides* fauna most closely at the familial level, having abundant Eodiscidae and Olenellidae, and rare Dorypygidae in common. He assigned this faunule to the "intermediate realm."

The youngest, *Pagetia connexa*, *Pagetides elegans*, and *Paedeumias–Bonnia* assemblages show more definite Pacific affinities (Rasetti and Theokritoff, 1967, p. 192–193).

Rasetti (1967, p. 24–26) discussed the biogeographic affinities of the Taconic Middle Cambrian faunules. All include Pacific province genera; some also include Acado–Baltic genera.

Discussion of the biogeographic affinities of the Taconic Upper Cambrian faunas does not seem justified pending their description.

Correlation of the *Elliptocephala asaphoides* Fauna

The presence of Acado–Baltic elements in the *Elliptocephala asaphoides* fauna suggests correlation with Acado–Baltic faunas that include *Serrodiscus speciosus* or closely related species (Rasetti, 1967, p. 16–17). On the other hand, owing to the rudimentary state of knowledge of the ranges of many trilobite families in the Lower Cambrian, the available paleontological evidence cannot at present be resolved to show the exact chronological relationships of the *E. asaphoides* fauna with the late Early Cambrian Pacific province faunas.

The *Olenellus* faunas of northwestern Vermont are known to range from the Cheshire Quartzite, and possibly even the underlying Dalton Formation (Thompson, 1967, p. 71), up into the Monkton Quartzite and Parker Slate. The Cheshire Quartzite is now correlated with the Bomoseen Graywacke (Cady, this volume; Zen, 1967, p. 49).

The Elliptocephala asaphoides fauna ranges from conglomerates in the upper part of the Mettawee Slate, which overlies the Bomoseen Graywacke, into the West Castleton Formation. Discussion of the age span of the *E. asaphoides* fauna is complicated by the fact that it occurs in conglomerates as well as in bedded limestones. Its occurrence in bedded limestones and, albeit at only few localities, in shales of the West Castleton Formation indicates that these strata must be dated within the age span of the fauna (Theokritoff, 1964, p. 183). The undoubted lowest stratigraphic occurrence of this fauna is in conglomerates at the top of the Mettawee Slate. Because these conglomerates cannot be older than the fauna, and furthermore because these conglomerates underlie strata that are dated within the age span of this fauna, the fossiliferous conglomerates in the Metawee Slate, as well as their enclosing strata, must also be dated within the age span of the fauna. Hence the strata in which this fauna is found represent its known age span. However, as its lowest occurrence is in conglomerates, it is possible that its age span is

greater and also includes underlying strata; the probability that the Brezee Phyllite falls within the age span of this fauna has been discussed above.

The *Elliptocephala asaphoides* fauna is succeeded by the *Acimetopus bilobatus* faunule, and that, in turn, by the *Pagetides elegans*, *Pagetia connexa*, and *Paedeumias–Bonnia* faunules which are probably the latest Early Cambrian assemblages in the Taconic sequence. I correlate the *Pagetides elegans* and synchronous assemblages with the *Pagetides–Olenellus* subfauna of the lower Parker Slate. Hence, the *E. asaphoides* fauna is probably older than the *Pagetides–Olenellus* subfauna. As the *Pagetides–Olenellus* subfauna includes important elements of the underlying *Olenellus* subfauna, there is no break at this point that could be correlated with the *E. asaphoides* fauna. Hence, the *E. asaphoides* fauna cannot be younger than the *Olenellus* subfauna. The age spans of the two probably overlapped to a greater or lesser degree.

Habitat of the Taconic Shelly Faunas

Lochman (1956, p. 1331, 1363) considered that the Pacific province faunas of the shelf sequence were indigenous to a shallow "continent-dominate" shelf with sand and carbonate, and some mud, deposition. She further considered that the *Elliptocephala asaphoides* fauna was indigenous to a northward-elongated basin situated in this shelf and opening southward and southeastward to the ocean; this basin was the site of deposition of the predominantly shale Taconic sequence. The supposed differences between these two environments appear to be mainly related to the proximity of the shelf to the craton, the greater depth of the basin, and its connection with a postulated northward-flowing tropical ocean current, allowing the influx of warmer water.

Citing the fact that the Taconic shelly faunas occur more commonly in limestones than in shales, Theokritoff (1964, p. 183–184) suggested that the faunas were optimally adapted to sites of carbonate, rather than mud, deposition. Bird and Theokritoff (in press) carried this idea further, suggesting that the organisms were indigenous to a sand-carbonate shelf lying west of the site of deposition of the Taconic sequence and that they were carried in turbidity flows and slides into the sites of dominantly mud deposition. Although the Taconic shelly fossils occur typically in limestone conglomerates and bedded limestones, *Elliptocephala asaphoides* and *Serrodiscus speciosus* have been collected from the shales at a few localities. These may represent individuals carried by currents into areas of mud deposition. Alternatively, these few species may have had a limited adaptability to a mud environment; it does not seem probable that any of these were swimmers, independent of bottom conditions, because all are incomparably more abundant in limestone than in shale. On the other hand, *Atops trilineatus*, usually considered to belong to the *E. asaphoides* fauna, is known almost exclusively from the shales, and extremely rarely from limestones (Rasetti, 1967, p. 97). It may be a species adapted to the basin rather than the shallow-water shelf environment.

The *Acimetopus bilobatus* faunule, and the *Pagetia connexa*, *Pagetides elegans*, and *Paedeumias–Bonnia* assemblages occur in essentially the same manner as the *E. asaphoides* fauna, in limestones enclosed in shales.

The Taconic shelly faunas were indigenous to the shelf on which the shelf sequence of western New England was deposited. At least in the Early Cambrian, they must have occupied this self synchronously with the Pacific province faunas known from this shelf in northwestern and west-central Vermont, and Dutchess County, New York. Because the organisms were carried from the west into the site of deposition of the Taconic sequence, it appears that the Pacific faunas occupied the inner margin of the shelf and the Taconic the outer margin. However, Taconic faunas are not known from any part of the shelf sequence as it is exposed now; this can probably be accounted for partly by the subsequent erosion of the eastern margin of the shelf and partly by the destruction of fossils in this sequence by progressively greater metamorphism toward the east.

Paleogeography

Throughout the Cambrian, the area of New England lay across a portion of the eastern margin of North America. In its western portion, there was a sand-carbonate shelf, covered by shallow water and sloping steeply in an easterly direction into the Appalachian geosyncline, a basin of mud deposition. In its eastern portion, there were non-volcanic islands such as those in eastern Massachusetts, New Brunswick, and Nova Scotia (see Rodgers, this volume, Fig. 10–3). Although mud deposition was dominant around these islands, lime was also deposited.

Slides and turbidity currents carried lime and limestone fragments from the sand-carbonate shelf into the basin sites. Subsidence of the outer margin of the shelf probably resulted in westward migration of the loci of such slides and turbidity currents, as well as the encroachment of basin sediments onto the older shelf sediments. Such an encroachment occurred at the close of the Early Cambrian in northwestern Vermont where the Parker Slate rests on the shelf sequence; it is dated by the presence of the Parker Slate. However, mud had not at that time *Olenellus—Pagetides* subfauna in the lower encroached onto the shelf in the latitude of Rutland because limestone conglomerates bearing the *Pagetides elegans* and synchronous faunas are found in the shales of the Taconic sequence.

The fact that the *Elliptocephala asaphoides* fauna was for a long time the only Cambrian fauna known from the Taconic sequence, suggests that conglomerates and beds with the *E. asaphoides* fauna are much more numerous than conglomerates and beds with the other faunas. I infer that during the late Early Cambrian, the margin of the sand-carbonate shelf stood closest to and at maximum elevation above the site of deposition of the Taconic sequence.

Middle Cambrian faunas older than the *Bathyuriscus—Elrathina* zone are so far unknown in the Taconic sequence. Although further search may lead to their discovery, the fact that they are not so far known suggests that if they occur, they occur, at best, at only few localities. If so, the outer margin of the sand-carbonate shelf may have generally subsided, even in southern Taconic latitudes.

The early Middle Cambrian may be represented by unusually thin sections in the basin; at Judson Point, the section between a horizon with *Serrodiscus speciosus* and *Atops trilineatus* in shale of the West Castleton Formation and a horizon with *Centropleura* in turbidites measures only approximately 85 feet (Bird and Rasetti, in press). Such thin sections may be due to unconformable overlap and stratigraphic convergence related to geanticlines (Cady, 1967, p. 62–64; this volume; *in* Zen, 1967, p. 47) or deposition at the foot of a starved shelf (Rodgers, this volume; Zen, 1967, p. 47).

Biogeography and Biostratigraphy

By the late Early Cambrian, hard-shelled organisms were established in the New England area. The sand-carbonate shelf was occupied by Pacific province faunas in northwestern Vermont, and farther south, by Pacific province faunas along its cratonic side and by the intermediate zone *Elliptocephala asaphoides* fauna along its outer margin. The basin adjacent to this part of the shelf may have been occupied by some members of the *Elliptocephala asaphoides* fauna, in particular by the Acado–Baltic *Atops trilineatus*. Acado–Baltic faunas with *Callavia* were established around the islands in eastern Massachusetts and elsewhere in the Maritime Provinces, as well as in shelf environments on the Fenno–Scandian craton (Henningsmoen, 1956; Opik, 1956).

It is not clear whether Acado–Baltic or any other faunas were established in parts of the basin distant from the shelf and islands. In north-central Maine, no crystalline Precambrian basement is exposed at the base of a thick Early Paleozoic sequence that includes volcanics; the lowest unit assigned to the Paleozoic sequence, named the Grand Pitch Formation, underlies strata with late Early or earliest Middle Ordovician fossils (Neuman, 1964; Neuman and Rankin, *in* Caldwell, 1966, p. 14; Whittington *in* Neuman, 1964) and includes gray, green, and red slate and siltstone lithologically most like strata referred to the Weymouth Formation in eastern Massachusetts. If the Grand Pitch Formation is Cambrian, it represents Cambrian eugeosynclinal sedimentation. If so, the absence from it of body fossils suggests that during the Cambrian, the eugeosyncline was not populated by significant numbers of hard-shelled organisms, or that conditions were not favorable for preservation.

At the close of the Early Cambrian, that part of the basin that encroached onto the shelf in northwestern Vermont was occupied by Pacific province faunas represented by those of the lower Parker Slate. At approximately the same time, the biofacies of that part of the outer margin of the shelf earlier occupied by the *Elliptocephala asaphoides* fauna had evolved to the Pacific type, although by the late Middle Cambrian, this outer shelf margin was again occupied by intermediate zone faunas.

In the Middle Cambrian of northwestern Vermont, the strongest Acado–Baltic affinity is seen in the presence of *Centropleura* and *Meneviella* in the St. Albans Slate.

In New England, during the Early and Middle Cambrian, Pacific province faunas occupied sites on the shelf and adjacent parts of the basin; intermediate zone faunas occupied sites on the outer margin of the shelf and, to a lesser extent, adjacent parts of the basin; Acado–Baltic faunas occupied sites around the islands in eastern Massachusetts. Thus, whatever factors may have determined biogeography, they do not appear to have been closely related to gross lithofacies or to tectonic regimen.

REFERENCES

Berry, W. B. N., 1961, Graptolite fauna of the Poultney Slate: Am. Jour. Science, v. 259, p. 223–228

Billings, M. P., 1929, Structural geology of the eastern part of the Boston basin: Am. Jour. Science, v. 218, p. 97–137

Bird, J. M., 1963. Generalities of the Lower Cambrian stratigraphy, Rensselaer County, New York, p. 118 *in* The Geological Society of America Abstracts for 1962: Geol. Soc. America Special Paper 73, 355 p.

————, *ed.*, 1963, Stratigraphy, structure, sedimentation, and paleontology of the southern Taconic region, eastern New York: Geol. Soc. America Guidebook for field trip 3, Albany, New York, 67 p.

————, and Rasetti, Franco, in press, Lower, Middle, and Upper Cambrian faunas in the Taconic sequence of eastern New York: Stratigraphic and biostratigraphic significance: Geol. Soc. America Spec. Paper 113.

————, and Theokritoff, George, in press, Mode of occurrence of fossils in the Taconic allochthon, *in* The Geological Society of· America Abstracts for 1966: Geol. Soc. America Special Paper

Burr, H. T., 1900, A new Lower Cambrian fauna from eastern Massachusetts: Am. Geologist, v. 25, p. 42–50

Cady, W. M., 1945, Stratigraphy and structure of west-central Vermont: Geol. Soc. America Bull., v. 56, p. 515–588

————, 1960, Stratigraphic and geotectonic relationships in northern Vermont and southern Quebec: Geol. Soc. America Bull., v. 71, p. 531–576

————, 1967, Geosynclinal setting of the Appalachian Mountains in southeastern Quebec and northwestern New England: *in* Clark, T. H., *ed.*, Appalachian Tectonics: Roy. Soc. Canada Special Pub. 10, p. 57–68

————, Albee, A. L., and Chidester, A. H., 1963, Bedrock geology and asbestos deposits of the upper Missisquoi Valley and vicinity, Vermont: U. S. Geol. Survey Bull. 1122–B, 78 p.

Caldwell, D. W., *ed.*, 1966, The Mount Katahdin region, Maine: Guidebook to 58th Ann. Mtg. of New England Intercollegiate Geol. Conf., Katahdin, Me., 64 p.

Clark, M. G., and Shaw, A. B., 1968, Paleontology of northwestern Vermont. XV. Trilobites of the Upper Cambrian Gorge Formation (lower part of bed 3): Jour. Paleontology v. 42, p. 382–396

————, in press, Paleontology of northwestern Vermont, XVI. Trilobites of the Upper Cambrian Gorge Formation (upper bed 3): Jour. Paleontology

Clark, T. H., 1923, New fossils from the vicinity of Boston: Boston Soc. Nat. Hist. Proc., v. 36, p. 473–485

————, 1924, The paleontology of the Beekmantown Series at Lévis, Quebec: Bull. Am. Paleontology, v. 10, no. 41, 134 p.

Cloud, P. E., and Nelson, C. E., 1966, Phanerozoic–Cryptozoic and related transitions: new evidence: Science, v. 154, p. 766–770

Dennis, J. G., 1964, The geology of the Enosburg area, Vermont: Vt. Geol. Survey Bull. 23, 56 p.

Doll, C. G., *ed.*, 1961, Vermont geologic map centennial: Guidebook to 53rd Ann. Mtg. of New England Intercollegiate Geol. Conf., Montpelier, Vt.

Doll, C. G., Cady, W. M., Thompson, J. B. Jr., and Billings, M. P., *compilers* and *editors*, 1961, Centennial geologic map of Vermont: Vermont Geol. Survey, Montpelier, Vt., scale 1:250,000

————, 1963, Reply to Zen's discussion of the Centennial geologic map of Vermont: Am. Jour. Science, v. 261, p. 94–96

Dowse, A. M., 1950, New evidence on the Cambrian contact at Hoppin Hill, North Attleboro, Massachusetts: Am. Jour. Science, v. 248, p. 95–99

Emerson, B. K., 1917, Geology of Massachusetts and Rhode Island: U. S. Geol. Survey Bull. 597, 289 p.

Fisher, D. W., 1956, The Cambrian System of New York State, p. 321–351 in Internat. Geol. Congress, 20th Mexico, 1956, El Sistema Cámbrico, su paleogeografía y el problema de su base, Symposium part 2: 762 p.

————, 1962, Correlation of the Cambrian rocks in New York State: N. Y. State Mus. and Science Service Geol. Survey, Map and Chart series: no. 2

————, 1964, Anatomy of the Past: The Empire State Geogram, v. 3, no. 1, p. 1–2

Foerste, A. F., 1889, The paleontological horizon of the limestone at Nahant, Mass.: Boston Soc. Nat. Hist. Proc., v. 24, p. 261–263

Grabau, A. W., 1900, Paleontology of the Cambrian terranes of the Boston basin: Boston Soc. Nat. Hist. Occasional Papers v. 4, pt. 3, p. 601–694

Henningsmoen, Gunnar, 1956, The Cambrian of Norway, p. 45–57: in Internat. Geol. Congress, 20th Mexico, El Sistema Cámbrico, su paleogeografía y el problema de su base, Symposium, part 1: 435 p.

Howell, B. F., 1937, Cambrian *Centropleura vermontensis* fauna of northwestern Vermont: Geol. Soc. America Bull., v. 48, p. 1147–1210

————, Shimer, H. W., and Lord, G. S., 1936, New Cambrian *Paradoxides* fauna from eastern Massachusetts: Geol. Soc. America Proceedings for 1935, p. 385

———— et al., 1944, Correlation of the Cambrian formations of North America: Geol. Soc. America Bull., v. 55, p. 993–1003

Hupé, Pierre, 1953, Sur les affinities des trilobites du Cambrien Inferieur Marocain: Le Paléozoique Nord-Africain et ses corrélations avec celui des autres régions du monde: Internat. Geol. Congress, 19th Algiers, comptes rendus, sec. 2, fasc. 2, p. 41–48

————, 1960, Sur le Cambrien Inférieur du Maroc: Late Pre-Cambrian and Cambrian stratigraphy: Internat. Geological Congress, 21st Copenhagen rept., pt. 8, p. 75–85

Hutchinson, R. D., 1962, Cambrian stratigraphy and trilobite faunas of southeastern Newfoundland: Geol. Survey Canada Bull. 88, 156 p.

Keith, Arthur, 1932, Stratigraphy and structure of northwestern Vermont: Washington Acad. Sci. Jour., v. 22, p. 357–379, 393–406

Kindle, C. H., and Tasch, Paul, 1948, Lower Cambrian fauna of the Monkton Formation of Vermont: Canadian Field-Naturalist, v. 62, p. 133–139

Knopf, E. B., 1962, Stratigraphy and structure of the Stissing Mountain area, Dutchess County, New York: Stanford University Publications, Geol. Sciences, v. 7, no. 1, 55 p.

Lochman, Christina, 1956, Stratigraphy, paleontology, and paleogeography of the *Elliptocephala asaphoides* strata in Cambridge and Hoosick quadrangles, New York: Geol. Soc. America Bull., v. 67, p. 1331–1396

————, and Wilson, J. L., 1958, Cambrian biostratigraphy in North America: Jour. Paleontology, v. 32, p. 312–350

Matthew, G. F., 1899, The Etcheminian fauna of Smith Sound, Newfoundland: Roy. Soc. Canada Trans., ser. 2, v. 5, sec. 4, p. 97–119

Neuman, R. B., 1964, Fossils in Ordovician tuffs in northeastern Maine: U. S. Geol. Survey Bull. 1181-E, 38 p.

Opik, A. A., 1956, Cambrian (Lower Cambrian) of Estonia, p. 97–126 in Internat. Geol. Congress, 20th Mexico, El Sistema Cámbrico, su paleogeografía y el problema de su base, Symposium, part 1: 435 p.

Osberg, P. H., 1965, Structural geology of the Knowlton-Richmond area, Quebec: Geol. Soc. America Bull., v. 76, p. 223–250

Rasetti, Franco, 1944, Upper Cambrian trilobites from the Lévis conglomerate: Jour. Paleontology, v. 18, p. 229–258

————, 1945, New Upper Cambrian trilobites from the Lévis conglomerate: Jour. Paleontology, v. 19, p. 462–478

————, 1946, Revision of some late Upper Cambrian trilobites from New York, Vermont, and Quebec: Am. Jour. Science, v. 244, p. 537–546

————, 1948, Lower Cambrian trilobites from the conglomerates of Quebec: Jour. Paleontology, v. 22, p. 1–24

————, 1952, Revision of the North American trilobites of the family Eodiscidae: Jour. Paleontology, v. 26, p. 434–451

————, 1963, Additions to the Upper Cambrian fauna from the conglomerate boulders at Lévis, Quebec: Jour. Paleontology, v. 37, p. 1009–1017

————, 1966, New Lower Cambrian trilobite faunule from the Taconic sequence of New York: Smithsonian Misc. Coll., v. 148, no. 9, 52 p.

————, 1967, Lower and Middle Cambrian trilobite faunas from the Taconic sequence of New York: Smithsonian Misc. Coll., v. 152, no. 4, 111 p.

————, and Theokritoff, George, 1967, Lower Cambrian agnostid trilobites of North America: Jour. Paleontology, v. 41, p. 189–196

Raymond, P. E., 1924, New Upper Cambrian and Lower Ordovician trilobites from Vermont: Boston Soc. Nat. Hist. Proc., v. 37, p. 389–466

————, 1937, Upper Cambrian and Lower Ordovician Trilobita and Ostracoda from Vermont: Geol. Soc. America Bull., v. 48, p. 1079–1146

Resser, C. E., 1937, Elkanah Billings' Lower Cambrian trilobites and associated species: Jour. Paleontology, v. 11, p. 43–54

————, and Howell, B. F., 1938, Lower Cambrian *Olenellus* zone of the Appalachians: Geol. Soc. America Bull., v. 49, p. 195–248

Rhodes, E. J., and Graves, W. H. Jr., 1931, A new Cambrian locality in Massachusetts: Am. Jour. Science, v. 222, p. 364–372

Schuchert, Charles, 1937, Cambrian and Ordovician of northwestern Vermont: Geol. Soc. America Bull., v. 48, p. 1001–1078

Shaler, N. S., and Foerste, A. F., 1888, Preliminary descriptions of North Attleboro fossils: Mus. Comp. Zool. Harvard Coll., Bull., Geol. Series, v. 2, p. 27–41

Shaler, N. S., Woodworth, J. B., and Foerste, A. F., 1899, Geology of the Narragansett Basin: U. S. Geol. Survey Mon. 33, 402 p.

Shaw, A. B., 1950, A revision of several Early Cambrian trilobites from eastern Massachusetts: Jour. Paleontology, v. 24, p. 577–590

__________, 1951, The paleontology of northwestern Vermont. I. New Late Cambrian trilobites: Jour. Paleontology, v. 25, p. 97–114

__________, 1952, Paleontology of northwestern Vermont. II. Fauna of the Upper Cambrian Rockledge Conglomerate near St. Albans: Jour. Paleontology, v. 26, p. 458–483

__________, 1953, Paleontology of northwestern Vermont. III. Miscellaneous Cambrian fossils: Jour. Paleontology, v. 27, p. 137–146

__________, 1954, Lower and lower Middle Cambrian faunal succession in northwestern Vermont: Geol. Soc. America Bull., v. 65, p. 1033–1046

__________, 1955a, The paleontology of northwestern Vermont. IV. A new trilobite genus: Jour. Paleontology, v. 29, p. 187

__________, 1955b, Paleontology of northwestern Vermont. V. The Lower Cambrian fauna: Jour. Paleontology, v. 29, p. 775–805

__________, 1957a, Paleontology of northwestern Vermont. VI. The early Middle Cambrian fauna: Jour. Paleontology, v. 31, p. 785–792

__________, 1957b, Paleontology of northwestern Vermont. VII. The Lower Cambrian fauna (corrections and addendum): Jour. Paleontology, v. 31, p. 812–814

__________, 1958, Stratigraphy and structure of the St. Albans area, northwestern Vermont: Geol. Soc. America Bull., v. 69, p. 519–567

__________, 1961, Cambrian of south-eastern and northwestern New England, p. 433–471 *in* Internat. Geol. Congress, 20th Mexico, 1956, Kembriyskaya sistema, yeye paleogeografiya i problema nizhney granitsy (The Cambrian system, its paleogeography and the problem of its lower boundary), Symposium, part 3, Western Europe, Africa, U.S.S.R., Asia, America: Moscow, Akad. Nauk U.S.S.R., 518 p.

__________, 1962a, Paleontology of northwestern Vermont. VIII. Fauna of the Hungerford Slate: Jour. Paleontology, v. 36, p. 314–321

__________, 1962b, Paleontology of northwestern Vermont. IX. Fauna of the Monkton Quartzite: Jour. Paleontology, v. 36, p. 322–345

__________, 1966a, Paleontology of northwestern Vermont. X. Fossils from the (Cambrian) Skeels Corners Formation: Jour. Paleontology, v. 40, p. 269–295

__________, 1966b, Paleontology of northwestern Vermont. XI. Fossils from the Middle Cambrian St. Albans Shale: Jour. Paleontology, v. 40, p. 843–858

__________, and Clark, M. G., 1968, Paleontology of northwestern Vermont. XIV. Type section of the Upper Cambrian Gorge Formation: Jour. Paleontology, v. 42, p. 374–381

Shimer, H. W., 1907, An almost complete specimen of *Strenuella strenua* (Billings): Am. Jour. Science, v. 173, p. 199–201, 319

Stone, S. W., and Dennis, J. G., 1964, The geology of the Milton quadrangle, Vermont: Vt. Geol. Survey Bull. 26, 79 p.

Tasch, Paul, 1949, A new fossil locality in the Lower Cambrian Monkton Formation of Vermont: Canadian Field-Naturalist, v. 63, p. 210–211

Theokritoff, George, 1959, Taconic sequence in northern Washington County, New York (abstract): Geol. Soc. America Bull., v. 70, p. 1686–1687

__________, 1964, Taconic stratigraphy in northern Washington County, New York: Geol. Soc. America Bull., v. 75, p. 171–190

Thompson, J. B. Jr., 1967, Bedrock Geology of the Pawlet Quadrangle, Vermont, part II: Eastern Portion: Vt. Geol. Survey, Bull. 30, p. 61–98

Walcott, C. D., 1884, On the Cambrian faunas of North America Preliminary studies: U. S. Geol. Survey Bull. 10, 72 p.

__________, 1886, Second contribution to the studies of the Cambrian faunas of North America: U. S. Geol. Survey Bull. 30, 369 p.

__________, 1890, The fauna of the Lower Cambrian or *Olenellus* zone: U. S. Geol. Survey 10th Ann. Rept., pt. 1, p. 509–763

__________, 1910, Cambrian geology and paleontology, no. 6—*Olenellus* and other genera of the Mesonacidae: Smithsonian Misc. Coll., v. 53, no. 6, 446 p.

__________, 1912, Cambrian Brachiopoda: U. S. Geol. Survey Mon. 51, pt. 1, 872 p., pt. 2, 363 p.

Wheeler, R. R., 1942, New mid-Cambrian ptychoparid: Am. Jour. Science, v. 240, p. 567–570

Whittington, H. B., 1957, Ontogeny of *Elliptocephala, Paradoxides, Sao, Blainia,* and *Triarthrus* (Trilobita): Jour. Paleontology, v. 31, p. 934–946

__________, 1966, Phylogeny and distribution of Ordovician trilobites: Jour. Paleontology, v. 40, p. 696–737

Zen, E-an, *ed.*, 1959, Stratigraphy and structure of west central Vermont and adjacent New York: Guidebook to 51st Ann. Mtg. of New England Intercollegiate Geol. Cong., Rutland, Vt., 87 p.

__________, 1961, Stratigraphy and structure at the north end of the Taconic range in west-central Vermont: Geol. Soc. America Bull., v. 72, p. 293–338

__________, 1967, Time and space relationships of the Taconic allochthon and autochthon: Geol. Soc. America Special Paper 97, 107 p.

Chapter 2

Ordovician Paleogeography of New England and Adjacent Areas Based on Graptolites

WILLIAM B. N. BERRY

INTRODUCTION

GRAPTOLITES HAVE LONG been known from the Lower Paleozoic succession in New England and adjacent areas of New York state and Canada. Recent mapping has resulted in many new Ordovician graptolite collections, which, although commonly isolated geographically and stratigraphically, may be used to date relatively precisely the rocks from which they come. The correlations that emerge have led to the recognition of some broad lithofacies patterns and changes in these patterns during the Ordovician.

GRAPTOLITE ZONES

In working with North American Ordovician graptolites, the author (Berry, 1960) recognized 15 graptolite zones (Table 2–1) in the Ordovician succession in the Marathon region, Texas. Studies of Ordovician graptolites in western Newfoundland (Kindle and Whittington, 1958), the Yukon, Canada (Jackson and Lenz, 1962; Jackson, 1966), the Great Basin (Ross and Berry, 1963), New York (Berry, 1962a; 1963a), and Quebec (Osborne and Berry, 1966; Osborne and Riva, 1966; Raymond, 1914; Clark, 1924), as well as unpublished data from the southern part of the Appalachians, have amplified knowledge of the species congregations by which each zone is identified and provided further data concerning the ranges of the species. The zonal succession itself has proved recognizable across North America. Because this is so, graptolite faunas collected in New England and adjacent areas have been compared and correlated with the congregations considered diagnostic of each zone and a zonal assessment made.

Berry's (1966) graptolite zones are time-stratigraphic units established following the procedures for zonation established by Oppel (1856–1858) in his work with Jurassic ammonites. The recognition of each zone is based upon a congregation of species identified by analyzing the overlapping ranges of species determined by collecting from many stratigraphic successions. Zone boundaries delimit time intervals, each zone defining a corresponding time span. That time span extends from the base of one zone, chosen at appearances of certain new species, to those appearances of species selected to designate the base of the next succeeding zone.

Ordovician graptolite-bearing rocks have been correlated with those bearing shelly fossils as indicated in Table 2–1 and discussed by Whittington (this volume). These correlations have been used in the delination of the lithofacies belts for four time intervals within the Ordovician (Figs. 2–1 to 2–4).

THE CANADIAN—ZONES 1–6 (FIGURE 2–1)

Shale Belt

Graptolites of Canadian age (Zones 1–6) have not been widely found in the New England states and adjacent regions; however, the area does include one of the world's best known successions of Early Ordovician graptolites in the Levis Shale which crops out across the St. Lawrence River from Quebec City. Study of this succession by Raymond (1914), Clark (1924), Osborne and Berry (1966), Osborne and Riva (1966), and Osborne (unpublished data) has demonstrated that a succession of shales and argillites was de-

TABLE 2-1. The North American Ordovician graptolite Zones and their correlation with the Series and Stages based on shelly fossils, after Ross and Berry (1963, Table 1) and Berry (*in* Bird, 1963). Correlation of the American and European Lower Ordovician shelly fossil Series, Stages, and Zones is discussed by Whittington in this volume.

ORDOVICIAN SHELLY FOSSIL SERIES AND STAGES	NORTH AMERICAN ORDOVICIAN GRAPTOLITE ZONES
Richmond	15 *Dicellograptus complanatus ornatus*
Maysville Eden	14 *Orthograptus quadrimucronatus*
Trenton	13 *Orthograptus truncatus intermedius*
Wilderness	12 *Climacograptus bicornis*
Porterfield	11 *Nemagraptus gracilis*
Ashby Marmor	10 *Glyptograptus euglyphus* (or *G.* cf. *G. teretiusculus*)
Whiterock	9 *Paraglossograptus etheridgei* 8 *Isograptus* 7 *Didymograptus bifidus*
Canadian Series	6 *Didymograptus protobifidus* 5 *Tetragraptus fruticosus* 3 & 4-branched 4 *Tetragraptus fruticosus* 4-branched 3 *Tetragraptus approximatus* 2 *Clonograptus-Adelograptus* 1 *Anisograptus* (with *Dictyonema flabelliforme* as a lower subzone)

posited continuously from the Late Cambrian through the Canadian into the Middle Ordovician. The graptolite zones listed in Table 2–1 as correlative with the Canadian have all been recognized in the Levis Shale.

Bulman (1950) described a zone 1 graptolite fauna from near Matane (the Matane Shale), northeast of the Levis Shale exposures. Recent collections from shales exposed in the vicinity of Matane include graptolites indicative of correlation within the span of graptolite zones 1 through 6 (F. F. Osborne, oral communication, 1965). The Cap Rosier Shale bears graptolites similar to those described by Bulman (1950) from the Matane Shale as well as some tetragraptids indicative of an age within the span of zones 3–6 (McGerrigle, 1950). Graptolite-bearing shales and siltstones correlative with the Canadian have been recorded by McGerrigle (1959) from other localities in the northern part of the Gaspé Peninsula.

South from Quebec, the Poultney Slate bears graptolites indicating that it has an age span similar to the Levis Shale (Berry, 1961; *in* Bird, 1963). The Schaghticoke and Deepkill Shales in eastern New York bear graptolites indicative of correlation with zone 1 and zones 5–7, respectively (Berry, 1962a). The Germantown Formation, bearing Late Cambrian and zone 1 graptolites, and the Stuyvesant Falls Formation, bearing graptolites indicative of the age span of zones 2–6, have been recognized in the southern part of the Taconic area (Fisher, 1962; Berry, *in* Bird, 1963).

The graptolites from these formations, all of which are dominantly shale and argillite units, indicate that fine-grained clastics were deposited in a relatively linear belt extending from the southern part of the Taconic area in eastern New York northward and northeastward to Cape Rosier. Deposition in much of this belt was essentially continuous from the Late Cambrian throughout the Canadian into the Middle Ordovician (zones 7–11).

Limestone Belt

Limestones and dolomites were deposited to the west of the Shale Belt during the Canadian. These carbonates have commonly been lumped in

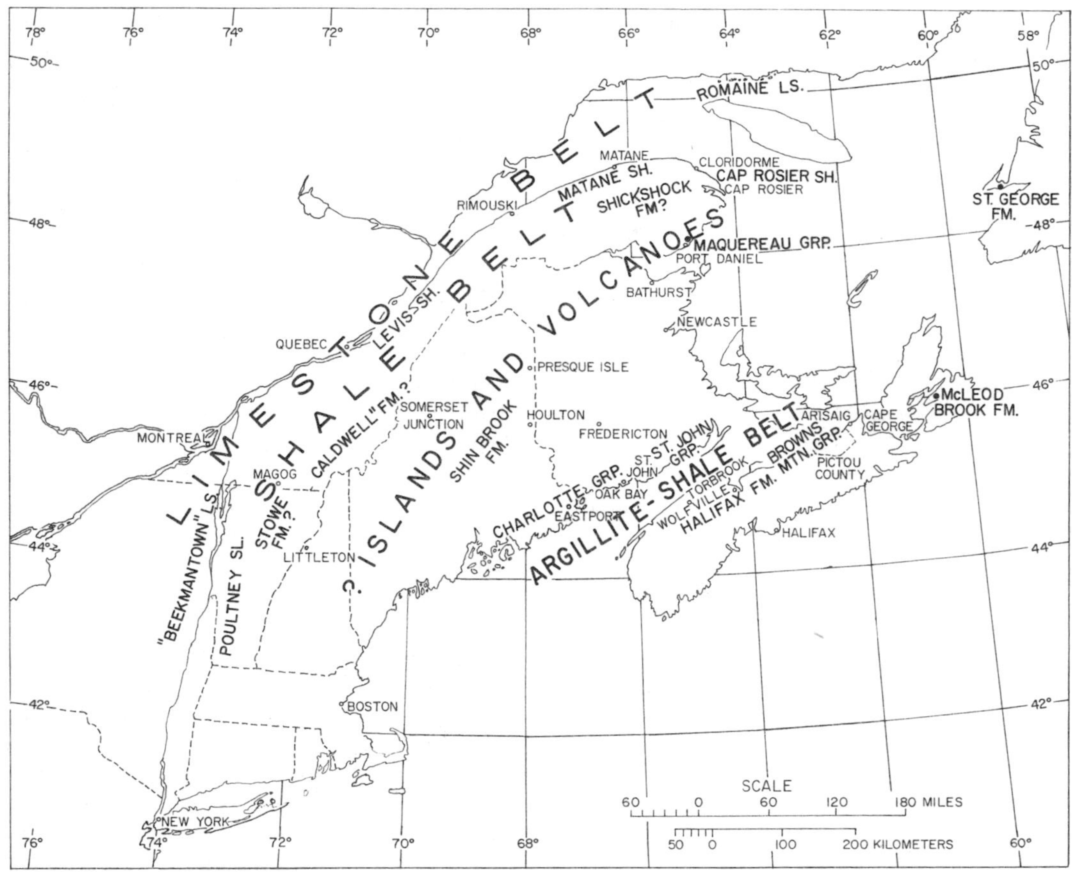

FIGURE 2–1. Canadian (zones 1–6) Lithofacies and Paleogeography. The Canadian Lithofacies and Paleogeographic patterns probably continued into the early Middle Ordovician Whiterock (at least zone 7) time.

the Beekmantown Group or referred to as Canadian carbonates.

Canadian-age carbonates have been recorded from southern Quebec (Clark, 1946; 1955) and from the Mingan Islands (the Romaine Limestone) (Twenhofel, 1926). The St. George Formation in western Newfoundland is dominantly carbonate and appears to be an extension of the carbonates that lay west of the Shale Belt during the Canadian. Cady (1960) discussed extension of the carbonates (here designated the Limestone Belt) through the Lake Champlain Valley area into adjoining parts of Quebec. Shaw (1958) discussed the trilobites from this belt in northwestern Vermont and also showed that the carbonates could be traced into Quebec.

The actual intergradation of the Canadian age carbonates with the shales and argillites of the same age has not been documented. It apparently occurs in a terrain that is highly faulted.

Some limestone breccias and conglomerates lie in the shales and argillites along the west side of the Shale Belt. These limestone boulders include those in the Levis Shale and the Mystic Conglomerate near Philipsburg, Quebec, and those in the Cow Head Group in western Newfoundland discussed by Kindle and Whittington (1958) and Whittington and Kindle (1963). The matrix shales and limestones of the Cow Head Group appear to be an extension of the Shale Belt. These blocks bear a dominantly trilobite fauna in contrast to the dominantly molluscan fauna of the Limestone Belt. These blocks and boulders may have been derived from local platforms within the Shale Belt or in the transition interval between the Limestone and Shale Belts (H. B. Whittington, oral communication, 1967).

Volcanoes and Islands Belt

Few fossils of Canadian age have been recovered from rocks exposed east from the Shale Belt to the costal regions of New England and New Brunswick. Following correlations suggested by Cady (1960), at least some of the Stowe Formation in northern Vermont (volcanics, quartzites, slates) and the Caldwell Formation in Quebec (including volcanic flows, slates, quartzites) may be Canadian in age. The Shickshock Formation (volcanics, arkose, quartzite) recognized by Crickmay (1932) in the Matapedia Valley in central Gaspé and the Maquereau Group (volcanics, quartzites, and arkoses) in the Chaleur Bay area near Port Daniel discussed by Alcock (1935) may also be, at least in part, Canadian in age. The Shin Brook Formation (Neuman, 1964; Neuman and Rankin, *in* Caldwell, 1966) bears brachiopods and trilobites that suggest a latest Canadian or earliest Middle Ordovician (early Whiterock) age (Neuman, 1964). The formation includes tuff, tuffaceous sandstone, flows, breccias, and conglomerate (Neuman, 1964). The predominance of volcanic rocks and the presence of arkoses, quartzites, breccias and conglomerates suggests (Fig. 2-1) that the terrain through central New England and adjacent parts of Quebec and New Brunswick was the site of various volcanic as well as non-volcanic islands. The islands served as sources for quartz- and feldspar-rich sediments during the Canadian. The presence of thick sequences of undated clastic and volcanic rocks lying stratigraphically beneath those of Middle Ordovician age in New Brunswick may support this suggestion.

Argillite-Shale Belt

Various argillites and shales exposed along the coastal regions of Maine and New Brunswick, as well as those on Cape Breton Island and Nova Scotia, bear graptolites indicative of correlation with the Canadian. Hutchinson (1952) noted the presence of *Dictyonema flabelliforme* (zone 1) and a slightly younger Ordovician fauna from the shales of the McLeod Brook Formation on Cape Breton Island. Crosby (1962) cited the occurrence in the Wolfville area of Nova Scotia of *Dictyonema flabelliforme* in the upper beds of the Halifax Formation (primarily shales), which has been recognized over much of the southern part of Nova Scotia. The Browns Mountain Group, which includes argillites, slates, and graywackes, and is exposed in Pictou County, Nova Scotia, is considered of possible Canadian age (McKerrow *in* Boucot et al., in press). Smith (*in* Poole, 1966, p. 6) noted that black shales of the Navy Island and Suspension Bridge Formations (the St. John Group) exposed near St. John, New Brunswick, bore *Dictyonema flabelliforme* and species of the genera *Tetragraptus* and *Didymograptus* commonly found in the span of zones 3–5. Smith (*in* Poole, 1966, p. 6) also

described a succession of slates, argillites and impure quartzites that are part of the Charlotte Group which is exposed south from St. John to the Eastport, Maine area. Beds assigned to the Charlotte Group exposed on Cookson Island in Oak Bay bear a clonograptid indicative of an early Canadian age (Smith, *in* Poole, 1966, p. 6). Smith (*in* Poole, 1966, p. 6) stated that the Charlotte Group probably included rocks of several ages but that the Cookson Island graptolite locality was the only known fossil locality in the group. Larrabee et al. (1965) described sediments similar to those of the Charlotte Group beneath Middle Ordovician graptolite-bearing rocks in the Grand Lake area of eastern Maine (Lat 45°00′–45°45′ N., Long. 67°30′–68°30′ W.), and noted the presence of the Charlotte Group in adjoining New Brunswick. Their data suggest the Charlotte Group may have been relatively widespread in eastern and coastal areas of Maine and New Brunswick.

The rocks that typify the McLeod Brook, Halifax, Browns Mountain, and Charlotte formations indicate that an Argillite–Shale Belt lay east of the Islands and Volcanoes Belt along the eastern part of New England and adjacent Canada during the Canadian. Sparse fossil evidence from this area prevents recognition of much of the geologic history of this belt.

EARLY MIDDLE ORDOVICIAN—ZONES 7–11 (Figure 2-2)

Shale Belt

Much of the time represented by graptolite zones 7 through 11 was apparently a time of widespread emergence and folding of most of New England and adjacent areas. However, evidence that deposition continued in at least a part of the Shale Belt exists in the record of Early through Middle Ordovician graptolites from the Levis Shale (zone 7, 9, 10, 11, and possibly zone 8). Graptolites from the upper part of the Deepkill Shale in New York (Berry, 1962a), from the Stuyvesant Falls Formation in New York (Berry *in* Bird, 1963), from the Stanbridge Formation in southern Quebec (Riva, 1966a), and from shales in the northern part of the Gaspé Peninsula (McGerrigle, 1959) are indicative of the *P. etheridgei* Zone (zone 9). Thus the graptolites of the Levis Shale, the zone 9 graptolites in other parts of the argillite-shale succession, and the apparent stratigraphic continuity of that succession suggest that deposition in the Shale Belt may have been continuous from the Canadian through the time span of zones 7 through 11.

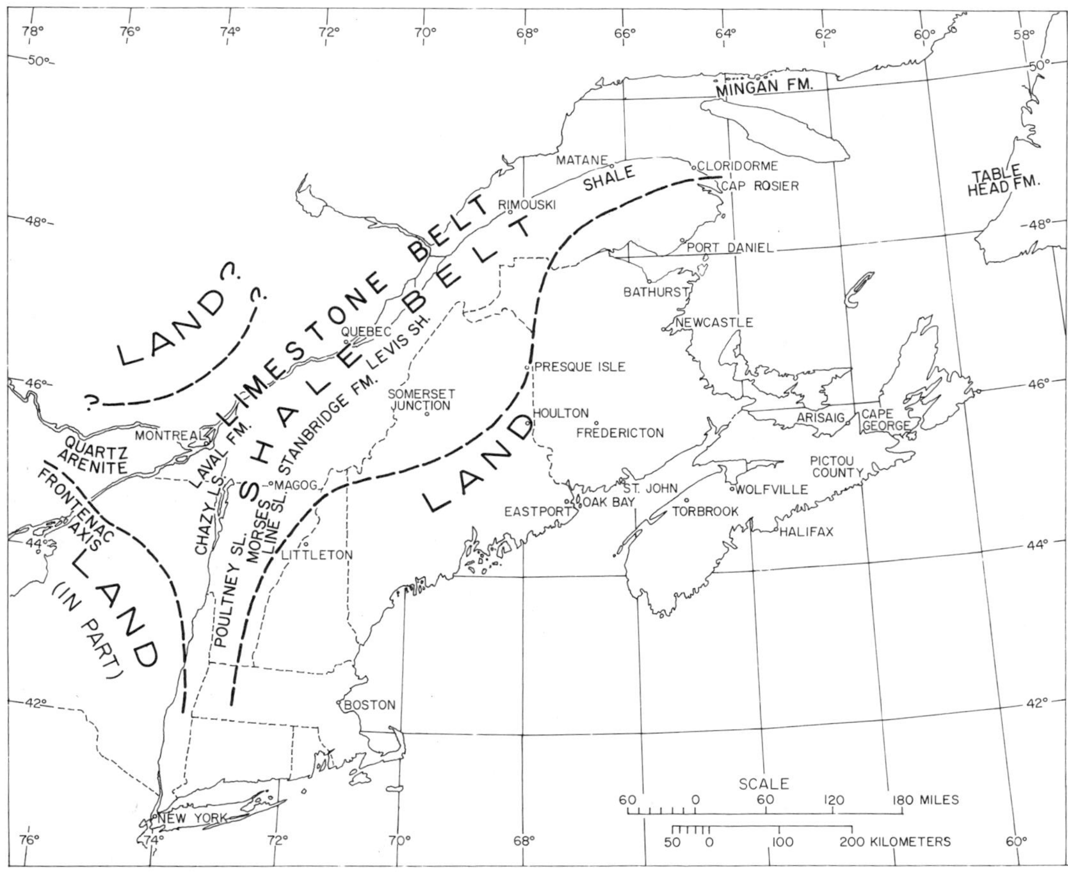

FIGURE 2–2. Early Middle Ordovician (zones 8–11) Lithofacies and Paleogeography. The patterns shown probably began to develop in post-early Whiterock (probably post-zone 7) time.

Limestone Belt

Carbonates included in the Chazy Group and correlative units were deposited west of the Shale Belt during most of the early Middle Ordovician. Hofmann (1963) discussed the Chazy and correlative units, indicating (Hofmann, 1963, Fig. 10) that rocks of this group crop out in a belt from the shores of Lake Champlain in Vermont north to an area extending from 53 kilometers east to 210 kilometers west of Montreal, and are present in the subsurface of the St. Lawrence valley. The carbonates of the Mingan Formation (Twenhofel, 1926) on the Mingan Islands are correlative with the Chazy Group. The presence of fossils of Whiterock age in boulders in the Mystic conglomerate (Clark and Eakins, this volume), and Flower's (1964, Fig. 53) correlation of the stratigraphically highest part of the Beekmantown Group (Providence Island Dolomite) with the Black Rock Formation in Arkansas which bears graptolites indicative of a possible zone 7 age (Berry, unpub-

lished data) suggests deposition of the Beekmantown carbonates continued into the early Middle Ordovician. These carbonate distributions indicate that a limestone belt lay west of the Shale Belt during at least a part of the time span of graptolite zones 7–11. Deposition in at least the western part of that belt was interrupted during late Whiterock (post-Providence Island Dolomite and pre-Chazy Group) as well as Ashby and Porterfield time (Cooper, 1956, correlation chart). As Cady (1960) indicated, the part of the early Middle Ordovician Limestone Belt that lies in northwestern Vermont and adjacent Quebec may have been the site of continuous carbonate deposition from the Canadian through the early Middle Ordovician. Zen (1967, p. 42, 64, Fig. 14) suggested that carbonate deposition may have been continuous through the early Middle Ordovician in local basins that were possibly grabens between submarine horsts within the Limestone Belt. Submarine erosion could have taken place over the horsts while deposition continued in the grabens.

Fisher (1954) demonstrated that an erosional surface with considerable relief developed upon carbonates of Canadian age in the Mohawk Valley in eastern New York, where units as young as Trenton (Barnveld age of Fisher, 1962) are locally in contact with Canadian carbonates. Cooper's (1956) brachiopod study indicated that the hiatus represented by this unconformity spans the Whiterock, Marmor, Ashby, and Porterfield. Land areas have been postulated (Fig. 2) northwest of Montreal and southwest from the Frontenac Axis after the studies of Hofmann (1963, Fig. 10) and Fisher (1954).

Folding, uplift, and erosion apparently took place over much of New England and adjacent Canada and New York state during the early Middle Ordovician inasmuch as some fossiliferous late Middle Ordovician formations, such as the Walloomsac Slate (Berry, *in* Bird, 1963, p. 28), unconformably overlie Canadian age units. Furthermore, units of late Middle Ordovician age, such as the Beauceville, Mictaw, and Moretown

(Fig. 2–3) include various types of conglomerates at the base. No fossils of definitely post-early Whiterock and pre-Wilderness age have been found east of the Shale Belt. The evidence for this area suggests that it underwent a period of tectonism, including folding and emergence of land, during at least a part of the early Middle Ordovician. The red shales and cherts at the base of the Normanskill Formation in New York may represent a redeposited fossil soil that developed on a part of such a land.

ZONE 12 TIME (Figure 2-3)

Beginning during zone 12 time, New England and adjacent parts of Canada became a relatively mobile area with platforms or ridges separating relatively deep troughs. Seas spread widely across terrain that had been land during the early Middle Ordovician. Many zone 12 graptolites have been found in New England and New Brunswick, permitting relatively precise correlation of the rocks bearing them.

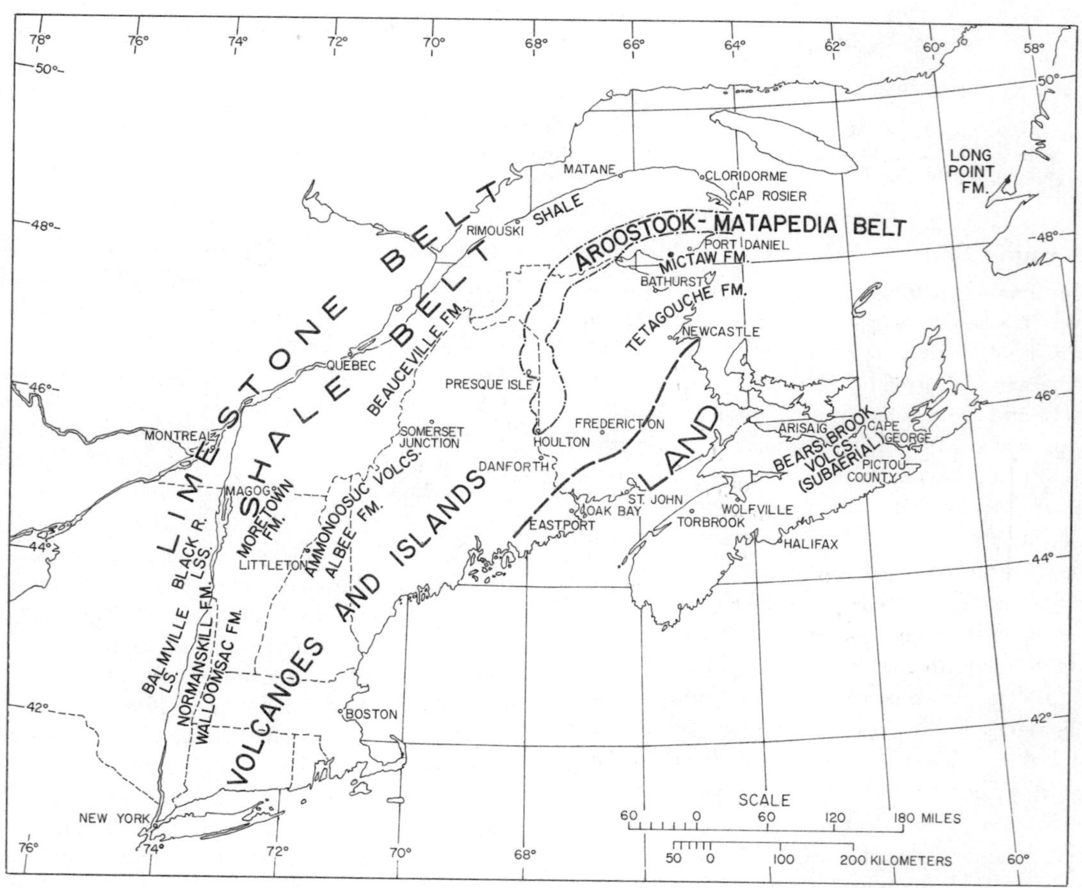

FIGURE 2–3. Zone 12 Lithofacies and Paleogeography.

Shale Belt

Deposition in the Shale Belt appears to have continued with little or no interruption from the early Middle into late Middle Ordovician, as suggested by the probable presence of zone 11 graptolites at two localities near Quebec (Osborne, and Riva, 1966; Riva, 1966b) and the proven presence of zone 12 graptolites in the sediments along the Etchemin River near Quebec (Riva, 1966b). These rocks may lie without hiatus on the youngest beds of the Levis Shale. Zone 12 graptolites are common in the Normanskill and Walloomsac Formations (Berry, 1962a; *in* Bird, 1963) in eastern New York and adjacent Vermont, the Magog and western exposures of the Beauceville Formations in Quebec (Berry, 1962b), and dark shales and argillites exposed near Rimouski, Quebec (Berry, unpublished data).

Limestone Belt

The shales and argillites of the Shale Belt grade laterally westward into the carbonates of the Black River Group. The Black River Group limestones have been traced through eastern New York (Fisher, 1962) northward into Quebec (Clark, 1946, 1955; Cady, 1960). The Balmville Limestone on the southern margin of the Taconic area in eastern New York and the Orwell Limestone on the northern margin of the Taconics in western Vermont, as well as correlative limestones east of the Taconics in eastern New York and adjacent Vermont, are also in part correlative with the Black River Group (Berry *in* Bird, 1963).

Cooper (1956, p. 12–13) indicated that the Long Point Formation in western Newfoundland is probably of Late Wilderness or slightly younger age. At least a part of the formation is thus probably zone 12 in age.

Volcanoes and Islands Belt

Shale Belt strata grade laterally eastward into thick sequences (approximately 1500–3000 meters) of argillites, graywackes, cherts, and various volcanic rocks. The westernmost exposures of the Beauceville Formation apparently have few if any volcanic flows and few tuffaceous beds, whereas the eastern exposures include many. The argillites and shales of the eastern exposures of the Beauceville Formation have yielded zone 12 graptolites. Billings (1937) described the Albee, Ammonoosuc and Partridge formations in west-central New Hampshire and considered them to be Ordovician in age. Since

his initial work, these units have been traced relatively widely in the northern part of New Hampshire and into adjacent parts of Maine (Billings, 1956; see Pavlides et al., Thompson et al., Naylor, and Green and Guidotti, this volume). Berry (*in* Harwood and Berry, 1967) identified zone 12 graptolites in a collection made by Harwood from rocks in west-central Maine that are the apparent lateral equivalents of the Partridge Formation. Zone 12 graptolites have also been found at many localities in northeastern Maine, as cited by Pavlides et al. (1964) and Neuman (1960). The Dunn Brook Formation near Houlton, Maine (Pavlides, 1964), which includes many volcanic rocks, as well as breccias, conglomerates, and slates, may be at least in part of zone 12 age. Zone 12 graptolites have also been found in the Shin Pond, Island Falls, Stacyville (Neuman and Rankin *in* Caldwell, 1966; Ekren, 1961), Millinocket Lake and Spider Lake (Hall, 1964; *in* Caldwell, 1966) quadrangles in north-central and northeastern Maine. Neuman (1963) cited shelly fossils and a graptolite, which may be indicative of zone 12 age, from tuffaceous mudstones and volcanic conglomerates west of Presque Isle, Maine. Larrabee et al. (1965) found zone 12 graptolites near Danforth, Maine. The Tetagouche Formation (or Group) (various volcanic rocks, slates, argillites, and sandstones), which has been mapped in the Bathurst–Newcastle area (Smith, 1957), in the Hayesville (Lat. 46°30′–46°50′N., Long. 66°30′–67°W.) and McNamee (Lat. 46°30′–46°45′N., Long. 66°15′–66°30′W.) map areas (Poole, 1960; 1963), and the Woodstock–Fredericton area (Anderson and Poole, 1959), bears zone 12 graptolites at many localities. Alcock (1935) mapped the Tetagouche and associated Fornier Group volcanics in the southern part of the Chaleur Bay region.

Near Somerset Junction, Maine, the Kennebec Formation bears brachiopods suggestive of a Middle Ordovician age (Boucot, 1961). The formation includes siltstones interbedded with massive felsite and rhyolite tuff. This formation may be of zone 12 age, or it may be somewhat older, as suggested by Neuman (this volume). If it is older, its presence could indicate that deposition in this Middle Ordovician Volcanoes and Islands Belt began somewhat earlier in some parts of it than in others. Cady (1960) suggested that at least some of the Beauceville Formation and related units rest without hiatus on older rocks. If this suggestion is supported by field relations, it would, with the evidence from the Kennebec Formation, indicate that deposition in a part of the Volcanoes and Islands Belt was continuous, or nearly so, from the Canadian into the late Middle Ordovician.

Alcock (1935) and Ayrton (1961) mapped the Mictaw Formation on the north side of Chaleur Bay in the vicinity of Port Daniel. Their studies indicated that the Mictaw Formation lies unconformably above the Maquereau and includes a basal conglomerate overlain by dark shale and graywacke. The Mictaw Formation bears graptolites indicative of zone 12 age and appears to lie on the margin of or perhaps outside of the Volcanoes and Islands Belt.

Aroostook–Matapedia Belt

The Volcanoes and Islands Belt is divided by a suite of rocks that includes slates, graywackes, conglomerates, siltstones, and shales, but not volcanics. This suite of rocks is the Chandler Ridge Formation of the Meduxnekeag Group (Pavlides et al., 1964). Other undated rocks of the same lithologic type and with the same stratigraphic relationship to the superjacent Carys Mills Formation of the Meduxnekeag Group are herein considered coeval with the Chandler Ridge Formation. These rocks occur in the same general belt through northeastern Maine and in New Brunswick as do the strata of the Carys Mills Formation. They and the Chandler Ridge Formation are thus considered to indicate the beginning of the Aroostook–Matapedia Belt. At least some of the Chandler Ridge Formation and the rocks considered here coeval with it are of probable zone 12 age because they conformably underlie the Carys Mills Formation, the basal part of which bears zone 13 graptolites.

Shales, slates, and phyllites without interbedded volcanics that lie east of the Aroostook–Matapedia Belt near Houlton, Maine, bear zone 12 graptolites. These rocks and the Mictaw Formation suggest the rocks of the Volcanoes and Islands Belt probably graded laterally into the rocks of the Aroostook–Matapedia Belt.

Eastern Land Area

A land mass apparently lay east of the Volcanoes and Islands Belt. The Bears Brook Group in Pictou County, Nova Scotia, includes volcanic rocks that were subaerially deposited on a part of this land (Dewey *in* Boucot et al., in press) and ignimbrites that contain lateritized surfaces (Dewey *in* Boucot et al., in press). Although not well dated, these rocks apparently range in age through much of the latter part of the Ordovician and so may be, in part, of zone 12 age. If some of the folded, undated rocks in eastern Nova Scotia prove to be zone 12 in age,

as proposed by J. F. Dewey (oral commun., 1967), then deposition may have taken place east of the suggested land mass during zone 12 time.

ZONE 13 TIME (FIGURE 2-4)

Shale Belt

The persistence of the Shale Belt into zone 13 time is demonstrated by the presence of zone 13 graptolites in the Magog and western parts of the Beauceville formations (Berry, 1962b) and in the subsurface in southern Quebec (Clark and Strachan, 1955). A land area in western Massachusetts and southwestern Vermont that probably began to rise in zone 12 time was a prominent feature early in zone 13 time. A submarine high may have developed as a northern extension of this land. Coarse-grained clastics (the Austin Glen Graywacke Member of the Normanskill Formation) were shed west from this land. Finer-grained clastics spread farther to the west, resulting in a westward gradation from coarse-grained clastics to somewhat finer grained clastics (the belt of siltstones and shales) to muds (the Utica and related shales).

During zone 13 time, some of the land area was apparently eroded down to a considerable degree because locally black shales succeed the coarse-grained clastics of the Austin Glen Member. During the time these black shales were being deposited, blocks that constitute the Taconic allochthon apparently slid from the submarine high west into the area of shale deposition. Locally the sliding appears to have taken place either after deposition of finer-grained materials resumed, or into areas where there was little graywacke in the Austin Glen Member. Locally, shales without slide blocks in them intervene between the graywackes of the Austin Glen and the shales with slide masses. Blocks do occur in many places, however, in the typical sediments of the Austin Glen.

Limestone Belt

Carbonates of early zone 13 age underlie the Utica and related shales in western Vermont and adjacent New York. The shales spread westward over the carbonates during zone 13 time with the result that the main body of the Trenton Group carbonates lies west of the Utica and related shales that bear zone 13 graptolites. Kay (1937) traced the lateral gradation of the shales into the carbonates. The distribution of the Trenton Group and related carbonates of zone 13 age in New York has been indicated by Fisher

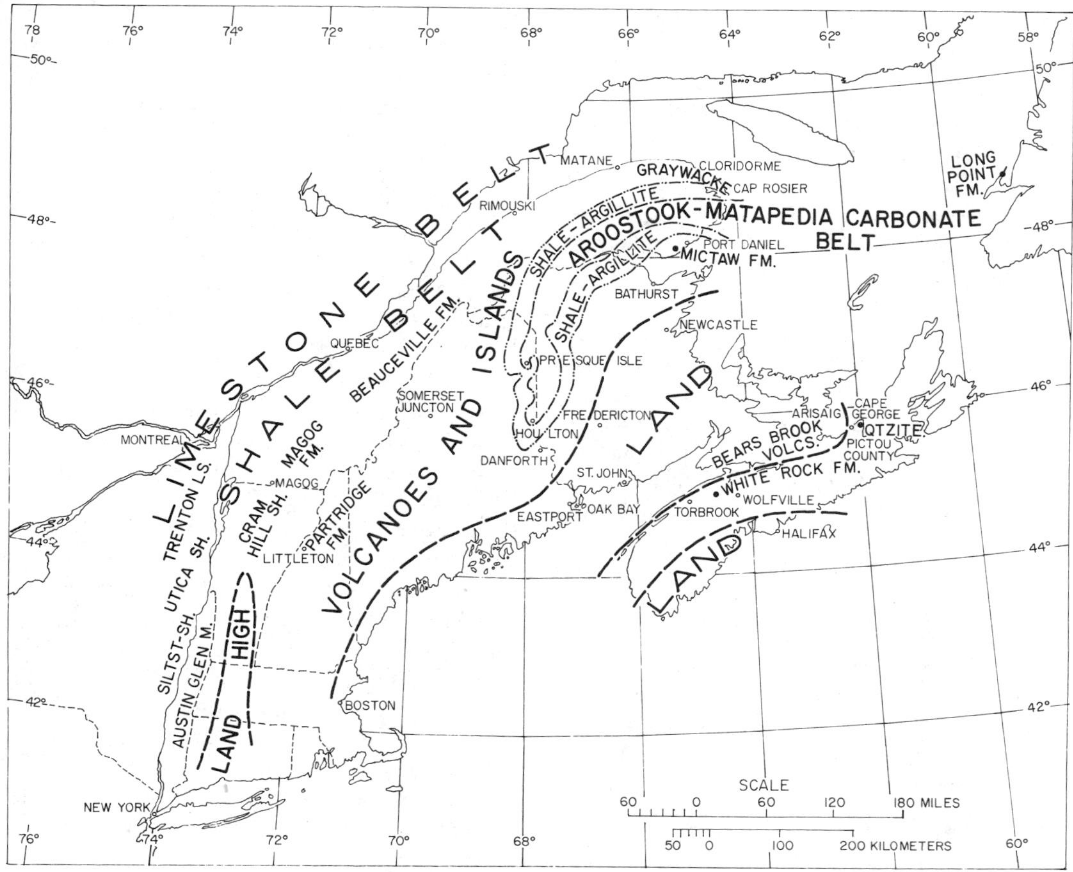

FIGURE 2–4. Zone 13 Lithofacies and Paleogeography. The relationships shown are for approximately early to middle zone 13 time.

(1962), and the rocks have been traced northward by Cady (1960) into Quebec, where Clark (1946, 1955) also recognized them.

Cooper (1956, p. 12–13) discussed the age of the Long Point Formation in western Newfoundland, and H. B. Whittington (oral communication, 1967) studied trilobites collected from beds probably belonging in the Long Point Formation. From these data, it appears that part of that formation may be of zone 13 age. Rodgers (1965) stated that the sediments of the Long Point were deposited after the western Newfoundland allochthon was emplaced, an event that somewhat preceded the emplacement of the Taconic allochthon.

Volcanoes and Islands Belt

The Volcanoes and Islands Belt continued to be a prominent feature during zone 13 time as evidenced by collections of zone 13 graptolites which have been obtained from the thick successions of volcanic rocks and associated argillites, cherts, slates, and graywackes that characterize the belt. Boucot et al. (1964), Hall (1964), Pavlides et al. (1964), and Neuman and Rankin (*in* Caldwell, 1966) cited occurrences of zone 13 graptolites in northeastern Maine and summarized the stratigraphy of the belt there. Some of the Partridge and Tetagouche Formations (see Fig. 2–3) may be zone 13 in age although no definitive zone 13 graptolite associations have been recorded from them.

Paul Enos (written communication, 1963) recognized a succession of graywackes near Cloridorme in the northern Gaspé Peninsula from which zone 13 graptolites were obtained. These graywackes were probably derived from a nearby source.

Eastern Land Area

A land area apparently continued to exist during zone 13 time east of the Volcanoes and Islands Belt. The volcanics of the Bears Brook Group probably include some rocks of zone 13 age.

Aroostook–Matapedia Carbonate Belt

The deposits of zone 13 age in the Aroostook–Matapedia Belt are characterized by gray limestones, calcareous siltstones, and tan weathering ankeritic beds. These rocks are included in the Meduxnekeag and Matapedia Groups. Zone 13 graptolites have been collected from the Carys Mills Formation (the Ribbon Rock Member of the Meduxnekeag Group of Pavlides et al., 1961). The dominantly carbonate rocks of the Carys Mills Formation may be traced in an embayment-like belt along the Aroostook–Matapedia anticlinorium in northeastern Maine (Pavlides et al., 1964), across western New Brunswick, and along the south side of the Gaspé Peninsula, Quebec (Pavlides and Berry, 1966, p. B53). The carbonates of this belt contrast with the volcanics and graywackes of the Volcanoes and Islands Belt. The rocks of the two belts apparently grade one into the other through a narrow band of argillites, siltstones, and shales, but without volcanics. The Mictaw Formation in the Port Daniel area, dominantly graywackes and shale with a decrease in grain size toward the west (Ayrton, 1961, p. 5), apparently lies in the area of gradation between the Aroostook–Matapedia and Volcanoes and Islands Belts. Zone 13 graptolites have been identified in collections from the Mictaw Formation.

White Rock Formation

The White Rock Formation has been mapped in the Torbrook (Smitheringale, 1960) and Wolfville (Crosby, 1962) areas in western Nova Scotia. The formation has not yielded age-diagnostic fossils but it lies stratigraphically above the Halifax Formation of Early Canadian age and beneath the Late Silurian (Early Ludlow) Kentville Formation. Both Crosby (1962, p. 23–24) and Smitheringale (1960, p. 15) noted that the contact between the White Rock and Halifax formations was apparently conformable. The White Rock Formation is dominantly quartzite, but includes slates in the Wolfville area (Crosby, 1962, p. 23) and some volcanic rocks (a "basalt member" and a "rhyolite member") in the Torbrook area (Smitheringale, 1960, p. 13–14).

The age of the White Rock Formation is in doubt. It may be in part Silurian in age, but Boucot et al. (1959) recorded some brachiopods suggestive of a late Middle or Late Ordovician age from quartzite boulders exposed on Cape George, Nova Scotia. Those quartzites could be zone 13 age. If they are, and if their deposition was related to the quartzites in the White Rock Formation, then at least some of the White Rock Formation may be zone 13 in age. The volcanics in the White Rock Formation may be coeval with some of the Bears Brook volcanics.

LATE ORDOVICIAN—ZONES 14 AND 15

During the Late Ordovician, the land area from which the Austin Glen sediments were derived may have been the source terrain for at least some of the Late Ordovician age coarse-grained clastics that comprise the Queenston Delta of New York State. Most of the Shale Belt and that part of the Volcanoes and Islands Belt west of the Aroostook–Matapedia Carbonate Belt became a land mass and a source for the clastic sediments of the Lorraine Group, the Pontgrave River Formation (gray shales) and Becanour River Formation (red sandstones and shales) in southern Quebec.

Deposition continued without interruption in the Aroostook–Matapedia Carbonate Belt from zone 13 time through the Late Ordovician into the Early Silurian, as shown by zone 15 and Early Silurian graptolies found in the limestones of that belt. The sediments of that belt continued to be dominantly carbonates.

Much of the terrain east from that Belt to western Nova Scotia may have been land. No Late Ordovician age fossils have been found in that area, with the possible exception of the brachiopods from Cape George (Boucot et al., 1959). The Bears Brook volcanic rocks and the apparently overlying Malignant Cove Conglomerates in Pictou County, Nova Scotia, indicate that land was present there. The Malignant Cove includes breccias and conglomerates of non-marine origin (Dewey and Ziegler *in* Boucot et al., in press).

As mentioned before, the White Rock Formation may include beds of Late Ordovician age. If so, an area of marine deposition lay east of the suggested land.

Small bodies of granodiorite, diorite, and related plutonic rocks were emplaced in the northern part of New Hampshire during the Late Ordovician. Billings (1937) named these rocks the Highlandcroft Plutonic Series.

HIGHLIGHTS OF THE ORDOVICIAN HISTORY

The Ordovician graptolite zones provide a reference net of time and related time–stratigraphic units by which correlations of the rock units in New England and adjacent parts of New York and Canada may be established. Although many successions are highly deformed, and only small and poorly preserved faunal collections can be made, comparison of each collection with the congregations that typify each zone has made them useful in working out a general Ordovician history of New England and adjacent areas.

From the maps (Figs. 2–1 to 2–4) and the discussion of correlations and lithologies, the Limestone, Shale, Volcanoes and Islands, and Aroostook–Matapedia Belts may be recognized as prominent features. The sediments of the Shale Belt were probably deposited in deeper water than those of the Limestone Belt and the gradation between them probably took place along an area of sea floor characterized by a significant slope. The quartz and feldspar content of many of the sediments of the Volcanoes and Islands Belt as well as the apparent shallow-water origin of some of these sediments suggests the presence of non-volcanic as well as volcanic islands. The coastal regions of New England, New Brunswick and Nova Scotia are relatively little known at present; strata included in the Charlotte and White Rock Formations may prove that deposition was more continuous during the Ordovician than has been indicated here.

Two tectonic events or intervals occurred in New England during the Ordovician. During the first, which occurred in the early Middle Ordovician, much of both the platform carbonates and the terrain east of the Shale Belt were folded and became emergent, and relief developed upon the lands formed at that time. During zone 13 through 15 time, various events took place, including the emplacement of the Taconic allochthon, the folding and emergence of lands in most of the Shale and Volcanoes and Islands Belts, the deposition of deltaic sediments westward from these lands, and the emplacement of the plutons related to the Highlandcroft magma series.

REFERENCES

Alcock, F. J., 1935, Geology of the Chaleur Bay region: Geol. Survey Canada, Mem. 183, 146 p.

Anderson, F. D., and Poole, W. H., 1959, Woodstock–Fredericton, York, Carleton, Sunbury, and Northumberland Counties, New Brunswick: Geol. Survey Canada Map 37–1959

Ayrton, W. G., 1961, Preliminary report on Chandler–Port Daniel area, Bonaventure and Gaspé–South Counties: Quebec Dept. of Nat. Resources, Rept. no. 447, 12 p.

Berry, W. B. N., 1960, Graptolite faunas of the Marathon region, West Texas: Univ. Texas Pub. 6005, 179 p.

————, 1961, Graptolite fauna of the Poultney Slate: Amer. Jour. Science, v. 259, p. 223–229

————, 1962a, Stratigraphy, zonation, and age of Schaghticoke, Deepkill, and Normanskill Shales,

eastern New York: Geol. Soc. America Bull., v. 73, p. 695–718

————, 1962b, On the Magog, Quebec, graptolites: Amer. Jour. Science, v. 260, p. 142–148

————, 1966, Zones and zones—with exemplification from the Ordovician: Amer. Assoc. Petroleum Geol. Bull., v. 50, p. 1487–1500

Billings, M. P., 1937, Regional metamorphism of the Littleton–Moosilauke area, New Hampshire: Geol. Soc. America Bull., v. 48, p. 463–566

————, 1956, The geology of New Hampshire; pt. 2, Bedrock geology: N. H. State Plan. and Devel. Comm., Concord, N. H., 203 p.

Bird, J. M., ed., 1963, Stratigraphy, structure, sedimentation and paleontology of the southern Taconic region, eastern New York: Guidebook for Field Trip 3, Geol. Soc. America 1963 Annual Meeting, Albany, New York, 67 p.

Boucot, A. J., 1961, Stratigraphy of the Moose River synclinorium, Maine: U. S. Geol. Survey Bull. 1111–E, p. 153–188

————, Fletcher, Raymond, and Griffin, John, 1959, Middle or Upper Ordovician in Nova Scotia [abs.]: Geol. Soc. America Bull., v. 70, p. 1572

————, Field, M. T., Fletcher, Raymond, Forbes, W. H., Naylor, R. S., and Pavlides, Louis, 1964, Reconnaissance bedrock geology of the Presque Isle quadrangle, Maine: Maine Geol. Survey Quad. Map Ser. no. 2, 123 p.

————, Dewey, J. F., Dineley, D. L., Fyson, W. K., Hickox, C. F., McKerrow, W. S., and Ziegler, A. M., in press, The geology of the Arisaig area, Antigonish County, Nova Scotia: Geol. Soc. America.

Bulman, O. M. B., 1950, Graptolites from the *Dictyonema* shales of Quebec: Quart. Jour. Geol. Soc. London, v. 106, p. 63–99

Cady, W. M., 1960, Stratigraphic and geotectonic relationships in northern Vermont and southern Quebec: Geol. Soc. America Bull., v. 71, p. 531–576

Caldwell, D. B., ed., 1966, The Mount Katahdin region, Maine; New England Intercollegiate Geological Conference Guidebook, 58th Annual Meeting, Shin Pond, Maine. 64 p.

Clark, T. H., 1924, The paleontology of the Beekmantown series at Levis, Quebec: Amer. Paleontology Bull., v. 10, 131 p.

————, 1946, Montreal area—Laval and Lachine Map areas: Quebec Dept. Mines Report 46, 159 p.

————, 1955, St. Jean-Beloeil area: Quebec Dept. Mines Report 66, 83 p.

————, and Strachan, Isles, 1955, Log of the Senigon well, southern Quebec: Geol. Soc. America Bull., v. 66, p. 685–698

Cooper, G. A., 1956, Chazyan and related brachiopods: Smithsonian Misc. Coll., v. 127, pt. 1, 1024 p.

Crickmay, G. W., 1932, Evidence of Taconic orogeny in Matapedia Valley, Quebec: Amer. Jour. Science, v. 24, p. 368–386

Crosby, D. G., 1962, Wolfville map-area, Nova Scotia: Geol. Survey Canada Mem. 325, 67 p.

Ekren, E. B., 1961, Volcanic rocks of Ordovician age in the Mount Chase Ridge, Island Falls quadrangle, Maine: U. S. Geol. Survey Prof. Paper 424–D, p. D43–46

Fisher, D. W., 1954, Lower Ordovician (Canadian) stratigraphy of the Mohawk Valley, New York: Geol. Soc. America Bull., v. 65, p. 71–96

__________, 1962, Correlation of the Ordovician rocks in New York State: N. Y. State Geol. Survey Map and Chart Series no. 3

Flower, R. H., 1964, The Nautiloid Order *Ellesmeroceratida* (Cephalopoda): N. Mex. State Bur. Mines and Min. Res., Inst. Mining and Tech. Mem. 12, 234 p.

Hall, B. A., 1964, Stratigraphy and structure of the Spider Lake quadrangle, Maine: Ph.D. Thesis, Yale University, 153 p.

Harwood, D. S., and Berry, W. B. N., 1967, Fossiliferous Lower Paleozoic rocks in the Cupsuptic quadrangle, west-central Maine: *in* U. S. Geol. Survey Prof. Paper 575-D, p. D16-D23

Hofmann, H. J., 1963, Ordovician Chazy Group in southern Quebec: Amer. Assoc. Petroleum Geol. Bull., v. 47, p. 270-301

Hutchinson, R. D., 1952, The stratigraphy and trilobite faunas of the Cambrian sedimentary rocks of Cape Breton Island, Nova Scotia: Geol. Survey Canada, Mem. 263, 124 p.

Jackson, D. E., 1966, Graptolitic facies of the Canadian Cordillera and Arctic Archipelago: A review: Can. Petroleum Geol. Bull., v. 14, p. 469-485

__________, and Lenz, A. C., 1962, Zonation of Ordovician and Silurian graptolites of northern Yukon, Canada: Amer. Assoc. Petroleum Geol. Bull., v. 46, p. 30-45

Kay, G. M., 1937, Stratigraphy of the Trenton Group: Geol. Soc. America Bull., v. 48, p. 233-302

Kindle, C. H., and Whittington, H. B., 1958, Stratigraphy of the Cow Head region, western Newfoundland: Geol. Soc. America Bull., v. 69, p. 315-342

Larrabee, D. M., Spencer, C. W., and Swift, D. J. P., 1965, Bedrock geology of the Grand Lake area, Aroostook, Hancock, Penobscot and Washington Counties, Maine: U. S. Geol. Survey Bull. 1201-E, 38 p.

McGerrigle, H. W., 1950, The geology of eastern Gaspé: Quebec Dept. Mines, Geol. Report 35, 168 p.

__________, 1959, Madelaine River area: Quebec Dept. Mines, Geol. Report 77, 50 p.

Neuman, R. B., 1960, Pre-Silurian stratigraphy in the Shin Pond and Stacyville quadrangles, Maine: U. S. Geol. Survey Prof. Paper 400-B, p. B166-B168

__________, 1963, Caradocian (Middle Ordovician) fossiliferous rocks near Ashland, Maine: U. S. Geol. Survey Prof. Paper 475-B, p. B117-B119

__________, 1964, Fossils in Ordovician tuffs, northeastern Maine, with a section on the Trilobita by H. B. Whittington: U. S. Geol. Survey Bull. 1181-E, 38 p.

Oppel, Albert, 1856-1858, Die Juraformation Englands, Frankreichs und des südwestlichen Deutschlands: Stuttgart, von Ebner and Seubert, 838 p.

Osborne, F. F., and Berry, W. B. N., 1966, Tremadoc rocks at Levis and Lauzon: Naturaliste Canadiene, v. 93, p. 133-143

__________, and Riva, John, 1966, Post-Levis beds of the Quebec Group at St. Apollinaire, Lotbiniere Co., P.Q.: Naturaliste Canadiene, v. 93, p. 145-151

Pavlides, Louis, 1964, The Hovey Group of northeastern Maine: U. S. Geol. Survey Bull. 1194-B, p. B1-B6

__________, Neuman, R. B., and Berry, W. B. N., 1961, Age of the "ribbon rock" of Aroostook County, Maine: U. S. Geol. Survey Prof. Paper 424-B, p. B65-B67

__________, Mencher, Ely, Naylor, R. S., and Boucot, A. J., 1964, Outline of the stratigraphic and tectonic features of northeastern Maine: U. S. Geol. Survey Prof. Paper 501-C, p. C28-C38

__________, and Berry, W. B. N., 1966, Graptolite-bearing Silurian rocks of the Houlton-Smyrna Mills area, Aroostook County, Maine: U. S. Geol. Survey Prof. Paper 550-B, p. B51-B61

Poole, W. H., 1960, Hayesville and McNamee map-areas, York, Northumberland, and Carleton Counties, New Brunswick: Geol. Survey Canada Paper 60-15, 10 p.

__________, 1963, Geology of Hayesville, New Brunswick: Geol. Survey Canada Map 6-1963

__________, ed., 1966, Guidebook, Geology of parts of Atlantic Provinces: The Geological Association of Canada and The Mineralogical Association of Canada, Guidebook for ann. mtg., Halifax, Canada, 155 p.

Raymond, P. E., 1914, The succession of faunas at Levis, P. Q.: Amer. Jour. Science, v. 38, p. 523-530

Riva, John, 1966a, Upper Levis graptolites from Cowansville, southern Quebec: Jour. Paleontology, v. 40, p. 220-221

__________, 1966b, New assemblage of Middle Ordovician graptolites from the Appalachian region, Quebec: Naturaliste Canadiene, v. 93, p. 153-156

Rodgers, John, 1965, Long Point and Clam Bank Formations, western Newfoundland: Geol. Assoc. Canada Proc., v. 16, p. 83-94

Ross, R. J., Jr., and Berry, W. B. N., 1963, Ordovician graptolites of the Basin Ranges in California, Nevada, Utah, and Idaho: U. S. Geol. Survey Bull. 1134, 177 p.

Shaw, A. B., 1958, Stratigraphy and structure of the St. Albans area, northwestern Vermont: Geol. Soc. America Bull., v. 69, p. 519-568

Smith, C. H., 1957, Bathurst-Newcastle map-area, New Brunswick: Geol. Survey Canada Map 1-1957

Smitheringale, W. G., 1960, Geology of Nictaux-Torbrook map-area, Annapolis and Kings Counties, Nova Scotia: Geol. Survey Canada Paper 60-13, 32 p.

Twenhofel, W. H., 1926, Geology of the Mingan Islands: Geol. Soc. America Bull., v. 37, p. 535-550

Whittington, H. B., and Kindle, C. H., 1963, Middle Ordovician Table Head Formation, western Newfoundland: Geol. Soc. America Bull., v. 74, p. 745-758

Zen, E-an, 1967, Time and space relationships of the Taconic allochthon and autochthon: Geol. Soc. America Special Paper 97, 107 p.

Chapter 3

Paleogeographic Implications of Ordovician Shelly Fossils in the Magog Belt of the Northern Appalachian Region*

ROBERT B. NEUMAN

INTRODUCTION

ORDOVICIAN BRACHIOPODS and other shelly fossils from a wide range of rock types at scattered localities from Maine to Newfoundland permit assignment of some specific events and details to Kay's (1951, fig. 1) graphic paleogeographic rendition of this region. Kay's subdivision of the Appalachians into two parts—the Champlain and Magog belts—corresponds to the distribution of dominantly carbonate rocks and detrital-volcanic rocks, respectively. This review is confined to the Magog belt southeast of the Green Mountain–Sutton Mountain anticlinorium and its presumed counterpart in Newfoundland, the Long Range (Williams, 1964, p. 1142).

European genera not generally found in North America characterize the fossil assemblages. Some forms are newly discovered, such as a new species of *Tritoechia* named for Professor Marland Billings and described herein. Other representative specimens are illustrated on the accompanying plates.

Although a few Ordovician fossils had been found in this area before World War II, the pace of geologic investigations in both the United States and Canada was much quickened after the war; most of the fossil discoveries have been made during this more recent work. These fossils are not easily found, as most of them occur in unlikely rocks such as tuff and conglomerate,

and some are considerably deformed. Because the fossils at many places are fragmentary, many large collections of rock were needed; much credit is therefore due the observant and persevering geologists whose collections make this review possible.

Pre-Silurian rocks form a minor part of the total exposed bedrock outcrop area of the Magog belt. They occur in anticlinal outcrop areas of different sizes, isolated from each other by intervening younger sedimentary rocks, plutonic rocks, and bodies of water. Only a small part of the pre-Silurian outcrop area is demonstrably of Ordovician age; the remainder is either assignable to the Precambrian or Cambrian, or is of uncertain age. Ordovician slate and shale containing graptolites but no brachiopods or other shelly fossils are more common than rocks that have shelly fossils; consequently, the area known to be underlain by Ordovician rocks that contain shelly fossils is an infinitesimally small percentage of the entire belt.

Ten localities are in the northern half of Maine; a few are known in New Brunswick, Quebec, and Newfoundland; and one poorly preserved collection has been found in Nova Scotia. Direct superposition of two zones was seen at only one place, Intricate Harbour, New World Island, Newfoundland. At a few places a sequence has been pieced together by geologic mapping, but most localities are isolated and their relative ages are inferred from their fossils. In addition, a few localities have yielded too few

---

* Publication authorized by The Director, U.S. Geological Survey.

35

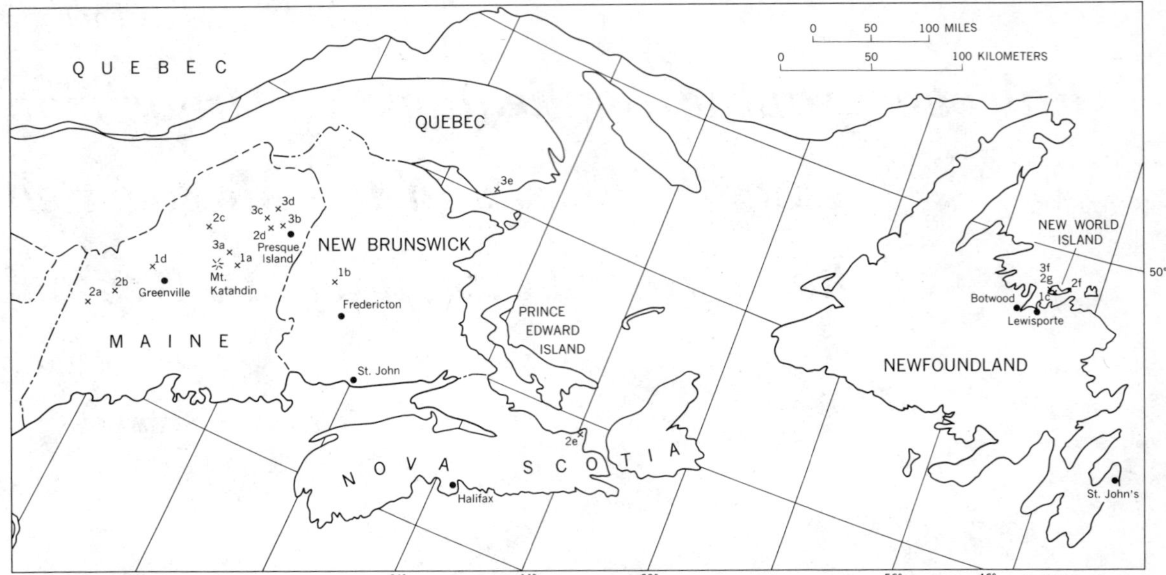

FIGURE 3–1. Index map of southeastern Canada and Maine showing localities of Ordovician shelly fossils in the central Magog belt.

specimens to furnish useful information, and no doubt some have been overlooked.

GENERAL COMPOSITION OF THE SHELLY FOSSILS

The assemblages of Ordovician shelly fossil genera in the Magog belt are more like those in Europe than they are like those elsewhere in North America. European subdivisions of the system are therefore more applicable than the North American ones. The older assemblages consist of unique associations of genera not known to coexist elsewhere, together with new genera not yet recorded from outside this area. Younger assemblages reflect the well-established trend toward cosmopolitanism with decreasing age in the Ordovician; nevertheless, in addition to some of the more cosmopolitan genera, other genera of both brachiopods and trilobites found here have not hitherto been found in North America.

In contrast with most North American assemblages but in common with many European ones, the number of genera at individual localities in the Magog belt is greater than at most other places in the Appalachians or in the midcontinent region, where relatively few genera are present throughout the Ordovician. Thus, correlations are more readily made with the Old World, even though many of these are uncertain, and few can be confirmed by associated graptolite faunas or by superposition of successive zones.

ROCKS OF ARENIG TO LLANDEILO AGE

Occurrences

Correlated with confidence. Shin Brook Formation, Shin Pond and Island Falls quadrangles, Penobscot and Aroostook Counties, Maine (1a of Fig. 1) (Neuman, 1964, 1967; Ekren and Frischknecht, 1967). Most of the formation is crystal tuff, with lesser amounts of flows, volcanic sandstone, and conglomerate. It unconformably overlies the Lower Cambrian(?) Grand Pitch Formation, and is overlain by metadiabase whose age is not well established.

	Britain		North America	Central Magog belt
			UPPER	Richmond	3
	Ashgill		Maysville	
			Eden	
ORDOVICIAN / UPPER	Caradoc		Barneveld	2
		MIDDLE	Wilderness	
			Porterfield	
	Llandeilo		Ashby	
ORDOVICIAN / LOWER	Llanvirn		Marmor	1
			Whiterock	
	Arenig	LOWER	Canadian	

FIGURE 3–2. Ordovician subdivisions in Britain and North America (after Whittington, 1966, Table 1) and equivalents in the central Magog belt.

TABLE 3-1.   Brachiopods in rocks of Arenig to Llandeilo age in Maine, New Brunswick, and Newfoundland. Localities plotted on Figure 1. "sp" indicates uncertain specific identification of specimens belonging to the genus.

	1a Shin Brook Formation, Maine	1b Hayesville and Napadogan quads., New Brunswick	1c New World Island, Newfoundland
Multispinula sp.			X
Productorthis mainensis Neuman	X	sp	X
Orthambonites robustus Neuman	X	X	sp
Valcourea sp.		X	X
Platystrophia sp.	X		
Tritoechia billingsi Neuman	sp	X	X
Platytoechia boucoti Neuman	X		sp
new genus aff. *Oslogonites*			X
Polytoechia sp.	X		
Porambonites sp.			X
Christiania sp.			X
Eoplectodonta sp.			X
syntrophid, new genus		X	
Eodalmanella sp.		X	X

Fossils in the Shin Brook Formation, in addition to brachiopods tabulated in Table 3-1, are: siliceous monaxon sponge spicules and root tufts; bryozoans—*Batotstoma* sp. and *Nicholsonella* sp.; gastropod—*Leseurilla* sp.; and trilobites—*Geragnostus* sp., *Ampyx* sp., *Nileus* sp., *Annamitella? borealis* Whittington, *Illaenus* (*s. l.*) *sp.*, *Raymondaspis* sp., *Hibbertia?* sp., and an undetermined miraspinid.

Unnamed formation, principally calcareous siltstone, Hayesville and Napadogan quadrangles, York County, New Brunswick (1b of Fig. 3–1) (Robb, 1870, p. 189–190; Poole, 1958, 1960, 1963). This unit is underlain by quartzite and slate similar to the Grand Pitch Formation, and is overlain by a heterogeneous unit consisting of graywacke, chert, red and green iron formation, volcanic rocks, and slate near the base that contains graptolites including *Climacograptus bicornis* (Hall) (W. H. Poole, written communication, 1967). Brachiopods (Table 3–1) are the most abundant fossils, associated with subordinate amounts of bryozoans and pelmatozoan debris.

Unnamed formation, Village Cove, New World Island, Twillingate quadrangle, Newfoundland (1c of Fig. 3–1) (Williams, 1963). Fossiliferous volcanic sandstone is part of Williams's (1963) map unit 4 consisting of volcanic and sedimentary rocks; relations of the fossiliferous rocks to the unfossiliferous rocks enclosing them are not known. A few trilobites, including *Annamitella?* sp. and a harpid, are associated with the varied brachiopod assemblage indicated in Table 3–1.

Provisionally correlated. *Kennebec Formation*, Brassua Lake quadrangle, Somerset County, Maine (1d of Fig. 3–1) (unpublished geologic quadrangle map; see Boucot, 1961, p. 183). The rock was described as ". . . massive felsite and rhyolite tuff . . ." It is underlain by rocks classified as undifferentiated sedimentary rocks of Cambrian or Ordovician age, but contact relations are unknown, and is overlain by undifferentiated strata of Silurian or Devonian age, probably unconformably (Boucot, 1961, p. 178). The following fossils were identified by Boucot (written communication, 1967): dasycladacean alga—*Mastopora* sp. (Boucot and Yochelson, 1966, p. A15); brachiopods—*Orthambonites?* sp., *Valcourea* sp., *Sowerbyites* sp.; gastropod—*Loxoplocus* (*Lophospira*) sp.; trilobite—*Isotelus?* sp.

Age Considerations

When the Shin Brook Formation and its fossils were originally described (Neuman, 1964), no correlatives in the region were known, and neither the brachiopods nor the trilobites (Whittington, *in* Neuman, 1964, p. E-25 to E-34) permitted a positive age assignment. At that time, as now, the Shin Brook fossil assemblage was found to consist of some forms known elsewhere in Lower Ordovician rocks (Arenig or Canadian), and others known elsewhere from the Middle Ordovician (Llanvirn to Caradoc; Marmor to Porterfield), but not older. The age assignment of the Shin Brook, arrived at by compromising the conflicting evidence, was considered to be either late Canadian or Whiterock by Neuman (1964, p. E24), or somewhat less equivocally Whiterock by Whittington (*in* Neuman, 1964, p. E34).

Four of the genera (and probably species) of the Shin Brook assemblage were found on New World Island, Newfoundland. Five of the 7 additional forms from New World Island (*Valcourea*, *Porambonites* with obscure cardinal process, *fide* Spjeldnaes, 1957, p. 217, *Christiania*, *Eoplectodonta*, and *Eodalmanella*) are characteristic of Llanvirn or younger rocks; *Tritoechia* and the undescribed genus similar to *Oslogonites* are more typical of the earlier Ordovician. Kay (1967, p. 588) considers this assemblage ". . . of late Chazyan (Valcourean) age similar to [that] near the top of the Whiterock Stage" in central Nevada." Kay (1967, p. 588) also reported that cephalopods, identified as ellesmeroceratids of probable Gasconade age by Flower, were found in limestone

PLATE 3–1.   Brachiopods from older Ordovician rocks on New World Island, Newfoundland. These and other fossils listed in the text occur together at the locality indicated by an F symbol on the geologic map of the Twillingate quadrangle (Williams, 1963), ½ mile (0.8 km) northwest of Village Cove. Specimens accessioned by the Geological Survey of Canada (GSC). Photographs by C. F. Buddenhagen IV.

(1) *Productorthis mainensis* Neuman: a–d, rubber replica of an external mold, dorsal, ventral, side, and posterior views, respectively, GSC 23294, × 2; a–g, silicified dorsal valve, external, posterior, and internal views, respectively, GSC 23295, ×2.5. Specimens of this species from this locality are better preserved than those from the Shin Brook Formation in Maine. Note the long carinate cardinal process and the absence of a chilidium.

(2) *Tritoechia billingsi* Neuman, new species: a, b, paratype, rubber replica from external mold, posterior view of ventral and dorsal valves, GSC 23297, ×2; c, d, holotype, internal mold and rubber replica of dorsal valve, GSC 23296, ×2; e, f, paratype, internal mold and rubber replica of ventral valve, GSC 23298, ×2; g–i, paratype, dorsal valve, rubber replica from external mold, internal mold, and rubber replica from internal mold, respectively, GSC 23299, ×2. The presence of a chilidium in this species distinguishes it from all others known to me. Ventribiconvex, parvicostellate shells having radially striated interareas, the dorsal about ⅛ the length of the ventral. Notothyrial platform supported above shell floor by median septum which extends about midway into shell; cardinal process short and bladelike in smaller specimens, but bulbous in larger ones. In small specimens chilidium covers only apical part of notothyrium, but in large ones, such as in Figure 2b, at least half of the notothyrium is covered. Unsupported stout brachiophores and associated shallow sockets extend outward from the margins of the notothyrium and merge with shell floor toward the posterolateral extremities of the shell. Ventral valve unexceptional for the genus, its interarea planar or slightly concave, and having an arched pseudodeltidium; no specimens at hand preserve evidence of the foramen, although presumably one perforated the apex of the pseudodeltidium. Dental plates of small specimens short and erect, but in larger ones they advance into the valve along the margins of the muscle field. Muscle impressions poorly differentiated, but anterior end of thickened muscle field of large specimens nearly straight.

The New Brunswick specimens also display a chilidium like these, but the single dorsal valve known from the Shin Brook Formation lacks the relevant part of the shell so that the presumption that it too is conspecific cannot be confirmed.

(3) *Valcourea* sp.; a,b, ventral valve, rubber replica of external mold, and internal mold, GSC 23300, ×2.

(4) *Christiania* sp.; a, ventral valve, internal mold, GCS 23301, ×2; b, dorsal valve, partially preserved silicified specimen and impression of exterior, GSC 23302, ×2.

(5) *Eodalmanella* sp.; a–d, articulated specimen, ventral and dorsal views of internal and external molds, GSC 23303, ×2. This genus has hitherto been recorded only from rocks of Llanvirn age in Czechoslovakia.

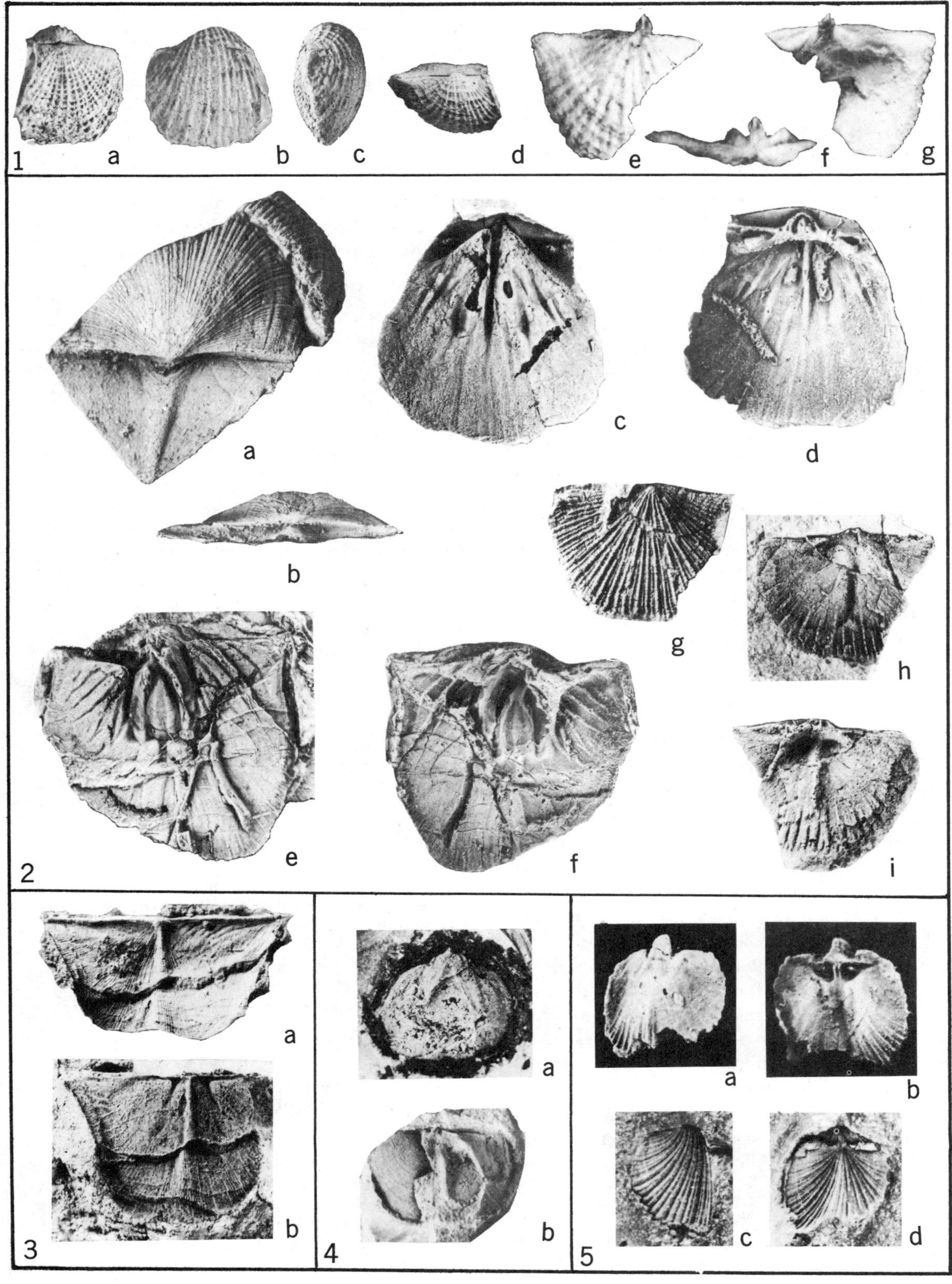
1
a
b
c
d
e
f
g
2
a
b
c
d
e
f
g
h
i
3
a
b
4
a
b
5
a
b
c
d

fragments incorporated in lava near the rocks bearing brachiopods, the only occurrence of rocks of this age bearing shelly fossils known in the Magog belt.

Three genera in the Hayesville and Napadogan map-areas are common to both the Shin Brook Formation and the New World Island assemblages, but the undescribed syntrophid belongs to a group most common in older rocks.

The few fossils known from the Kennebec Formation do not aid in the resolution of the age problem.

Independent evidence from graptolites is not yet available. Rocks bearing graptolites that might be the equivalent of the shelly rocks are known in the central Magog belt only at one place, Snooks Arm, Newfoundland (lat. 55°42′ W., long. 49°51′ N.) (Snelgrove, 1931, p. 488; Ruedemann, 1947, p. 59), but shelly fossils have not been reported there. In the vicinity of the New Brunswick brachiopod localities, slate bearing graptolites of early Caradoc age overlies the rocks bearing the brachiopods (Poole, 1963).

Thus firm correlation of these rocks will have to await further evidence. They will probably prove to be of Whiterock or only slightly younger age, correlative with the Llanvirn or Llandeilo Series; an age as young as Caradoc seems unlikely.

Paleogeographic Interpretation

Three of the four occurrences under consideration have yielded fossils from rocks of dominantly volcanic composition. The stratigraphy of the associated sequences is not well known in all the areas; that of the Shin Brook Formation is best documented (Neuman, 1964). The formation consists of 1,000 feet (350 meters) or more of tuff, tuffaceous sandstone and conglomerate, and lava flows; clasts of terrigenous material occur at the base of the formation. Thus the Shin Brook apparently represents a segment of a volcanic island and its shore, as the Kennebec Formation and the rocks on New World Island probably do too. The rocks at the New Brunswick localities, containing fragments of quartzite but not of volcanic material, also appear to have been deposited in shallow water, but the effects of volcanism are not evident there.

Slate probably of the same age as these rocks has been identified in the central part of the Magog belt only at the Snooks Arm, Newfoundland, locality already referred to. Along the eastern and western margins of the belt are other occurrences of such slate, but in entirely different stratigraphic associations.

Occurrence of brachiopod genera common to these fossil assemblages indicates marine connections between them and suggests that the shoal water in which the organisms lived merged with a common seaway. Interpretation of the nature of these connections is limited by the general absence of rocks of this age. The Snooks Arm rocks may be a remnant of offshore, deeper water equivalents, but the absence of evidence of inferred connections suggests their removal by post-Shin Brook, pre-Caradoc uplift and erosion.

ROCKS OF CARADOC AGE

Occurrences

Dixville Formation?, Cupsuptic quadrangle, Franklin and Oxford Counties, Maine (2a of Fig. 3–1) (Harwood and Berry, 1967). Dark-gray graywacke containing cherty rock fragments and slate chips as well as quartz and feldspar, bearing fragments of shelly fossils, is interbedded with dark-gray slate bearing specifically indeterminate specimens of *Climacograptus* and *Dicellograptus*. Shelly fossils are nearly obliterated by deformation, and unidentifiable except for ventral valve of a leptestinid brachiopod, either *Bilobia* or *Diambonia*.

Unnamed formation, Brassua Lake quadrangle, Somerset County, Maine (2b of Fig. 3–1) (E. V. Post, written communication, 1962). Gray feldspathic graywacke containing calcareous siltstone pebbles appears to be in a small erosion remnant above gray slate of Cambrian or earlier Ordovician age, and below calcareous siltstone of Silurian or Devonian age. Brachiopods in the calcareous siltstone pebbles and in graywacke are: *Glyptorthis* sp., *Platystrophia* sp., and *Leptellina* sp.

Unnamed sequence, Spider Lake and Churchill Lake quadrangles, Piscataquis County, Maine (2c of Fig. 3–1) (Hall, *in* Caldwell, 1966, p. 42–50, and written communication, 1960). Ordovician rocks overlie complexly deformed gray slate and siltstone correlated with the Lower Cambrian? Grand Pitch Formation of the Shin Pond area. At the base is conglomeratic, dark-gray graywacke with fragmentary brachiopods; this is overlain successively by mafic volcanic rocks, a complex unit of volcanic breccia, bedded chert, sandstone and conglomerate, slate that contains *Climacograptus bicornis*, a black slate unit that contains *Orthograptus truncatus* var. *intermedius*, and by a unit consisting of felsitic tuff and chert containing severely deformed brachiopods. The Ordovician rocks are overlain unconformably by Silurian sedimentary and volcanic rocks and younger units.

Fossils in the conglomeratic graywacke in the lower part of the sequence at several localities include the following brachiopod genera: *Nicolella*, *Glyptorthis*, *Skenidioides*, dalmanellid indet., *Rhynchotrema*, *Sowerbyella*, *Leptellina*, *Diambonia*, and *Catazyga*.

Fossils in the felsitic tuff in the upper part of the sequence include heliolitid corals and the following brachiopod genera: *Hesperorthis*, dalmanellid indet., *Eoplectodonta*, and *Diambonia*.

Unnamed formation, Ashland quadrangle, Aroostook County, Maine (2d of Fig. 3–1) (Neuman, 1963). Exposures, confined to the floor of a single gravel pit, include dark-gray, tuffaceous and conglomeratic mudstone, granule and boulder conglomerate, and green concretionary mudstone and siltstone. Fossil list is revised and supplements that of 1963; selected species illustrated on Plate 3–2, Figs. 5–7 show their remarkable preservation.

Corals—*Catenipora* sp., unidentified rugose and tabulate corals; brachiopods—*Cyrtonotella* sp., *Nicolella* cf. *N. actoniae* (Sowerby), *Boreadorthis* sp., *Chaulistomella* sp., *Glyptorthis* sp., *Skenidioides* sp., *Howellites* sp., *Triplecia* cf. *T. insularis* (Eichwald), *Oxoplecia* sp., *Catazyga* sp., *Chonetoidea* sp., *Christiania* sp., *Leptellina* sp., *Diambonia* sp., *Sampo* sp., *Eoplectodonta* sp., *Ptychoglyptus* sp., and *Leptaena* sp.; graptolite—*Amplexograptus* sp.

Unnamed formation, Cape George, Antigonish County, Nova Scotia (2e of Fig. 3–1) (Boucot and others 1959). Hematitic, calcareous quartzite occurs in blocks in fault breccia and in boulders in Carboniferous conglomerate, but is not known elsewhere. Brachiopods, revised from study of the collections, are: *Kjaerina* sp., rhynchonellid, indet., and an indeterminate atrypacean.

Unnamed formation, Cobbs Arm, New World Island, Twillingate quadrangle, Newfoundland (2f of Fig. 3–1) (Williams, 1963). Impure limestone, part of a unit that consists of volcanic rocks, argillite, siltstone, and sandstone as well as limestone, contains the brachiopods *Hesperorthis* sp., *Valcourea* sp., *Refinesquina* sp., and *Cyclospira* sp.

Unnamed formation on the northwest shore of Intricate Harbour, New World Island, Twillingate quadrangle, Newfoundland (2g of Fig. 1) (Williams, 1963). Gray, fossiliferous, pebbly siltstone containing the fossils listed here is overlain by 75 feet (25m) of polymict boulder conglomerate which is in turn overlain by a much more pebbly mudstone containing Ashgill-age fossils that are listed on p. 44.

Corals: *Catenipora* sp. and unidentified rugose and tabulate corals; brachiopods—*Boreadorthis* sp., *Glyptorthis* sp., *Ptychopleurella* sp., *Platystrophia* sp., *Atelelasma?* sp., *Howellites* sp., *Christiania* sp., *Diambonia* sp., *Sampo* sp., *Eoplectodonta* sp., and *Leptaena* sp.; and unidentified trilobite fragments.

Age Considerations

The Caradoc age of six of these seven occurrences can be deduced from their brachiopods with reasonable confidence (The exception is the Nova Scotia locality, 2e, whose fossils are few and poorly preserved.) Their relative ages within the Caradoc can also be estimated.

The Cobbs Arm, Newfoundland, assemblage is unlike any known elsewhere in the northern Appalachians, but its genera suggest a Porterfield or Wilderness age, equivalent to the lower Caradoc.

Graptolites of Berry's (1960) zones 12 and 13 associated with the brachiopod-bearing graywacke in the Spider Lake–Churchill Lake area, Maine, establish a middle Caradoc age for these rocks; presumably the similar graywackes with similar fragmentary fossils in northwestern Maine are of about the same age.

The tuffaceous beds in the Spider Lake–Churchill Lake area, the rocks in the Ashland gravel pit, and beds at Intricate Harbour, Newfoundland, that contain the fossils listed here are

probably of younger Caradoc age. Although the generic suite of the rich assemblage at Ashland is similar to that of the Girvan district, Scotland (Williams, 1962), specific determinations of the Maine fossils have not yet been made; generic comparisons are not significant because many genera range through more than one part of the Girvan succession. Most of the genera of the Intricate Harbour suite listed here are the same as those at Ashland. Inasmuch as the beds at Intricate Harbour that yielded these fossils are conformably succeeded by beds yielding fossils of probable Ashgill age, listed below, it seems reasonable to estimate that the Caradoc part here, and thus at Ashland, represents a segment of late Caradoc time.

Paleogeographic Interpretation

The absence of early Caradoc-age rocks through most of the Magog belt, also noted by Berry (this volume), suggests that most of the area stood above sea level at this time. This interpretation is consistent with post-Shin Brook uplift and erosion that is deduced from the distribution of the older Ordovician rocks.

The single occurrence of rocks assigned to the early Caradoc is a crossbedded, impure limestone on New World Island; purer limestone (Williams, 1963, p. 28; Kay, 1967, p. 591–592) that contains fossils which have not yet proved identifiable occurs nearby. Such limestone is unique in the Ordovician of the Magog belt, and its presence indicates that if there was a general withdrawal of the sea or emergence of the land in early Caradoc time, at least a small part of the area remained covered by shallow water.

Middle Caradoc seas were apparently deep, judging from the widespread occurrence of black slate and chert (Berry, this volume). The interbedded graywacke contains large proportions of siliceous volcanic fragments together with scattered fragmentary brachiopods and other fossils which indicate their derivation from local volcanic islands. Debris from the erosion of such islands incorporating the remains of shallow water benthonic organisms was probably swept into the adjacent deep sea by turbidity currents. Although the location of islands and their extent is not yet known, they seem to have been most abundant in northwestern and north-central Maine where the graywackes are known and where Ordovician volcanic rocks are abundant. Study of the stratigraphy of the graywackes and their sedimentary structures may permit delineation of the islands.

Plate 3–2. Caradoc and Ashgill age brachiopods from Maine and Newfoundland. Photographs by R. H. McKinney.

(1) *Retrorsirostra* sp.; a,b, dorsal valve, internal mold and rubber replica made from external mold, USNM 159951, ×1.5. USGS loc. 3889-CO at Pond Pitch, East Branch of the Penobscot River, Traveler Mountain quadrangle, Maine.

(2) *Cryptothyrella* sp., mold of ventral valve, GSC 23527, ×2. Northwest shore of Intricate Harbour, Twillingate quadrangle, Newfoundland.

(3) *Schizophorella* sp.; a,b, dorsal valve, internal mold and rubber replica, USNM 159952, ×2. USGS loc. 3889-CO.

(4) *Eochonetes* sp.; a,b, dorsal valve, internal mold and rubber replica, GSC 23528, ×2; c–f, ventral valve, internal and external molds and rubber replicas, respectively, GSC 23529, ×2. Same locality as Figure 2. The canals that pierce the interarea are seen as bridges of matrix in 4c.

(5) *Boreadorthis* sp.; a–c, ventral valve, internal mold, rubber replica, and part of rubber replica made from external mold, respectively, USNM 159953, ×1.5; d,e, dorsal valve, rubber replica and internal mold, USNM 159954, ×1.5. USGS loc. 4112-CO, floor of gravel pit 5.6 miles east of Ashland, Maine.

(6) *Nicolella* cf. *N. actoniae* (Sowerby); a,b, ventral valve, rubber replica and internal mold, USNM 159955, ×1.5; c–f, dorsal valve, external and internal molds, and rubber replicas, respectively, USNM 159956, ×1.5. USGS loc. 4112-CO.

(7) *Howellites* sp.; a,b, ventral valve, internal mold and rubber replica, USNM 159957, ×2; c–f, dorsal valve, internal mold, rubber replica, external mold and rubber replica, respectively, USNM 159958, ×2. USGS loc. 4112-CO. Fossils from this locality are very well preserved permitting the observation of such details as punctae which are preserved as a mat of hairlike spines of matrix on internal and external molds.

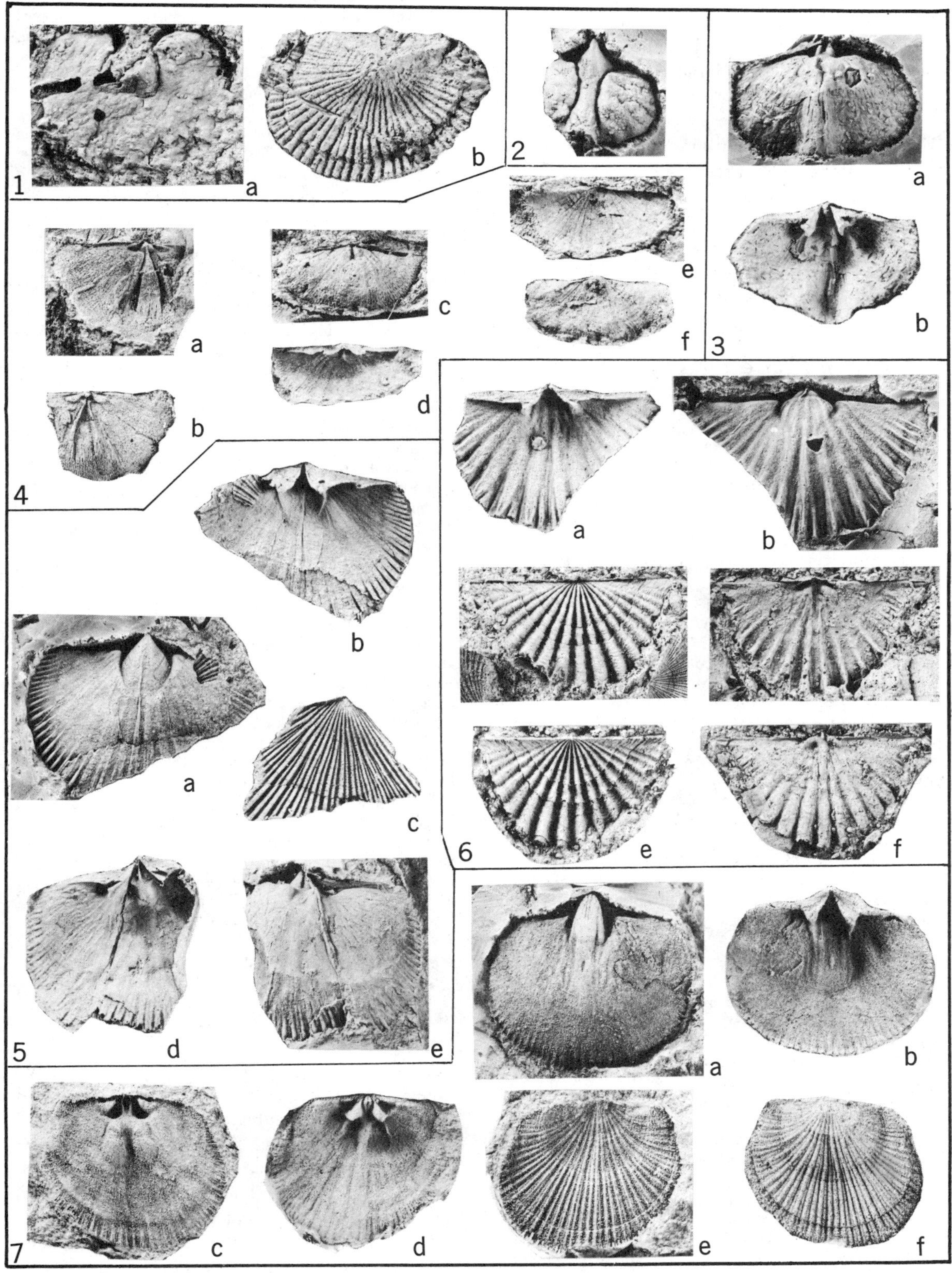

The rocks here regarded as of late Caradoc age are considerably different from the middle Caradoc-age rocks just described. The fossils occur in a wide range of rock types, all of which have abundant volcanic constituents, but none of which displays evidence of turbidity currents or similar kinds of redeposition. Graptolites characteristic of the late Caradoc (zone 15 of Berry, 1960) are known in the Magog belt only in argillaceous limestone in the Aroostook-Matapedia anticlinorium (Berry, this volume); their scarcity seems best explained by postulating that thin deposits that might have contained them have been removed by post-late Caradoc erosion.

ROCKS OF ASHGILL AGE

Occurrences

Correlated with confidence. *Unnamed formation*, Traveler Mountain and Shin Pond quadrangles, Penobscot County, Maine (3a of Fig. 3–1) (Neuman, 1967). A succession of varied polymict conglomerate, graywacke and siltstone crops out in a belt about 4,000 feet (1.2 km) wide in the small area between Pond Pitch and Haskell Rock Pitch of the East Branch of the Penobscot River and immediate vicinity.

The fossils from several localities through this belt are: corals (identified by W. A. Oliver, Jr., written communication, 1961)—*Catenipora* sp., *Acidolites* sp. and petraiid corals, genera and species undetermined; brachiopods—*Orthambonites* sp., *Nicolella* sp., *Hesperorthis* sp., *Boreadorthis* sp., *Glyptorthis* sp., *Ptychopleurella* sp., *Retrorsirostra* sp. (Plate 3–2, Fig. 3–1), *Schizophorella* sp. (Plate 3–2, Fig. 3–3), *Horderleyella* sp., *Laticrura* sp., *Reuschella* sp., *Vellamo* sp., *Triplecia* sp., *Oxoplecia* sp., *Leangella* sp., *Sampo* sp., *Eoplectodonta* sp., *Strophomena* sp., *Christiania* sp., *Leptaena* sp., *Fardenia* sp., *Rhychotrema* sp., and *Catazyga* cf. *C. anticostiensis* (Billings) (see Boucot *in* Boucot and others, 1964, p. 108–111); gastropod (identified by E. L. Yochelson, oral communication, 1963), *Pterotheca* sp.; trilobites (identified by H. B. Whittington, written communication, 1963), *Sphaerocoryphe* cf. *S. thompsoni* Begg and *Diacalymene* sp.

Pyle Mountain Argillite, Presque Isle quadrangle, Aroostook County, Maine (3b of Fig. 3–1) (Boucot and others, 1964). The rock is olive-colored argillite, greenish-gray where unweathered, underlain without structural contrast by volcanic rocks with interbedded dark chert and slate bearing graptolites of Berry's (1960) zone 13; overlain by the Frenchville Formation of Silurian (late Llandovery) age but the nature of the contact is equivocal (Boucot and others, 1964, p. 32). Fossils (occurrences at individual localities are listed by Boucot and others, 1964, p. 21–22) are: brachiopods (preliminary determinations by Neuman)—*Skenidioides* sp., *Leangella* sp., *Sowerbyella* sp., an undetermined plectambonitid, *Christiania* sp., and an undetermined leptaenid; trilobites (listed by Boucot and others, 1964, p. 21, from identification by Whittington)—*Trinodus* sp., *Remopleurides* sp., ?*Symphysops* sp., *Tretaspis* sp., *Novaspis* sp., *Dionide* sp., ?*Raphiophorus* sp., *Pseudosphaerexicus* sp., *Dindymene* sp., *Platylichus* sp., and *Carmon* sp.

White Head Formation, Gaspé County, Gaspé Peninsula, Quebec (3e of Fig. 3–1) (Schuchert and Cooper, 1930; Cooper and Kindle, 1936, and accompanying papers; P. J. Lespérance, written communications, 1964–1967). The formation consists of thin-bedded, fine- to coarse-grained, more or less argillaceous limestone, and calcareous siltstone; its stratigraphic relations to enclosing formations are uncertain. Fossils from the White Head are described in the papers cited above. Study of recent brachiopod collections by Neuman, and of trilobite collections by Lespérance (written communication, 1967), largely confirms Schuchert and Cooper's correlation of the White Head with the Ashgill Series. According to Lespérance, however, one collection from this formation yielded Caradoc graptolites.

Unnamed formation on the northwest shore of Intricate Harbour, New World Island, Twillingate quadrangle, Newfoundland (3f of Fig. 1) (Williams, 1963). Fossils occur in a pebbly mudstone interbed in coarse conglomerate that immediately overlies the less pebbly mudstone containing Caradoc fossils already listed (loc. 2g); brachiopods—*Retrorsirostra* sp., *Dalmanella* sp., *Sampo* sp., *Eochonetes* sp. (Plate 3–2, Fig. 4), *Leptaena* sp., and *Cryptothyrella* sp. (Plate 3–2, Fig. 2).

Provisionally correlated. *Unnamed formation* at Ashland, Ashland quadrangle, Aroostook County, Maine (3c of Fig. 3–1) (Collection of David Roy, Mass. Inst. of Technology, 1966, assigned U. S. National Museum collection no. 13030). Shelly fossils are in a thin bed of gray pebbly mudstone included in a sequence of siltstone. Poorly preserved graptolites indicating an age in the range of Berry's (1960) zone 12 to 15 occur in slate a few feet beneath the pebbly mudstone (Berry, written commun., 1967). The siltstone is overlain by the Frenchville Formation at a discordant contact that is probably an angular unconformity. Brachiopods are: *Skenidioides* sp., ?*Horderleyella* sp., *Hirnantia* sp., an undetermined enteletacean, *Leptaena* sp., *Cryptothyrella* sp., and *Plectothyrella* sp.

Unnamed formation, 6.5 miles (10 km) northeast of Ashland, Ashland quadrangle Aroostook County, Maine (3d of Fig. 3–1) (collection of David Roy, assigned U.S.N.M. collection no. 13031). The fossiliferous rock is a gray conglomeratic siltstone containing pebbles of volcanic rocks in a sandstone sequence that apparently grades upward into the Frenchville Formation. Brachiopods are: *Hirantia* sp., *Triplecia* sp., *Oxoplecia* sp., *Cryptothyrella* sp., and *Plectothyrella* sp.

Age Considerations

The recognition of Ashgill-age assemblages and their discrimination from those of Caradoc age has been considerably aided by several recent papers (Temple, 1965; Bassett et al., 1966; Marek and Havlíček, 1967). From these and earlier works it is apparent that Caradoc- and Ashgill-age rocks have many brachiopod genera in common, and many species of the common genera are closely related. A few brachiopod genera, however, have not yet been found in beds older than Ashgill; their presence led to the Ashgill age assignment of most of the assemblages listed above.

Thus, *Retrorsirostra*, *Schizophorella*, *Hirnantia*, and *Fardenia* are the key elements of the assemblage from Penobscot County, Maine, in assigning it an Ashgill age; this assignment is confirmed by the trilobite, *Sphaerocoryphe* cf. *S. thompsoni.* Most of the genera found there, however, are common in beds of Caradoc age, and some range well into the Lower Silurian.

Similarly, the *Eochonetes* as well as the *Retrorsirostra* and *Cryptothyrella* date the New World Island assemblage. This date is consistent with the occurrence of these fossils in rocks in sequence above those containing a late Caradoc assemblage.

Neither of these assemblages, however, can be correlated with any of the established subdivisions of the Ashgill Series. The fossils at several different levels through a wide outcrop belt in Penobscot County suggests that more than one subdivision may be represented, but these have not yet been distinguished.

The trilobites of the Pyle Mountain Argillite have not yet been intensively studied. Whittington (*in* Boucot and others, 1964, p. 21) noted that most of the genera occur also in the Upper Ordovician rocks of Poland. Kielan examined some of the Maine specimens briefly, and commented on the probable specific identity of some forms with Polish species that she had studied (Kielan, 1960). All the genera from the Pyle Mountain occur only in rocks dated as lower or middle Ashgill in Poland; upper Ashgill equivalents of the *Dalmanitina* beds have not been identified in Maine.

The stratigraphic succession within the White Head Formation on the Gaspé Peninsula and its correlation are under study by P. J. Lespérance (written communication, 1967) and his colleagues. Although ages from late Caradoc through late Ashgill have been inferred from paleontologic evidence, complex structure has destroyed direct evidence that might be obtained from superposed strata. The Caradoc graptolites occur in a structurally disjunct body of shale; some of the more certainly Ashgill-age trilobite assemblages are assigned by Lespérance to the early or middle part of that epoch, and assemblages containing *Dalmanitina* are assigned to its later part. Other assemblages, as well as those found in the Matapedia Group farther southwest, have fossils that do not permit discrimination between Caradoc or Ashgill ages, and some limestone mapped as White Head contains fossils of Silurian (lower to upper Llandovery) age.

The two localities in the Ashland quadrangle, Maine, have yielded assemblages that may be of early Llandovery rather than Ashgill age. Discrimination between these is difficult (Boucot and Johnson, 1964, p. 2), but the Ashland collections lack the genera most indicative of a Silurian age, and are thus provisionally assigned to the Ashgill.

Paleographic Interpretations

The wide range of rock types and fossil assemblages of Ashgill age indicates that paleogeography of the Magog belt consisted of a complex pattern of contrasting elements that in places changed rapidly through the epoch. The variety of rocks containing Ashgill-age fossils along the East Branch of the Penobscot River, Maine, ranging from fine-grained calcareous siltstone to boulder conglomerate, indicates rapidly changing local tectonic controls. By contrast, the thin-bedded calcareous rocks of the White Head Formation on the Gaspé Peninsula are contiguous with similar rocks as old as late Caradoc and as young as Silurian (late Llandovery) (Pavlides and Berry, 1966, p. B53, Lespérance, written communication, 1967); thus the calcareous rocks of the Aroostook–Matapedia belt (Pavlides and others, 1964) represent a stable paleogeographic element that changed little through a long span of time.

A problem similar to those already posed for older, erratically distributed marine Ordovician rocks is repeated here: although marine interconnections are implied by the fossils, the connections are not preserved. Whereas coarse-grained rocks of Ashgill age may have been connected by thin, fine-grained deposits that have been completely removed, they seem more likely to have accumulated in local, tectonically controlled troughs that were arms of a sea whose main body presumably lay to the southeast. This suggestion is derived from consideration of the limited occurrence of Lower Silurian (upper Llandovery) conglomerate above the Ashgill-age rocks in the East Branch of the Penobscot River area (Neuman, 1967). These Silurian rocks probably represent such an arm; the main body of this sea is recorded by a broad outcrop belt of contemporaneous Silurian rocks on the southeast that extends for many miles along the northeast-southwest structural trend. A comparable seaway of Ashgill age is suggested by the calcareous rocks of the Aroostook–Matapedia belt in which the Ordovician rocks plunge beneath present erosion levels in eastern Maine (Pavlides and others, 1964, p. C30), but may re-emerge farther southwest (Ekren and Frischknecht, 1967, p. 15) and then continue an unknown distance along strike below the surface. Local tectonic control of sedimentation was also suggested for these rocks in Newfoundland (Kay, 1967, p. 589).

The White Head Formation of the Matapedia Group and the Carys Mills Formation of the Meduxnekeag Group (Pavlides, *in* Cohee and West, 1966) seem to record the floor of a segment of the sea that persisted from Middle Ordovician (Caradoc) time through most if not all of the Early Silurian. Pavlides (in press) suggested that these rocks in Maine might be comparable to the "calcareous flysch" of the Carpathian Mountains of Poland (Dzulynski and others, 1959, p. 1095–1096). However, the limestone beds of the "calcareous flysch" are graded calcarenites that differ from "normal flysch" only in the chemical composition of the constituent particles of the sandstone layers; such graded calcarenites are rare in the Carys Mills. Although the Carys Mills possesses abundant flute casts, convolute structures, and occasional burrows, these features are common in several sedimentary environments; furthermore, the argillaceous layers between limestone beds are lighter-colored than those normally associated with acknowledged "flysch" deposits. In addition, the White Head Formation at the northeastern end of the Aroostook–Matapedia outcrop belt displays few if any of these distinctive sedimentary structures, and its thin-bedded argillaceous limestones are typical of shallow water deposition.

The calcareous rocks of the Aroostook–Matapedia belt seem more likely to have been deposited on a platform within an elongate trough. Deeper troughs flanking the platform may have served as traps for most of the terrigenous clastic debris originating from adjacent sources, probably islands. Trapped debris of this kind may be represented by the Chandler Ridge Formation (Pavlides, *in* Cohee and West, 1966, p. A54–A55), which lies at the western border of the outcrop of calcareous rocks, although similar sandstone forms thin units at several places across the belt (Pavlides, 1965, p. 19).

The lithologic contrast between the Pyle Mountain Argillite and the conglomeratic rocks to the west in the Ashland quadrangle provisionally assigned an Ashgill age are consistent with such an interpretation, especially if they prove to be contemporaneous. If so, the conglomeratic rocks may represent entrapped coarse debris in contrast to the fine-grained Pyle Mountain.

SUMMARY

Abundant and varied fossils of marine benthonic organisms imply that the rocks that contain them were deposited in shallow to moderately deep water rather than at abyssal depths. Consideration of the composition of such rocks, together with their geologic and spatial relations to the contrasting graptolitic facies and to unfossiliferous equivalents, yields some acceptable paleogeographic interpretations, while others are prohibited. Although the suggestions offered here are based on but a few bits of evidence, this evidence suggests that the main outlines of the principal paleogeographic features appear to be significantly different from one to another of three phases of the Ordovician, and tectonic events seem to have separated these phases.

The oldest Ordovician rocks in the central Magog belt (Arenig to Llandeilo age) include shelly deposits in volcanic ejecta which record the presence of volcanic islands from north-central Maine to northern Newfoundland. Graptolitic facies of this age, known only in northern Newfoundland, may represent the interconnections between islands, now largely removed. A Middle Ordovician tectonic event is implied, during which the thinner deposits and much of the volcanic material were removed by subaerial erosion, leaving only small and isolated remnants.

Rocks containing lower Cardoc fossils are virtually absent from the area. Middle Caradoc rocks, largely dark pelites containing graptolites, are widespread, but graywacke having large amounts of volcanic debris and scattered shells is interbedded with pelite at several places in northwestern Maine. Such relations seem most likely to record sediment derived from volcanic islands and swept into deeper water. Thus marine connections between islands are better demonstrated in this segment of the Ordovician than in earlier or later parts. The known distribution of middle Caradoc graywacke probably represents but a fraction of their actual distribution; however, present knowledge indicates that they are more restricted than are the earlier Ordovician shell-bearing volcanic rocks. Although this difference in distribution may reflect the distribution of volcanoes in these parts of the Ordovician, the picture will remain unsatisfactory until the widespread unfossiliferous volcanic rocks attributed to the Ordovician can be ascribed to a specific part of the system. The study of the distribution and composition of fossiliferous rocks of volcanic derivation may help in better discriminating these less well-dated rocks.

The oldest rocks of the Aroostook–Matapedia belt of calcareous rocks are of middle Caradoc (zone 13 of Berry, 1960) age, suggesting that this major feature, which persisted well into the Silurian, came into being at this time. Although the shallow-water origin of the rocks in this belt is questioned (Pavlides, in press), the rocks are conspicuously different from those that surround them, both in their composition and in their temporal and areal continuity. Thus the Aroostook–Matapedia belt represents a stable paleogeographic element whose origin seems to date from about the time of sliding of the Taconic allochthon (Zen, 1967, p. 68) ; the tectonic significance of this coincidence, however, remains to be appraised.

Other paleogeographic realinements, perhaps coincidental with the Taconic orogeny, seem more evident from the kinds of rocks dated as late Caradoc and Ashgill, and from the general absence of pelitic facies of these ages. Interconnections linking late Caradoc and Ashgill shelly rocks may have been through ephemeral arms of the sea extending from the Aroostook–Matapedia belt onto the axial part of the Magog belt. Tectonic control of the coarse-grained rocks bearing the shelly fossils is suggested by their conglomeratic nature, and by the coincidence of such Ashgill-age rocks with very similar rocks of Llandovery age that overlie them in one small area in Maine. Thus the major elements of Early Silurian paleogeography seem to have been established in Late Ordovician time.

REFERENCES

Bassett, D. A., Whittington, H. B., and Williams, Alwyn, 1966, The stratigraphy of the Bala district, Merionethshire: Geol. Soc. London, Quart, Journ., v. 122, p. 219–271

Berry, W. B. N., 1960, Graptolites faunas of the Marathon region, west Texas: Texas Univ. Pub. 6005, 179 p.

Boucot, A. J., 1961, Stratigraphy of the Moose River synclinorium, Maine: U. S. Geol. Survey Bull. 1111-E, p. 153–188

————, Fletcher, Raymond, and Griffin, John, 1959, Middle or Upper Ordovician in Nova Scotia [abs]: Geol. Soc. America Bull., v. 70, p. 1572

————, Field, M. T., Fletcher, Raymond, Forbes, W. H., Naylor, R. S., and Pavlides, L., 1964, Reconnaissance bedrock geolgy of the Presque Isle quadrangle, Maine: Maine Geol. Survey Quad. Map Ser. no. 2, 123 p.

————, and Johnson, J. G., 1964, Brachiopods of the Ede Quartzite (lower Llandovery) of Norderön, Jämtland: Univ. Uppsala Paleont. Inst. Pub. 51 (Geol. Inst. Bull., v. 42), 11 p.

————, and Yochelson, E. L., 1966, Paleozoic Gastropoda from the Moose River synclinorium, northern Maine: U. S. Geol. Survey Prof. Paper 503-A, 20 p.

Caldwell, D. W., 1966, ed., The Mount Katahdin region, Maine: New England Intercol. Geol. Conf., 58th ann. mtg., guidebook for field trips, Shin Pond, Me., 61 p.

Cohee, G. V., and West, W. S., 1966, Changes in stratigraphic nomenclature by the U. S. Geological Survey, 1965: U. S. Geol. Survey Bull. 1244-A, 60 p.

Cooper, G. A., and Kindle, C. H., 1936, New brachiopods and trilobites from the Upper Ordovician of Percé, Quebec: Jour. Paleontology, v. 10, p. 348–372

Dzulynski, Stanislaw, Ksiazkiewicz, Marion, and Kuenen, P. H., 1959, Turbidites in flysch of the Polish Carpathian mountains: Geol. Soc. America Bull., v. 70, p. 1089–1118

Ekren, E. B., and Frischknecht, F. C., 1967, Geological-geophysical investigations of bedrock in the Island Falls quadrangle, Aroostook and Penobscot Counties, Maine: U. S. Geol. Survey, Prof. Paper 527, 36 p.

Harwood, D. S., and Berry, W. B. N., 1967, Fossiliferous lower Paleozoic rocks in the Cupsuptic quadrangle, west-central Maine: U. S. Geol. Survey Prof. Paper 575-D, p. D16–D23

Kay, Marshall, 1951, North American geosynclines: Geol. Soc. America Mem. 48, 143 p.

————, 1967, Stratigraphy and structure of northeastern Newfoundland bearing on drift in North Atlantic: Am. Assoc. Petroleum Geologists Bull., v. 51, p. 579–600

Kielan, Zofia, 1960, Upper Ordovician trilobites from Poland and some related forms from Bohemia and Scandinavia: Palaeontologia Polonica, no. 11, 198 p.

Marek, Ladislav, and Havlícek, Vladmír, 1967, The articulate brachiopods of the Kosov Formation: Ustred ustavu geol. Vestnik, v. 42, p. 275–284

Neuman, R. B., 1963, Caradocian (Middle Ordovician) fossiliferous rocks near Ashland, Maine: U. S. Geol. Survey Prof. Paper 475-B, p. B117–119

————, 1964, Fossils in Ordovician tuffs, northeastern Maine: U. S. Geol. Survey Bull. 1181-E, 38 p.

————, 1967, Bedrock geology of the Shin Pond and Stacyville quadrangles, Maine: U. S. Geol. Survey Prof. Paper 524-I, 37 p.

Pavlides, Louis, 1965, Geology of the Bridgewater quadrangle, Aroostook County, Maine: U. S. Geol. Survey Bull. 1206, 70 p.

———— in press, Stratigraphy and facies relationships of the Carys Mills Formation of Silurian and Ordovician age of northeast Maine: U. S. Geol. Survey Bull. 1264

————, Mencher, Ely, Naylor, R. S., and Boucot, A. J., 1964, Outline of the stratigraphic and tectonic features of northeastern Maine: U. S. Geol. Survey Prof. Paper 501-C, p. C28–C38

————, and Berry, W. B. N., 1966, Graptolite-bearing Silurian rocks of the Houlton-Smyrna Mills area, Aroostook County, Maine: U. S. Geol. Survey Prof. Paper 550-B, p. B51–B61

Poole, W. H., 1958, Napadogan map-area, 1:63,360: Canada Geol. Survey, Prelim,. Ser. map 11–1958

————, 1960, Hayesville and McNamee map-areas, York, Northumberland, and Carleton Counties, New Brunswick: Canada Geol. Survey Paper 60–15, 10 p.

————, 1963, Hayesville, New Brunswick, 1:63,360: Canada Geol. Survey Map 6–1963

Robb, Charles, 1870, On [the geology of] a part of New Brunswick: Geol. Survey Canada Rept. of Progress, 1866–1869, p. 173–209

Ruedemann, Rudolf, 1947, Graptolites of North America: Geol. Soc. America Mem. 19, 652 p.

Schuchert, Charles, and Cooper, G. A., 1930, Upper Ordovician and Lower Devonian stratigraphy and paleontology of Percé, Quebec: Am. Jour. Sci., 5th ser., v. 20, p. 161–176, 265–288, 365–392

Snelgrove, A. K., 1931, Geology and ore deposits of the Betts Cove-Tilt Cove area, Notre Dame Bay, Newfoundland: Canadian Inst. Mining Metall. Bull. 228, p. 477–519

Spjeldnaes, Nils, 1957, The Middle Ordovician of the Oslo region, Norway, 9: brachiopods of the family Porambonitidae: Norsk geol. tidsskr., v. 37, p. 217–228

Temple, J. T., 1965, Upper Ordovician brachiopods from Poland and Britain: Acta palaeont. Polonica, v. 10, p. 379–427

Whittington, H. B., 1966, Phylogeny and distribution of Ordovician trilobites: Jour. Paleontology, v. 40, p. 696–737

Williams, Alwyn, 1962, The Barr and lower Ardmillan Series (Caradoc) of the Girvan district, south-west Ayrshire, with descriptions of the Brachiopoda: Geol. Soc. London Mem. 3, 267 p.

Williams, Harold, 1963, Twillingate map-area, Newfoundland: Canada Geol. Survey Paper 63–36, 30 p.

__________, 1964, The Appalachians in northeastern Newfoundland—a two-sided symmetrical system: Am. Jour. Sci., v. 262, p. 1137–1158

Zen, E-an, 1967, Time and space relationships of the Taconic allochthon and autochthon: Geol. Soc. America Spec. Paper 97, 107 p.

Chapter 4

Zonation and Correlation of Canadian and Early Mohawkian Series

H. B. WHITTINGTON

INTRODUCTION

ROCKS OF CANADIAN and Early Mohawkian age in North America exhibit contrasting carbonate and dark shale facies, the faunas being dominantly shelly and graptolitic respectively. Correlation between these facies depends upon their interdigitation, but such interbedding rarely occurs throughout the entire sequence. Rather, one facies crops out in an area which is geographically or structurally separated from exposures of the other. A further complication is that, within the carbonate facies, correlations based on particular groups of fossils have been worked out over only a limited area. For example, while brachiopods (Ulrich and Cooper, 1938; Cooper, 1956) and cephalopods (Flower, 1964) have been studied in geographically widespread regions, lettered zones based on trilobites from sequences in Utah (Ross, 1951; Hintze, 1953) have been found readily applicable only northwards into western Canada (Aitken and Norford, 1967). Correlations based on gastropods (Yochelson and Bridge, 1957) have been made in the central and southern Appalachians and southern mid-continent. The relations between these correlations are problematical, and the faunas of the northern Appalachians, in which part of the type area for the Canadian Series lies, are unfortunately poorly known. It is hardly surprising that the National Research Council's correlation chart of the Ordovician System (Twenhofel, 1954) gives no stages for the Canadian Series. Much progress has been made in understanding the faunal succession in the dark shale of the Levis (Osborne and Berry, 1966) and Deepkill (Berry, 1962) Shales, a com-

plete graptolitic zonal succession has been described from West Texas (Berry, 1960), and much new information obtained from the Canadian Cordillera (Jackson, 1966).

In western Newfoundland (Kindle and Whittington, 1958), rocks which constitute part of the type Canadian show interbedding of the carbonate and dark shale facies. Preliminary studies of the Canadian faunas, combined with detailed work on early Mohawkian faunas, allows a re-examination of correlations in North America and their relation to recently-proposed stages (Table 4–1). New information is incorporated in a survey of correlations of these stages over the North Atlantic region (Table 4–2). As these latter correlations show, part of the Tremadoc Series of Britain may be equivalent to an early part of the Canadian Series. British practice places the Cambrian–Ordovician boundary at the top of the Tremadoc, whereas North American usage places it at the base of the Canadian. I have adhered to North American practice in employing the terms Cambrian and Ordovician.

I am indebted to Drs. W. B. N. Berry, V. Jaanusson, R. B. Neuman, A. R. Palmer and E-an Zen for their helpful criticism of my manuscript.

DEFINITION OF SERIES AND STAGES

Canadian Series

Proposed by Dana (1874, p. 214) for rocks in eastern Canada, particularly those near Quebec (essentially the Quebec Group of Logan, 1863), these rocks were considered to be of about the same age as the "Calciferous sand-rock" and

TABLE 4–1. Faunas of the Cow Head Group and their correlation. Numbers of boulder beds and shales from Kindle and Whittington, 1958, fig. 3. Lettered trilobite zones of Utah after Ross (1951) and Hintze (1953). Numbered graptolite zones after Berry (1960).

| Britain | North America | | Cow Head Group, Western Newfoundland | | Utah | Graptolites |
Series	Series	Stages	Trilobites in boulders	Graptolites in shales	Ross, Hintze	Berry
LLANVIRN	MOHAWKIAN	Whiterock			N	9
			14: Ampyx, Edymionia cf. schucherti, Ischyrophyma, Ischyrotoma twenhofeli, Remopleurides, Nileus, Bathyurellus nitidus, Uromystrum, Illaenus spp., Raymondaspis, Kawina arnoldi, ? Kawina sp. ind., Apatolichas jukesi.		N	
				13: Isograptus, Trigonograptus.		8
			12: Geragnostus, Selenoharpes, Ampyx, Stegnopsis, asaphids, aff. Eorobergia, Remopleurides, Telephina, Ischyrotoma, Lloydia, Nileus cf. scrutator, aff. Benthamaspis, ? Bathyurina, Illaenus, Pseudomera cf. barrandei, Ceraurinella, Kawina.		M	
						6, 7
ARENIG	CANADIAN	Cassinian		11. T. fruticosus, Goniograptus	H – L	4, 5
			9, 10: Geragnostus, alsataspidid, Ampyx, aff. Praepatokephalus, aff. Remopleuridiella, asaphid, Shumardia, Bolbocephalus, Petigurus, aff. Benthamaspis, Bathyurellus, Illaenus, Colobinion. n. sp.			
		Jeffersonian		8f, 9: Tetragraptus approximatus, Clonograptus spp.		3
			8e. Geragnostus, Ampyx, alsataspidid, Selenoharpes, cf. Glaphurina, cf. Carolinites, Psalikilus, aff. Remopleuridiella, Protopresbynileus, cf. Ogygiocaris, cf. Benthamspis, Petigurus?, illaenid, scutelluid, pliomerid.		G	
			8a: agnostids, Ampyx, shumardid, Ischyrophyma?, isocolid, cf. Glaphurina Protopresbynileus, komaspidid, cf. Apatokephalus, Remopleurides, cf. Benthamaspis, Bolbocephalus, Petigurus?, illaenid, scutelluid, Pliomeroides?, odontopleurid			
		Demingian			E, F	
TREMADOC		Gasconadian	7: Lloydia, Leiostegium, Onchonotus, Pilekia Broom Pt.: Symphysurina, Hystricurus, Pseudagnostus spp. cf. Shaw, 1951, Plethometopus, Highgatella?, aff. Euloma, Ulrichaspis, Levisaspis, Bienvillia, Parabolinella?		D	
					C	2
					B	1
				Dictyonema of flabelliforme type		
			6: Hungaia magnifica fauna			

Chazy Limestone of the Champlain Valley. Present practice excludes rocks of Chazy (and Whiterock) age from the series. Type sequences of the Canadian thus include the little understood "Beekmantown" of southern Quebec, part of the Levis Shale, the St. George Formation and part of the Cow Head Group of western Newfoundland. Many authors have, however, taken the sequence in southeast Missouri and north Arkansas as a standard, including as highest Canadian the Smithville and Black Rock Formations. The latter contains brachiopod genera (Ulrich and

TABLE 4–2. Correlation with North American stages of sequences in North America, northern Europe and the North Atlantic. The column for Newfoundland shows the position of Logan's lithological division H in the St. George Formation, the approximate positions of the lower (L) middle (M) and upper (U) parts of the Table Head Formation, and on the right the positions of the numbered graptolite zones of Berry (1960) and the lettered trilobite zones of Ross (1951) and Hintze (1953), based on the Cow Head Group and Table Head Formation. Columns for Britain, Sweden and Norway after Jaanusson (1960); trilobite zones of the Volkhov and Kundan Stages after V. Poulsen (1965). Lasnam is an abbreviation for Lasnamägi Stage.

Stages	Ozark	Oklahoma	Newfoundland	Britain	Sweden	Norway	Greenland	Spitsbergen
Ashby	Buffalo River Group	Tulip Creek		Llandeilo	Viruan — Uhaku	4aα; Ogygiocaris		
Marmor		McLish			Viruan			
Whiterock		Oil Creek	Table Head U — 9: Glossogr; N; Table Head M — 8: Isograptus; L, M; Table Head L — ?	Llanvirn	Aseri, Lasnam			? un-named unit
					Kunda — gigas, obtusicauda, raniceps, expansus	3c	?	?
Cassinian	Black Rock; Smithville; Powell	Joins; West Spring Creek	7: bifidus; J, K; 6: protobifidus; H, I; 4,5: fruticosus	hirundo / Arenig / extensus	Volkhov — lepidurus; ?; lata; stigmata	3b	? Nunatami	Upper Oslobreen
Jeffersonian	Cotter; Theodosia; Rich Fountain	Creek; Kindblade	H (St. George) — 3: approximatus; G	Arenig / extensus	Latorp — Billingen; Hunneberg		Narhval Sound, Cape Weber	Middle Oslobreen
Demingian	Roubidoux	Cool Creek	E, F		Oelandian			
Gasconadian	Gasconade	McKenzie Hill	2: Anisograptus; B – D; 1: Dictyonema, Staurograptus	Tremadoc	Tremadoc		Cape Clay, Cass Fjord	Lower Oslobreen
Trempealeau								

Cooper, 1938) also present in the Fort Cassin Formation (highest Canadian of New York and Vermont), and the graptolites *Didymograptus bifidus* and *D. artus* (Ruedemann, 1947, Pl. 54, Figs. 10–12). These graptolites characterize zone 7 of Berry (1960) and range into beds placed in the overlying Whiterock Stage. Shelly faunas of Black Rock age have not been recognised in the western States, so Ross (1964) uses trilobite zone J (*Pseudocybele*) as indicative of the highest Canadian. Zone J faunas are unknown in the Ozark or Appalachian regions, and until the faunas of the Fort Cassin Formation are better known it will remain difficult to define the upper limit of the Canadian. Gasconade and equivalent faunas are apparently better understood, and the lower limit, zone 1 of Berry (1960) or zone B of Ross (1951) and Hintze (1953), is widely recognisable.

Stages within the Canadian Series were proposed by Flower (1957, p. 18; 1964, pp. 17–19) who regarded the Canadian cephalopod faunas as so distinctive from those of the remainder of the Ordovician as to justify regarding the Canadian as a separate system. Neither brachiopod, trilobite, or graptolite faunas seem to be so different in the post-Canadian from the Canadian as to justify such a major subdivision.

Whiterock Stage

This stage was introduced by Cooper and Cooper (*in* Cooper, 1956, pp. 7–8, 126–127) for a group of brachiopod and gastropod zones in Nevada considered to be post-Canadian and pre-Marmor (Chazy). It is thus the earliest stage of the Mohawkian. Trilobites from these zones (except the highest or *Rhysostrophia* zone) have been described, and correlation of the stage discussed (Whittington and Kindle, 1963), showing that it embraces zones L, M, and N of Utah (Ross, 1951; Hintze, 1953) and the Table Head Formation of Newfoundland. These correlations are supported by Flower's (1964) work on cephalopods, and the stage is used in the same way by Ross (1964). Kay (1962, fig. 2) has claimed that zones L–N in Utah and Nevada, and the Table Head Formation, are of the same age as the Day Point Formation, lowest in the Chazy Group, New York. Studies of the brachiopods (Cooper, 1956, p. 28) and trilobites (Whittington, 1965, p. 292; F. C. Shaw, personal communication, 1967) do not support this claim, and emphasise that the faunas of zones L–N are older than any known in the Chazy Group.

The lower limit of the Whiterock Stage is here placed as the base of zone L, but Cooper (1956, p. 130) includes the underlying zone K in this stage. The placement of the uppermost zone, of *Rhysostrophia*, has also been disputed (Ross, 1964) but I have no new information on this matter.

ORDOVICIAN SEQUENCES IN WESTERN NEWFOUNDLAND AND THEIR CORRELATION WITHIN NORTH AMERICA

In western Newfoundland two distinct rock sequences occur, the distribution, thickness and faunas of which have recently been described (Whittington and Kindle, in press). The autochthonous sequence includes the St. George and Table Head Formations, the other (presumably allochthonous) sequence embraces part of the Cow Head Group. In the following paragraphs the faunas of these rocks are considered in some detail.

St. George Formation

These limestones and dolomites overlie Cambrian rocks, without visible break, though there may be a disconformity between the two systems. The earliest fossils found are cephalopods of ellesmeroceratid type, typically Gasconade in age (Flower, 1964), and the brachiopod *Finkeln-burgia*, characteristic of the early Canadian. Higher in the formation are the faunas of lithological division H of Logan (1863), best seen on the outer side of Port au Choix peninsula and

at Cape Norman at the tip of the northern peninsula (Fig. 4–1). This division may be 200 feet (60 m) or more in thickness, and yields the trilobites *Petigurus nero* Billings, 1865, *Bolbocephalus*, four species of bathyurids described by Billings, 1865 (*abruptus*, *marginiatus*, *timon*, and *caudatus*), a bathyurid allied to *Benthamaspis* and referable to the species *gibberulus* Billings, 1865, a carolinitid, an asaphid of *Isoteloides* type, and a pliomerid. At Cape Norman these trilobites are accompanied by the gastropod operculum *Ceratopea billingsi* Yochelson, 1964, who considers that these beds should be correlated with the middle third of the Kindblade Formation of Oklahoma. At Port au Choix, in the mid-part of division H, are opercula similar to *C. tennesseensis*, and presumably of high Kindblade age. From a similar horizon Cumming (1967) has collected graptolites which include *Clonograptus flexilis*, a species which occurs in Newfoundland with *Tetragraptus approximatus* and at slightly lower horizons. Between 50–80 feet (15–25 m) below the top of the St. George Formation on Port au Port peninsula the gastropods *Ceratopea unguis*, *C. keithi* and *Maclurites? odenvillensis* have been collected, suggesting a high West Spring Creek or Smithville equivalent (Yochelson, written communication, 1965).

The faunas of the St. George thus suggest that the formation ranges through the entire Canadian Series. Certain trilobites of lithological division H (*Petigurus*, *Bolbocephalus*, and the *caudatus* type of pygidium) are known in the Fort Cassin Formation of Vermont (Whittington, 1953). In

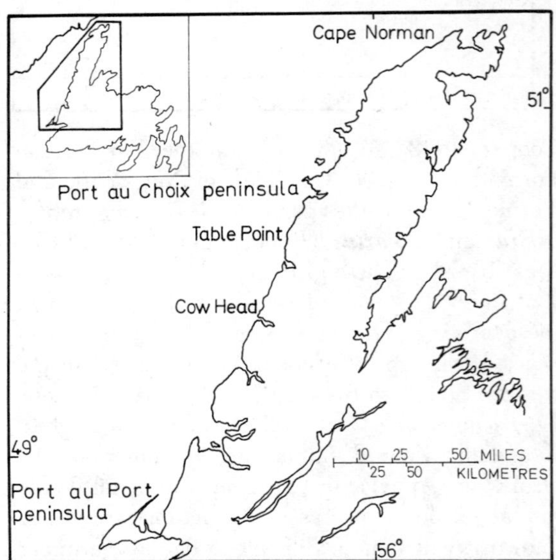

FIGURE 4–1. Principal localities which yield Ordovician fossils in western Newfoundland.

Missouri *Bolbocephalus* ranges from low Jeffersonian (Rich Fountain) to low Cassinian (Powell), and is present in zone G in Utah (Hintze, 1953). Correlation of the Upper Canadian in Table 4–2 is based on these trilobites, supported by occurrences of *Ceratopea* (Yochelson and Bridge, 1957); cephalopods from the upper half of the St. George are also Jeffersonian and Cassinian in aspect (Flower, 1964, p. 151). It is suggested in Table 4–2 that the Upper Canadian (Jeffersonian and Cassinian) is equivalent to part of graptolite zone 3, zones 4–6, and the lower part of zone 7. This is based on the graptolites in division H and the occurrence of *Didymograptus patulus* (Ruedemann, 1947, pl. 55, fig. 22; pl. 56, fig. 17) about 150 feet (45 m) below the top of the St. George (bed 10 of Schuchert and Dunbar, 1934, p. 58). This latter species ranges through zones 4–6 of Berry (1960). In Oklahoma (Decker, 1941) *D. protobifidus* of zone 6 occurs in the West Spring Creek Formation, and in the Ozark region graptolites of zone 7 are recorded from the Black Rock Formation (Ruedemann, 1947, pp. 326–327).

Table Head Formation

These limestones (Whittington and Kindle, 1963) disconformably overlie the St. George Formation, and contain trilobite, brachiopod and cephalopod faunas which are not known elsewhere in bedded strata in the Appalachians, but only in boulders of the Mystic Conglomerate of Quebec (Whittington, 1965, p. 291). Black shales interbedded with limestones which yield middle Table Head trilobites, and which overlie these limestones, contain a graptolite fauna (Whittington and Kindle, 1963; Morris and Kay, 1966) of zone 9 of Berry (1960); the upper Table Head Shales contain graptolites of the same zone. The characteristic trilobites and brachiopods of the Table Head occur not only in Utah and Nevada but also in British Columbia (Norford, 1966).

Cow Head Group

The Ordovician rocks of this group are approximately 500 feet (150 m.) thick, and include thin bedded limestones with some black shale partings, and limestone conglomerates (Kindle and Whittington, 1958). Graptolites were obtained from some of the thin black shale partings; these show the approximate position of the graptolite zones, but because of their very localized occurrence in the section they do not show the lower and upper limits of each zone. No addition to the graptolite species previously recorded is here made. Preliminary lists of trilobite genera obtained from fossiliferous boulders in the conglomerates were given by Kindle and Whittington (1958). Since then many new collections have been made and while the trilobite faunas have not been described in detail, new identifications which supersede previous ones are given in Table 4–1. The conglomerate layers are numbered and lettered as in the earlier account. The list of trilobites from a particular conglomerate layer combines assemblages found in several boulders, but many of those recorded do occur in a single boulder. There is considerable overlap between the assemblages of individual boulders, so that each list can fairly be assumed to be of a single fauna. It is remarkable that no "out of place" boulders have been found, except one Upper Cambrian assemblage previously recorded (Kindle and Whittington, 1958, p. 324) from a boulder in bed 8e. This particular boulder was obtained early in our investigations, and may have been mislabelled.

On the north shore of Cow Head peninsula, in the highest part of conglomerate band 6, boulders have been obtained containing many of the typical elements of the *Hungaia magnifica* fauna so well known from the boulders near Levis, Quebec; those obtained in Newfoundland include the zone species. This conglomerate is overlain by thin bedded limestones with shale partings which have yielded *Dictyonema* and *Staurograptus*.

Five miles (8 km) south of Cow Head peninsula is Broom Point (Kindle and Whittington, 1959). On the north side of the point a series of conglomerates contains boulders which yield the *H. magnifica* fauna and are overlain by thin-bedded limestones containing *Dictyonema*. Professor O. M. B. Bulman considers this graptolite to be of *flabelliforme* type, and probably of early Tremadoc age, though a slightly younger age cannot be excluded. Directly overlying these Broom Point limestones is a boulder bed which has yielded the trilobites listed in Table 4–1. The agnostids and the species of *Symphysurina*, *Highgatella?*, and *Parabolinella?* are like those in the fauna from Vermont described by Shaw (1951), and the species of *Ulrichaspis* resembles one from a boulder of unknown association in the Levis region (Rasetti, 1945). *Levisaspis* occurs in the Lower Ordovician boulder fauna described by Rasetti (1943). *Hystricurus* is typical of beds with *Symphysurina* in Utah and Nevada. It therefore appears that this Broom Point fauna is equivalent in age to zone B of Utah (Hintze, 1953) and Colorado (Berg and Ross, 1959), though the characteristic asaphid *Bellefontia* has not been seen in Newfoundland. Boulders in conglomerate 7 on Cow Head yield some of these same genera and in addition those listed in Table 4–1. This assemblage, though it lacks *Kainella*, suggests zone D of Utah (Hintze, 1953) and Colorado (Berg and Ross, 1959) and is also like that in a boulder at Levis (Rasetti, 1943). This part of the column belongs within the Gasconadian Stage of Flower (1964, pp. 17–18, 23).

Considerable faunas (which contain a number of new genera that have not been seen elsewhere in Newfoundland) have been obtained from boulders in conglomerates 8a and 8e on the south shore of Cow Head peninsula. There is a resemblance between trilobites of conglomerate 8a and those from Nevada described by Ross (1958), but the latter fauna contained a number of new and unnamed forms and its stratigraphic position is uncertain. However, *Bolbocephalus* is present in conglomerate 8a and also in division H of the St. George and the Jeffersonian of Missouri. *Petigurus?*, cf. *Benthamaspis* and cf. *Carolinites* are like forms present in the St. George Formation, division H. The asaphid *Protopresbynileus* is a genus first described from Utah (Hintze, 1953), where it ranges from zones G–J. The faunas of conglomerates 8a–8e seem to be approximately of the same age as the Jeffersonian of Missouri, zone G in Utah, and some part of division H of the St. George. Thin limestones directly overlying conglomerate 8e contain shale partings which yielded *T. approximatus* and species of *Clonograptus*, so that the trilobite faunas of 8a–8e lie within the later part of the *Clonograptus* zone (zone 2 of Berry) and the early part of the succeeding *approximatus* zone 3 of Berry.

Conglomerates 9 and 10 on the south shore of Cow Head and on The Ledge have yielded the trilobites listed (Table 4–1). *Bolbocephalus, Petigurus*, aff. *Benthamaspis gibberulus*, and *Bathyurellus* are present in division H of the St. George Formation. Other genera in these boulders are not known in the St. George Formation, and *Illaenus* is first present in the upper part of the lower Table Head. *Colobinion* is a pliomerid first described from the boulder at Lower Head; fragments of similar white limestone with the same fauna occur in conglomerate 14 at The Ledge. Thus the assemblage from conglomerates 9 and 10 has elements of division H but also elements that suggest younger faunas transitional to the Table Head. The graptolites in Bed 11 include *Tetragraptus fruticosus*, and suggest that the fauna of conglomerates 9 and 10 lies either within this zone or in the highest part of the underlying zone of *T. approximatus*. The

only trilobite in conglomerates 9 and 10 in common with the Utah sequence is *Bolbocephalus* of zone G. There is no sign in the boulders at Cow Head (nor in the St. George Formation) of the trilobites of zones H–J of Utah—for example the asaphids *Trigonocerca, Trigonocercella, Lachnostoma*, and *Ptyocephalus*, the bathyurid *Goniotelina*, and the pliomerids *Cybelopsis, Pseudocybele, Hintzeia*, and *Protopliomerella*. Since *Bolbocephalus* is present in both division H in Newfoundland and zone G in Utah, it appears that these faunas of zones H to J of Utah are younger than those of division H in Newfoundland. This would further imply that at least part of zone G, and zones H–J are all Cassinian in age, a slightly different correlation from that of Flower (1964, p. 23). It is curious that the characteristic asaphid *Ptyocephalus* (Whittington, 1948), found widely in the higher Canadian of the west, the Smithville of Arkansas, and boulders in the Mystic Conglomerate of Quebec, has not been found in either sequence in western Newfoundland. Conversely, not only is the typical assemblage of the Fort Cassin Formation unknown in the sequence in Utah, but the work of Dr. L. F. Braithwaite (written communication, 1967) has not revealed the presence of either *Tetragraptus approximatus* or *T. fruticosus* in the Utah sequence. One explanation is that there may be a non-sequence in the Utah succession, somewhere about the top of zone G and below zone H, where the Fort Cassin faunas and the graptolite zones mentioned belong. An alternative explanation may be that the tribolites from the Utah sequence have all been obtained as silicified specimens from limestones dissolved in acid, and that collections made by breaking the limestones may reveal trilobite faunas that are not silicified and have so far been overlooked.

Limestone conglomerate 12, exposed at The Ledge and Jim's Cove on Cow Head peninsula, has yielded trilobites typical of both the lower and middle Table Head. Unusual is the presence of a pygidium of *Lloydia*, which appears to represent a different species from that described from a boulder which may have come from Lower Head (Whittington, 1963, p. 65–66). The present specimen suggests that the range of this genus may be much greater than is at present assumed. From shale partings in limestones overlying conglomerate 12 come the graptolites *Trigonograptus* and *Isograptus*, characteristic of Berry's zone 8.

Limestone conglomerate 14 crops out on either side of Deep Cove on the south west side of Cow Head, and contains boulders yielding trilobites typical of the middle Table Head, as well as large white boulders containing some of the genera of the Lower Head boulder. At Table Point the upper Table Head shales contain graptolites of Berry's zone 9. This evidence, combined with that at Cow Head, suggests that the lower and middle Table Head spans zone 8 and ranges into at least the lower part of zone 9. The lower Table Head may range down into graptolite zone 7, for in Oklahoma (Cooper, 1956, p. 118–119) the characteristic *Didymograptus artus* occurs in the Joins Formation with Whiterock brachiopods, and in Utah *D. bifidus* occurs in zone M.

Correlations within North America of shelly and graptolite zones are summarised in Tables 4–1 and 4–2. Correlations within the Gasconadian and Jeffersonian stages seem reasonably well established. Equivalents of the Demingian on the basis of either trilobites or graptolites are problematical, but the characteristic *Lecanospira* fauna is widely recognised (Sando, 1957; Flower, 1964). There appears room for debate on the exact equivalents of the lower and middle parts of the Cassinian, but the general outlines of the correlation, based on a variety of faunal elements, are clear. These correlations are in agreement with those given by Berry (1960, p. 19, 23, table 2) who used some different lines of evidence.

CORRELATIONS WITHIN THE NORTH ATLANTIC REGION (Figure 4-2)

Greenland

The following notes are based on work by Cowie *in* Raasch, 1961; Cowie and Adams, 1957; C. Poulsen, 1927, 1937, 1951; and Troelsen, 1950. Upper Cambrian strata are not known in Greenland, and the earliest Ordovician is the Cass Fjord Formation, evidently Gasconadian in age since it contains *Symphysurina, Hystricurus* and other trilobites. The overlying Cape Clay Formation also contains *Symphysurina* and appears to belong within the same stage. Overlying formations in northwest Greenland, the Poulsen Cliff Shale and the Nygaard Bay Limestone, contain few fossils and their correlation is uncertain. The Cape Weber Limestone, however, yields a fauna typical of lithological division H of western Newfoundland including *Bolbocephalus, Petigurus* and other bathyurids, *Carolinites* and *Isoteloides*. The occurrence of *Raymondaspis* and *Remopleurides* suggests the presence of genera like those in conglomerate 8e of western Newfoundland. In eastern Greenland, the Cape Weber is overlain by the Narhval Sound Formation, which contains no fossils that appear to be diagnostic for correlation, except that Yochelson (1964) records a species of *Ceratopea* similar to that from division H in northern Newfoundland. The age therefore appears to be Cassinian and not younger as has been suggested. On the other hand, the Heimbjerge Formation is apparently much younger, the fauna having the aspect of the Barnveld Stage of the Trenton.

In northwestern Greenland the Nunatami Formation overlies the Cape Weber, and includes beds which yield *Didymograptus* of a type found in the highest *fruticosus* zone or slightly younger beds (W. B. N. Berry, oral communication, May, 1967). Trilobites in the overlying gastropod and ostracod limestones contain *Bolbocephalus*, species of *Goniotelina* and the pliomerid *Cybelopsis*. The latter two genera are typical of zone J in Utah, and the strata may be slightly younger than division H in Newfoundland. The lower part of the Centrum Limestone of northeast Greenland appears to be of the same age as the Nunatami Formation.

Spitsbergen

I have re-examined the material from Ny Friesland described by Gobbett and Wilson (1960), and as their work shows, the Lower Oslobreen Formation contains *Hystricurus* and ellesmeroceratids indicative of the Gasconadian Stage. The Middle Oslobreen Formation contains *Bolbocephalus, Bathyurina, Bathyurellus*, and *Hystricurus*, suggestive of an age approximately the same as division H of northern Newfoundland. No trilobites were recovered from the Upper Oslobreen Formation. I have also examined limestones collected by the Cambridge University Expedition in 1965 from northern Ny Friesland, which contain a raphiophorid, *Carolinites, Nileus*, an asaphid, and a cybelid. This assemblage is suggestive of the Table Head Formation and is the youngest Ordovician so far discovered in the island.

Rocks of Canadian age farther south, in Sørkapp land, were described by Major and Winsnes (1955). These include cephalopods and the gastropod *Ceratopea*, the latter indicative of Upper Canadian age.

Ellesmere Land

Fossils from eastern Ellesmere Land have been described by C. Poulsen (1946) and further comments on the region given by Troelsen (1950). The fossils indicate that both the Cape Weber and Nunatami formations are present, yielding faunas that are similar to those known from Greenland.

Western Ireland

The following lists give my new identifications of trilobites from west of Lough Mask (Reed, 1945, which see for 1909 and 1910 references), and the Sedgwick Museum (group A) catalogue numbers:

Shangort Beds (ashy grits):
Ischyrotoma sp. 10430
Niobe? ornatus (Reed, 1945). 10418
asaphid gen. ind. 10440
Nileus (not *Symphysurina* as Reed, 1945, p. 64) sp. 10397, 10417, 10465
Uromystrum glensaulensis (Reed, 1910). 10410–2
Acidiphorus sp. 10381, 10413, 10429
Annamitella? reynoldsi (Reed, 1945). 10414
Illaenus of *I. consobrinus* type. 10464
Illaenus of *I. marginalis* type. 10577
Ectenonotus connemaricus (Reed, 1909) (= *octocostatus* Reed, 1909). 10382–4, 10392–5, 10405, 10415, 10419, 10420, 10427–8, 10431, 10454, 10462, 10480.
Colobinion? shangortensis (Reed, 1945). 10380, 10391 (pygidia only)
Kawina sp. 10421

Tourmakeady Beds (fine-grained pink and grey mottled limestone).
Telephina hibernicus (Reed, 1909). 10398–401
Nileus sp. 1043
Illaenus weaveri Reed, 1909. 10386, 10389–90.
Illaenus cf. *tumidifrons*. 10385, 10387, 10488–9.
Illaenus cf. *consobrinus*. 10456–7
Kawina divergens Reed, 1945. 10396
Cydonocephalus? sp. 10399
odontopleurid. 1043

The resemblance between these trilobites and those of the Whiterock Stage of North America was previously commented upon (Whittington, 1963, p. 20), but the revised lists greatly strengthen this resemblance. *Ischyrotoma, Acidiphorus, Ectenonotus, Kawina* and *Cydonocephalus* are otherwise known only from North America, and *Uromystrum* from North America and eastern Greenland. The pygidium here referred to *Annamitella?* is another possible record of this widespread genus, which has been recognised in ashy strata in northern Maine of presumed Whiterock age (Whittington *in* Neuman, 1964). Thirty miles (48 km) northeast of Lough

Mask, near Charlestown, cherty shales interbedded with volcanic rocks have yielded Table Head type graptolites, *Isograptus* and *Glossograptus*, together with *Oncograptus* (Cummins, 1954).

Durness Limestone, Scotland

Brief notes on these rocks, and a section, are given by Gobbett and Wilson (1960), but though the presence of 'Canadian' faunas is indicated, such faunas have never been described. I have observed a pygidium of *Petigurus* in the collections of the Geological Survey & Museum, London, and examples of *Ceratopea* in the collections of the Grant Institute of Geology, Edinburgh. These fossils suggest a relationship to North American and Greenland faunas.

Trondheim Area, Norway

Fossils from the Hølonda Limestone, which occurs in a thick sequence of conglomerates, sandstones and shales in this region, were described by Strand (1948). I have re-examined these fossils through the kindness of Professor L. Størmer, and they include a pygidium of *Acidiphorus* type, a pliomerid, an asaphid, and a pygidium which appears to belong to *Niobe*. These fragmentary and poorly preserved fossils may belong within the time interval represented by the Table Head Formation. Beneath the Hølonda Limestone is the Bogo Shale, which has yielded a graptolite fauna (Blake, 1962; Skevington, 1963a) which includes species of *Isograptus, Glossograptus* and *Hallograptus* suggestive of Berry's zones 8 and 9.

Southern Norway, Sweden and Denmark (Bornholm)

In their considerations of correlations of graptolite sequences, Skevington (1963b) and Jackson (1964) have maintained that the *bifidus* zone of North America (zone 7 of Berry, 1960) is equivalent to the middle part of the *extensus* zone of the Arenig of Britain. Correlations of the Swedish and Norwegian sequences with British graptolite zones have been given by Tjernvik (1956) and Jaanusson (1960). In addition, Dr. V. Jaanusson informs me (personal communication, 17 Oct., 1967) that *Tetragraptus approximatus* is recorded from Sweden close to the base of the Hunneberg substage, and that *T. fruticosus* ranges to the top of the Billingen substage. The Skevington–Jackson correlation implies that North American zone 7 should be equated approximately with the Billingen substage of Sweden (cf. Table 4–2). The Table Head faunas, which cer-

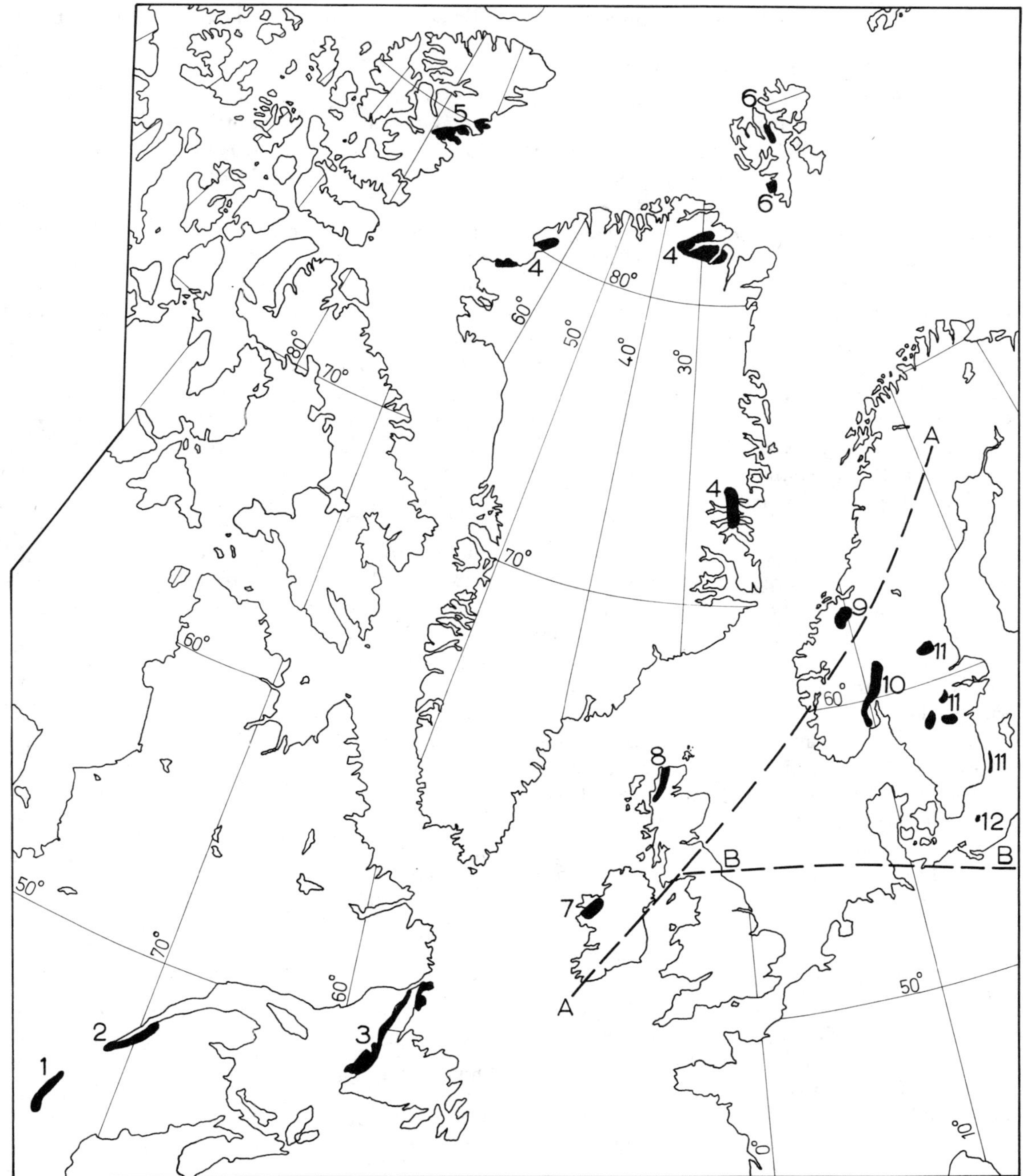

FIGURE 4–2. Areas where fossiliferous Ordovician sequences of Canadian and early Mohawkian age are present in the North Atlantic and Baltic regions. Fit of continents after Bullard, Everett and Smith (1965). 1, northern New York State and southern Quebec. 2, Levis, Quebec, shales and boulder beds. 3, western Newfoundland. 4, Greenland. 5, Ellesmereland. 6, Spitsbergen. 7, western Ireland. 8, northwestern Scotland. 9, Trondheim area, Norway. 10, Oslo district, Norway. 11, central and southern Sweden. 12, Bornholm, Denmark. Areas west of line A–A have shelly faunas of bathyurid type; those between line A–A and B–B of asaphid type; south of line B–B and east of line A–A, in southeastern Ireland, central Britain and central and southern Europe, the shelly faunas are of a third or southern type (Whittington, 1966).

tainly embrace zone 8 and may begin at the top of zone 7, would thus have to be correlated with the Volkhov Stage. There are some genera in common between the Table Head and an early Volkhov Stage fauna from Bornholm described by V. Poulsen (1965), but these are all long-ranging (*Geragnostus, Harpides, Selenoharpes, Ampyx, Niobe, Nileus* and *Raymondaspis*) and the species are dissimilar. On the other hand, the Table Head (Whittington, 1965) is characterised by a profusion of illaenid species, quite like those from the *raniceps* to *gigas* beds of the Kundan Stage in Sweden. The species of *Niobe* in the Table Head is remarkably similar to that from the Kundan Stage, while in the early part of the Ogygiocaris Series of Norway such Table Head genera as *Telephina, Miracybele* and *Bronteopsis* are represented by similar species. The appearance of these unrelated trilobite groups in beds of lower Llanvirn age in Norway and Sweden may be at about the same time as in North America, so that I argue for the correlation given in Table 4–2. This implies that the Chazy (Marmor-Ashby) faunas are roughly the same age as the Uhaku Stage. This seems a more likely alternative than the Skevington–Jackson view, which implies that the Chazy faunas would have to be equated with much of the Kundan Stage as well as the Aseri, Lasnamägi and Uhaku stages. Strong similarities between shelly faunas of particular horizons of the Canadian to Whiterock interval with those of beds in the Baltic region have not yet been discovered, so correlations are necessarily tentative.

Central Britain, Central and Southern Europe

Shelly faunas of Tremadoc to Llandeilo age in this region are almost entirely different from those of North America (Whittington, 1966), and the column for Britain in Table 4–2 is based on correlations between British and Baltic graptolite zones given by Tjernvik (1956) and Jaanusson (1960). The British zones are brought, by the arguments given in the preceding section, into a position that is in agreement with correlations of British and American graptolite zones advocated by Berry (1967).

Correlations and Faunal Provinces

Successive faunas of the Canadian and Whiterock may be correlated over a vast area which lies west of the line A–A in Figure 4–2, a single, distinctive faunal province termed bathyurid (Whittington, 1966). East of this line in the Baltic region (10–12, Fig. 4–2) are different faunas termed the asaphid type. In central Britain and southwards into central Europe are faunas of a third or southern type. The lines A–A and B–B (Fig. 4–2) between these faunal provinces do not represent barriers to migration, for certain genera of both trilobites and graptolites occur over the whole area. Such trilobite genera, however, are stratigraphically long-ranging and a small proportion of any particular fauna, so that their significance to correlation is debatable. Factors which may have determined the extent, duration and nature of these faunal provinces are discussed elsewhere (Whittington, 1966). They began to break down in Whiterock and succeeding time, at an increasing rate. A particular fit of continents around the North Atlantic (Bullard, et al., 1965) is used as a base for Figure 4–2, and has the effect of reducing the areal extent of the North American province in this region. This effect is minor on the size of the entire province, however, for it extended across southern and western North America to northeastern USSR and the Siberian platform.

REFERENCES

Aitken, J. D., and Norford, B. S., 1967, Lower Ordovician Survey Peak and Outram Formations, southern Rocky Mountains of Alberta: Bull. Canadian Petrol. Geol., v. 15, p. 150–207

Berg, R. R., and Ross, R. J., 1959, Trilobites from the Peerless and Manitou formations, Colorado: Jour. Paleont., v. 33, p. 106–119

Berry, W. N., 1960, Graptolite faunas of the Marathon region, west Texas: Bur. Econ. Geology, Univ. Texas, Pub. no. 6005, 179 p.

__________, 1962, Stratigraphy, zonation and age of Schaghticoke, Deepkill and Normanskill shales, eastern New York: Geol. Soc. America Bull., v. 73, p. 696–718

__________, 1967, Comments on correlation of the North American and British Lower Ordovician: Geol. Soc. America Bull., v. 78, p. 419–427

Billings, Elkanah, 1861–65, Palaeozoic fossils, vol. 1: Geol. Surv. Canada, 426 p.

Blake, D. M., 1962, A new Lower Ordovician graptolite fauna from the Trondheim region: Norsk Geol. tiddskr., v. 42, p. 223–238

Bullard, E., Everett, J. E., Smith, A. Gilbert, 1965, The fit of the continents around the Atlantic: Phil. Trans. Royal Soc., London, v. 258, p. 41–51

Cooper, G. A., 1956, Chazyan and related brachiopods: Smithson. Misc. Coll., v. 127, pts. 1 and 2, 1245 p.

Cowie, J. W., and Adams, P. J., 1957, The geology of the Cambro–Ordovician rocks of central east Greenland. Part 1. Stratigraphy and Structure: Medd. Grønland, v. 153, no. 1, 193 p.

Cumming, L. M., 1967, *Clonograptus* from the St. George Formation, Newfoundland: Geol. Surv. Canada, Paper 67–1, part B, Report of Activities, p. 61–63.

Cummins, W. A., 1954, An Arenig volcanic series near Charlestown, Co. Mayo: Geol. Mag., v. 91, p. 102–104

Dana, J. D., 1874, Reasons for some of the changes in the subdivisions of geological time in the new edition of Dana's Manual of Geology: Amer. J. Sci., Ser. 3, v. 8, p. 213–216

Decker, C. E., 1941, *Didymograptus protobifidus* in North America: Jour. Paleont., v. 15, p. 362–365

Flower, R. H., 1957, Studies of the Actinoceratida: State Bureau Mines and Mineral Res., New Mexico, Mem. 2, 73 p.

————, 1964, The nautiloid order Ellesmeroceratida (Cephalopoda): State Bureau Mines and Mineral Res., New Mexico, Mem. 12, 234 p.

Gobbett, D. J., and Wilson, C. B., 1960, The Oslobreen Series, Upper Hecla Hoek of Ny Friesland, Spitsbergen: Geol. Mag., v. 97, p. 441–460

Hintze, L. F., 1953, Lower Ordovician trilobites from western Utah and eastern Nevada: Utah Geol. Mineral. Surv. Bull. 48, 249 p.

Jaanusson, V., 1960, Graptoloids from the Ontikan and Viruan (Ordov.) Limestones of Estonia and Sweden: Geol. Instit. Uppsala, Bull., v. 38, p. 289–366

Jackson, D. E., 1964, Observations on the sequence and correlation of Lower and Middle Ordovician graptolite faunas of North America: Geol. Soc. America Bull., v. 75, p. 523–534.

————, 1966, Graptolite facies of the Canadian Cordillera and Arctic Archipelago: A Review: Bull. Canadian Petrol. Geol., vol. 14, p. 469–485

Kay, M., 1962, Classification of Ordovician Chazyan shelly and graptolite sequences from Central Nevada: Geol. Soc. America Bull., v. 73, p. 1421–1429

Kindle, C. H., and Whittington, H. B., 1958, Stratigraphy of the Cow Head region, western Newfoundland: Geol. Soc. America Bull., v. 69, p. 315–342

————, 1959, Some stratigraphic problems of the Cow Head area in western Newfoundland: New York Acad. Sci., Trans., Ser. II, v. 22, no. 1, p. 7–18

Logan, W. E., 1863, Geology of Canada: Montreal, Dawson Brothers, xxvii + 983 p.

Major, H., and Winsnes, T. S., 1955, Cambrian and Ordovician fossils from Sørkapp Land, Spitsbergen: Skr. Norsk Polarinstit., no. 106, 47 p.

Morris, R. W., and Kay, M., 1966, Ordovician graptolites from the Middle Table Head Formation at Black Cove, Port au Port, Newfoundland: Jour. Paleont., v. 40, p. 1223–1229

Neuman, R. B., 1964, Fossils in Ordovician tuffs, northeastern Maine: U. S. Geol. Surv., Bull. 1181–E, 38 p.

Norford, B. S., 1966, Ordovician and Silurian stratigraphy of the Southern Rocky Mountains of Canada: Geol. Soc. America, Abstracts for 1965, Spec. Pap. 87, p. 118.

Osborne, F. F., and Berry, W. B. N., 1966, Tremadoc rocks at Levis and Lauzon: Naturaliste Canadien, v. 93, p. 133–143

Poulsen, C., 1927, The Cambrian, Ozarkian and Canadian faunas of northwest Greenland: Medd. Grønland, v. 70, p. 233–343

————, 1937, On the Lower Ordovician faunas of East Greenland: Medd. Grønland, v. 119, no. 3, 72 p.

————, 1946, Notes on Cambro–Ordovician fossils collected by the Oxford University Ellesmere Land Expedition, 1934–5; Geol. Soc. London, Quart. Jour., v. 102, p. 299–337

————, 1951, The position of the East Greenland Cambro–Ordovician in the palaeogeography of the North Atlantic region: Medd. Dansk Geol. Foren., v. 12, p. 161–2

Poulsen, V., 1965, An early Ordovician trilobite fauna from Bornholm: Medd. Dansk geol. foren., v. 16, p. 49–113

Raasch, G. O., Editor, 1961, Geology of the Arctic, vol. 1, Toronto: Univ. Toronto Press, 732 p.

Rasetti, F., 1943, New Lower Ordovician trilobites from Levis, Quebec: Jour. Paleont., v. 17, p. 101–104

————, 1945, New Upper Cambrian trilobites from the Levis conglomerate: Jour. Paleont., v. 19, p. 462–478

Reed, F. R. C., 1945, Revision of certain Lower Ordovician faunas from Ireland. 1. Trilobites: Geol. Mag., v. 82, p. 55–66

Ross, R. J., 1951, Stratigraphy of the Garden City Formation in northeastern Utah and its trilobite faunas: Peabody Mus. Nat. Hist., Bull. 6, 161 p.

————, 1958, Trilobites in a pillow–lava of the Ordovician Valmy Formation, Nevada: Jour. Paleont., v. 32, p. 559–570

————, 1964, Middle and Lower Ordovician Formations in southernmost Nevada and adjacent California: U. S. Geol. Surv., Bull. 1180–C, 101 p.

Ruedemann, R., 1947, Graptolites of North America: Geol. Soc. America, Memoir 19, 652 p.

Sando, W. J., 1957, Beekmantown Group (Lower Ordovician) of Maryland: Geol. Soc. America, Mem. 68, 161 p.

Schuchert, Charles, and Dunbar, C. O., 1934, Stratigraphy of western Newfoundland: Geol. Soc. America, Memoir 1, 123 p.

Shaw, A. B., 1951, The paleontology of northwestern Vermont. 1. New late Cambrian trilobites: Jour. Paleont., v. 25, p. 97–114

Skevington, D., 1963a, A note on the age of the Bogo shale: Norsk geol. tiddskr., v. 43, p. 257–260

————, 1963b, A correlation of Ordovician graptolite-bearing sequences: Geol. Fören. Förhandl., v. 85, p. 298–319

Strand, T., 1948, New trilobites from the Hølonda limestone (Trøndheim region, southern Norway): Norsk. Geol. tidsskr., v. 27, p. 74–88

Tjernvik, T. E., 1956, On the early Ordovician of Sweden. Stratigraphy and fauna: Geol. Instit. Uppsala, Bull., v. 36, p. 107–284

Troelsen, J. C., 1950, Contributions to the geology of northwest Greenland, Ellesmere Island and Axel Heiberg Island: Medd. om Grønland, v. 149, 86 p.

Twenhofel, W. H., Chairman, 1954, Correlation of the Ordovician formations of North America: Geol. Soc. America Bull., v. 65, p. 247–298

Ulrich, E. O., and Cooper, G. A., 1938, Ozarkian and Canadian Brachiopoda: Geol. Soc. America, Spec. Pap. 13, 323 p.

Whittington, H. B., 1948, A new Lower Ordovician trilobite: Jour. Paleont., v. 22, p. 567–572

————, 1953, North American Bathyuridae and Leiostegiidae (Trilobita): Jour. Paleont., v. 27, p. 647–678

————, 1963, Middle Ordovician trilobites from Lower Head, western Newfoundland: Mus. Comp. Zoology, Harvard Univ., Bull., v. 129, p. 1–118

————, 1965, Trilobites of the Ordovician Table Head Formation, western Newfoundland: Mus. Comp. Zoology, Harvard Univ., Bull., v. 132, no. 4, p. 275–441

————, 1966, Phylogeny and distribution of Ordovician trilobites: Jour. Paleont., v. 40, p. 696–737

__________, and Kindle, C. H., 1963, Middle Ordovician Table Head Formation, western Newfoundland: Geol. Soc. America, Bull., v. 74, p. 745–758

__________, and Kindle, C. H., in press, Cambrian and Ordovician stratigraphy of western Newfoundland. Proceedings of Gander (Newfoundland) International Conference on "Stratigraphy and structure bearing on continental drift in the North Atlantic Ocean". August, 1967, publication as Memoir, Amer. Assoc. Petrol. Geol.

Yochelson, E. L., 1964: the early Ordovician gastropod *Ceratopea* from East Greenland: Medd. Grønland, v. 164, no. 7, 10 p.

__________, and Bridge, J., 1957, The Lower Ordovician gastropod *Ceratopea:* U. S. Geol. Surv., Prof. Pap. 294–H, p. 281–304

Chapter 5

Stratigraphic Evidence for the Taconic Orogeny in the Northern Appalachians[*]

LOUIS PAVLIDES, A. J. BOUCOT,

AND W. B. SKIDMORE

INTRODUCTION

The Taconic orogeny, as generally defined for eastern North America in most text books, is the orogenic event that occurred at or near the end of the Ordovician. The classical terrane of the Taconic is mostly the folded Appalachian chain that extends from the Delaware Water Gap, Pennsylvania, northward to the Maritime Provinces of Canada. This report briefly reviews and evaluates the stratigraphic evidence for the Taconic unconformity in this region, exclusive of Newfoundland.[†]

H. D. Rogers (1856, p. 178–180; 1858, p. 784–786) was the first geologist to recognize clearly the regional magnitude of the Taconic break; an unnamed unconformity that he described as extending from the Gaspé Peninsula of Quebec to the Hudson River Valley of New York and that separates Ordovician (approximately "Matinal" of Rogers) rocks from overlying younger rocks. Dana (1895, p. 386) named the orogeny associated with this unconformity "the Taconic mountain making crisis" and he ascribed to it the formation of the Taconic Mountains and other uplifts in New Brunswick and Nova Scotia as well as the general folding of the Ordovician strata that are unconformably overlain by Silurian or younger rocks. Since then, this diastrophic event has been variously described as

the "Taconic" Disturbance, Orogeny, or Revolution. Some geologists prefer the name "Taconian" to avoid confusion with the Taconic "klippe" problem, whose type locality is also the Taconic Range of New York and Vermont. Despite this unfortunate synonymy of terms preference is given to the term "Taconic" in this report, following Dana's nomenclature of 1895.

While the concept of the Taconic orogeny and a resultant regional unconformity was being developed by Rogers and Dana for the Appalachians northward from Otisville, N. Y., controversy developed as to the presence of such an unconformity in Pennsylvania (White, 1882; Chance, 1882; Lesley, 1892). However, by the early part of the 20th century, many of the Silurian and Ordovician sections exposed in the wind and water gaps of Pennsylvania were accepted as being unconformable and the Taconic orogeny became accepted as a major orogenic event extending from southern Pennsylvania through Maritime Canada (Schuchert, 1916). A major challenge to the regional extent and significance of the Taconic orogeny was published by T. H. Clark in 1921. Clark essentially tried to discredit the stratigraphic evidence for a Taconic unconformity at many of the then classic localities of the northern Appalachians. He suggested that the "Taconic" angular discordances described in parts of Quebec and elsewhere in Maritime Canada may be the result of disharmonic folding followed by thrusting. By this hypothesis, the less competent rocks of the Ordovician were closely folded and eventually overridden by the more openly folded, competent

<hr>

[*] Publication authorized by the Director, U. S. Geological Survey.

[†] Since the preparation of this report, Rodgers (1967) has reviewed the evidence for all orogenies in the Appalachian belt, including the Taconic orogeny in Newfoundland.

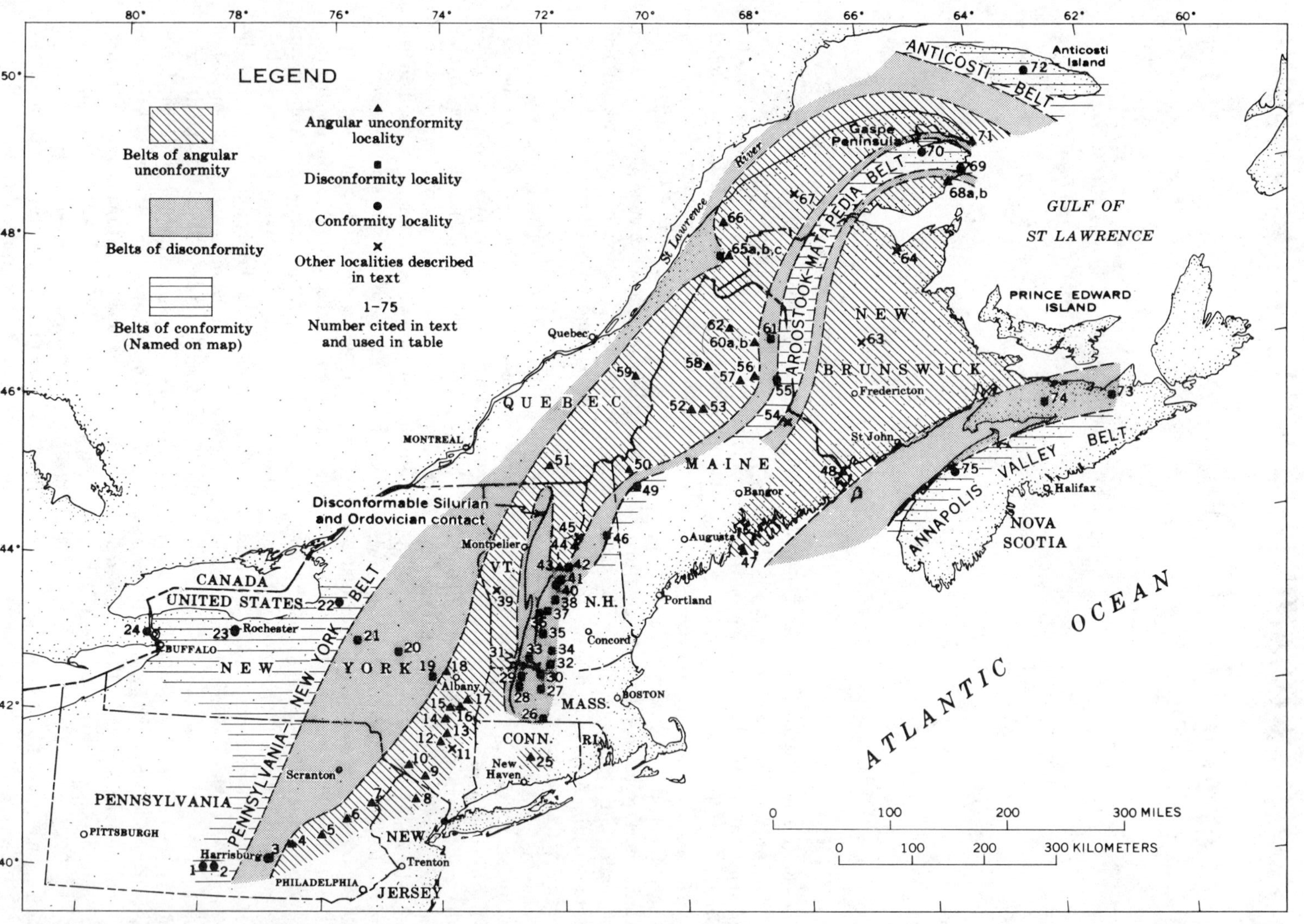

Figure 5–1. Distribution of unconformities ascribed to the Taconic orogeny and their relationship to areas of conformity (uninterrupted sedimentation) during this orogeny.

rocks (generally sandstones) of the Silurian. Clark proposed to limit the Taconic orogeny to a local event confined to eastern New York, essentially to an area between the unconformities at Otisville and Rondout, N. Y. Clark's paper had a marked, albeit short-lived, influence on geologic thinking; Schuchert (1925; and *in* Pirsson and Schuchert, 1924, p. 243–244) apparently accepted Clark's evaluation. However, reevaluation of the evidence for the Taconic unconformity by many investigators at various places in the Appalachians soon reestablished its validity as a regional diastrophic event, and it has been more or less uncontested to this day. Problems still exist, however, as to (1) the geologic time-span in which the Taconic orogeny occurred; (2) whether it was a single, synchronous event or a series of discontinuous but related events; and (3) whether it was a geographically widespread or restricted event. This report attempts to gain an insight into these problems by evaluating the various types of unconformities ascribed to the Taconic orogeny and the hiatuses represented at the unconformities, and their areal distribution.

ACKNOWLEDGMENTS

The helpful suggestions of many geologists during the preparation of this report are greatly appreciated. We are also grateful to colleagues who generously allowed use of their unpublished data as written communications. Special thanks are due Wallace M. Cady, John M. Aaron, and Donald W. Fisher who critically reviewed this report.

UNCONFORMITIES ASCRIBED TO THE TACONIC OROGENY

Numerous angular unconformities and disconformities throughout the central and northern Appalachians have been cited as evidence for the Taconic orogeny. The more important and characteristic of these are located on Figure 5–1 and listed in Table 5–1 under the major headings of angular unconformity and disconformity. Nonconformities are not considered in this paper. Only a minimum of description is given for the listed unconformities; more detailed accounts of these localities can be obtained from the included selected references. Formation names and the ages of the rocks above and below the unconformities may differ from the original nomenclature and age assignment of the cited references; many of the ages of the Silurian units are those established by Berry and Boucot (in press). Figure 5–2 indicates the correlation between European and American sections and also indi-

cates the magnitude of the hiatus at each described unconformity.*

In a general way, areas of Taconic unconformity and disconformity occur in belts that parallel areas of conformity wherein there is no Taconic break between Ordovician and Silurian strata (Fig. 5–1). For example, between the areas of conformity of Anticosti Island and of the Aroostook–Matapedia belt is a belt characterized by angular unconformity. On the northwest and southeast sides of the Arrostook–Matapedia belt are areas of probable Silurian and Ordovician disconformity; still farther to the southeast, in the Chaleurs Bay region, is an area of angular unconformity. This area may extend southward as a belt and include the coastal regions of southern New Brunswick and adjacent Maine. Farther to the southeast, in the Arisaig–Cobequid Mountains region of Nova Scotia (loc. 74), a disconformity may separate Ordovician and Silurian strata, and southeast of the Arisaig–Cobequid Mountains region is the Annapolis Valley area of Nova Scotia (loc. 75), an area of possible conformity between Ordovician and Silurian rocks.

Unconformities generally pass into disconformities and eventually into conformable sequences in progressing away from land areas outwards into the basin of sedimentation. This geometric pattern has been followed in constructing parts of Figure 5–1, on the assumption that although physical evidence is not now available to demonstrate the existence of a particular conformable or disconformable relationship, it did or does exist.

As will be subsequently described, the Taconic orogeny occurred at different times, at different places. For purpose of definition and on the basis of regional evaluation of Taconic unconformities in the Appalachians, it is suggested that the term Taconic orogeny be limited to the diastrophic events that took place in the central and northern Appalachians in the time span from Trenton (Graptolite Zones 12–13) in the Ordovician up to Llandovery C_3–C_5 in the Silurian. Of necessity, Figure 5–1 generalizes the physical limits of the belts of unconformity and conformity into composite belts of tectonism for the entire time interval of the Taconic orogeny. Indeed, the actual boundaries of these belts are not precisely known. Figure 5–1 only suggests the general distribution of such belts during the entire time span

---

* For convenience, unconformities described in this report generally pertain to a specific place where a break can be demonstrated; actually many such individual localities represent the same regional unconformity.

of the Taconic; the configuration of these belts may have been quite different at various times during the time span of the Taconic orogeny.

ANGULAR UNCONFORMITIES

Western Belt

New York, New Jersey, and Pennsylvania. One of the classic Taconic angular unconformities occurs near Otisville, N. Y. (loc. 10, Table 5–1, and Figs. 5–1 and 5–2)* where the Shawangunk Conglomerate of Silurian (Llandovery) age† rests on more steeply dipping and truncated Normanskill Shale of Late Ordovician age (Graptolite Zone 14). Southwestward from the Otisville area, the Shawangunk Conglomerate continues through New Jersey along Kittatinny Mountain and then into Pennsylvania where it forms Blue Mountain and is known as the Tuscarora Sandstone. Blue Mountain is cut by a series of water gaps that expose essentially the same Silurian and Ordovician contact between Delaware and Schuylkill Gaps (locs. 7 and 5), as near Otisville. Whether the angular discordance at some of these Pennsylvania localities, and particularly at Schuylkill Gap (loc. 5), is an unconformity or a fault has been debated for many years. At Schuylkill and some of the other gaps, some local fault movement may have taken place at or near the contact between the Tuscarora and the Martinsburg Shale, but the regional relationships suggest that the contact is essentially an angular unconformity.

The unconformity at Swatara (loc. 4) and Lehigh (loc. 6) Gaps, Pa., differs from that in the Schuylkill (loc. 5) and Delaware (loc. 7) Gaps to the east in that the Bald Eagle Conglomerate of Grabau (1909), that here may be of Maysville–Richmond age, intervenes between the Tuscarora and the Martinsburg. The Martinsburg below the Bald Eagle at Swatara Gap contains fossils of middle(?) Eden age (Stose, 1930; and Willard and Cleaves, 1939, p. 1180). In contrast, at Schuykill Gap, the Martinsburg below the Tuscarora contains fossils as young as early Maysville in age (Stose, 1930).

In the Green Pond outlier of New Jersey (loc. 8) and New York (loc. 9) the Silurian Green Pond Conglomerate (Wenlock age) rests with angular unconformity on rocks that range in age from Precambrian to Middle Ordovician.

---

* Hereafter, for brevity, only locality numbers will be cited in the text with the understanding that these localities can be identified by the reader in Table 5–1 and Figures 5–1 and 5–2.

† Ages of the units bracketing the unconformities generally refer to those strata immediately adjacent to the contact.

In New York the lithology of the rocks above the Taconic unconformity change from conglomerate and sandstone at Otisville and in the Rondout Valley area (loc. 12) to limestone in the Hudson River Valley to the east. At Rondout (loc. 13) the tilted beds of the Normanskill Shale of Middle Ordovician (Graptolite Zone 13) age, that have been beveled to an irregular erosion surface containing small channels, are overlain with angular unconformity by the Silurian Rondout Limestone of Pridoli age. Farther north (locs. 16–18) where the Manlius Limestone (Gedinnian age) is above the unconformity, the hiatus is much larger. At Becraft Mountain (loc. 16) furthermore, the post-unconformity rocks lie across the trace of a Taconic thrust (Fisher and others, 1962). However, at Mount Ida (loc. 17), about 2 miles northeast of Becraft Mountain, the autochthonous Manlius unconformably overlies the allochthonous Lower Ordovician and Upper Cambrian rocks (Fisher and others, 1962). However, because the allochthon was emplaced during the Middle Ordovician, the hiatus here spans Middle Ordovician to Manlius time.

New England, Quebec, and New Brunswick. The recognition of angular unconformities and disconformities (described below) of Taconic age in much of New England (exclusive of Maine) depends on the recognition or correlation of a stratigraphy synthesized from widely spaced localities. The pre-Taconic marker beds are generally the Ammonoosuc Volcanics and the overlying Partridge Formation, both probably of Middle Ordovician age (Billings, 1956, p. 96). The age of these units is based on lithologic similarity to the Cram Hill Formation (Cram Hill Volcanic Member of the Missisquoi Formation of Doll and others, 1961) which Currier and Jahns (1941) traced from northern Vermont into the Magog Slate (Clark, 1934, p. 80) or the Beauceville Formation of Quebec, that contains graptolites (zones 12 and 13) of Middle Ordovician age (Berry, 1962). In central Vermont the Cram Hill grades southward into the Barnard Volcanic Member of the Missisquoi of Doll and others (1961). The Barnard extends southward into Massachusetts as the Hawley Formation.

Additional support for a Middle Ordovician age for the Ammonoosuc Volcanics and Partridge Formation is the presence of zone 12 graptolites in a black slate unit in the Cupsuptic quadrangle of Maine (loc. 50) described by Harwood and Berry (1967). This slate is considered the correlative of both the Beauceville Formation and the Dixville Formation of Green (1964) in New Hampshire: the Dixville, in turn, is considered the equivalent of both the Partridge Formation and the Ammonoosuc Volcanics (Green, 1964, p. 61–62).

Post-Taconic marker beds include the Clough Quartzite and the Fitch Formation, both of Silurian age, which overly the Ammonoosuc Volcanics or Partridge Formation or both units at various localities. The Clough contains fossils of late Llandovery (C_3–C_6) age (Boucot and Thompson, 1963) at the north, inverted end of the Skitchewaug nappe (loc. 36) in Vermont (Thompson, 1954, p. 37–41). However, Boucot (unpublished data) now considers that the type Clough is of Ludlow age whereas the widespread Clough of the Bronson Hill anticline is of Llandovery C_3–C_6 age. Fossils similar to those at locality 36 also occur within the Skitchewaug nappe at Bernardston, Mass. (loc. 31). The Fitch Formation of early Ludlow age (Boucot and Thompson, 1963) is dated from collections made at Fitch Farm (loc. 45). However, the Fitch Formation of the Bronson Hill anticline may be of Wenlock and Ludlow age (Berry and Boucot, in press).

Among the angular unconformities involving the above stratigraphy, is the one demonstrated by the map relationships at the south end of the Walker Mountain syncline (loc. 44) in western New Hampshire (Billings, 1956). Here, a syncline of Partridge in the Ammonoosuc Volcanics (locally intruded by granite) is overlain by the Clough Quartzite. At Piermont Mountain, on the west side of Owls Head dome (loc. 43) there is also an angular discordance; the Ammonoosuc Volcanics dip 10 degrees steeper than the overlying Clough and the two formations have a 30 degrees difference in strike (L. R. Page, written communication, 1967). The southernmost angular unconformity in the New England states is in south-central Connecticut at Great Hill, near Portland (loc. 25). Here the Great Hill Formation of Rosenfeld and Eaton (1958) is believed to overlie the Collins Hill Formation of Rodgers, Gates, and Rosenfeld (1959) with angular discordance, a relationship established by regional mapping. The Great Hill Formation is correlated with the Clough Quartzite of New Hampshire of Llandovery C_3–C_6 age; the Collins Hill Formation is believed to be the equivalent of the Cram Hill and Partridge Formations of Vermont and New Hampshire that are considered to be of Middle Ordovician age.

Farther north, in the Cupsuptic quadrangle of Maine (loc. 50), map relationships also indicate that a polymict conglomerate overlies with angular unconformity the slate unit containing graptolites of Middle Ordovician (Graptolite Zone 12) age, described above. The overlying polymict conglomerate is considered equivalent to a similar conglomerate, about 6 miles northeast of and on strike with it, which has yielded brachiopods of late Llandovery age (U. S. Geol. Survey, 1965, p. A74), in the range of C_3–C_5. In the Lake Memphremagog (loc. 51) and Lake Etchemin (loc. 59) areas, map relationships also indicate the presence of angular unconformities. Near Lake Etchemin, the Beauceville Formation of Middle Ordovician (Graptolite Zones 12 and 13) age is overlain by the Cranbourne Formation of Ludlow age. The "Bolton lavas" of the Memphremagog area are now believed to be folded with the Magog Formation (Ambrose, 1957, p. 162) of Trenton age (Berry, 1962) and overlain by the Peasley Pond Conglomerate of Wenlock(?) age.

The angular unconformities in northern Maine include exposed contacts as well as unconformities deduced from map patterns. About 1 mile from the west side of Caribou Lake (loc. 53), an unnamed Silurian unit which is locally conglomeratic at its base, contains brachiopods of Llandovery C_3–C_5 age. The Silurian rocks here truncate the upturned edges of graywacke and slate beds that are part of an unnamed pre-Silurian unit (Dennis Erinakes, written communication, 1966). The age of these pre-unconformity rocks is not known, and it is not certain whether they were deformed by the Taconic or an older orogeny, or by more than one orogeny. Similar uncertainties characterize the angular discordance at Lobster Lake (loc. 52).

In the Island Falls (loc. 56) and Shin Pond (loc. 57a) quadrangles, map relationships indicate that the complexly deformed Grand Pitch Formation of Cambrian(?) age is overlain with angular discordance by conglomerate and sandstone of late Llandovery C_3–C_5 age. The fossiliferous rocks of Early, Middle, and Late Ordovician age of this region (Neuman, 1967) are herein inferred to rest also with angular unconformity beneath the rocks of late Llandovery age (loc. 57b). Regional relationships also indicate that an angular unconformity is present in the Spider Lake region (loc. 58) and the Pennington anticline area (loc. 62).

At an unconformity that is no longer exposed in Ashland (loc. 60), in northern Maine, the Silurian Frenchville Formation of Llandovery C_3–C_5 age (dated by fossils in nearby areas) unconformably overlies an unnamed siltstone unit that has yielded fossils of probable Late Ordovician age. David N. Roy and Ely Mencher (written communication, 1966), measured a maximum discordance of about 24 degrees in strike and 37 degrees in dip between beds on opposite

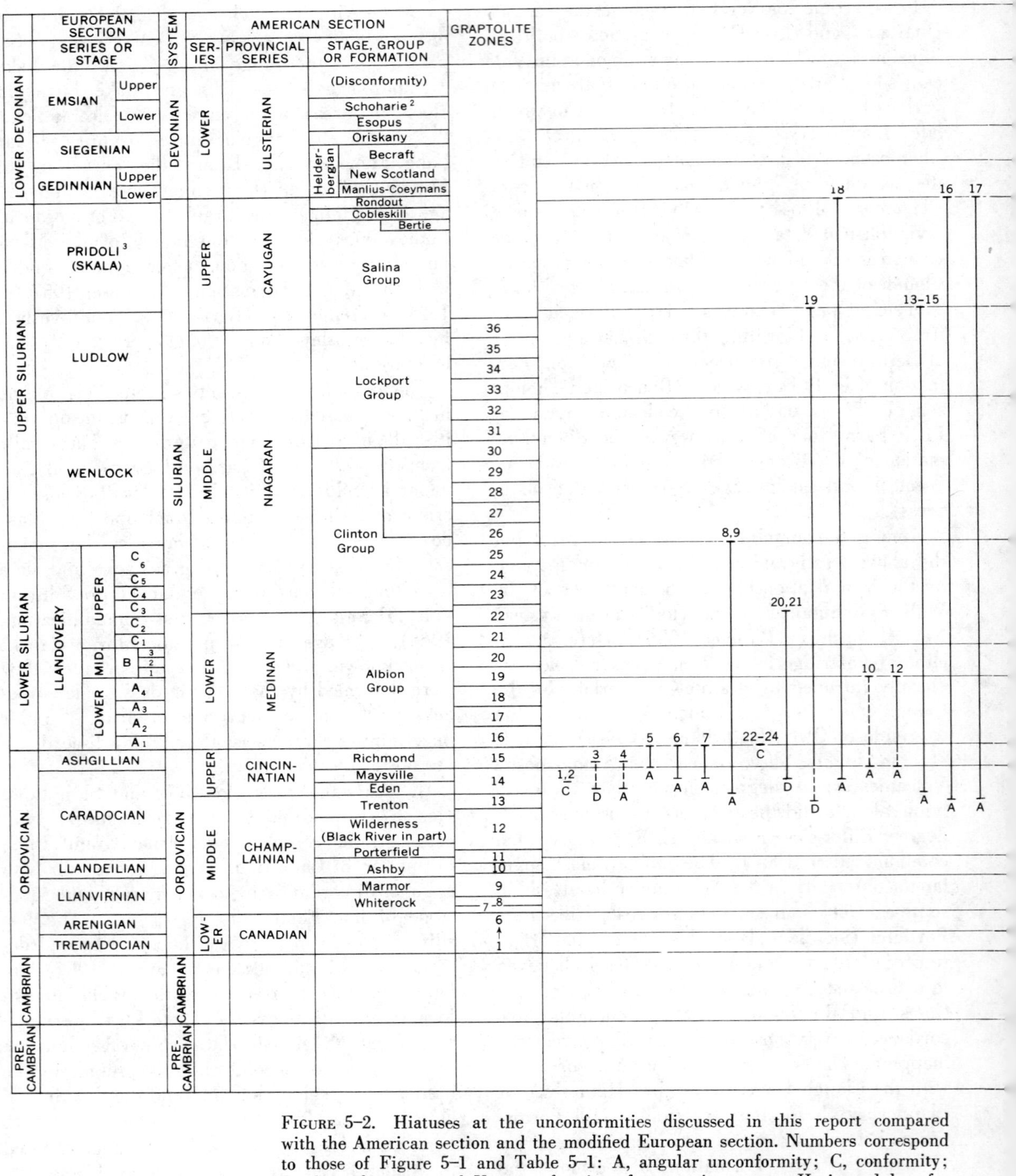

EUROPEAN SECTION			SYSTEM	AMERICAN SECTION			GRAPTOLITE ZONES
SERIES	SERIES OR STAGE			SER-IES	PROVINCIAL SERIES	STAGE, GROUP OR FORMATION	
LOWER DEVONIAN	EMSIAN	Upper	DEVONIAN	LOWER	ULSTERIAN	(Disconformity)	
		Lower				Schoharie [2]	
						Esopus	
	SIEGENIAN					Oriskany	
						Becraft (Helder-bergian)	
	GEDINNIAN	Upper				New Scotland (Helder-bergian)	
		Lower				Manlius-Coeymans (Helder-bergian)	
UPPER SILURIAN	PRIDOLI [3] (SKALA)		SILURIAN	UPPER	CAYUGAN	Rondout	
						Cobleskill	
						Bertie	
						Salina Group	
	LUDLOW			MIDDLE	NIAGARAN	Lockport Group	36
							35
							34
							33
							32
							31
	WENLOCK						30
							29
							28
							27
						Clinton Group	26
							25
LOWER SILURIAN	LLANDOVERY — UPPER	C₆		LOWER	MEDINAN	Albion Group	24
		C₅					23
		C₄					22
		C₃					21
		C₂					20
		C₁					19
	LLANDOVERY — MID.	B 3					18
		B 2					17
		B 1					16
	LLANDOVERY — LOWER	A₄					
		A₃					
		A₂					
		A₁					
ORDOVICIAN	ASHGILLIAN		ORDOVICIAN	UPPER	CINCIN-NATIAN	Richmond	15
	CARADOCIAN					Maysville	14
						Eden	
				MIDDLE	CHAMP-LAINIAN	Trenton	13
						Wilderness (Black River in part)	12
	LLANDEILIAN					Porterfield	11
						Ashby	10
	LLANVIRNIAN					Marmor	9
						Whiterock	7-8
	ARENIGIAN			LOW-ER	CANADIAN		6
	TREMADOCIAN						↑
							1
CAMBRIAN			CAMBRIAN				
PRE-CAMBRIAN			PRE-CAMBRIAN				

Unconformity markers (read left to right across the graptolite-zone range):

1,2 — C
3 — D
4 — A
5 — A
6 — A
7 — A
8,9 — A
22–24 — C
20,21 — D
19 — D
18 — A
10 — A
12 — A
13–15 — A
16 — A
17 — A

FIGURE 5-2. Hiatuses at the unconformities discussed in this report compared with the American section and the modified European section. Numbers correspond to those of Figure 5-1 and Table 5-1: A, angular unconformity; C, conformity; D, disconformity; and X, unconformity of uncertain nature. Horizontal bar for

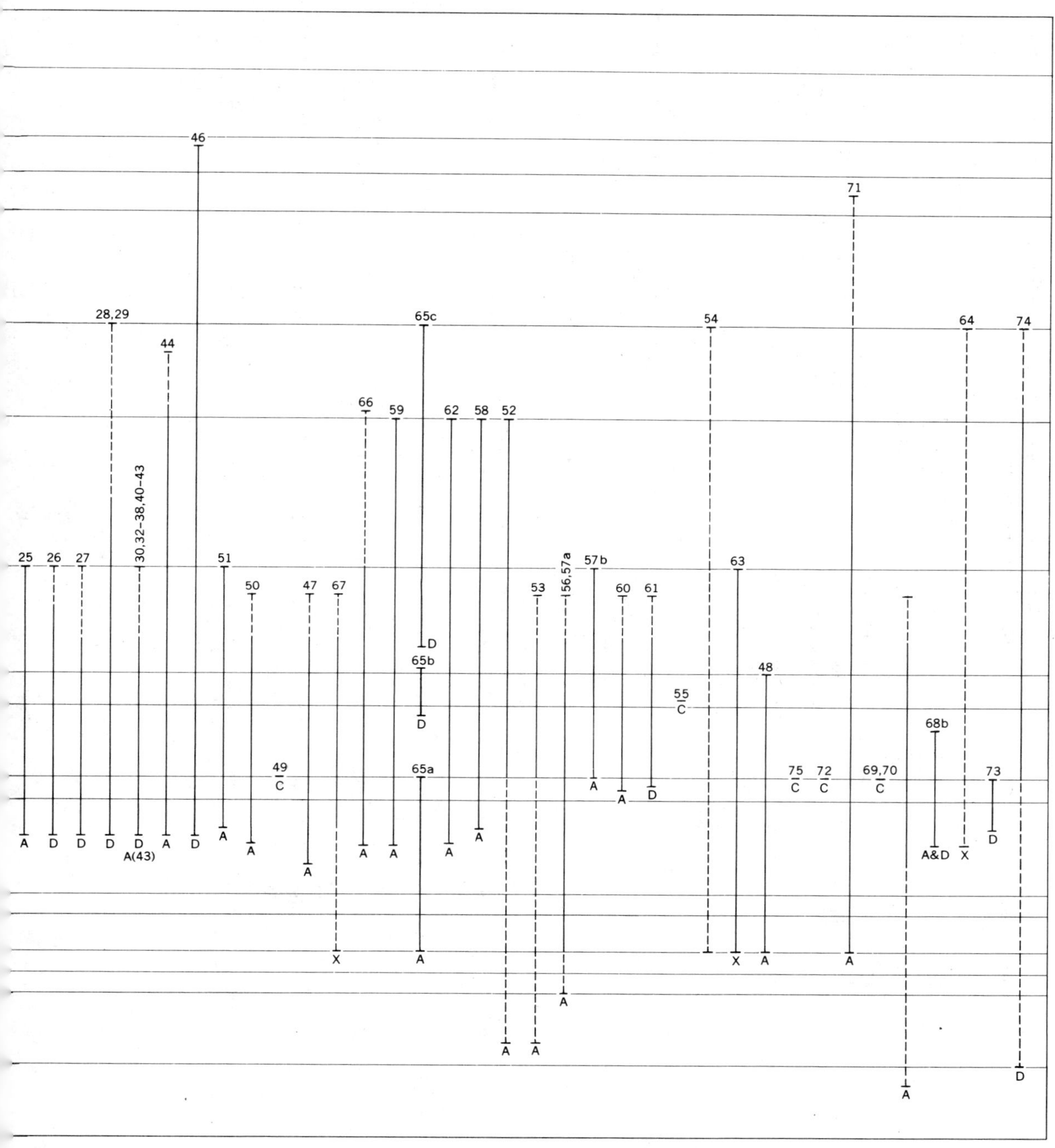

conformable contacts is placed at major lithologic breaks. Insofar as possible, unconformities and conformities have been plotted in groups that form various sections across the strike of the regional structure.

TABLE 5-1. Unconformities related to the Taconic orogeny. Figure 5-2 contains comparision of American and British terms used in age columns. Silurian ages indicated by asterisk are taken from Berry and Boucot (in press). Names or ages accepted in this report may not correspond to the usage of the references cited.

Map No.	Underlying beds			Overlying beds			Selected references
	Unit	Lithology	Age	Unit	Lithology	Age	
				A. Angular unconformities			
4	Lower shale member of Martinsburg Shale	shale	middle(?) Eden	Bald Eagle Con- glomerate of Grabau (1909)	coarse conglomerate	Maysville– Richmond(?)	Willard and Cleaves (1939) and Stose (1930)
5	Shochary Sandstone Member (of Willard and Cleaves, 1939) of Martinsburg Shale	arkosic sandstone	early Maysville	Tuscarora Sandstone	fine grained con- glomerate	*Llandovery	Willard and Cleaves (1939) and Stose (1930)
6	Lower member of Martinsburg Shale	sandstone and shale	pre-Maysville(?)	Bald Eagle Con- glomerate of Grabau (1909) or Tuscarora Sandstone	coarse conglomerate	Maysville– Richmond(?) or Silurian	Willard and Cleaves (1939) and Stose (1930)
7	Martinsburg Shale	sandstone and shale	pre-Maysville(?)	Tuscarora Sandstone Shawangunk Conglomerate	coarse conglomerate and sandstone	*Llandovery	Schuchert (1916), Miller (1926), Stose (1930)
8	1. Kittatinny Lime- stone	1. limestone	1. Late Cambrian and Early Ordovician	Green Pond Con- glomerate	conglomerate	* Wenlock	Kümmel and Weller (1902)
	2. crystalline rocks	2. gneiss	2. Precambrian				
9	Snake Hill Forma- tion	shale	Middle Ordovician (Graptolite Zone 13)	Green Pond Con- glomerate	conglomerate	*Wenlock	Fisher (1960, 1962)
10	Normanskill Shale	sandy shale	Late Ordovician (Graptolite Zone 14)	Shawangunk Con- glomerate	conglomerate	*Middle Llandovery	Schuchert (1916); Fisher (1960, 1962)
12	Austin Glen Mem- ber Normanskill Shale	sandstone and shale	Middle Ordovocian (Graptolite Zone 13)	Shawangunk Con- glomerate	shale, sandstone and conglomerate	*Middle Llandovery	Brown (1914), Fisher and others (1962) and Fisher (1962)
13–15	Austin Glen Member of Normanskill Shale	sandstone, gray- wacke, and shale	Middle Ordovician (Graptolite Zone 13)	Wilbur Member of Roundout Lime- stone	limestone	*Pridoli	Van Ingen and Clark (1903), Schuchert (1916), and Fisher (1962); Schuchert and Longwell (1932); Chadwick (1913)

16	1. Blue Hill facies (of Fisher, 1962) of Normanskill Shale; east side of fault 2. Mt. Merino Member (of Ruedemann, 1942) of the Normanskill Shale; west side of fault	1. red shale and green chert 2. black shale and chert	1. Middle Ordovician 2. Middle Ordovician	Manlius Limestone	limestone	Early Gedinnian	Schuchert and Longwell (1932), Fisher and others (1962), Zen (1967) and Fisher (1960, 1962 and written comm. 1967)
17	Unnamed unit	shale	Upper Cambrian and Lower Ordovician	Manlius Limestone	limestone	Early Gedinnian	Fisher and others (1962)
18	Indian Ladder Formation	shale	Eden	Manlius Limestone	limestone	Early Gedinnian	Fisher (1962, and written comm., 1967)
25	Collins Hill Formation	schist	Middle Ordovician	Great Hill Formation	conglomerate and quartzite	Late Early Silurian (Llandovery C_3–C_6)	Rosenfeld and Eaton (1958)
43	Ammonoosuc Volcanics	metavolcanic rocks	Middle Ordovician	Clough Quartzite	quartzite and conglomerate	*Llandovery C_3–C_6	L. R. Page (written comm., February 1967)
44	1. Ammonoosuc Volcanics 2. Partridge Formation	1. metavolcanic rocks 2. slate	Middle Ordovician	Clough Quartzite	quartzite and conglomerate	*Ludlow	Billings (1956)
47	North Haven Greenstone	metamorphosed diabase, trachyte, and syenite	Middle Ordovician(?)	Ames Knob Formation	conglomerate, silty limestone, and shale	*Llandovery C_3–C_6 to Pridoli, or early Gedinnian	Smith and others (1907), Beecher and Dodge (1892)
48	Dark Argillite Division of the Charlotte Group	slate	Early Ordovician (Arenig)	Oak Bay Conglomerate	conglomerate	*Late Llandovery	Smith *in* Poole (1966)
50	Unnamed unit	slate	Middle Ordovician (Graptolite Zone 12)	Unnamed unit	polymict conglomerate	Late Llandovery C_3–C_5	Harwood and Berry (1967)
51	Magog (Beauceville) Formation and "Bolton lavas"	greenstone and slate	Middle Ordovician (Graptolite Zones 12 and 13)	Peasley Pond Conglomerate	conglomerate	*Wenlock(?)	Ambrose (1957)
52	Unnamed unit	phyllite	Cambrian or Ordovician	Big Claw Red-Bed Member of Lobster Lake Formation	quartzite, slate and conglomerate	Ludlow	Boucot (1961)

(continued)

TABLE 5-1 (continued)

Map No.	Underlying beds			Overlying beds			Selected references
	Unit	Lithology	Age	Unit	Lithology	Age	
	A. Angular unconformities (continued)						
53	Unnamed unit	graywacke and slate	Pre-Silurian	Unnamed unit	quartzose conglomerate and limestone	Late Llandovery C_3–C_5	Dennis Erinakes (written comm. Jan. 1966)
56	Grand Pitch Formation	quartzite and phyllite	Cambrian(?)	Unnamed unit	sandstone and siltstone	Late Llandovery C_3–C_5	Ekren and Frischknecht (1967)
57a	Grand Pitch Formation	quartzite and phyllite	Cambrian(?)	Unnamed unit	conglomerate and sandstone	Late Llandovery C_3–C_5	Neuman (1967)
57b	Unnamed units	volcanic and sedimentary rocks	Ordovician (up to Ashgillian)	Unnamed unit	conglomerate and sandstone	Late Llandovery C_3–C_5	Neuman (1967)
58	Unit with no published name	dolorite, mafic tuff, and volcanic breccia	Middle Ordovician (Graptolite Zones 12-13 ?)	Unit with no published name	mafic and felsic volcanic rocks	Ludlow	Hall (1964)
59	Beauceville Formation	slate	Middle Ordovician (Graptolite Zones 12-13)	Cranbourne Formation	siltstone and limestone	Ludlow	Boucot and Drapeau (in press), Berry (this volume)
60	Unnamed unit	siltstone	Ashgill	Frenchville Formation	conglomerate and sandstone	Llandovery C_3–C_5	Mencher, Ely and Roy, David C. (written comm., December, 1966)
62	Unnamed unit	volcanic rocks	Middle Ordovician (Graptolite Zone 12)	Unnamed unit	conglomerate, limestone and siltstone	Ludlow	Pavlides and others (1964)
65a	Quebec Group	slate and quartzite	Ordovician	Cabano Formation	sandstone and conglomerate	Early Llandovery	Lespérance (1960)
66	Quebec Group	slate	Middle Ordovician (Graptolite Zone 12)	Robitaille Formation	siltstone	Wenlock or Early Ludlow	Lajoie (1962), Berry (this volume)
68a	Maquereau Group	volcanic rocks graywacke, and slate	pre-Middle Ordovician (pre-Mictaw Group)	Clemville Formation	sandstone and limestone	*Llandovery C_3–C_5	Alcock (1935); Ayrton (1961)
68b	Mictaw Group	shale, graywacke, and conglomerate	Middle Ordovician (Graptolite Zones 12 and 13)	Weir Formation	siltstone, and sandstone	*Llandovery A_3–A_4	Ayrton (written comm., 1964)
71	"Cape Rosier beds"	shale, limestone and conglomerate	Early Ordovician	Roncelles Member of St. Alban Formation	pebbly limestone, limestone	*Pridoli(?)– Early Gedinnian	McGerrigle (1950), Cumming (1959)

				B. Disconformity			
3	Martinsburg Shale	shale	Eden	Bald Eagle Conglomerate of Grabau (1909)	coarse red conglomerate	Maysville(?)–Richmond(?)	Willard and Cleaves (1939) and Stose (1930)
19	Schenectady Formation	sandstone, graywacke, and shale	Middle Ordovician (Graptolite Zone 13)	Brayman Shale	shale	Late Silurian	Fisher and others (1962), Fisher and Rickard (1953)
20	Frankfort Shale	shale	Late Ordovician	Oneida Conglomerate	quartz pebble conglomerate	Llandovery C_3–C_5	Fisher (1960), Rickard (1964)
21	Frankfort Shale	shale	Eden	Oneida Conglomerate	conglomerate and sandstone	Llandovery C_3–C_5	Dale (1953), Fisher (1960)
26	1. Ammonoosuc Volcanics	1. quartz-feldspar gneiss and amphibolite	1. Middle Ordovician	Fitch Formation	calc-silicate granulite	*Wenlock(?)	Peter Robinson (written comm. 1966)
	2. Partridge Formation	2. sulfidic mica schist and gneiss	2. Middle Ordovician				
27	Partridge Formation	sulfidic mica schist and gneiss	Middle Ordovician	Clough Quartzite	quartzite and conglomerate	Late Llandovery	Peter Robinson (written comm. 1966)
28 & 29	Hawley Formation	amphibolites, feldspar schist and gneiss, graphitic schists and quartzite	Middle Ordovician	Goshen Formation	phyllite and cyclically layered phyllite and quartzite	*Wenlock–Ludlow	Balk (1946); N. L. Hatch (written comm., Feb., 1967)
30, 32-38, 40-42	Ammonoosuc Volcanics or Partridge Formation	metavolcanic rocks, slate and quartzite; gneiss, schist	Middle Ordovician	Clough Quartzite	quartzite and conglomerate	Late Llandovery C_3–C_6	Billings (1956), Boucot and Thompson (1963), Peter Robinson (written comm., 1966)
46	Ammonoosuc Volcanics	metavolcanic rocks, slate and quartzite	Middle Ordovician	Littleton Formation	slate, quartzite, and metavolcanic rocks	Early Devonian	Billings (1956)
61	Pyle Mountain Argillite	argillite	Ashgillian	Frenchville Formation	conglomerate and sandstone	Late Llandovery C_3–C_5	Boucot and others (1964)
65b	Cabano Formation	sandstone and conglomerate	Early Llandovery to Late Llandovery C_1–C_2	Pointe Aux Trembles Formation	shale and conglomerate	Late Llandovery C_1–C_2	Lespérance (1960) and Berry and Boucot (in press)
65c	Pointe Aux Trembles Formation	shale and conglomerate	Late Llandovery C_1–C_2	Robitaille Formation	mudstone	Wenlock to Ludlow	Lespérance (1960) and Berry and Boucot (in press)
73	Unnamed unit	greenstone and felsite	Late Ordovician(?)	Beechhill Cove Formation	sandstone	*Early Llandovery	A. J. Boucot (unpub. data)
74	Unnamed unit	undivided sedimentary and volcanic rocks	Cambrian or Ordovician	Arisaig Group	shale and mudstone	*Early Llandovery (?)	A. J. Boucot (unpub. data)

(continued)

TABLE 5-1 (continued)

Map No.	Underlying beds			Overlying beds			Selected references
	Unit	Lithology	Age	Unit	Lithology	Age	
C. Unconformities of uncertain nature							
54	Unnamed unit	slate, shale, silt-stone and tuff	Middle or Late Ordovician	Unnamed unit	conglomerate	Silurian	Larrabee and Spencer (1963), Larrabee and others (1965)
63	Unnamed unit	graywacke and greenstone	Middle and Late Ordovician	Unnamed unit	grit conglomerate, slate and gray-wacke	*Wenlock(?)	Poole (1963)
64	Tetagouche Group	slate and meta-volcanic rocks	Middle Ordovician (Graptolite Zone 12)	Unnamed unit of Chaleur Bay Group	conglomerate, shale, graywacke and volcanic rocks	Silurian	Smith (1957)
67	Quebec Group	phyllite, sandstone and meta-volcanic rocks	Early to Middle Ordovician	Awantjish Forma-tion	shale	*Late Llandovery C_3-C_5	Beland (1960)
D. Conformities							
1 & 2	Juniata Formation	sandstone	Richmond	Tuscarora Sand-stone	sandstone and con-glomerate	Llandovery	Willard and Cleaves (1939)
22	Oswego Sandstone	gray sandstone	Late Ordovician	Grimsby Sandstone as used by Fisher (1954)	red sandstone	Llandovery	Fisher (1954, 1960, 1962 and written comm., 1967)
23	Queenston Shale	shale	Richmond	Grimsby Sandstone as used by Fisher (1954)	red sandstone	Llandovery	Fisher (1954, 1960 and written comm., 1967)
24	Queenston Shale	shale	Richmond	Whirlpool Sand-stone	white sandstone	Llandovery	Grabau (1901), Fisher (1954, 1960, and written comm., 1967)
49	Unnamed unit	metasandstone and metashale	Ordovician	Rangeley Con-glomerate of Smith (1923)	conglomerate and sandstone	Llandovery	Moench (written comm., 1967)

	Formation	Lithology	Age	Formation	Lithology	Age	Reference
55	Carys Mills Formation	impure limestone, calcareous slate and siltstone	Middle Ordovician (Graptolite Zone 13) to Llandovery (Graptolite Zone 19)	Smyrna Mills Formation	slate, quartzite and conglomerate	Llandovery (Graptolite Zone 19) to Ludlow (Graptolite Zone 32)	Pavlides and Berry (1966)
69	White Head Formation	limestone and shale	Middle Ordovician to Clinton	Mont Alexander Group	siltstone, limestone and volcanic rocks	Late Llandovery–Ludlow	Skidmore (1965) and this report
70	White Head Formation	limestone and shale	Middle Ordovician to Llandovery	Burnt Jam Brook Formation	shale and siltstone	Late Llandovery (Graptolite Zone 22–23)	McGerrigle (1950), Cumming (1954)
72	Vauréal Formation	limestone, intraformational conglomerate and shale	Late Ordovician	Ellis Bay Formation	limestone and shale	Early Llandovery, at least in part	Twenhofel (1927), Boucot and Johnson (1967)
75	Halifax Formation	argillite	Early Ordovician (Tremadoc)	White Rock Formation	quartzite and slate	Silurian	Taylor (1965)

sides of the unconformity. A pebbly layer in the probable Ordovician siltstone unit yielded from its matrix the brachiopods *Cryptothyrella* and *Plectothyrella* which were identified by Boucot and assigned an Ashgillian age (an early Llandovery age is possible but considered less likely). This break probably spans part of Ashgill through late Llandovery C_3–C_5 time.

Angular unconformity is also demonstrable at several localities in Quebec. Discordance is apparent in outcrops in the Chénier–Bédard area (loc. 66), and at Cap-des-Rosiers Cove (loc. 71), and from geologic map relationships in the Cabano area (loc. 65a).

Eastern Belt

The data for an eastern belt of angular unconformities are less abundant than for the western belt. Nonetheless its existence is strongly suggested although the generalization of including these unconformities in a single belt may be modified when additional information becomes available.

The southernmost angular unconformity exposed in this belt is at Ames Knob (loc. 47) in the Penobscot Bay region of Maine. Smith and others (1907) believed that elsewhere in the area the North Haven Greenstone is more highly folded than the overlying Ames Knob Formation, suggesting Taconic deformation. They considered the greenstone to be of Cambrian(?) age, but a Middle Ordovician age now appears more likely (Boucot, unpublished data). A hiatus of similarly large magnitude is represented by exposures on Cookson Island in New Brunswick (loc. 48).*

In central and northern New Brunswick two additional unconformities of uncertain nature (locs. 63 and 64), but believed to be angular unconformities (described elsewhere), are also included in this eastern belt, along with the Port Daniel area (locs. 68a,b) of Quebec. The Port Daniel area is classical for its unconformities, only two of which may pertain to the Taconic orogeny. Near the coast, the Clemville Formation of Llandovery C_3–C_5 age overlies the folded Maquereau Group of pre-Middle Ordovician age (loc. 68a). The upper age limit of the Maquereau is fixed by the age of the Mictaw Group

* Since the preparation of this report R. L. Brown (written comm., 1967) and A. A. Ruitenberg (1967, p. 87) have concluded that the Cookson Island unconformity is a faulted disconformity rather than an angular unconformity. Such an interpretation of this unconformity would necessitate placing the inferred contact between the belts of angular unconformity and disconformity northwest of locality 48 rather than southeast of it as now shown on Figure 5–1.

which unconformably overlies it and has yielded Trenton age graptolites (zones 12 and 13). Thus, the Clemville–Maquereau unconformity may telescope Taconic and older orogenies. Farther north in the Port Daniel area (loc. 68b), the Mictaw is overlain with possible local angular unconformity by the Weir Formation of Llandovery A_3–A_4 age (Ayrton, 1967, p. 36–39). Laterally, however, the Weir and Mictaw appear to be structurally conformable and the angular unconformity thus becomes a disconformity along strike.

DISCONFORMITIES

Pennsylvania and New York

At the Taconic angular unconformity of southeastern Pennsylvania the folded Martinsburg Shale is overlain by younger clastic rocks of either the Bald Eagle Conglomerate of Grabau (locs. 4 and 6) or the Tuscarora Sandstone (loc. 5). At the Susquehanna Gap (loc. 3), however, the Martinsburg, which here contains fossils no younger than Eden age, is conformably overlain by the Bald Eagle (Willard and Cleaves, 1939, p. 1178–1179). The Bald Eagle here lacks the *Orthorhynchula* zone of Maysville age that occurs near its base (Pierce, 1966, p. 25) in the region of localities 1 and 2. At Susquehanna Gap it grades transitionally upward into the Juniata Formation (Richmond) which in turn grades upward into the Tuscarora. Hence, the Taconic break here is the contact between the Martinsburg and the Bald Eagle; it is probably a disconformity that spans part of late Maysville and possibly some of early Richmond time.

Several disconformities are also present in New York. The westernmost at Clinton (loc. 21) and Vanhornesville (loc. 20), span late Ordovician (Eden) to Llandovery C_3–C_5 time. At Schoharie (loc. 19) the hiatus of the disconformity is much larger; spanning Middle Ordovician (Graptolite Zone 13) up to Late Silurian time.

Western New England and Adjacent Quebec

Central Vermont and adjacent Quebec, Massachusetts, and Connecticut. The Silurian and Ordovician boundary trending southward through Vermont and western Massachusetts is considered to represent a disconformity related to the Taconic orogeny. On the west side of this contact are the coeval Cram Hill Formation (northern Vermont), the Barnard Volcanic Member of the Missisquoi (central and southern Vermont), and the Hawley Formation (Massachusetts). In Vermont the rocks above (east of) the Silurian–Ordovician boundary consist of the Shaw Mountain Formation (Doll and others, 1961) of probable late Llandovery to Ludlow age (Boucot and Thompson, 1963) and the Northfield Slate of Silurian age (Berry and Boucot, in press). To the south, in Massachusetts, the Goshen Formation is considered the on-strike equivalent of the Northfield Slate of Vermont and hence is dated as Silurian in age. Near Lake Memphremagog, Quebec (loc. 51), the Peasley Pond Conglomerate overlies the Magog with angular unconformity. Thus, the disconformable Ordovician–Silurian contact here changes northward to one of angular discordance.

Mantled domes of the Connecticut Valley. Disconformity between Ordovician and Silurian strata is present in the series of mantled domes and inverted nappes that occur along the Connecticut River Valley of New England (locs. 26–38, 40–42, and 46). The en echelon domes and folds of the Bronson Hill anticline (Billings, 1956, p. 109, 122; also described by Thompson and others, this volume), commonly have a core of crystalline rock and are mantled by either the Ammonoosuc Volcanics or the Partridge Formation of Middle Ordovician age. These two formations are in turn disconformably overlain by Silurian strata of either the Clough Quartzite of late Llandovery (C_3–C_6) age or the Fitch Formation of Wenlock(?) age. The Bronson Hill anticline (more properly anticlinorium) continues southward from New Hampshire into Massachusetts and Connecticut where it is also defined by a series of en echelon mantled domes (locs. 26, 27, 30, and 32). These are similar, in most respects, to those of New Hampshire in that the metavolcanic rocks that mantle these domes are correlated with the Ammonoosuc and Partridge Formations of New Hampshire. The various quartzite, conglomerate, and calc-silicate rocks that overlie these metavolcanic rocks are correlated with the Clough Quartzite and Fitch Formation of New Hampshire and Vermont.

Maine, Quebec, and Nova Scotia

Disconformities are also present elsewhere in the northern Appalachians. At the Castle Hill anticline (loc. 61) the contact is not exposed and the nature of the unconformity is inferred from the map relationships. In the Cabano area two disconformities, in addition to the Cabano Formation–Quebec Group angular unconformity (loc. 65a) described earlier, are also present. A local and small break (loc. 65b) occurs within the interval of Llandovery C_1–C_2 where the Cabano Formation of Llandovery C_1–C_2 age is apparently overlain disconformably by the Pointe

Aux Trembles Formation also of Llandovery C_1–C_2 age (Lespérance, 1960). About 30 miles to the northeast this disconformity is absent and the two formations are conformable. The youngest disconformity in this region (loc. 65c) is the contact between the Pointe Aux Trembles and the overlying Robitaille Formation of Wenlock–Ludlow age. Less well documented disconformities occur to the east in Nova Scotia. The basal Beechhill Cove Formation of the Arisaig Group, of early Llandovery age, rests with apparent disconformity upon poorly dated unnamed volcanic rocks, in Antigonish County (loc. 73). These underlying volcanic rocks are thought to be probably of Late Ordovician age (Boucot, unpublished data) but adequate paleontologic information relative to their age is absent. In the Glencoe region of Pictou County (Maehl, 1961) the lower Llandovery Glencoe Quartzite rests with apparent disconformity upon pre-Silurian strata. In northern Nova Scotia, including the Cobequid Mountain region (loc. 74) all of the Arisaig Group except the Beechhill Cove Formation has been recognized. The relationship suggests a disconformity between rocks of early Llandovery age and older Paleozoic rocks. Northwest of the Arisaig region, however, there is a possibility of an angular break between some part of the pre-Silurian and the lower Llandovery.

UNCONFORMITIES OF AN UNCERTAIN NATURE

Descriptions of several unconformities in the northern Appalachians fail to specify whether they are angular or disconformable and their map relations are generally too equivocal to make this distinction. However, based on regional relationships and the degree of deformation of the Ordovician rocks as compared to that of the nearby Silurian rocks, it is concluded that the unconformities at Hayesville (loc. 63) and Bathurst (loc. 64), New Brunswick, and at St. Cleophas, Quebec (loc. 67), are angular. The unconformity at Danforth, Maine (loc. 54), is interpreted as a disconformity.

CONFORMITIES

Pennsylvania and New York

A conformable sequence of Ordovician and Silurian rocks occurs at East Tuscarora Mountain (loc. 1) and at East Blue Mountain (loc. 2). Here the Martinsburg Shale is gradationally overlain by the Bald Eagle Conglomerate of Grabau, which contains an *Orthorhynchula* zone near its base (described earlier) and hence is probably of Maysville age. The Bald Eagle is gradationally overlain by the Juniata Formation (Richmond) which in turn grades up into the Tuscarora Sandstone (Willard and Cleaves, 1939, p. 1175–1177). Hence, Taconic diastrophism, as demonstrated by the disconformity at Susquehanna Gap (loc. 3), and angular unconformities in the gaps to the east (locs. 4 to 7), had no effect in this part of south-central Pennsylvania. The same is true for western New York (locs. 22–24).

Anticosti Island, Quebec

One of the classic sections of uninterrupted sedimentation from the Ordovician into the Silurian is the carbonate sequence on Anticosti Island (loc. 72). Originally, Twenhofel (1927, p. 15) assigned the Ellis Bay Formation to his "Gamachian" subdivision of the Ordovician and considered it to be the youngest Ordovician unit on Anticosti Island. Reevaluation of some of the brachiopods within the Ellis Bay by Boucot and Johnson (1967), however, indicates that at least part of this unit is of early Llandovery rather than Ordovician age. It is probable, therefore, that the Ordovician and Silurian boundary here is between the underlying Vauréal Formation and the Ellis Bay or within the Ellis Bay.

Aroostook-Matapedia Belt

Regional relationships, supplemented by local, paleontologically documented sections at each end of the Aroostook–Matapedia belt (Fig. 5–1), suggest that the impure limestones of this belt that are of Middle Ordovician through Early Silurian age, grade upward into more clastic younger Silurian rocks without any physical interruption. The Middle Ordovician to Lower Silurian rocks consist of impure, fine-grained calcareous rocks (chiefly quartzose silty limestone and calcareous quartzite) interlayered with slate and locally with more clastic rocks. These are the core rocks of the Aroostook–Matapedia anticlinorium (Pavlides and others, 1964) to which different formation names have been applied at different places. In Maine, they are the Carys Mills Formation (formerly the ribbon rock member of the Meduxnekeag Formation) and have been traced as such into adjoining western New Brunswick (Pavlides, in press). Across northern New Brunswick parts of Map Unit 5 of Rose and others (1962) and Map Unit 15 of Potter (1964) are considered to be coextensive with the Carys Mills (Pavlides, unpublished data) as well as the Matapedia Group of Alcock (1935) in the Chaleur Bay region of New Brunswick. Along the southern part of the Gaspé Peninsula the temporal equivalents of these formations are the impure limestones of the Matapedia Group of the usage of Béland (1958, 1960) in the Matapedia

and Oak Bay areas of Quebec. These limestones strike eastward to Percé, Quebec, and are included in the Pabos and White Head Formations of the Matapedia Group (Series) of the usage of McGerrigle (1950).

The St. John River anticline (McGerrigle, 1950) exposes a narrow strip of limestones of the Matapedia Group about 10 miles north of the main outcrop belt of the Pabos and White Head Formations, and these rocks are included within the Aroostook–Matapedia belt (loc. 70) of this report.

The Carys Mills Formation in Maine includes beds that range in age from Middle Ordovician, in the interval of *Orthograptus truncatus* var. *intermedius* (zone 13), up through Graptolite Zone 19, which is near the base of B_1 of the middle Llandovery. Near Smyrna Mills (loc. 55) graptolites of zone 19 occur both near the top of the Carys Mills Formation and near the base of the more clastic Smyrna Mills Formation (which ranges up into the Ludlow); the contact between the two formations is gradational (Pavlides and Berry, 1966). Hence, the Taconic unconformity is absent here. Wherever the Carys Mills contact with overlying rocks has been mapped in Maine and western New Brunswick, it appears to be one of stratigraphic conformity, except where it is a fault (Pavlides, in press), and the assumption, therefore, by analogy with locality 55, is that it is conformable.

The contiguous limestones of the Matapedia Group, including the White Head Formation, to the north in this belt, also are in apparent conformable or fault contact with younger rock and have yielded fossils generally of Late Ordovician (Ashgill) age. In the Percé area (loc. 69) at the east end of the Gaspé Peninsula, the White Head Formation had been dated as Richmond (Alcock, 1935, p. 25–26), but recent fossil collections from various localities within the White Head in this area have yielded trilobites, graptolites, and ostracodes which indicate that this formation ranges from Middle Ordovician (Caradoc) to Clinton of the Silurian. These rocks have a greater age range and include younger rocks than the Carys Mills Formation of Maine with which they are correlated. Among the key Silurian genera are the trilobites *Denckmannites* and *Protoscutellum*, identified by P. J. Lespérance of the University of Montreal (written communication, 1963), and the ostracodes *Bolbineossia* cf. *B. billingsi* and "*Bythocypris*" identified by M. J. Copeland of the Geological Survey of Canada (written commun., 1964). There is no evidence for a stratigraphic or structural break within the White Head in the Percé area, although the

White Head is in fault contact here with the younger Mont Joli Formation of Gedinnian age.

The Silurian faunas within the upper part of the White Head have been traced about 25 miles westward from the Percé area. From there westward, lithologically similar limestone continues in a strike belt along the north edge of the Matapedia Group, but the age of this belt of rocks has not been established by fossils. This belt (loc. 69) is overlain by siltstone, limestones, shales, and volcanic rocks of the Mont Alexander Group of Llandovery C_3 or younger age. The contact along this belt is not exposed, but map relationships suggest it is one of stratigraphic conformity.

The limestones of the Matapedia Group in the St. John River anticline (loc. 70) have yielded a *Streptis* of early Llandovery age (A. D. Wright, written communication, 1964) indicating that these rocks are also coeval, in part, to the Silurian part of the White Head near Percé. These limestones here (loc. 70) are overlain by shales of the Burnt Jam Brook Formation of late Llandovery age (Cumming, 1959, p. 25–27). Although the contact in this region is not exposed, map relationships also suggest that it is one of conformity.

Rangeley Quadrangle, Maine

An area of probable conformity from Middle Ordovician into Silurian times is also present in northwestern Maine (loc. 49) according to R. H. Moench (written communication, 1967). At the base of the stratigraphic section are rocks assigned to the Dixville Formation, which is probably correlative to an unnamed unit to the northwest (loc. 50) dated as Middle Ordovician, including the interval of *Climacograptus bicornis* (zone 12) (Harwood and Berry, 1967). Within the Rangeley quadrangle these rocks are successively overlain by units of: (a) interbedded metashale and metagraywacke; (b) interlaminated and locally calcareous metasandstone and metashale; (c) the Rangeley Conglomerate of Smith (1923); and (d) thick conglomerate metashale. According to Moench, the Rangeley and overlying conglomeratic metashale are of Early Silurian age, based on fossils in similar rocks found in a nearby area (U.S. Geol. Survey, 1965, p. A74). Basal metasandstone and subordinate metashale of the Rangeley Conglomerate conformably overlie unit b, which with unit a, and the Dixville, comprise the Ordovician rocks of the Rangeley region. To the south the coarse clastics of the Rangeley intertongue with a thick unit of metashale of the same age which also conformably overlies unit b. The sequence from Dixville through Rangeley, as well as the younger Silurian and Devonian rocks, are considered by Moench to be a conformable and continuous sedimentary sequence free of unconformities, although locally interrupted by faults.

Nova Scotia

The Kentville to Yarmouth region of the Annapolis Valley of Nova Scotia (loc. 75) is generally an area unaffected by Taconic diastrophism, as determined from the relationship of the Halifax Formation of Ordovician age to the White Rock Formation of Silurian age (not to be confused with the lower Middle Ordovician Whiterock stage of the American section, Fig. 5–2). The Halifax consists dominantly of slate and contains graptolites of Tremadoc age (Crosby, 1962, p. 18–21). The White Rock is more complex. Near Kentville (Wolfville area) it is dominantly slate with interbeds of quartzite in its upper and lower parts. Towards the southwest, it is thicker and contains abundant, thick sections of volcanic rocks. The White Rock does not contain fossils usable for establishing its age. However, it lies conformably below, and in gradational contact with the fossiliferous Kentville Formation of Ludlow age and is hence considered to be of Silurian age. The White Rock and Halifax Formations are in gradational and conformable contact except at Cape St. Mary, Nova Scotia, where rhyolite of the White Rock lies with angular discordance on the Halifax (Taylor, 1965, p. 17), although the contact here may be a fault. In general, the Silurian and Ordovician boundary in this part of Nova Scotia has been placed at the contact of the White Rock and Halifax, but there is no convincing evidence that this boundary does not actually lie within one or the other of the two formations. In any event, the conformable and gradational nature of the contact between the White Rock and Halifax, except for the local unconformity near Cape St. Mary, suggests that these rocks were deposited without any significant interruption in sedimentation. The only effect of the Taconic orogeny was the change to a more detrital and volcanic regimen in parts of the basin. This is reflected in the change of lithology in progressing upward from the Halifax into the White Rock, and along strike within the White Rock itself.

NATURE AND TIME SPAN OF THE TACONIC OROGENY

Nature and Distribution of Diastrophism

Evaluation of the Taconic unconformities indicates that the Taconic orogeny was characterized by different types and degrees of tectonism at different places. Although the data are sparse and in part equivocal, certain tentative conclusions can be drawn concerning the distribution of diastrophically active and inactive belts during the interval of Taconic orogeny. In general, belts of folding and faulting of Taconic or earlier age (angular unconformities) locally parallel belts of uplift without, or with only local and very minor Taconic folding (disconformities) which at some places adjoin belts unaffected directly by Taconic diastrophism (conformities). The magnitude of the hiatus at the unconformities also increases away from the belts of conformity (Fig. 5–2). This is interpreted to mean, for example, that the Aroostook–Matapedia belt was situated in a part of a basin of marine deposition that remained unaffected by the Taconic diastrophism, although such diastrophism occurred near its margins. The only record of tectonism within the basin is the change in sedimentation that occurred as deformation and uplift started near the margins. Thus, in about early middle Llandovery time (Graptolite Zone 19) the regimen of sedimentation changed from essentially calcareous rocks (Carys Mills Formation) to a more clastic suite (Smyrna Mills Formation).

A generally similar distribution of conformity, bordered successively by disconformity, and then angular unconformity, in progressing from a basin of deposition away from its margin, is present in Pennsylvania and New York. The conformable sections in south-central Pennsylvania (locs. 1 and 2) are succeeded to the east by the disconformity at Susquehanna Gap (loc. 3). Farther east, angular unconformity first appears at Swatara Gap (loc. 4) where only a small angular discordance is present. The succeeding unconformities to the east (locs. 5 to 6) are more pronouncedly angular. The hiatus, in general, also increases landward from the sedimentation basin (Fig. 2), being greatest in the Green Pond outlier of New Jersey (loc. 8) and New York (loc. 9). A similar pattern is present in New York where the conformable sections (locs. 22–24) are succeeded eastward by disconformities (locs. 19–21), and eventually by unconformities (locs. 15–18). The hiatus of the unconformities also generally increases eastward. This general increase in the magnitude of the hiatus is clearly related to the local paleogeography; the eastern areas represent unconformities with greater hiatus as a result of longer exposure to subaerial erosion. Marine onlap eastward in New York is also demonstrated by the unconformities at Rondout (loc. 13) where the Rondout Formation (mostly dolomite) of Pridoli

age unconformably overlies the pre-Taconic beds, whereas at Becraft Mountain (loc. 16) and Mt. Ida (loc. 17) the basal Rondout is absent and the Manlius Limestone overlies the pre-Taconic beds.

The disconformities (uplift) along the Bronson Hill anticline in southern New England are in the last tectonic belt now definable to the east in this part of the Appalachians. It is only conjectural whether additional tectonic belts ever lay to the east of this belt of disconformities.

The very sparse data in Nova Scotia may also represent different belts of conformity and unconformity (Fig. 5–1). These include the probable conformity in the Annapolis Valley, the disconformity at Beechhill Cove, at Glencoe (loc. 73) and possibly in the Cobequids (loc. 74). Farther west are the angular unconformities at Cookson Island, New Brunswick (loc. 48), and in Penobscot Bay, Maine (loc. 47). The presumed angular unconformities near Hayesville (loc. 63) and the Bathurst area (loc. 64) also most likely belong to the belt of angular unconformities to which Cookson Island and Penobscot Bay belong.

Obviously, Anticosti Island (loc. 72) must have been in an anorogenic belt of uninterrupted deposition with tectonism occurring to the south as along the belt of angular unconformities to which the one at Cap-des-Rosiers Cove (loc. 71) belongs. The pattern elsewhere suggests that a belt of disconformity intervened between the Anticosti belt and the belt of angular unconformities (Fig. 5–1).

In general, therefore, the paleogeographic and tectonic features of the northern Appalachians during the interval of the Taconic orogeny include more or less linear basins of deposition not directly affected by diastrophism (the Pennsylvania–New York belt, Anticosti Island, the Aroostook–Matapedia belt, and southern Nova Scotia of Figure 5–1) bounded on one or both sides by tectonically active areas wherein some folding, faulting and uplift occurred (belts of angular unconformity). Peripheral to these highly active diastrophic belts there locally occurred areas of less intense tectonism wherein chiefly uplift took place (disconformities). The tectonically active border lands may, in part, have been a series of islands, uplifted during different times within the interval of the Taconic orogeny.

Temporal Relationships

In general, the Taconic orogeny has been considered as occurring at the end of the Ordovician. It has been obvious for some time, however, that this is too restrictive a time span for this event.

By the definition of Dana (1895), the Taconic Mountains and nearby areas are the type locality of the Taconic orogeny, and the temporal relationships there are pertinent to any evaluation of the time span of this orogeny. The thrust slices associated with the emplacement of the Taconic klippe of Vermont and New York have been cited as the diastrophic events associated with the Taconic orogeny at or near the close of the Ordovician in this region (Kay, 1942, p. 1618). Zen, however, has demonstrated that in the area between localities 11 and 39 (Figs. 5–1) the Taconic allochthon consists of a series of thrust slices that probably were emplaced at somewhat different times within the Middle Ordovician (Zen, 1967, p. 68–69; this volume). Uplift to the east, in the Green Mountain area, is believed to have resulted in the detachment and eventual sliding westward of these allochthonous slices. The emplacement of the allochthon in the Normanskill sea (Graptolite Zone 13) is believed to have been preceded by normal longitudinal faults that produced a submarine horst and graben topography earlier in Trenton time (Graptolite Zone 12) (Zen, this volume). The episode of faulting preceding the emplacement of the allochthon is considered as the key diastrophic event by Zen. The lithologic nature of the Normanskill Shale underlying the Taconic allochthon changes in aspect in zone 13 time, probably as a result of uplift of a more easterly terrane with "Taconic sequence" facies (eugeosynclinal) owing to Taconic diastrophism in the area (Zen, 1967, p. 66).

The inception of the Taconic orogeny in its type area therefore is herein considered to have occurred in Trenton times and involves two events: an initial event of faulting (Graptolite Zone 12) was immediately succeeded by the start of the emplacement of the allochthonous slices of the Taconic klippe by gravity sliding off the Green Mountain anticlinorium to the east. This emplacement occurred during later Trenton times (Graptolite Zone 13) when the allochthonous blocks slid into the Normanskill sea and became incorporated in the Normanskill Formation.

The Taconic unconformities to the south, in Pennsylvania, suggest a somewhat younger age for the inception of Taconic diastrophism there. The presence of folded Martinsburg containing early Maysville fossils below the Tuscarora at Schuykill Gap (loc. 5) dates the Taconic orogeny in Pennsylvania as occurring in the interval of late Maysville to Richmond time or in the interval of Graptolite Zones 14 (in part) and 15.

The Taconic unconformities of New England and particularly northern Maine are of both local and regional significance in dating Taconic diastrophism. The unconformities at Ashland (loc. 60) and Castle Hill (loc. 61), Maine, are considerably younger (in the interval of Ashgill to pre-late Llandovery C_3–C_5) than in the type area of the Taconic orogeny (Trenton).

Other unconformities in the northern Appalachians are less useful for dating Taconic diastrophism because of the large and, in places, uncertain nature of the hiatus they represent. Among these, for example, are the unconformities at Cookson Island (loc. 48) and Cap-des-Rosiers Cove (loc. 71). Indeed some of these unconformities may telescope younger and older orogenies as well as one or more pulse of Taconic diastrophism. In general, however, most of these unconformities are not incompatible with dating the start of the Taconic orogeny as not later than Middle Ordovician time.

Bench marks useful for establishing an upper age limit to Taconic diastrophism in the northern Appalachians are the widely distributed and generally well-dated quartzites and conglomerates of about Llandovery C_3–C_5 age, that occur above the Taconic unconformity in New England and adjacent Canada. In northern Maine, the Frenchville Formation, and its temporal equivalents southwestward towards New Hampshire, unconformably overlie Ordovician or older rocks (locs. 50, 52, 53, 56, 57, 60, and 61). Farther south in New England, and particularly in New Hampshire, the Clough Quartzite or its equivalents mostly overlie Middle Ordovician formations such as the Ammonoosuc Volcanics and Partridge Formation (locs. 26–30, 32–38, and 40–44). Thus, the Clough-Frenchville horizons of Llandoverian C_3–C_6 and C_3–C_5 ages set the upper temporal limits for Taconic diastrophism in New England. However, tectonic activity in Maine did not altogether cease after Frenchville time and resumed during the climactic, Middle Devonian, Acadian orogeny. The Salinic break (Boucot, 1962) in the northern Appalachians is characterized by the absence of the Pridoli through lower Gedinnian and may be a local tectonic episode between the Taconic and Acadian orogenies. Some of the Taconic unconformities overlain by post-lower Gedinnian rocks may indeed telescope the Taconic and Salinic events.

In similar manner the observation that the Taconic orogeny seems to have started at about Trenton time does not exclude the possibility that earlier tectonic events such as the Blountain disturbance in the southern Appalachians (Kay,

1942, p. 1632), and the Cambrian to Early Ordovician orogeny of Maine and possibly Quebec (Pavlides and others, 1964, p. 634), named the Penobscot (Neuman, 1967), may have been precursors if not directly related to the tectonism herein described as Taconic. In addition, the Vermontian disturbance named by Kay (1942) appears to coincide in place and time with the initial phase of the Taconic movements recognized by Zen (this volume).

SUMMARY AND CONCLUSIONS

The type locality of the Taconic orogeny indicates that Taconic diastrophism there occurred during Trenton time (Graptolite Zones 12 and 13). Information is absent in the type area as to when Taconic diastrophism may have ceased. Regional considerations, however, suggest that the Silurian Clough and Frenchville Formations of Llandovery C_3–C_6 and C_3–C_5 ages in New England can conveniently be used to define the close of the Taconic: major diastrophism is absent in the northern Appalachians following this time until the inception of the Acadian orogeny in early Middle Devonian time. Of necessity, this definition excludes all pre-Trenton unconformities from the Taconic orogeny. It is recognized, however, that these earlier unconformities probably are related to the regional episodic tectonism that occurred in the northern Appalachians from Cambrian up into the Taconic orogeny as defined herein. Indeed the evidence at hand suggests that diastrophism was episodic during the Taconic orogeny and included different degrees and types of tectonism at different places. In general, however, three tectonic pulses can be distinguished during Taconian times:

(1) An initial phase in western Vermont and eastern New York involved faulting, folding, and uplift (the emplacement of the Taconic allochthon) during Trenton (Graptolite Zones 12–13) time.

(2) A somewhat later phase that marked the inception of Taconic diastrophism in Pennsylvania during late Maysville and Richmond time or approximately in the interval of Graptolite Zones 14 (upper part) and 15, involved both folding and uplift.

(3) A terminal phase that can be bracketed no closer than between Ashgill and Llandovery C_3–C_5, in the interval of Graptolite Zones 22–24, in northern Maine and no closer than Trenton to Llandovery C_3–C_6 (or the interval of Graptolite Zones 22–25, in part) elsewhere in New England. Near Port Daniel, Quebec (loc. 68b), however,

the terminal Taconic break is bracketed as occurring between Trenton and early Llandovery A_3–A_4 time. The northern Maine unconformities (locs. 60 and 61) can be more closely dated as occurring within the interval of Graptolite Zone 19 (base of middle Llandovery B_1), the interval of the gradational boundary from the Carys Mills up into the Smyrna Mills at locality 50. This change from an older regime of carbonate deposition with subordinate amounts of clastic rocks to one of greater clastic rock content and lesser amounts of carbonate probably reflects penecontemporaneous tectonism marginal to the Aroostook–Matapedia belt, as in the Ashland (loc. 60) and Castle Hill (loc. 61) areas.

Minor episodic pulses of diastrophism within this terminal phase probably include the unconformities of late Llandovery C_1–C_2 age (loc. 65b) and Llandovery C_1–C_2 to Wenlock–Ludlow age (loc. 65c) of the Cabano, Quebec area.

It is concluded, therefore, that the age span of the Taconic orogeny should be defined to include the diastrophic events encompassed between Middle Ordovician (Trenton) and pre-Llandovery C_3–C_5 times; and as characterized by at least three pulses of diastrophism. Tectonism associated with these pulses of Taconic orogeny occurred in belts characterized by an eastern and western area of angular unconformity; these are flanked by parallel belts of disconformity which in some places bound belts of conformity or belts of uninterrupted sedimentation free of diastrophism. Three such belts of conformity are recognized in the northern Appalachians. Two of these belts, that in Nova Scotia and the Aroostook–Matapedia belt of Maine through the Gaspé Peninsula, reflect the Taconic diastrophism that occurred in peripheral areas by the change in the regime of sedimentation within these belts with the inception of Taconic tectonism.

REFERENCES

Alcock, F. J., 1935, Geology of Chaleur Bay region: Canada Geol. Survey Mem. 183, 146 p.

Ambrose, J. W., 1957, The age of the Bolton lavas, Memphremagog district, Quebec: Naturaliste Canadien, v. 84, p. 161–170

Ayrton, W. G., 1967, Chandler-Port Daniel area: Quebec Dept. of Nat. Resources, Geol. Rept. no. 120, 41 p.

Balk, Robert, 1946, Gneiss dome at Shelburne Falls, Mass.: Geol. Soc. America Bull. v. 57, p. 125–160

Beecher, C. E., and Dodge, W. W., 1892, On the occurrence of Upper Silurian strata near Penobscot Bay, Maine: Am. Jour. Sci., 3d ser., v. 43, p. 412–418

Béland, Jacques, 1958, Preliminary report on the Oak Bay area: Quebec Dept. Mines, Prelim. Rept. 375, 12 p.

————, 1960, Preliminary report on Rimouski-Matapedia area: Quebec Dept. Mines, Prelim. Rept. 430, 18 p.

Berry, W. B. N., 1962, On the Magog, Quebec, graptolites: Am. Jour. Sci., v. 260, no. 2, p. 142–148

————, and Boucot, A. J., in press, Silurian of North America: Geol. Soc. America

Billings, M. P., 1956, The geology of New Hampshire; pt. 2, Bedrock geology: Concord, New Hampshire State Plan. and Devel. Comm., 203 p.

Boucot, A. J., 1961, Stratigraphy of the Moose River synclinorium, Maine: U. S. Geol. Survey Bull. 1111–E, 188 p.

————, 1962, Appalachian Siluro-Devonian *in* Coe, Kenneth, *ed.* Some aspects of the Variscan fold belt: Inter-University Geol. Cong., 9th, Exeter, Manchester Univ. Press, p. 155–163

————, and Drapeau, Georges, in press. The Siluro-Devonian rocks of Lake Memphremagog and their correlatives in the Eastern Townships, Quebec: Quebec Dept. of Nat. Resources, Spec. Paper no. 1.

————, and Johnson, J. G., 1967, Silurian and Upper Ordovician atrypids of the genera *Plectatrypa* and *Spirigerina*: Norsk Geolo. Tidsskrift, v. 47, pt. 1, p. 70–101

————, Field, M. T., Fletcher, Raymond, Forbes, W. H., Naylor, R. S., and Pavlides, Louis, 1964, Reconnaissance bedrock geology of the Presque Isle quadrangle, Maine: Maine Geol. Survey Quad. Map Ser. no. 2, 123 p.

————, and Thompson, J. B., Jr., 1963, Metamorphosed Silurian brachiopods from New Hampshire: Geol. Soc. America Bull., v. 74, p. 1313–1334

Brown, T. C., 1914, The Shawangunk conglomerate and associated beds near High Falls, Ulster County, New York: Am. Jour. Sci., 4th ser., v. 37, p. 464–474

Chadwick, G. H., 1913, Angular unconformity at Catskill [abs.]: Geol. Soc. America Bull., v. 24, p. 676

Chance, H. M., 1882, Special survey of the Delaware Water Gap: Pennsylvania Geol. Survey, 2d, Rept. G6, p. 335–363

Clark, T. H., 1921, A review of the evidence for the Taconic revolution: Boston Soc. Nat. History Proc., v. 36, no. 3, p. 135–163

————, 1934, Structure and stratigraphy of southern Quebec: Geol. Soc. America Bull., v. 45, p. 1–20

Crosby, D. G., 1962, Wolfville map-area, Nova Scotia: Canada Geol. Survey Mem. 325, 67 p.

Cumming, L. M., 1959, Silurian and Lower Devonian formations in the eastern part of Gaspé Peninsula, Quebec: Canada Geol. Survey Mem. 304, 45 p.

Currier, L. W., and Jahns, R. H., 1941, Ordovician stratigraphy of Central Vermont: Geol. Soc. America Bull., v. 52, p. 1487–1512

Dale, N. C., 1953, Geology and mineral resources of the Oriskany quadrangle (Rome quadrangle): New York State Mus. Bull. no. 345, 197 p.

Dana, J. D., 1895, Manual of geology, 4th ed.: New York, American Book Co., 1087 p.

Doll, C. G., Cady, W. M., Thompson, J. B., Jr., Billings, M. P., 1961, *compilers* and *editors*, Centennial geologic map of Vermont: Vermont Geol. Survey, scale 1:250,000

Ekren, E. B., and Frischknecht, F. C., 1967, Geological-geophysical investigations of bedrock in the Island Falls quadrangle, Aroostook and Penobscot, Maine: U. S. Geol. Survey Prof. Paper 527, 36 p.

Fisher, D. W., 1954, Stratigraphy of the Medinan group, New York and Ontario: Am. Assoc. Petroleum Geologists Bull., v. 38, no. 9, p. 1979–1996

————, 1960, Correlation of the Silurian rocks in New York State: New York State Mus. and Sci. Services Geol. Survey Map and Chart Ser., no. 1

————, 1962, Correlation of the Ordovician rocks in New York State: New York State Mus. and Sci. Services Geol. Survey Map and Chart Ser., no. 3

————, and Rickard, L. V., 1953, Age of the Brayman Shale: New York State Mus. Circ. 36, 14 p.

————, Isachsen, Y. W., Rickard, L. V., Broughton, J. G., and Offield, T. W., *compilers* and *editors*, 1962, Geologic map of New York, 1961: New York State Mus. and Sci. Service; Geol. Survey Map and Chart series no. 5, 5 sheets, scale 1:250,000

Grabau, A. W., 1901, Guide to the geology and paleontology of Niagara Falls and vicinity: New York State Mus. Bull. 7, 284 p.

————, 1909, Physical and faunal evolution of North America during Ordovicic, Siluric, and early Devonic time: Jour. Geology, v. 17, p. 209–252

Green, J. C., 1964, Stratigraphy and structure of the Boundary Mountain anticlinorium in the Errol quadrangle, New Hampshire-Maine: Geol. Soc. America Spec. Paper 77, 78 p.

Hall, B. A., 1964, Stratigraphy and structure of the Spider Lake quadrangle, Maine: PhD thesis, Yale Univ. 153 p.

Harwood, D. S., and Berry, W. B. N., 1967, Fossiliferous Ordovician and Silurian rocks in the Cupsuptic quadrangle, west-central Maine: U. S. Geol. Survey Prof. Paper 575D, p. D16–D23

Kay, Marshall, 1942, Development of the northern Allegheny synclinorium and adjoining regions: Geol. Soc. America Bull., v. 53, no. 11, p. 1601–1657

Kümmel, H. D., and Weller, Stuart, 1902, The rocks of the Green Pond Mountain region: New Jersey Geol. Survey, Ann. Rept., p. 1–51

Lajoie, Jean, 1962, Preliminary report on Chénier-Bédard area, Rimouski County: Quebec Dept. Nat. Resources Prelim. Rept. 493, 8 p.

Larrabee, D. M., and Spencer, C. W., 1963, Bedrock geology of the Danforth quadrangle, Maine: U. S. Geol. Survey Geol. Map GQ–221.

————, Spencer, C. W., and Swift, D. J. P., 1965, Bedrock geology of the Grand Lake area, Aroostook, Hancock, Penobscot, and Washington Counties, Maine: U. S. Geol. Survey Bull. 1201–E, p. E1–E38

Lesley, J. P., 1892–5, A summary description of the geology of Pennsylvania [in part by D'Invilliers and others]: Pennsylvania Geol. Survey, 2d, Final Report, v. 1, p. 1–1915, v. 2, p. 2153–2188, v. 3, p. 2189–2588

Lespérance, P. J., 1960, The Silurian and Devonian rocks of the Temiscouata Region, Quebec: Ph.D. thesis, McGill University, 264 p. with Appendices p. A–RR

Maehl, R. H., 1961, The older Paleozoic of Pictou County, Nova Scotia: Nova Scotia Dept. Mines Mem. no. 4, 112 p.

McGerrigle, H. W., 1950, The geology of eastern Gaspé: Quebec Dept. Mines, Geol. Surveys Br. Geol. Rept. 35, 168 p.

Miller, B. L., 1926, Taconic folding in Pennsylvania: Geol. Soc. America Bull., v. 37, p. 497–511

Neuman, R. B., 1967, Bedrock geology of the Shin Pond and Stacyville quadrangles, Maine: U. S. Geol. Survey Prof. Paper 524–I, p. 1–37

Pavlides, Louis, in press, Stratigraphic and facies relationships of the Carys Mills Formation of Silurian and Ordovician age of northeast Maine: U. S. Geol. Survey Bull. 1264

————, and Berry, W. B. N., 1966, Graptolite-bearing Silurian rocks of the Houlton-Smyrna Mills area, Aroostook County, Maine: U. S. Geol. Survey Prof. Paper 550–B, p. B51–B61

————, Mencher, Ely, Naylor, R. S., and Boucot, A. J., 1964, Outline of the stratigraphic and tectonic features of northeastern Maine: U. S. Geol. Survey Prof. Paper 501–C, p. C28–C38

Pierce, K. L., 1966, Bedrock and surficial geology of the McConnellsburg quadrangle, Pennsylvania: Pennsylvania Geol. Survey Atlas 109a, 111 p.

Pirsson, L. V., and Schuchert, Charles, 1924, A textbook of geology. Part II, Historical Geology, 2d Revised Ed.: New York, John Wiley and Sons, Inc., 724 p.

Poole, W. H., 1963, Hayesville, New Brunswick: Canada Geol. Survey Map 6–1963

————, ed., 1966, Geology of parts of Atlantic Provinces: Geol. Assoc. Canada and Mineralog. Assoc. Canada Guidebook, 1966, Hailfax, Nova Scotia, 155 p.

Potter, R. R., 1964, Geology, Upsalquitch Forks: Canada Geol. Survey Map 14–1964

Rickard, L. V., 1964, Correlation of the Devonian rocks in New York State: New York State Mus. and Sci. Service Geol. Survey Map and Chart Ser., no. 4

Rodgers, John, 1967, Chronology of tectonic movements in the Appalachian region of eastern North America: Am. Jour. Sci., v. 265, p. 408–427

————, Gates, R. M., and Rosenfeld, J. L., 1959, Explanatory text for preliminary geologic map of Connecticut, 1956: Connecticut Geol. and Nat. History Survey Bull. no. 84, 64 p.

Rogers, H. D., 1856, On the correlation of the North American and British Paleozoic Strata: Rept. of 26th mtg. of the British Assoc. (Cheltenham) p. 175–186

————, 1858, The Geology of Pennsylvania; a government survey: Philadelphia, 2 v., 586 p. and 815 p.

Rose, B., Anderson, F. D., Greenshields, R. A., Johnston, J. M., and MacGregor, I. D., 1962, Geology, Tobique, New Brunswick: Canada Geol. Survey Prelim. Ser. Map 37–1962, scale 1:126,720

Rosenfeld, J. L., and Eaton, G. P., 1958, Stratigraphy, structure, and metamorphism in the Middle Haddam quadrangle and vicinity, Connecticut: New England Intercollegiate Geol. Conf. Guidebook 50th Ann. Mtg., 8 p.

Ruedemann, Rudolf, 1942, Cambrian and Ordovician geology of the Catskill quadrangle, in Pt. 1 of Geology of the Catskill and Kaaterskill quadrangles: New York State Mus. Bull. 331, p. 7–188

Ruitenberg, A. A., 1967, Stratigraphy, structure and metallization, Piskahegan-Rolling Dam area (northern Appalachians, New Brunswick, Canada): Ph. D. thesis, Leiden University, p. 79–120

Schuchert, Charles, 1916, Silurian formations of southeastern New York, New Jersey, and Pennsylvania: Geol. Soc. America Bull., v. 27, p. 531–554

————, 1925, Significance of Taconic orogeny: Geol. Soc. America Bull., v. 36, p. 343–350

————, and Longwell, C. R., 1932, Deformation of the Hudson Valley region, New York: Am. Jour. Sci., 5th ser., v. 23, p. 304–326

Skidmore, W. B., 1965, Honorat-Reboul area, Bonaventure County: Quebec Dept. Nat. Resources, Geol. Rept. 107

Smith, C. H., 1957, Bathhurst-Newcastle map-area, New Brunswick: Canada Geol. Survey Map I–1957

Smith, E. S. C., 1923, The Rangeley conglomerate [Maine]: Am. Jour. Sci., 5th ser., v. 5, p. 147–154

Smith, G. O., Bastin, E. S., and Brown, C. W., 1907, Description of the Penobscot Bay quadrangle [Maine]: U. S. Geol. Survey Geol. Atlas, Folio 149, 14 p.

Stose, G. W., 1930, Unconformity at the base of the Silurian in southeastern Pennsylvania: Geol. Soc. America Bull., v. 41, p. 629–658

Taylor, F. C., 1965, Silurian stratigraphy and Ordovician-Silurian relationships in southwestern Nova Scotia: Canada Geol. Survey Paper 64–13, 24 p.

Thompson, J. B., Jr., 1954, Structural geology of the Skitchewaug Mountain area, Claremont quadrangle Vermont-New Hampshire: New England Intercollegiate Geol. Cong. 46th (Dartmouth College) Guidebook p. 35–41

Twenhofel, W. H., 1927, Geology of Anticosti Island: Canada Geol. Survey Mem. 154, 481 p.

U. S. Geological Survey, 1965, Geological Survey Research 1965: U. S. Geol. Survey Prof. Paper 525–A, 376 p.

Van Ingen, Gilbert, and Clark, P. E., 1903, Disturbed fossiliferous rocks in the vicinity of Rondout, N. Y.: New York State Mus. Bull. no. 69, p. 1176–1227

White, I. C., 1882, Geology of Pike and Monroe Counties: Pennsylvania Geol. Survey, 2d, Rept. G6, 407 p.

Willard, Bradford, and Cleaves, A. B., 1939, Ordovician-Silurian relations in Pennsylvania: Geol. Soc. America Bull., v. 50, p. 1165–1198

Zen, E-an, 1967, Time and space relationships of Taconic allochthon and autochthon: Geol Soc. America Spec. Paper 97, 107 p.

Chapter 6

Silurian and Devonian of the Northern Appalachians

A. J. BOUCOT

INTRODUCTION

DESPITE THE OBLITERATING effects of deformation and metamorphism, many Silurian and Early Devonian fossil localities are now known in the Northern Appalachian region. Over 2000 collections have been accumulated in the area between the Hudson River and northern Newfoundland since 1948, when my studies began. The following list of fossil localities from my file in Pasadena will serve to illustrate the volume of data and their geographic distribution:

Massachusetts 3; New Hampshire 3; Vermont 1; Maine: coastal area 58, Aroostook Co. 124, Franklin Co. 8, Oxford Co. 1, Penobscot Co. 31, Piscataquis and Somerset Counties 370; northern New Brunswick 37, southern New Brunswick 37; Newfoundland 24; Nova Scotia: Antigonish Co. 166, Cape Breton Island 2, Cobequid Mountain region 14, Annapolis Valley region 3, Pictou Co. region 149; Quebec: Anticosti Island 48, Eastern Townships 8, Gaspé 300, Montreal 1, Matapedia–Temiscouata region 179.

In an area as extensive as the northern Appalachians region, the normal procedure, when one attempts to unravel lithofacies and paleogeographic relationships on the basis of isolated outcrop areas, is to plot the physical data for each lithologic unit on a suitable base map, and then to correlate the strata on the basis of the relationships as they emerge from the plot. This procedure is not satisfactory for the northern Appalachian region, however, because the area is extremely complex, not only stratigraphically but also structurally. For instance, rapid lithofacies changes may make physical correlation doubtful, and metamorphism makes the original nature of some rock units uncertain. Because of these complications, a more productive procedure is to plot on a map correlatable fossil localities which belong to a particular fossil zone or to closely related zones and then note the nature of rock distributions at those localities. In this manner, a lithofacies map and accompanying paleogeographic interpretation may be developed; in addition, scattered information derived from fossil collections that came from poorly mapped or unmapped areas can be used to advantage.

The paleogeographic conclusions discussed in this paper are derived primarily from considerations of such lithofacies and paleogeographic maps. For the Silurian Period, the maps have been presented by Berry and Boucot (in preparation), and for the Early Devonian, by Boucot and Johnson (1967). Because of limitations of space, it will be impossible to present here the primary data or sources of information. Rather, this summary paper, specifically requested by the editors of this volume in the interest of chronostratigraphic continuity, will be concerned with interpretations and conclusions of the author and his coworkers; the reader is referred to the cited publications for details and for references to the literature. Correlation of rock units is shown on the accompanying correlation charts which should be studied in conjunction with the paleogeographic maps.

The Silurian and Devonian history of the northern Appalachian region can be summarized as follows:

1. A Taconic orogeny. This affected some areas strongly, resulting in angular unconformities; it affected some other areas only as general

TABLE 6–1. Correlation chart for the Silurian of the northern Appalachian region.

Region groupings across the columns: **GASPÉ** (columns 1–3), **EASTERN TOWNSHIPS** (columns 4–7), **CONN. R. VALLEY** (columns 8–9), **NORTHERN** [Maine] (columns 10–16). Diagonal hatching in the chart marks intervals that are missing or not represented.

No.	Locality	Pridoli	Ludlow	Wenlock	Upper Llandovery	Middle Llandovery	Lower Llandovery	Ashgillian or Older
1	Forillon / Griffin Cove R.	St. Alban Fm.; Griffin Cove River Fm.; Conglomerate	—	—	—	—	—	Cape Rosier Shale
2	Metapedia Valley Region to Mt. Albert Region	St. Leon Formation	St. Leon Formation	Sayabec Limestone	Val Brillant Quartzite; Awantjish Formation	—	—	Quebec Group
3	Cap Chat River	St. Leon Formation	St. Leon Formation	Missing (Fault)	Val Brillant Quartzite; Jupiter Formation	Missing (Fault)	Missing (Fault)	Middle Ordovician
4	Lake Temiscouata Region	Mt. Wissick Group (Lac Croche Formation; Undifferentiated Sayabec & St. Leon Formations; Robitaille Formation)			Pointe Aux Trembles Formation		Cabano Formation	Quebec Group
5	Lake Etchemin Area	Cranbourne Formation	Cranbourne Formation	Cranbourne Formation	—	—	—	Beauceville Formation
6	Lake Aylmer Region	Lake Aylmer Formation	Lake Aylmer Formation	Lake Aylmer Formation	—	—	—	Beauceville Formation
7	Lake Memphremagog	—	Sargent Bay Limestone; Glenbrooke Slate	Peasley Pond Conglomerate	—	—	—	Magog Formation
8	East Side Connecticut Valley – N.H.-Mass. (type area near Littleton N.H.)	—	Fitch Formation; Clough Fm	Fitch Formation; Clough Formation	Clough Formation			pre-Silurian Formations
9	West Side Connecticut Valley – Vt.-Mass.	Waits River Limestone	Waits River Limestone; Northfield Slate	Shaw Mtn. Formation	—	—	—	Cram Hill Formation
10	Chesuncook Area	—	Un-named Siltstones	Chesuncook Limestone	Un-named Conglomerate	Chesuncook Limestone	Chesuncook Limestone	pre-Silurian Complex
11	Northern Shin Pond and Stacyville Quads.	—	Un-named Siltstones	Un-named Calcareous Siltstone and Limestone	Un-named Red Cong.; Un-named Conglomerate on Kimball & Lane Brooks			Ashgillian Age Bed
12	Southern Shin Pond, Stacyville, & Sherman Falls Quads.	—	Un-named Limestone	Allsbury Formation	Frenchville Fm; Un-named Conglomerate	Allsbury Formation	Allsbury Formation	Wissataquoik Chert
13	Spider Lake Quad.	—	East Branch Group	East Branch Group	—	—	—	Blind Brook Formation
14	Lobster Lake Area	—	Lobster Lake Fm.	Lobster Lake Fm.	—	—	—	pre-Silurian Complex
15	Moose River Synclinorium	Fox's Camp Fm.	Hardwood Mountain Formation	Hardwood Mountain Formation	—	—	—	Basement Complex
16	Cupsuptic & Kennebago Lake Quads.	Un-named Volcanics	Un-named Volcanics	Un-named Unit	Un-named Unit	Un-named Unit	Un-named Unit	Un-named Ordovician

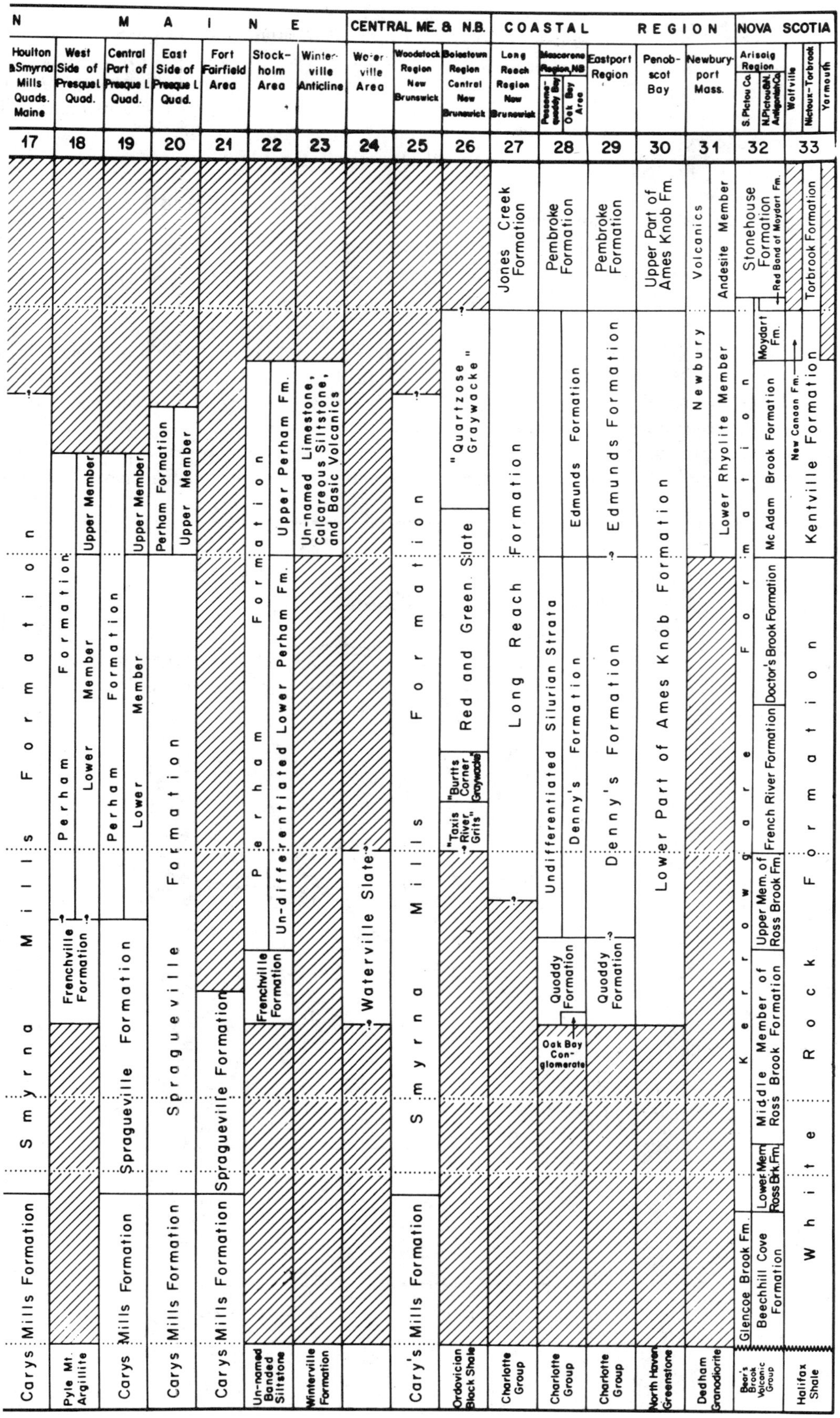

Stratigraphic correlation chart (columns 17–33).

Regional groupings (top header):
- **N   M A I N E   E** — columns 17–24
- **CENTRAL ME. & N.B.** — columns 25–26
- **C O A S T A L   R E G I O N** — columns 27–31
- **NOVA SCOTIA** — columns 32–33

Column headings:

17	Houlton & Smyrna Mills Quads. Maine
18	West Side of Presque I. Quad.
19	Central Part of Presque I. Quad.
20	East Side of Presque I. Quad.
21	Fort Fairfield Area
22	Stockholm Area
23	Winterville Anticline
24	Waterville Area
25	Woodstock Region New Brunswick
26	Boiestown Region Central New Brunswick
27	Long Reach Region New Brunswick
28	Mascarene Region, NB — Passamaquoddy Bay / Oak Bay Area
29	Eastport Region
30	Penobscot Bay
31	Newbury-port Mass.
32	Arisaig Region — S. Pictou Co. / N. Pictou & N. Antigonish Co.
33	Wolfville / Nictaux-Torbrook / Yarmouth

Stratigraphic units by column (top to bottom):

- **17:** Carys Mills Formation — Smyrna Mills Formation — Carys Mills Formation
- **18:** Perham Formation (Upper Member / Lower Member) — Frenchville Formation — Pyle Mt. Argillite
- **19:** Perham Formation (Upper Member / Lower Member) — Spragueville Formation — Carys Mills Formation
- **20:** Perham Formation (Upper Member / Lower Member) — Spragueville Formation — Carys Mills Formation
- **21:** Spragueville Formation — Carys Mills Formation
- **22:** Perham Formation / Frenchville Formation — Un-named Banded Siltstone
- **23:** Upper Perham Fm. / Un-named Limestone, Calcareous Siltstone, and Basic Volcanics / Un-differentiated Lower Perham Fm. — Winterville Formation
- **24:** Waterville Slate
- **25:** Smyrna Mills Formation — Cary's Mills Formation
- **26:** "Quartzose Graywacke" / Red and Green Slate / "Taxis River Grits" / "Burtts Corner Graywacke" — Ordovician Black Shale
- **27:** Jones Creek Formation — Long Reach Formation — Charlotte Group
- **28:** Pembroke Formation — Edmunds Formation — Undifferentiated Silurian Strata / Denny's Formation — Quoddy Formation — Oak Bay Conglomerate — Charlotte Group
- **29:** Pembroke Formation — Edmunds Formation — Denny's Formation — Quoddy Formation — Charlotte Group
- **30:** Upper Part of Ames Knob Fm. — Lower Part of Ames Knob Formation — North Haven Greenstone
- **31:** Newbury Volcanics (Andesite Member / Lower Rhyolite Member) — Dedham Granodiorite
- **32:** Stonehouse Formation — Moydart Fm. (Red Band of Moydart Fm.) — McAdam Brook Formation — Doctor's Brook Formation — French River Formation — Upper Mem. of Ross Brook Formation — Middle Member of Ross Brook Formation — Lower Mem. Ross Brk. Fm. / Glencoe Brook Fm. — Beechhill Cove Formation — Bear's Brook Volcanic Group
- **33:** Torbrook Formation — New Canaan Fm. — Kentville Formation — White Rock Formation — Halifax Shale

uplift, resulting in disconformities; it failed to affect yet other areas at all, resulting in perfect conformity and continuous sedimentation (see Pavlides and others, this volume).

2. A Silurian interval during which seas gradually encroached upon and completely or almost completely covered those areas previously affected by the Taconic orogeny.

3. An interval in the Silurian and earliest Devonian, the so-called Salinic disturbance, during which a few relatively minor areas were affected by uplift.

4. An Early Devonian interval, during which marine sedimentation was widespread to the northwest of a coastal New England and New Brunswick belt, extending well up onto the Canadian Shield, and also to the southeast in the Annapolis Valley region.

5. The Acadian orogeny. Following the Early Devonian and Eifelian marine sedimentation, which covered most of the northern Appala-

chians, this orogeny affected virtually the whole region. Record of marine sedimentation is completely absent from the northern Appalachian region for the remainder of the Devonian period.

Figures 6–1 through 6–6 indicate shoreline and lithofacies relationships that existed from the time of the Taconic orogeny up through various faunal time zones of the Silurian and Early Devonian.

LITHOFACIES RELATIONSHIPS

Silurian

From a lithofacies point of view, the distribution of Silurian (Figs. 6–1, 6–2, 6–3) rock types can be broken down as follows:

1. Carbonate rocks, typical of a platform that extended from Anticosti to Ontario and New York, grade to the east or southeast into a restricted belt of argillaceous rocks (calcareous mudstone belt of Fig. 6–2) which are too limited in some areas to plot on the maps. This belt in

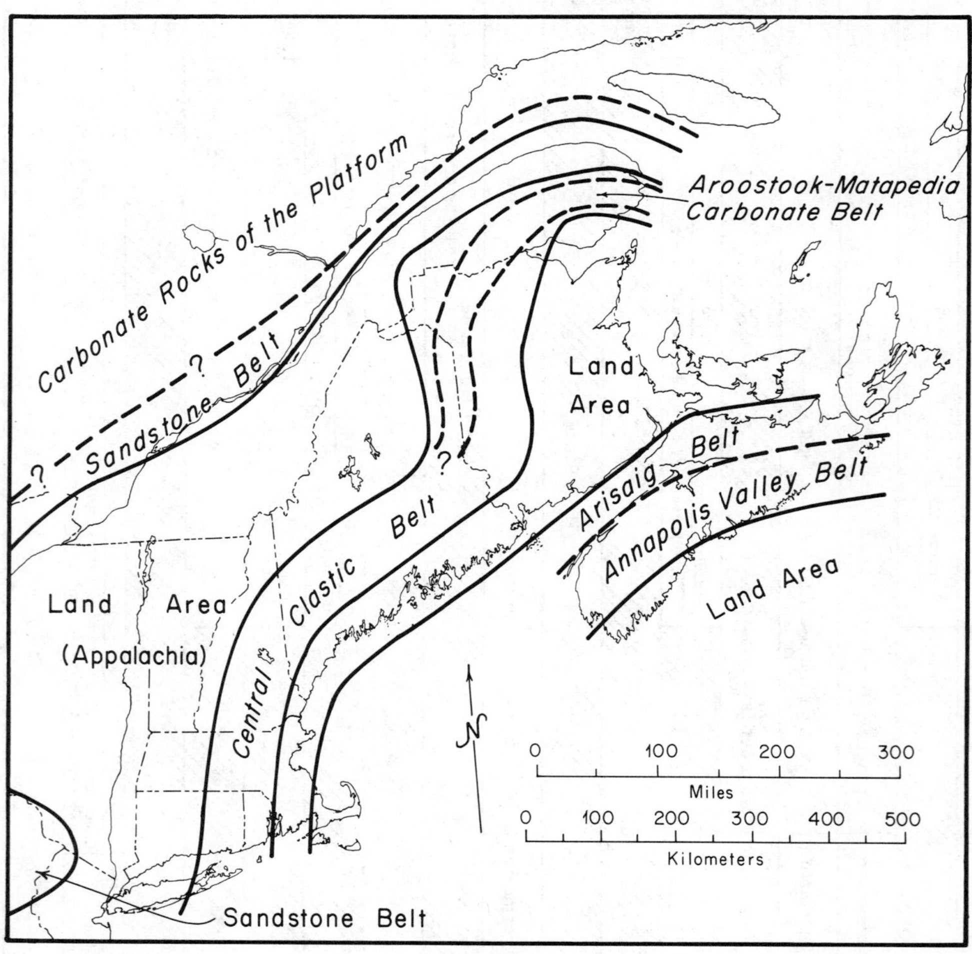

FIGURE 6–1. Lithofacies and paleogeography of the lower Llandovery Stage in the northern Appalachian region.

turn grades rapidly to the east or southeast into non-marine rocks of Medina aspect (the sandstone belt of Figs 6–1, 6–2, 6–3).

2. The Medina rocks (sandstone belt) were presumably in turn replaced to the east by a major land area, the so-called Appalachia of many writers (Dunbar, 1960, p. 108); no Silurian sedimentary record is known from here. On the east and southeast side of this major land area, marine sedimentation occurred during the time interval between Early Llandovery and Ludlow; immediately adjacent to this area of marine sedimentation, occasional remnants of non-marine beds of a variety of ages are preserved. The marine sedimentation in this belt was primarily of the clastic type and commonly included very coarse material.

3. To the east of the northeast end of this belt of marine sedimentation, and adjacent to it, a relatively narrow strip developed in the Aroostook–Matapedia anticlinorium region (Figs. 6–1, 6–2), in which calcareous flysch (Aroostook–Matapedia carbonate belt) was deposited through the interval of the Taconic orogeny and into the Llandovery Stage of the Silurian; the precise time of termination in the Llandovery depends on geographic location. In this belt, deposition of calcareous flysch was succeeded by deposition of non-calcareous, fine- to medium-grained sediments through the remainder of the period (Fig. 6–3).

4. Geographically associated with the calcareous flysch belt south and east of Appalachia, there was a second belt (Figs. 6–1 through 6–5) of marine sediments, characterized by clastic sedimentation. In contrast to the marine belt (calcareous mudstone belt of Figs. 6–2, 6–3) adjacent to Appalachia, this belt (central clastic belt) lacks abundant shelly faunas; is characterized by much greater known stratigraphic thicknesses, abundant graywackes and thick argillaceous sandstones; and is the locus of the thickest sequence of Silurian sediment in the northern Appalachian region.

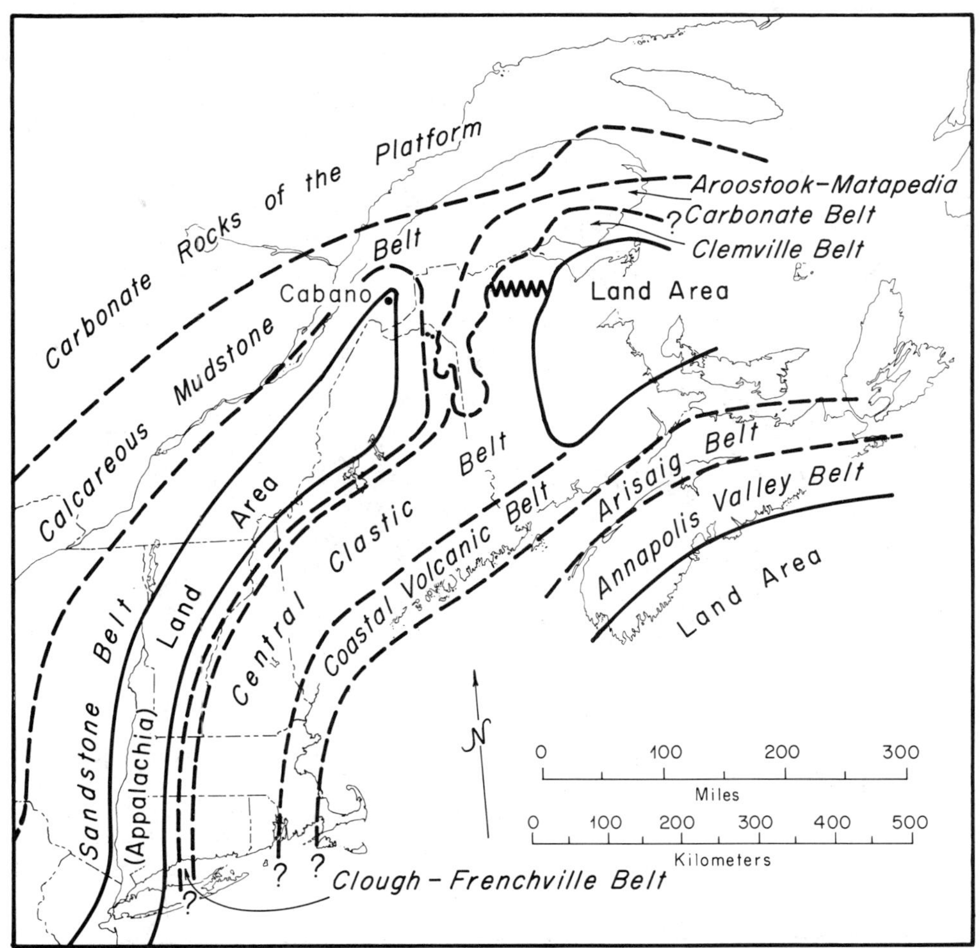

FIGURE 6–2. Lithofacies and paleogeography of the upper Llandovery Stage (C_4–C_5) in the northern Appalachian region.

5. Adjacent to and east and northeast of this last belt are scattered remnants of rocks that suggest the former presence of a belt of relatively shallow water clastic debris adjacent to a second land area. This second land area was already present in the earlier Llandovery (Fig. 6–1), and became a site of intensive volcanism, largely in the marine environment, from the middle part of the Upper Llandovery through the remainder of the Silurian and into the Early Devonian time (Figs. 6–2, 6–3). Marine deposition in this volcanic belt continued until some time in the Early Gedinnian (not shown; see Boucot and Johnson, 1967), when the area became non-marine (Figs. 6–4, 6–5, 6–6).

6. Southeast of this coastal belt of land and marine volcanism is the Arisaig belt, of terrigenous deposition. During the entire Silurian, from the Early Llandovery through the Pridoli and extending into the Early Gedinnian time (Figs. 6–1, 6–2, 6–3), this area was occupied by deposition of marine mudstone and siltstone. The deposits are never very thick, and are characterized by relatively shallow marine faunas.

7. The Arisaig belt grades to the southeast into an area of non-marine Medina-type sediments (Figs. 6–1, 6–2) in the Annapolis Valley region during the Llandovery and Wenlock intervals. Southeast of this Annapolis Valley belt, the presence of an additional land area during the Llandovery and Wenlock is inferred. By Ludlow time (Fig. 6–3), however, a graptolitic facies is present in the Annapolis Valley region and during the Early Devonian graptolitic, terrigenous type rocks are replaced by shelly sandstones, mudstones and siltstones interbedded with sedimentary iron ores.

Early Devonian

Early Devonian lithofacies distributions continue the pattern existing toward the end of the Silurian, even though there is no evidence for the presence of the so-called Appalachia in the northern Appalachian region at this later time.

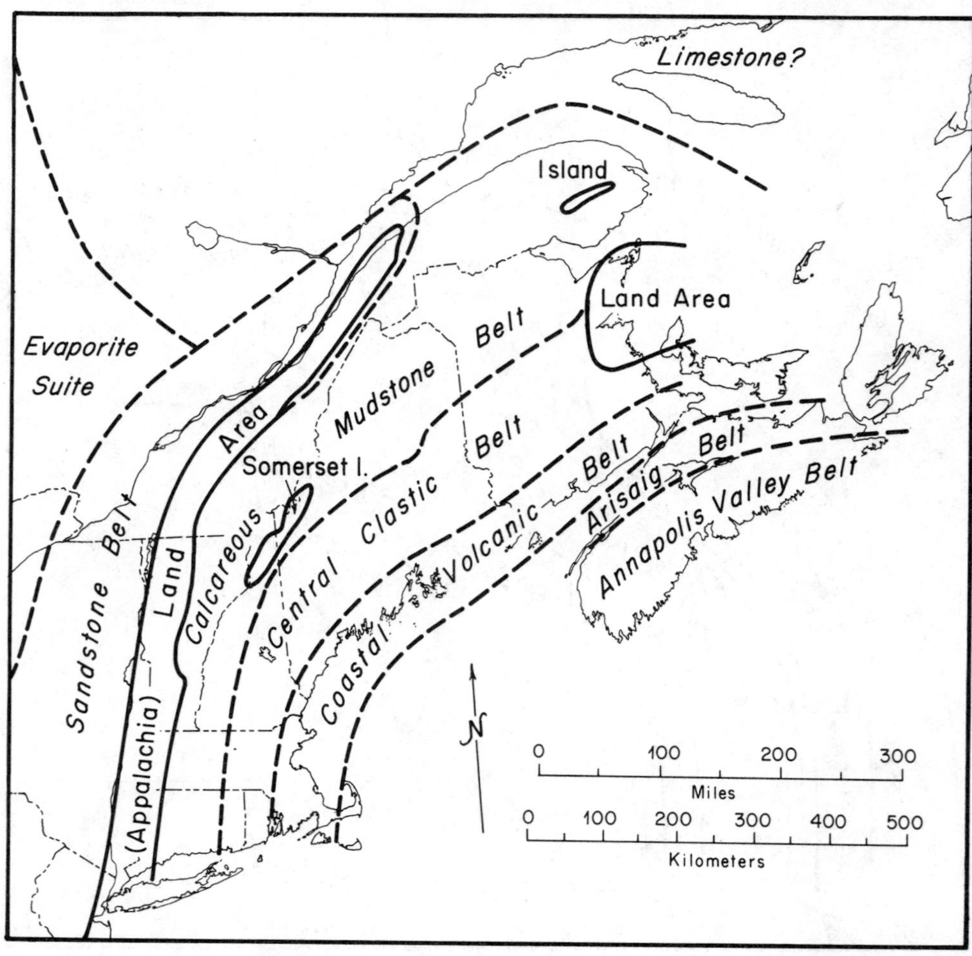

Figure 6–3. Lithofacies and paleogeography of the Ludlow Stage in the northern Appalachian region.

Lithofacies relationships for the Early Devonian are best known during the Oriskany and Schoharie time intervals; the relationships for other Early Devonian time intervals are essentially extrapolations from these better known intervals or from the preceding Silurian. Therefore, description of relationships will be made only for the Middle and Late Helderberg, Oriskany, and Schoharie time intervals.

The Middle and Late Helderberg time interval (essentially Kalkberg–New Scotland through Becraft) is characterized by the emergence of a major land area (Fig. 6–4) at the sites of the marine Arisaig belt and the coastal volcanic belt that existed from Late Llandovery into Early Gedinnian times. The northwestern part of this land area is characterized by abundant volcanic rocks, whereas the southeastern part, restricted insofar as is known to Nova Scotia, is free of volcanic rocks and contains little more than red beds. During this time interval, the northern limit of Appalachia was in southeast New York and northern New Jersey; the land supplied quartz sand to the west. In the area north of Appalachia, limestone of the platform type occurs to the west and is known as far as Montreal, whereas a calcareous siltstone and limestone belt occurs to the east and grades southeasterly into a belt of shale and siltstone. This last belt includes within it a region characterized by thick volcanic deposits and also an island where the basement complex was exposed (Somerset Island).

During the Oriskany time interval (Fig. 6–5), platform-type carbonate and arenaceous rocks are well documented in the area to the west and northwest. This belt is followed to the southeast by a relatively narrow belt of calcareous rocks, containing an abundant, relatively shallow-water marine fauna. Southeast of this belt of shallow-water marine carbonate rocks, the most widespread stratigraphic unit known in the northern Appalachian region occurs. This unit consists of cyclically-banded slate and argillaceous sandstone, variously called the Fortin (which is fossiliferous; see Boucot and Johnson, 1967), Temiscouata, Seboomook, Meetinghouse, Littleton, and Gile Mountain, among other names (see Table 6–2; see also Thompson et al., this volume; Green and Guidotti, this volume).

In northern Maine, within this area of cyclically-banded slate and argillaceous sandstone, are scattered areas of massive argillaceous sandstones, almost all of which are associated at least in part with relatively shallow-water faunas, indicating proximity to major or minor islands (see Rankin, this volume). Boucot and Johnson (1967) have shown the location of approximately five of these, including Somerset and Lobster Islands. Because of absence of detailed mapping and fossil collecting, our knowledge for the Early Devonian in the area southeast of the belt of cyclically-banded sediments is extremely sparse. However, along the coastal region of New England and southern New Brunswick, as well as in adjacent areas in the Arisaig belt of Nova Scotia, we have excellent evidence for the presence during the Oriskany time interval of non-marine environment, probably accompanied by volcanism. In the Annapolis Valley belt, the Oriskany time interval was recorded by marine sedimentation. It is critical to note here that the marine faunas of the Annapolis Valley belt are of Old World Province type, rather than the Appalachian Province type present in most of the northern Appalachian region.

Lithofacies and paleogeographic relationships during the Schoharie time interval Fig. 6–6) continue those of the Oriskany and Helderberg times. In northern Maine and extending northeasterly into the Gaspé region, however, the shoreline migrated considerably, to the northwest in Maine and to the northwest and west in Gaspé. No significant changes are apparent for coastal Maine, New Brunswick, or Nova Scotia, where the previous paleogeographic arrangements appear to have persisted. The chief novelty is the absence of the cyclically-banded shale and sandstones that characterize the Seboomook and the bulk of the Littleton, although the type Littleton Formation in the Connecticut River valley region is actually of Schoharie age. The type Littleton thus differs both in age (Schoharie instead of Oriskany) and in rock type from the bulk of the rocks mapped under that name. In the Whitefield Quadrangle, New Hampshire, fossiliferous Littleton Formation of Schoharie age rests directly upon pre-Silurian beds, but the relationships are difficult to evaluate. Obviously, a land mass existed here in pre-Schoharie time, but its extent is still unknown and may or may not represent a major feature. As yet we have no evidence for the existence of Schoharie age, cyclically-banded sediments in the Seboomook and other named rock units that have, in the past, been correlated with the Littleton.

Later Devonian Relations

Devonian rocks of post-Early Devonian age are very restricted in the northern Appalachian region. Known occurrences include the Mapleton Sandstone in Aroostook County, Maine, of Late Middle Devonian age; the Perry Formation in

TABLE 6–2. Correlation chart for the Lower and Middle Devonian of the northern Appalachian region.

The left-hand column lists the standard stages, from top to bottom: Hamilton, Onondaga, Schoharie, Esopus, Oriskany, Middle & Upper Helderberg, Lower Helderberg. The regional columns are numbered 1–15. Diagonally ruled (hachured) areas indicate no record; they are left blank below.

Stage	1 Forillon	1 Griffin Cove R.	2 Matapedia Valley Region to Mt. Albert Region	3 Cap Chat River
Hamilton				
Onondaga	Battery Point Formation (?)			
Schoharie	Battery Point Formation	York River Sandstone	York River Sandstone	
Esopus	Battery Point Formation	York River Sandstone	York River Sandstone	
Oriskany	Grande Greve Limestone — Lower Unit	Grande Greve Limestone — Upper Unit / Lower Unit	York River Formation	
Middle & Upper Helderberg	Cap Bon Ami Formation	Cap Bon Ami Formation	Cap Bon Ami Formation	
Lower Helderberg	Cap Bon Ami Formation	Cap Bon Ami Formation	Cap Bon Ami Formation	

Stage	4 Lake Temiscouata Region	5 Lake Etchemin Area	6 Lake Aylmer Region	7 Lake Memphremagog	8 East Side Connecticut Valley — N.H.-Mass.	9 West Side Connecticut Valley — Vt.-Mass.
Hamilton					Type area near Littleton N.H.	
Onondaga	Temiscouata Slate / Touladi Limestone	St. Juste Group / Famine Limestone	Compton Group	Mountain House Wharf Limestone		
Schoharie					Littleton Formation	Formation
Esopus					Littleton Formation	Gile Mountain Formation
Oriskany					Littleton Formation	Gile Mountain Formation
Middle & Upper Helderberg					Littleton Formation	Gile Mountain Formation
Lower Helderberg					Littleton Formation (Waits R. Ls.)	Gile Mountain Formation (Waits R. Ls.)

Stage	10 Chesuncook Area	11 Northern Shin Pond and Stacyville Quads.	12 Southern Shin Pond Stacyville & Is. Falls Quads.	13 Spider Lake Quad.	14 Lobster Lake Area	15 Moose River Syncline north
Hamilton						
Onondaga						
Schoharie		Trout Valley Formation				Tomhegan (?)
Esopus	Traveler Rhyolite	Traveler Rhyolite	Traveler Rhyolite			Kineo
Oriskany	Seboomook Formation — Sandstone	Seboomook Formation / Matagamon Formation	Matagamon Formation	Seboomook Formation — Sandstone	Seboomook Formation / Tarratine Formation	Seboomook Formation
Middle & Upper Helderberg	Seboomook Formation	Seboomook Formation		Seboomook Formation	Seboomook Formation	Seboomook Formation
Lower Helderberg	Seboomook Formation	Seboomook Formation		Seboomook Formation (Un-named Volcanics & Sediments)	Seboomook Formation	Seboomook Formation (Parker Bog Fm; Beck... Limestone)

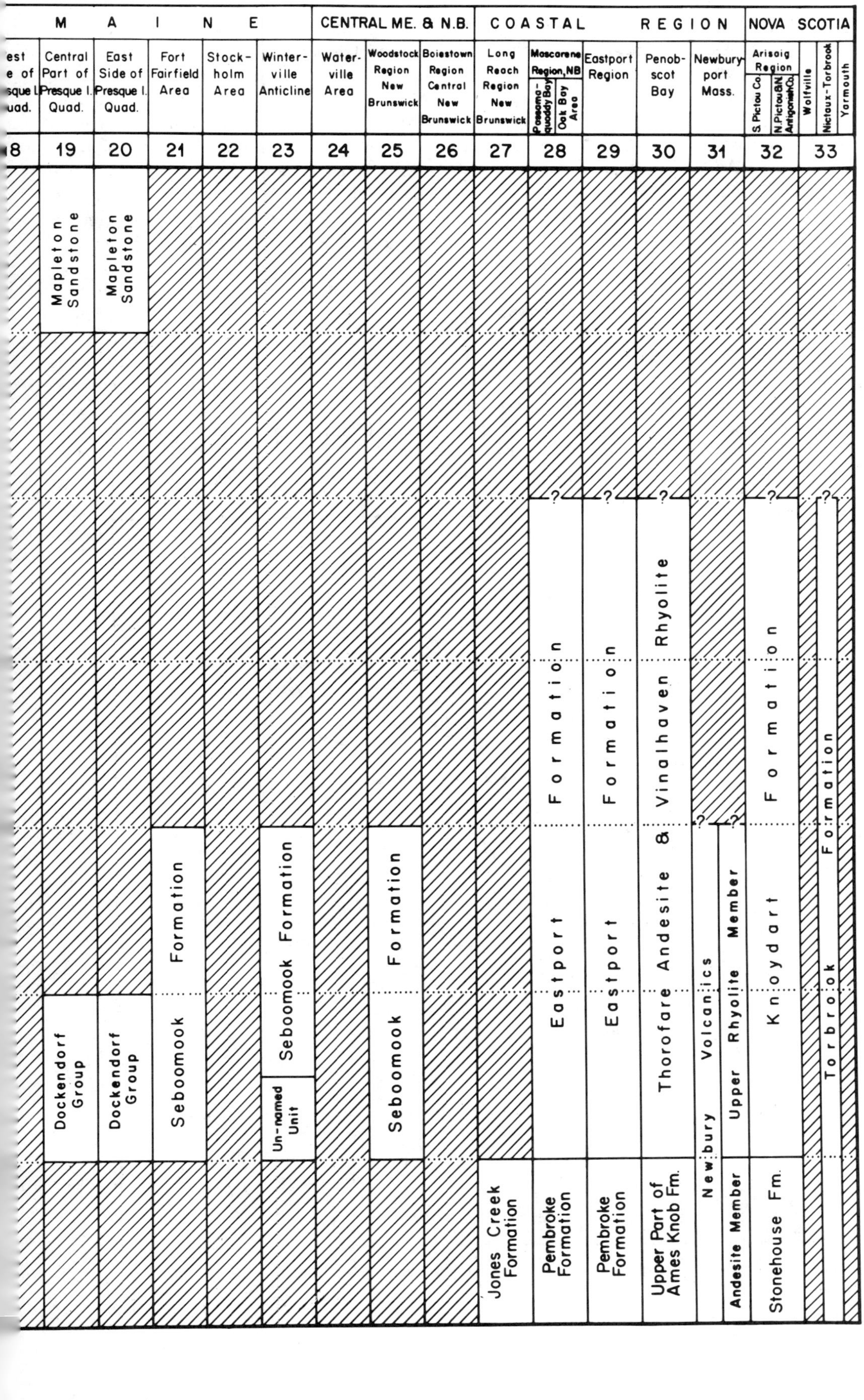

MAINE						CENTRAL ME. & N.B.			COASTAL			REGION		NOVA SCOTIA	
West Side of Presque I. Quad.	Central Part of Presque I. Quad.	East Side of Presque I. Quad.	Fort Fairfield Area	Stockholm Area	Winterville Anticline	Waterville Area	Woodstock Region New Brunswick	Boiestown Region Central New Brunswick	Long Reach Region New Brunswick	Mascarene Region, NB (Passamaquoddy Bay / Oak Bay Area)	Eastport Region	Penobscot Bay	Newburyport Mass.	Arisaig Region (S. Pictou Co. / N. Pictou & Antigonish Co.)	Wolfville / Nictaux-Torbrook / Yarmouth
18	19	20	21	22	23	24	25	26	27	28	29	30	31	32	33

Formation labels shown in the chart columns:

- Column 19: Mapleton Sandstone (upper); Dockendorf Group (lower)
- Column 20: Mapleton Sandstone (upper); Dockendorf Group (lower)
- Column 21: Seboomook Formation
- Column 23: Seboomook Formation; Un-named Unit
- Column 25: Seboomook Formation
- Column 27: Jones Creek Formation
- Column 28: Eastport Formation; Pembroke Formation
- Column 29: Eastport Formation; Pembroke Formation
- Column 30: Vinalhaven Rhyolite; Thorofare Andesite & Vinalhaven Rhyolite; Upper Part of Ames Knob Fm.
- Column 31: Newbury Volcanics; Upper Rhyolite Member; Andesite Member
- Column 32: Formation; Knoydart Formation; Stonehouse Fm.
- Column 33: Torbrook Formation

southeastern Maine and adjacent New Brunswick, of Late Devonian age; various named and unnamed units in the Chaleur Bay region, of both post-Helderberg Early Devonian and of Middle and Upper Devonian age; and the Bay du Nord Series in southern Newfoundland, of Early or Middle Devonian age.

Marine Middle Devonian rocks were unrecognized in the northern Appalachians until the results summarized by Boucot and Johnson (1967) were presented. The present conclusions are that fossiliferous marine rocks, chiefly limestones, of Eifelian age are present in the Lake Memphremagog area (Mountain House Wharf Limestone), St. George area (Famine Limestone), and Lake Temiscouata area (Touladi Limestone). These rocks have been folded and cleaved. The Mountain House Wharf Limestone is faulted against the pre-Silurian; the Famine Limestone is interbedded at its top with calcareous slate of the overlying St. Juste Group (Boucot, unpublished field data); and the contact of the Touladi with the overlying Temiscouata Slate is not exposed, but could be either conformable or faulted. These relationships indicate to me that the Seboomook-like slates extending from Lake Temiscouata (Temiscouata Slate) southwest (Compton Group and St. Juste Group) to Lake Memphremagog, and thence southward down the Connecticut River Valley (Gile Mountain Formation) are of early Middle Devonian (Eifel) age rather than Late Early Devonian, as has been previously concluded. There is no suggestion for the presence in the northern Appalachian region of Devonian marine strata of post-Middle Devonian age.

The above data for the Middle Devonian indicate that the Acadian orogeny is a mid-Middle Devonian rather than Late Lower to Early Middle Devonian event, at least in the Lake Temiscouata–Connecticut Valley region. The occurrence of fossiliferous Givetian mudstone of the Hamilton type at Montreal (Boucot and Johnson, 1967) places a northwestern limit in this region for the effect of the Acadian orogeny. According to my

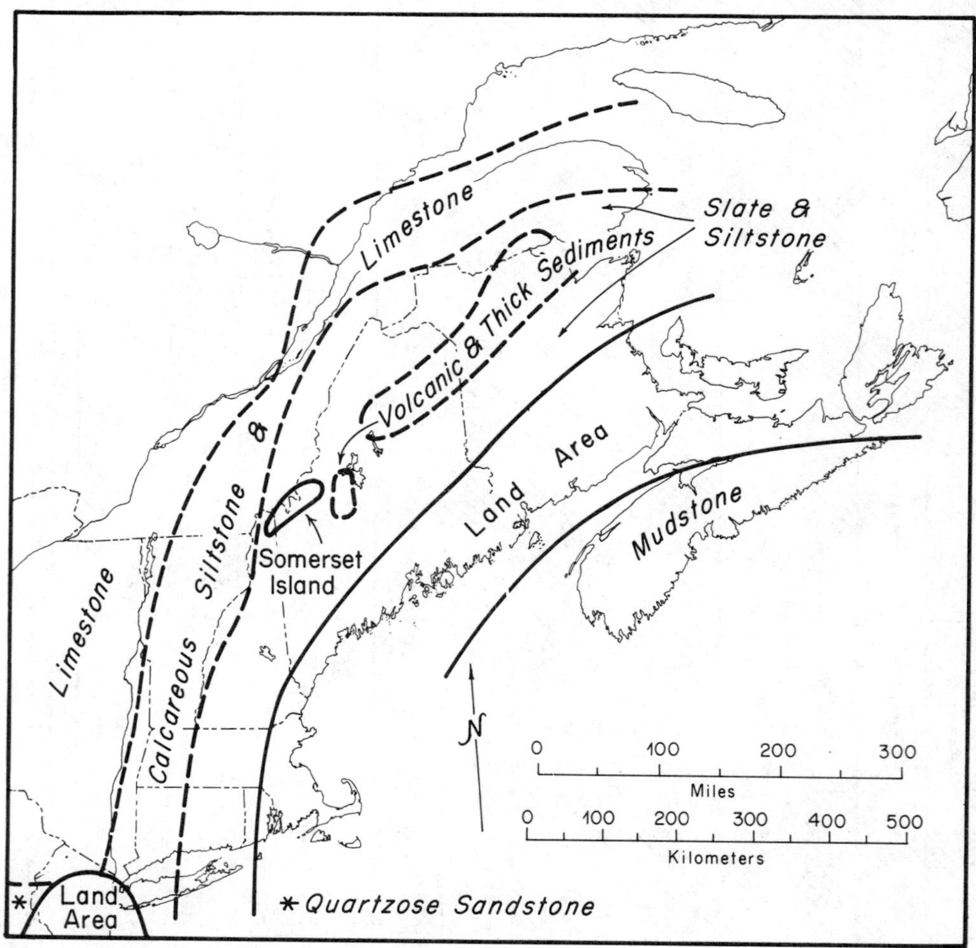

FIGURE 6–4. Lithofacies and paleogeography of the middle and upper Helderberg Stage in the northern Appalachian region.

interpretation, the major lithofacies change in the north-central Appalachian region, between the Onondaga Limestone (Eifel age) and the overlying clastic rocks of the Hamilton Group coincides with the onset of Acadian orogeny. The various radiometric dates from the northern Appalachian region for the "Acadian" igneous rocks must now be viewed in the light of this new stratigraphic information.

CONCLUSIONS

The interrelationships of the paleogeography and lithofacies distributions to both preceding and subsequent structural and metamorphic trends is of great interest. It can be stated categorically that the areas of most intensive regional metamorphism, in western and southwestern Maine, New Hampshire, Vermont and adjacent parts of Massachusetts, Connecticut, and Rhode Island, lie astride the lithofacies and paleogeographic trends of the Silurian and Lower Devonian. Major structural trends, including fold axes and fault traces, do parallel the general trends of the lithofacies and paleogeography in most areas. Because most of the individual structural trends persist for only relatively short distances, gradual divergences of original sedimentary and shoreline trends from structural trends are very difficult to decipher. However, there is clearly no major overall divergence between them that is comparable with the divergence between trends of regional metamorphism and trends of structure and lithofacies-paleogeographic distribution.

The present distribution of large intrusive bodies of Acadian, i.e. Middle Devonian age, and the correlation between their location and the lithofacies trends as well as shoreline trends is nearly random. Acadian intrusive bodies are concentrated in southwestern Nova Scotia, in the Katahdin region of north-central Maine, in the coastal region of Maine and adjacent New Brunswick, and, of course, in the highly metamorphosed terrains of western New England,

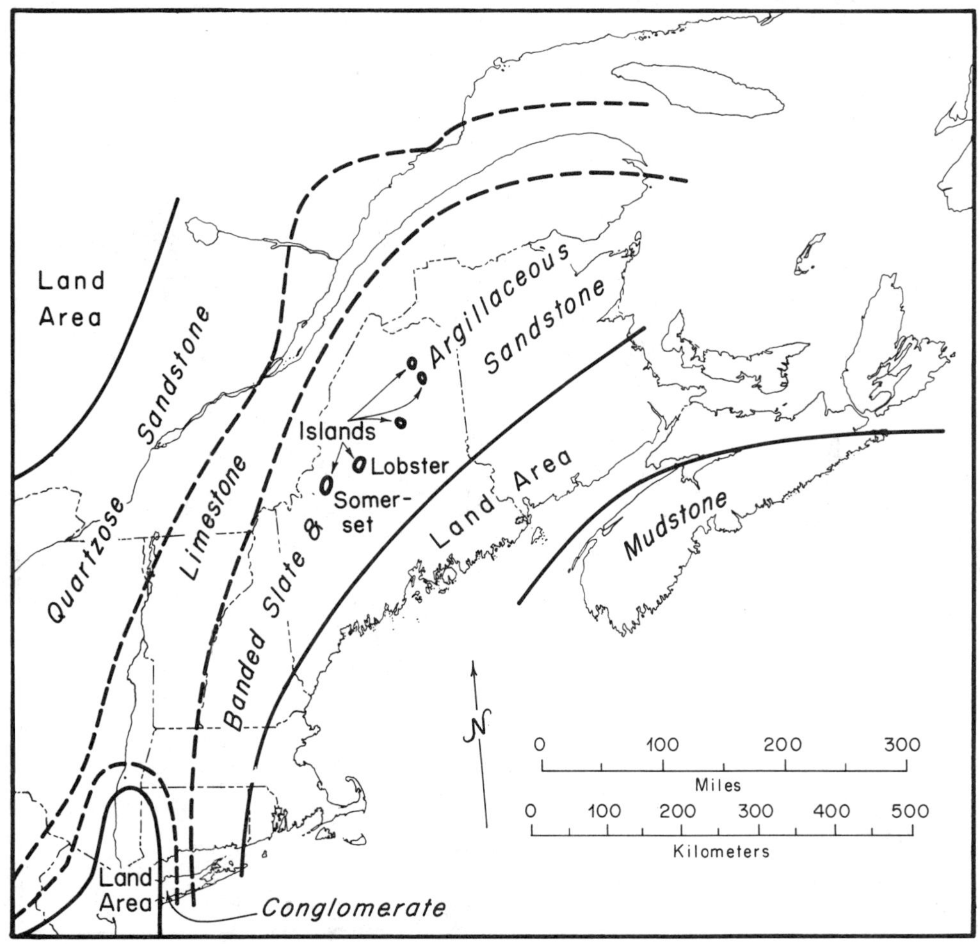

FIGURE 6–5. Lithofacies and paleogeography of the Oriskany Stage in the northern Appalachian region.

where intrusive bodies are present in great abundance (see Page, this volume). However, the intrusive bodies seem to occur in different sedimentary belts, and no pattern is discernible.

The distribution of land areas during the Silurian and Devonian displays a certain degree of consistency. The western land mass, Appalachia, persisted throughout the Silurian, but after the Silurian it seems to have disappeared as a positive element within the northern Appalachian region and exerted no further effect upon the lithofacies pattern of the Early Devonian. The land mass inferred for southeastern New Brunswick and coastal Maine persisted throughout the Silurian either as a land mass or, later, as a site of volcanic activity. It probably extended up into the Chaleur Bay region during part of the Llandovery, and possibly even during the later Silurian. During the Early Devonian (Gedinnian), it was still a site of volcanism in the coastal region, accompanied by marine sedimentation, but then became part of a larger land area during the remainder of the Lower Devonian. The underlying controls for these inferred relationships among sedimentary patterns, shoreline and land distribution, areas of intrusive rocks, structural trends, and areas of regional metamorphism are not yet understood. Whether these phenomena are genetically related to one another is a question that cannot yet be answered by data from the northern Appalachian region.

REFERENCES

Berry, W. B. N., and Boucot, A. J., (in press), Silurian of North America: Geological Society of America

Boucot, A. J. and Johnson, J. G., 1967, Paleogeography and correlation of Appalachian Province Lower Devonian sedimentary rocks: Tulsa Geological Society Digest, v. 35, p. 35–87

Dunbar, C. O., 1957, Historical Geology [2nd ed.]: New York: John Wiley & Sons, 500 p.

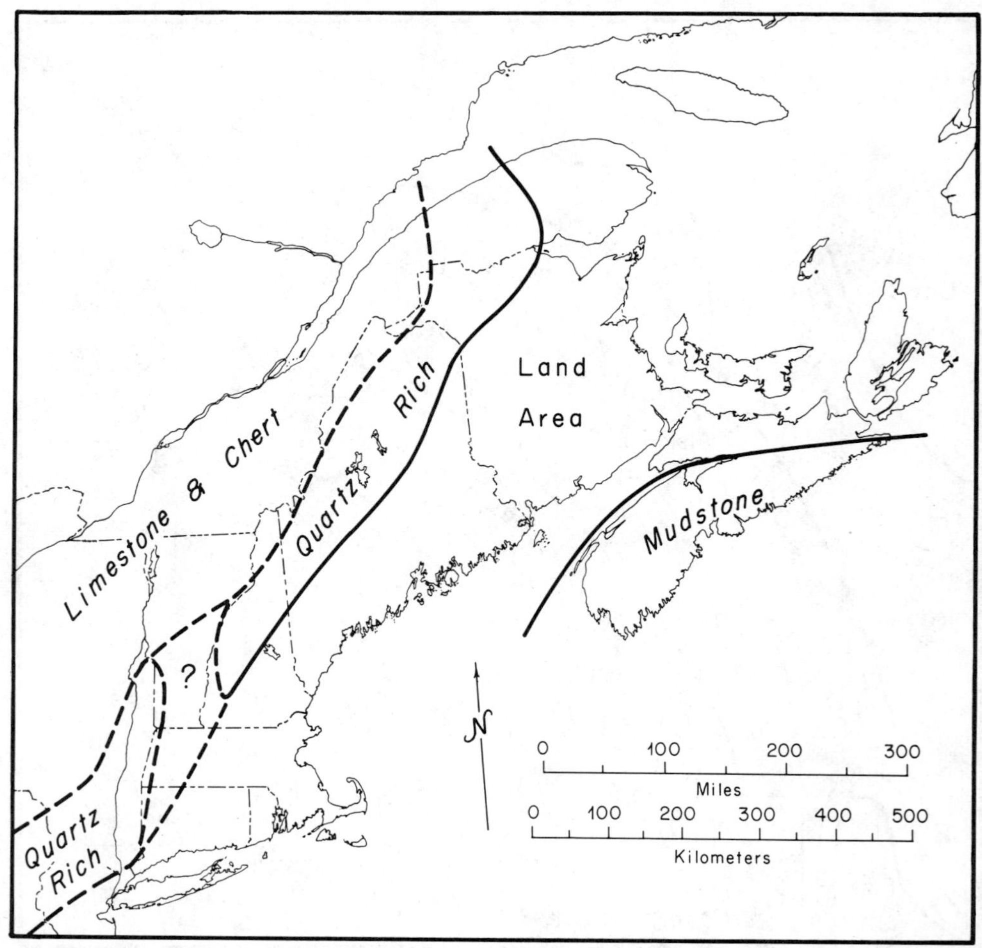

Figure 6–6. Lithofacies and paleogeography of the Schoharie Stage in the northern Appalachian region.

Chapter 7

Post-Acadian Rifts and Related Facies, Eastern Canada

EDWARD S. BELT

INTRODUCTION

UNDERSTANDING OF THE sedimentary facies of Carboniferous rocks in eastern Canada has improved over the past ten years to the extent that the geometry of those facies can be shown to be the result of source areas uplifted along high-angle faults. These faults are visible today and can be traced for many tens of kilometers. The structural picture emerging from the new data is that of a rift valley or complex graben system that developed in Late Devonian time and functioned as a rift until mid-Pennsylvanian time. The rift framework of deposition was revived during Late Triassic time.

This report is a review of what has been published to date on eastern Canadian Carboniferous geology, with the addition of unpublished field data by the author and unpublished miospore dates by Canadian Survey palynologists on many of the non-marine facies. It is hoped that the several hypotheses presented herein will stimulate added interest in the Late Paleozoic geology of eastern Canada.

ACKNOWLEDGMENTS

The data included in this report resulted from field work supported by the Geological Society of America, Inc. (grant-in-aid, 1960, 1961), the National Science Foundation (GP-2858, 1964–1966; GA-499, 1966–1967), and the Nova Scotia Research Foundation (1956–1961). Their support is gratefully acknowledged.

M. S. Barss, J. de Boer, A. J. Boucot, W. K. Fyson, N. L. Hatch, and D. C. McGregor have provided helpful critical reviews of this report. The views expressed here are entirely the writer's, however.

Special appreciation is expressed to Gregory W. Webb for the many hours of fruitful discussion on structural and stratigraphic problems in eastern Canada and for his unpublished manuscript (Webb, in press), from which he allowed me to obtain information.

The following members of the Geological Survey of Canada also allowed me use of unpublished information: M. S. Barss (Coal Research Section reports of 1961 and 1967); D. G. Benson, information from his field mapping in eastern mainland Nova Scotia; A. K. Goodacre, geophysical data; R. D. Howie, information on boreholes in Prince Edward Island; and D. C. McGregor (Paleontology Section, rept. FL-13-1962-DCM). In addition, I wish to thank R. E. Sheridan and Charles Drake (in press) for unpublished seismic data in the Cabot Strait.

Finally, I wish to express special appreciation to Marland P. Billings, without whose encouragement I might never have started to study eastern Canadian geology.

LATE DEVONIAN AND CARBONIFEROUS SECTION

The Carboniferous and Late Devonian strata described here have been previously subdivided into the rock units shown in Table 7–1. The equivalent time-rock units, each based on the strata included in the type section of the rock unit, are proposed here to extend an earlier precedence (Belt, 1965, p. 793–795). Eastern Canadian time-rock terms rather than European time-rock terms are used throughout this report because intercontinental correlation of portions of the Canadian sequence is still very tentative.

Major revisions are currently being made.[*] Table 7-1, compiled from the most recent published information (after Hacquebard, 1960; Barss and others, 1963; Globensky, 1967), shows no major hiatus within the Carboniferous section although those authors do not report the presence of Namurian B and C, Early Viséan, and Late Westphalian A within the sparsely fossiliferous nonmarine strata of eastern Canada. Sedimentologic study by the writer indicates continuous deposition from one dated horizon within the Carboniferous section to the higher dated ones.

The term "Horton Group" in Table 7-1 includes strata within the rift that are older than strata of the type Horton (Bell, 1960) and are thus, strictly speaking, pre-Hortonian. The pre-Hortonian time-rock term, informally proposed

[*] Recent unpublished work by Roger Neves and the author indicates that much of the Cansoan strata in the Antigonish basin, central Nova Scotia, is of Late Viséan age.

here, includes rocks of Late Devonian age because the lowest formation of the Horton Group in the rift of southeastern New Brunswick, the Memramcook Formation, contains probable Late Devonian miospores at its base (and Early Mississippian miospores at a higher stratigraphic level, according to D. C. McGregor, Geol. Surv. Canada, Paleo. Section, rept. FL-13-1962-DCM). For this reason the time-rock term pre-Hortonian could possibly include very Late Devonian-Early Mississippian rock units such as the Perry, Terrenceville, and Great Bay de l'Eau Formations and the Miguasha Group (Table 7-1) that are found outside the rift. It surely includes the Cape John (Baird and Coté, 1964, p. 515) and the Fisset Brook (Kelley and Mackasey, 1965, p. 4) Formations of very Early Mississippian age within the rift.

The term "Pictou Group" here includes strata within the rift that are younger than strata of the type Pictou (Hacquebard, 1960), and are

TABLE 7-1. **Provisional correlation of eastern Canadian time-rock units with those in Europe. All Canadian time-rock terms used here are defined from the type section of the rock unit. Post-Pictouan and pre-Hortonian are informal terms.**

North America	Europe		Eastern Canada		
EARLY PERMIAN			post-Pictouan		
	Stephanian			Pictou Group	
PENNSYLVANIAN	Westphalian	D C B	Pictouan		
			Cumberlandian	Cumberland Group	
		A	M Riversdalian		Port Hood Fm. / Boss Point Fm. / Mabou Group / Bonaventure Fm. / Cannes de Roche Fm. / Barachois Group / Deer Lake Group
	Namurian		M Cansoan *		
LATE MISSISSIPPIAN	Viséan	Late Middle Early	M Windsorian *	Windsor Group	
EARLY MISSISSIPPIAN	Tournaisian		M Hortonian *	Horton Group	Piskahegan Group
			M pre-Hortonian **		Cape John & Fisset Brook Formations
LATE DEVONIAN	Famennian & Frasnian		*		Perry, Great Bay de l'Eau & Terrenceville Formations & Miguasha Group

**, widespread volcanism; *, local volcanism; M, metamorphism locally within rift. Where several rock units appear within a time-rock unit, they are combined with the symbol "&" or presented at right angles to the rest of the table to avoid time significance.

thus post-Pictouan. The post-Pictouan time-rock term is here proposed informally to include the youngest strata of the Pictou Group in eastern Canada. These strata occur on Prince Edward Island and are of Early Permian and Stephanian age (Barss and others, 1963).

FUNDY BASIN RIFT

Carboniferous deposits in eastern Canada (Fig. 7–1) are thick and highly deformed in a narrow region in northern Nova Scotia, central Newfoundland, and southeastern New Brunswick (Fig. 7–2), but are thin, undeformed, uniform in facies and widespread outside this narrow region (Fig. 7–3, "thin cover"). The region of thickly developed strata was termed the Fundy Basin by Bell (1958, p. 4), who visualized the area surrounding it as having been broad uplands throughout much of Carboniferous time (Bell, 1958, p. 5). Kay (1951, Pl. 14, p. 56) proposed that the Fundy Basin is a geosyncline (viz. epieugeosyncline) in which the sedimentary fill thins gradually towards the margins.

Appendices A, B, and C and Figures 7–2 and 7–3 present the evidence that the Fundy Basin is, on the contrary, a complex rift valley (graben system) bounded by high-angle faults or sharp downflexures and surrounded by relatively stable platforms, rather than an epieugeosyncline. The corroborating information is drawn from the literature, the writer's observations in eastern Canada (Belt, 1965; 1968; in press) and data

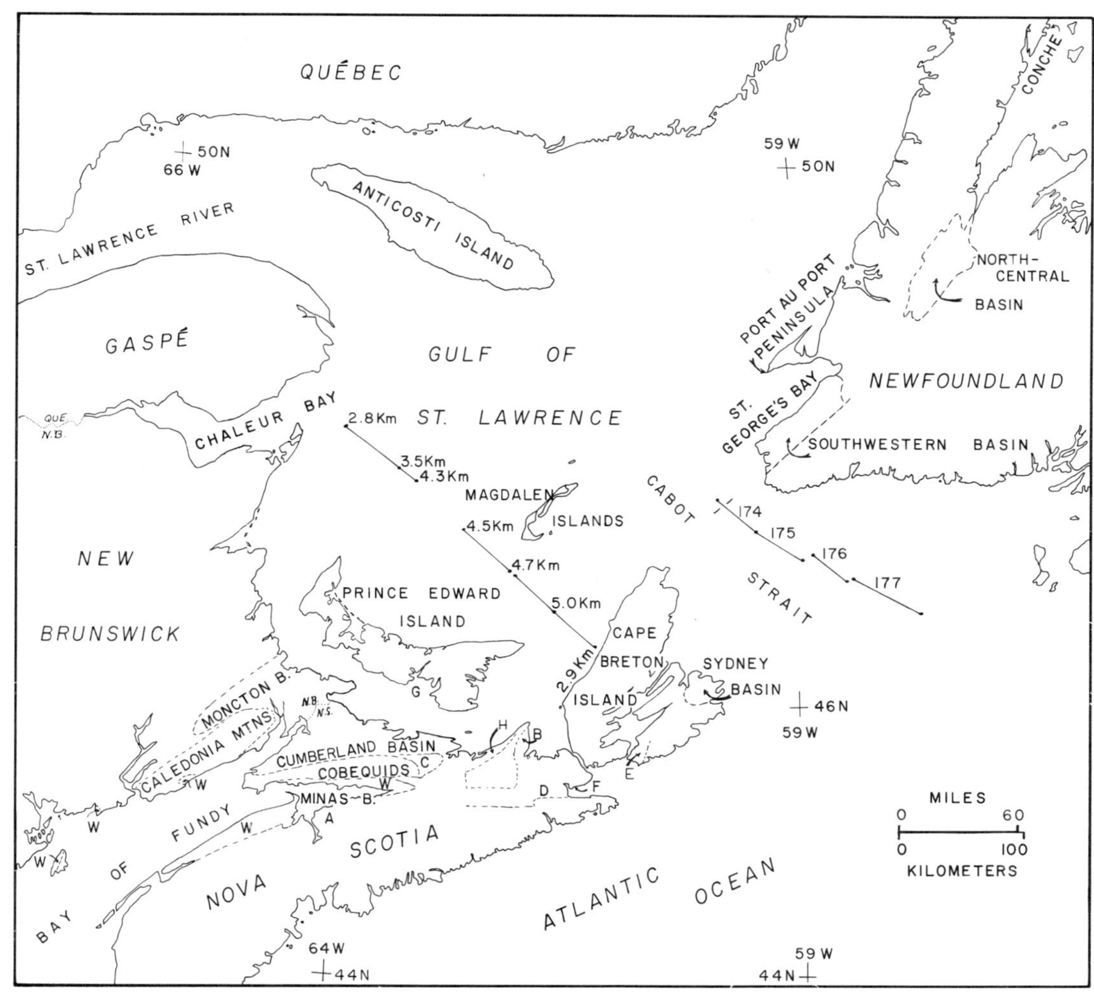

FIGURE 7–1. Index map of eastern Canada. Thickness (in kilometers) of stratigraphic cover at seismic stations west of Magdalen Islands from Goodacre and Nyland (1966, Fig. 5). Locality A, type Horton-Windsor sections; B, Fisset Brook Formation near Cape St. George; C, Scotsburn anticline; D, Fisset Brook Formation in Guysborough County; E, region between Grand River and St. Peters Inlet; F, occurrence of Triassic strata; G, borehole in Hillsborough Bay; H, Merigomish structural basin; W, occurrence of Triassic strata.

from Webb (in press) for southeastern New Brunswick.

Faults and Thickness of Units

The published thickness data (Appendix A) are based on measured sections and boreholes too few to indicate precisely how rapidly the thicknesses of rock units change when the units cross major faults that at least partly bound the rift today (Fig. 7–3). The following regions, where the oldest strata in the basin (Horton Group) are faulted against the pre-Carboniferous basement, are critical for the basin-margin problem; in each area more than 1500 meters (5000 feet) of strata are found within a few miles of the basin-side of a high-angle fault, whereas no Early Mississippian (Hortonian and pre-Hortonian) strata are found on the up-thrown side.

(1) The Price Brook section, Westmorland Co., New Brunswick, (Stewart, 1939) (Fig. 7–1, western Moncton basin, northern side), has approximately 900 meters (3000 feet) of upper Horton Group exposed at the surface, with up to an additional 3000 meters (10,000 feet) buried (Gussow, 1953).

(2) County Harbour River and Desbarres Sinclair Brook, Guysborough Co., Nova Scotia (Schiller, 1961) (Fig. 7–1, loc. D), show over 1500 meters (5000 feet) of deformed Horton Group exposed at the surface.

(3) In the region between Grand River and St. Peters Inlet, on southern Cape Breton Island (Weeks, 1964) (Fig. 7–1, loc. E), over 3000 meters (10,000 ft.) of deformed Horton Group are poorly exposed at the surface northwest of the pre-Carboniferous rocks.

(4) In the north-central basin of Newfoundland (Fig. 7–1), over 3000 meters (10,000 feet) of deformed Anguille Group (Early Mississippian, equivalent to the Horton Group) are exposed between pre-Carboniferous platform regions in a zone 19 km (12 miles) wide (Baird, 1959).

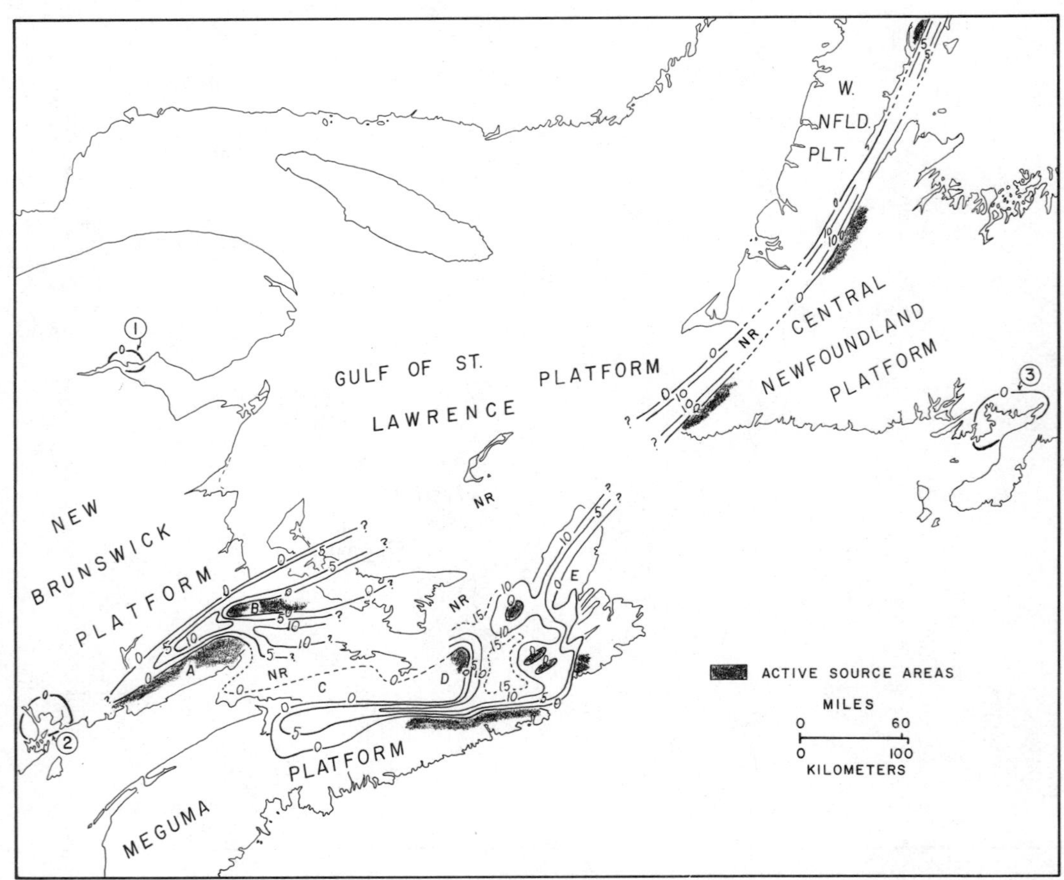

FIGURE 7–2. Fundy Basin rift as defined by a reconstructed isopachous map of the Horton Group and its equivalents. Locality A is the Caledonia Arch; B, the Westmorland Arch; C, the Cobequid Arch; D, the Antigonish Massif; E, the Cape Breton Highlands Massif. Contours in thousands of feet; contour interval 1500 meters (5,000 feet) except in Newfoundland. NR means not reported. Numbers refer to areas of Late Devonian platform deposits mentioned in text.

Other faults (Fig. 7–3, locs. 5, 6, 9), active during a different time than Hortonian, show similar relationships to facies and their thickness.

Because of the rapid facies change from fine-grained sandstone and shale to fanglomerate next to the faults, the above data cannot be interpreted to signify that post-Carboniferous high-angle faults dropped Horton Group strata into a graben, and that the original Early Mississippian depositional basin was a geosyncline with gradually thinning margins that have since been eroded away on the upthrown blocks.

Faults and Sedimentary Facies

The facies data and relation of facies to major faults are outlined in Appendix B. The lateral nonmarine facies variations locally delineate both the margin of the Fundy Basin and the isolated or semi-isolated source areas within the basin (Fig. 7–2, locs. A to E). In general, the Carboniferous rocks immediately adjacent to the pre-Carboniferous basement rocks are fanglomerates locally carrying boulders several meters across. Over a distance of a few kilometers (in some cases meters) coarse sandstone grades toward the center of the local depositional basins into fine-grained sandstone and then into shale, which becomes dominant in the section. The deposits in the center of the basin consist almost entirely of shale having a few thin beds of fine-grained sandstone and nonmarine calcilutite and dololutite. The marine facies of the Windsor

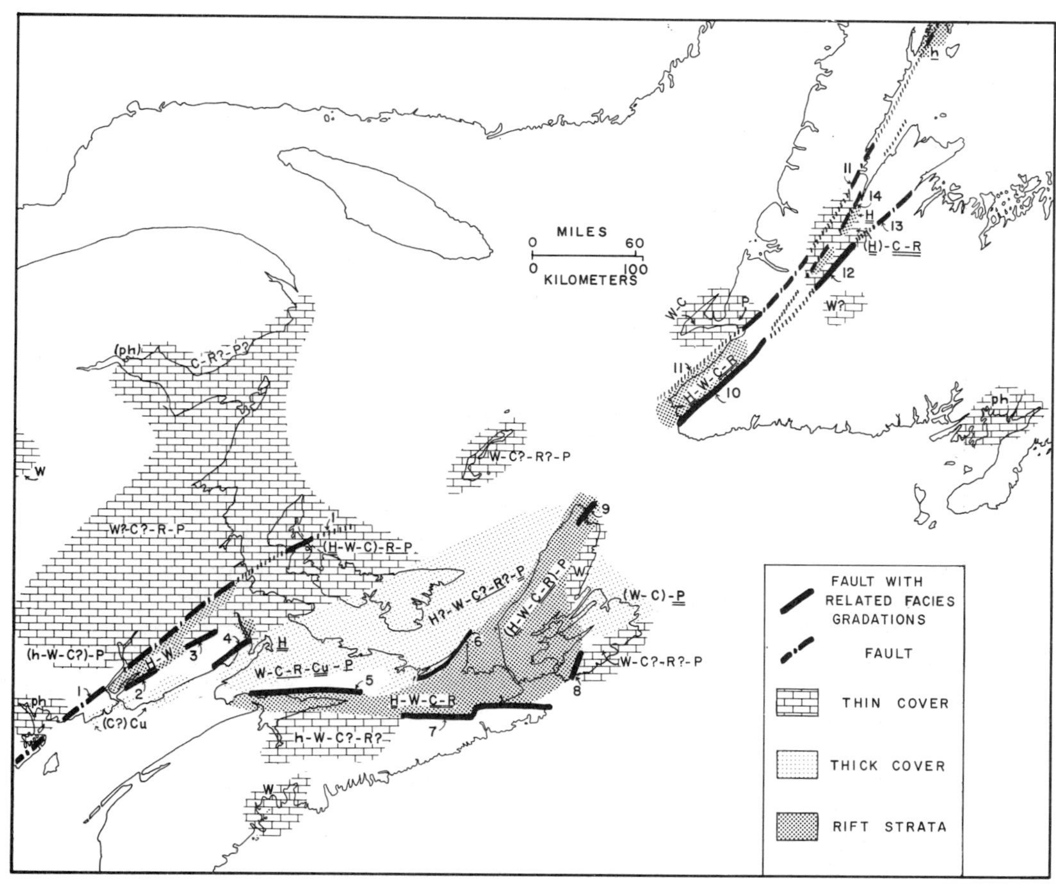

FIGURE 7–3. Carboniferous tectonic map of eastern Canada. "Thin cover" represents less than 1500 meters (5,000 feet) of undeformed strata, "thick cover" represents more than 1500 meters (5,000 feet) of undeformed strata; "rift strata" represents thick, highly deformed strata. H is Hortonian and pre-Hortonian; h is Hortonian; ph is pre-Hortonian (Late Devonian where shown); W is Windsorian, C is Cansoan, R is Riversdalian, Cu is Cumberlandian, P is Pictouan. Units in brackets are unconformably overlain by units to right of bracket symbol. Absence of underscoring means unit is thin, single underscoring means unit is moderately thick; double underscoring means unit is excessively thick. Numbered fault localities: 1, the Lubec–Belleisle; 2, the Clover Hill; 3, the Pollett–Caledonia; 4, the Harvey–Hopewell; 5, the Cobequid; 6, the Hollow; 7, the Guysborough; 8, the Grand River; 9, the Aspy; 10, the Long Range; 11, the St. George's Bay–Taylors Brook; 12, the Grand Lake; 13, the Birchy Lake; 14, the Birchy Ridge.

Group show little influence of the rift margin, but pass into fanglomerate facies near positive blocks active within the rift (Fig. 7–2, locs. A, D, E and other smaller ones on Cape Breton Island).

Deformation and Metamorphism of Rift Strata

The platforms and the Fundy Basin differ in degree of deformation and metamorphism (Appendix A, item 4). The heavily stippled "rift strata" of Figure 7–3 represent highly folded, faulted, and locally mildly metamorphosed rocks, in contrast with the surrounding undeformed or gently deformed and unmetamorphosed platform strata (Fig. 7–3, "thin cover"). Where the "thin cover" exists on the platform, few examples of unconformity are found between the various rock units within the sequence; but in some areas, the "thin cover" unconformably overlaps greatly deformed older strata in the Fundy Basin; two examples of this appear on Figure 7–3.

The first, in southeastern New Brunswick, shows "thin cover" southeast of the Belleisle Fault (Fig. 7–3, loc. 1). Riversdalian and Pictouan strata there overlie Mississippian strata (Horton and Windsor Groups) with a marked angular unconformity. To the northwest of the Belleisle Fault (northwest of the Moncton structural basin, Fig. 7–1), minor discordances exist between the Windsor and younger strata in the platform sequence (Muller, 1951; Anderson and Poole, 1959), although in some areas the sequence appears conformable (van de Poll, 1967a). The second region is in north-central Newfoundland (Fig. 7–1) where Hortonian strata are unconformably overlain by Cansoan and Riversdalian strata that also overlap onto the platforms on either side. The marked difference in degree of deformation within the rift itself occurs between the older "rift strata" of pre-Hortonian, Hortonian, Windsorian, Cansoan, and Riversdalian age, and the "thick cover" (Fig. 7–3) that is principally made up of Cumberlandian and Pictouan strata. The older strata below the thick cover are generally not exposed; exceptions are in western mainland Nova Scotia through the cores of steep anticlines (Cumberland structural basin, Fig. 7–1) and in eastern Cape Breton Island beneath the unconformable Pictou Group (Sydney structural basin, Fig. 7–1).

Nature and Continuity of the Faults

Some of the major longitudinal faults in eastern Canada (Fig. 7–3) have been dated as Carboniferous in age (Appendix A, item 5) either because of associated sedimentary facies or because they have been unconformably overlain by younger Carboniferous strata whilst cutting older Carboniferous strata. Appendix B cites evidence for dip-slip movement on many of those faults (most dip *into* the rift), and Appendix C cites evidence that horizontal movements have also occurred on some of them. Probable connections between various faults are indicated on Figure 7–3 by dotted lines. The only faults which apparently show no associated Carboniferous facies changes are the Lubec–Belleisle Fault (Fig. 7–3, loc. 1) and the St. George's Bay–Taylors Brook Fault (Fig. 7–3, loc. 11), which are both unconformably overlapped by Cansoan and Riversdalian strata not cut by the faults. Evidence of fault movement in Early Mississippian time is discussed later in the report; however, the obvious similarity in structure and age of the two faults makes it probable that they represent the two ends of a fault system that passed through the Gulf of St. Lawrence slightly to the south of the Magdalen Islands, where it probably did not function as the rift margin but as a fault zone within the Gulf of St. Lawrence Platform.

The seismic data listed in Appendix A (item 6) are not conclusive enough to determine whether the Long Range Fault (Fig. 7–3, loc. 10) joins the Aspy Fault (loc. 9), the Hollow Fault (loc. 6), the Harvey-Hopewell Fault (loc. 4), or all three. A connection with the Aspy Fault seems least likely because the Aspy cannot be traced through Cape Breton Island (Belt, 1965, Pl. 1). A connection with the Hollow or the Harvey–Hopewell Faults (or both) seems more likely because they each lie along the boundary between two of the tectonic units shown in Figure 7–3; they surround the region of "thick cover" on the northwest and the east.

Fundy Basin Boundaries

The Fundy Basin boundaries are well defined by Figures 7–2 and 7–3 except in the Gulf of St. Lawrence, particularly south of the Magdalen Islands region. Discovering the obscured basin boundaries depends on determining the thickness and facies of pre-Pictouan strata (i.e. "rift strata") within that area.

The only surface exposures of pre-Pictouan strata on the Gulf of St. Lawrence Platform occur in the Magdalen Islands. Coarse Fluvial Facies (see Belt, 1965, p. 792–793 for definition) of probable Cansoan and Riversdalian age (Fig. 7–3) lie stratigraphically between Pictouan and Windsorian strata. The Windsor Group there is unusual in that it contains basalt lava flows (elsewhere found only at Arisaig, Fig. 7–1, loc. H) in addition to the more typical marine limestone, evaporites, and nonmarine redbeds. No Horton strata are reported from the Magdalen Islands, but no boreholes have penetrated beneath the Windsor Group that apparently have intruded the Pictouan blanket deposits in diapiric anticlines (Sanschagrin, 1964). The seismic profile that passes just to the west of the islands (Fig. 7–1) indicates platform thicknesses of strata between the islands and Cape Breton Island (Goodacre and Nyland, 1966, Fig. 5; Sheridan and Drake, in press), and thus the isopach trends of Horton strata that might be projected between New Brunswick and Newfoundland (Fig. 7–2) would have to terminate approximately where shown.

Pictou Group strata (mostly of post-Pictouan age) are exposed throughout Prince Edward Island. Boreholes in the western part of that island have penetrated the "thin cover" (Fig. 7–3), and Horton and Windsor rocks have been identified in a trough northwest of a basement high termed the Westmorland Arch (Fig. 7–2, northwest of loc. B, as projected into Prince Edward Island). The Fundy Basin margin northwest of the Horton–Windsor trough in western Prince Edward Island is probably a fault, an extension of the Belleisle Fault (Fig. 7–3, loc. 1). A well-defined platform succession lies within a few kilometers to the northwest of the trough. The evidence for the fault and the platform is as follows: (a) the Imperial Porthill Well No. 1, where Windsor strata unconformably overlie the basement on the northwest side of the fault (Fig. 7–3, loc. 1), and (b) the Imperial MacDougal Well No. 1 where more than 1200 meters (4000 feet) of Horton strata conformably underlie Windsor strata on the southeast side of the fault (Howie and Cumming, 1963, Table 1, ref. no. 36 and 37; Howie, 1966, p. 39–42). The two wells are approximately 8 kilometers (5 miles) apart. Other wells in Prince Edward Island show that the Westmorland Arch already existed before the Pictouan overlap. South of the arch, a well at locality 6 (Fig. 7–1) penetrated 2700 meters (8800 feet) of rocks of Coarse Fluvial Facies (post-Pictouan age at the top; Barss and others, 1963) which are conformable above Windsor marine and evaporite strata, but did not penetrate the rocks older than Windsor. The Westmorland Arch may pinch out to the northeast to become another isolated basement block (as shown in Belt, 1965, Fig. 1), or it may widen to the east and northeast to become a somewhat faulted southern margin of the Gulf of St. Lawrence Platform (as shown in Fig. 7–4a). The present

seismic data, though scanty, favors the latter hypothesis.

Conclusions

The evidence presented shows that the tectonic style and facies of the Fundy Basin have many similarities to features found in the East African rift system (McConnell, 1967, and references therein; Charles Downie, oral communication, 1967) as well as the rift troughs of Late Triassic age in eastern North America. Lacustrine, fluvial, and fanglomerate facies characterize an African rift which may show faults in pairs determining a graben in one region, one-sided fault troughs in another region, and a down flexed rift margin elsewhere. Horst-like structures of basement blocks within the rift basin are not uncommon in the East African rifts, and several periods of rift movement are recorded, the earliest being Precambrian in age and the latest Plio–Pleistocene.

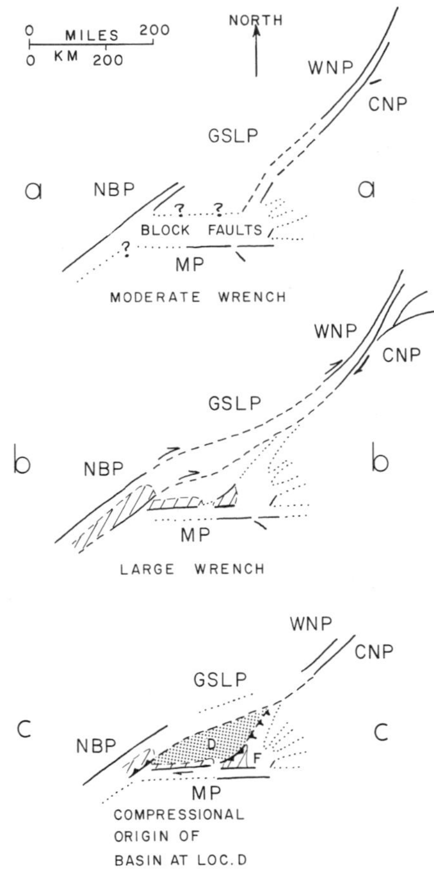

FIGURE 7–4. Hypotheses proposed for late Paleozoic rifts in eastern Canada. Faults accurately placed with respect to each other. MP, Meguma Platform; NBP, New Brunswick Platform; GSLP, Gulf of St. Lawrence Platform; CNP, Central Newfoundland Platform; WNP, Western Newfoundland Platform.

The Fundy Basin rift is here defined as bounded on the north and northwest by the Lubec–Belleisle and the St. George's Bay–Taylors Brook Faults, on the east by the Long Range–Grand Lake Faults, and on the south by the Guysborough Fault. Quite possibly the rift, at least in its early stages, assumed the pattern shown in Figure 7–4a, with an extension down the present Bay of Fundy and another from mainland Nova Scotia through central Newfoundland. It is probable that the southern Gulf of St. Lawrence and eastern Prince Edward Island were initially the site of a platform (i.e. did not receive Early Mississippian rift strata), although later rift faults may have cut through that portion of the platform between southeastern New Brunswick and southwestern Newfoundland (Fig. 7–4b).

LATE·PALEOZOIC AND EARLY MESOZOIC GEOLOGIC HISTORY

The post-Acadian geologic history of eastern Canada cannot be firmly established until we know the precise age of the widespread nonmarine rocks of pre-Hortonian age (see Table 7–1) in order to date the first occurrence of rifting and relate it to deposition on the surrounding platforms. Details of how the rift formed and what possible stress systems may have caused it will not be known until the precise outlines of the rift at various stages of its development are better defined. The following account simply suggests some possible alternatives to stimulate interest in these problems.

Acadian Orogeny

The Acadian Orogeny left its imprint on much of eastern Canada in the form of folded and faulted Middle Devonian and older rocks, lowgrade (locally high-grade) regional metamorphism and large granitic intrusions (Neale and others, 1961; see Boucot, this volume). Although granite intrusions of a later date are recorded from strata as young as pre-Hortonian (Belleoram granite, Newfoundland; Wanless and others, 1967, p. 108; Anderson, 1965) and Cansoan (cuts West Beach Formation, southeastern New Brunswick; Wanless and others, 1966, p. 94–95; Alcock, 1938, p. 31–35), the granites presumably reached a peak of intrusive activity during the middle Devonian (Fairbairn and others, 1960, Pl. 1; Poole and others, 1964, Fig. 4; Wanless and others, 1966, Fig. 1) after which they were exposed by erosion before the deposition of the Late Devonian and Early Mississippian strata (Table 7–1). Outside the rift,

several periods of Devonian deformation are recorded as unconformities, but the most intense uplift is late Early Devonian in age (Late Emsian, Boucot and others, 1964, Table 7, p. 93–99). The youngest Devonian strata involved in the Acadian Orogeny within the Fundy Basin rift occur in the Scotsburn anticline (Fig. 7–1, loc. C) and have tentatively been dated by miospores as Middle Devonian (Kelley, 1965, p. 125–126). Early Late Devonian strata of the Miguasha Group (Alcock, 1935; Williams and Dineley, 1966) of southern Gaspé (Table 7–1; Fig. 7–2, loc. 1) were gently folded before the deposition of the Bonaventure and Cannes de Roche Formations of probable Cansoan age, and those movements may be a northerly record of the rifting of the Fundy Basin to the south.

Rift Valley Development and Sedimentation

The age of the earliest deposit within the rift that can be said to have resulted from the rift itself has been dated as of probable Late Devonian age. D. C. McGregor of the Geol. Survey of Canada (Paleo. Section, Rept. FL-13-1962-DCM) has analyzed the Memramcook Formation in its type section between the Caledonia Arch and the Westmorland Arch, southeastern New Brunswick (Fig. 7–2, locs. A and B); he indicates this early age on the basis of miospores from a sample near the base of the section and an Early Mississippian age of miospores from samples stratigraphically higher in the same section.

Outside the rift, local basins developed during early pre-Hortonian time in southern Gaspé (Fig. 7–2, loc. 1), southwestern New Brunswick, eastern Maine (Fig. 7–2, loc. 2), and south-central Newfoundland (Fig. 7–2, loc. 3). These deposits are all dated as Late Devonian (Smith and White, 1905; Bradley, 1962; Williams and Dineley, 1966), although some recent workers suggest that an Early Mississippian age (Smith, *in* Poole, 1966, p. 7) is more probable for the Perry Formation of eastern Maine–southwestern New Brunswick. The writer is inclined to date all these formations (Perry, Terrenceville, Great Bay de l'Eau, basal Memramcook, and Miguasha) as early pre-Hortonian, roughly equivalent in time, although the age eventually decided for each may show they are not precisely time-equivalent. None of these platform pre-Hortonian rock units exceeds 900 meters (3000 feet) in thickness; more typically, less than 300 meters (1000 feet) is recorded. None is metamorphosed or more than gently deformed except where faults locally cut them. They are considered unrelated to a rift framework of deposition.

During Late Devonian and/or Early Mississippian time, therefore, a rift was produced in eastern Canada. Rifting was accompanied by approximately 600 meters (2000 feet) of rhyolite, dacite, and andesite lava interbedded with nonmarine sediments in the rift on Cape Breton Island and eastern mainland Nova Scotia (Fisset Brook Formation; Kelley and Mackasey, 1965; Schiller, 1961, unit #5; D. G. Benson, oral communication, 1967). Approximately 300 meters (1000 feet) of andesite and basalt lava, porphyry, and tuff are interbedded with redbeds of the Perry Formation (Bastin and Williams, 1914) and the Piskahegan Group (van de Poll, 1967a; MacKenzie, 1964, unit #16) of the western New Brunswick Platform (see also Muller, 1951, p. 11–14). At no other time was volcanism so widespread, although thin basalts are reported from rocks as young as Windsorian (Sanschagrin, 1964; with another occurrence at Arisaig, Fig. 7–1, loc. H) and Cansoan (MacKenzie, 1964, unit #20; Alcock, 1938, p. 34–35) in the vicinity of transcurrent faults (Hollow, Belleisle–St. George's Bay) active during those times. Ten meters (30 feet) of trachyte lava are reported from one locality of the Pictouan section of the central New Brunswick Platform (Muller, 1951, p. 22–23).

Outside the regions of volcanism just mentioned, the oldest deposits consist entirely of nonmarine conglomerate and coarse sandstone with interbeds of shale. The conglomerates on the platforms were basally derived, and the average rock texture decreases upsection. Only one occurrence of tectonic fault breccia is known, in the Perry Formation, Midjik Bluff, eastern Passamaquoddy Bay (Cumming, 1967, fig. 2). Within the rift, basal Horton conglomerates were formed; but in younger Horton strata, rift-margin conglomerates, as yet undated, presumably grade laterally into the finer-grained strata found farther away from the faults. Much better dating of the facies gradations is available for the younger (Cansoan and Riversdalian) deposits (Belt, 1965; in press).

The main east-west portion of the Fundy Basin rift must have been a region of "Basin and Range" topography in pre-Hortonian and Hortonian time, but this pattern was changed when the Windsorian sea entered eastern Canada and covered platform and rift alike. Only a few of the basement blocks within the rift (Fig. 7–2, A, D, E, and three others shown on Cape Breton Island) continued to shed fanglomerates into the Windsorian sea. Evaporites developed over much of eastern Canada during the dying phases of the marine invasion; these are highly deformed within the rift (e.g., see Evans, 1967) whereas they are undeformed outside the rift except locally where diapirically intruded into overlying strata. These evaporite sediments are followed within the rift, in Nova Scotia, by fine-grained lacustrine (Hastings Formation, West Bay Formation; Belt, 1965) and fine-grained fluvial (Maringouin Formation, Middleborough Formation, Pomquet Formation; Belt, 1965) facies of the Mabou Group, and by the Coarse Fluvial Facies on the platforms outside the rift.

In Newfoundland, on the other hand, a major period of erosion occurred during the late Windsorian in the north–central Newfoundland basin (Fig. 7–1), while the Long Range Fault actively shed upper Windsorian fanglomerates into the southwestern Newfoundland basin (Belt, in press). During the Cansoan, the Long Range Fault became more active in the southwest basin (Barachois Group, especially Searston Beds; Riley, 1962) and a second depositional basin (Deer Lake Group; Belt, in press) developed in the north-central part of the island. Deposition of coarse fluvial facies with the appearance of a few coal swamps continued into Riversdalian time in both Newfoundland basins. In both depositional basins, the southeastern side of the basin was the active fault side (Appendix B), indicating vertical movement on the Long Range–Grand Lake Faults during middle Carboniferous time.

A major unconformity that occurs in southeastern New Brunswick during early Cansoan or Windsorian time between the Riversdalian and pre-Cansoan rift strata (Fig. 7–3) correlates well with the late Windsorian period of erosion in north-central Newfoundland. This suggests a possible synchroneity of movement between the Lubec–Belleisle Fault (Fig. 7–3, loc. 1) in southeastern New Brunswick and the St. George's Bay–Taylors Brook Fault (Fig. 7–3, loc. 11) in Newfoundland and would necessitate a connection across the Gulf of St. Lawrence Platform. Both of those major faults cut the Horton Group strata but show no evidence of basin-margin fanglomerates (Fig. 7–3), and both were overlapped by middle Carboniferous strata (Hopewell Group of possible late Cansoan and Riversdalian age in New Brunswick; Deer Lake Group of Cansoan and Riversdalian age in Newfoundland). No angular discordance exists within the Hortonian and Windsorian succession of southwestern Newfoundland (Baird and Coté, 1964), but the mapping in the Anguille Mountains is poorly known.

Further confirmation of the possible connection of the two major fault zones results from the comparison of Webb's (in press) contention that approximately 160 kilometers (100 miles) of dextral strike-slip movement took place on the major faults in southeastern New Brunswick (Fig. 7–3, locs. 1, 2, 3, 4, and others not shown; Gussow, 1953) with this writer's analysis of the amount and sense of movement on the northeast–trending fault system in Newfoundland (Appendix C, item 5).

The direction of dip of the major longitudinal faults (Fig. 7–3) can be observed only in a few localities in eastern Canada; where known, they dip towards the rift (see Appendix A, item 5). A typical example is the Long Range Fault zone of Newfoundland (Fig. 7–3, loc. 10) which shows a 35 to 45 degree dip towards the rift in the coastal exposure at its southwestern end, an 80 degree dip towards the rift a few kilometers northeast (Stevenson's Brook), and a vertical dip at other exposures nearby (Phair, 1949, p. 72–88). Thus the dip of the fault zone (up to 100 meters wide) is not constant, a feature common in major transcurrent faults. In addition, none of the dips of the major faults in eastern Canada is known below the surface. Consequently, the following comments on the possible stress system that produced the rift are necessarily speculative.

The combination of structural, sedimentological, and tectonic events that are concentrated in the area here termed the Fundy Basin strongly suggests a genetic uniformity of orgin through the rift system. The evidence of age movements on the component faults of the rift is clearly demonstrated by the stratigraphic and facies record (Appendix B, C). The northeast–trending faults in Newfoundland and southeastern New Brunswick suggest shear directions, and the evidence available indicates strike-slip movement on them during Early Mississippian and later times (see Appendix C, items 4, 5). In contrast, east-west faults, specifically the Guysborough Fault (Fig. 3, loc. 7), indicate that no wrench movement has occurred on it since pre-Hortonian time, if any occurred at all, because unfaulted Hortonian and Windsorian strata overlap its western extension (see Belt, 1965, Pl. 1). On the other hand, some recent evidence (Eisbacher, 1967) indicates dextral wrench movement on the east-west Cobequid Fault (Fig. 7–3, loc. 5) as late as post-Riversdalian. This movement is much younger (Early Cumberlandian; evidence discussed later) and it apparently did not affect the older Guysborough Fault system to the south; the stress was absorbed exclusively by the meta-

morphosed Horton strata to the north (Schiller's 1961 map unit #7A).

The northeast-southwest trending fault systems of the Fundy rift were probably shear directions, and the east-west trending fault systems were probably either tensional or compressional fault directions (Fig. 7–4a). Fyson (1966) presents evidence for east-west compression during the later stages of Acadian deformation in the Meguma Platform. Some of the secondary folds are related to this deformation and pre-date Acadian granite because they are cut by the granite, but some were produced after the period of the mid-Devonian intrusions. Present evidence (Appendix C, items 4, 5; Webb, in press) suggests that the northeast-southwest alignment of faults (Fig. 7–3) could be due to dextral shear during east-west compression following the granite intrusion. The portion of the rift in Nova Scotia north of the Meguma Platform would then be the result of tension faulting from the same compression. Dextral movement of a few kilometers on the northeast-southwest system would open up the east-west tension graben.

We may entertain alternative hypotheses for the tectonic events of late Hortonian through Pictouan time (Figs. 7–4a, 7–4b, 7–4c). Figure 7–4a postulates that the Westmorland Arch formed part of the Gulf of St. Lawrence Platform and that the system of faults in Newfoundland did not join with the one in southeastern New Brunswick, but was en echelon and parallel to it. The figure further postulates that the east-west region between the Gulf of St. Lawrence Platform (as shown) and the Meguma Platform was a complex graben containing several block faults. The basic fault pattern of Figure 7–4a might have been produced by an east-west compression during Late Devonian time and could have continued to operate through at least Riversdalian (Middle Pennsylvanian) time. The Newfoundland and southeastern New Brunswick portions of the fault system could have been produced by moderate shearing from an east-west compression. The "shear-grabens," nevertheless, received thousands of meters of Carboniferous sediment.

The hypothesis exemplified by Figure 7–4a fits the known facts that a major graben received thousands of meters of sediment during pre-Hortonian, Hortonian, Windsorian, Cansoan, and Riversdalian time; that the graben in Nova Scotia shows an east-west southern margin and is at least 80 kilometers (50 miles) wide; that in New Brunswick and Newfoundland the graben has a northeast-southwest trend and is less than 16 kilometers (10 miles) wide; and that the region surrounding the graben shed sediments into the graben. On the other hand, if we accept Webb's (in press) argument that large dextral movement took place on most major faults during Late Hortonian and pre-Riversdalian time, the hypothesis of Figure 7–4a would be untenable after Early Hortonian time; the faults would have to link up across Cabot Strait in some such manner as that suggested in Figure 7–4b.

Under the hypothesis of Figure 7–4b, an early east-west compression produced a graben system, much like the one shown in Figure 7–4a, during Late Devonian time. This pattern lasted from the Late Devonian through the Late Hortonian, after which additional compression was so severe that the southeastern New Brunswick faults sheared across the Gulf of St. Lawrence Platform to join with the Newfoundland shear system. The evidence in support of this hypothesis is the coincidence of dates of movement on the New Brunswick and the Newfoundland faults (Appendix A, item 5; Appendix C) with accompanying unconformities and fanglomerates, and the lack of evidence for comparable movement (no unconformities) within the rift of Nova Scotia. The Guysborough Fault (Fig. 7–3, loc. 7) ceased to operate by Late pre-Hortonian time because Hortonian strata (Horton Group, type section) were deposited across its western end, and the fault has not subsequently cut them. It is possible, however, that the Guysborough Fault passes westward into a monoclinal downflexure; pre-Hortonian strata are not found south of the trend, but were deposited north of it. Later strata (Cansoan and Riversdalian age) indicate a facies change across the rift margin in this area (Belt, 1965, p. 798–799, Fig. 5), although no rift-margin fanglomerates were developed. The apparent synchroneity of Late Hortonian–pre-Riversdalian movements at the southwestern and northeastern end of the rift, separated by a distance of 680 kilometers (420 miles), is difficult to explain in terms of the apparent lack of comparable evidence for movements within the rift elsewhere. Figure 7–4b offers one possible hypothesis.

Under the hypothesis of Figure 7–4c, the Cumberland–Pictou basin (Fig. 7–4c, loc. D) was produced during Cumberlandian and Pictouan time by compression directed northwest-southeast rather than by a continuation of the east-west direction of the earlier compression. Recent evidence from pre-Carboniferous basement structures in the Cobequid Arch (Fig. 7–1) as well as the Cobequid Fault zone (Fig. 7–3, loc. 5) indicates post-Riversdalian dextral movement along approximately east-west fractures in excess of 1 kilometer (Eisbacher and Kelley, 1966; Eisbacher, 1967). This was postulated by Eisbacher (Eisbacher and Kelley, 1966, Fig. 1) to have resulted from northwest-southeast directed maximum compression that most probably occurred at least initially during the period of mid-Pennsylvanian movements.

Mid-Pennsylvanian Movements

Some time during the Late Riversdalian and the Early Cumberlandian, a deep trough developed in northwestern Nova Scotia (Fig. 7–3, "thick cover") and received thousands of meters of the type Cumberland Group strata. Here the Cumberland Group is conformable with the underlying Riversdalian strata, but elsewhere within the trough the Cumberland Group (e.g. Cobequid Arch, south side of Cumberland Basin; southeast of Caledonia Arch, St. John region, New Brunswick) lies unconformably upon older strata. Cumberlandian strata is unknown from other regions of eastern Canada (although reported south of the Cobequid Arch; see Belt, 1965, Pl. 1), and its absence is marked by a major unconformity at the appropriate horizon. This unconformity is prominent in central Nova Scotia and in Cape Breton Island (Fig. 7–3, "rift strata") beneath the Pictouan strata. It may also occur on the platforms in Nova Scotia, Newfoundland, Magdalen Islands, and Gaspé, but without adequate fossil and structural evidence within generally coarse, unfossiliferous strata there, it has not been recognized. However, it is well defined on the New Brunswick Platform (Anderson and Poole, 1959; Muller, 1951).

The youngest metamorphosed rocks in eastern Canada are of Riversdalian age. These reach a slate grade, and are confined to the rift southeast of the Cobequid Arch and west of the Antigonish Massif, a region known as the Stellarton Gap area (see Belt, 1965, Pl. 1, loc. 24). Nowhere else are strata of Riversdalian age even mildly metamorphosed. Metamorphosed Cansoan strata are found in the Stellarton Gap area and on the west side of the Bay of Fundy near St. John (Mispeck Group). The writer attributes the metamorphosed Cansoan and Riversdalian strata in Nova Scotia to the Late Riversdalian–Early Cumberlandian movements; these produced the dextral shearing within the 16 kilometer (10 miles) wide Cobequid Arch, as well as along the Cobequid Fault (Eisbacher, 1967), and at the same time uplifted the arch so that early Cumberlandian conglomerates were shed to the south (Weeks, 1948: Bass River sheet, unit #14; Londonderry sheet, unit #11) and to the east (New Glasgow conglomerate; Bell, 1940, p. 18–21). The movement must have occurred partly along the Hollow Fault that curves to the northeast and partly along the fault bounding the southern side of the Antigonish Massif and from thence eastward (see Schiller, 1961, unit #7A) into the Atlantic Ocean (Fig. 7–4d). The metamorphosed Mispeck Group is undated within the Lower Carboniferous and may be as old as Hortonian in age.

The Pictouan strata above the widespread unconformity are undeformed except where diapirically intruded by Windsor evaporites on Cape Breton Island and Magdalen Islands (Belt, 1965, Pl. 1), the strata contrast markedly with the highly folded and faulted strata within the rift. Much of the pronounced folding of pre-Pictouan strata in Cape Breton Island and mainland Nova Scotia (Belt, 1965, Pl. 1) is probably the result of pre-Pictouan movements that occurred during Cumberlandian, and possibly slightly earlier (Riversdalian) time.

Pictouan Depositional Framework

Deposition in the Cumberland–Pictou basin continued from Cumberlandian time into Pictuoan time without a break, but the site of maximum subsidence shifted from northwestern Nova Scotia to southeastern Prince Edward Island (see Howie and Cumming, 1963, Fig. 7). The southeastern limit of the deeper part of this Pictouan basin was approximately the position of western Cape Breton Island. Pictouan strata from the Prince Edward Island basin may once have joined across Cape Breton Island to the Sydney basin (Fig. 7–1), but the evidence has been eroded away (Belt, 1965, Pl. 1; see also Fig. 7–3). In southeastern New Brunswick, paleocurrent analysis by van de Poll (1966) indicates a strong source from the Caledonian Arch and a much weaker one from part of the uplifted Moncton basin (Kingston Uplift; van de Poll, 1966, Fig. 6). A third source was from central New Brunswick beyond the present region of platform strata (Fig. 7–3; van de Poll, 1967b, Fig. 2); a shallow basin may have existed on the central New Brunswick Platform at that time.

The part of the post-Pictouan strata assigned to the Lower Permian on Prince Edward Island (Barss and others, 1963) contains at least one unconformity (Frankel, 1966) suggesting increased structural instability towards the end of the Paleozoic. The succession of vertebrate fossils within the Pictouan and post-Pictouan strata indicate an increasing aridity of the climate (Langston, 1963).

Rejuvenation of Rift Framework

Minor folding and faulting of Pictouan and older strata presumably took place in Middle and Late Permian time, but no deposits of this age are reported in eastern Canada (see Frankel, 1966, p. 30–31). Triassic deposits (both redbeds and basaltic lava) are widespread in the western part of the Fundy Basin (Fig. 7–1, loc. W); they overlie pre-Pictouan strata with an angular unconformity and were deposited in a rift with high-angle fault boundaries located in approximately the same positions as in Carboniferous time (Klein, 1962, 1963). Sporadic outcrops of Triassic strata are found as far east as the north shore of Chedabucto Bay (Fig. 7–1, loc. F), Nova Scotia (Stevenson, 1959a, 1960). No Triassic sedimentary rocks have been reported outside the boundaries of the Carboniferous Fundy Basin. Bradley (1962, p. 52) suggests that the northeast-trending graben and associated dikes near Terrenceville, south-central Newfoundland, are possibly of Triassic age, although no strata are preserved that could have been deposited in the graben.

The Triassic graben or hinged trough (hinge to the southeast) located in the region generally occupied by the Bay of Fundy today implies the possibility of a maximum horizontal compressional stress directed northeast-southwest. There is no present evidence to support such a stress direction other than the orientation of the basin itself.

Note in proof: An excellent summary of the geologic evolution of eastern Canada has just appeared: Poole, W. H., 1967, Tectonic evolution of Appalachian region of Canada: Geol. Assoc. Canada, Sp. Paper 4, p. 9–51.

REFERENCES

Alcock, F. J., 1935, Geology of Chaleur Bay region: Geol. Survey Canada Mem. 183, 146 p.

__________, 1938, Geology of Saint John region, New Brunswick: Geol. Survey Canada, Mem. 216, 65 p.

__________, 1949, Geological map of the Maritime Provinces: Geol. Survey Canada Map 910A, with explanatory notes

Anderson, F. D., 1965, Belleoram, Newfoundland: Geol. Survey Canada, Map 8–1965, with marginal notes

__________, and Poole, W. H., 1959, Woodstock-Fredericton, New Brunswick: Geol. Survey Canada, Map 37–1959, with marginal notes

Baird, D. M., 1959, Geology of Sandy Lake (west half), Newfoundland: Geol. Survey Canada, Map 47–1959, with marginal notes

__________, 1966, Carboniferous rocks of the Conche-Groais Island area, Newfoundland: Canadian Jour. Earth Sciences, v. 3, p. 247–257

__________, and Coté, P. R., 1964, Lower Carboniferous sedimentary rocks in southwestern Newfoundland and their relations to similar strata in western Cape

Breton Island: Canadian Mining and Metallurgical Bull., v. 57, p. 509–520

Barss, M. S., Hacquebard, P. A., and Howie, R. D., 1963, Palynology and stratigraphy of some upper Pennsylvanian and Permian rocks of the Maritime Provinces: Geol. Survey Canada, Paper 63–3, 13 p.

Bastin, E. S., and Williams, H. S., 1914, Description of the Eastport quadrangle, Maine: U. S. Geol. Survey Atlas, Eastport Folio (no. 192), 15 p.

Bell, W. A., 1940, The Pictou Coalfield, Nova Scotia: Geol. Survey Canada, Mem. 225, 160 p.

————, 1958, Possibilities for occurrence of petroleum reservoirs in Nova Scotia: Halifax, Nova Scotia Dept. Mines, 177 p.

————, 1960, Mississippian Horton group of type Windsor-Horton district, Nova Scotia: Geol. Survey Canada Mem. 314, 112 p.

Belt, E. S., 1965, Stratigraphy and paleogeography of Mabou Group and related middle Carboniferous facies, Nova Scotia, Canada: Geol. Soc. America Bull., v. 76, p. 777–802

————, 1968, Carboniferous continental sedimentation, Atlantic Provinces, Canada, in G. deV. Klein, ed., 1966, Symposium on continental sedimentation: Geol. Soc. America, Spec. Paper 106, p. 127–176

————, in press, Newfoundland Carboniferous stratigraphy and its relation to the Maritimes and Ireland: in Marshall Kay, ed., 1967, Symposium on the North Atlantic—geology and continental drift, Gander meeting, Am. Assoc. Petroleum Geologists, Memoir 12

Benson, D. G., 1962, Hopewell, Nova Scotia: Geol. Survey Canada, Map 3–1962, with marginal notes

————, 1964, Lochaber, Nova Scotia: Geol. Survey Canada, Map 58–1963, with marginal notes

Boucot, A. J., Field, M. T., Fletcher, Raymond, Forbes, W. H., Naylor, R. S., and Pavlides, Louis, 1964, Reconnaissance bedrock geology of the Presque Isle Quadrangle, Maine: Maine Geol. Survey, Quadr. Mapping Ser. No. 2, 123, p.

Bradley, D. A., 1962, Gisborne Lake and Terrenceville map-areas, Newfoundland (1 M/15 and 10): Geol. Survey Canada, Mem. 321, 56 p.

Cameron, H. L., 1966, The Cabot Fault zone: in, G. D. Garland, 1964, Symposium on continental drift, Roy. Soc. Canada, Sp. Publ. no. 9, p. 129–149

Cumming, L. M., 1967, Geology of the Passamaquoddy Bay region, Charlotte County, New Brunswick (21 B, 21 G, parts of): Geol. Survey Canada, Paper 65–29, 36 p.

Eisbacher, G. H., 1967, Tectonic analysis in the Cobequid Mountains, Nova Scotia, Canada: Ph.D. thesis, Princeton Univ., 108 p. (University Microfilms MIC 67–13,489)

————, and Kelley, D. G., 1966, Tectonic studies in the Cobequid Mountains, Nova Scotia: Maritime Sediments, v. 2, p. 180–183

Evans, Robert, 1967, The structure of the Mississippian evaporite deposit at Pugwash, Cumberland County, Nova Scotia: Econ. Geology, v. 62, p. 262–273

Fairbairn, H. W., Hurley, P. M., Pinson, W. H., Jr., and Cormier, R. F., 1960, Age of the granitic rocks of Nova Scotia: Geol. Soc. America Bull., v. 71, p. 399–414

Frankel, Lawrence, 1966, Geology of southeastern Prince Edward Island: Geol. Survey Canada, Paper 66–14, p. 330–344

Fyson, W. K., 1966, Structures in the Lower Paleozoic Meguma Group, Nova Scotia: Geol. Soc. America Bull., v. 77, p. 931–944

Geol. Survey Canada, 1967, Index to map sheets, aeromagnetic series, Ottawa, sheet #1

Gillis, J. W., 1964, Geology of northwestern Pictou County, Nova Scotia, Canada: Ph.D. thesis, The Pennsylvania State Univ., 130 p.

————, 1965, Port aux Basques (11 0) Map-area: Geol. Survey Canada, Paper 65–1, report of activities 1964, p. 133–135

Globensky, Yvon, 1967, Middle and Upper Mississippian conodonts from the Windsor Group of the Atlantic Provinces of Canada: Jour. Paleo., v. 41, p. 432–448

Goodacre, A. K., and Nyland, E., 1966, Underwater gravity measurements in the Gulf of St. Lawrence: in, G. D. Garland, ed., 1964, Symposium on continental drift, Roy. Soc. Canada, Sp. Publ. no. 9, p. 114–218

Gussow, W. C., 1953, Carboniferous stratigraphy and structural geology of New Brunswick, Canada: Am. Assoc. Petroleum Geologists Bull., v. 37, p. 1713–1816

Hacquebard, P. A., 1960, A summary of Carboniferous stratigraphy and paleontology of the Maritime Provinces of Canada: 4th Internat. Cong. of Carboniferous Stratigraphy and Geology (Heerlen, 1958), Comptes Rendus, v. 1, p. 233–235

Howie, R. D., 1966, Catalogue of well samples from Nova Scotia, New Brunswick, Prince Edward Island and Newfoundland at the Geological Survey of Canada, Ottawa: Geol. Survey Canada, Paper 65–40, 43 p.

————, and Cumming, L. M., 1963, Basement features of the Canadian Appalachians: Geol. Survey Canada Bull. 89, 18 p.

Kay, Marshall, 1951, North American geosynclines: Geol. Soc. America Mem. 48, 144 p.

————, and Colbert, E. H., 1965, Stratigraphy and life history: New York: John Wiley & Sons, Inc., 736 p.

Kelley, D. G., 1958, Mississippian stratigraphy and petroleum possibilities of central Cape Breton Island, Nova Scotia: Canadian Inst. Mining and Metallurgy Trans., v. 61, p. 175–185

————, 1965, Cobequid Mountains: in Report of activities: Field, 1964, Geol. Survey Canada, Paper 65–1, p. 125–127

———— and Mackasey, W. O., 1965, Basal Mississippian volcanic rocks in Cape Breton Island, Nova Scotia: Geol. Survey Canada, Paper 64–34, 10 p.

Klein, G. deV., 1962, Triassic sedimentation, Maritime Provinces, Canada: Geol. Soc. America Bull., v. 73, p. 1127–1146

————, 1963, Regional implications of Triassic paleocurrents, Maritime Provinces, Canada: Jour. Geology, v. 71, p. 801–808

Langston, Wann, Jr., 1963, Fossil vertebrates and the Late Paleozoic red beds of Prince Edward Island: Natural Mus. Canada, Bull. 187, 36 p.

MacKenzie, G. S., 1964, Hampstead, New Brunswick: Geol. Survey Canada, Map 1114A, with marginal notes

McConnell, R. B., 1967, The East African rift system: Nature, v. 215, p. 578–581

McConnell, R. K., Jr., and McTaggart-Cowan, G. H., 1963, Crustal seismic refraction profiles: Inst. Earth Sciences, Univ. Toronto Sci. Rept. no. 8, p. 39–40

McGerrigle, H. W., 1950, the geology of eastern Gaspé: Quebec Dept. Mines, Geol. Surveys Branch, Rept. 35, 168 p.

Muller, J. E., 1951, Geology and coal deposits of Minto and Chipman map-areas, New Brunswick: Geol. Survey Canada, Mem. 260, 40 p.

Neale, E. R. W., 1963a, Dingwall, Nova Scotia: Geol. Survey Canada, Map 1124A, with marginal notes

———, 1963b, Pleasant Bay, Nova Scotia: Geol. Survey Canada, Map 1119A, with marginal notes

———, Béland, Jacques, Potter, R. R., and Poole, W. H., 1961, A preliminary tectonic map of the Canadian Appalachian region based on age of folding: Canadian Inst. Mining Metallurgy, Trans., v. 54, p. 687–694

Phair, George, 1949, Geology of the southwestern part of the Long Range, Newfoundland: Ph.D. thesis, Princeton Univ., 165 p. (University Microfilms, microfilm no. 11, 217)

Poole, W. H., ed., 1966, Geology of parts of Atlantic Provinces: Geol. Assoc. Canada and Mineralogical Assoc. Canada, Guidebook, Halifax, Nova Scotia, 155 p.

———, Kelley, D. G., and Neale, E. R. W., 1964, Age and correlation problems in the Appalachian region of Canada: in, F. F. Osborne, 1963, Symposium on geochronology in Canada, Roy. Soc. Canada, Sp. Publ. no. 8, p. 61–84

Riley, G. C., 1962, Stephenville map-area, Newfoundland: Geol. Survey Canada Mem. 232, 72 p.

Sanschagrin, Roland, 1964, Magdalen Islands: Quebec Dept. Natural Resources, Geol. Rept. 106, 58 p.

Schiller, Edward, 1961, Guysborough, Nova Scotia: Geol. Survey Canada, Map 27–1961, with marginal notes

Sheridan, R. E., and Drake, Charles, in press, Seaward extension of the Canadian Appalachians: Canadian Jour. Earth Sciences

Shroder, J. F., Jr., 1963, Stratigraphic and tectonic history of the Moncton Group of nonmarine redbeds of New Brunswick, Canada: M. S. thesis, Univ. Mass., 81 p.

Smith, G. O., and White, David, 1905, The geology of the Perry basin in southeastern Maine: U. S. Geol. Survey, Prof. Paper 35, 291 p.

Stevenson, I. M., 1959a, Chedabucto Bay, Nova Scotia: Geol. Survey Canada map 3–1959, with marginal notes

———, 1959b, Shubenacadie and Kennetcook map-areas, Colchester, Hants, and Halifax counties, Nova Scotia: Geol. Survey Canada Mem. 302, p. 41

———, 1960, New occurrences of Triassic sedimentary rocks in Chedabucto Bay area, Nova Scotia: Geol. Soc. America Bull., v. 71, p. 1807–1808

Stewart, J. S., 1939, Alward Brook, New Brunswick: Geol. Survey Canada, Map 605A, with marginal notes

van de Poll, H. W., 1966, Sedimentation and paleocurrents during Pennsylvanian time in the Moncton basin, New Brunswick: New Brunswick Dept. Natural Resources, Rept. Investigation 1, 33 p.

———, 1967a, Carboniferous volcanic and sedimentary rocks of the Mount Pleasant area, New Brunswick: New Brunswick Dept. Natural Resources, Rept. Investigation 3, 52 p.

———, 1967b, Pennsylvanian sedimentation in the Central Basin of New Brunswick: New Brunswick Dept. Natural Resources, Info. Circular 67–1, p. 3–5

Wanless, R. K., Stevens, R. D., Lachance, G. R., and Rimsaite, J. Y. H., 1966, Age determinations and geological studies, K-Ar isotopic ages, report 6: Geol. Survey Canada, Paper 65–17, p. 94–95

———, Stevens, R. D., Lachance, G. R., and Edmonds, C. M., 1967, Age determinations and geological studies, K-Ar isotopic ages, report 7: Geol. Survey Canada, Paper 66–17, p. 108

Webb, G. W., 1963, Occurrence and exploration significance of strike-slip faults in southern New Brunswick, Canada: Am. Assoc. Petroleum Geologists Bull., v. 47, p. 1904–1927

———, in press, Paleozoic wrench faults in Canadian Appalachians: in Marshall Kay, 1967 ed., Symposium on the North Atlantic—geology and continental drift, Gander meeting: Am. Assoc. Petroleum Geologists, Memoir 12

Weeks, L. J., 1948, Londonderry and Bass River map-areas, Colchester, and Hants counties, Nova Scotia: Geol. Survey Canada Mem. 245, 86 p.

———, 1964, St. Peters, Nova Scotia: Geol. Survey Canada, Map 1083A, with marginal notes

Williams, Harold, 1967, Geology, Island of Newfoundland: Geol. Survey Canada, Map 1231A

Williams, B. P. and Dineley, D. L., 1966, Studies on the Devonian Strata of Chaleur Bay, Quebec: Maritime Sediments, v. 2, p. 7–10

APPENDIX A

Evidence That The Fundy Basin Is A Complex Rift Valley (Graben)

1. *Thickness data.* Isopachous maps by Kay (1951, Fig. 14, p. 57); Kay and Colbert (1965, Fig. 13–1, p. 265); and Howie and Cumming (1963, Figs. 3, 4, 5, 6, 7, 9, p. 6–8) show a region of greater thickness within the Fundy Basin than in the surrounding regions but do not indicate the abrupt thickness change at the margin of the basin, described in the text.

2. *Completeness of section.* Cumberland Group and its time-rock equivalent (Cumberlandian) have not been recognized outside the Fundy Basin. Horton Group (as defined in the text) is reported outside the Fundy Basin only on the central Meguma Platform (Fig. 7–3) where it comprises only the younger half of all Horton Group strata. Lower Horton Group (pre-Hortonian of Lower Mississippian and upper Devonian age) is confined to the Fundy Basin as here defined.

3. *Facies data.* The facies of sedimentary rocks adjacent to the major faults bounding the Fundy Basin or major faults bounding basement blocks within the rift suggest fault scarps rather than gently sloping basin margins (Appendix B). The Cannes de Roche Formation (Coarse Fluvial Facies of eastern Gaspé) has recently been dated as middle Cansoan (M. S. Barss, Coal Research Section Rept. 2-67-MSB), and the Barachois Group (Coarse Fluvial Facies of Port au Port Peninsula, Newfoundland, Fig. 7–1) as lower Cansoan (M. S. Barss, written communication,

1965). Elsewhere, Cansoan strata are presumed to be present on the Gulf of St. Lawrence Platform (Magdalen Islands; Sanschagrin, 1964) and on the Meguma Platform (Stevenson, 1959b), but the evidence is structural rather than paleontologic. Coarse fluvial facies of the Deer Lake Group on the Western Newfoundland Platform have also recently been dated as of Cansoan age (M. S. Barss, Coal Research Section Rept. 2-67-MSB). These coarse facies are all laterally equivalent to finer-grained rift facies of the same age.

4. *Degree of deformation and metamorphism.* Strata within the Fundy Basin are more highly deformed than on the surrounding platforms (Alcock, 1949; McGerrigle, 1950; Belt, 1965; Williams, 1967). This concept was portrayed by Neale and others (1961) in their Figure 2, and applies to Carboniferous strata older than Pictouan and Cumberlandian, i.e., those strata affected by the pre-Pictouan orogeny. Thus as shown in Figure 7–3, Windsorian, Cansoan, and some Riversdalian strata in the Fundy Basin ("rift strata") are highly deformed and unconformable beneath the Pictouan rocks, whereas those same strata on the platforms ("thin cover") are undeformed and conformable beanth the Pictouan. Horton strata on the Meguma Platform are only locally deformed by Triassic movements whereas those in the rift are everywhere highly deformed.

In contrast, rocks of "thick cover" (Fig. 7–3) are apparently structurally concordant from the Windsorian through the Pictouan. Pre-Hortonian through Riversdalian strata are selectively metamorphosed within the Fundy Basin (table 7–1). The metamorphic grade is generally very low, never exceeding slate grade, and seems to be associated with regions (some 16 km, or 10 miles wide) of intense deformation. Unmetamorphosed strata of the same age can be found nearby outside the regions of intense deformation (see Schiller, 1961, and compare units 7A and 7B across the fault separating the two).

5. *Major longitudinal faults.* Some major faults bound the Fundy Basin; others bound active source blocks within the basin. All faults defined here as major can be traced as individual fault zones (up to 100 meters wide) or closely parallel fault systems that extend 80 to 200 kilometers (50 to 120 miles) along strike. In southeastern New Brunswick, Cape Breton Island, and Newfoundland (Fig. 7–3), they trend northeast–southwest. In Nova Scotia, their trend is east-west, except for the Hollow Fault (Fig. 7–3, loc. 6). Evidence for vertical movement on most

of these faults (shown in Fig. 7–3 by the solid fault symbol) is presented in detail in Appendix B. Other faults (e.g. Lubec–Belleisle, Fig. 7–3, loc. 1; St. George's Bay–Taylors Brook, Fig. 7–3, loc. 11) are not characterized by associated fanglomerate facies, and may have had only strike-slip movement (Appendix C). Some of the faults that show vertical movement also underwent strike-slip movement. Well-preserved slickensides on one of the fault planes of the Long Range Fault plunge 42 degrees down the dip to the northeast (Stevenson's Brook; Phair, 1949, p. 80). Thus the basin northwest of the fault went down to the northeast as the pre-Carboniferous basement block southeast of the fault went up to the southwest. Lineations on tectonic boudins within the Cobequid Fault zone plunge 20 degrees down the dip to the west (Eisbacher, 1967). Thus the basin south of the fault went down to the west as the pre-Carboniferous basement block north of the fault went up to the east. A great many very careful measurements by Eisbacher on a wide variety of other tectonic features in the Cobequid Arch support this sense of movement. The ages of the faults on Figure 7–3 are given in the following list (number, name, and age is in the order of presentation):

1. Lubec–Belleisle, cuts lower Horton Group strata (as old as Late Devonian); unconformably overlain by Riversdalian strata.

2. Clover Hill, produced facies gradations of Horton age (Hortonian and older Early Mississippian); later reactivated.

3. Pollett–Caledonia, produced facies gradations of Horton age; later reactivated.

4. Harvey–Hopewell, cuts Windsorian and Cansoan strata; unconformably overlain by Riversdalian strata; nearby parallel fault produced facies gradations of Windsorian and Cansoan age.

5. Cobequid, cuts Riversdalian strata; presumed covered at east end by Cumberlandian strata; produced facies gradations of Cansoan, Riversdalian, and Cumberlandian age; reactivated at west end later. The few exposures of this fault zone indicate dips from 40 to 70 degrees south, towards the depositional basin (Eisbacher, 1967).

6. Hollow, cuts upper Windsorian strata; overlain at west end by Cumberlandian strata; produced facies gradations of upper Windsorian age. Fault zone, marked by a topographic low, is nowhere exposed.

7. Guysborough, cuts lower Horton Group; covered at west end by upper Horton Group and Windsor Group strata; facies gradations of Hortonian age. Fault zone, not exposed, presumed to dip north towards depositional basin (Schiller, 1961).

8. Grand River, cuts lower Horton Group and Windsor Group; facies gradations of Hortonian age possibly present.

9. Aspy, cuts Horton strata; facies gradations of Cansoan and Windsorian age (Neale, 1963a, 1963b; Belt, 1965) may be related to fault. Neale's map (1963b) indicates a normal component with a minimum of 600 meters (2000 ft) of displacement.

10. Long Range, cuts lower Horton Group strata; produced facies gradations of Late Windsorian, Cansoan, and Riversdalian age. Fault zone dips 35 to 45 degrees into the basin (northwest) at its southwestern exposure, but dips of 80 degrees northwest and vertical dips are recorded at other localities within 30 kilometers (20 mi.) from that locality along strike to the northeast (Phair, 1949, p. 72–88). A strike-slip component is observed on slickensides (Phair, 1949, p. 80).

11. St. George's Bay–Taylors Brook, cuts lower Horton Group strata (Anguille Group) and is unconformably overlain by Cansoan strata in central Newfoundland (Belt, in press). Extension of Taylors Brook Fault to the northeast, along the eastern side of the Great Northern Peninsula of Newfoundland, is overlain by Late Hortonian (Cheverie equivalent) strata in the Conche area (Baird, 1966). The Taylors Brook Fault is basically a series of parallel faults, those to the northwest with slight vertical or horizontal displacement, those to the southeast with increasing amount of displacement (Brian Lock, oral communication, 1967). These faults dip into the basin (southeast), one exposure indicating approximately 60 degrees (George R. Heyl, oral communication, 1967).

12. Grand Lake, cuts lower Horton Group strata; produced facies gradations of sures of this fault zone commonly indicate a vertical dip, but one exposure Cansoan and Riversdalian age. Expo-

(Connors Brook) indicates a 75 to 80 degree dip towards the pre-Carboniferous basement (possibly a high-angle reverse component).

13. Birchy Lake, cuts lower Deer Lake Group; unconformably overlain by Late Riversdalian strata.

14. Birchy Ridge, cuts lower Horton Group; unconformably overlain by Cansoan strata.

Other faults (e.g. the Hampden Fault of north-central Newfoundland; Belt, in press), cut all Carboniferous strata, show no effect on Carboniferous facies, and are not unconformably overlain by Carboniferous strata. Faults of this category are not shown in Figure 7–3. Faults overlapped by all of the Carboniferous strata (e.g. major pre-Cansoan faults in central Gaspé) are likewise not shown in Figure 7–3. The numerous faults between the Belleisle and Clover Hill–Pollet–Caledonia (Gussow, 1953; Webb, 1963) could not be shown on a map of the scale of Figure 7–3.

6. *Geophysical data.* Seismic, gravity, and aeromagnetic surveys of parts of the Gulf of St. Lawrence made during the past 10 years are summarized by McConnell and McTaggart–Cowan (1963), Goodacre and Nyland (1966), and the Geological Survey of Canada (1967). Two seismic profiles are shown on Figure 7–1. The profile west of the Magdalen Islands (Goodacre and Nyland, 1966, Fig. 5, p. 120–121) shows a gradual southeastward thickening of the sedimentary cover which is interrupted adjacent to western Cape Breton Island by a sudden decrease from 5.0 km to 2.9 km (Fig. 7–1). This break is well-documented in aeromagnetic and bottom-contour trends (Cameron, 1966) and is interpreted by the writer to lie along the boundary between the "thick cover" and the "rift strata" of Figure 7–3.

The other seismic profile (Sheridan and Drake, in press) lies within the Cabot Strait. The results have been interpreted by Sheridan and Drake as showing a cover of coastal plain sediments over basement rocks that thins from station 177 to 175, a major fault at station 174, and an estimated 1800 meters (6000 feet) of sedimentary rock, postulated to be Carboniferous in age, above the basement northwest of the fault. Unfortunately, no station was made northwest of station 175: the profile of Sheridan and Drake does not continue across the possible extension of the St. George's Bay Fault south of the Magdalen Islands, but their results are compatible with an extension of the Long Range Fault across the Cabot Strait.

APPENDIX B

Facies Evidence for Carboniferous Dip-Slip Component on Faults Related to Fundy Basin

Blocks within the Fundy Basin

A. COBEQUID ARCH (FIG. 7–2, LOC. C)

1. Millsville conglomerate, Windsorian and Cansoan, (Fig. 7–1, loc. C) grades south into lacustrine shales and east into fine-grained fluvial shales both of the same age (Belt, 1965). These facies are associated with the Loganville Fault (Gillis, 1964).

2. Middleborough Formation, Cansoan (Fig. 7–1, Cumberland structural basin) contains lenses of coarse sandstone and conglomerate in a predominantly red shale formation. Coarse material was most likely derived from Cobequid Arch (now the Cobequid Mountains, Fig. 7–1) south of the basin.

3. West Bay Formation, Cansoan (Fig. 7–1, Minas structural basin) contains a few thin coarse sandstone beds in an otherwise predominantly lutite sequence. Coarse material derived from Cobequid Arch a few miles north of the Minas basin. Arch uplifted along Cobequid Fault (Fig. 7–3, loc. 5).

4. Unnamed fanglomerates, forming the basal part of the Cumberland and/or Pictou section (Weeks, 1948; Belt, 1965, Pl. 1, 8 kilometers (5 miles) northeast of loc. 3) in the Minas structural basin, were derived from the Cobequid Arch 300 meters (1000 feet) to the north. Cobequid Arch uplifted along Cobequid Fault (Fig. 7–3, loc. 5).

5. Unnamed conglomerates of probable Pennsylvanian age within the Cobequid Fault zone (Fig. 7–3, loc. 5), Londonderry area (Weeks, 1948; Belt, 1965, Pl. 1, 13 kilometers (8 miles) northeast of loc. 3), and Minas structural basin, were derived from the Cobequid Arch a few meters north of the fault zone. Ruptured pebbles and twisted boudins have been used to determine strike-slip movement in fault zone (Eisbacher and Kelley, 1966; Eisbacher, 1967).

B. CALEDONIA ARCH (FIG. 7–2, LOC. A)

1. Hopewell conglomerates, Windsorian and Cansoan, southeast of the Caledonia Arch, grade southeast away from the basement block into fine-grained red fluvial facies of the same age (Webb, 1963). Harvey–Hopewell Fault (Fig. 7–3, loc. 4) cuts the fanglomerates, but is believed to parallel an earlier fault scarp that produced the fanglomerates.

2. Weldon Formation, middle Hortonian, southeastern Moncton structural basin (Fig. 7–1), contains the Peek Creek Conglomerate member, found in contact with pre-Carboniferous rocks of the Caledonia Arch which formed the provenance area for the fanglomerate (Shroder, 1963). The fanglomerate facies grade northward away from the basement block into fine-grained fluvial sandstones of the North River Member of the Weldon Formation, and finally into the red siltstone and shale member of the Weldon Formation (lacustrine facies). Peek Creek Fault (Fig. 7–1, near N.E. end loc. 3) post-dates fanglomerates, but is believed to parallel an earlier fault scarp that produced them.

3. Hillsborough Formation conglomerates, upper Hortonian, of the southern Moncton structural basin grade northward, away from the Caledonia Arch, into the Steeves Creek Marlstone Member of the Hillsborough Formation, and is interpreted as a lacustrine deposit (Shroder, 1963). High-angle fault present between basement and fanglomerates.

C. WESTMORLAND ARCH (FIG. 7–2, LOC. B)

Hillsborough Formation conglomerates, upper Hortonian, between Caledonia Arch and Westmorland Arch shows clasts derived from the Westmorland Arch (Shroder, 1963). High-angle fault present between basement block and fanglomerates.

D. ANTIGONISH MASSIF (FIG. 7–2, LOC. D)

Unnamed conglomerate, upper Windsorian, Merigomish structural basin (Fig. 7–1, loc. H) on northwest side of the Antigonish Massif, grades laterally away from the basement block into fine-grained fluvial strata with interbedded marine limestone. Hollow Fault (Fig. 7–3, loc. 6) cuts these strata but is believed to parallel an earlier fault scarp that produced them.

E. CAPE BRETON HIGHLANDS MASSIF (FIG. 7–2, LOC. E)

The middle member of the Hastings Formation, lower Cansoan (Belt, 1965, p. 790–791) on northern Cape Breton Island, contains coarse sandstone and conglomerate believed to be derived from the basement block a few hundred meters to the east. A nearby major fault (Aspy, Fig. 7–3, loc. 9) is believed to parallel the fault responsible for uplift.

F. OTHER CAPE BRETON ISLAND SOURCE AREAS

The Grantmire and St. Ann's Formations, Windsorian (Belt, 1965, p. 780, Pl. 1), in eastern Cape Breton Island are locally derived from pre-Carboniferous basement but are apparently not related to high-angle faults. Other examples of Windsorian and Hortonian age fanglomerates associated with basement blocks are mentioned in Kelley (1958). Three such source areas are shown in western Cape Breton Island, Figure 7–2. Some of these fanglomerates may be related to activities on high-angle faults.

Platforms bordering the Fundy Basin

A. Unnamed red boulder-bearing mudrock of the Horton Group north of the Guysborough Fault (Fig. 7–3, loc. 7, western end) presumably passes northward into fine-grained facies of the same age (Benson, 1962; 1964; and oral communication, 1967). Boulders as much as 30 cm (1 foot) across in the fanglomerate a few meters north of the fault are lithologically identical with the basement rock south of the fault.

B. Coarse Horton Group conglomerates north of the Chedabucto Fault (Fig. 7–3, loc. 7, eastern end) are reported (Schiller, 1961) to grade northward within a few hundred meters into fine-grained facies of the same age.

C. Barachois Group fanglomerates (Cansoan and Riversdalian age, Belt, in press) of the Southwestern Newfoundland basin (Fig. 7–1) are closely related to the Long Range Fault (Fig. 7–3, loc. 10). Clasts in the conglomerate were derived from the basement rocks southeast of the fault, and 1.3-meter (4-foot) boulders grade to coarse sandstone within a kilometer from the fault.

D. Unnamed conglomerate and coarse sandstone of the undifferentiated Deer Lake Group faulted against the Grand Lake Fault (Fig. 7–3, loc. 12) contain clasts derived from the basement rocks southeast of the fault. Grain size diminishes within a few kilometers away from the fault (Belt, in press).

APPENDIX C

Evidence for Strike-Slip Movement on Faults and along Cleavage Planes Related to the Fundy Basin

1. Cobequid Fault (Fig. 7–3, loc. 5). Eisbacher (1967) presents evidence of ruptured pebbles and oriented boudins in the Cobequid Fault zone near Londonderry, Nova Scotia. Stress orientation derived from the evidence at seven localities indicates that a dextral strike-slip movement occurred in post-Riversdalian time. The Cumberlandian New Glasgow conglomerate possibly overlaps the eastern end of the fault (see Appendix A, item #5).

2. Harvey–Hopewell Fault (Fig. 7–3, loc. 4). Webb (1963; in press) presents evidence that the toe of a Windsorian and Cansoan alluvial fan was cut off and sinistrally faulted to the northeast for 16 kilometers (10 miles). Pebbles and oriented stringers within fault zone support strike-slip nature of movement. Fault is unconformably overlain to the northeast by the Boss Point Formation (Riversdalian, Belt, 1965, Pl. 1).

3. Cleavage folds and kinks in the Meguma Group (pre-Carboniferous) of the Meguma Platform. Fyson (1966) presents a scheme of primary, secondary, and tertiary folding and strike-slip micro-movement related to the cleavage in the Lower Paleozoic Meguma Group of central Nova Scotia. The total strike-slip movement is great when the whole platform is considered. Some movement preceded Acadian granite intrusions; some movement followed the intrusives.

4. Lubec Fault, Eastport district, Maine (Fig. 7–3, southwestern portion, loc. 1). Bastin and Williams (1914) recognized dextral movement along this fault, traceable for 19 kilometers (12 miles) in southeastern Maine on the shores of the Bay of Fundy and this has recently been re-examined by Cumming (1967). This same fault is traced for an additional 185 kilometers (115 miles) in New Brunswick, where it becomes the Belleisle Fault west of the Moncton basin (Fig. 7–3, loc. 1); dextral movement is proposed for the Belleisle portion of the fault (Webb, 1963, p. 1918–1923; Cumming, 1967, Fig. 3). Fault cuts lower Horton Group and is unconformably overlain by the Boss Point and Petticodiac strata (Riversdalian ? and Pictouan) at its northeastern end in New Brunswick and Prince Edward Island (Fig. 7–3).

5. Precambrian anorthosites and gabbros in Newfoundland. Belt (in press) has proposed that 260 kilometers (160 miles) of dextral strike-slip movement might have occurred on the Long Range–Grand Lake Fault system (Fig. 7–3, locs. 10 and 12, and others not shown to the northeast) and the St. George's Bay–Taylors Brook Fault system (Fig. 7–3, loc. 11) so as to split a single anorthosite-gabbro body into the three parts seen today (Baird, 1959; Riley, 1962; Gillis, 1965). Movement may account for the strong structural deformation of the Horton Group strata, containing slivers of basement, unconformably overlain

by relatively undeformed Cansoan and Riversdalian strata in the North-central Newfoundland basin (Fig. 7–1).

6. Several faults in southeastern New Brunswick (Clover Hill Fault, Fig. 7–3, loc. 2; also Prosser Mountain Fault, not shown) are postulated to have resulted from dextral strike-slip movement, a few kilometers in extent, that brought slivers of pre-Carboniferous basement rock northeast into the Moncton depositional basin north of the Caledonia Arch (Webb, in press).

7. Aspy Fault of northern Cape Breton Island (Fig. 7–3, loc. 9) may have a large dextral strike-slip component according to Neale (1963b). A large zone of cataclasis and chloritization within the pre-Carboniferous basement complex uplifted along the fault is reminiscent of the mylonite schist zone in the Cobequid Arch near the Cobequid Fault (Eisbacher, 1967).

STRATIGRAPHIC AND STRUCTURAL RELATIONSHIPS IN THE OUTER OR NORTHWESTERN BELT

ᴧes of Origin and Deformation of Bedrock in the Manhattan Prong[*]

LEO M. HALL

INTRODUCTION

METAMORPHIC ROCKS in the Manhattan Prong of southeastern New York State (Fig. 8–1) have long been subdivided into five formations: Yonkers Gneiss, Fordham Gneiss, Lowerre Quartzite, Inwood Limestone and Manhattan Schist (Merrill, 1890, 1896). These general rock types were recognized prior to their definition as formations (Gale, 1839; Mather, 1839, 1843; Dana, 1880) and since definition, all except the Lowerre Quartzite have been consistently recognized as mapping units by field workers in this region. The ages of these formations have been interpreted in various ways by previous workers. Some believe that all of the rocks are Precambrian (Berkey, 1907, p. 377; Berkey and Rice, 1919); others interpret them to be partly Precambrian and partly Lower Paleozoic (Merrill, 1896; Balk, 1936; Isachsen, 1964). Prucha (1956) and Scotford (1956) conclude that these rocks represent a continuous depositional history and both of them favor Paleozoic as the time of deposition but do not rule out Precambrian. An attempt is made here to solve this stratigraphic puzzle by relating the results of detailed field studies in the White Plains area (Fig. 8–1 and Fig. 8–2) to our knowledge of the stratigraphy and structure of rocks farther north in western New England and eastern New York.

ACKNOWLEDGMENTS

The New York State Museum and Science Service, Geological Survey financed my work in the Manhattan prong and I am thankful for the stimulation provided by John Broughton and Yngvar Isachsen. Many colleagues and students provided stimulating discussions in the field. John Rodgers and E-an Zen, in particular, were very helpful in aiding my understanding of the regional geologic relations and this manuscript has been improved as a result of their comments and corrections. I am also indebted to Kurt Lowe who took the time to introduce me to Manhattan Prong geology in the fall of 1960 and to John Prucha who spent several days in the field to show me many of the results of his detailed field work in the Manhattan Prong and Hudson Highlands.

STRATIGRAPHY OF THE WHITE PLAINS AREA

General Statement

All the bedrock in the White Plains area (Fig. 8–1 and Fig. 8–2) has been subjected to at least sillimanite-grade metamorphism. Except for minor pegmatitic dikes and veins, the rocks are gneisses, amphibolites, schists, marble, and quartzites of sedimentary or volcanic origin.

Bedrock in the Manhattan Prong can be divided into three general categories. First is an assemblage of rocks, here referred to as a gneiss complex, consisting predominantly of the gneisses and amphibolites in the Yonkers Gneiss and Fordham Gneiss. Second is the sequence Lowerre Quartzite, Inwood Marble, and Manhattan Schist, here referred to collectively as the cover rocks. Third are the undivided schists, gneisses, and amphibolites which are the continuation of the

[*] Publication authorized by the Assistant Commissioner for the New York State Museum and Science Service.

Hartland Formation of western Connecticut (Rodgers et al., 1959).

Gneiss Complex

Yonkers Gneiss. The Yonkers Gneiss (Merrill, 1890, p. 388) consists of well-foliated biotite and/or hornblende-quartz-feldspar gneiss. It is most commonly pinkish or orange although it is bluish-purple on local freshly blasted exposures. Dark greenish-black amphibolite is common but amounts to less than 5 per cent of the entire rock unit.

Fordham Gneiss. The Fordham Gneiss (Merrill, 1890, p. 389) has been subdivided (Hall, 1966) into five members in the White Plains area (Fig. 8–2). Although the relative ages of these members are uncertain, they are described in what is now interpreted to be their proper age sequence beginning with the oldest.

FORDHAM-A—Variety distinguishes Fordham-A. This member consists of interbedded brown weathering garnet-biotite-quartz-feldspar gneiss, rusty weathering garnet-biotite-

quartz-plagioclase gneiss, and minor amounts of various other gneisses.

FORDHAM-B—Gray garnet - biotite - quartz - feldspar gneiss locally containing sillimanite is the prevalent rock; in some places it contains spectacular poikilitic garnets up to eight inches (20 cm) in diameter. Amphibolite is characteristically interbedded with the gray garnet gneiss.

FORDHAM-C—This member is characterized by uniformity. It consists primarily of gray biotite - hornblende - quartz - feldspar gneiss, commonly with pale green epidote grains. White or pink quartz-feldspar layers and lenses that are locally pegmatitic are abundant in the gray gneiss. Amphibolite is common and pinkish biotite and/or hornblende-quartz-feldspar gneiss is locally prominent.

FORDHAM-D—This very distinct rock unit is composed mainly of rusty weathering sillimanite-lavender garnet-biotite-quartz-feldspar gneiss that commonly contains pyrite and/or graphite. Some siliceous biotite

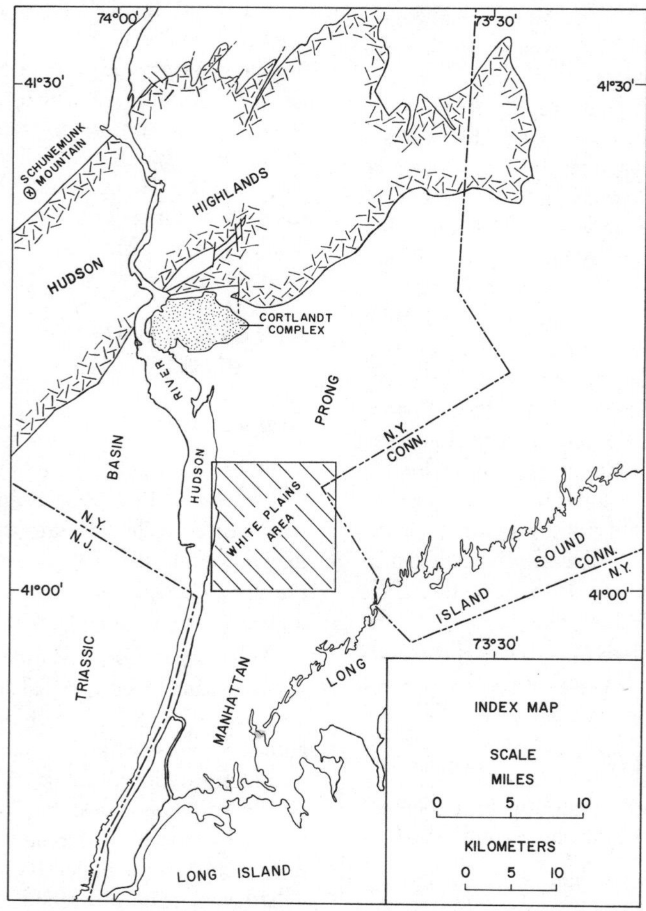

FIGURE 8–1. Index map for the White Plains area.

gneiss and quartzite are locally interbedded with the rusty gneiss.

FORDHAM-E—This member is predominantly siliceous biotite-quartz-plagioclase gneiss interbedded with gray garnet-biotite-quartz-feldspar gneiss, brown to rusty weathering biotite-quartz-feldspar gneiss and amphibolite.

Cover Rocks

Lowerre Quartzite. The Lowerre was named and described as "a stratum of thinly bedded quartzite" at the base of the Inwood Marble by Merrill (1896, p. 26). In the White Plains area, interbedded tan or buff feldspathic quartzite, micaceous quartzite and vitreous quartzite occupy this stratigraphic position (Fig. 8–2). Some of the feldspathic quartzite beds in the Lowerre have a granulitic texture. Beds are 2 inches (5 cm) to 6 feet (2 m) thick, and fractures more or less perpendicular to bedding are common. The Lowerre Quartzite has a maximum known thickness of 40 feet (12 m) in the White Plains area

but is absent in many places in the area (Fig. 8–2).

The status of the Lowerre Quartzite as a valid stratigraphic unit has been in dispute (Berkey, 1907, p. 367; Prucha, 1956, 1959, p. 1166; Norton and Giese, 1957; Norton, 1959). Sparse distribution of exposures of Lowerre Quartzite in the Manhattan Prong and the presence of abundant feldspar in many of the beds, leading to the interpretation that these beds are sheared granitic gneiss at the top of the Fordham Gneiss, are the two main lines of evidence that have been used to discredit the stratigraphic significance of the Lowerre. A paucity of Lowerre outcrops is not denied but the fact that the same interbedded assemblage of rocks described above occurs in the same stratigraphic position, at the base of the cover rocks, in each locality where it is exposed (Fig. 8–2) indicates the stratigraphic importance of these rocks. The Lowerre Quartzite is in contact with several members of the Fordham Gneiss (Fig. 8–2); hence the explanation that the Lowerre is sheared gneiss in the Ford-

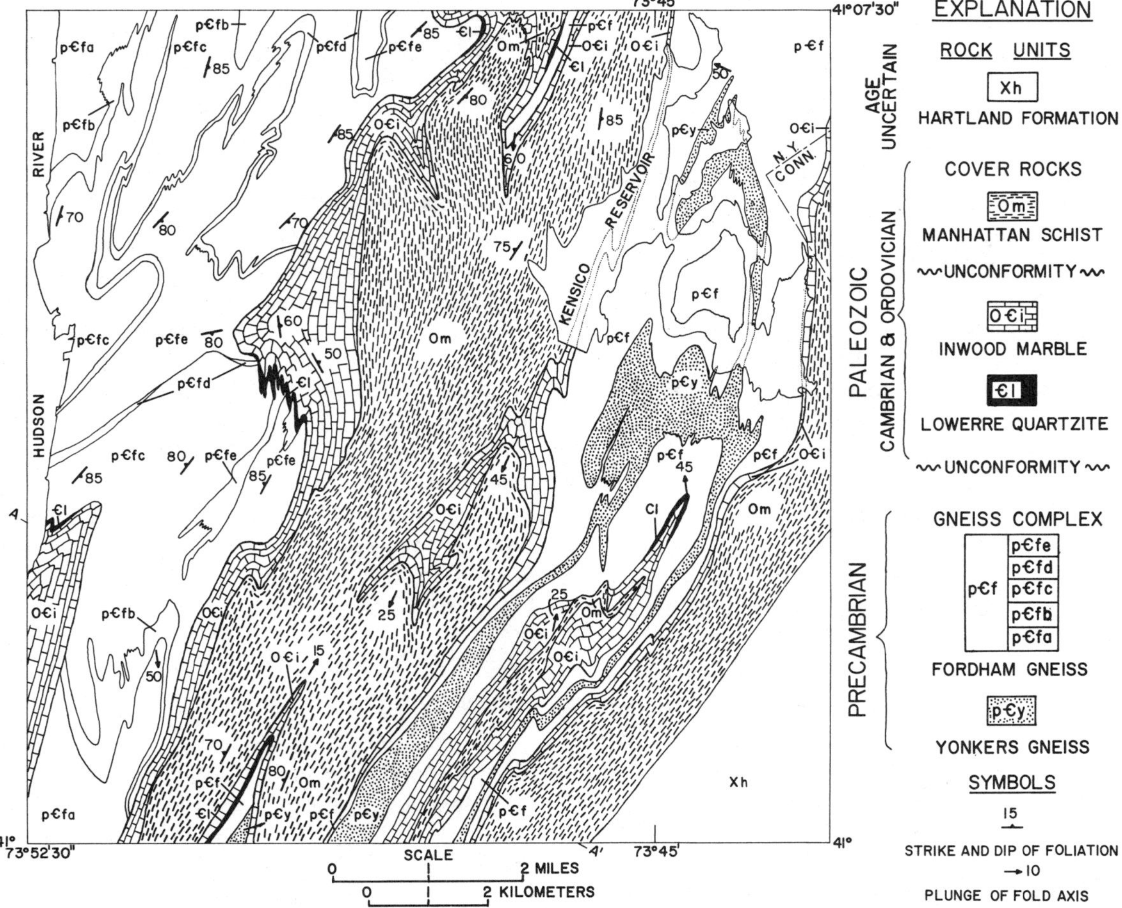

FIGURE 8–2. Generalized geologic map of the White Plains area.

ham is unlikely because the shearing of different types of gneiss seemingly would produce a variety of rocks instead of the consistently similar assemblage that is found and mapped as Lowerre Quartzite in the White Plains area. It is clear that the Lowerre Quartzite is present in the White Plains area and that it is a distinct and important part of the stratigraphic sequence in the Manhattan Prong.

Inwood Marble. The Inwood Marble (Merrill, 1890, p. 390) is composed preodminantly of clean dolomite marble and clean calcite marble with smaller amounts of calcite-dolomite marble, tan to brown weathering granular calcite marble, tan weathering calc-schists and granular dolomitic siliceous rocks as well as some tan mica schist. The maximum apparent thickness of the Inwood Marble is 2000 feet (600 m) in the White Plains area. The Inwood Marble has been divided into five members, not shown separately on Figure 8–2, in the White Plains area. The following brief description of each member is in order of sequence from the base.

INWOOD–A—White, gray, or blue-gray dolomite marble composes member A of the Inwood.

INWOOD–B—Interbedded white, gray, buff, or pinkish dolomite marble, tan and reddish-brown calc-schist, purplish-brown or tan siliceous calc-schist, and gray calcite-dolomite marble are common. Bedding is one-half inch (1 cm) to 3 feet (1 m) thick. The upper and lower contacts of Inwood–B are gradational.

INWOOD–C—White or blue-gray clean dolomite marble in beds 6 inches (15 cm) to 8 feet (2.4 m) thick are characteristic; in some places member C is massively bedded and it is locally fetid. Irregular pods and lenses of quartz may be present.

INWOOD–D—This consists of thin-bedded tan dolomite marble, gray calcite marble and tan calc-schist.

INWOOD–E—Gray or white calcite marble that commonly weathers tan is characteristic of member E.

Units B and C are the most commonly exposed members of the Inwood. There are some places where, apparently as a result of sedimentary facies changes, one or more of the Inwood members were not deposited. Such lensing in and out is apparently common for Inwood members D and E.

Manhattan Schist. This assemblage of schists, schistose gneisses and amphibolites was named by Merrill (1890, p. 390). Three members of the Manhattan have been recognized and mapped in the White Plains area but are not individually shown on Figure 8–2 due to the limitation of scale. These three members are here briefly described beginning with the oldest.

MANHATTAN-A—Manhattan A is a predominantly gray or dark-gray fissile sillimanite-garnet-muscovite-biotite schist with interbedded white calcite marble at the base. It should be noted that these marble beds have heretofore been considered part of the Inwood and that this is a redefinition of the basal Manhattan Schist. The schist is locally rusty or brown on the weathered surface. The entire unit has an apparent maximum thickness of 900 feet (300 m).

MANHATTAN-B—Dark greenish-gray or black amphibolite makes up member B. This unit is discontinuous, being absent or only a few feet (meters) thick in many places but as much as 100 feet (30 m) thick elsewhere.

MANHATTAN-C—Brown weathered garnet-muscovite-biotite schist or schistose gneiss characterizes Manhattan C. These rocks are typically feldspar rich and commonly contain sillimanite nodules. Bedding is distinguished with difficulty in most exposures but siliceous beds are prominent in some places.

Hartland Formation

The name Hartland Formation refers to an assemblage of various schists, gneisses and amphibolites in western Connecticut (Rodgers et al., 1959) and southeastern New York. In the White Plains area, the formation is characterized by brownish-weathering, gray feldspathic garnet-muscovite-biotite schist with local sillimanite and/or kyanite and by white, gray or pinkish muscovite - biotite - quartz - feldspar gneiss that locally contains sillimanite. Granitic pegmatites bearing coarse muscovite and biotite are common within the Hartland Formation in the White Plains area.

DEPOSITIONAL HISTORY

Gneiss Complex

The Yonkers Gneiss is granitic in composition. Merrill (1890, p. 388) originally suggested that the Yonkers is a metamorphosed arkose but subsequently decided that it represents an intrusive granite (Merrill, 1896, p. 29; Merrill and others, 1902, p. 5). In map pattern, the Yonkers Gneiss appears to be a normal stratigraphic unit within the gneiss complex (Fig. 8–2). Consequently, the Yonkers is here interpreted to be a metamorphosed rhyolite or arkose.

Metamorphosed clastic sedimentary rocks compose most of the gneisses in the Fordham Gneiss. The amphibolites in the Fordham are believed to represent metamorphosed mafic volcanics. Granitic gneisses in the Fordham, similar to the Yonkers Gneiss in composition and appearance, are believed to represent metamorphosed rhyolitic volcanics.

The gneiss complex, then, is interpreted as a predominantly clastic and volcanic eugeosynclinal sequence that has undergone high grade metamorphism.

Cover Rocks

The Yonkers Gneiss and each member of the Fordham Gneiss is truncated along the cover-rock contact (Fig. 8–2). Truncation of the Yonkers and Fordham at the Inwood contact is also indicated by Merrill in the vicinity of Bryn Mawr on the geologic map of the Harlem quadrangle (Merrill and others, 1902). The stratigraphic discontinuity at this contact is interpreted as a major angular unconformity. Interlayered marble and gneiss are found in some places along this contact elsewhere in the Manhattan Prong and this interlayering has been interpreted as evidence that the Fordham Gneiss and Inwood Marble are part of a continuous depositional sequence (Prucha, 1956; Scotford, 1956). In order to reconcile the presence of the unconformity indicated by mapping in the White Plains area with this evidence it is suggested that the interlayering is not the result of primary gradational deposition but was produced by some other process, perhaps folding or sedimentary reworking.

The Lowerre Quartzite represents metamorphosed arkosic and clean quartz sand deposits that accumulated on a weathered erosion surface developed on the basement gneiss complex. Part of the explanation for the scattered distribution of the Lowerre Quartzite in the Manhattan Prong is that it was deposited on an irregular topographic surface with high areas shedding clastic debris into local low areas. This is interpreted as an onlap unconformity (Fig. 8–3); by the time the high areas were inundated, clastic deposition had given way almost entirely to clean carbonate deposition. Consequently, there are many places where no clastic sands were deposited on the basement gneiss complex. Similar onlap relationships are known in many other places; the Precambrian gneisses of the Adirondacks, for example, are unconformably overlain by the Potsdam Sandstone in many places, but

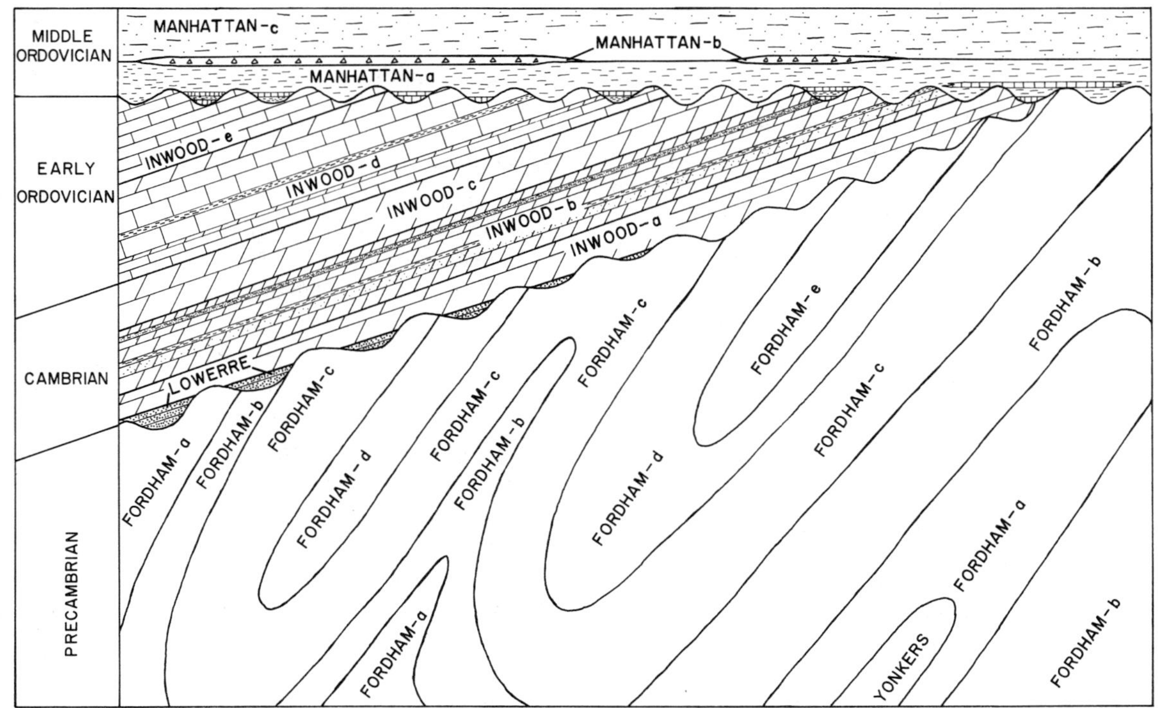

FIGURE 8–3. Diagrammatic illustration of the interpreted stratigraphic history of the Manhattan Prong.

elsewhere younger carbonate rocks overlap the Potsdam and are in contact with the basement gneisses (Fisher et al., 1962).

The Inwood Marble was deposited as a sequence of clean carbonate rocks with minor amounts of interbedded calcareous shales and sands. A sedimentary accumulation such as this is typical of the carbonate bank or shelf environment postulated by Rodgers (this volume) for eastern North America during parts of the Cambrian and Ordovician.

From place to place, the basal rocks of the Manhattan Schist are in contact with each member of the Inwood Marble and locally even with the gneiss complex (Fig. 8–2). This discordance between the base of the Manhattan and the older cover rocks as well as the gneiss complex is interpreted as another angular unconformity (Fig. 8–3) in the stratigraphy of the Manhattan Prong. Because the calcite marble beds in Member A of the Manhattan have previously been interpreted as beds of Inwood Marble in the Manhattan Schist, it has formerly been concluded that the change from Inwood deposition to Manhattan deposition was gradual and occurred during continuous sedimentation (Prucha, 1956). Such an interpretation was very reasonable at the time. However, we now know that the Inwood Marble is made up of several members that occur in a recognizable stratigraphic sequence. The marble interbedded with schist in member A of the Manhattan is not a part of this sequence of units in the Inwood; instead it and the schist are in contact with all members of the Inwood in various places and thus truncate the Inwood.

Accumulation of interbedded shale and limestone that marked the early deposition of member A of the Manhattan gave way to predominantly shale deposition. This deposition was apparently followed by volcanic eruption of mafic lavas represented by the amphibolite of member B. This was apparently followed by deposition of member C of the Manhattan, consisting of feldspathic shales, siltstones and sandstones.

Hartland Formation

The schists in the Hartland Formation were originally deposited as a group of shales, siltstones and sandstones. The light colored gneisses may well have originally been volcanic rocks of granitic composition whereas the amphibolites suggest original mafic volcanics.

In the White Plains area the Hartland Formation is in contact with the easternmost portion of the Manhattan Schist (Fig. 8–2). This contact is very difficult to locate in the White Plains area because of the similarity of the schists in both formations and the lack of characteristic light colored gneisses in the Hartland Formation near the contact. The preliminary geologic map of Connecticut (Rodgers and others, 1959) shows that, in southwestern Connecticut, the Hartland is in contact with Inwood Marble or Fordham Gneiss as well as Manhattan Schist.

One might speculate that the Hartland, at least in part, represents an eastward eugeosynclinal-facies equivalent to the Lowerre-Inwood-Manhattan miogeosynclinal facies. The contact between the Hartland and Manhattan Prong rocks may be a fault, thus explaining the contact of the Hartland with different formations from place to place in southwestern Connecticut and southeastern New York. A solution to this "Hartland Problem" has yet to be demonstrated but a firmer explanation should result from work now in progress.

PROPOSED REGIONAL CORRELATION AND AGE

Mather (1843), Dana (1880, p. 371), Merrill (1896, p. 23; Merrill and others, 1902) and Balk (1936) are among the many geologists who have proposed that the cover rocks of the Manhattan Prong correlate with the Cambrian and Ordovician clastic and carbonate sequence north of the Hudson Highlands in eastern New York and westernmost New England (Fig. 8–1 and regional geologic map, this volume). This suggested correlation has been based largely on the gross similarity of the sandstone-limestone-shale sequence both in the Manhattan Prong and in the Cambrian and Ordovician terranes north of the Hudson Highlands. Detailed stratigraphic subdivision of the cover rocks in the Manhattan Prong adds a great deal more resolving power to the stratigraphy and permits a more precise correlation than was previously possible.

Rock units of known age that are the presumed correlatives of the rock units in the Manhattan Prong are presented in the correlation table (Fig. 8–4). The unconformity at the base of the cover rocks marks the base of Paleozoic rocks in the Manhattan Prong so that the Lowerre Quartzite correlates with the Lower Cambrian Poughquag Quartzite as well as the fossiliferous Dalton and Cheshire farther north. The Yonkers–Fordham gneiss complex is therefore Precambrian, but these gneisses are not necessarily coeval with the gneisses in the Hudson Highlands or other terranes of known Precambrian rocks that underlie the Poughquag, Dalton, or Cheshire in eastern New York and western New England.

The Inwood Marble and the lower part of the Manhattan Schist are remarkably similar in lithology, as well as in detailed sequence, to the rocks that succeed the Poughquag and Cheshire in New York and New England. All these rocks were deposited on the extensive carbonate bank referred to by Rodgers (this volume). The Cambrian deposits comprise a sequence of predominantly clean dolomite limestones with some argillaceous and arenaceous beds toward the middle and particularly massive dolomite limestone beds at the top. The beginning of Ordovician deposition was marked by calcite limestones throughout the region of the carbonate bank.

The unconformity at the base of the Manhattan Schist is the same as the widespread Middle Ordovician unconformity elsewhere in New York (Fisher, 1962b) as well as New England (Cady, 1945, p. 560; Thompson, *in* Zen 1959, p. 72; Zen, 1967, p. 40). On this basis, it is concluded that at least the lower part of the Manhattan, member A, is Middle Ordovician. This correlation is strengthened by the recent discovery of pelmatazoan columnals at Verplanck Point (Ratcliffe and Knowles, 1968) on the east side of the Hudson River near the west edge of the Cortlandt Complex (white area at the west edge of the Cortlandt Complex on Fig. 8–1). These fossils are in a marble bed that is interlayered with dark gray schist, an assemblage of rocks that is identical to the lower part of member A of the Manhattan Schist. There is little question that the fossil bearing horizon and the lower part of member A of the Manhattan are stratigraphically equivalent.

On the basis of regional geologic relations as well as lithologic similarity, member B and member C of the Manhattan may correlate with the Waramaug Formation that is locally in contact with Precambrian basement in western Connecticut (Rodgers and others, 1959). It is thus possible that member B and member C of the Manhattan are older than member A of the Manhattan and at least part of the Inwood Marble. If this hypothesis turns out to be valid, it would necessitate a major thrust fault to explain the present spatial relationship between these rocks. Such a thrust fault would be similar to one farther north, in New York and New England, between the Taconic allochthon and the autochthon; a fault which is not sharply defined everywhere even in the region to the north (Zen, 1967, p. 13).

AGE / LOCATION	MANHATTAN PRONG — THIS REPORT	DUTCHESS COUNTY, NEW YORK — FISHER, (1962a, 1962b); KNOPF, (1962)	WESTERN MASSACHUSETTS AND CONNECTICUT — ZEN AND HARTSHORN, (1966); ZEN, (1967)	WESTERN VERMONT — DOLL AND OTHERS, (1961); ZEN, (1961)
MIDDLE ORDOVICIAN	C-MEMBER (MANHATTAN SCHIST)	WALLOOMSAC FORMATION	WALLOOMSAC FORMATION	IRA FORMATION
	B-MEMBER			
	A-MEMBER / INTERBEDDED MARBLE AND SCHIST	BALMVILLE LIMESTONE	INTERBEDDED LIMESTONE AND SCHIST	WHIPPLE MARBLE MEMBER
EARLY ORDOVICIAN	E-MEMBER (MARBLE)	COPAKE LIMESTONE (WAPPINGERS GROUP)	UNIT-G (STOCKBRIDGE FORMATION)	CHIPMAN FORMATION
		ROCHDALE LIMESTONE		BASCOM FORMATION
	D-MEMBER	HALCYON LAKE	UNIT-F	CUTTING DOLOMITE
			UNIT-E	SHELBURNE FORMATION
			UNIT-D	
CAMBRIAN	C-MEMBER (INWOOD)	BRIARCLIFF DOLOMITE (STOCKBRIDGE GROUP ≅ WAPPINGERS GROUP)	UNIT-C	CLARENDON SPRINGS DOLOMITE
				DANBY FORMATION
	B-MEMBER	PINE PLAINS FORMATION	UNIT-B	WINOOSKI DOLOMITE
				MONKTON QUARTZITE
	A-MEMBER	STISSING DOLOMITE	UNIT-A	DUNHAM DOLOMITE
	LOWERRE QUARTZITE	POUGHQUAG QUARTZITE	CHESHIRE QUARTZITE	CHESHIRE QUARTZITE
			DALTON FORMATION	DALTON FORMATION
PRECAMBRIAN	FORDHAM GNEISS / YONKERS GNEISS	GNEISS	BERKSHIRE MASSIF	MOUNT HOLLY COMPLEX

FIGURE 8–4. Proposed correlation of the stratigraphic subdivisions in the Manhattan Prong.

The regional correlation of the Hartland Formation is very tenuous at present. As stated above, the Hartland may be a facies equivalent of the Lowerre-Inwood-Manhattan. If so, the Hartland would also correlate with the Cambrian and Ordovician rocks, described by Hatch and others (this volume) on the east limb of the Berkshire anticlinorium in Massachusetts and Connecticut. No attempt was made to include the Hartland Formation in the correlation table (Fig. 8–4) because of the extreme uncertainty of its correlation.

DEFORMATIONAL HISTORY

Gneiss Complex

Evidence that the Yonkers and Fordham were subjected to at least one episode of intense metamorphism and isoclinal folding prior to the deposition of the Lower Paleozoic cover rocks is clear from the map pattern in the western part of the White Plains area (Fig. 8–2). The axial surfaces of the isoclinal folds in the gneiss complex were very likely deformed by refolding before the cessation of Precambrian deformation. Structural details in the gneiss complex have not yet been completely worked out but a disparity in orientation and style exists between some of the minor structures, as well as major structures, in the gneiss complex and those in the cover rocks. To be sure, many of the structures in the gneiss complex are parallel to those in the cover rocks. A complete understanding of the Precambrian deformation pattern will be difficult to obtain because of the extremely intense refolding subsequently imposed on the gneiss complex during the Paleozoic.

Cover Rocks and Gneiss Complex

The intensity and style of deformation that occurred after deposition of the Inwood and before deposition of member A of the Manhattan is uncertain, but the pre-Manhattan rocks were rotated enough to allow the erosion surface, on which the lower Manhattan was deposited, to be bevelled down to the basement gneiss complex. This is apparent from the fact that the Manhattan Schist is in contact with the gneiss complex in several places (Fig. 8–2).

The geologic map of the White Plains area (Fig. 8–2) and the diagrammatic structure section (Fig. 8–5) display a complex fold pattern involving both the gneiss complex and the cover rocks. A refolded fold is outlined by the contact at the base of the cover rocks in the eastern part of the White Plains area. A similar structural relationship farther north in the Manhattan Prong has been described by Scotford (1956). It is particularly important to note that the cover rocks plunge northward and structurally beneath the gneiss complex in the region south of the Kensico Reservoir (Fig. 8–2). This defines an antiform with younger rocks in the core and older rocks on the flanks (Fig. 8–5). The general structure, then, is interpreted as a refolded isoclinal fold that might be thought of as a tightly refolded nappe; the synformal hinge on the east side of the antiform (Fig. 8–5) is the arch bend (Billings, 1954, p. 41) and the hinge of the antiform is the crest. It is also apparent (Figs. 8–2 and 8–5) that the axial surface of the antiform has been deformed by smaller scale north plunging dextral folds.

Three varieties of folds, in addition to pre-Manhattan tilting, are recognizable in the cover rocks: early isoclinal folds, later isoclinal folds and latest dextral folds. The basement gneiss complex, of course, has additional folds. Although each variety of fold did not necessarily result from a separate orogeny, the sequential relationship between them is evident. The earliest stage of folding of the cover rocks resulted in the development of isoclinal, possibly recumbent, folds involving both basement gneiss and cover rocks as a unit. Refolding of these isoclinal folds took place as the deformation continued and finally the more open dextral folds, of shorter wavelength, developed. All of these folds were formed after deposition of the Manhattan Schist and their development was probably synchronous with the sillimanite-grade metamorphism. The high metamorphic grade is consistent with the fact that the gneisses, carbonates and shale behaved with similar plasticity during the folding.

DATES OF THE DEFORMATIONS

At least three separate pre-Mesozoic episodes of rock deformation took place in the Manhattan Prong. The earliest, which involved intense folding and high grade metamorphism occurred sometime in the Precambrian. The second, an apparently mild deformation, took place in the Middle Ordovician. Dating the third deformation, which was extremely intense, is somewhat problematical. Based on the regional correlation, it is concluded that this deformation and metamorphism occurred in or later than the Middle Ordovician. It is therefore possible that the Taconic orogeny, the Acadian orogeny, or both of these orogenies were major deformational events in the Manhattan Prong.

Isotopic mineral ages in the Manhattan Prong have been studied by Long (1961; also Long and Kulp, 1962). Long reported evidence that the Manhattan Prong underwent a widespread thermal event 360 million years ago; he believes this thermal event was the result of "mild pervasive reheating" (Long, 1961, p. 402) and not intense metamorphism which he believes took place 470 million years ago (Long, 1961, p. 406). The reason for Long's interpretation is that the Cortlandt Complex (Fig. 1), a mafic igneous intrusive dated at 435 million years (Long, 1961), crosscuts isoclinally folded rocks but is itself little deformed (Long and Kulp,

1962, p. 982). Bucher (*in* Creagh, 1948, p. 38) believed that the Cortlandt Complex intruded the Manhattan Prong during the deformation and metamorphism. Ratcliffe (1967) concluded that the emplacement of the Cortlandt Complex occurred after the main Paleozoic deformation in the Manhattan Prong because the complex truncates refolded isoclinal folds in the country rocks. According to Long (1961) and Ratcliffe (1967), then, the major episode of deformation must have taken place well before the Acadian orogeny if the 435 million year age identified for the Cortlandt Complex is correct. Thus the 470 million year metamorphic event identified by

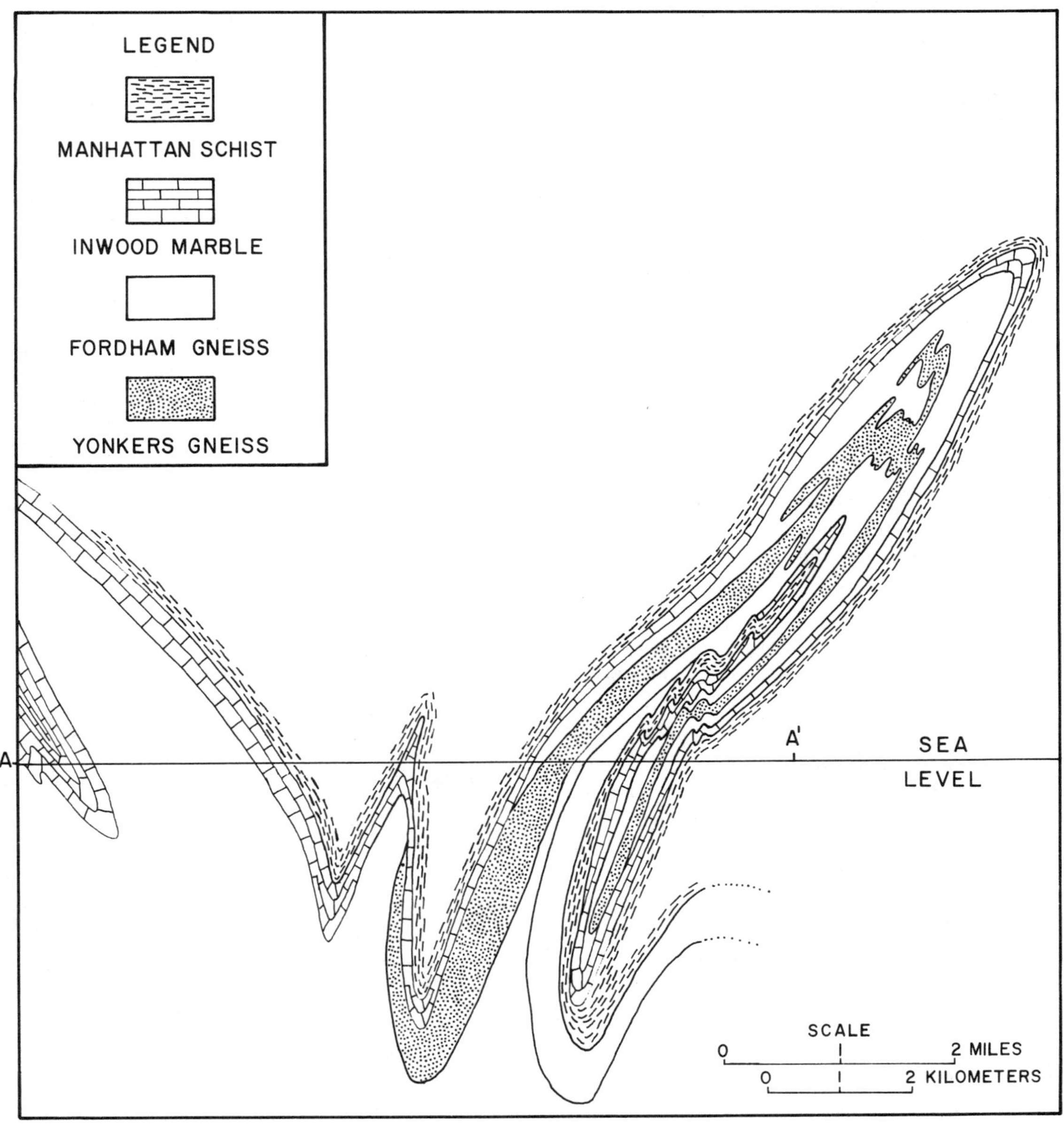

FIGURE 8–5. Generalized and diagrammatic structure section through the White Plains area.

Long (1961) dates the Taconic orogeny in the Manhattan Prong and indicates that, as suggested by Lowe and Schaffel (*in* Lowe, 1958, p. 12), this orogeny is of major importance in the Manhattan Prong.

The widespread geographic distribution of mineral dates at or near 360 million years in the Manhattan Prong (Long, 1961), north of the Hudson Highlands in Dutchess County (Long, 1962) and in northwestern Connecticut (Zen and Hartshorn, 1966, p. 3) suggests that the Acadian orogeny affected the rocks of the Manhattan Prong. The presence of folded and very slightly metamorphosed Silurian and Devonian rocks in the nearby vicinity of Schunemunk Mountain northwest of the Hudson Highlands (Fig. 8–1) as well as in other areas of deformed Silurian and Devonian rocks north and west of the Manhattan Prong (Chadwick, 1944) enforces this suggestion of Devonian tectonic activity. It is difficult to reconcile these facts with the evidence for the Taconic orogeny cited above and the dating of Paleozoic tectonic events in the region is not yet completely understood.

SUMMARY AND DISCUSSION

The bedrock of the Manhattan Prong represents Cambrian and Ordovician miogeosynclinal rocks that were deposited on a Precambrian basement complex. An area on a large carbonate shelf that extended over a vast area of the Appalachian geosyncline during most of the Cambrian and Early Ordovician was the site of deposition of most of the rocks of that age in the Manhattan Prong.

In the Middle Ordovician or possibly the late Early Ordovician, gentle tilting marked the first effect of the Taconic orogeny in the Manhattan Prong. Evidence for deformation at this time is cited by Cady (1945) in the Champlain Valley of Vermont, by Thompson (*in* Zen, 1959, p. 72) in the vicinity of Dorset, Vermont and by Zen (Zen and Hartshorn, 1966) in western Massachusetts and western Connecticut. A major tectonic episode occurred in the Manhattan Prong after the Middle Ordovician and it was during this event that the peak of Paleozoic metamorphism was realized. Evidence now at hand, although inconclusive, favors the Taconic orogeny as the major tectonic episode.

REFERENCES

Balk, Robert, 1936, Structural and petrologic studies in Dutchess County, New York. Part 1, Geologic structure of sedimentary rocks: Geol. Soc. America Bull., v. 47, p. 685–774

Berkey, C. P., 1907, Structural and stratigraphic features of the basal gneisses of the Highlands: New York State Museum Bull. 107, p. 361–378

———— and Rice, Marion, 1919, Geolgy of the West Point quadrangle, New York: New York State Museum Bull. 225–226, 152 p.

Billings, M. P., 1954, Structural geology (2nd ed.): New York, Prentice-Hall, Inc., 514 p.

Cady, W. M., 1945, Stratigraphy and structure of west-central Vermont: Geol. Soc. America Bull., v. 56, p. 515–587

Chadwick, G. H., 1944, Geology of the Catskill and Kaaterskill quadrangles. Part 2, Silurian and Devonian geology, with a chapter on glacial geology: New York State Museum Bull. 336, 251 p.

Creagh, Agnes, 1948, *ed.*, Geol. Soc. America, 61st Annual Meeting, Guidebook of Excursions, New York, New York, 135 p.

Dana, J. D., 1880, On the geological relations of the limestone belts of Westchester County, New York: Am. Jour. Sci., 3rd ser., v. 20, p. 21–32, 359–375, and 450–456

Doll, C. G., Cady, W. M., Thompson, J. B. Jr., and Billings, M. P., *compilers* and *editors*, 1961, Centennial geologic map of Vermont: Montpelier, Vt., Vt. Geol. Survey, scale 1:250,000

Fisher, D. W., 1962a, Correlation of the Cambrian rocks in New York State: New York State Museum Map and Chart Series, No. 2

————, 1962b, Correlation of the Ordovician rocks in New York State: New York State Museum Map and Chart Series, No. 3

————, Isachsen, Y. W., Rickard, L. V., Broughton, J. G., and Offield, T. W., *compilers* and *editors*, 1962, Geologic map of New York, 1961: Albany, N. Y., New York State Museum Map and Chart Series No. 5, 5 sheets, scale 1:250,000

Gale, L. D., 1839, Geological survey of the island of New York: Appendix C, Third Annual Rept. of the First Geological District of the State of New York, p. 177–199

Hall, L. M., 1966, Some stratigraphic relationships within the New York City group in Westchester County, New York [abs.]: Geol. Soc. America Special Paper 87, p. 69

Isachsen, Y. W., 1964, Extent and configuration of the Precambrian in northeastern United States: New York Acad. Sci. Trans., ser. 2, v. 26, p. 812–829

Knopf, E. B., 1962, Stratigraphy and structure of the Stissing Mountain area, Dutchess County, New York: Stanford Univ. Publications, Geol. Sciences, v. 7, No. 1, 55 p.

Long, L. E., 1961, Isotopic ages from northern New Jersey and southeastern New York: Annals of the New York Academy of Sciences, v. 91, p. 400–407

————, 1962, Isotopic age study, Dutchess County, New York: Geol. Soc. America Bull., v. 73, p. 997–1006

———— and Kulp, J. L., 1962, Isotopic age study of the metamorphic history of the Manhattan and Reading Prongs: Geol. Soc. America Bull., v. 73, p. 969–996

Lowe, K. E., 1958, *ed.*, Field Guide Book: New York State Geol. Assn. Thirtieth Ann. Mtg., Peekskill, New York, 52 p.

Mather, W. W., 1839, Third Annual Report on the Geology of the First Geological District of the State of New York: Albany, New York, p. 69–134

————, 1843, Geology of New York, Part 1, Comprising the Geology of the First Geological District: Albany, New York, 653 p.

Merrill, F. J., 1890, On the metamorphic strata of southeastern New York: Am. Jour. Sci., v. 39, p. 383–392

———, 1896, The geology of the crystalline rocks of southeastern New York: New York State Museum Ann. Rept. No. 50, Appendix A, p. 21–31

———, Darton, N. H., Hollick, Arthur, Salisbury, R. D., Dodge, R. E., Willis, Bailey, and Pressey, H. A., 1902, Description of the New York City District: U. S. Geol. Survey, Geologic Atlas of the United States, New York City Folio, No. 83, 19 p.

Norton, M. F., 1959, Stratigraphic position of the Lowerre Quartzite: New York Acad. Sci., Annals, v. 80, 1148–1158

——— and Giese, R. F., 1957, Lowerre Quartzite problem: Geol. Soc. America Bull., v. 68, p. 1577–1580

Prucha, J. J., 1956, Stratigraphic relationships of the metamorphic rocks in southeastern New York: Am. Jour. Sci., v. 254, p. 672–684

———, 1959, Field relationships bearing on the age of the New York City Group of the Manhattan Prong: New York Acad. Sci., Annals, v. 80, p. 1159–1169

Ratcliffe, N. M., 1967, Paleozoic post-tectonic plutonism at Stony Point, New York [abs.]: Geol. Soc. America Program for Northeastern Section Meeting, Boston, p. 51

——— and Knowles, R. R., 1968, Fossil evidence from the "Manhattan Schist-Inwood Marble" sequence at Verplanck, New York [abs.]: Geol. Soc. America Program for Northeastern Section Meeting, Washington, p. 48

Rodgers, John, Gates, R. M., and Rosenfeld, J. L., 1959, Explanatory text for preliminary geologic map of Connecticut, 1956: Conn. State Geol. and Natural History Survey Bull. 84, 64 p.

Scotford, D. M., 1956, Metamorphism and axial plane folding in the Pound Ridge area, New York: Geol. Soc. America Bull., v. 67, p. 1155–1198

Zen, E-an, 1959, ed., Stratigraphy and structure of west-central Vermont and adjacent New York: Guidebook to 51st Ann. Mtg. of New England Intercollegiate Geol. Conf., Rutland, Vt., 87 p.

———, 1961, Stratigraphy and structure at the north end of the Taconic Range in west-central Vermont: Geol. Soc. America Bull., v. 72, p. 293–338

———, 1967, Time and space relationships of the Taconic allochthon and autochthon: Geol. Soc. America Special Paper 97, 107 p.

——— and Hartshorn, J. H., 1966, Geologic map of the Bashbish Falls quadrangle, Massachusetts, Connecticut and New York: U. S. Geol. Survey Geol. Quad. Map, GQ-507, explanatory text, 7 p.

Chapter 9

Nature of the Ordovician Orogeny in the Taconic Area*

E-AN ZEN

INTRODUCTION AND HISTORICAL SURVEY

AMERICAN GEOLOGY was still in its infancy when Mather (1838, p. 166) at Becraft Mountain in eastern New York and Rogers (1838, p. 37) in eastern Pennsylvania simultaneously observed that certain lower Paleozoic rocks have been tilted and overlain unconformably by younger rocks. Dana (1863, p. 227) showed that this hiatus is significant for the Appalachian belt as a whole and fixed its stratigraphic position as between the "Lower Silurian" (i.e. Ordovician) and the "Upper Silurian" (Silurian of current usage). Dana (1895, p. 526) later described this hiatus more fully, stating that at the end of the "Lower Silurian" "mountain-making finally ensued, producing, among its effects, the Taconic Mountain Range along western and northwestern New England."

This hiatus became attributed to the Taconic orogeny. It is commonly supposed to mark the close of the Ordovician Period in the northern Appalachian region. The significance of this hiatus has been discussed by many authors; the survey studies of Clark (1921), of Schuchert (1925, 1930) and of Schuchert and Longwell (1932) merit special mention.

The present paper will describe the pre-Silurian diastrophic relations within the area now occupied by the Taconic sequence (the areas labeled LT, HT, and also the area labeled S between the latitudes of Ticonderoga and Poughkeepsie on Figure 9–1), which includes the

Taconic Range itself. A fresh look at the evidence for the early Paleozoic diastrophism in the Taconic area is desirable first of all because this is the type area of the Taconic orogeny. Moreover, recent detailed mapping in western New England and eastern New York has made it possible to reconstruct an apparently coherent sequence of events that occurred here during Cambrian and Ordovician times, and to suggest some causal relationships among these events, whose grand finale was the emplacement of the Taconic allochthon (surely the mountain-making event referred to by Dana in 1895). The structural history of the Taconic sequence, therefore, is of considerable intrinsic interest.

Lingering uncertainties on the nature and age of the Taconic orogeny in the northern Appalachian region stem in part from the nature of the evidence adduced for the orogeny. A summary, in chronological order, of the major events in early Paleozoic time in the Taconic area, therefore, should help to clarify the situation. Such a discussion will be the major purpose of this paper; a brief outline will be given here by way of introduction. In Cambrian and Early Ordovician time, the site now occupied by the Taconic sequence was relatively tranquil, and carbonate rocks were laid down in a shelf environment. At the end of Early Ordovician time, a system of longitudinal (north-south) high angle faults developed. During this event and shortly thereafter, the sediments being deposited west of the present Green Mountains became rapidly and increasingly argillaceous owing to uplifts in the area of the present Green Mountain Range and farther east, where rocks of the Taconic sequence

---

* Publication authorized by the Director, U. S. Geological Survey.

were depositing. Within this Taconic sequence, these same tectonic activities were registered by disrupted, local areas of erosion and deposition, although the sedimentary facies was not affected.

The carbonate-shale interface retreated westward across the shelf in late Middle Ordovician (Trenton) time, resulting in a regionally widespread onlap which at many places is an unconformity. Continued uplift to the east led to increasingly coarser clastic sediments in the former shelf area; these sediments, in late Trenton time, began to include a wildflysch-type conglomerate which marked the beginning of wholesale westerly transport of the Taconic rocks in a series of submarine slides. The emplacement of these slides, which constitute the Taconic allochthon, appeared to culminate the episode of intense but superficial Ordovician tectonic activity.

A gap in the stratigraphic record followed. The next sedimentary record here was the deposition, within and around the Taconic allochthon, of Late Silurian marine sediments. These rocks now rest with sharp angular unconformity upon the older rocks. The Silurian and younger rocks do not possess the early structures of the older rocks; however, all the rocks deformed as a unit during the Devonian Acadian orogeny, when they were also regionally metamorphosed.

Much of this protracted and complex sequence of events has been regarded, at one time or another, as evidence for the Taconic orogeny. Thus, for instance, Dana (1895, p. 527–531) spoke in this context of the deformation that gave rise to the Taconic Mountains, including folding and faulting of rocks that underlie the Taconic Range as well as the "slate belt" (Dale, 1899) west of it. Many geologists, however, cited as evidence for the orogeny the unconformity between the deformed Middle Ordovician and older strata, on one hand, and the Upper Silurian and younger rocks, on the other; this unconformity has been recorded extensively in the Appalachian belt (for summary, see Pavlides and others, this volume), but the classical exposures, at Becraft Mountain near Hudson, New York, and at Rondout, New York (Schuchert and Longwell, 1932), as well as an excellent exposure at Mount Ida, in Claverack, New York (Craddock, 1957, p. 691) are within the Taconic region.

Yet a third type of evidence cited for an Ordovician diastrophic event is the change, in Middle Ordovician time, of the nature of the youngest preserved rocks in the Taconic area, from carbonate to shale. Schuchert (1943, p. 50) suggested that "these muds came from the east and are indicative of the rising of the Taconic Mountains"; Kay (1942, p. 1612) likewise noted that the change of sedimentation signified land rise in the source area to the east, but referred the diastrophism to the Vermontian disturbance, whose genetic relation to the Taconic orogeny was not specified.

Recent field work sheds light on the significance and mutual relations of these events. They will be discussed next.

ACKNOWLEDGMENTS

Professor M. P. Billings early aroused my interest in Taconic geology while I was a beginning graduate student. J. M. Bird, P. H. Osberg, John Rodgers, George Theokritoff, O. T. Tobisch, and W. S. White searchingly reviewed the manuscript; their numerous helpful suggestions for its improvement are gratefully acknowledged. C. S. Zen and Priestley Toulmin III improved my writing style significantly.

BRIEF SUMMARY OF TACONIC GEOLOGY

The Taconic sequence of Cambrian(?), Cambrian, Lower Ordovician and Middle Ordovician shales and graywackes is partly autochthonous and partly allochthonous. The autochthonous part is restricted temporally to post-Lower Ordovician rocks and areally to the south and west peripheral areas of the Taconic sequence. The allochthonous Taconic sequence is divisible into six structurally distinct thrust slices; the emplacement of the two lowest (early) slices has been dated as late Trenton (*Orthograptus truncatus* var. *intermedius* zone of Berry's Marathon, Texas, succession; Berry, 1960) whilst the emplacement of the highest, latest slice, which occupies the Taconic Range, occurred before the onset of the main phase of the Acadian orogeny and has been assigned tentatively to the end of the Middle Ordovician orogenic episode (Zen and Ratcliffe, 1966).

In order to deduce the geologic history of the area, it is necessary to know the paleogeographic arrangements of the rock sequences preceding the Middle Ordovician diastrophism. Therefore, the allochthonous rocks must first be restored to their depositional site. Most probably, this site was between the eugeosynclinal east Vermont sequence and the miogeosynclinal sequence, over the basement rocks of the Green Mountains in Vermont and the Berkshire Highlands in Massachusetts and Connecticut, as well as over the area between the two massifs, now occupied by Paleozoic rocks (Fig. 9–1). The detailed reasons for this choice of site, from sedimentologic, stratigraphic, structural, and geometric ("room") points of view, have been discussed elsewhere (Zen, 1967, p. 56).

One decisive factor in favor of this choice of site is the facies relations in northern Vermont and adjacent Quebec between Lake Champlain and the Green Mountains, where the relative distribution of sedimentary facies of the lower Paleozoic rocks has been fairly well preserved despite many thrust faults (Fig. 9–2; Cady, 1960; this volume). Here, the Cambrian and Ordovician rocks lying above the Champlain Thrust and the Hinesburg Thrust occur geographically between the eugeosynclinal sequence typical of the area east of the Green Mountains–Berkshire Highlands massifs (see Doll et al., 1961; Hatch et al., this volume) and the miogeosynclinal sequence typical of the autochthonous rocks of the Taconic region, and consist of a relatively thin, transitional, largely clastic sequence (see, for instance, the descriptions of Shaw, 1958; Cady, 1960; Doll et al., 1961; Dennis, 1964). Carbonate rocks become more important in the western part of the transition zone, towards the miogeosynclinal rocks. At the base of this transitional sequence is a graywacke, the Pinnacle Formation, that thins eastward to merge into the finer grained argillaceous and quartzofeldspathic beds of the Underhill Formation (Doll et al., 1961).

These features of the transitional sequences of northern Vermont are shared by the Taconic sequence if account is taken of tectonic shuffling of the stratigraphy in the latter. Thus, like the Pinnacle, the Rensselaer Graywacke is a basal Paleozoic wedge that thins eastward into phyllites and schists; as a whole, the western parts of the Taconic sequence contain more carbonate rocks than do the eastern parts. In addition, the Taconic sequence and the transitional sequence of northern Vermont have detailed lithostratigraphic correspondences, and the scanty faunal evidence is consistent with the time-correlation of the comparable rock units. It seems beyond reasonable doubt that the transitional rocks of northern Vermont are the on-strike, autochthonous (or parautochthonous) continuation of the Taconic sequence. In this paper, the relations in northern Vermont will be taken as a template for the original relations that prevailed in the area farther south in New England.

The Bascom Formation contains a significant proportion of argillaceous rocks. The Brownell Mountain Phyllite Member of the Bascom (of Doll et al., 1961; see also Cady, 1960, p. 539, footnote 7), appearing just below the Hinesburg thrust near Mechanicsville, Vermont, at the extreme eastern limit of exposures of the miogeosynclinal sequence, is a calcareous phyllite. The correlative of the Bascom in the Taconic sequence is the Poultney Slate, which differs from the Brownell Mountain Phyllite Member or other, thinner lentils of phyllite in the Bascom chiefly by its lack of carbonate. In northern Vermont the Bascom-correlative rocks are absent across the Green Mountain anticlinorium through erosion, and so direct tracing of the Bascom-correlative units eastward cannot be done. Nevertheless, there is little doubt that the argillites in the Bascom were derived from the east, and that the now-eroded Bascom-Poultney contact, like the Winooski–West Castleton contact, was one of lateral facies change.

The early Paleozoic east-west facies change in northern Vermont, from argillite to carbonate, has been long recognized. Cady (1960) summarized this facies contrast, and showed (1960, Pl. 3) that it migrated forth and back during this time interval. According to Cady, the westward transgression of argillite, as recorded in the Bascom Formation, marked the beginning of the final retreat of the carbonate facies as a whole in this area and, by implication, in the latitude of the Taconic sequence as well.

Rodgers (this volume) suggests that this contact between the two lithofacies was the outer limit of the shallow carbonate bank, such that argillaceous rocks were laid down in deeper water to the east. This interpretation contrasts with the concept of Shaw (1958, p. 534; 1961, p. 445) who suggested that an east-west submarine topographic high (the Milton axis) separated the trough into two sedimentary basins. Zen (1967, p. 46) envisioned the thick eugeosynclinal sequence as proximal slope deposits on the east flank of the geosyncline, and the miogeosynclinal rocks as shallow platform deposits on its west flank; in this reconstruction, the Taconic rocks represent the distal deposits from both flanks, a "starved section." Bird and Theokritoff (1966) suggested that the shelly faunas found in Cambrian and Lower Ordovician carbonate rocks of the Taconic sequence were for the most part detrital, brought into their present sites from contemporary living areas on the carbonate shelf to the west by submarine slumps and turbidite deposition. This idea may fit with the reconstructions of Rodgers and Zen better than with that of Shaw (see also Cady, quoted by Zen, 1967, p. 47); however, the fact that sedimentary rocks characteristic of the Taconic sequence do appear within the miogeosynclinal sequence shows that, on the whole, the difference in water

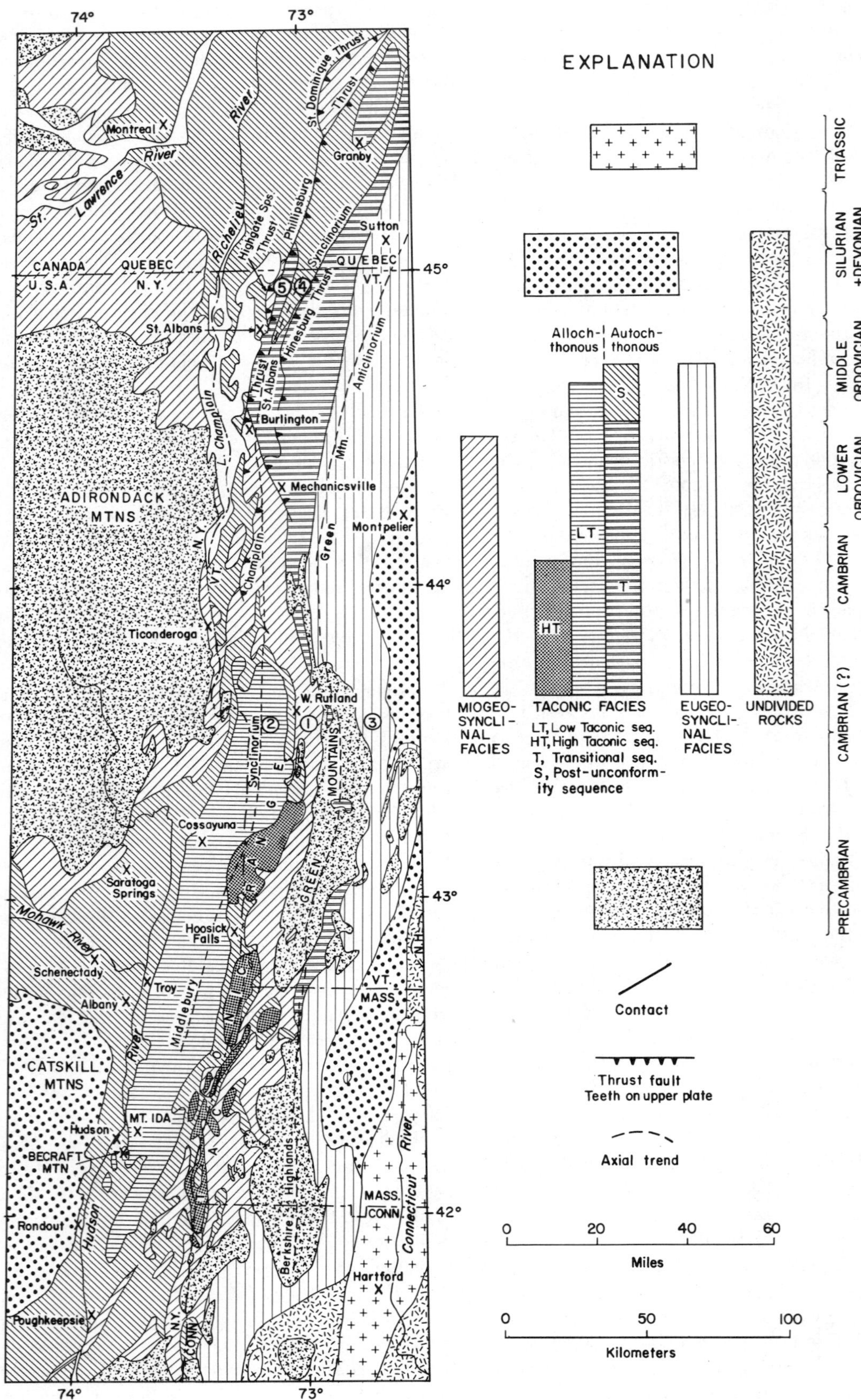
74°
73°
Montreal X
St. Lawrence River
Richelieu River
Highgate Sps
St. Dominique Thrust
Granby X
Thrust
Phillipsburg
St. Albans Thrust
Hinesburg Thrust
Sutton X
Sutton Synclinorium
CANADA   QUEBEC   QUE BEC
U.S.A.   N.Y.   VT.
45°
5 4
St. Albans X
Anticlinorium
St. Albans
Thrust
Burlington X
Green Mtn.
X Mechanicsville
ADIRONDACK
MTNS
Montpelier X
L. Champlain
N.Y.
VT.
Champlain Thrust
44°
Ticonderoga
W. Rutland X
2 1
3
Synclinorium
Cossayuna X
R A N G E
GREEN MOUNTAINS
Saratoga
Springs X
43°
Mohawk River
Hoosick
Falls X
Schenectady X
Troy X
VT.
MASS.
Albany X
Middlebury
Hudson River
N.H.
CATSKILL
MTNS
Hudson X
MT. IDA
X
A C A D I A N
Berkshire Highlands
MASS.
CONN.
42°
BECRAFT
MTN
Rondout X
Connecticut River
N.Y.
CONN.
Hartford
X
Poughkeepsie X
74°
73°

EXPLANATION

Alloch- Autoch-
thonous thonous
S
LT
T
HT
MIOGEO-   TACONIC FACIES   EUGEO-   UNDIVIDED
SYNCLI-   SYNCLI-   ROCKS
NAL   LT, Low Taconic seq.   NAL
FACIES   HT, High Taconic seq.   FACIES
           T, Transitional seq.
           S, Post-unconform-
           ity sequence

TRIASSIC
SILURIAN +DEVONIAN
MIDDLE ORDOVICIAN
LOWER ORDOVICIAN
CAMBRIAN
CAMBRIAN (?)
PRECAMBRIAN

Contact

Thrust fault
Teeth on upper plate

Axial trend

0   20   40   60
Miles

0   50   100
Kilometers

depth was neither profound nor irreversible at least during parts of early Paleozoic time, although a topographic break presumably existed at other times to cause the development of a sharp facies contrast.

Development of High-Angle Faults

At the end of Early Ordovician time, tectonic events began in the area of miogeosynclinal deposition.

The first event was the development of an extensive system of high-angle, longitudinal faults. The movement directions of these faults are not known, but the stratigraphic throws are as much as several thousand feet, as Trenton rocks are locally juxtaposed with rocks as old as Precambrian (see Doll et al., 1961). The dip component must have been sufficient to develop significant submarine or even subaerial relief.

Both normal and reverse faults exist within the

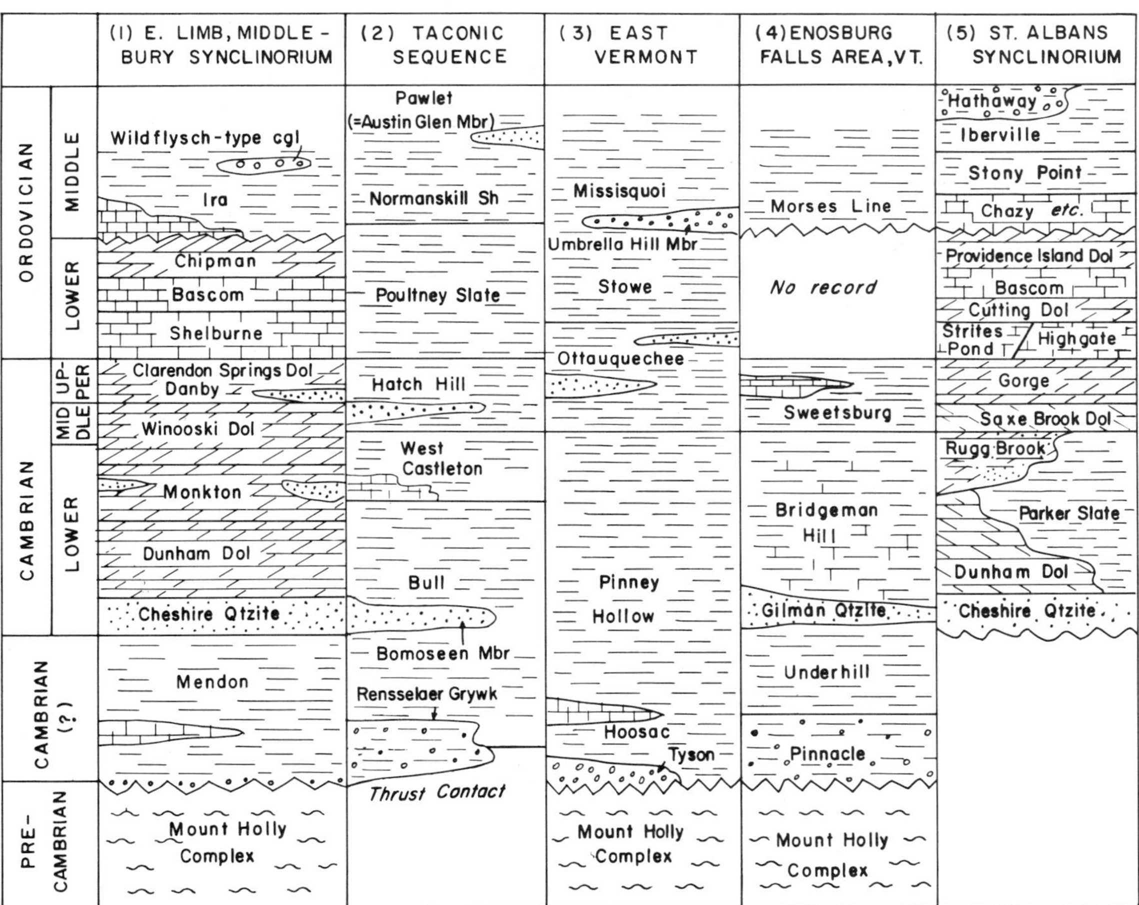

FIGURE 9–2. Correlation chart for western New England and eastern New York. The relative height assigned to each rock unit does not necessarily bear any relation to the thickness of that unit. The units are formations except where otherwise specified. Sources of data: Column 1, Doll et al., 1961; Zen, 1967. Column 2, Zen, 1967. Column 3, Cady, 1960; Doll et al., 1961. Column 4, Cady, 1960; Doll et al., 1961; Dennis, 1964. Column 5, Hawley, 1957; Shaw, 1958; Cady, 1960; Doll et al., 1961; Stone and Dennis, 1964. Nomenclature and age designations in columns 1, 3, 4, and 5 do not necessarily follow the usage of the U. S. Geological Survey.

system. On the whole, east of the axis of the Middlebury synclinorium, the individual fault blocks tend to tilt to the east, and the converse west of the axis (Rodgers, 1937, p. 1584; Megathlin, 1938, p. 106, Fig. 3; Cady, 1945, p. 570; Zen, 1967, p. 71).

The dating of the faults is indirect. On the east limb of the Middlebury synclinorium, most of the faults cut at least the basal strata of the black slate of middle Trenton age (Ira Formation, correlative of most of the Normanskill Shale). However, the trace of at least one major fault is overlapped by outcrops of the Ira Formation (Thompson, *in* Zen, 1959). Thompson interpreted the relations to mean that the fault here is older than at least part of the Ira. Therefore, if the faults all belong to the same system, their development must be a Middle Ordovician event.

The system of faults may have produced a horst-and-graben topography within the east limb of the Middlebury synclinorium. The following relations support this idea:

1. The base of the Middle Ordovician section is an important unconformity within much of the area on the east limb of the Middlebury synclinorium, where this younger section rests on rocks as old as the Precambrian. However, within short distances from outcrops showing such unconformable relations, the Middle Ordovician rocks may be found in apparent continuous section with the Lower Ordovician rocks (for examples, see Zen, 1961, p. 309; similar interpretation is feasible for the area surrounding the West Rutland marble quarries, Zen, 1964). Zen (1967, p. 42) suggested that erosion took place on the horst, leading to the later unconformable overlap of the younger rocks; continuous deposition, however, occurred within the graben so that uninterrupted sections were preserved here.

2. In southwestern Massachusetts and adjacent Connecticut and New York, the contacts between the Middle Ordovician rocks and the underlying units are locally marked by the former presence of bodies of bog iron ore that have been mined out (see Mather, 1838, p. 182, for a contemporary account and interpretation of these deposits). In the same general area, the basal unit of the Middle Ordovician rocks is locally a highly ferruginous, phyllitic limestone (Zen and Hartshorn, 1966). In west-central Vermont, the uppermost beds of the Lower Ordovician carbonate units are locally hematite-bearing (Zen, 1961, p. 321). All these iron-rich deposits may represent weathering residua of the carbonate rocks developed over the horst, more or less transported and diluted by other sediments.

3. In West Rutland, lenses of a vitreous quartzite occur in Ira black slate. Zen (1964, p. 67) suggested that these quartzite lenses represent sand derived from the disintegration of the vitreous Lower Cambrian Cheshire Quartzite, which is now exposed a short distance to the east across a fault belonging to this Ordovician fault system. This interpretation implies the former existence of local relief, perhaps as a result of the fault movement.

The age of movement of the faults on the west flank of the Middlebury synclinorium has been much debated. Megathlin (1938, p. 118) suggested that the movement was post-Taconic orogeny, whereas Swinnerton (1932, p. 413) concluded that the evidence is equivocal. On the other hand, Rodgers (1937, p. 1585) and Quinn (1933, p. 133) pointed out that these high-angle faults are not known to cut the thrust faults to the east, including the faults at the base of the Taconic allochthon, and so proposed the high-angle faults to be older. Zen (1967) suggested that the faulting was Middle Ordovician in age. This age is supported by the detailed relation of one fault near Saratoga Springs, New York, described by Fisher and Hanson (1951, p. 808), and by Platt's (1960, p. 88, 94) interpretation of the relations near Cossayuna, New York. These data indicate that high-angle faulting was active in that part of the west limb of the Middlebury synclinorium during the emplacement of the Taconic allochthon.

In contrast, Potter (*in* Bird, 1963, p. 61) showed that near Hoosick Falls, New York, high-angle faults cut the allochthonous rocks. This evidence, however, does not necessarily contradict the other cited relations, for high angle faulting could have continued after the allochthon was emplaced.

Development of an Unconformity

A second aspect of diastrophic activity in the Middlebury synclinorium during Middle Ordovician time is the development of a regional unconformity. Only a few features of this unconformity need our attention here.

The first feature is the already-discussed discontinuous nature of the unconformity. Combined with the dating of the longitudinal faults on both flanks of the Middlebury synclinorium as Middle Ordovician, this characteristic of the unconformity has been taken to mean that the post-unconformity rocks were laid down in a horst-and-graben terrane created by an active system of faults.

The second feature pertains to the nature of

the sediments at or just above the unconformity. The presence of iron-ore bodies or iron-rich beds along this horizon, probably representing residua preserved in situ or redeposited and incorporated in other types of sedimentary material, has been mentioned. The fault blocks have been uplifted locally above sea level; eventually, however, they were dropped down again, to be covered by fresh argillaceous sediments.

TECTONIC EVENTS IN THE AUTOCHTHON WITHIN THE TACONIC AREA

Stratigraphic Correlations

No pre-Middle Ordovician Paleozoic diastrophism is known to have occurred within the area of miogeosynclinal sedimentation. The sediments consist mainly of carbonate and clean quartzite; apart from the quartzofeldspathic basal Cambrian Mendon Formation, the two formations that contain notable argillaceous components are the Lower to Middle Cambrian Winooski Dolomite (for supporting data, see Herz, 1958; Knopf, 1962, p. 25–29; Zen, 1964, p. 32; Zen and Hartshorn, 1966) and the middle Lower Ordovician Bascom Formation (for supporting data, see Herz, 1958; Doll et al., 1961, and references cited; Zen, 1964, p. 36; Zen and Hartshorn, 1966). The source of the argillaceous components in the Winooski is uncertain. However, the temporally correlative unit of the Winooski within the Taconic sequence is at least partly the West Castleton Formation (Fig. 9–2), which was deposited farther to the east. The West Castleton is a black slate that contains interbeds of calcareous sandstone, limestone, dolostone, and limestone or dolostone conglomerate. In northern Vermont, the correlative and lithic equivalents of the West Castleton are the Parker Slate and the Bridgeman Hill Formation (of Doll et al., 1961; see also Fig. 9–2 and Zen, 1967); these units also contain bedded carbonate as well as carbonate conglomerates resembling those in the West Castleton, which become more extensive westward (Shaw, 1958; Doll et al., 1961; Stone and Dennis, 1964; Dennis, 1964). These relations support the idea that the argillaceous component of the Winooski was derived from the east, and conversely, the carbonate component in the West Castleton was derived from the west.

Basal conglomerate or similar coarse clastic beds are very rare along this unconformity. Bucher (1957, p. 668) described pebbles of Precambrian rocks in limestone of Trenton age near Peekskill (73° 55.4′W, 41° 17.3′N), New York,

and also suggested (p. 665) that the "Rysedorph conglomerate" represents a basal conglomerate; however, Bucher's interpretation for the Rysedorph is now known to be invalid (Zen, 1967, p. 87). On the other hand, conglomerate of the Umbrella Hill Member of the Missisquoi Formation (of Doll et al., 1961) in central Vermont and similar rocks in adjacent Quebec at the same stratigraphic level (Béland, 1957; Riordon, 1957; P. H. Osberg, written communication, 1967) may represent the same stratigraphic break in the eugeosyncline (Doll et al., 1961).

On the whole, the basal beds above the unconformity are clean to argillaceous, massive or thin-bedded blue-gray limestone, whose ages are early to middle Middle Ordovician, ranging from Chazy to early Trenton (Fisher, 1954; Doll et al., 1961; Zen, 1964, p. 41; 1967, p. 43); the rocks are locally fossiliferous, containing rugose corals, mollusks, brachiopods, trilobites, and bryozoa (Cady, 1945, p. 552–558; Thompson, *in* Zen, 1959, p. 75; Hewitt, 1961, p. 39; Potter, *in* Bird, 1963, p. 62; Zen, 1964, p. 39–42). The strata immediately above the unconformity thus locally indicate organism-related calcareous sedimentation in a clear, shallow, probably warm sea with little or no clastic sediment supply. This regime does not seem to differ greatly from those of the immediately preceding Early Ordovician environment.

The real break in the pattern of sedimentation appears higher in the section, where carbonate deposits give place to argillaceous strata. In individual outcrops, this transition is commonly gradual, and is shown by the increased content of pelitic material in the carbonate beds, and by interbedding of shaly and calcareous strata (Cady, 1945, p. 558; Zen and Hartshorn, 1966). As a whole, the zone of limestone-shale transition moved westward with time (Zen, 1967, p. 44), to result in an eastward wedging out of the carbonate facies.

In the context of the areal distribution of sedimentary facies, the Middle Ordovician unconformity marks an onlap of marine argillaceous sediments over the carbonate shelf, much like the facies intertonguing that occurred during earlier epochs in the same area. The Middle Ordovician event is distinguished by the scale of the onlap and by the complications caused by penecontemporaneous faulting.

The enormous quantity of clayey and silty material represented by the postunconformity shaly formations (the Normanskill and its correlative units) in the Taconic region suggests considerable uplift of the source area. According to the facies relationships cited above, the source area of these sediments was eastward; this agrees with the scantily studied current-direction indications of the "Martinsburg Formations" (i.e. Normanskill Shale) in the Hudson valley area immediately southwest of the Taconic allochthon (McBride, 1962).* The source area presumably was in part the eugeosynclinal deposits, and in part the Taconic sequence; these sources consisted of sedimentary rocks, in contrast with the coarsely crystalline source of the Cambrian(?) and Early Cambrian deposits.

The older, noncarbonate strata of the post-unconformity rocks are generally fine-grained argillites. The fine grain size may reflect partly the soft, unconsolidated state of the source material, and partly the gentle bottom slope down which the sediments were carried. The uplift at the source area seemed to have been gradual and steady rather than sudden and cataclysmic; the westward migration of the carbonate-shale interface might have resulted partly from a westward migration of the source area itself as the uplift gradually became more intense and as local sedimentary sources appeared. Graywacke beds, with grains in the coarse sand to fine pebble range and interpreted as turbidite beds (Bird, *in* Bird, 1963), first appeared in the upper part of the Normanskill and its correlative beds. The uplift of the source area became so pronounced in late Trenton time that progressively larger pieces of the Taconic sequence began to slide into the sedimentary basin, heralding the impending emplacement of the Taconic allochthon itself.

Emplacement of the Taconic Allochthon

The following features relating to the emplacement of the Taconic allochthon (Zen, 1967) are pertinent to the present discussion: (1) The allochthon consists not of a single slice, but probably of six distinct slices; (2) the earliest emplaced slices contain the youngest part of the stratigraphic section, and the latest, the oldest; (3) immediately or a short distance below the two earlier slices, there is a discontinuous zone

* Middleton (1965) arrived at a different conclusion on the current direction and source of material. However, interpretation of Middleton's data is uncertain because he did not distinguish bottom features according to the different types of scouring action, with implied possibility of different directions, nor did he attempt to rotate the highly deformed strata back to the original position. As the beds are commonly multiply deformed, their rotation back must not only be along correct axes, but must also follow a chronologically correct sequence; for most of the Normanskill rocks in this area this cannot yet be done.

of wildflysch-type conglomerate, whose matrix is marine Normanskill Shale or its correlative and equivalent units, and whose pebbles, chips and blocks are dominantly rocks of the Taconic sequence in addition to intraformationally derived material. These features show that the emplacement of at least these older slices of the allochthon was a surficial affair in a marine environment; the presence of chips and blocks of Taconic rocks within the conglomerate indicates that the slices, approaching the site of repose, had their toes in sea water but with the tops possibly above sea level. It seems certain that the movement of the slices of the allochthon was by near-surface gravity sliding; the emplacement could best be regarded as an incidental effect of the Middle Ordovician uplift in the area of deposition of the Taconic rocks.

Fossils in the matrix of the wildflysch-type conglomerate date the emplacement of the early slices to the *Orthograptus truncatus* var. *intermedius* zone (zone 13) of Berry (1960; see Zen, 1967, p. 68). The emplacement of the later slices cannot be dated with certainty. A late Middle Ordovician age is possible (Zen and Ratcliffe, 1966), but an early Acadian age cannot be excluded. Because the rocks in the later slices behaved more as brittle material during their emplacement, the movement might have occurred in a subaerial environment as the trough became filled by the older slices and by normal, particulate sedimentation.

It seems proper to regard the Middle Ordovician sedimentary onlap of quartzo-argillaceous deposits onto the carbonate shelf as the central event of the diastrophism of this time. The unconformity at or near the base of this onlap was the result of interaction of the onlap with a system of evolving high-angle, longitudinal faults; in this respect the onlap differed qualitatively from previous Paleozoic onlaps in this area. The emplacement of the Taconic allochthon was the culmination of the uplift and onlap; though its effects were spectacular, it probably did not cause additional major diastrophic events.

TECTONIC EVENTS IN THE ALLOCHTHON

Within the Taconic allochthon, available evidence points toward continual sedimentation between the earliest Paleozoic (Cambrian?) and the Middle Ordovician, extending at least into the *Climacograptus bicornis* zone (zone 12) of Berry, 1960 (see Berry, 1962, p. 713; Zen, 1967, p. 29), which immediately preceded the emplacement of the Taconic allochthon.

The time of emplacement of the Taconic allochthon corresponded to the time of deposition of the uppermost part of the Normanskill Shale (upper part of the Austin Glen Member). The equivalent and correlative unit of the Austin Glen Member in the northern part of the allochthon is the Pawlet Formation, whose base is also in the *C. bicornis* zone (Berry, 1962, p. 713). The Pawlet locally rests on allochthonous rocks of various ages, including Early Cambrian (Zen, 1964, p. 27–28). Such an unconformity, developed shortly before the allochthon arrived at its present site, without much doubt reflects the beginning of the movements that led to the gravity sliding of the rocks. Indeed, the Pawlet Formation might have been deposited partly on a moving substrate and partly even after the allochthon had come to rest.

During the emplacement of the allochthon, rocks of the Taconic sequence were deformed into isoclinal, recumbent folds of various sizes, and were also thrust faulted. At least within the older slices (which encompass the younger strata), this deformation took place when the rocks were still soft. This idea is supported by the presence of the wildflysch-type conglomerate, already discussed, and also by the fact that blocks of Taconic slates within the conglomerate, so far as is known, possess only a single regionally prevalent, "Acadian" cleavage, showing that these blocks were not yet cleaved at the time of their incorporation into the conglomerate.

All the slices of the allochthon show discontinuous masses of carbonate rocks along their soles, that have been interpreted as tectonic slivers (Zen, 1967). These rocks are commonly upper Lower Ordovician, though Middle Ordovician rocks are also present. The slivers further support the idea of near-surface movement because they could not have been very deeply buried in late Trenton time. Indeed, they indicate that areas of exposed carbonate rocks persisted, either on the Trenton sea floor or on the land nearby.

RECORD OF LATER PRE-ACADIAN EVENTS

The next record of events in the Taconic area is represented by the deposition of the Upper Silurian and Lower Devonian rocks found at Becraft Mountain and the adjacent Mount Ida near Hudson, New York, as well as west of the Hudson River (for regional summary, see Schuchert and Longwell, 1932; Pavlides et al., this volume). At Becraft Mountain, the Silurian and Devonian rocks straddle the contact between the autochthon and the allochthon (Zen, 1967, p.

56), and do not exhibit evidence of having been subjected to the same early deformation that affected the Taconic sequence.

Nonetheless, deformation of the Taconic sequence almost certainly continued during the period between the emplacement of the allochthon and the deposition of the Silurian and Devonian rocks. Hall (this volume) suggests that this deformation may have involved intense isoclinal folding in the Manhattan prong area. Pavlides et al. (this volume) show that even farther south, in the Lehigh Gap area, this deformation involved rocks that belong to Berry's graptolite zone 14. The record, however, is not preserved in the Taconic area.

DISCUSSIONS

Within the region now underlain by the Taconic sequence as well as within the confines of the Taconic allochthon, the evidence demonstrates that the major diastrophic events—high angle faulting, gravity sliding, isoclinal folding, unconformity, sedimentary overlap—all took place within a relatively short time interval in the Middle Ordovician.

The ultimate cause of this Middle Ordovician diastrophism is unknown. The records of the various events, with the exception of the longitudinal faults, indicate near-surface phenomena; however, it seems clear that uplift of land and deposition of sediments in the magnitude involved must entail the participation of the deeper part of the crust. The longitudinal faults may have resulted from keystone collapse of the shelf area (Zen, 1967, p. 71), an event that was coupled with the uplift of the area to the east (indeed, the area of the Green Mountains–Berkshire Highlands massifs might have been large horsts). Noting the coincidence of a deep negative Bouguer anomaly with the Taconic allochthon and of a sharp, linear positive anomaly with the Precambrian massifs (see Diment, this volume), Zen (1967, p. 72) proposed that the coupled subsidence and uplift might have been the result of a deep seated transfer of material. South of the Vermont–Massachusetts line, the positive anomaly becomes less sharp and is displaced to the east of the Berkshire massif; however, the west front of the massif there is now known to be a zone of extensive subhorizontal movement (Ratcliffe, 1965), so that the lack of coincidence could be the result of a later deformation.

In the literature, there is no generally accepted designation for the Middle Ordovician diastrophic event. Part of the problem lies in the fact that the definition of the Taconic orogeny refers both to the pre-Silurian-Devonian unconformity, and to the formation of the structure of the Taconic Range. The latter event may be Middle Ordovician in age, but the lack of record cannot be used to rule out categorically another Late Ordovician or Early Silurian event in this area.

The name "Vermontian disturbance" (Kay, 1942), describing at least the sedimentary onlap phase of the Middle Ordovician event, can be expanded to include perhaps the entire sequence of Middle Ordovician events described in this paper; the name "Taconic orogeny" can be used as a more general term, including the Vermontian disturbance as an early episode. There is some risk, however, that the main phase of the Taconic orogeny elsewhere in the Appalachian belt might prove to be totally unrelated to the diastrophism affecting the Taconic rocks; the Taconic orogeny thus might be devoid of representation in the type area.

OTHER SIMILAR DIASTROPHIC EVENTS IN THE NORTHERN APPALACHIAN REGION

Along the shores of Lake Champlain in northwestern Vermont, a wildflysch-type conglomerate, called the Hathaway Formation, was described by Hawley (1957). This unit is latest Trenton in age (Hawley, 1957, p. 58). The boulders include carbonate rocks, chert, and graywacke, and Hawley (1957, p. 68) concluded that the rock is a submarine slide breccia deformed while the rock was soft.

The carbonate blocks can be matched with lithic types of the shelf and of the transitional sequences within the Phillipsburg and Highgate Springs slices (Cady, 1960). The chert may be at least in part intraformational. However, the graywacke and possibly some other rock types must have been derived from the rocks above and to the east of the Hinesburg thrust (David Hawley, oral commun., 1967). The Hinesburg thrust therefore may have already been active in Middle Ordovician time; the rocks above this fault may have extended farther west than their present limit. The Hinesburg thrust was probably active during the Acadian orogeny (Cady, 1960, p. 564), which doubtless modified its configuration and appearance; however, in its primitive form, the Hinesburg thrust was perhaps genetically closely comparable with the gravity slide surfaces of the Taconic allochthon, the main difference being that here detached slices were not developed.

A Middle Ordovician submarine slide breccia or conglomerate is known in Quebec City (Os-

borne, 1956, p. 163, 189, and references cited therein). This rock is very similar to the wild-flysch-type conglomerate in the Taconic area, and may well bear a similar relation to gravity tectonism. The "Sillery formation" of the Granby, Quebec area (Cady, 1960, p. 540) lithically resembles the Taconic rocks. These rocks may have slid to the present position by gravity tectonics (Clark and Eakins, this volume); if so the Middle Ordovician St. Germain complex (Clark, 1947, p. 15) may be an associated wildflysch-type unit. In view of these relations and of the recent decipherment of a major Middle Ordovician gravity sliding episode in western Newfoundland (Rodgers and Neale, 1963; Cumming, 1966; Brückner, *in* Poole, 1966), also associated with wild-flysch-type conglomerates, it seems possible that the kind of diastrophism described in this paper had a wider regional extension than has hitherto been realized.

REFERENCES

Béland, Jacques, 1957, St. Magloire and Rosaire—St. Pamphile areas: Quebec Dept. Mines, Geol. Rept. 76, 49 p.

Berry, W. B. N., 1960, Graptolite faunas of the Marathon region, west Texas: Texas Univ. Pub. 6005, 179 p.

————, 1962, Stratigraphy, zonation, and age of Schaghticoke, Deepkill, and Normanskill Shales, eastern New York: Geol. Soc. America Bull., v. 73, p. 695–718

Bird, J. M., ed., 1963, Stratigraphy, structure, sedimentation and paleontology of the southern Taconic region, eastern New York: Geol. Soc. America Guidebook for field trip 3, Albany, New York, 67 p.

————, and Theokritoff, George, 1966, Mode of occurrence of fossils in the Taconic allochthon [abs.]: Geol. Soc. America, Northeastern Sec., 1966 Ann. Mtg., Philadelphia, Program, p. 12

Bucher, W. H., 1957, Taconic klippe—a stratigraphic-structural problem: Geol. Soc. America Bull., v. 68, p. 657–674

Cady, W. M., 1945, Stratigraphy and structure of west-central Vermont: Geol. Soc. America Bull., v. 56, p. 515–587

————, 1960, Stratigraphic and geotectonic relationships in northern Vermont and southern Quebec: Geol. Soc. America Bull., v. 71, p. 531–576

Clark, T. H., 1921, A review of the evidence for the Taconic Revolution: Boston Soc. Nat. History Proc., v. 36, p. 135–163

————, 1947, Summary report on the St. Lawrence Lowlands south of the St. Lawrence River: Quebec Dept. Resources, Prelim. Rept. no. 204, 18 p.

Craddock, J. C., 1957, Stratigraphy and structure of the Kinderhook quadrangle, New York, and the "Taconic Klippe": Geol. Soc. America Bull., v. 68, p. 675–724

Cumming, L. M., 1966, Tectonics and stratigraphy of the west coast of Newfoundland [abs.]: Geol. Soc. America Special Paper 87, p. 40–41

Dale, T. N., 1899, The slate belt of eastern New York and western Vermont: U. S. Geol. Survey Ann. Rept. 19, pt. 3, p. 153–300

Dana, J. D., 1863, Manual of Geology: 1st ed.: Theodore Bliss and Co., Philadelphia: 798 p.

————, 1895, Manual of Geology: 4th ed.: American Book Co., New York: 1088 p.

Dennis, J. G., 1964, The geology of the Enosburg area, Vermont: Vermont Geol. Survey Bull. 23, 56 p.

Doll, C. G., Cady, W. M., Thompson, J. B., Jr., and Billings, M. P., *compilers* and *editors*, 1961, Centennial geologic map of Vermont: Vermont Geol. Survey, Montpelier, scale 1:250,000

Fisher, D. W., 1954, Lower Ordovician (Canadian) stratigraphy of the Mohawk valley, New York: Geol. Soc. America Bull., v. 65, p. 71–96

————, and Hanson, G. F., 1951, Revisions in the geology of Saratoga Springs, New York, and vicinity: Am. Jour. Sci. v. 249, p. 795–814

Hawley, David, 1957, Ordovician shales and submarine slide breccias of northern Champlain valley in Vermont: Geol. Soc. America Bull., v. 68, p. 55–94

Herz, Norman, 1958, Bedrock geology of the Cheshire quadrangle, Massachusetts: U. S. Geol. Survey Geol. Quad. Map GQ 108

Hewitt, P. C., 1961, The geology of the Equinox quadrangle and vicinity, Vermont: Vermont Geol. Survey Bull. 18, 83 p.

Kay, G. M., 1942, Development of the northern Allegheny synclinorium and adjoining regions: Geol. Soc. America Bull., v. 53, p. 1601–1657.

Knopf, E. B., 1962, Stratigraphy and structure of the Stissing Mountain area, Dutchess County, New York: Stanford Univ. Pubs., Geol. Sci., v. 7, no. 1, 55 p.

Mather, W. W., 1838, Report of the Geologist of the 1st Geological District of the State of New York: New York Geol. Survey 2nd Ann. Rept., Albany, N. Y., p. 121–184

McBride, E. F., 1962, Flysch and associated beds of the Martinsburg Formation (Ordovician), central Appalachians: Jour. Sed. Petrology, v. 32, p. 39–91

Megathlin, G. R., 1938, Faulting in the Mohawk Valley: New York State Mus. Bull. 315, p. 85–122

Middleton, G. V., 1965, Paleocurrents in Normanskill graywackes north of Albany, New York: Geol. Soc. America Bull., v. 76, p. 841–844

Osborne, F. F., 1956, Geology near Quebec City: Naturaliste Canadien, v. 83, p. 157–223

Platt, L. B., 1960, Structure and stratigraphy of the Cossayuna area, New York: Yale University, thesis, 126 p.

Poole, W. H., *ed.*, 1966, Geology of parts of Atlantic Provinces: Geol. Assoc. Canada and Mineralog. Assoc. Canada guidebook, Halifax, Nova Scotia, 155 p.

Quinn, A. W., 1933, Normal faults of the Lake Champlain region: Jour. Geology, v. 41, p. 113–143

Ratcliffe, N. M., 1965, Bedrock geology of the Great Barrington area, Massachusetts: Pennsylvania State University, Ph.D. thesis, 213 p.

Riordon, P. H., 1957, Evidence of a pre-Taconic orogeny in southeastern Quebec: Geol. Soc. America Bull., v. 68, p. 389–394

Rodgers, John, 1937, Stratigraphy and structure in the upper Champlain Valley: Geol. Soc. America Bull., v. 48, p. 1573–1588

————, and Neale, E. R. W., 1963, Possible "Taconic" klippen in western Newfoundland: Am. Jour. Sci., v. 261, p. 713–730

Rogers, H. D., 1838, Second annual report on the geological exploration of the State of Pennsylvania: Harrisburg, Pa., 93 p.

Schuchert, Charles, 1925, Significance of Taconic orogeny: Geol. Soc. America Bull., v. 36, p. 343–350

————, 1930, Orogenic times of the northern Appalachians: Geol. Soc. America Bull., v. 41, p. 701–724

————, 1943, Stratigraphy of the eastern and central United States: John Wiley and Sons, Inc., New York, 1013 p.

————, and Longwell, C. R., 1932, Paleozoic deformation of the Hudson valley region, New York: Am. Jour. Sci., 5th Ser. v. 23, p. 305–326

Shaw, A. B., 1958, Stratigraphy and structure of the St. Albans area, northwestern Vermont: Geol. Soc. America Bull., v. 69, p. 519–568

————, 1961, Cambrian of south-eastern and northwestern New England, p. 433–471 *in* Internat. Geol. Congress, 20th Mexico, 1956, Kembriyskaya sistema, yeye paleogeografiya i problema nizhney granitsy (The Cambrian system, its paleogeography and the problem of its lower boundary), Symposium, part 3, Western Europe, Africa, U.S.S.R., Asia, America: Moscow, Akad. Nauk U.S.S.R., 518 p.

Stone, S. W., and Dennis, J. C., 1964, The geology of the Milton quadrangle, Vermont: Vermont Geol. Survey Bull. 26, 79 p.

Swinnerton, A. C., 1932, Structural geology in the vicinity of Ticonderoga, New York: Jour. Geology, v. 40, p. 402–416

Zen, E-an, *ed.*, 1959, Stratigraphy and structure of west-central Vermont and adjacent New York: Guidebook for field trip, New England Intercollegiate Geol. Conference, Rutland, Vermont, 85 p.

————, 1961, Stratigraphy and structure at the north end of the Taconic Range in west-central Vermont: Geol. Soc. America Bull., v. 72, p. 293–338

————, 1964, Stratigraphy and structure of a portion of the Castleton quadrangle, Vermont: Vermont Geol. Survey Bull. 25, 70 p.

————, 1967, Time and space relationships of the Taconic allochthon and autochthon: Geol. Soc. America Special Paper no. 97, 107 p.

———— and Hartshorn, J. H., 1966, Geologic map of the Bashbish Falls quadrangle, Massachusetts, Connecticut, and New York: U. S. Geol. Survey Geol. Quad. Map GQ 507, with 5 p. explanatory text

————, and Ratcliffe, N. M., 1966, A possible breccia in southwestern Massachusetts and adjoining areas, and its bearing on the existence of the Taconic allochthon: U. S. Geol. Survey Prof. Paper 550-D, p. D39–D46

The Eastern Edge of the North American Continent during the Cambrian and Early Ordovician

JOHN RODGERS

INTRODUCTION

THIS ARTICLE presents the speculation that the boundary between the contrasting Cambro-Ordovician carbonate and clastic facies in eastern North America was an abrupt declivity, like the margins of the present Bahama Banks, that it is exposed in two places (northwestern Vermont and Lancaster County, Pennsylvania) and that it was roughly localized by and hence roughly records the eastern edge of the North American continent at that time, all the terrain farther east having been deep water, probably diversified by volcanic and non-volcanic island arcs.

It is not a new idea that the great sheet of Cambro-Ordovician carbonate rocks in the eastern and central United States records a carbonate bank far vaster than, but in many respects like, today's Great Bahama Bank; the idea was clearly stated by Cloud and Barnes (1948, p. 79–109) for the Ellenburger group in Texas and implied for the rest of the region. The further idea that contemporaneous pelitic deposits in the western part of the Appalachians represent a deeper water facies was worked out by Lochman (1956, p. 1362–1367, Pl. 10) for the Taconic slate mass and has been adopted, although with different structural implications, by most workers in those rocks. What may be new in this article is the idea that an abrupt contact between the two facies may still be visible in certain areas; this idea came to me in 1965, during a field trip in northwestern Vermont with John M. Bird and E-an Zen, and was further elaborated in 1967 during a field trip in Lancaster County, Pennsylvania, with Donald U. Wise and Jacob Freedman; I am very grateful to these gentlemen for their critical discussion of the idea in the field. MM. John M. Bird, Arthur J. Boucot, Robert B. Neuman, Allison R. Palmer, George Theokritoff, Harry B. Whittington, and E-an Zen have kindly read and criticized the manuscript for me; I wish to thank them heartily, even if I have not accepted all their suggestions.

THE CAMBRO-ORDOVICIAN CARBONATE SEQUENCE

Along most of the western side of the Appalachian region from Newfoundland to Alabama one finds a thick sequence of carbonate rocks, ranging in age from Early Cambrian to Middle Ordovician. These rocks are exposed throughout the Valley and Ridge province of the Central and Southern Appalachians, especially in the Great Valley along the southeast side of that province, and also in the northeastern continuation of the province in the Hudson and Champlain Valleys and again in western Newfoundland. They are not known in place in the Appalachian part of Quebec, except close to the Vermont border, but they must once have been present there, for fragments of them are found as pebbles in lime-conglomerate and lime-breccia beds in pelitic rocks of the same age all along the south side of the St. Lawrence estuary from Quebec to Gaspé. Representative thicknesses of the carbonates (estimated from the references cited) are:

Area (reference)	Thickness	
	in meters	in feet
Western Newfoundland (Whittington and Kindle, in press)	1,200	3,900
West-central Vermont (Cady, 1945)	1,900	6,250
Central Pennsylvania (Butts, 1939)	2,350	7,700
Northern Virginia (Butts, 1940)	3,000	10,000
East Tennessee (Rodgers, 1953)	2,600	8,500
Alabama (Butts, 1926)	2,900	9,500

The stratigraphy within this sequence is remarkably consistent from Newfoundland to Alabama. Thus its Lower Cambrian base, exposed all along the southeast margin of the Valley and Ridge province and continuations, rests conformably on a basal Cambrian clastic group of greatly varying thickness, and its Middle Ordovician top grades upward and eastward everywhere into the basal parts of westward-advancing clastic wedges composed of black graptolite shale below and shale with graywacke above. Clastic materials are also present at several levels within the carbonate sequence. A particularly persistent clastic zone (possibly associated with evaporites in the southern Appalachians—Rodgers, 1968) lies at about the Lower-Middle Cambrian boundary from Newfoundland (Hawke Bay and March Point quartzites) to Alabama (Rome formation), shale invades the Middle Cambrian and lowest Upper Cambrian in several areas, clear quartz sand appears in the Upper Cambrian almost throughout the chain, and residual clastic materials (mud and fragments of chert) appear in many places above a fairly persistent disconformity between Lower and Middle Ordovician strata. Almost all these clastic materials within the carbonate sequence seem to have come from the northwest, although the clastic wedges above, and perhaps the clastic materials below, were clearly derived from the southeast.

The thicknesses cited above all lie within the Appalachian geosyncline; to the west the sequence thins rapidly, and the underlying basal clastics become younger—lower Upper Cambrian in the Adirondack Mountains and across much of the central interior of the United States, as young as Lower Ordovician in Ontario. Generally, in rocks of any one age, dolostone predominates to the west and northwest, limestone to the east and southeast, but at certain levels dolostone penetrates far to the southeast and at others limestone to the northwest.

Both the carbonate rocks and the intercalated clastic beds show abundant evidence of deposition in well aërated shallow water—very common cross-bedding, intraformational ("edgewise") lime-conglomerate and lime-breccia, erosional disconformities (especially toward the northwest), occasional reefy structures and stromatolites, and moderately abundant shelly fauna. Comparison with present-day carbonate deposits strongly suggests (e.g. Cloud and Barnes, 1948) that the sediments were formed on the surface of a carbonate bank not unlike the present Great Bahama Bank or other banks and shelves around the Gulf of Mexico or those around northern Australia. If so, the bank must have been immense, at least in Late Cambrian and Early Ordovician time, for its deposits still form a virtually continuous sheet from Vermont and Alabama westward to Minnesota and Texas, and they reappear in western Newfoundland, the eastern Cordillera, the Canadian Arctic Islands, northern and eastern Greenland, and perhaps Spitzbergen and northwest Scotland. The distribution of clastic materials within the sequence suggests, however, a continuing source of detritus somewhere within the central portion of the continent, probably roughly the area of the present Canadian Shield and the Sioux Arch. Thickness relations and the distribution of dolostone and limestone indicate that the southeastern or Appalachian margin (and probably the western, Cordilleran, and northern, Innuitian, margins as well) subsided much more rapidly than the rest, shallow water being maintained there because the rate of deposition kept pace with the subsidence, as during the Cenozoic in such areas as Florida, the Bahamas, and Yucatán. Only in the Middle Ordovician did subsidence finally outstrip carbonate deposition, at least along the eastern edge of the bank where black shale and then graywacke encroached over it from rising uplifts to the east.

THE CAMBRO-ORDOVICIAN CLASTIC OR PELITIC SEQUENCE

The carbonate rocks just described can also be followed a short distance eastward into the metamorphic part of the Appalachain chain where they appear as marble, in several places forming commercial marble belts such as that of western Vermont and Massachusetts. In general, however, Cambro-Ordovician rocks in the western part of the metamorphic Appalachians are clastic, now chiefly schist, which in certain belts and at certain stratigraphic levels contain abundant volcanics, now largely greenstone or amphibolite.

Furthermore, several large outlying masses of less metamorphosed or unmetamorphosed rocks of the same facies appear west of the metamorphic belt, surrounded by or northwest of belts of the carbonate sequence and containing fossils that prove their contemporaneity to the carbonates. The best known body of such rocks is the Taconic slate mass along the border between New York and New England; other large masses occur in western and northern Newfoundland and in eastern Pennsylvania. The structural position of these outlying masses has been or is being vehemently debated (cf. Lochman, 1956; Bucher, 1957; Rodgers and Neale, 1963; Zen, 1967; and see discussion below).

The structural complexity of these rocks and the metamorphism that most of them have undergone have impeded stratigraphic and sedimentary analysis. Within the metamorphic belt, as in eastern Vermont where the succession is now fairly well understood, the clastic sequence appears to have been immensely thick and to have included much graywacke as well as shale and, at several levels, abundant relatively mafic volcanics. In the less metamorphosed outlying masses, on the other hand, graywacke appears only near the bottom and top of the sequence, while the middle part is mainly shale and quite thin; volcanics are important in Newfoundland, locally significant in Pennsylvania, but rare in the Taconic region. Some of the shale is fairly calcareous, and beds of lime-conglomerate and lime-breccia are intercalated through much of the sequence, especially along the western margins of the outlying masses. Several studies (e.g. Ruedemann, 1901; Ross, 1949; Kindle and Whittington, 1958) have shown that the fragments in these beds are pieces of the carbonate sequence; some fragments can be matched almost exactly with their units, even to fossil content. The Cow Head breccias of western Newfoundland are particularly spectacular examples; blocks the size of box cars occur at several stratigraphic levels, shown by fossils in both fragments and matrix to range from the Middle Cambrian to the Middle Ordovician (Kindle and Whittington, 1958, 1959). In many areas, clean quartzite or carbonate-cemented quartz sandstone beds are moderately common in the Upper Cambrian part of the sequence, both metamorphic and non-metamorphic; for example, quartz-sandy beds are intercalated in the sequence of Cow Head breccias at the level of the Upper Cambrian.

The paleogeographic significance of the clastic sequence described above has also been less clear than that of the carbonate sequence but is now being elucidated by several workers (e.g. Lowman, *in* La Fleur, 1961; Bird, *in* Bird, 1963; Zen, 1967, and this volume; Cady, this volume; Enos, in press). The shale and shaly limestone both tend to be dark and to contain pyrite, suggesting relatively stagnant water, and ankerite, especially in the silty layers. Evidence of turbidity currents is abundant where the metamorphism has been minor, and the metagraywackes and metavolcanics of the more metamorphosed belts suggest deep-water troughs between volcanic islands. Beds of lime-sandstone associated with the lime-breccias show graded bedding, and both rock types can be interpreted as debris brought down into deep water by slumps and turbidity currents from the carbonate bank where the carbonate sequence was being contemporaneously deposited; the coarseness of the debris in breccias such as those at Cow Head demands that in those places the edge of the bank be only a few hundred meters away. The clean quartz sand in the Upper Cambrian part of the sequence could likewise have been transported from the bank, where similar clean quartz sand was spreading widely at that time.

The fossils in the pelitic sequence appear to support the inference that it was deposited in relatively deeper, perhaps poorly aërated water. Trilobites in the Cambrian are largely different from those in the shallow-water carbonate sequence, except that fragments of the latter trilobites appear in lime-breccia and lime-sandstone beds (Lochman, 1956; Bird and Rasetti, in press; Bird and Theokritoff, 1967; Theokritoff, this volume). In the Ordovician strata, shelly fossils are rare and mostly fragmental, whereas graptolites and a few other probably planktonic fossil types predominate (cf. Ruedemann, 1934).

With a few possible exceptions discussed below, the boundary between the two Cambro-Ordovician sequences has not been observed. Moreover, no evidence of a gradation between them has been found, unless one so interprets the appearance of limy shale and lime-breccia beds along the western sides of the outlying masses of the clastic sequence. The paleogeographic comparisons attempted above suggest the reason; thus, the edges of present-day carbonate banks such as the Bahama Banks are notably abrupt, the bottom dropping off into the surrounding deep water with slopes that may average 50% and that are locally vertical for hundreds of meters. The Bahama Banks have been growing since the Cretaceous, yet apparently their margins have not shifted very much since that time

(Newell, 1955). Considerable parts of the eastern boundary of the North American Cambro-Ordovician carbonate bank may have been further fixed by contemporaneous down-to-basin normal faulting of the kind that appears to characterize the early geosynclinal history of several mountain belts (e.g., Trümpy, 1960, for the Alps), but this boundary runs, throughout its known extent in the Appalachians, in a belt that has since been powerfully deformed and in good part metamorphosed, so that the original relations have been seriously obscured. The very strong contrast in competence between the two different sequences undoubtedly helped to localize thrusting and décollement at their abrupt contact; the older normal faults, where they existed, might thus have been converted into thrusts, covering over the transition zone, if any, and juxtaposing the two facies as we find them today.

The eastern margin of the restricted deep-water basin or basins is even less clear, being hidden in the metamorphic core of the Appalachian chain. In western New England, volcanic rocks become increasingly abundant eastward, reaching a climax along the course of the present Bronson Hill anticlinorium and its northeastward prolongation through northern Maine (see Hatch et al., this volume; Thompson et al., this volume; Naylor, this volume; Berry, this volume; Rankin, this volume), a belt also characterized by irregular overlaps or unconformities and by an incomplete section. Perhaps this belt was then a volcanic island arc, helping to restrict circulation in the basins to the west between the arc itself and the carbonate bank—one can find rough parallels to this situation in the East Indies today.

EXPOSURES OF THE BANK EDGE

The original eastern boundary of the carbonate bank may be preserved, little altered, in a few places where its course was not determined by faulting and ran oblique to later structural trends. One such place may be in northwestern Vermont, in the Milton area some 70 kilometers south of the Quebec border, within the Champlain or Rosenberg thrust sheet between the Champlain thrust fault and a zone of connected thrust faults (variously called Hinesburg, Oak Hill, Arrowhead, and St. Albans) along the west flank of the Green Mountain anticlinorium. The carbonate sequence forms this thrust sheet from west-central Vermont north to the vicinity of the Lamoille River near Milton (Cady, 1945; Stone and Dennis, 1964), whereas in the St. Albans area to the north (Shaw, 1958) the succession consists largely of gray to black slate, some silty

and ankeritic, some very calcareous, or of thin-bedded dark argillaceous limestone, with numerous lenticular bodies of lime-conglomerate and lime-breccia and also with some beds of dolomite-sandstone and dolomite-conglomerate. Massive carbonate (dolostone, in good part dolomite-breccia, dolomite-conglomerate, and dolomite-sandstone, commonly ankeritic) occurs only at the base of the succession above the basal clastics. Much of the transition area between carbonate and slate facies is hidden under Pleistocene deposits of the Lamoille River.

From these relations, Cady (1945, 1960) and Stone and Dennis (1964) inferred intertonguing of shale and carbonate, whereas Shaw (1958, p. 531–534) interpreted the area as a positive axis, commonly emergent, between a carbonate basin to the south and a clastic basin to the north, for he considered the lime-conglomerate lenses to be evidence of subaërial erosion, as had most earlier workers. These lenses are interbedded, however, only in the shale, many features of which suggest deep-water deposition, whereas the carbonate section, only a few kilometers to the south, has all the usual features of the shallow-water carbonate sequence. The limy beds in the slate sequence are either fragmental (locally graded) or simply dark fine-grained limy shale or argillaceous limestone; the latter lack shallow-water features, containing instead thin even beds of sand, ankeritic silt, or limy silt, and fossils less like the faunas of the bank than of the typical pelitic facies. The local representative of the Lower to Middle Cambrian clastic zone in the carbonate sequence (Monkton quartzite) disappears close to where the carbonate sequence ends, its place taken by a dark slate containing trilobites (Parker slate).

As an alternative explanation of these facts (Fig. 10–1), I therefore suggest that, but for the glacial deposits, we would here see the original edge of the carbonate bank crossing the thrust sheet diagonally, tipped up on edge by the thrusting and bevelled off by erosion. In the deeper water north of the bank margin, lime-rubble and lime-sand repeatedly slumped or were washed into the basin where mud, some limy, some carbonaceous, was being deposited.

Another possible exposure of the facies boundary is in Lancaster County, Pennsylvania. Here the east-west Chickies-Welsh Mountain anticline separates the Ordovician and younger Cambrian rocks of the county into two synclinoria, both severely deformed and at least slightly metamorphosed (Jonas and Stose, 1926, 1930; Knopf and Jonas, 1929; Stose and Jonas, 1933); the youngest rocks crossing the crest of the anticline are Lower or Middle Cambrian (Ledger dolostone). In the northern synclinorium the standard carbonate sequence is exposed, but in the southern it is largely replaced by thin-bedded, dark, very argillaceous limestone, the Conestoga limestone, severely deformed and altered and lacking fossils. This limestone rests on various Lower Cambrian carbonate units, which are successively older to the south (D. U. Wise, 1967, written communication) until it reaches the top of the basal Cambrian clastics (Antietam formation) near the south margin of the synclinorium. This overlapping contact has naturally been interpreted as an unconformity (see references above), and certain lime-breccia beds near it have been considered basal conglomerates. They are not basal, however, but occur up in the lower part of the thin-bedded limestone, whose basal beds, in some places at least, are black, finely laminated shale or phyllite, some of it limy or ankeritic.

Here again an alternative explanation seems possible (Fig. 10–2). The Conestoga limestone can be interpreted as a deep-water deposit contemporaneous with instead of later than the more normal shallow-water carbonates to the north; its apparently transgressive overlap thus becomes the encroachment of deep-water over shallow-water deposits, as subsidence of part of the outer margin of the carbonate bank became so rapid that

shallow-water deposition could not keep up but was defeated and cut off. After further subsidence had brought this part of the former bank surface into relatively deeper water, deep-water mud began to accumulate on top of the older shallow-water deposits, and slumps from the edge of the carbonate bank, still not far away, produced the lime-breccias in the lower part of the Conestoga limestone. Furthermore, within the sequence of Lower Cambrian carbonate units in the northern part of the southern synclinorium, over which the Conestoga limestone overlaps, is a unit of shale and limestone, the Kinzers formation, rather unlike the purer dolostone units above and below but enough like the main body of the Conestoga to have been confused with it (cf. successive interpretations of a single cross-section of these rocks north of Wrightsville, Penna., given by Stose and Jonas, 1933, Fig. 6, and Stose and Stose, 1944, Fig. 8). On the alternative explanation, the Kinzers formation can be thought of as a tongue of the Conestoga facies intercalated into the more normal bank carbonates, like the Parker slate of Vermont of which, incidentally, it is the exact correlative.

ORGINAL COURSE OF THE EDGE OF THE CARBONATE BANK

In the absence of severe deformation and tectonic transport, the original course of the bank edge could be deduced from the present distribu-

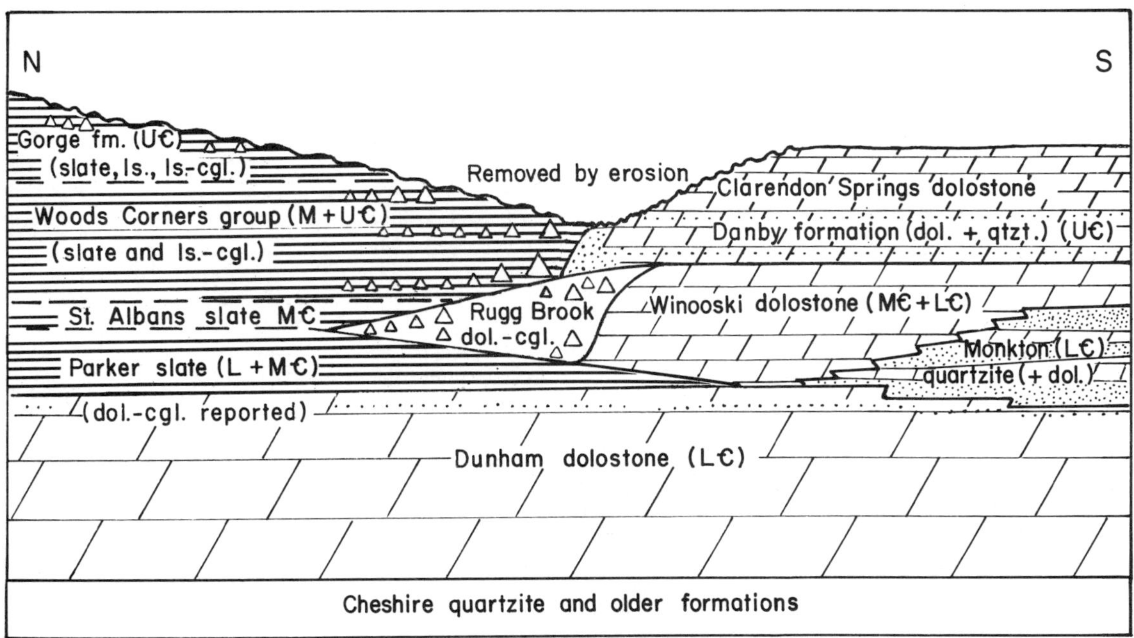

FIGURE 10–1. Stratigraphic diagram, along strike, of Cambrian strata near Milton in northwestern Vermont, to illustrate interpretation of Cambrian dark slates as deep-water deposits.

tion of the two facies. As mentioned above, however, outlying masses of the pelitic facies occur west of or are surrounded by belts of the contemporaneous carbonate facies; in many areas indeed they appear to overlie the carbonates. These masses can be interpreted in two main ways, and the interpretation one selects will greatly influence how one draws the bank edge; Figure 3 is drawn in accordance with the allochthonous interpretation, (2) below, which I accept.

(1) The outlying masses may be the deposits in deep-water embayments well within the carbonate bank, somewhat like the present Tongue of the Ocean in the Great Bahama Bank (Lochman, 1956, Pl. 10). The present structural relations would be the result of squeezing the incompetent shales in these embayments out over the carbonate banks around them.

(2) The outlying masses may be allochthonous, having slid into their present location during the Middle Ordovician, when this part of the carbonate bank sank—we know it did, relatively at least, because the carbonate rocks there are overlain by Middle Ordovician black shale and then by eastward-derived graywacke turbidites. Uplifted source areas from which the masses could have slid are indicated both by these sediments and by erosional unconformities beneath Middle Ordovician rocks to the east (Zen, 1967, and this volume). To be sure, the evidence has been considerably obscured by later

uplift in or west of the supposed source areas, forming the great anticlinoria cored by Precambrian rocks that lie just east of the Valley and Ridge province and its equivalents—the northern Long Range of Newfoundland, the Green Mountains, the Berkshire Highlands, the Hudson-New Jersey Highlands, and the Blue Ridge—and also by metamorphism possibly contemporaneous with their uplift.

The inferred bank edge seems everywhere to lie not too far from the axes of these anticlinoria, generally to the east of them. In Newfoundland, its course should be through White Bay and under the Carboniferous basins that cross the western part of the island. In Quebec, it must have lain somewhere between the rocks that now face each other across the estuary of the Saint Lawrence; perhaps it is now completely buried beneath the great thrust fault, Logan's Line, that bounds the deformed Appalachians in this sector, bringing the pelitic facies within sight of the Canadian Shield. If the above interpretation of the relations in northwestern Vermont is correct, the bank edge lay west of the Green Mountain anticlinorium at the international border but crossed over to it near Milton. In southern Vermont its course must have followed the present anticlinorium or lain slightly to the east, coming into western Massachusetts along the present Berkshire uplift, which lies en echelon to the Green Mountains. In southwestern Connecticut

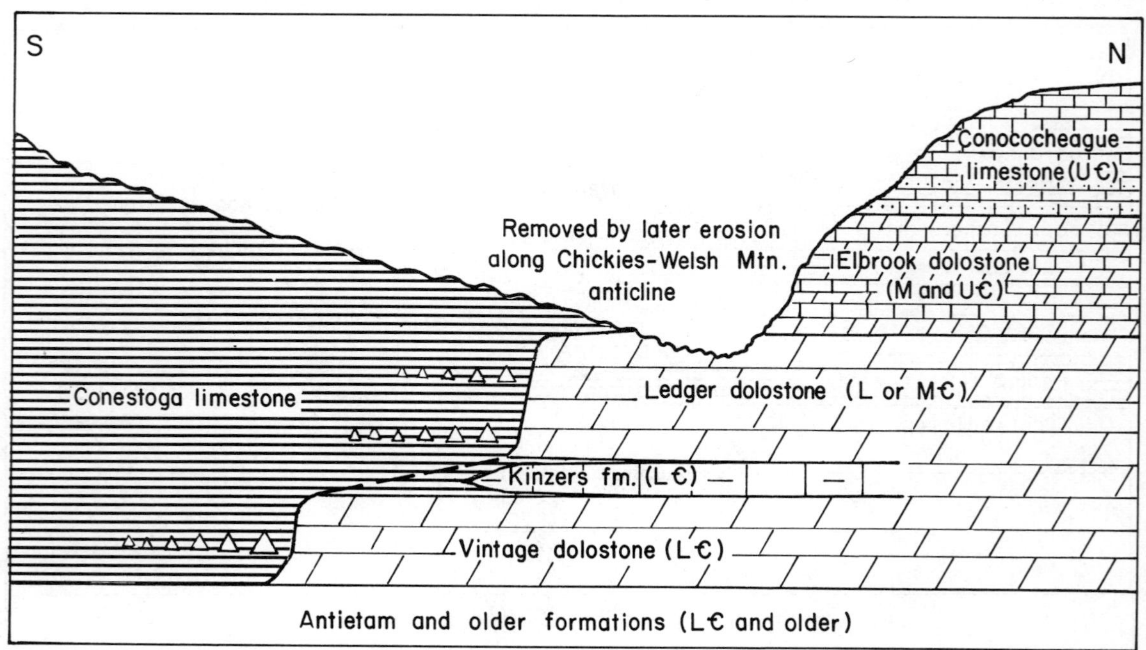

FIGURE 10–2. Stratigraphic diagram, across strike, of Cambrian strata in Lancaster County, Pennsylvania, to illustrate interpretation of Conestoga limestone as deep-water deposit.

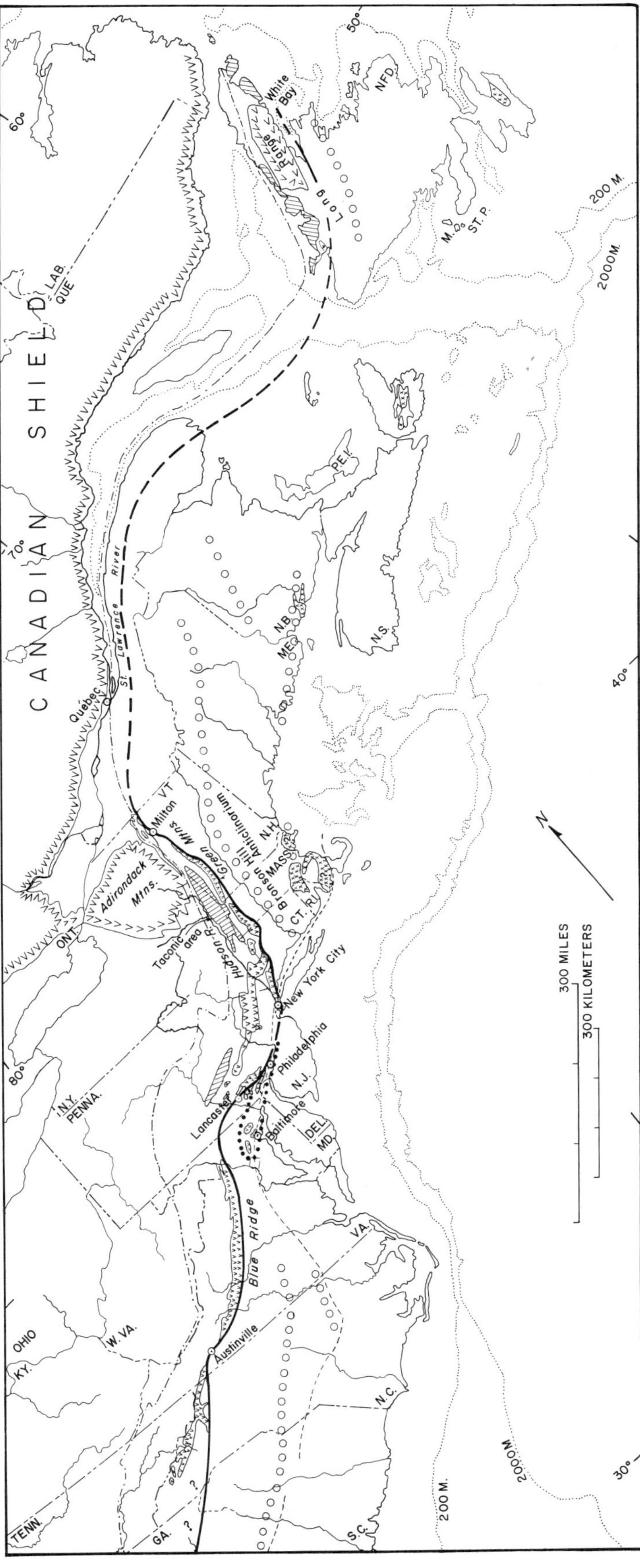

FIGURE 10–3. Eastern edge of North American continent during the Cambrian and Early Ordovician, according to concepts developed in this article. Heavy line, presumed continental border (dashed where concealed); heavy dotted line, extension to include Baltimore gneiss. Check marks, edge of exposures of billion-year-old basement. Dash-dot line, northwestern limit of Valley and Ridge province and its northeastern extension. Ruled pattern, outlying mass of Cambro-Ordovician pelitic sequence ("Taconic" masses). Circles, axes of zones of maximum Paleozoic volcanism. Crosses, exposure of rocks supposed to be about 580 million years old (Avalonian orogeny). Light dashed line, overlap of Coastal Plain sediments.

and southeastern New York, the Inwood marble represents the Cambro-Ordovician carbonate sequence (Hall, this volume); hence the bank edge must swing farther and farther away from the present location of the Hudson Highlands and probably lies along the tectonic line that in this region separates the schists variously called the Berkshire, Waramaug, and Manhattan from the Hartland formation. Hidden beneath Coastal Plain sediments from New York City to Philadelphia, it should then cross the Honeybrook uplift of Precambrian rocks in southeast Pennsylvania into Lancaster County. In Maryland and northern Virginia it would lie near the line of the present Blue Ridge anticlinorium, for the Frederick limestone just to the east appears to resemble the Conestoga limestone of Lancaster County. At Austinville in southwest Virginia, a peculiar facies of the Lower Cambrian, with a near duplicate of the Kinzers formation and the Parker slate, crops out in a slice under the thrust fault at the northwest foot of the Blue Ridge (Currier, 1935; Stose and Jonas, 1939); probably the bank edge passed very near here. Elsewhere in the southern Appalachians, its course is uncertain, for the ages and depositional environments of the various marble belts in the Piedmont province are unknown. One might consider as platform carbonates the Sylacauga marble belt in Alabama, the Murphy marble belt in Georgia and North Carolina, and even the local carbonate slivers along the Brevard zone in the same states; if so, the bank edge would there lie just east of the Blue Ridge anticlinorium and its southwestward prolongation.

From Newfoundland at least to Pennsylvania, then, the edge of the carbonate bank seems to follow the Precambrian anticlinoria in which are found the easternmost exposures of billion-year-old continental basement in the northern Appalachians; this close relation suggests the possibility that the edge of the bank may itself have been localized by the original eastern edge of this basement, i.e., what was then the eastern edge of sialic crust. In other words, the drop-off from bank to deep water may have approximately marked the true edge of the Cambro-Ordovician North American continent, the crust beyond being oceanic, as it is today off the Florida, Bahama, and Yucatán Banks (see Diment, this volume). If so, virtually all that part of the present American continent east of the drop-off must have been converted into continent later, by the several orogenic episodes that affected the northern Appalachians during the Paleozoic (Rodgers, 1967). Along the present Atlantic sea-

board, Precambrian rocks are found again in southeastern Newfoundland, Cape Breton Island in Nova Scotia, southern New Brunswick, and probably eastern Massachusetts and adjacent Rhode Island, but here they are apparently very late Precambrian, the granitic rocks giving ages of about 580 million years, barely half as old as the basement rocks in the nearest part of the North American platform. One might therefore suggest that the process of changing the Appalachian belt into continent began in the late Precambrian Avalonian orogeny on the ocean side of the belt, thus laying the foundations for a Paleozoic non-volcanic island arc out beyond the volcanic one.

So little is known of the main body of the Piedmont province south of the Potomac River that no test of these speculations is possible there. The Baltimore gneiss of Maryland, however, is known to represent billion-year-old sialic basement, and here at least, therefore, the continent must have extended farther than the bank edge delineated above. Moreover the Baltimore gneiss is overlain, over basal clastics, by a carbonate unit, the Cockeysville marble, including some dolostone (Choquette, 1960), perhaps suggesting a carbonate bank this far southeast, but its sedimentary features are obliterated and its age is quite uncertain. A lower Paleozoic age has always been one possibility, strengthened by the general resemblance of stratigraphy and structure northeastward along strike past Philadelphia to the vicinity of New York City, where the Paleozoic age of the Inwood marble is now secure, but Hopson (1964) has given excellent reasons for assigning a late Precambrian age to the Cockeysville and associated formations. As to the older sialic basement, one might rationalize it as an original projection of the continent, enclosing the Conestoga limestone basin of Lancaster County like the Tongue of the Ocean (as shown by the dotted line on Fig. 10–3), or one might suggest that wrench or thrust faulting has displaced an originally simple continental margin.

REFERENCES

Bird, J. M., ed., 1963, Stratigraphy, structure, sedimentation and paleontology of the southern Taconic region, eastern New York: Geol. Soc. America Ann. Mtg. 1963, New York, Gdbk. trip 3, Albany, N. Y., 67 p.

————, and Rasetti, Franco, in press, Lower, Middle, and Upper Cambrian faunas in the Taconic sequence of eastern New York: stratigraphic and biostratigraphic importance: Geol. Soc. America Bull.

————, and Theokritoff, George, 1967, Mode of occurrence of fossils in the Taconic allochthon [abs.]: Geol. Soc. America Spec. Paper 101, p. 248

Bucher, W. H., 1957, Taconic klippe: a stratigraphic-structural problem: Geol. Soc. America Bull., v. 68, p. 657–673

Butts, Charles, 1926, The Paleozoic rocks [of Alabama]: Alabama Geol. Survey Spec. Rept. 14, p. 41–230

————, 1939, Description of the Tyrone quadrangle, Pennsylvania: Penna. Geol. Survey, 4th ser., A–96, 118 p.

————, 1940, Geology of the Appalachian Valley in Virginia: Virginia Geol. Survey Bull. 52, 568 p.

Cady, W. M., 1945, Stratigraphy and structure of west-central Vermont: Geol. Soc. America Bull., v. 56, p. 515–587

————, 1960, Stratigraphic and geotectonic relationships in northern Vermont and southern Quebec: Geol. Soc. America Bull., v. 71, p. 531–576

Choquette, P. W., 1960, Petrology and structure of the Cockeysville marble (pre-Silurian) near Baltimore, Maryland: Geol. Soc. America Bull., v. 71, p. 1027–1052

Cloud, P. E., Jr., and Barnes, V. E., 1948, The Ellenburger group of central Texas: Univ. Texas (Bur. Econ. Geology) Pub. no. 4621, 473 p.

Currier, L. W., 1935, Zinc and lead region of southwestern Virginia: Virginia Geol. Survey Bull. 43, 122 p.

Enos, Paul, in press, Cloridorme formation, Middle Ordovician flysch, northern Gaspé Peninsula, Quebec: Geol. Soc. America Bull.

Hopson, C. A., 1964, The crystalline rocks of Howard and Montgomery Counties: Maryland Geol. Survey, Howard and Montgomery Counties, p. 27–215

Jonas, A. I., and Stose, G. W., 1926, Geology and mineral resources of the New Holland quadrangle, Pennsylvania: Pennsylvania Geol. Survey, 4th ser., A–178, 40 p.

————, and Stose, G. W., 1930, Geology and mineral resources of the Lancaster quadrangle, Pennsylvania: Pennsylvania Geol. Survey, 4th ser., A–168, 106 p.

Kindle, C. H., and Whittington, H. B., 1958, Stratigraphy of the Cow Head region, western Newfoundland: Geol. Soc. America Bull., v. 69, p. 315–342

————, and Whittington, H. B., 1959, Some stratigraphic problems of the Cow Head area in western Newfoundland: New York Acad. Sci. Trans., ser. 2, v. 22, p. 7–18

Knopf, E. B., and Jonas, A. I., 1929, Geology of the McCalls Ferry-Quarryville district, Pennsylvania: U. S. Geol. Survey Bull. 799, 156 p.

LaFleur, R. G., ed., 1961, Guidebook to field trips: N. Y. State Geol. Assoc. 33rd ann. mtg., Troy, N. Y., 95 p.

Lochman, Christina, 1956, Stratigraphy, paleontology, and paleogeography of the *Elliptocephala asaphoides* strata in Cambridge and Hoosick quadrangles, New York: Geol. Soc. America Bull., v. 67, p. 1331–1396

Newell, N. D., 1955, Bahamian platforms: Geol. Soc. America Spec. Paper 62, p. 303–315

Rodgers, John, 1953, Geologic map of East Tennessee with explanatory text: Tennessee Division of Geology Bull. 58, pt. 2, 168 p.

————, 1967, Chronology of tectonic movements in the Appalachian region of eastern North America: Am. Jour. Sci., v. 265, p. 408–427

————, 1968, The Pulaski fault, and the extent of Cambrian evaporites in the central and southern Appalachians: [Cloos volume]

Rodgers, John, and Neale, E. R. W., 1963, Possible "Taconic" klippen in western Newfoundland: Am. Jour. Sci., v. 261, p. 713–730

Ross, M. H., 1949, Source and correlation of the Deepkill conglomerates [abs.]: Geol. Soc. America Bull., v. 60, p. 1973

Ruedemann, Rudolf, 1901, Trenton conglomerate of Rysedorph Hill, Rensselaer County, New York, and its fauna: New York State Mus. Bull. 49, p. 3–114

————, 1934, Paleozoic plankton of North America: Geol. Soc. America Mem. 2, 141 p.

Shaw, A. B., 1958, Stratigraphy and structure of the St. Albans area, northwestern Vermont: Geol. Soc. America Bull., v. 69, p. 519–567

Stone, S. W., and Dennis, J. G., 1964, The geology of the Milton quadrangle, Vermont: Vermont Geol. Survey Bull. 26, 79 p.

Stose, A. J., and Stose, G. W., 1944, Geology of the Hanover-York district, Pennsylvania: U. S. Geol. Survey Prof. Paper 204, 84 p.

Stose, G. W., and Jonas, A. I., 1933, Geology and mineral resources of the Middletown quadrangle, Pennsylvania: U. S. Geol. Survey Bull. 840, 86 p.

————, and Jonas, A. I., 1939, A southeastern limestone facies of Lower Cambrian dolomite in Wythe and Carroll Counties, Virginia: Virginia Geol. Survey Bull. 51, p. 1–30

Trümpy, Rudolf, 1960, Paleotectonic evolution of the central and western Alps: Geol. Soc. America Bull., v. 71, p. 843–907

Whittington, H. B., and Kindle, C. H., in press, Cambrian and Ordovician stratigraphy of western Newfoundland: [to appear in] Proc. Gander (Newf.) Intl. Conf., Am. Assoc. Petroleum Geologists Mem.

Zen, E-an, 1967, Time and space relationships of the Taconic allochthon and autochthon: Geol. Soc. America Spec. Paper 97, 107 p.

Chapter 11

The Lateral Transition from the Miogeosynclinal to the Eugeosynclinal Zone in Northwestern New England and Adjacent Quebec[*]

WALLACE M. CADY

INTRODUCTION

THE COMPLICATED stratigraphic relations between the eugeosynclinal and miogeosynclinal zones in northwestern New England and adjacent Quebec (Fig. 11–1) are shown in restored sections of the lower and middle Paleozoic (Fig. 11–2, henceforth referred to only by mention of the sections—A–A' to F–F'). This region is on the northwest side of the northern part of the Appalachian Mountain belt and in the nearby Hudson, Champlain, and St. Lawrence Valleys, and is adjoined on the northwest by the North American craton whose basement rocks are exposed in the Canadian Shield and Adirondack Mountains.

The Appalachians coincide mainly with the early and middle Paleozoic orthogeosyncline (Kay, 1951, Pl. 1), which was a mobile belt, mostly of subsidence, locally and briefly of stillstand, and less briefly of general uplift. The orthogeosyncline (Cady, 1950, p. 784) includes a broad eugeosynclinal zone, a miogeosynclinal zone to the northwest, several geanticlines, and, after the early Paleozoic, a quasi-cratonic belt (Fig. 11–3). An exogeosyncline (Cady, 1967,

p. 65; Kay, 1951, p. 17–20) is superimposed on the northwestern part of the orthogeosyncline and on adjacent parts of the craton.

The eugeosynclinal deposits, whose maximum thickness is more than 50,000 feet or 15,000 m average at least three times as thick as rocks of the same age in the miogeosynclinal zone. The terms "eugeosynclinal *zone*" and "miogeosynclinal *zone*" emphasize the fact that the eugeosynclinal and miogeosynclinal rocks merge laterally within the orthogeosyncline, at facies transitions whose geographic position changed with time; also that they are not necessarily the contents of discrete and time-persistent geosynclinal troughs (Cady, 1960, p. 557–559; 1967, footnote 1, p. 62; see also Kay, 1951, p. 67; Stille, 1940a, p. 653–654, 656).

ACKNOWLEDGMENTS

The author is indebted to more than half of those contributing to this volume and to twice as many others, totaling over 60 geologists, for many kinds of assistance in connection with the studies upon part of which this review is based. They are being acknowledged individually in a more general paper now in press. Critical review of the manuscript of this paper, by K. B. Ketner, R. H. Moench, George Theokritoff, and E-an Zen, is gratefully acknowledged.

[*] Publication authorized by the Director, U. S. Geological Survey.

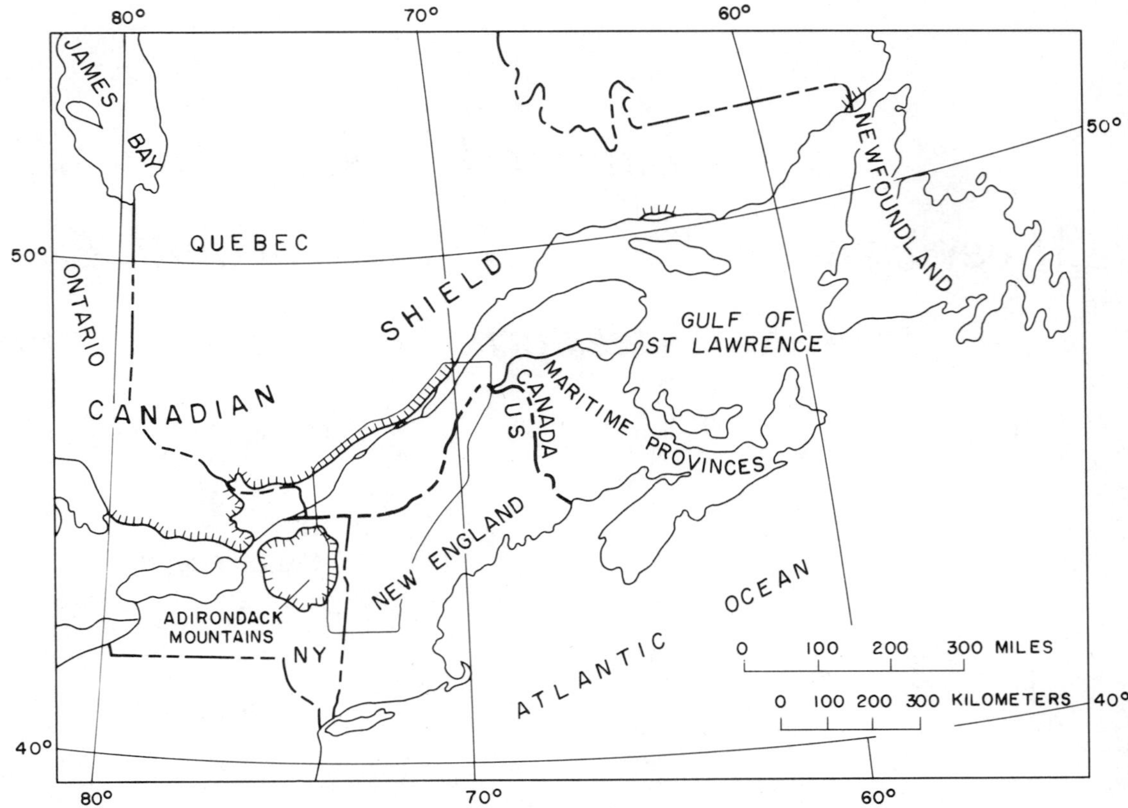

FIGURE 11–1. Index map outlining region of northwestern New England and adjacent Quebec.

ROCKS

Both the eugeosynclinal and miogeosynclinal zones contain variously metamorphosed lower and middle Paleozoic marine bedded rocks that lie unconformably upon the Precambrian basement complex.

The eugeosynclinal rocks are slate, phyllite, schist, metaquartzite, granofels, partly foliate granitic rock, calc-silicate rock, gneiss, greenstone, amphibolite, and metarhyolite—metamorphic derivatives of interbedded shale and graywacke (collectively semipelites), calcareous siltstone, and mafic to felsic volcanic rocks. The volcanic rocks, especially mafic volcanics, are the critical earmarks of the eugeosynclinal zone (Stille, 1940a, p. 14–15; 1940b, p. 8–9; 1950, p. 22–24), which is extrapolated into thick sections dominated by pelites and semipelites that, though barren of volcanics, are otherwise lithically and sequentially similar to sections that contain volcanics.

The miogeosynclinal rocks include marble, quartzite, and slate obviously derived from limestone and dolomite, quartz sandstone, and siltstone and shale that near the margin of the craton are unmetamorphosed. Pelitic and semipelitic rocks dominate the upper deposits of both the miogeosynclinal and eugeosynclinal zones and lap northwestward over the geanticlines and the margin of the craton. The uppermost of these, mainly in the miogeosynclinal zone, grade upward into similar rocks in the exogeosyncline (Sect. B–B′, C–C′) that before they were eroded lapped far onto the craton.

STRUCTURAL RELATIONS

Three geanticlines are recognized, by unconformable overlap and stratigraphic convergence of bedded rock units toward their axes (Cady, 1967, p. 62). The geanticlines were relatively stable tracts of slowed deposition, nondeposition and local erosion during episodic general uplifts of the northwestern half of the orthogeosynclinal belt, and of relatively mild deformation during subsequent orogenic folding. Two of the geanticlines coincide with gravity highs (Diment, 1953; Fitzpatrick, 1959), apparently produced by warping and/or faulting (Diment, 1956; this volume) of a dense crustal layer left structurally high, possibly while adjoining geosynclinal troughs subsided. The lower Paleozoic Vermont–Quebec geanticline is northwest of the axis of the middle Paleozoic (Devonian Acadian) Green Mountain–Sutton Mountain anticlinorium in northwestern Vermont and nearby parts of Quebec, but it swings more into line with the anticlinorium to the northeast in Quebec and crosses it to the south in southern Vermont (Fig. 11–3). The lower Paleozoic Stoke Mountain geanticline coincides with the middle Paleozoic (Acadian) Stoke Mountain anticlinal tract on the southeast limb of the Green Mountain–Sutton Mountain anticlinorium in Quebec and northern Vermont. The lower and middle Paleozoic Somerset geanticline nearly coincides with the northern axial anticline of the middle Paleozoic (Acadian) Bronson Hill–Boundary Mountain anticlinorium. Despite their near coincidence with the anticlinoria the geanticlines, left as structural highs after subsidence of adjoining and intervening geosynclinal troughs, are not to be confused with the anticlinoria produced later by folding under lateral compression. They probably tend to coincide because the early structures guided the different later movement.

Unconformities indicate episodes of stillstand in the miogeosynclinal zone and general uplift of the northwestern part of the orthogeosyncline, followed by erosion and tectonic denudation of the geanticlines in both the eugeosynclinal and miogeosynclinal zones and by repeated folding within the geosynclinal deposits. (See sect. C–C′, D–D′, E–E′, F–F′). The unconformities are at several levels, (1) within the lower Paleozoic especially in Middle Ordovician rocks (Albee, 1957; Ambrose, 1957, p. 164; Cady, 1945, p. 537–539, 560; 1960, p. 564; Cady et al., 1963, p. 26; Lamarche, 1965, p. 60–61, 144–150; Riordon, 1954, p. 8; 1957; Zen, 1964, Pl. 2, sect. A–A′, B–B′; see also Wanless et al., 1966, p. 79; St. Julien, 1963a, p. 30, 32, 125, 143–144, 193–195; 1965; Zen, 1964, p. 28–30; 1967, p. 40–44; this volume), (2) most extensive between the lower and middle Paleozoic marking the Taconic disturbance (Cady, 1960, p. 563; Clark, 1921; Pavlides et al., this volume), and (3) within the middle Paleozoic, especially beneath the Lower Devonian (Boucot, 1961, p. 159, 170; this volume).

Unconformities marking the Devonian Acadian orogeny, responsible for the principal folding that accounts for most of the map patterns and that involved the Precambrian basement as well as the geosynclinal deposits, have been eroded from the region of this review, but one has been reported in northeastern Maine that is referred to the Acadian (Pavlides et al., 1964, p. C33, C37).

There, various steeply dipping and closely folded and cleaved Silurian and Lower Devonian units are overlain unconformably by probable Middle Devonian strata (Mapleton Sandstone) that are themselves gently flexed in an open syncline. An interesting problem in this connection concerns whether the tight folds beneath the unconformity are, as their style suggests, of a geosynclinal fold regime (sect. C–C′, E–E′, F–F′) earlier than that of the larger, more open basement folds such as that of the Green Mountain–Sutton Mountain anticlinorium. If so, the open syncline of Middle Devonian strata may very well be of the late regime, which would place the final pulse of the Acadian orogeny in the late Middle Devonian, or possibly later. Alternatively, the tight folds southeast of the Green Mountain–Sutton Mountain anticlinorium, in Maine, are of the late regime and mark the final pulse of the Acadian orogeny—owing their style to the possibility that the rocks had not been deformed and thus made competent, in earlier regimes (R. H. Moench, oral communication, 1967).

LATERAL TRANSITION AND CORRELATION

Lateral transition from the eugeosynclinal zone northwestward into the miogeosynclinal zone is best described in relation to the three geanticlines and the geosynclinal troughs between and adjoining the geanticlines. The transition zone is in the vicinity of the Vermont–Quebec geanticline in the lower Paleozoic rocks, but moved southeast of the Vermont–Quebec geanticline and the Stoke Mountain geanticline in the middle Paleozoic (Fig. 11–3; Cady, 1967, p. 59–61). It is marked by change in both sedimentary facies (Logan, 1862, p. 323–325) and total thickness of the rocks, and, where geanticlines intervene, by thinning of sections through unconformable overlap and stratigraphic convergence. An alternative explanation for stratigraphic convergence in the zone of transition is that the section is "starved," being too far from sources of clastic sediments, and at the foot of a bank or shelf (Rodgers, this volume; Zen, 1967, p. 47; this volume). This evidence must be weighed with other evidence, including that provided by unconformable overlap (Cady, 1968), but makes no difference with the facts of lateral transition and correlation.

LOWER PALEOZOIC

In the lower Paleozoic rocks the zone of transition coincides with the Vermont–Quebec geanticline, except to the north in Quebec where it trends to the northwest of the geanticline (Fig.

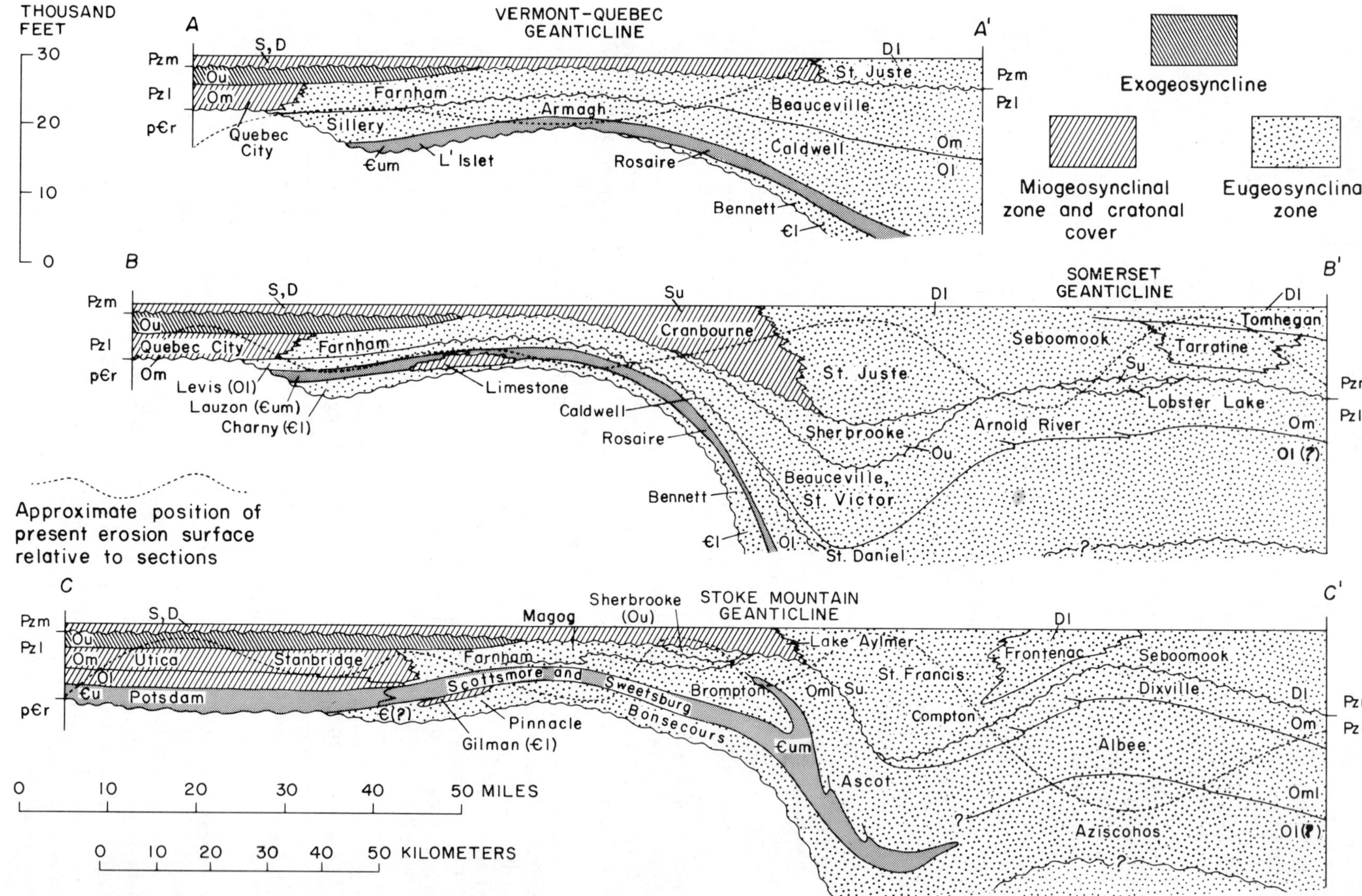
THOUSAND FEET
30
20
10
0
Exogeosyncline
Miogeosynclinal zone and cratonal cover
Eugeosynclinal zone
VERMONT-QUEBEC GEANTICLINE
A
A'
Pzm
Pzl
p€r
S,D
Ou
Om
Quebec City
Farnham
Sillery
€um
L'Islet
Armagh
Rosaire
Bennett
€l
Dl
St. Juste
Beauceville
Caldwell
Om
Ol
Pzm
Pzl
SOMERSET GEANTICLINE
B
B'
Pzm
Pzl
p€r
S,D
Ou
Quebec City
Om
Levis (Ol)
Lauzon (€um)
Charny (€l)
Farnham
Limestone
Caldwell
Rosaire
Bennett
€l
Ol
St. Daniel
Su
Cranbourne
St. Juste
Sherbrooke
Ou
Beauceville,
St. Victor
Dl
Seboomook
Arnold River
Su
Tarratine
Tomhegan
Lobster Lake
Om
Ol (?)
?
Pzm
Pzl
Approximate position of
present erosion surface
relative to sections
STOKE MOUNTAIN GEANTICLINE
C
C'
Pzm
Pzl
p€r
S,D
Ou
Om
Ol
€u
Utica
Potsdam
Stanbridge
Farnham
€(?)
Scottsmore and
Sweetsburg
Bonsecours
Pinnacle
Gilman (€l)
€um
Ascot
Brompton
Magog
Sherbrooke (Ou)
Lake Aylmer
Oml Su
St. Francis
Compton
Dl
Frontenac
Seboomook
Dixville
Albee
Aziscohos
?
?
Dl
Pzm
Om
Pzl
Oml
Ol(?)
0   10   20   30   40   50 MILES
0   10   20   30   40   50 KILOMETERS

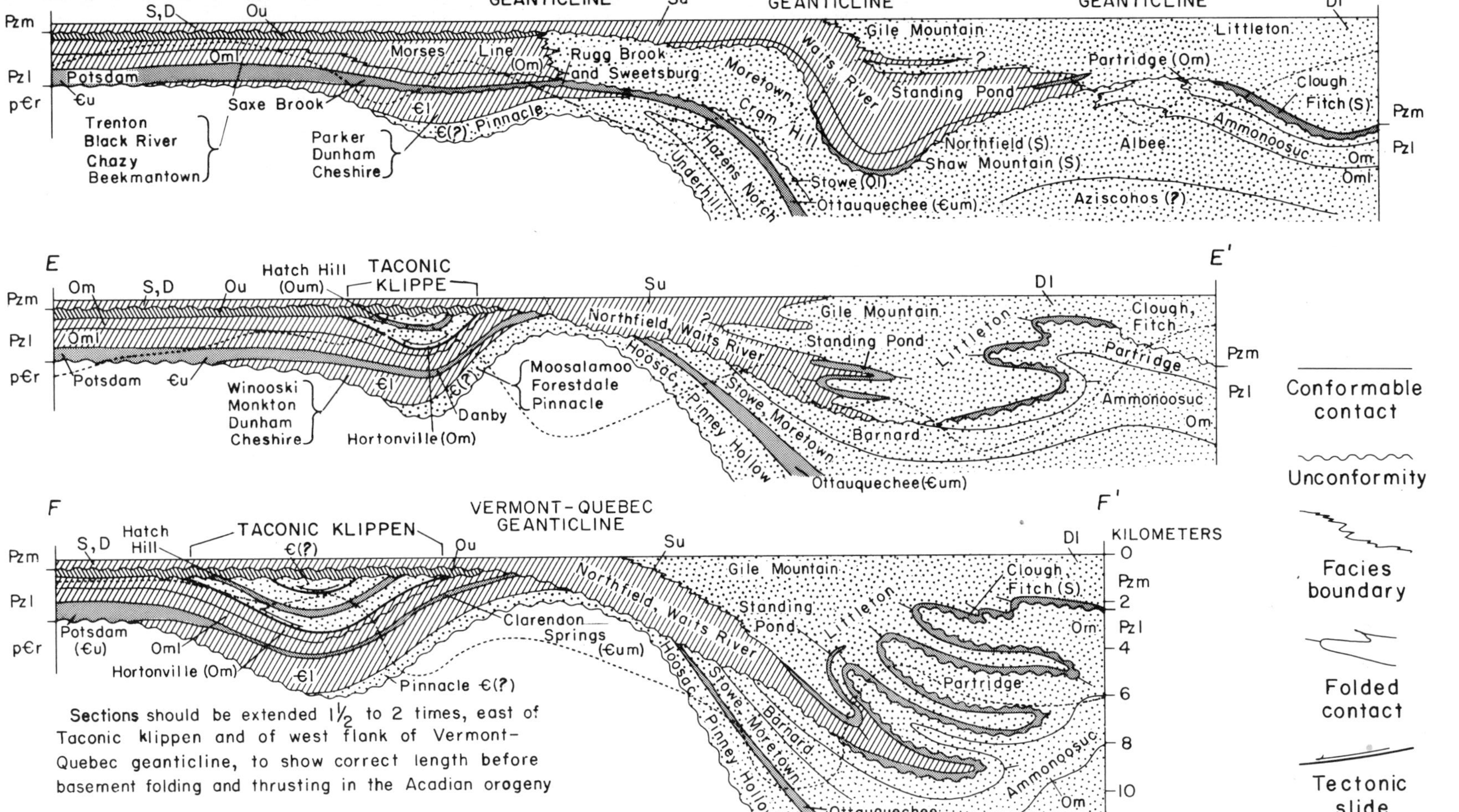

Figure 11–2. Restored sections of the lower (Pzl) and middle (Pzm) Paleozoic of northwestern New England and adjacent Quebec, showing distribution of eugeosynclinal and miogeosynclinal zones and of critical and major stratigraphic units (rank is indicated only in text), before the final episode of Devonian Acadian orogeny. Devonian (D) rocks: Lower (Dl). Silurian (S) rocks: Upper (Su). Ordovician rocks: Upper (Ou); Upper and Middle (Oum); Middle (Om); Middle and Lower (Oml); Lower (Ol). Cambrian (€) rocks: Upper (€u); Upper and Middle (€um); Lower (€l). Precambrian (p€r) rocks.

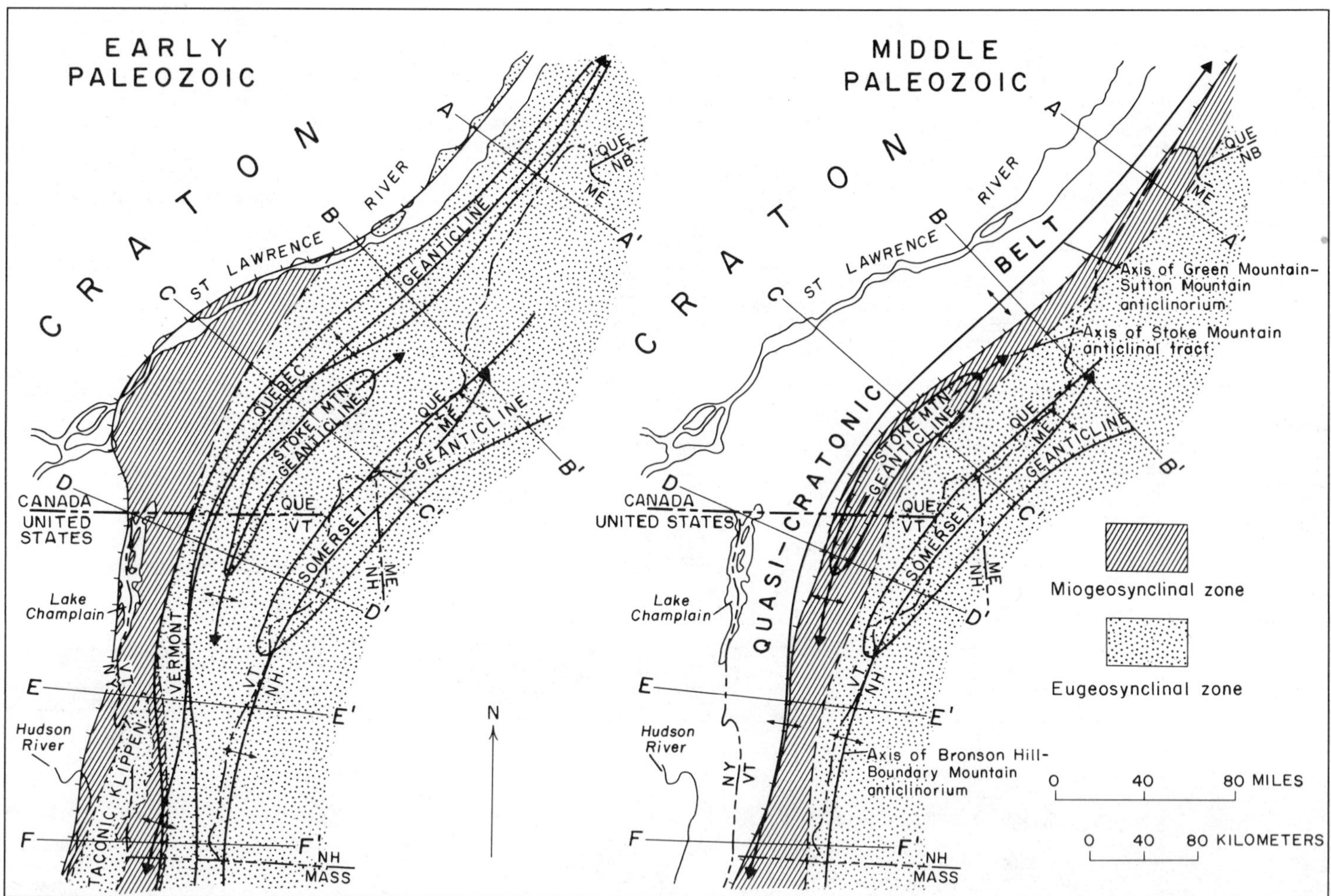

FIGURE 11-3. Paleotectonic maps of the early and middle Paleozoic of northwestern New England and adjacent Quebec showing location of the restored sections.

11–3), and all but part of the Middle Ordovician of the miogeosynclinal zone apparently pinches out between the eugeosynclinal zone and the craton (Cady, 1960, p. 557–558). Stratigraphic correlation between the eugeosynclinal and miogeosynclinal zones is clearest in Quebec, where the slates of the fossiliferous Middle and Upper Cambrian Sweetsburg Formation (Doll et al., 1961; Shaw, 1966a, p. 275; 1966b, p. 845–846), plainly in a miogeosynclinal section at its western extremity (sect. D–D'), are traced northward and eastward into the eugeosynclinal zone (sect. C–C') across the north-plunging axis of the Green Mountain–Sutton Mountain anticlinorium (Fig. 11–3) (Osberg, 1956) and then southward into similar rocks of the Ottauquechee Formation (sect. D–D', E–E', F–F') (Doll et al., 1961, sect. A–A'). To the south and up plunge, especially in Vermont, all but the lowest Cambrian(?) is eroded from axial anticlines of the anticlinorium (sect. C–C', D–D'—note that the anticlines correspond to "swales" in the plots of the present erosion surface). The Sweetsburg is also traced northeastward in Quebec, on both limbs of the anticlinorium and in both flanks of the geanticline, into the Rosaire Group and equivalent L'Islet Formation (sect. A–A', B–B') (Cady, 1960, p. 548–549), where to the northwest especially it includes and is associated with massive quartz sandstone (Béland, 1962, p. 14; Dresser, 1912, p. 14–22) that suggests transition to the miogeosynclinal zone (sect. A–A'). Limestone (sect. B–B') is possibly an unexposed facies of the Rosaire and rocks at lower stratigraphic levels in the Vermont–Quebec geanticline, perhaps the source of massive blocks found to the northwest in the Lower Ordovician Lévis Formation (Osborne, 1956, p. 185–187, 197–198).

Other units in the vicinity of the Vermont–Quebec geanticline, above the Sweetsburg and Ottauquechee and useful in correlation between the eugeosynclinal and miogeosynclinal zones, are Middle Ordovician slates—the Morses Line Formation (sect. D–D') (Shaw, 1958, p. 553–554) in the miogeosynclinal zone and the Magog Slate (sect. C–C') (Berry, 1962) in the eugeosynclinal zone. The Morses Line is traced northeast into the miogeosynclinal Stanbridge Formation (Clark, 1934, p. 6–7; Riva, 1966) and the eugeosynclinal Farnham Formation (sect. C–C'); physical connection with the Magog is eroded at the axial anticline of the Green Mountain–Sutton Mountain anticlinorium (Cady, 1960, p. 550–554; Osberg, 1956; see also Dresser, 1912, p. 27–28; 1913, p. 25; Dresser and Denis, 1944, p. 404).

Beneath the Sweetsburg and Ottauquechee, the Cambrian(?) Pinnacle Formation, Cambrian Forestdale Marble, and Moosalamoo Phyllite in the eugeosynclinal zone and the Lower Cambrian Cheshire Quartzite, Dunham Dolomite, and Winooski Dolomite in the miogeosynclinal zone, on the west flank of the Vermont–Quebec geanticline, grade laterally into the eugeosynclinal Hoosac and Pinney Hollow Formations on the east flank (sect. E–E', F–F'). The Lower Cambrian units are eroded in Vermont (Doll et al., 1961), but show gradational and intertonguing relationships in Quebec, where they connect with the Bonsecours Formation on the east flank of the geanticline (Osberg, 1965, Pl. 1).

Despite the fact that the lower Paleozoic rocks are largely eroded from the Green Mountain–Sutton Mountain anticlinorium (Fig. 11–3) in Vermont, thereby obliterating the zone of transition from miogeosynclinal to eugeosynclinal zone in much of the southern part of the Vermont–Quebec geanticline (sect. E–E', F–F'), it appears that the record is partly preserved to the west in the Taconic klippen (Fig. 11–3) (Doll et al., 1963, p. 95; Theokritoff, this volume; Zen, 1961, p. 331–333) whose rocks probably slid westward from the geanticline in the Middle Ordovician (Cady, 1968; Zen, 1967, p. 66–69; this volume). The Middle(?) and Upper Cambrian Hatch Hill Formation in the western part of the klippen (Theokritoff, 1964, p. 178–179) is much like both the eugeosynclinal Ottauquechee and the miogeosynclinal and eugeosynclinal Sweetsburg Formation, and contains "rotten" weathering sandstones similar to those of the Middle Cambrian Scottsmore (sect. C–C') and Rugg Brook (sect. D–D') Formations (Dennis, 1964, p. 30; Shaw, 1958, p. 536; Stone and Dennis, 1964, p. 42), which lie immediately beneath the Sweetsburg in and west of the Vermont–Quebec geanticline in northwestern Vermont and neighboring Quebec. Below the Hatch Hill in the klippen is the phyllitic Lower Cambrian St. Catherine Formation (not labeled in Figure 11–2) which is much like the schistose eugeosynclinal Underhill Formation (Doll et al., 1961; Theokritoff, 1964, p. 175–176; Zen, 1961, p. 301–302) in the geanticline in northern Vermont (sect. D–D') and like the eugeosynclinal Pinney Hollow Formation (Doll et al., 1963, p. 95) on its east flank in southern Vermont (sect. E–E', F–F'); the St. Catherine sandstones, notably some in the Bomoseen Graywacke (not labeled in Figure 11–2), are like the miogeosynclinal Lower Cambrian Cheshire Quartzite (Dennis, 1964, p. 26–28; Stone and Dennis,

1964, p. 25–28) or upper part of the Gilman Quartzite (Clark, 1934, p. 9) in the west flank of the geanticline in Vermont and Quebec (sect. C–C′, D–D′). The Brezee Formation and the Rensselaer Graywacke at the base of the sections in the klippen (also not labeled in Figure 11–2) probably correlate with the Cambrian(?) Hoosac and Pinnacle (≈ Dalton Formation of Doll and others, 1961) Formations (sect. E–E′, F–F′) respectively in the east and west flanks of the geanticline (Doll et al., 1963, p. 95; Zen, this volume).* Above the Hatch Hill Formation at the top of the sections in the klippen is the slaty Middle Ordovician Pawlet Formation (not labeled in Figure 11–2) comparable to the eugeosynclinal Magog and Farnham, and to the miogeosynclinal Morses Line and Stanbridge discussed above.

Correlation within the lower Paleozoic eugeosynclinal and miogeosynclinal zones is complicated chiefly by differences in terminology, although some significant facies changes and structural barriers to correlation are also encountered.

The principal facies transition within the eugeosynclinal zone is lengthwise of the geosynclinal trough between the Vermont–Quebec and Stoke Mountain geanticlines near the international boundary, where granofels and slate of the Lower(?) and Middle Ordovician Moretown and Cram Hill Members (sect. D–D′) of the Missisquoi Formation (Doll et al., 1961) pass northward into the volcanics and slates respectively of the Brompton Formation and Magog [Slate] (sect. C–C′) (St. Julien, 1961b, p. 4–8; 1963b, p. 4–8). Another significant facies relationship, to the west and lower in the section, is shown by the carbonaceous Lower Cambrian Hazens Notch Formation (sect. D–D′), which grades southward into the Pinney Hollow Formation (sect. E–E′) westward into the Underhill Formation (sect. D–D′), and northward into the Bonsecours Formation (sect. C–C′) (Doll et al., 1961; Osberg, 1965, Pl. 1; see also Rickard, 1965, Fig. 1, p. 525). Volcanics (not shown in Figure 11–2) within the Hazens Notch, notably the Belvidere Mountain Amphibolite Member (Doll et al., 1961) near its top and traceable into other volcanics in adjoining facies, assist in correlation. Similarly, the Lower Ordovician Stowe Formation in Vermont (sect. D–D′) contains volcanics that may be traced northward into the lower part of the Brompton Formation in Quebec (sect. C–C′) and the Middle(?) Ordovician

Coburn Hill Volcanic Member of the Missisquoi Formation (not shown on Figure 11–2; sect. D–D′—in the Moretown and Cram Hill Members) continues into the upper part of the Brompton Formation (equivalent to Bolton Group of Cooke, 1951; see also Ambrose, 1957; Cady et al., 1963, Table 1). The correlation of the Cambrian and Lower and Middle Ordovician Bonsecours, Sweetsburg, Brompton, Ascot, and Magog Formation (sect. C–C′) (Berry, 1962; St. Julien, 1961a, p. 4–7; 1961b, p. 4–8; 1963b, p. 4–8; St. Julien and Lamarche, 1965, p. 4–9) with the Bennett, Rosaire, and Caldwell Groups and the St. Daniel, Beauceville, and St. Victor Formations (sect. B–B′) (St. Julien, 1967, p. 46–47), farther northeast on the southeast flank of the Vermont–Quebec geanticline, is more nearly a semantic than a facies problem. The latter section is followed by the only Upper Ordovician rocks known east of the Vermont–Quebec geanticline—those of the Sherbrooke Formation (Lamarche, 1965, p. 155–157; St. Julien, 1965). In the Somerset geanticline there is a similar problem of longitudinal correlation, northeastward from the Middle Ordovician Ammonoosuc Volcanics and Partridge Formation (sect. D–D′) in New Hampshire into the Dixville Formation (sect. C–C′) in New Hampshire and Maine, and the Arnold River Formation (sect. B–B′) in Maine and Quebec (Green, 1964, p. 23–32; 68–71; Hatch, 1963, p. 12–18, 76; Harwood and Berry, 1967).

The principal structural barriers to transverse correlation within the lower Paleozoic eugeosynclinal zone are Acadian folds and faults northwest of the Vermont–Quebec geanticline (Fig. 11–3) in Quebec, and the broad Acadian Connecticut Valley-Gaspé synclinorium (not labeled in Figure 11–3) between the Vermont–Quebec and Stoke Mountain geanticlines and the Somerset geanticline. The folds and faults present a fairly serious obstacle to correlation between the Cambrian and Ordovician Bonsecours, Gilman, Scottsmore, Sweetsburg, Brompton, and Farnham Formations in the Vermont–Quebec geanticline (sect. C–C′) with the Charny, Lauzon, Lévis, and Quebec City Formations to the northwest near the craton (sect. B–B′) (Osborne, 1956, p. 173–188; see also Béland, 1967; Hubert, 1967). The Connecticut Valley–Gaspé synclinorium, on the other hand, is much less of a barrier because of the striking lithic and sequential similarities of the granofels, volcanic rocks, and slate of the Lower(?) and Middle Ordovician of the Moretown, Barnard Volcanic, and Cram Hill Members of the Missisquoi Formation in the

---

* Zen (1967, p. 95–97) considers the Brezee to be autochthonous Middle Ordovician.

east flank of the Vermont–Quebec geanticline to the Albee Formation, and the Ammonoosuc and Partridge Formations in the Somerset geanticline (sect. D–D′, E–E′, F–F′) (Billings, 1956, p. 94–97; Cady, 1960, p. 550–553; Doll et al., 1961; White, 1946, p. 131–133), and of the Ascot and Magog Formations in the Stoke Mountain geanticline to the Albee and Dixville in the Somerset geanticline (sect. C–C′). Correlation of the schists of the Aziscohos Formation at the base of the known section in the Somerset geanticline (sect. C–C′) with the Middle and Upper Cambrian Ottauquechee and the Lower Ordovician Stowe on the east flank of the Vermont–Quebec geanticline (sect. D–D′) seems likely on the basis of both lithology and sequence (Green, 1964, p. 65–67), though their respective exposures are widely separated by the overlying rock in the Connecticut Valley–Gaspé synclinorium. The transverse correlation of the Sillery Formation and Armagh Group in the northwest flank of the Vermont–Quebec geanticline in Quebec with the Lower Ordovician Caldwell Group in the southeast flank (sect. A–A′) seems entirely a semantic problem (Béland, 1962, p. 12; Dresser, 1912, Map 34A).

The principal facies complications in the miogeosynclinal zone are in the geosynclinal trough west of the Vermont–Quebec geanticline in northwestern Vermont, where the Lower Cambrian Monkton Quartzite (sect. E–E′; not indicated in sect. F–F′) grades laterally northward into the top of the Dunham Dolomite and base of the Winooski Dolomite and the Parker Slate appears at about the same level a little farther north (sect. D–D′; see also Rodgers, this volume). Also the Upper(?) Cambrian Danby Formation and its equivalent the Upper Cambrian Potsdam Sandstone (sect. E–E′) passes north into the Rugg Brook and Sweetsburg Formations (sect. D–D′), which in turn, yet farther north near the international boundary, pass into the Saxe Brook Dolomite (sect. D–D′); to the south it passes into the lower part of the Clarendon Springs Dolomite and top of the Winooski Dolomite (sect. F–F′) (Cady, 1960, p. 538–539, 571–572; Doll et al., 1961). Thrust faults and erosion separate the slates of the Middle Ordovician Morses Line (sect. D–D′) and Stanbridge (sect. C–C′) Formations from the shales of the Cumberland Head, Stony Point, Iberville, and Hathaway Formations (not labeled in Figure 11–2) and the Utica Shale to the west and the Hortonville Slate (sect. E–E′, F–F′) to the south (Cady, 1960, Pl. 3); the lower part of the Stanbridge (Riva, 1966) and also the Morses Line probably correlate with lower Middle Ordovician limestones to the west, in the upper part of the limestone sections containing the Beekmantown, Chazy, Black River, and Trenton Groups (indicated only in sect. D–D′).

MIDDLE PALEOZOIC

The zone of transition from the eugeosynclinal zone to the miogeosynclinal zone in the middle Paleozoic (Cady, 1960, p. 558–559) is in the geosynclinal trough between the Vermont–Quebec and Stoke Mountain geanticlines and the Somerset geanticline (Fig. 11–3), a geosyncline with which the Devonian Acadian Connecticut Valley–Gaspé synclinorium (not labeled in Figure 11–2 or 11–3) closely coincides. Correlation can be carried by west-east comparisons of sequences of similar stratigraphic units (sect. D–D′, E–E′, F–F′), although because of the synclinorium the units cannot be traced across the geosyncline. Thus the Silurian and Devonian Shaw Mountain, Northfield, Waits River, and Gile Mountain Formations to the west are nearly matched by the section Clough, Fitch, and Littleton Formations to the east—both sequences of quartzites and quartz conglomerate, limestone, pelite and semipelite, and sandstone (Billings, 1956, p. 94–97; Cady, 1960, p. 551–553; White, 1946, p. 131–133). In making these correlations the thin Silurian Clough and Fitch Formations, whose protoliths were quartz sandstone and limestone respectively, may be considered genetically as both eastern extremities and as outliers of the middle Paleozoic miogeosynclinal zone, although they are in sequences with thick sections of eugeosynclinal rocks above and below that produce the overall aspect of the eugeosynclinal zone.

The transition is obscured in the upper units, because they are thicker and less distinctive and also because they are exposed in a belt of complex recumbent folding and doming near the axis of the Connecticut Valley–Gaspé synclinorium (Doll et al., 1961). It is further complicated both by lack of primary features indicating tops of beds, such as would make it possible to distinguish facies tongues from recumbent folds (note query in sect. E–E′) (Chang, Ern, and Thompson, 1965, p. 54–62; Ern, 1963, p. 70–75; Goodwin, 1962; 1963, p. 98–101; Woodland, 1965, p. 112–117), and by overlap extinction of at least the Shaw Mountain, Clough, and Fitch Formations by the upper units on the northwest flank of the Somerset geanticline (sect. C–C′, D–D′) (Cady, 1960, p. 551–552). The correlation of the upper rocks is partly tied, however, to tracing of the intermittently exposed mafic Standing Pond Volcanic Member of the Upper Silurian and Lower Devonian Waits River Formation and of its probable correlatives within the Devonian Gile Mountain Formation (sect. D–D', E–E', F–F'). The Standing Pond extends principally north-northeast parallel to the geosynclinal trough, and pinches out laterally westward into interbedded crystalline limestones and slates of the Waits River Formation near the east edge of the miogeosynclinal zone and eastward into the interbedded schist and metaquartzite of the Gile Mountain and its correlative, the Littleton Formation (see Doll et al., 1961; also Johannson, 1963, p. 72; Myers, 1964, p. 19–21).

Correlation within the middle Paleozoic eugeosynclinal and miogeosynclinal zones is complicated by differences in terminology and by facies changes. This is especially true near and north of the international boundary, where the Shaw Mountain, Northfield, Waits River, and Gile Mountain Formations become the Peasley Pond Conglomerate, Glenbrooke Shale, and Sargent Bay Limestone (not labeled in Figure 11–2) and the Lake Aylmer Group northwest of the Stoke Mountain geanticline (sect. C–C') (Cady, 1960, p. 551; Duquette, 1959, p. 252–253) and the St. Francis Group to the southeast (sect. C–C') (Duquette, 1959, p. 253–255; St. Julien and Lamarche, 1965, p. 8–9); and the Gile Mountain Formation (sect. D–D') becomes the Compton Formation (sect. C–C') and grades into the Lower Devonian mafic volcanic Frontenac Formation and a slate, the Seboomook Formation in the vicinity of the Somerset geanticline (Green, 1964, p. 76). The Seboomook and succeeding Lower Devonian units in the Somerset geanticline, including the Tarratine and Tomhegan Formations, form a section (sect. B–B') that unconformably overlies the Upper Silurian Hardwood Mountain (not shown on Figure 11–2) and Lobster Lake Formations (Boucot, 1961). The St. Francis and Seboomook become the St. Juste Group to the northeast along the southeast flank of the Vermont–Quebec geanticline (sect. A–A', B–B'). The Cranbourne Formation (Boucot and Drapeau, in press) is a northwestern facies of the St. Juste.

REFERENCES

Albee, A. L., 1957, Bedrock geology of the Hyde Park quadrangle, Vermont: U. S. Geol. Survey Geol. Quad. Map GQ–102

Ambrose, J. W., 1957, The age of the Bolton lavas, Memphremagog district, Quebec: Naturaliste Canadien, v. 84, p. 161–170

Béland, J. R., 1962, Ste-Perpétue area, Kamouraska and L'Islet counties [Quebec]: Quebec Dept. Nat. Resources, Geol. Surveys Br., Geol. Rept. 98, 22 p.

————, 1967, Contributions from systematic studies of minor structures in the southern Quebec Appalachians, in Clark, T. H., ed., Appalachian Tectonics: Royal Soc. Canada Spec. Pub. 10. p. 48–56

Berry, W. B. N., 1962, On the Magog, Quebec, graptolites: Am. Jour. Sci., v. 260, p. 142–148

Billings, M. P., 1956, The geology of New Hampshire, Part II—Bedrock geology: New Hampshire Plan. Devel. Comm., 203 p.

Boucot, A. J., 1961, Stratigraphy of the Moose River synclinorium, Maine: U. S. Geol. Survey Bull. 1111–E, p. 153–188

————, and Drapeau, Georges, in press, Silurian-Devonian rocks of Memphremagog Lake and the correlatives in the Eastern Townships, Quebec: Quebec Dept. Nat. Resources

Cady, W. M., 1945, Stratigraphy and structure of west-central Vermont: Geol. Soc. America Bull., v. 56, p. 515–587

————, 1950, Classification of geotectonic elements: Am. Geophys. Union Trans., v. 31, p. 780–785

————, 1960, Stratigraphic and geotectonic relationships in northern Vermont and southern Quebec: Geol. Soc. America Bull., v. 71, p. 531–576

————, 1967, Geosynclinal setting of the Appalachian Mountains in southeastern Quebec and northwestern New England, in Clark, T. H., ed., Appalachian Tectonics: Royal Soc. Canada Spec. Pub. 10, p. 57–68

————, 1968, Tectonic setting and mechanism of the Taconic slide: Am. Jour. Sci., v. 266

————, Albee, A. L., and Chidester, A. H., 1963, Bedrock geology and asbestos deposits of the upper Missisquoi Valley and vincinity, Vermont: U. S. Geol. Survey Bull. 1122–B, 78 p.

Chang, Ping Hsi, Ern, E. H., Jr., and Thompson, J. B., Jr., 1965, Bedrock geology of the Woodstock quadrangle, Vermont: Vermont Geol. Survey Bull. 29, 65 p.

Clark, T. H., 1921, A review of the evidence for the Taconic revolution: Boston Soc. Nat. History Proc., v. 36, p. 135–163

————, 1934, Structure and stratigraphy of southern Quebec: Geol. Soc. America Bull., v. 45, p. 1–20

Cooke, H. C., 1951, [Geological map] Magog-Weedon, Quebec: Canada Geol. Survey Map 994A

Dennis, J. G., 1964, The geology of the Enosburg area, Vermont: Vermont Geol. Survey Bull. 23, 56 p.

Diment, W. H., 1953, A regional gravity survey in Vermont, western Massachusetts, and eastern New York: Ph.D. thesis, Harvard Univ., 176 p.

————, 1956, Regional gravity survey in Vermont, western Massachusetts, and eastern New York [abs.]: Geol. Soc. America Bull. v. 67, p. 1688

Doll, C. G., Cady, W. M., Thompson, J. B., Jr., and Billings, M. P., Compilers and Editors, 1961, Centennial geologic map of Vermont: Montpelier, Vermont, Vermont Geol. Survey, scale 1:250,000

————, 1963, Reply to Zen's discussion of the Centennial geologic map of Vermont: Am. Jour. Sci., v. 261, p. 94–96

Dresser, J. A., 1912, Reconnaissance along the National Transcontinental Railway in southern Quebec: Canada Geol. Survey Mem. 35, 42 p.

————, 1913, Preliminary report on the serpentine and associated rocks of southern Quebec: Canada Geol. Survey Mem. 22, 103 p.

————, and Denis, T. C., 1944, Geology of Quebec: Vol. II, Descriptive geology: Quebec Dept. Mines, Geol. Rept. 20, 544 p.

Duquette, Gilles, 1959, Le groupe de Québec et le groupe de Gaspé près du lac Weedon: Naturaliste Canadien, v. 86, p. 243–263

Ern, E. H., Jr., 1963, Bedrock geology of the Randolph quadrangle, Vermont: Vermont Geol. Survey Bull. 21, 96 p.

Fitzpatrick, M. M., 1959, Gravity in the Eastern Townships of Quebec: Ph.D. thesis, Harvard University, 135 p.

Goodwin, B. K., 1962, An alternate interpretation for the structure of east-central and northeastern Vermont: Pennsylvania Acad. Sci. Proc., v. 36, p. 200–207

————, 1963, Geology of the Island Pond area, Vermont: Vermont Geol. Survey Bull. 20, 111 p.

Green, J. C., 1964, Stratigraphy and structure of the Boundary Mountain anticlinorium in the Errol quadrangle, New Hampshire-Maine: Geol. Soc. America Spec. Paper 77, 78 p.

Harwood, D. S., and Berry, W. B. N., 1967, Fossiliferous lower Paleozoic rock in the Cupsuptic quadrangle, west-central Maine: U. S. Geol. Survey Prof. Paper 575–D, p. D16–D23

Hatch, N. L., 1963, The geology of the Dixville quadrangle, New Hampshire: New Hampshire Dept. Resources and Econ. Development Bull. 1, 81 p.

Hubert, Claude, 1967, Tectonics of part of the Sillery Formation in the Chaudière-Matapédia segment of the Québec Appalachians, in Clark, T. H., ed., Appalachian Tectonics: Royal Soc. Canada Spec. Pub. 10, p. 33–40

Johansson, W. I., 1963, Geology of the Lunenburg-Brunswick-Guildhall area, Vermont: Vermont Geol. Survey Bull. 22, 86 p.

Kay, Marshall, 1951, North American geosynclines: Geol. Soc. America Mem. 48, 143 p.

Lamarche, R. Y., 1965, Géologie de la Région de Sherbrooke, comté de Sherbrooke, Quebec: D. Sc. thesis, Laval Univ. 291 p.

Logan, W. E., 1862, Considerations relating to the Quebec Group, and the upper copper-bearing rocks of Lake Superior: Am. Jour. Sci., 2d. ser., v. 33, p. 320–327

Myers, P. B., Jr., 1964, Geology of the Vermont portion of the Averill quadrangle, Vermont: Vermont Geol. Survey Bull. 27, 69 p.

Osberg, P. H., 1956, Stratigraphy of the Sutton Mountains, Quebec; key to stratigraphic correlation in Vermont [abs.]: Geol. Soc. America Bull., v. 67, p. 1820

————, 1965, Structural geology of the Knowlton–Richmond area, Quebec: Geol. Soc. America Bull., v. 76, p. 223–250

Osborne, F. F., 1956, Geology near Quebec city: Naturaliste Canadien, v. 83, p. 157–223

Pavlides, Louis, Mencher, Ely, Naylor, R. S., and Boucot, A. J., 1964, Outline of the stratigraphy and tectonic features of northeastern Maine: U. S. Geol. Survey Prof. Paper 501–C, p. C28–C38

Rickard, M. J., 1965, Taconic orogeny in the western Appalachians: experimental application of microtextural studies to isotopic dating: Geol. Soc. America Bull., v. 76, p. 523–535

Riordon, P. H., 1954, Preliminary report on Thetford Mines-Black Lake area, Frontenac, Megantic, and Wolfe counties [Quebec]: Quebec Dept. Mines, Mineral Deposits Br., Prelim. Rept. 295, 23 p.

————, 1957, Evidence of a pre-Taconic orogeny in southeastern Quebec: Geol. Soc. America Bull., v. 68, p. 389–394

Riva, John, 1966, Upper Lévis graptolites from Cowansville, southern Quebec: Jour. Paleontology, v. 40, p. 220–221

St. Julien, Pierre, 1961a, Preliminary report on Fraser Lake area, Shefford and Stanstead counties [Quebec]: Quebec Dept. Mines, Mineral Deposits Br., Prelim. Rept. 439, 11 p.

————, 1961b, Preliminary report on Lac Montjoie area, Sherbrooke, Richmond and Stanstead counties [Quebec]: Quebec Dept. Nat. Resources, Mineral Deposits Br., Prelim. Rept. 464, 14 p.

————, 1963a, Geologie de la region Orford-Sherbrooke [Quebec]: D. Sc. thesis, Laval Univ., 369 p.

————, 1963b, Preliminary report on Saint Elie D'Orford area, Sherbrooke and Richmond counties [Quebec]: Quebec Dept. Natural Resources, Mineral Deposits Br., Prelim. Rept. 492, 14 p.

————, 1965, [Geological Map] Orford-Sherbrooke area, Richmond, Sherbrooke, Shefford, Brome, Stanstead, and Compton counties [Quebec]: Quebec Dept. Natural Resources, Mineral Deposits Service, Map 1619

————, 1967, Tectonics of part of the Appalachian region of southeastern Quebec southwest of the Chaudière River, in Clark, T. H., ed., Appalachian Tectonics: Royal Soc. Canada. Spec. Pub. 10, p. 41–47

————, and Lamarche, R. Y., 1965, Geology of Sherbrooke area, Sherbrooke county [Quebec]: Quebec Dept. Natural Resources, Mineral Deposits Service, Prelim. Rept. 530, 34 p.

Shaw, A. B., 1958, Stratigraphy and structure of the St. Albans area, northwestern Vermont: Geol. Soc. America Bull., v. 69, p. 519–567

————, 1966a, Paleontology of northwestern Vermont. X. Fossils from the (Cambrian) Skeels Corners Formation: Jour. Paleont., v. 40, p. 269–295

————, 1966b, Paleontology of northwestern Vermont. XI. Fossils from the Middle Cambrian St. Albans Shale: Jour. Paleont., v. 40, p. 843–858

Stille, Hans, 1940a, Einführung in den Bau Amerikas: Berlin, Gebrüder Borntraeger, 717 p.

————, 1940b, Zur Frage der Herkunft der Magmen: Preuss. Akad Wiss. Abh. 1939, no. 19, 31 p.

————, 1950, Der "subsequente" Magmatismus: Deutsche Akad. Wiss. Berlin, Kl. Math. u. Naturw. 1950, Abh. Geotektonik 3, 25 p.

Stone, S. W., and Dennis, J. G., 1964, The geology of the Milton quadrangle, Vermont: Vermont Geol. Survey Bull. 26, 79 p.

Theokritoff, George, 1964, Taconic stratigraphy in northern Washington county, New York: Geol. Soc. America Bull., v. 75, p. 171–190

Wanless, R. K., Stevens, R. D., Lachance, G. R., and Rimsaite, J. Y. H., 1966, Age determinations and geological studies. Report 6, K–Ar isotopic ages: Canada Geol. Survey Paper 65–17, 101 p.

White, W. S., 1946, Structural geology of the Vermont portion of the Woodsville quadrangle: Ph.D. thesis, Harvard Univ., 166 p.

Woodland, B. G., 1965, The geology of the Burke quadrangle, Vermont: Vt. Geol. Survey Bull. 28, 151 p.

Zen, E-an, 1961, Stratigraphy and structure at the north end of the Taconic Range in west-central Vermont: Geol. Soc. America Bull., v. 72, p. 293–338

————, 1964, Stratigraphy and structure of a portion of the Castleton quadrangle, Vermont: Vermont Geol. Survey Bull. 25, 70 p.

————, 1967, Time and space relationships of the Taconic allochthon and autochthon: Geol. Soc. America Spec. Paper 97, 107 p.

Chapter 12

The Stratigraphy and Structure of the Sutton Area of Southern Quebec

T. H. CLARK AND P. R. EAKINS

INTRODUCTION

THE SUTTON AREA is an important part of the western Appalachian mountain system of southern Quebec because the type localities for many formations of regional importance were first described there by Clark in 1934. A quadrangle map of most of the area, largely based upon Clark's work, was published by the Geological Survey of Canada, as part of Paper 63–34, containing a brief description of the main geological features and much additional structural data (Eakins, 1964). The area extends into the southeastern corner of the adjoining Lacolle quadrangle, which covers most of the important carbonate sequence of the Philipsburg succession. This quadrangle was mapped by Clark and McGerrigle in 1928–30 but has yet to be published. Because this paper deals with the deformed Appalachian terrain, no attention will be paid to the foreland part of the Sutton area west of Logan's Line.

The history of work in the Sutton area and the Appalachian region in Southern Quebec as a whole is long, involved, and extensive, dating back to reports by Sir William Logan. A review of this history, of the many stratigraphic and structural complications, and of the problems of correlation of stratigraphic units on a regional basis—often involving the Sutton area as the central part because of Clark's early formational descriptions—is presented in a monumental review article by Cady (1960). Aspects of Sutton geology discussed by Cady need not be reviewed here.

Of more recent works, particular reference must be made to the work of Rickard, who carried out a three-year study (1957–60) of the eastern half of the area under discussion, and who has adduced important evidence on the tectonics of this section of the Appalachians, both through personal communications to the authors (Eakins in particular) and through publications (1961; 1965). In addition, Roth (1965) and Williams (1966) have completed detailed studies of parts of the Sutton area here considered, under a grant to Eakins from the Geological Survey of Canada.

The major and most interesting features in the Sutton area which the authors intend to illustrate here are as follows:

1. A wide range of stratigraphic change and variety within the area, involving the eugeosynclinal–miogeosynclinal boundary in this section of the Appalachian geosyncline (Cady, 1960; this volume).

2. The influence of lithology and depth of burial on the style and variety of deformation in the various stratigraphic and structural units.

3. The amazing structural complexity in a terrain whose gross structure is, in general, simple.

No attempt will be made to give detailed stratigraphic descriptions or correlations, either on a local or regional basis, because they have already been adequately covered elsewhere. Basic stratigraphic descriptions accompany the explanation of Eakins' (1964) map.

MAIN ELEMENTS OF THE SUTTON AREA

The Sutton area as here treated extends geographically from the eastern shores of Lake

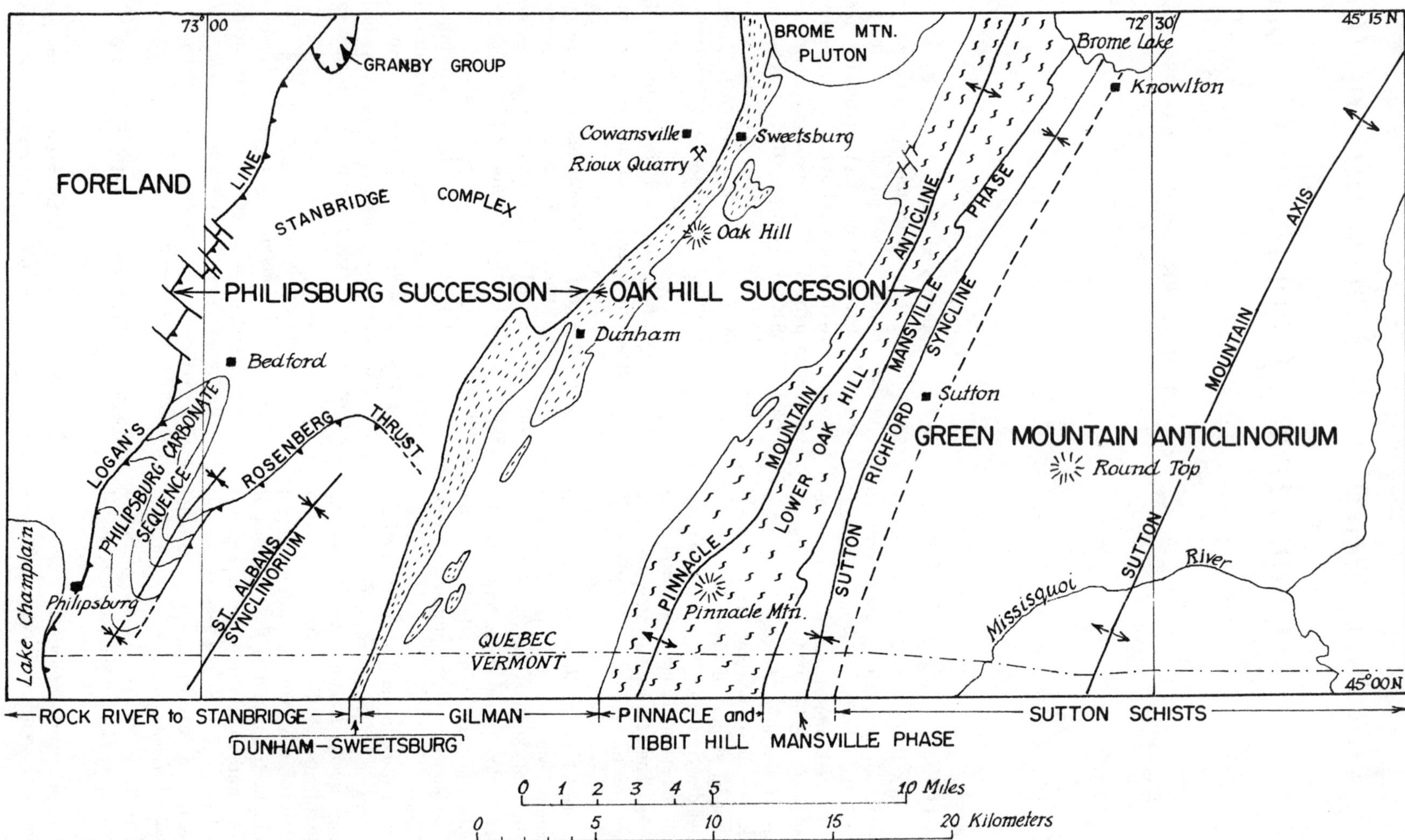

FIGURE 12–1. Map of the Sutton area showing the distribution of structural and stratigraphic units (after Clark, 1934, Eakins, 1964, and Rickard, 1965). For cross section along international boundary, see figure 12–2.

Champlain to the chain of summits known as the Sutton Mountains (see Fig. 12–1). It is bounded on the south by the Canada–United States international boundary, separating Quebec and Vermont, and on the north by latitude 45°15'.

Geologically the area is bounded on the west by the major thrust fault zone, sometimes known as Logan's Line, which separates rocks of the folded geosynclinal assemblage from those of the relatively undeformed foreland of the St. Lawrence Lowlands. It is bounded on the east by the axis of the Green Mountain–Sutton Mountain anticlinorium. West of this anticlinorial axis only lower Paleozoic rocks exist, whereas to the east Silurian and Devonian rocks are present as well.

The stratigraphy of the Sutton area involves many units, shown in Table 12–1. A facies change from a miogeosynclinal dolomite-quartzite-shale sequence in the west to a typical eugeosynclinal shale-greywacke-volcanic sequence in the east is displayed within the area.

The various structural-stratigraphic units of the Sutton map-area occur in north-northeast trending belts which from west to east are (Figs. 12–1 and 12–2):

1. The Foreland (St. Lawrence Lowland succession; not considered in this paper)
2. The St. Albans synclinorium (Philipsburg succession)
3. The Pinnacle Mountain anticline (Oak Hill succession)
4. The Sutton–Richford syncline (Mansville phase of the Oak Hill succession)
5. The Green Mountain–Sutton Mountain anticlinorium (Sutton Schists)

The St. Albans synclinorium (Philipsburg succession)

The westernmost structural and stratigraphic unit of the Appalachian system in the Sutton area consists of a broad synclinorial belt of slates with at least two thrust slices of Upper Cambrian to Lower Ordovician carbonate rocks on its western flanks (Clark, 1934; Shaw, 1958). Its western boundary is Logan's Line, known to the south in Vermont chiefly as the Champlain fault, and exposed within this area only on the shores of Lake Champlain at Philipsburg. The stratigraphic succession in the westernmost, or Philipsburg, slice is given by Clark (1934), but no formal definitions of the formations have appeared in print. Above the Rock River Formation, probably of Upper Cambrian age, there follows a thick sequence of carbonate rocks, of which the Ordovician Strites Pond, Wallace Creek, and Morgan Corner Formations crop out within the Sutton area, while the others occur only in the neighboring Lacolle map-area. The beds are disposed in a shallow syncline (Eakins, 1964). The formations of the Upper Cambrian to Upper Ordovician Philipsburg slice are distinct from contemporaneous Foreland formations as to lithology, fossils and thicknesses, and can best be considered as developments within the miogeosyncline. The Mystic Conglomerate, or Breccia, contains limestone blocks of Chazyan ages, with faunas distinct from those of the foreland, and is interpreted as the result of disruption consequent upon submarine sliding of carbonates deposited farther east. The uppermost formation is the Stanbridge Slate of Middle and Late Ordovician age, a widespread development of cleaved shales with limestone lenses, having affinities with the Trenton, probably with the Utica and certainly with the Lorraine. These are certainly miogeosynclinal deposits, and probably were continuous with the shales of the Stony Point and Iberville Formations of the foreland.

Although the thicknesses of the carbonate rocks (Strites Pond through Mystic) are well known, and total nearly 2,500 feet (800 m), the thickness

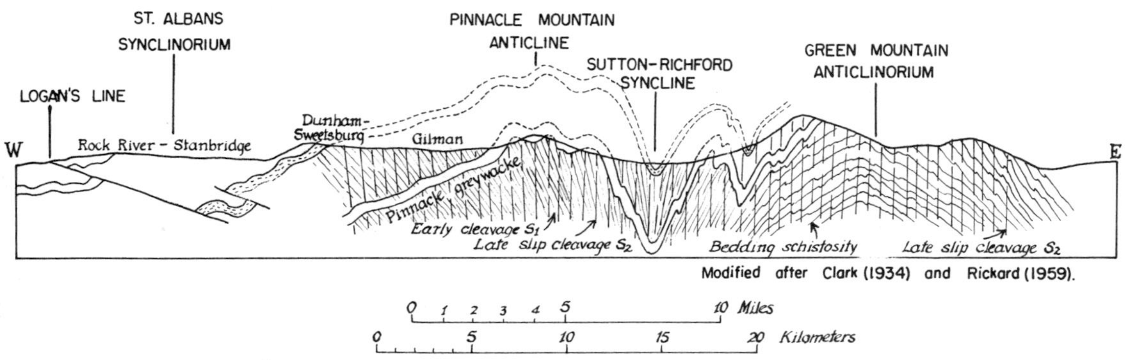

FIGURE 12–2. Cross section of the Sutton area along the international boundary.

TABLE 12-1. Formations exposed in the Sutton area. Members shown in small type.

STANDARD GROUPS ETC.	FORELAND		GRANBY SLICE	ST. ALBANS SYNCLINORIUM PHILIPSBURG-ROSENBURG SLICES		PINNACLE ANTICLINE OAK HILL SEQUENCE	SUTTON-RICHFORD SYNCLINE	SUTTON MOUNTAIN ANTICLINE SUTTON SCHISTS
LORRAINE	NICOLET RIVER	ST. GERMAIN COMPLEX						
TRENTON-UTICA	IBERVILLE STONY POINT LACOLLE			STANBRIDGE (FORMATION & COMPLEX)	MORSE'S LINE			
CHAZY	LAVAL			MYSTIC				
BEEKMANTOWN	BEAUHARNOIS		GRANBY GROUP ("SILLERY")	BASSWOOD CREEK COREY ST. ARMAND SOLOMONS CORNER LUKE HILL NAYLOR LEDGE HASTINGS CREEK MORGAN CORNER WALLACE CREEK STRITES POND				
UPPER AND MIDDLE CAMBRIAN	MARCH POTSDAM			ROCK RIVER	ROCK RIVER SWEETSBURG SAXE BROOK	SWEETSBURG Vail	SUTTON SCHISTS	OTTAUQUECHEE
LOWER CAMBRIAN						DUNHAM Scottsmore Oak Hill GILMAN Gilman phyllite West Sutton WHITE BROOK PINNACLE Call Mill Tibbit Hill	MANSVILLE PHASE ALONG WESTERN BORDER	BONSECOURS
PRECAMBRIAN								

of the highly deformed Stanbridge is unknown. Eakins' (1964) use of the term St. Germain Complex to include Clark's Stanbridge Formation and its correlate, the Morse's Line Formation in Vermont, seemed justified by the similarities in the structural and stratigraphic complexities of the slates both east and west of Logan's Line. However, it appears preferable to restrict the term St. Germain Complex to its original meaning—the deformed slates west of the bounding fault (Logan's Line: see Clark, 1964) and to revert to the use of Stanbridge Formation or Complex for the deformed beds east of the fault. From a structural viewpoint the Stanbridge presents a fairly homogeneous assemblage of pelites and carbonates which, except at a very few localities, yield no fossils upon which an age can be based. Nevertheless, they have a structural tenor which, independent of their stratigraphical position, allows them to be grouped as the Stanbridge Complex.

A small tongue of red Sillery Slate in the northwestern corner of the Sutton area is the southernmost extension of the Granby Group, composed largely of red slates and greenish gray sandstones and grits. These, it is believed, were deposited between their present position and the Green Mountain–Sutton Mountain anticlinorium; subsequent to the uplift of the anticlinorium, they moved westward by gravity sliding, deforming among other formations the incompetent Ordovician shales and thereby giving rise to the Stanbridge Complex, over which they came to lie as a veneer. The source of the sediments of the Granby Group is unknown. The large, fresh-looking, angular feldspar grains indicate rapid disintegration of the parent rock and short transportation before deposition. A Cambrian to Early Ordovician age is as likely as any, in the absence of paleontological evidence. If their sliding were responsible for the soft-sediment contortion of the Stanbridge slates, they must predate the latest Ordovician. The association of similar rocks near Quebec City with graptolite-bearing shales (Lévis) indicates that there they are no younger than Early Ordovician.

The structure within the western carbonate areas is relatively simple and consists of anticlines and synclines inclined or slightly overturned to the west and complicated by minor subsidiary thrusts. Throughout the rest of the belt, which is underlain by slates and occasional beds of limestone, outcrops are of poor quality, scattered, and few in number, with insufficient evidence for a structural analysis at the present time. Numerous small-scale folds are present throughout the slate terrain, with many different plunges and cleavage orientations; the rocks have obviously undergone several deformations. The complicated structure of this belt of slates is well illustrated by the rocks within the Rioux Quarry.

The following account of the geology of this quarry is based upon work by Williams (1966) and Eakins (unpublished). The quarry is in a succession of layers of grey impure limestone about 2 feet (0.6 m) thick, interbedded with fine-grained dark grey to black graphitic shales. Early Middle Ordovician graptolites have been collected from shales from one section of the quarry (Riva, 1966). The beds have been thrown into a number of broad, open folds plunging gently to the north-northeast and are cut by several faults. A well-developed crenulation cleavage (Rickard, 1961) is present as an axial-plane foliation related to these folds. Detailed study reveals an extremely complicated structural mélange, reflecting the following sequence of events:

1. Sliding and slumping, mainly westward, of thick units of water-saturated sediments shortly after deposition.

2. Development of an early calcite-filled conjugate joint system in the limestone beds.

3. Formation of quartz-filled tension fractures and gashes associated with movements along early faults.

4. Compression normal to the regional fold axis, resulting in the development of a crenulation cleavage and open flexural-flow folds (see below). This compression constituted the main deformational phase, which was accompanied by additional conjugate jointing developed in limestone beds.

5. Four additional minor structural events, whose chronology is uncertain:

    a. development of a few ill-defined kink zones, affecting cleavage orientation;

    b. intrusion of a 2-foot basic dyke presumably associated with the Monteregian intrusive rocks of Cretaceous age;

    c. further cross jointing;

    d. development of a young fault with gouge.

The development of the early folds that are penecontemporaneous with deposition was probably synchronous with the early folding in the Pinnacle Mountain anticlinorial zone to the east, involving the Oak Hill succession. In the Oak Hill rocks such early folds as developed have in general the same structural form and face west, but associated with them is a well-developed early slaty cleavage. In the Rioux Quarry, the later upright open folds with the well-developed crenulation cleavage are undoubtedly coeval with the late open upright folds and crenulation cleavage of the Oak Hill succession to the east.

The Pinnacle Mountain Anticline (Oak Hill succession)

This belt contains the classical stratigraphic succession of Clark (1934). It has been designated the Enosburg Falls anticline in Vermont (Doll et al., 1961), and the Tibbit Hill anticline in Quebec (Rickard, 1965), but it is felt here that the designation Pinnacle Mountain anticline is more appropriate because of the outstanding topographic feature represented by that mountain. In Vermont, the rocks of the Pinnacle Mountain belt are separated from those of the St. Albans synclinorium by a well-defined thrust and klippen zone. In the Sutton area only local thrusting, if any, characterizes the contact zone.

The Pinnacle Mountain anticlinal zone, underlain by the Oak Hill succession, is bounded on the east by the eastern limit of the main Tibbit Hill outcrop belt and on the west by the eastern margin of the Stanbridge Formation. The eastern margin of this zone is a conformable contact against a remarkable miniature Oak Hill succession, designated the Mansville phase of the Oak Hill succession by Clark (1934). At the western margin, the actual Oak Hill–Stanbridge contact is nowhere seen in the field; to the south in Vermont the contact is a major thrust plane, but this thrust appears to die out in the Sutton area and the Middle and Upper Ordovician Stanbridge rocks may simply overlie the Cambrian Oak Hill disconformably.

The Oak Hill beds exposed in the Sutton area have been described by Clark (1936) and need not be redescribed here. Largely for convenience in mapping, Eakins has suggested, with the concurrence of Clark, a grouping of Clark's ten formations into five, relegating the remaining five to member status (see Table 12–1). Earlier Clark had suggested a grouping of the formations into pelite-psammite-carbonate cycles. The Oak Hill formations can be followed for more than 50 miles (80 km), both to the north and the south of the Sutton area.

The presence of these formations in the Mansville Phase and in part in the Sutton Schists to the east shows their very widespread distribution over the central and eastern parts of the area. The differentiation of miogeosynclinal, eugeosynclinal and Mansville phases seems to have existed very early, for it is likely that Oak Hill rocks were the earliest geosynclinal sediments to be deposited within the Sutton area.

Obscure algal forms and brachiopods belonging to the genus *Kutorgina* have been found in the Gilman Formation, and fragments of olenellid trilobites and *Hyolithes* come from the Dunham Dolomite. Certainly the middle part of the series is Early Cambrian in age; probably the lower part is as well, but the upper part (Sweetsburg) may be of Middle or even Upper Cambrian age.

The Oak Hill rocks have clearly undergone two major deformations. The first deformation formed folds with a slaty cleavage as axial plane foliation (Fig. 12–2). These early folds are open and upright in what is now exposed as the basal Pinnacle Mountain section and become progressively more inclined westward, until overturning occurs and axial planes dip moderately eastward in the Sweetsburg section of the present Oak Hill outcrop band. The westward fanning is consistent along the length of the Oak Hill succession in the Sutton area. At least minor thrust faults developed in the western section during this early deformation; many folds have steep, short west limbs and long, more gently dipping east limbs.

A later deformation produced a new series of upright open folds across the belt with the development of a second axial plane cleavage (Rickard, 1961, 1965) (Fig. 12–3). The early and late cleavages are not readily distinguishable in the eastern part of the belt along the Pinnacle Mountain anticlinal axis, where the later deformation was more intense and strikes and dips of cleavages are more nearly parallel; but they are readily separable to the west where there is a strong divergence in dip. The later cleavage is not uniformly developed across the belt and is certainly not as strongly developed to the west as to the east.

The response of the different Oak Hill lithologies to deformation is noteworthy: each lithology—lava, greywacke, mudstone, siltstone, sandstone and dolomite—has been affected differently. Overall, a marked variation exists in the amount of structural detail and complexity in the different lithological units.

Kink or knick layering of cleavage is observed in the Gilman Formation at widely scattered localities (Fig. 12–4). These kink fold layers are only readily discernible in first class exposures such as road cuts and other artificial openings: observations are too few and far between to permit any statements about their distribution or significance.

The Tibbit Hill lava varies from a weakly foliated, nearly featureless greenstone to a strongly foliated schist. Although amygdules, both undeformed and highly stretched, are common, other primary volcanic features such as flow contacts, pillows, and flow layering are remarkable for their absence. Primary features such as bedding and crossbedding are well pre served, although often highly deformed, in the exposures of the Pinnacle greywacke. On the other hand, the primary features of the Gilman were so strongly affected by deformation that bedding and fold forms are hard to find. Cleavage has completely disrupted the bedding to form a phyllitic rock that contains lenticles of quartzitic material representing the remains of these sandy intercalations.

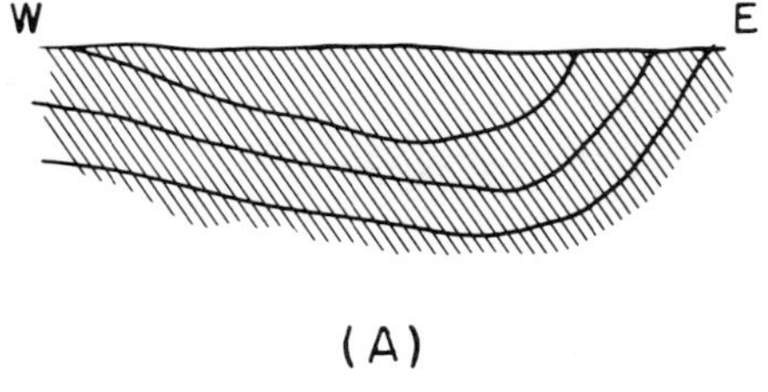

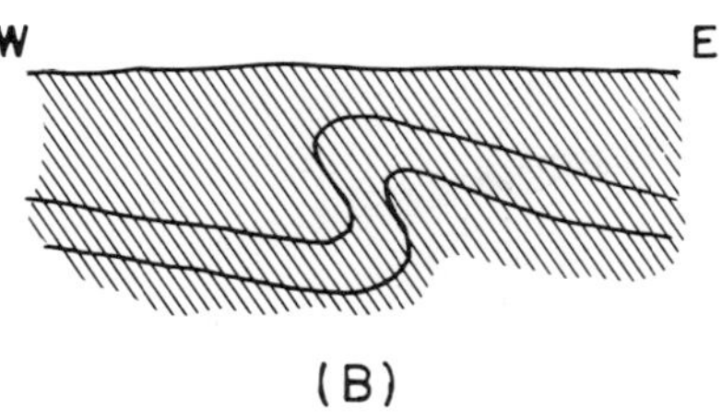

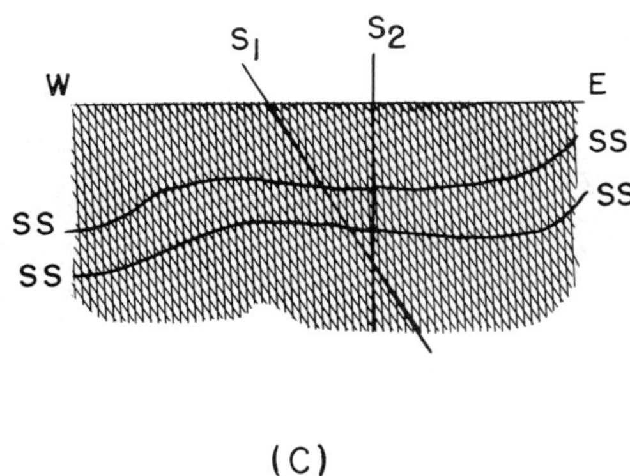

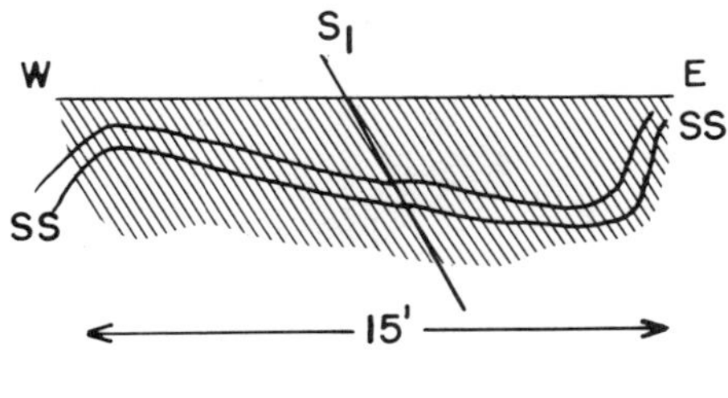

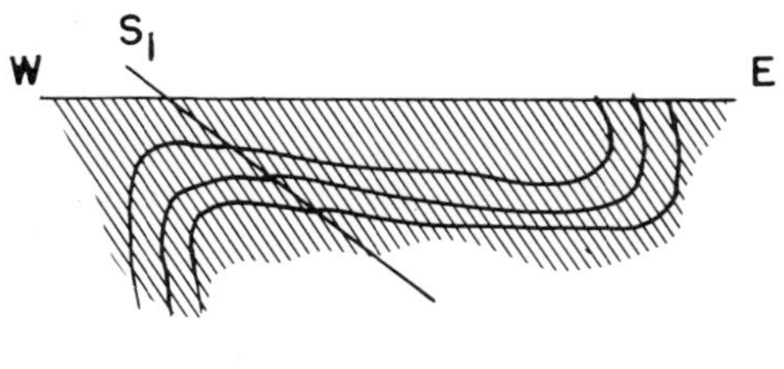

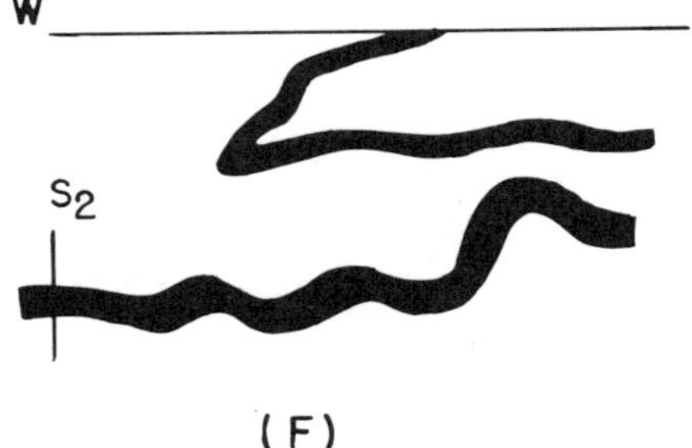

FIGURE 12–3. Different styles of early folds. (A), small syncline with earlier cleavage exposed on front face, N45°W, 90°, south of Cowansville in calcareous slates of the Stanbridge Formation. (B), small overturned fold in Stanbridge Formation north of Cowansville with earlier east-dipping cleavage. (C), earlier (S_1) and later (S_2) cleavages in Sweetsburg Formation north of Cowansville. (D), right section of early fold in Stanbridge Formation north of St. Ignace, plunging 40° S40°E. (E), overturned folds in Sweetsburg Formation east of Meigs. Folds plunge 10° S20°W. (F), early fold in Stanbridge Formation in north wall of Rioux Quarry.

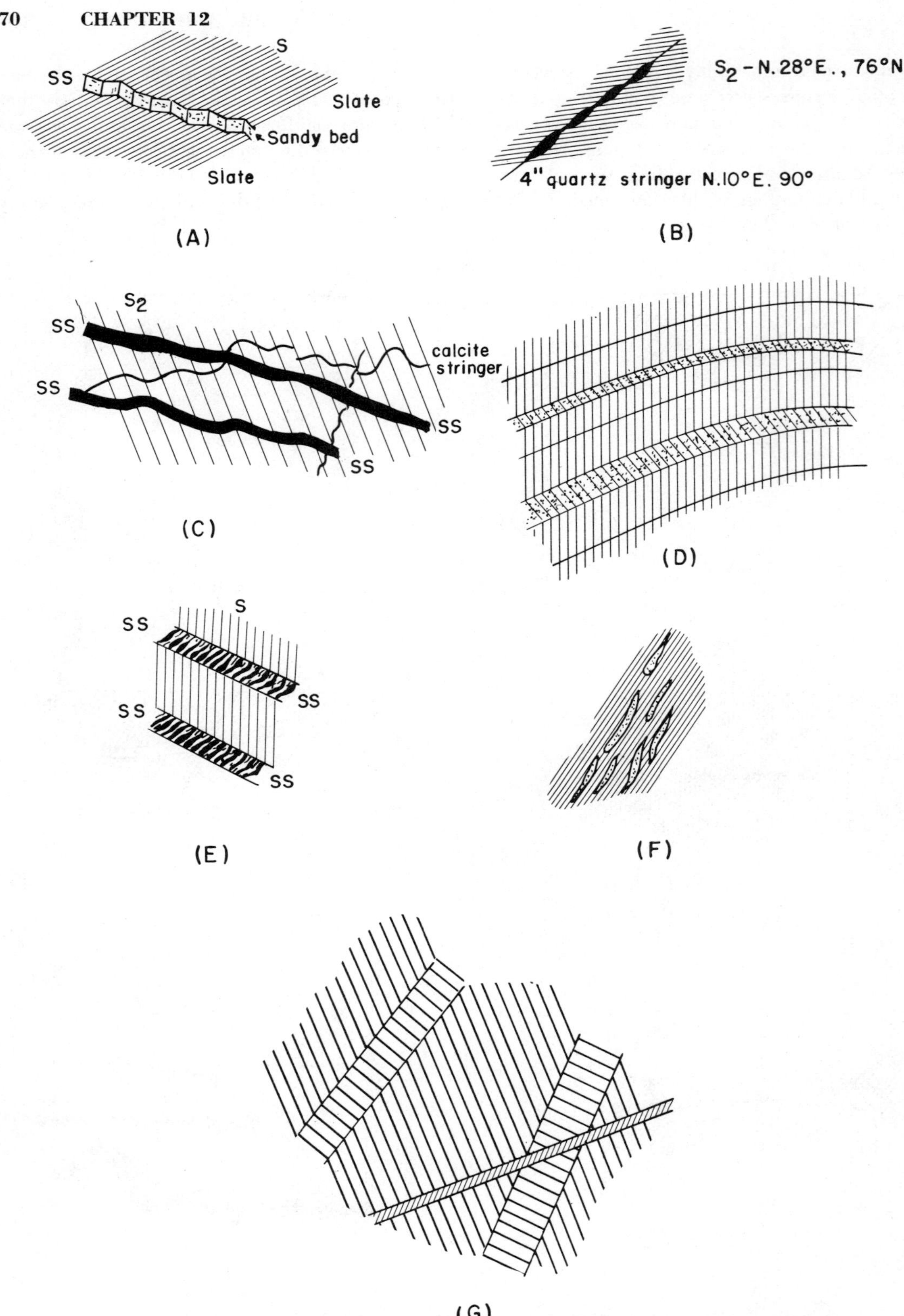

FIGURE 12–4. Different styles of cleavages. (A), variations in cleavage with lithology in Gilman Formation. (B), quartz stringer in Gilman Formation disrupted by later cleavage. (C), pre-late cleavage calcite stringer in Stanbridge Formation, Rioux Quarry. (D), variations in attitude of cleavage with lithology in Gilman Formation. (E), quartzite bands in Gilman Formation broken into lenticular fragments by cleavage. (F), Gilman "lenticulite" quartzite fragments in phyllite. (G), knick-fold bands in Gilman slate north of Frelighsburg.

The Sweetsburg folded in a flexural flow manner, with the sandy beds bent concentrically and the interbedded shales strongly squeezed to accommodate the more "competent" sandy layers.

The dolomites readily recrystallized under deformation and are massive except for well-developed joint and tension–gash systems.

The Sutton–Richford syncline (Mansville phase of Oak Hill succession)

The structure and stratigraphy of this belt is the least known and understood of the Sutton area, not only because of its structural complexity, but because of a frustrating lack of outcrop along the Sutton valley. The western boundary with the Tibbit Hill lava is clear enough, but the eastern is indeterminate. Where the Mansville Phase passes eastward into the Sutton Schist the attitude of bedding begins to change from steep, as in the Sutton-Richford tight synclinal fold zone, to gentler dips on the flanks of the Green Mountain–Sutton Mountain anticlinorium.

The Mansville phase of the Oak Hill Group is found within and to the north and south of the Sutton area in discontinuous exposures. It is a series of beds that reproduces, often in faithful detail, the succession of the Oak Hill Group from the Tibbit Hill Schist to the shales lying above the Dunham Dolomite, but the formations are thinned to from 5%–10% of the thicknesses of the normal Oak Hill succession (e.g. Pinnacle greywacke has a thickness of 400 feet (130 m) in the normal succession and only 30 feet (10 m) in the Mansville phase). Metamorphism is of low grade, but affects all units. Differences in thickness and to some extent in lithological character occur along the belt of outcrop, but the order of the Oak Hill stratigraphic succession is maintained. Thus there is always hematite at the base of the West Sutton Slate, both in the normal Oak Hill succession and in corresponding beds of the Mansville phase. This and other features leave no doubt as to the equivalence of the Mansville phase and the normal phase of the Oak Hill succession in everything but metamorphism and unit-thicknesses.

Outcrops of the Mansville phase show extremely complex structure. Minor folds, crenulations, mineral lineations, small refolded folds, etc., are plentiful and plunge in many directions. The double deformational pattern of the Oak Hill succession can, however, be discerned and mapped, as can other features as yet unexplained.

The amazing thinning of the Oak Hill succession in the Mansville Phase is difficult to explain on either a sedimentary or a tectonic basis.

Clark favors tectonic thinning, Eakins sedimentary. According to the latter view it represents the sequence developed in a geanticlinal high separating the mio- from the eugeosynclinal sections of the geosyncline where sedimentation was thin but persistent, while the actual interfingerings of the mio- and eugeosynclinal facies shifted eastward and westward across it (Cady, this volume). The subsequent intense deformation could, of course, have further contributed to the thinning.

The Green Mountain–Sutton Mountain Anticlinorium (Sutton Schists)

Both flanks of the Green Mountain–Sutton Mountain Anticlinorium are occupied by the Sutton Schists. This schist belt is recognizable southward along the Green Mountain anticlinorium (Camel's Hump Group, Hazen's Notch Formation of Vermont) and for a few score miles north of the Sutton area (Bonsecours Formation). Sericitic, graphitic, and quartzitic schists composed dominantly (98–100%) of quartz, muscovite and chlorite, in that order, form the bulk of the outcrops, with quartzite, dolomitic marble (White Brook?), greenstone (Tibbit Hill?) and albite gneiss making up minor parts. The rocks belong to the greenschist metamorphic facies. It has long been suspected that the Sutton Schists are the eastern correlatives of the Oak Hill succession (cf. Osberg, 1965), probably representing, principally, thickened Gilman and Pinnacle formations.

Although most of the Sutton Schists are clearly metamorphosed sedimentary rocks, no clear stratigraphic succession can yet be made out.

The Green Mountain–Sutton Mountain anticlinorium is a giant arch of doubly deformed schists with an overall north-northeastward gentle plunge. Many parasitic folds on the flanks of the arch show a gentle reverse plunge to the south as well as to the north. A prominent crenulation cleavage forming an axial plane foliation accompanied the major folding which produced the arch.

Associated with the arching are parasitic or "drag" folds of the normal type. In addition, "drags" showing an abnormal sense of movement as illustrated in the accompanying sketch (Fig. 12–5) are associated with and resulted from movements along the crenulation cleavage. As pointed out by Rickard (1961) and Cady (1965, personal communication), the abnormal drag fold movements are a normal consequence of compressional up-arching.

FIGURE 12–5. Normal "drag" folds in competent beds formed by flexural-slip folding accompanied by enlarged views of crenulation cleavage slip and micro-folding showing abnormal "drag" movements (c = clockwise; cc = counterclockwise); after Rickard, 1961.

An early phase of folding imposed a schistosity parallel or nearly parallel to lithological layering. Folds of this earlier phase are best preserved in the quartzite layers where the early schistosity can be demonstrated to be an axial plane foliation. The early folds and a well-developed associated mineral lineation plunge east and west away from the later Green Mountain–Sutton Mountain arch. The early east-west trending folds are set at right angles to the main Acadian fold pattern. Cady, however, points out (Cady et al., 1963; also personal communication) that such early crossfolds may develop in saddles in the basement topography.

RESUME OF GEOLOGICAL HISTORY

Our interpretation of the geologic history of the Sutton area, greatly influenced by the papers of Korn and Martin (1959) and Harland and Bayley (1958), can be summarized as follows:

1. The Sutton area straddles the miogeosynclinal basin and adjacent edge of the eugeosynclinal trough of the Appalachian geosyncline. The earliest known sediment, of Early Cambrian age (Oak Hill), was associated with basic lava flows (Tibbit Hill) and consisted largely of clastics spread over both mio- and eugeosynclines. As the basins deepened, pelitic sediments prevailed in the eastern and central parts, resulting in great thicknesses of deposits later transformed into parts of the Sutton Schists, and, farther west, into the Sweetsburg Formation (Middle or Upper Cambrian). Still farther west, in clearer and shallower seas, out of reach of the clastics poured into the eugeosyncline, carbonate sedimentation prevailed during Early Ordovician time (Philipsburg succession), to be followed by the widespread accumulation of pelites (Stanbridge Formation) which persisted through the Middle and Late Ordovician.

2. Compression and uplift along the Green Mountain–Sutton Mountain axial zone formed early folds and schistosity in deeply buried sediments (the present Sutton Schists). In the more deeply buried Oak Hill succession regular west-facing, inclined to overturned folds with a pronounced western *regard* or *vergenz* were formed, with a well-developed slaty cleavage as an axial plane foliation, while the relatively unconsolidated younger Stanbridge Complex rocks slid and slumped westward. These slump folds developed in a more chaotic fashion because of a lighter cover load. Uplift along the axis caused gravitational sliding towards the west of more surficial sedimentary accumulations (e.g. the "Sillery" or Granby Group), overriding and deforming the Oak Hill succession and the rocks farther west, which at the same time were themselves folded by westward slumping and possibly by directed pressures. Farther west, competent carbonate sequences were folded and bodily thrust westward over the Foreland.

3. Pronounced compression produced the Green Mountain–Sutton Mountain anticlinorium, the tight Richford–Sutton syncline, the Pinnacle Mountain anticline, and extended westward into the Stanbridge Complex. A strong crenulation cleavage accompanied this deformation as an axial plane foliation. Kink layers formed in some formations towards the end of this deformation or at a later date.

Rickard (1965) has adduced arguments that both deformational events are Taconic in age, whereas other workers maintain that the events are Taconic and Acadian respectively.

REFERENCES

Cady, W. M., 1960, Stratigraphy and geotectonic relationships in northern Vermont and southern Quebec: Geol. Soc. Amer. Bull., v. 71, p. 531–576

————, Albee, A. L., and Chidester, A. H., 1963, Bedrock geology and asbestos deposits of the Upper Missisquoi Valley and vicinity, Vermont: U. S. Geol. Survey Bull. 1122–B, 78 p.

Clark, T. H., 1934, Structure and stratigraphy of Southern Quebec: Geol. Soc. Amer. Bull., v. 45, p. 1–20

————, 1936, A Lower Cambrian Series from Southern Quebec: Roy. Can. Inst. Trans., v. 21, pt. 1, p. 135–151

————, 1964, Upton Area: Department of Natural Resources, Province of Quebec, Geol. Rept. 100, p. 10–11

Doll, C. G., Cady, W. M., Thompson, J. B., Jr., and Billings, M. P., *compilers* and *editors*, 1961, Centennial geological map of Vermont: Vermont Geol. Survey, Montpelier. Scale 1:250,000

Eakins, P. R., 1964, Sutton Map-Area, Quebec: Geol. Surv. Canada Paper 63–34, 3 p.

Harland, W. H. and Bayley, M. B., 1958, Tectonic regimes: Geol. Mag., v. 95, p. 89–104

Korn, H. and Martin, H., 1959, Gravity tectonics in the Naukluft mountains of South West Africa: Geol. Soc. Amer. Bull., v. 70, p. 1047–1077

Osberg, P. H., 1965, Structural geology of the Knowlton-Richmond area, Quebec: Geol. Soc. Amer. Bull., v. 76, p. 223–250

Rickard, M. J., 1961, A note on cleavages in crenulated rocks: Geol. Mag., v. 98, p. 324–332

————, 1965, Taconic orogeny in the Western Appalachians: experimental application of microtextural studies to isotope dating: Geol. Soc. Amer. Bull., v. 76, p. 523–536

Riva, John, 1966, Upper Levis graptolites from Cowansville, southern Quebec: Jour. Paleontology, v. 40, p. 220–221

Roth, H., 1965, A structural study of the Sutton Mountains, Quebec: Ph.D. thesis, McGill University, 139 p.

Shaw, A. B., 1958, Stratigraphy and structure of the St. Albans area, northwestern Vermont: Geol. Soc. Amer. Bull., v. 69, p. 519–567

Williams, F. M. G., 1966, Structural studies in the Rioux Quarry, Cowansville, Quebec: M. Sc. thesis, McGill University, 95 p.

STRATIGRAPHIC AND STRUCTURAL RELATIONSHIPS IN THE CENTRAL BELT

Stratigraphy and Correlation of the Rocks on the East Limb of the Berkshire Anticlinorium in Western Massachusetts and North-Central Connecticut*

NORMAN L. HATCH, JR., ROBERT W. SCHNABEL, AND STEPHEN A. NORTON

INTRODUCTION

THE AGE of the rocks on the east limb of the Berkshire anticlinorium in southwestern Massachusetts and Connecticut has long been a source of speculation and discussion. No fossils have been found in these rocks that occur in a north-trending belt between the Precambrian rocks of the Berkshire Highlands on the west and the Triassic rocks of the Connecticut River valley on the east (Fig. 13–1). Dating this sequence of rocks must therefore depend on lithic correlation or lateral continuity with rocks of known age in nearby areas in Vermont (Doll et al., 1961) and western New Hampshire (Billings, 1956). This correlation is made difficult, however, by lateral facies changes and by structural complications, especially south of Blandford, Massachusetts (Fig. 13–1).

One objective of the mapping program of the U.S. Geological Survey, done in cooperation with the Massachusetts Department of Public Works and the Connecticut Geological and Natural History Survey, is to trace the reasonably well dated units of eastern Vermont (Doll et al., 1961) south across Massachusetts to Connecticut and thus to

* Publication authorized by the Director, U. S. Geological Survey.

establish the lithic correlation and probable age of the stratified metamorphic rocks of north-central Connecticut.

Hatch and Norton, working in part with A. H. Chidester and P. H. Osberg, have been tracing units southward from Vermont, while Schnabel has been working northward from north-central Connecticut in the same belt of rocks. Detailed mapping at a scale of 1:24,000 is not yet complete, but reconnaissance mapping in the intervening area permits correlation of some units in the stratigraphic sequences. The purpose of this paper is to present these correlations along with tentative correlations for the remaining units, and to propose age assignments for the units of north-central Connecticut (Fig. 13–2).

We wish to thank A. H. Chidester and P. H. Osberg for their contributions both to the mapping and to the stratigraphic nomenclature for western Massachusetts.

STRATIGRAPHIC SECTION OF WESTERN MASSACHUSETTS

The lithologic units currently being studied in northwestern Massachusetts are briefly characterized below. More detailed descriptions of these units are available elsewhere (Chidester et al., 1967; Hatch, 1967; Norton, 1967; and Hatch

177

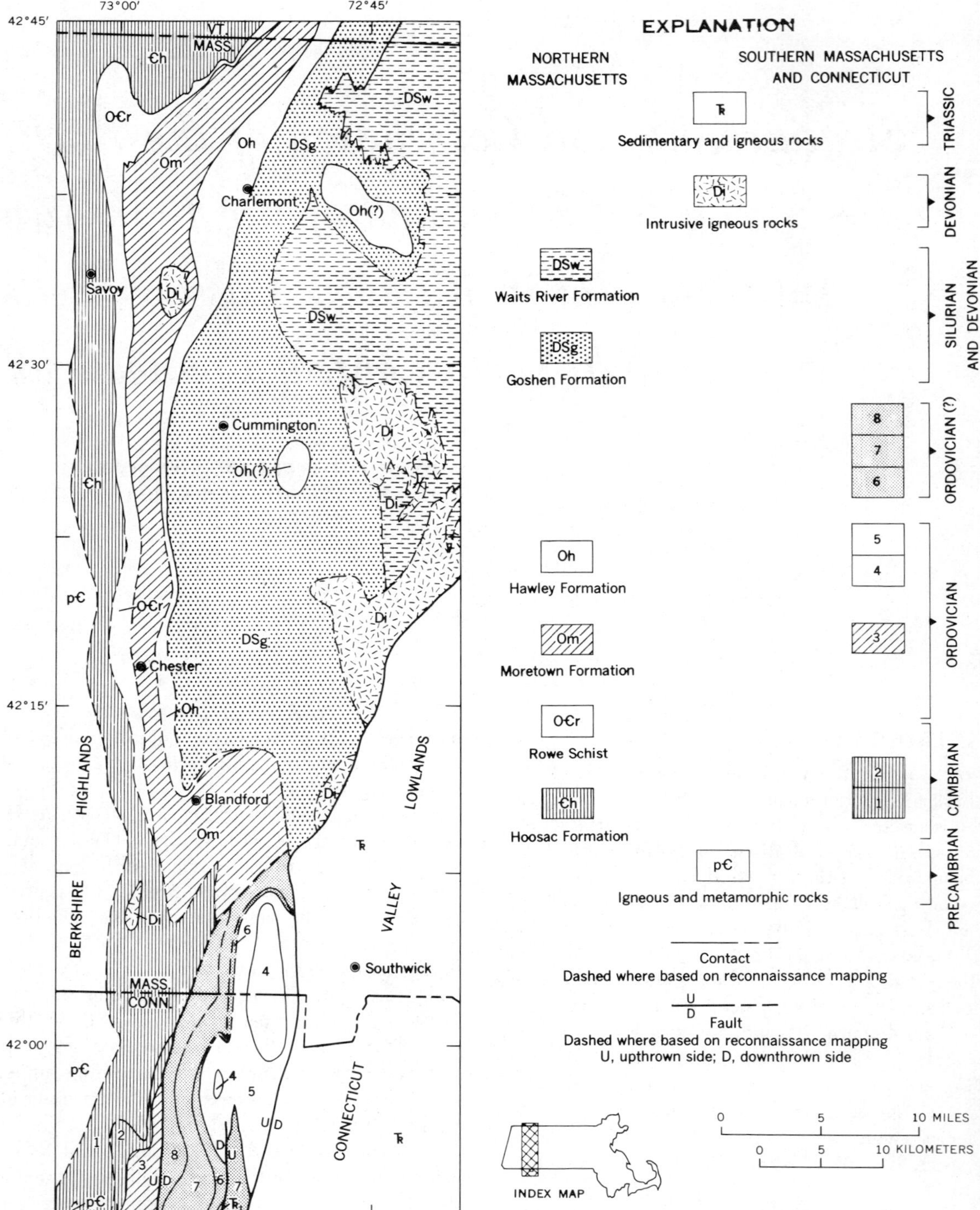

FIGURE 13-1. Geologic map of part of western Massachusetts and adjacent parts of Connecticut and Vermont. North–south correlations indicated by patterns are based primarily on lithologic similarity and only secondarily on assumed time equivalence.

and Hartshorn, in press). The emphasis here is on the lateral lithologic changes in the rock units as they are traced south from Vermont to the vicinity of Blandford (Fig. 13–1).

Hoosac Formation

The Hoosac Formation in western Massachusetts is predominantly gray and brown, medium-grained quartz-albite-biotite-muscovite gneiss and schist with, locally, layers a few hundred to 1,000 feet (60 to 300 m) thick of garnet schist, quartz-muscovite-calcite schist, and pebbly quartz-feldspar granulite* and gneiss. The thickness of the Hoosac in northwestern Massachusetts is generally about 4,000 to 6,000 feet (1200 to 1800 m). The formation in northwestern Massachusetts is continuous with the Lower Cambrian Hoosac Formation of Vermont.

Rowe Schist

The Rowe Schist (Hatch and others, 1966) comprises three distinctive associations of rocks that have been mapped as unnamed lithologic members. They are (1) pale-green, fine-grained quartz-muscovite-chlorite schist characteristic of the Pinney Hollow and Stowe Formations of Vermont; (2) gray, fine-grained carbonaceous muscovite-paragonite-quartz-biotite-chlorite schist, gray quartzites, and gray-brown, thin-laminated, granular quartz-biotite schist, all characteristic of the Ottauquechee Formation of Vermont; and (3) amphibolites and plagioclase-hornblende-chlorite-epidote schists and gneisses, some of which were mapped as Chester Amphibolite by Emerson (1898, 1917). The formation consists of 2 to 11 thin, lenticular map units representing these three rock associations, all of which are present at least as far south as Blandford (Fig. 13–1). Although the formation ranges in thickness from 300 to 5,000 feet (90 to 1500 m), it is about 2,000 to 3,000 feet (600 to 900 m) thick in most of the area (Chidester et al., 1967; Hatch et al., 1966).

A layer about 1,000 feet (300 m) thick of carbonaceous schist, quartzite, and quartz-biotite schist at the base of the Rowe Schist has been traced from near Savoy (Fig. 13–1) southward across Massachusetts. The contact of this unit with the underlying Hoosac Formation is sharp in the northern part of the State; to the south, however, the two units are interbedded over a greater and greater stratigraphic interval. A few miles

---

* Granulite is used in its textural sense in this paper, and means a metamorphic rock composed of interlocking grains of equidimensional minerals.

north of the Connecticut State line, the entire Hoosac Formation is interlayered with graphitic schist so that it is impossible to define a boundary between the two units. Present evidence is insufficient to rule out the possibility that this interlayering may be in part tectonic. The Rowe is continuous with the Pinney Hollow, Ottauquechee, and Stowe Formations of Vermont. Although this Vermont sequence, as such, cannot be traced across Massachusetts, the rocks characteristic of these three Vermont formations are recognized to within about 10 miles (16 km) north of the Connecticut State line.

Moretown Formation

The Moretown Formation in Massachusetts consists predominately of light-gray-green to beige, fine- to medium-grained quartz-plagioclase-muscovite-biotite-chlorite-garnet schist, granular schist, and granulite. The "pinstripe" structure described by Cady (1956) and others is recognized in the formation throughout western Massachusetts. A zone of more micaceous schist about 1,000 feet (300 m) thick at the top of the formation has been traced 25 miles (40 km) south from Vermont. The Moretown is 4,000 to 8,000 feet (1200 to 2400 m) thick in western Massachusetts; it is correlative with the Moretown Member of the Missisquoi Formation of Doll et al. (1961) in Vermont.

Hawley Formation

The Hawley Formation in western Massachusetts, as redefined by Hatch (1967), consists of amphibolites, feldspar schists and gneisses, and black and gray carbonaceous schists and quartzites. It contains distinctive 1-inch- to 2-foot-thick beds of thinly laminated, pink to gray, hard quartz-garnet granulite, locally with cummingtonite; this rock, called "coticule" by Emerson (1898, p. 174), is generally within, or near, zones or beds of carbonaceous schist and quartzite. A few thin beds of this distinctive rock are also present locally in the upper part of the Moretown Formation. In the northern half of Massachusetts, the formation is predominantly metamorphosed volcanic rocks, but it passes laterally into predominantly carbonaceous schists and quartzites in the vicinity of Chester (Fig. 13–1). The Hawley Formation ranges in thickness from about 5,000 feet (1500 m) near Charlemont (Fig. 13–1), at what may have been the center of an old shield volcano, to about 600 feet (180 m) southwest of Cummington where the volcanic pile is very thin. The volcanic pile also thins to the north (Hatch and Hartshorn, in press; Doll et al., 1961) and

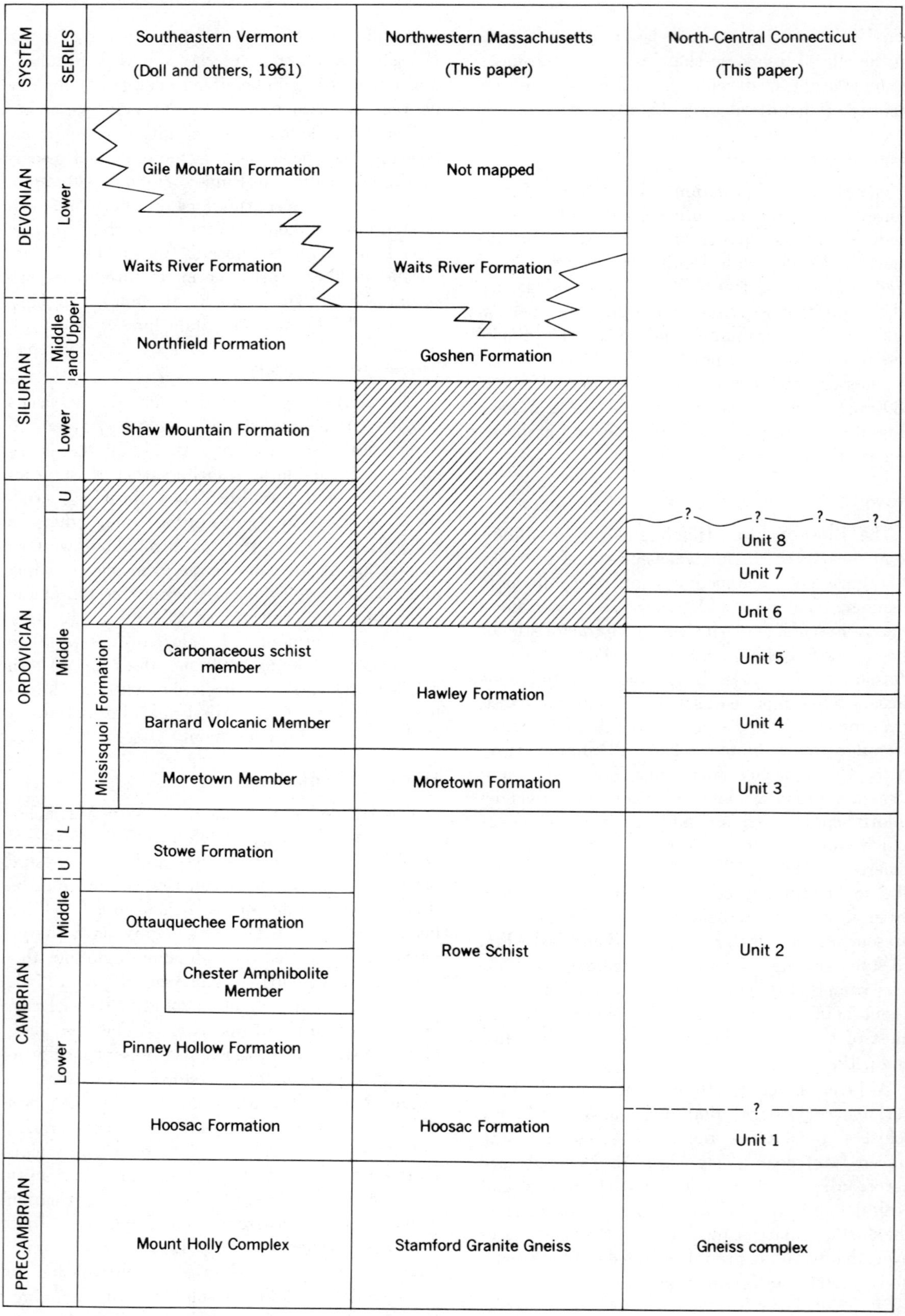

SYSTEM	SERIES	Southeastern Vermont (Doll and others, 1961)	Northwestern Massachusetts (This paper)	North-Central Connecticut (This paper)
DEVONIAN	Lower	Gile Mountain Formation	Not mapped	
DEVONIAN	Lower	Waits River Formation	Waits River Formation	
SILURIAN	Middle and Upper	Northfield Formation	Goshen Formation	
SILURIAN	Lower	Shaw Mountain Formation		
ORDOVICIAN	U			Unit 8
ORDOVICIAN				Unit 7
ORDOVICIAN				Unit 6
ORDOVICIAN	Middle	Missisquoi Formation — Carbonaceous schist member	Hawley Formation	Unit 5
ORDOVICIAN	Middle	Missisquoi Formation — Barnard Volcanic Member	Hawley Formation	Unit 4
ORDOVICIAN	Middle	Missisquoi Formation — Moretown Member	Moretown Formation	Unit 3
ORDOVICIAN	L	Stowe Formation		
CAMBRIAN	U	Stowe Formation		
CAMBRIAN	Middle	Ottauquechee Formation	Rowe Schist	Unit 2
CAMBRIAN	Lower	Chester Amphibolite Member	Rowe Schist	Unit 2
CAMBRIAN	Lower	Pinney Hollow Formation		
CAMBRIAN	Lower	Hoosac Formation	Hoosac Formation	Unit 1
PRECAMBRIAN		Mount Holly Complex	Stamford Granite Gneiss	Gneiss complex

FIGURE 13–2. Chronostratigraphic correlation of the rock units of southeastern Vermont, northwestern Massachusetts, and north-central Connecticut.

pinches out in the vicinity of Randolph, Vermont (Doll et al., 1961). The formation may pinch out, or have been eroded away, northeast of Blandford.

The Hawley is continuous and correlative with both the Barnard Volcanic Member and the carbonaceous schist and quartzite member of the Missisquoi Formation of Doll et al. (1961) in Vermont.

Goshen Formation

The Goshen Formation, as redefined by Hatch (1967), consists of gray, carbonaceous quartz-muscovite-biotite-garnet-chlorite schist with local staurolite, cyclically bedded gray schist and quartzite, and thicker bedded (1 to 2 feet (30 to 60 cm) gray quartzite containing a few beds of brown-weathering carbonate granulite, generally less than 10 inches (25 cm) thick. In the vicinity of the Vermont State line, the Goshen is predominantly massive gray schist, but it passes southward into a sequence of thin, cyclically bedded gray schist and quartzite near Charlemont (Fig. 13-1). Farther south, these cyclically bedded rocks, and to a lesser extent thicker bedded quartzites, characterize the formation. These thicker bedded quartzites, which immediately overlie the cyclically bedded part of the Goshen south from Cummington (Fig. 13-1), resemble the quartzites of the Gile Mountain Formation of Vermont (Doll et al., 1961). The Goshen ranges in thickness from 2,500 to 7,000 feet (750 to 2100 m) in northwestern Massachusetts (Hatch, 1967). It is continuous with the Northfield Formation of Vermont (Doll et al., 1961), although the Goshen Formation south of Charlemont is lithologically unlike the Northfield.

Waits River Formation

The Waits River Formation in western Massachusetts consists predominantly of gray schists that are similar to the schists of the Goshen Formation, but are interbedded with beds of punky, brown-weathering quartz-carbonate granulite more than a foot (30 cm) thick. The presence of these granulite beds is the sole basis for distinguishing the Waits River from the Goshen. In accordance with this definition, a few miles south of the Vermont–Massachusetts line, about 2,500 feet (750 m) of schists of the Goshen Formation passes southward into an equivalent thickness of rocks of the Waits River Formation, and, within another few miles, back again into Goshen Formation.

The Silurian and Devonian rocks of Massachusetts are lithologically similar to the North-field, Waits River, and Gile Mountain Formations of Vermont (Fig. 13-2), but their stratigraphic sequence is different. The thicker bedded quartzites in the Goshen Formation south of Cummington apparently are, at least in part, stratigraphically below the Waits River Formation, whereas similar rocks in the Gile Mountain Formation in Vermont generally overlie the Waits River Formation (Doll et al., 1961).

The top of the Waits River is not exposed within the area of Figure 13-1, but an estimated 2,500 (750 m) feet of Waits River rocks has been mapped northeast of Charlemont (Fig. 13-1). The formation is continuous with the Waits River Formation in Vermont.

Summary of the Section near Blandford, Massachusetts

The Paleozoic formations of northwestern Massachusetts have been traced, in part by detailed mapping and in part by reconnaissance, to a few miles south of Blandford. The section there, from west to east, expressed in terms of the units being mapped in northern Massachusetts, is:

Lithology	Thickness		Formation
	(feet)	(meters)	
Gray, carbonaceous cyclically bedded schists	5000+	1500+	Goshen
Gray and black carbonaceous schist and quartzite with minor amphibolite and feldspar schist	2000±	600±	Hawley
Gray-buff quartz-feldspar granulite and quartz-feldspar-mica schist	5000±	1500±	Moretown
Green quartz-muscovite-chlorite schist with interbedded amphibolites	1000±	300±	Rowe
Brown and gray quartz-albite-biotite gneiss interbedded with gray carbonaceous schist	5000±	1500±	Hoosac+ lower part of Rowe
Pink granitic gneisses	?		

STRATIGRAPHIC SECTION IN NORTH-CENTRAL CONNECTICUT AND SOUTHWEST MASSACHUSETTS

The stratified rocks in north-central Connecticut and southwest Massachusetts have been separated into eight units, as yet unnamed (Fig. 13-2). The oldest, Unit 1, rests unconformably upon Precambrian rocks to the west. Units 1, 2, and 3 are lithologically similar to some of the formations described above from northern and central Massachusetts. Units 4 through 8 crop out around the domes west and southwest of Southwick (Fig. 13-1); only part of this sequence, which is in fault contact with Triassic rocks on the east, can be correlated with formations of northern and central Massachusetts. Stanley (1964) has mapped part of the area immediately south of Figure 13–1, but his stratigraphic column and the one described below are different because of different interpretations of the geologic structure.

Unit 1

Unit 1 consists of well-bedded, gray, fine-grained quartz-plagioclase-biotite-muscovite schist with scattered beds of calc-silicate rock and quartzite. Many beds of biotite rich schist contain abundant megacrysts of plagioclase as much as ¼ inch (6 mm) in diameter. The unit is characterized and distinguished by the presence of a rind, ½ to 1 inch (1.2 to 2.5 cm) thick, of brown, rusty-weathered rock on most outcrops. Megascopic kyanite or sillimanite is present locally. Unit 1 is estimated to be about 6,000 feet (1800 m) thick.

Rocks of Unit 1 have been mapped only along the west side of the area adjacent to the Precambrian rocks. They may be traced northward into rocks mapped as Hoosac Formation in central and northern Massachusetts and Vermont and are thus assigned an Early Cambrian age.

Unit 2

Unit 2 is composed of moderately well bedded, fine- to medium-grained muscovite-biotite-quartz-plagioclase schist that weathers a dark-yellowish brown to reddish brown. The unit is characterized by muscovite porphyroblasts that average about ⅛ inch (3 mm) in long dimension and are randomly oriented relative to the schistosity. The contact with Unit 1 is gradational. Unit 2 is 2,000 to 3,000 feet (600 to 900 m) thick.

Unit 2 is lithologically correlated with the upper part of the Hoosac Formation and the lower part of the Rowe Schist of central and northern Massachusetts. Although Unit 2 contains none of the pale-green quartz-muscovite-chlorite schist characteristic of the Rowe, it may include time equivalents of all of the Rowe. Thus, Unit 2 probably includes rocks of Early, Middle, and possibly Late Cambrian age. Units 1 and 2 are also tentatively correlated with the Waramaug Formation of Rodgers et al. (1956) at the type locality.

Unit 3

Unit 3 consists of very thinly laminated, fine-grained, gray quartz-plagioclase-muscovite-biotite granulite and schist locally containing a "pinstripe" structure identical to that described by Cady (1956) in Vermont. Layers and lenses of amphibolite are scattered throughout the unit, which also contains many thin beds of quartz-garnet granulite (similar to the "coticule" of Emerson, 1898, p. 174) in the southern part of the area of Figure 13–1. The unit is characterized and distinguished by the nonrusty weathering and the scarcity of aluminum silicate minerals. The contact with Unit 2 is generally marked by a distinctive bed of thinly laminated amphibolite characteristic only of Unit 3. Unit 3 is 0 to 5,000 feet (0 to 1500 m) thick, but the top of the unit is faulted out in the area.

Unit 3 is correlated with the Moretown Formation of Massachusetts and is thus assigned a Middle Ordovician age.

Unit 4

Unit 4, which forms the core of the two domes west and southwest of Southwick (Fig. 13–1), consists of coarse-grained, locally garnetiferous amphibolite and felsic gneiss in beds 1 inch to 20 feet (2.5 cm to 6 m) thick. The upper part of the unit contains many 1- to 6-inch (2.5 to 15 cm) thick beds of quartz-garnet granulite similar to the rock described in Unit 3. Amphibolite predominates at the margins, but the cores of the domes are so poorly exposed that little is known of the character of the rocks, although the alternation of felsic and mafic layers suggests a volcanic origin. Although only about 2,000 feet (600 m) of rocks of Unit 4 is exposed around the domes, projections into the unexposed central part of the domes, also assumed to be underlain by Unit 4, give an estimated total thickness of 5,000 feet (1500 m).

Unit 4 is correlated with the lower part of the Hawley Formation to the north and is thus assigned a Middle Ordovician age.

Unit 5

Unit 5, which surrounds the two domes west of Southwick, consists predominantly of massive, gray and brown, slightly carbonaceous, rusty-weathering quartz - plagioclase - muscovite - biotite schist. The lower part of Unit 5 around the larger dome west of Southwick is characterized by many thin beds of quartz-garnet granulite; the middle part is characterized by abundant thin lenses of conspicuous zoisite-bearing amphibolite. Neither rock was seen around the smaller dome to the southwest. Local mineralogical and textural grading, interpreted as original graded bedding within the carbonaceous schist, indicates that Unit 5 stratigraphically overlies Unit 4; the contact is sharp and is the top of the highest amphibolite bed. Unit 5 is about 3,000 feet (900 m) thick.

Unit 5 is correlated with the upper part of the Middle Ordovician Hawley Formation in Massachusetts.

Unit 6

Unit 6 is composed of calc-silicate rock, amphibolite, quartzite, and schist and forms a narrow belt north and west of the domes west of Southwick. Although it varies in composition, the unit nearly everywhere contains beds of calc-silicate rock which make it distinctive and mappable. The contact with Unit 5 appears to be gradational by interlamination. Unit 6 thickens from a few tens of feet (6 or 8 m) at the north to a maximum of about 2,000 feet (600 m) at the south edge of the area.

This unit has no known equivalent in the Massachusetts section to the north.

Unit 7

Unit 7 consists of very distinctive, extremely coarse grained, gray, nonrusty-weathering quartz-plagioclase-muscovite-biotite schist. Quartz lenses as much as 1 foot (30 cm) long are common throughout the unit. Garnet in crystals as much as 1 inch (2.5 cm) in diameter, kyanite, which locally constitutes as much as 30 percent of the rock in blades as much as a foot (30 cm) long, and sillimanite are important minerals in this unit west of the southern dome; kyanite and sillimanite decrease in abundance to the north. Unit 7 is 4,000 feet (1200 m) thick in the area of Figure 13–1.

This unit has no obvious correlative in the Massachusetts section to the north.

Unit 8

The uppermost unit mapped in the southern part of the area, Unit 8, is a heterogeneous sequence of schists, gneisses, granulites, and, locally, amphibolites. The unit is poorly exposed, and distinctive lithologies within it cannot be traced more than a few hundred feet (60 m). The only distinctive rock characteristic of the whole unit is a soft, friable, quartz-mica schist. In the southernmost part of the area of Figure 13–1, the upper part of the unit is characterized by relatively abundant thin layers of quartz-garnet granulite identical with the quartz-garnet granulite in Units 3, 4, and 5. Graded bedding is preserved in many places and indicates that Unit 8 lies stratigraphically above Unit 7. It is at least

3,000 feet (900 m) thick in the area of Figure 15–1.

Unit 8 has no obvious correlative in the Massachusetts section to the north.

CORRELATION OF THE NORTHWEST MASSACHUSETTS AND NORTH-CENTRAL CONNECTICUT SECTIONS

If the correlations of Units 1, 2, 3, 4, and 5 presented above are accepted, the problem of correlating the northwest Massachusetts and north-central Connecticut stratigraphic sections becomes a problem of relating Units 6, 7, and 8 to the Massachusetts section. The area is an excellent example of the interdependence of stratigraphy and structure; each possible stratigraphic correlation with northern Massachusetts leads to a particular structural or facies relationship. The correlation and consequent structural interpretation favored by the authors is shown on Figures 13–1 and 13–2 and is outlined below. The principal alternative interpretation is discussed briefly.

Relict graded bedding and channeling indicate that Units 6, 7, and 8 overlie Units 4 and 5. Because Units 6, 7, and 8 do not resemble any of the Silurian or Devonian Formations of northern Massachusetts or southern Vermont, the northern Connecticut-southernmost Massachusetts section must include a sequence of rocks (Units 6, 7, and 8) younger than the Middle Ordovician Hawley Formation and probably older than the Lower Silurian Shaw Mountain Formation, which is not present in Massachusetts. Because of their general lithologic character and stratigraphic setting, these units are tentatively considered to be Middle Ordovician in age.

Regional correlations and sedimentary structures indicate that Units 1, 2, and 3 became progressively younger to the east. If the conclusions of the preceding paragraph are valid, Units 4 through 8 form a sequence that dips west and gets progressively younger westward, and the contact between the older east-facing rocks and the younger west-facing rocks must be a fault (Fig. 13–1). Schnabel has mapped a fault in northernmost Connecticut between Units 3 and 8, and it is extrapolated northward across Figure 13–1. Although geometrically this fault could be either a west-dipping thrust, an east-dipping thrust, or a normal fault, field evidence indicates that it probably is a westward dipping reverse fault, at least in the southern part of the area of Figure 13–1.

The absence of Units 6, 7, and 8 from the northwestern Massachusetts section can be explained in many ways. Lateral facies changes between the sites of deposition of the rocks on the two sides of the fault could explain the relations if the net slip on the fault is more than a few miles. The relations could also be explained if more detailed field work shows that the movement on the fault was pre-Goshen and the subsequent pre-Goshen period of erosion cut relatively deeper stratigraphically into the western block. Nondeposition of Units 6, 7, and 8 on the west block, or concealment of these units in the west block by onlap of younger Goshen rocks, would also explain these relations. Whatever the exact structural nature of this fault, its regional importance is shown by the fact that rocks comparable to Units 6, 7, and 8 have not been reported north of the fault at least as far north as the Quebec border in Vermont (Doll et al., 1961), whereas they apparently persist for many miles to the south on the south side of the fault. They may correlate in part with the upper part of the Partridge Formation in New Hampshire (Billings, 1956).

The simplest alternative interpretation for the correlation of the rocks of the northern Massachusetts and northern Connecticut sections makes the basic assumption that the gneisses in the cores of the domes west of Southwick are Precambrian and that a simple syncline is present between the domes and the Precambrian rocks to the west. This is the interpretation followed by Emerson (1898, 1917). Units 1 and 2 on the west limb of the syncline would in this case be equivalent to Units 5 through 8 on the east limb, and all would correlate with the Hoosac Formation and Rowe Schist to the north. Unit 3 would occupy the center of the syncline and would correlate with the Moretown to the north. This interpretation is not favored by the writers because the rocks on the west "limb" simply do not look like the rocks on the east "limb," and the magnitude and abruptness of the east-west facies changes required by this interpretation make it much less reasonable than the other.

Other alternative correlations involve very complex structures, chiefly large recumbent folds. None of these seem to have a particular advantage over the adopted interpretation, and most involve correlation of lithologically dissimilar units as well as overturning of units in areas where primary sedimentary structures indicate that the units are right-side-up.

REFERENCES

Billings, M. P., 1956, The geology of New Hampshire, Part II, Bedrock geology: Concord, N. H., New Hampshire State Plan. and Devel. Comm., 203 p.

Cady, W. M., 1956, Bedrock geology of the Montpelier quadrangle, Vermont: U. S. Geol. Survey Geol. Quad. Map GQ–79

Chidester, A. H., Hatch, N. L., Jr., Osberg, P. H., Norton, S. A., and Hartshorn, J. H., 1967, Geologic map of the Rowe quadrangle, Massachusetts-Vermont: U. S. Geol. Survey Geol. Quad. Map GQ–642

Doll, C. G., Cady, W. M., Thompson, J. B., Jr., and Billings, M. P., *compilers* and *editors*, 1961, Centennial geologic map of Vermont: Montpelier, Vermont Geol. Survey, scale 1:250,000

Emerson, B. K., 1898, Geology of old Hampshire County, Massachusetts, comprising Franklin, Hampshire and Hampden Counties: U. S. Geol. Survey Mon. 29, 790 p.

————, 1917, Geology of Massachusetts and Rhode Island: U. S. Geol. Survey Bull. 597, 289 p.

Hatch, N. L., Jr., 1967, Redefinition of the Hawley and Goshen Schists in western Massachusetts: U. S. Geol. Survey Bull. 1254–D, 16 p.

————, Chidester, A. H., Osberg, P. H., and Norton, S. A., 1966, Redefinition of the Rowe Schist in northwestern Massachusetts: U. S. Geol. Survey Bull. 1244–A, p. A33–A35

Hatch, N. L., Jr., and Hartshorn, J. H., (in press), Geologic map of the Heath quadrangle, Massachusetts-Vermont: U. S. Geol. Survey Geol. Quad. Map GQ–735

Norton, S. A., 1967, Geology of the Windsor quadrangle, Massachusetts: U. S. Geol. Survey open-file report, 210 p.

Rodgers, John, Gates, R. M., Cameron, E. N., and Ross, R. J., Jr., 1956, Preliminary geological map of Connecticut: Hartford, Conn., Connecticut Geol. and Nat. History Survey

Stanley, R. S., 1964, The bedrock geology of the Collinsville quadrangle: Connecticut Geol. and Nat. History Survey Quad. Rept. no. 16, 99 p.

Chapter 14

Garnet Rotations Due to the Major Paleozoic Deformations in Southeast Vermont

JOHN L. ROSENFELD

The beliefs which we have most warrant for, have no safeguard to rest on, but a standing invitation to the whole world to prove them unfounded.
John Stuart Mill, On Liberty

INTRODUCTION

Problem

SINCE THEIR INITIAL RECOGNITION, apparently by Flett (*in* Peach et al., 1912, p. 111, 178), rotated garnets have increasingly been recognized* as intriguing objects containing potentially important information for geologic interpretation of areas of highly deformed metamorphic rocks (see, for example, Gilluly, 1938). Their scant use in solving geologic problems seems to have resulted largely from the absence of adequate methods of study. The abundance of rotated garnets in most of southeast Vermont and the existing ideas on structural evolution of that region caused me to develop such methods in order to test these ideas in an area astride one of the main belts of the northern Appalachians. The methods, developed in 1959, will be discussed in extensive detail elsewhere (Rosenfeld, in preparation) and here in sufficient detail only to enable the reader to follow the discussion. The primary idea that prompted this investigation was the attractive hypothesis of Thompson (1950, p. 146; *in* Billings et al., 1952, p. 21), based largely on use of minor folds, that the mantled gneiss domes of the region (Fig. 14–1) and their attendant structural features are largely the result of a single upthrusting, possibly caused by buoyant forces. Secondarily I hoped to reconcile the apparent conflict of that hypoth-

---

* Many of the principal references to this topic are contained in a paper by Spry (1963).

esis with a very different hypothesis proposed almost simultaneously by White and Jahns (1950, p. 214–219) to explain a nearby "cleavage arch" having many structural features in common with the gneiss domes.

Principal Results

The methods developed enabled determination in considerable detail of the magnitudes, directions, and sequence of movements relative to the metamorphism. A common explanation for the structures discussed by Thompson (1950) and White and Jahns (1950) resulted from the discovery, based on evidence at many localities in Siluro–Devonian strata of two periods of pre-Triassic garnet rotation about non-parallel axes within the schistosity, that upthrust of the domes or cleavage arches (here diastrophic event II) followed major subhorizontal flow toward the west-southwest (diastrophic event I) centered within the Waits River Formation (lithologic units and their chronologic relations are described in Table 14–1). Major recumbent folds formed concurrently with that flow. Evidence of a pre-Silurian (Taconic?) period of garnet rotation and concomitant metamorphism is also found in the Cambrian Pinney Hollow Formation (see also Albee, this volume).

ACKNOWLEDGMENTS

In the study itself, principal acknowledgment goes to my collaborator of long standing, J. B. Thompson, Jr., for permission to use unpublished field data in the area of the Chester Dome and for many stimulating arguments in and out of the field. I owe a very large debt to my colleague,

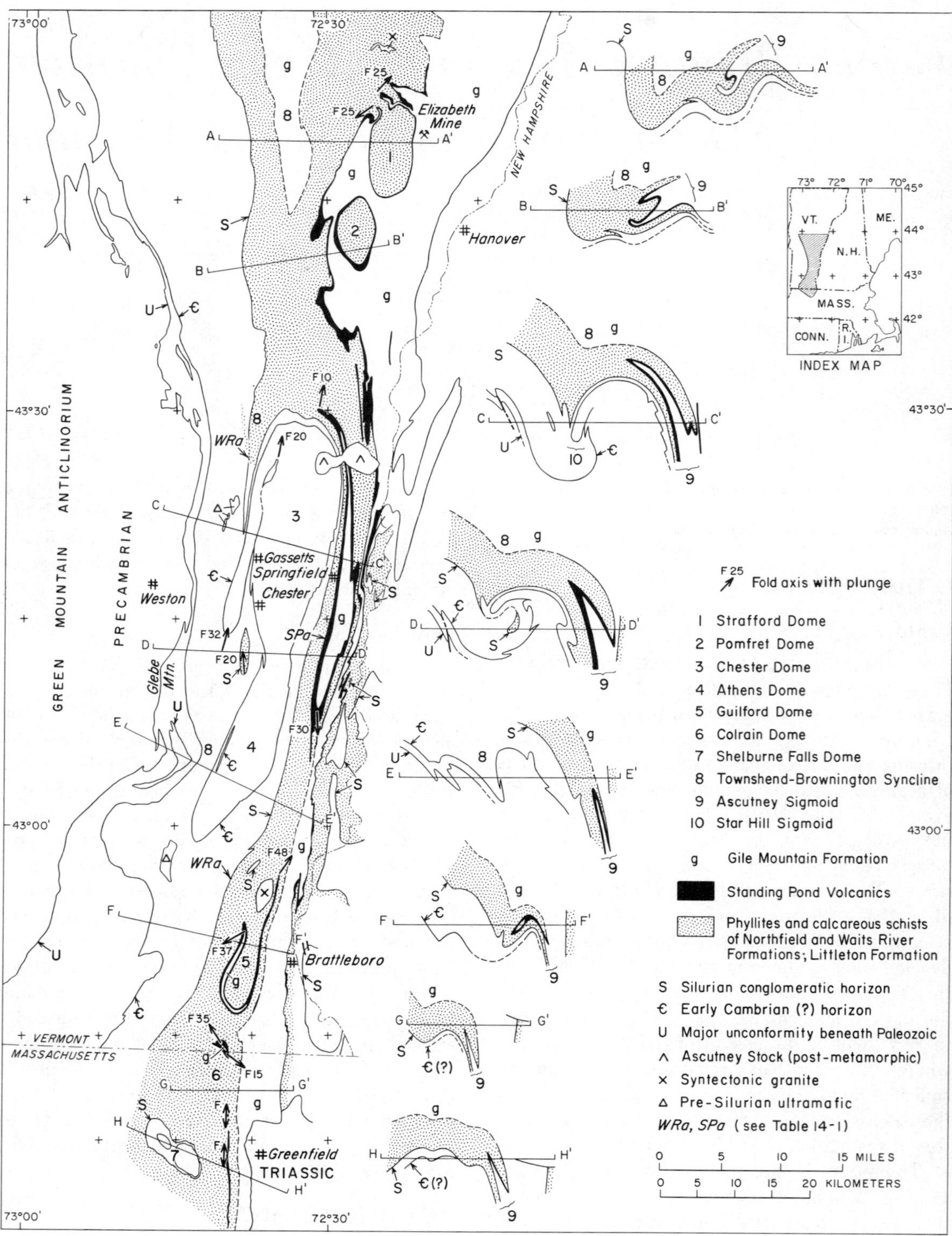

FIGURE 14-1. Geologic map and interpretative cross-sections, southeast Vermont and adjacent Massachusetts. Formation boundaries largely from Doll et al., 1961, and Segerstrom, 1956a, b.

R. L. Shreve, for numerous challenging discussions of the problems considered here and for a thorough and helpful review of this manuscript. M. K. Hubbert and the late E. S. Larsen, Jr., provided indispensible support and encouragement. O. T. Tobisch, T. E. Tullis, and W. S. White provided helpful comments on this paper in manuscript. E. C. Bowman, John Gieling, John Matthews, F. R. Boyd, William Shepard, C. J. Venuto, R. D. Cotter, and E-an Zen provided valuable assistance in the field. I have also benefited from discussions with and information and encouragement from A. J. Boucot, W. F. Brace, W. M. Cady, A. H. Chidester, J. M. Christie, G. P. Eaton, James Gilluly, J. F. Hays, H. C. Heard, P. F. Howard, J. B. Lyons, G. J. F. MacDonald, W. R. Muehlberger, R. S. Naylor, Peter Robinson, John Rodgers, J. W. Skehan, S. J., and W. S. White. I have benefited from the services of: John de Grosse (thin sections); Mrs. Peter Kurtz (drafting); James Groom and L. S. Weymouth (photography); Mrs. E. F. Petersen (typing).

Since 1956 I have received, in support of this study, financial assistance from Wesleyan University and the University of California and a fellowship from the John Simon Guggenheim Memorial Foundation. During the academic year, 1963–1964, The Department of Geological Sciences at Harvard University provided me with facilities and congenial surroundings for the pursuit of this study. I acknowledge all of this support with gratitude.

I thank, in advance, those who, in visiting the localities within the area studied, help to preserve for others the many fine exposures of rotated garnets that depend on weathering and erosion for their visibility. One such exposure in the Saxtons River just west of the village of the same name has recently been irrevocably damaged, apparently with explosives, by mineral collectors interested only in the large garnets present independently of their structural context. Collectors may obtain excellent rotated garnets from highway cuts at locality S15b (Fig. 14–4) opposite the Townshend Dam in the West River Gorge.

To guidance by Marland Billings in the early stages of this work and his encouragement in its later stages must be credited much of any merit it may have.

INITIAL WORK ON ROTATED GARNETS IN VERMONT

This included field and laboratory measurements of rotated garnets, started in 1948, at a large number of localities during the course of geologic mapping of the Athens Dome and part of the Chester Dome (Fig. 14–1). The garnets were broken approximately perpendicular to the fold axis in the surrounding rock if this was visible or perpendicular to the strike if no fold axis was visible. The maximum apparent rotation of the included surface of schistosity or compositional banding passing through the center of the garnet was measured from the continuation

TABLE 14-1. Condensed Chronologic Table of Metamorphosed Rocks (See Doll et al., 1961, for more details)

Geologic Age	Unit or Feature	Lithology
Devonian(?)	New Hampshire Plutonic Series	Late synkinematic granitic rocks
Devonian*	Gile Mountain Formation, Littleton Formation (?)*	Quartzo-feldspathic schist, graphitic schist, some calcareous
Siluro-Devonian	Standing Pond Volcanics	Chiefly amphibolites, greenschists of volcanic origin
	~UNCONFORMITY (?) (SPa)⇒	
Siluro-Devonian	Northfield and Waits River Formations, Littleton Formation (?)* ⇐(WRa)⇒	Graphitic calcareous and non-calcareous schist
Silurian	Shaw Mountain Formation	Quartz conglomerate, porphyritic volcanics
	~UNCONFORMITY~	
Late Ordovician	Ultramafic intrusives	Dunite, serpentinite, steatite
Early Cambrian to Mid-Ordovician	Pinney Hollow through Missisquoi Formations	Heterogeneous schists, hornblende gneisses and amphibolites
Late Precambrian to Early Cambrian	Cavendish through Hoosac Formations	Augen gneiss, conglomerate gneiss, albitic and paragonitic schist, dolomite
	~MAJOR UNCONFORMITY~	
Precambrian	Mt. Holly complex	Assorted gneisses, granites, schists, amphibolites, and marbles

*The direction of facing across the Standing Pond Volcanics is still uncertain. I have followed Chang *et al.*, 1965, in elevating this unit to formational status. Implications of the alternative possibilities are discussed in the latter part of the paper.

of the same surface outside of the garnet.* The first results (Rosenfeld, 1954, p. 128) seemed to be in accord with Thompson's hypothesis, which had previously been based only partly on rotated garnets and primarily on the rotational senses derived from common drag folds† associated with the domes. Detailed study of the garnets, however, gave a more complicated picture than that implied by this hypothesis. Garnets from the base of the Waits River Formation, where it strikes west-northwest on the northern part of the Chester Dome, commonly implied large rotations opposite from those expected. Also, garnets in Cambro-Ordovician units from the east flank of the Chester Dome near the line of section C–C′ of Figure 14–1 showed two stages of rotation: an inner, earlier, counterclockwise‡ and an outer, later, clockwise stage, commonly less conspicuous and with less rotation. Thus, drag folds were not telling the whole story; and the rotated garnets had been both incompletely and improperly interpreted in the initial work.

Meanwhile continued geologic mapping in southern Vermont and adjacent Massachusetts elicited the interesting map pattern of Figure 14–1. Particularly noteworthy were the large isoclinal folds involving the Standing Pond Volcanics, and the large isoclinal fold on the west limb of the Chester Dome (the "Star Hill Sigmoid," identified by 10, section C–C′, Fig. 14–1). The pattern of these folds apparently cannot be explained by simple upthrust in the centers of the domes.

The confusing pattterns of the garnets and the geologic map necessitated a new approach to kinematic analysis of the area.

NEW METHODS FOR STUDY OF ROTATED GARNETS

Although it can be shown from application of the Curie symmetry principle (Paterson and Weiss, 1961, p. 859) that the garnets rotated about axes that lay within the schistosity (Rosenfeld, in preparation), it does not follow that the two rotations in a doubly-rotated garnet could be

assumed *a priori* to have been about the same axis within the schistosity. With different axes, the later rotation would have carried the early axis, as preserved in the garnet, out of the schistosity, thereby transforming it into a *relict rotational axis*. The early rotation might therefore be unclear or misinterpreted in garnets observed using the methods of the initial study. Simple tests showed that the two axes of rotation are generally not parallel within the schistosity and that the relict rotational axis is commonly at a high angle to the schistosity. Clearly, if the garnets were to be used effectively for tectonic analysis, the first ingredient in a new approach was the development of better procedures for determination of axes, senses, amounts, and sequence of rotations recorded by garnets in each specimen to be studied. These procedures (Rosenfeld, in preparation) depend on the fact that there were only two important rotational stages about non-parallel axes within the schistosity during the last major regional metamorphism. Examples of the observational information upon which these methods are based are illustrated for two occurrences in Figures 14–2 and 14–3 (Figure 14–4 gives positions of localities cited in text and figures.)

In the interest of continuity, one may abstract from the fairly involved procedures and notation used in analysis of the rotations of the garnets those elements that are pertinent to the geological discussion. Let there be a surface of bedding schistosity, S, within which a garnet nucleates, grows, and simultaneously rotates relative to S about an axis constantly oriented within S. This axis is called b_1. The amount of rotation about b_1 of the included schistosity at the center of the garnet relative to S is called Ω_1. There will also be a sense of rotation determined from the pattern within the inner part of the garnet that grew with rotation Ω_1. After this first rotation, let the garnet continue to grow concurrently with rotation about a second axis within S not parallel to the first. The second axis within S is called b_2. The amount of this second rotation about b_2, Ω_2, equals the small circle rotation about b_2 that carries the early axis, b_1, out of S and transforms it into the observable relict rotational axis. The orientation of b_2 is not ordinarily easily determined from the pattern within a garnet, but is readily obtained from small neighboring drag folds. The sense of rotation about b_2 is obtained from both the drag folds and the pattern of the included schistosity in the outer part of the garnet (Fig. 14–3). The sense of rotation of garnets about b_2 reverses across the axis of the Town-

---

* In all cases in which both schistosity and banding were observed in the same garnet, they are parallel to one another. In this paper no distinction will be made between these two types of surface; the term "schistosity" will be used for either or both.

† Here defined as that kind of minor fold whose cross-section is a sigmoid flexure (i.e. an S-shaped curve), with middle limb shorter than either of the remaining limbs, and from which a shear sense is equated to the rotational sense of the short limb.

‡ When viewed looking northward. This convention will be used in the remaining text unless otherwise stated.

shend–Brownington Syncline (Fig. 14–1); it seems clear from this fact that b_2, Ω_2, and the present attitudes of S developed concurrently with that fold (and also with the formation of the domes and the Green Mountain Anticlinorium). Therefore, relative to a *geographic frame of reference*, b_1 is transformed by the folding accompanying rotation Ω_2 into a new orientation b'_1. To obtain the orientation of b_1 in the same geographic frame of reference, folds formed concurrently with Ω_2 must be "unfolded." The axial surface of the Townshend–Brownington Syncline dips steeply, and its axis plunges gently. Therefore, it seems reasonable to suppose that the Siluro–Devonian strata affected by this structure to the east and west were sub-horizontal before its formation (except in the hinges of early recumbent isoclinal folds). The orientation of the early axis of rotation, b_1 (referred to a geographic frame of reference), is obtained by performing that small circle rotation of b'_1 about the only horizontal axis within S in its present attitude, the direction of strike, through the angle, Ω_s (commonly the angle of dip for upright beds), that would restore S to the horizontal position considered to represent its approximate attitude at the end of rotation Ω_1. Ω_1, Ω_2, b'_1 and b_1 are not directly observable, but they are obtained by graphical analysis based on the above sequence.

One may contest the validity of the "unfolding" about the strike and, perhaps, propose a better operation. Some will say that b_2 parallels the hinge lines of the late folds and that "unfolding" should have been about b_2, but there are serious objections to both suggestions that will be discussed at length in a future publication. More importantly, because of the generally gentle plunges of b_2, "unfolding" of S about b_2 to, say, the attitude of gentlest dip will not lead to conclusions much different from those resulting from the simple transformation used.

IMPORTANCE OF STRATIGRAPHIC AND STRUCTURAL POSITION OF ROTATED GARNETS

The second feature of the new approach is based on contrasts in the rotational behavior of garnets in various structural positions (e.g., east or west flank of dome, long or short limb of drag fold, position with respect to boudinage in stratum) and on their stratigraphic position (e.g. the major (early) rotations for garnets in the Standing Pond Volcanics are opposite in sense to rotations in older units on the east limb of the Chester and Athens Domes (Fig. 14–1)).

Refined determinations were therefore made at many localities along two horizons at which good rotated garnets had been observed and in which the garnets showed this contrasting behavior: the base of the Northfield–Waits River sequence (henceforth horizon *WRa*) and the horizon of the Standing Pond Volcanics (horizon *SPa*) immediately adjacent to the Waits River Formation.

DIRECT RESULTS BASED ON ROTATED GARNETS

Figure 14–4 summarizes the principal results derived from the rotated garnets in Siluro-Devonian strata. In Figure 14–5, cross-sections of rotated garnets are projected into appropriate positions on structure section C–C′ of Figure 14–1 so as to provide a graphical picture of the tectonic sequence. Study of the rotated garnets makes a number of points apparent.

Two Major Tectonic Events Imprinted on Siluro-Devonian Rocks

In the region now occupied by the Chester and Athens Domes, isogradic surfaces (Rosenfeld, 1961, p. 15) for various reactions producing garnet traversed the rocks now exposed early enough to record in the garnets two major phases of diastrophism involving Siluro-Devonian and older rocks. The common continuity of and absence of sharp creases in the schistosity between the centers of the garnets and their exteriors comprise evidence that there was no long hiatus between these two events.

Earlier Deformation Much Stronger Than Later

The earlier of these two events (I) involved much greater deformation than the later (II), even ignoring the unknown deformation that may have occurred before garnet nucleation. Comparison of Figures 14–4a and 14–4b shows that rotations connected with event II are absent at many localities where garnets record event I, particularly in the Standing Pond Volcanics south of locality T9e. Figure 14–6 shows a garnet at locality T10b, 1.25 miles (2.01 km) south of locality T9e, that records only an early rotation of 180°. There the late rotation is recorded only as a crinkle in the surrounding rock not visible in the photograph.

The numerous large values of the early rotation (Ω_1) indicate that great strain must have accompanied event I. Experimentally confirmed deductions relating the rotation to various measures of simple shear and other types of deforma-

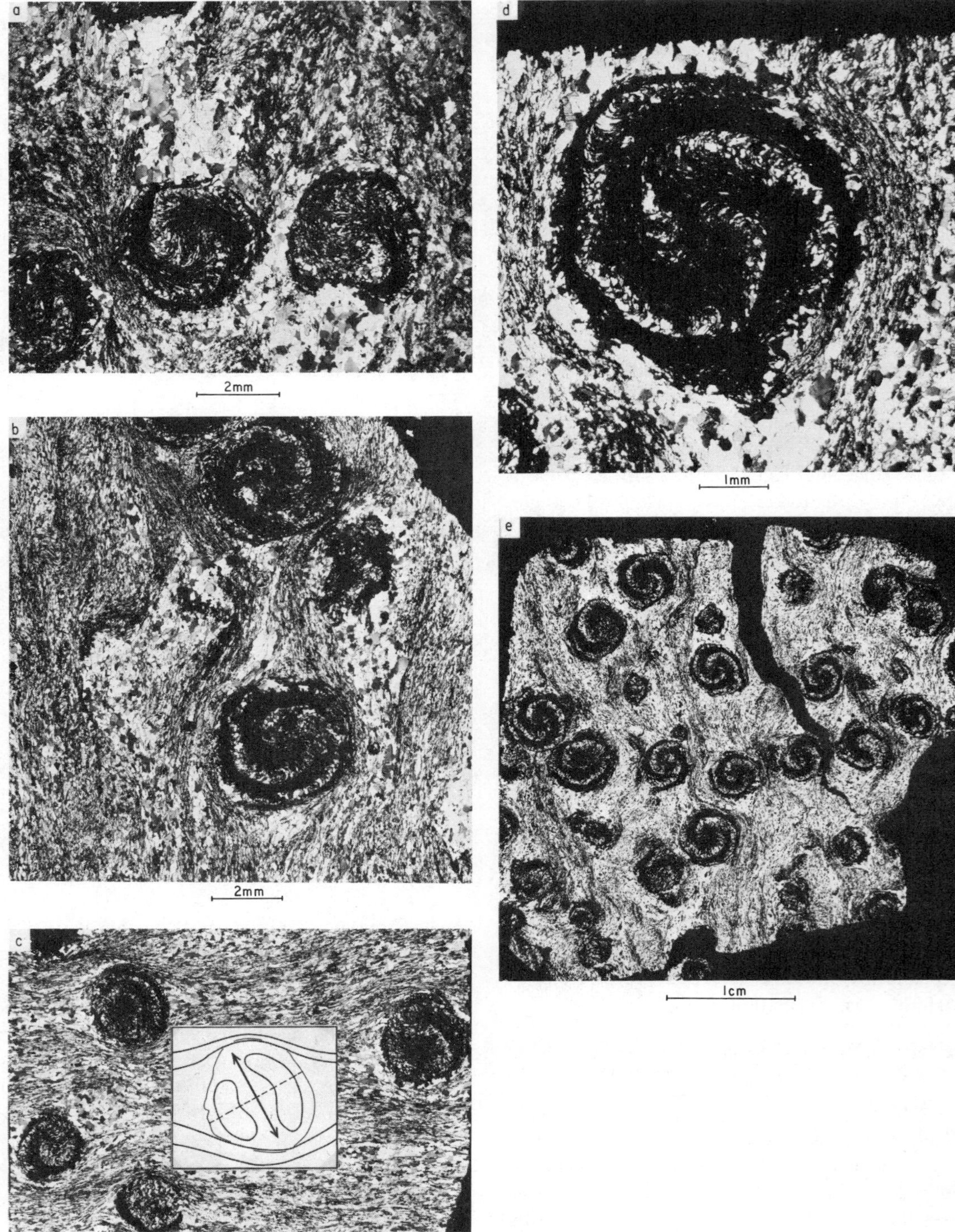

Figure 14–2. Photomicrographs of garnetiferous calcareous schist at locality S41e (see fig. 14–4); garnets up to 5 mm across: (a) Plane of schistosity (S), looking east. Apparent rotation indicates that relict rotational axis is not parallel to S. (b) Plane normal to b_2, view northerly along b_2. Note small drag fold parallel to b_2 in upper left showing clockwise rotation, garnet with counter-clockwise (early) rotation in lower center, and garnet with non-typical clockwise (early) rotation caught between two other garnets at top center (this non-typical example is illustrated to emphasize that, where garnets are crowded, analysis must be based on examination of a large number of garnets). (c) View downward on representative plane normal to S containing relict rotational axis for specimen. A plane of this kind, due to a certain amount of scatter of relict axes for individual garnets, normally has a small proportion of the garnets showing apparent rotations, approximately equally distributed as to sense, and a considerably larger proportion showing no apparent rotation. One reliable way to determine the orientation of the relict rotational axis for the specimen is to find two non-parallel planes of this kind. The orientation of their intersection is that of the relict rotational axis. Another way is to utilize the symmetry of the traces of schistosity surfaces for individual garnets on plane surfaces showing no apparent rotation. The relict rotational axis of an individual garnet is the normal to a plane of symmetry within this set of surfaces in the garnet. The appearance of this plane of symmetry on the plane surface of no apparent rotation is a line of symmetry (illustrated in inset by dashed line for representative traces of schistosity surfaces for garnet in upper left). The average direction of the normals to these lines (cf. arrow in inset) is the direction of the relict rotational axis of the specimen. The geometry of the included surfaces of schistosity is erroneously treated by Spry (1963, p. 212), who considers them to be cylindrical. There is only one surface of schistosity within a garnet that has rotated about a single axis during growth to which a straight line may be applied (in the mathematical sense), the surface of schistosity passing through the center of nucleation of the garnet. That straight line is unique and is the relict rotational axis. (d) Section normal to relict rotational axis viewed in a northeasterly direction; maximum *apparent* counter-clockwise rotation measured in this plane for another garnet at S41e is 458°. (e) Section with same orientation as (d) showing several rotated garnets.

tion will be discussed in a separate paper (Rosenfeld, in preparation). Of particular interest is the relationship between the amount of simple shear (γ) in the enclosing rock and the rotation (Ω) of a spherical garnet, $\gamma = 2\Omega$, where Ω is given in radians. Thus, to the extent that the deformation can be considered as approximating simple shear, the greatest first-stage rotation (Ω_1) deduced, 625° (10.91 radians) at locality S54a, implies an amount of simple shear γ equal to 21.82. A unit sphere subjected to a homogeneous deformation of this magnitude would become an ellipsoid with axes in the ratio, $[(\Omega^2+1)^{1/2}+\Omega]:1:[\Omega^2+1)^{1/2}-\Omega] = 21.87:1:0.0457$. The axis of maximum stretch would be inclined to the plane of schistosity (interpreted as a shearing plane) at an angle, $\alpha' = \frac{1}{4}\pi-\frac{1}{2}\tan^{-1}\Omega = 4.8°$. The axial ratios resulting from event I would be greater and the angle α' would be smaller if some deformation with the same sense occurred before nucleation of the garnet and if stretching occurred parallel to the schistosity. In contrast the second-stage rotation (Ω_2) is only 105° (1.79 radians), giving $\gamma = 3.58$, axial ratios $= 3.84:1:0.261$, and $\alpha' = 14.6°$.

Later Deformation Associated With Development of Domes and the Green Mountain Anticlinorium at Horizon *WRa.*

Thompson's hypothesis of upthrust for the domes seems to provide a partial explanation for event II. It does not account for the observed clockwise rotations at horizon *WRa* between the axis of the Townshend–Brownington Syncline and the crest of the Chester Dome (Figs. 14–1 and 14–5). It does explain the clockwise rotations on the east limb of the Chester and Athens Domes.

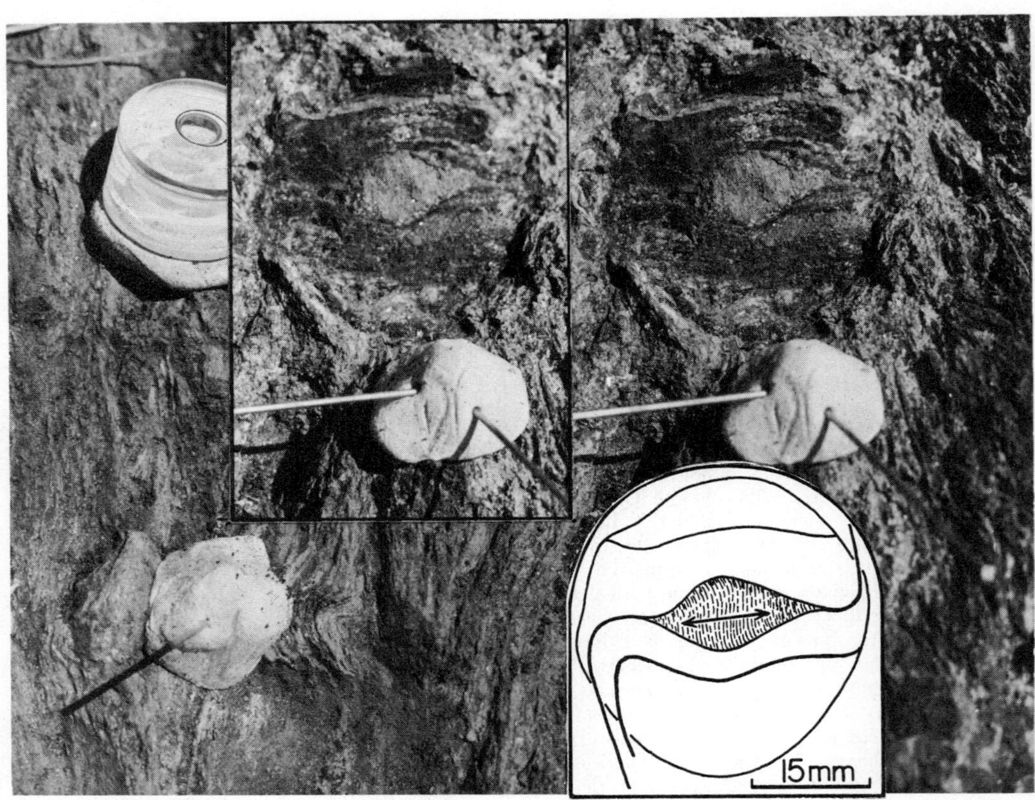

FIGURE 14–3. Doubly rotated garnet and small drag fold parallel to b₂ at locality T9e. The photographic inset provides stereoscopy for the garnet. This garnet has ruptured along a formerly ankeritic (now limonitic due to weathering) compositional band (also a surface of schistosity) passing through its center; and rupture along this band, indicated in line drawing inset by dashed pattern, therefore exposes a major segment of the relict rotational axis, indicated by arrow in line drawing (see explanation for figure 14–2c). The trace of the same band bends down on the left side of the garnet and up on the right, reflecting the same late clockwise rotation about b₂ during growth shown in the small fold. The lengths of straight wire (diam.: 0.0325 in. = 0.82 mm) from left respectively parallel b₂, the relict rotational axis, and the normal to the relict schistosity at the center of the garnet.

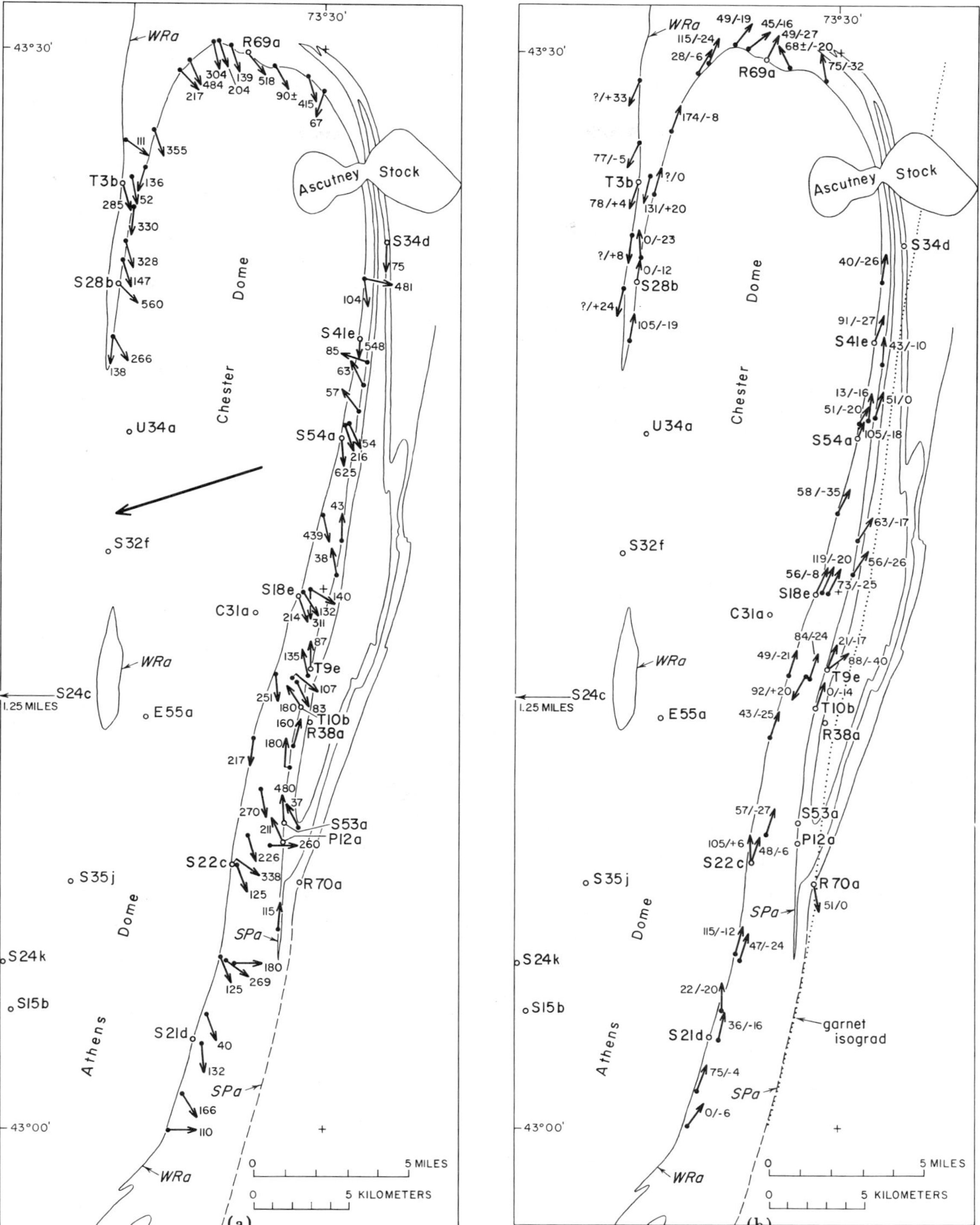

FIGURE 14-4. Maps showing localities referred to in text and garnet rotations associated with diastrophic events I and II in Siluro-Devonian strata: (a) Plot of axes of first stage rotation (b_1), giving senses of rotation (clockwise as viewed in direction of arrow) and amounts of rotation about the axes in degrees (Ω_1). (b) Plot of axes of second stage rotation (b_2), giving senses of rotation (clockwise as viewed in direction of arrow), amounts of rotation about the axes in degrees (Ω_2), bearings, and vertical angles from horizontal (figure after slash, $+$ if arrow points above horizontal, $-$ if below). Plot (a) ignores changes of position due to translation resulting from both events I and II; plot (b) ignores same for event II. Large arrow indicates direction of subhorizontal flow of event I.

As the sense of the late rotation at *WRa* reverses across the Townshend–Brownington Syncline, use of the conventional theory of drag-folding seems to imply that some crustal shortening accompanied development of the Green Mountain Anticlinorium and the Chester and Athens Domes as expressed at that horizon. Figure 14–4b shows the general nature of the rotations that developed at and between horizons *WRa* and *SPa* during event II. A flow pattern consistent with the late rotations at horizon *WRa* is indicated by the streamlines in Figure 14–11. Minor structural evidence, not presented here, supports an increased role for late upthrust in the genesis of the domes at deeper horizons.

Earlier Deformation Associated With Westward Subhorizontal Flow in Waits River Formation

Event I is most easily interpreted as a major subhorizontal westward flow before doming centered in the Waits River Formation. The existence of this flow is shown by the opposite senses of early rotation for the garnets at the two horizons most extensively sampled: counterclockwise at and near horizon *WRa* and clockwise at and near horizon *SPa*. Approximate horizontality of the strata during the early rotation is suggested by the fact that graphical restoration of the relict rotational axes to the horizontal by reversal of Ω_2 and Ω_s results in a surprising degree of paral-

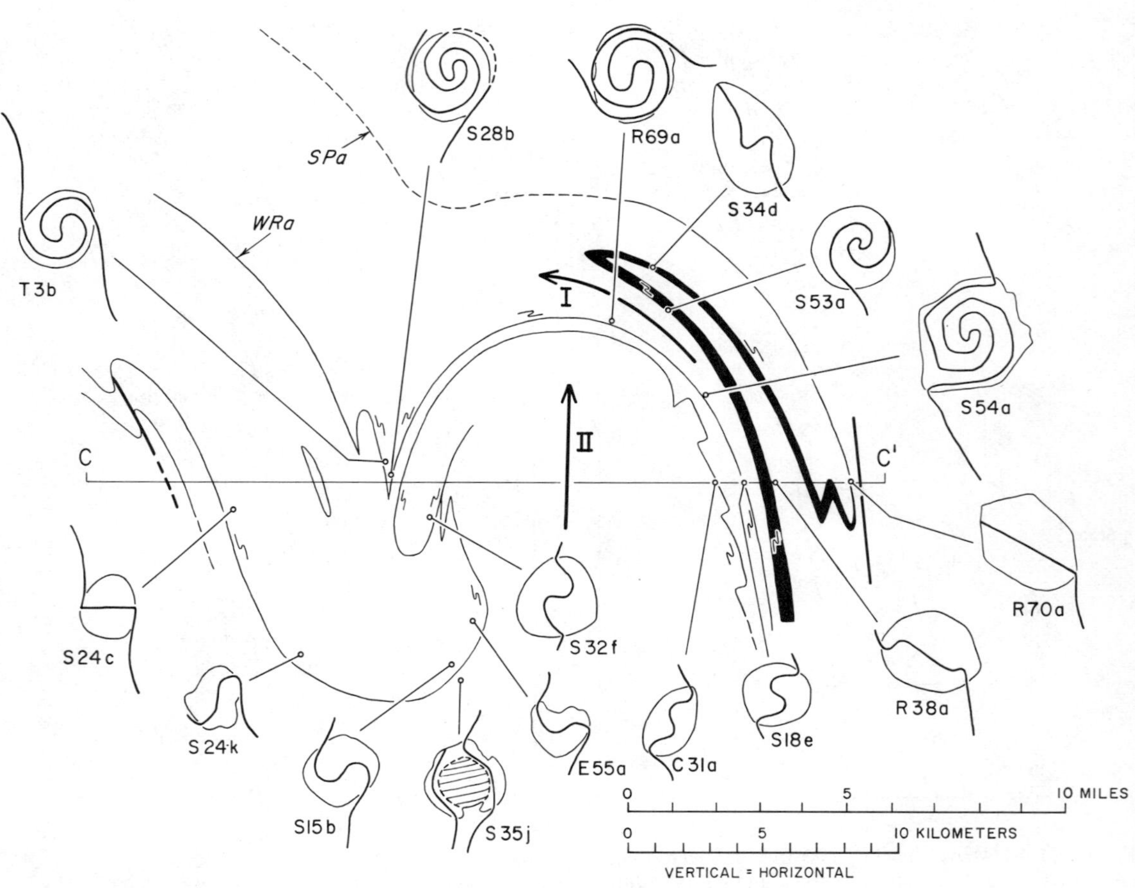

FIGURE 14–5. Patterns of garnets projected into cross-section C–C' of Figure 14–1. Sections of garnets are normal to the relict rotational axis where garnets grew during event I, and are viewed northerly along that axis. Garnets S24c and R70a are sectioned normal to the only axis present, b_2 for the latter and possibly b_2 for the former. Drag fold patterns are those observed to be conspicuous in field without regard to time of generation. Most represent event II. Scale of patterns variable. Figure 14–7 is a photomicrograph of garnet S35j.

lelism of the early rotational axes, as illustrated in Figure 14–4a. This orderly pattern over such a sizable area is most easily explained if the strata were relatively undeformed in their original near-horizontal attitude at the start of event I. Also transverse upthrust doming caused by buoyancy is more easily explained if the strata were approximately horizontal rather than highly inclined at the initiation of doming.

Early Deformation Associated With Development of Large Sigmoid Isoclinal Recumbent Folds

Event I, as recorded in the garnets, is clearly connected with the large sigmoid isoclinal folds adjacent to the domes. The axes of first-stage rotation (Fig. 14–4a), corrected for dip and the second rotation (b'_1), are approximately parallel to the axes of these folds. If the large sigmoids are interpreted as giant flexure-slip drag folds, they indicate the same kinematic picture as that implied by the small rotated garnets excepting those on the short limbs of the sigmoids (cf. loc. S34d).

Recognition of event I casts new light on the map pattern of the Standing Pond Volcanics

FIGURE 14–6. Schist from locality T10b containing large garnet separated along ankeritic band passing through its center. The relict rotational axis is parallel to b'_1 (i.e. $\Omega_2 = 0°$). Small rubber model simulates general form of surface of schistosity passing through centers of rotated garnets. The double-ended arrow (26 mm long) on the model is parallel to b'_1 of the garnet. Curved lines on model are schematic simulations of plane sections of the included schistosity normal to (dotted) and oblique to (solid) the axis of a garnet that has rotated during growth. Ankerite in garnet gives way to coarse fascicular hornblende outside garnet.

within the area. What had appeared to represent separate isoclinal folds associated with the domes now appears to be much more reasonably interpreted as a single giant sigmoid fold (= Strafford Village "syncline" and Grannyhand "anticline" of White and Jahns, 1950, p. 211), extending at least 100 miles (160 km) in an approximately north-south direction, whose map pattern has been modified by the culminations associated with development of seven domes (Fig. 14–1). This fold is here named the Ascutney Sigmoid after the conspicuous and well known Mount Ascutney Stock that transects it. The orientations of the axes of minor folds on the exposed hinges of this sigmoid and their shear senses are entirely consistent with this interpretation (Fig. 14–1; Fig. 14–5).

The Star Hill Sigmoid on the west limb of the Chester Dome probably dies out in a rather short distance somewhere over the Athens Dome. Otherwise it should manifest itself in the outcrop pattern on the southeast side of this dome. The small synclinal trough containing Siluro-Devonian rocks to the west of the waist between the Chester and Athens Domes just south-southwest of Chester is tentatively interpreted here as the detached (i.e. boudinaged) lower hinge of this sigmoid.

Early and Late Deformations Associated With Rotations About Different Axes

The early and late rotational axes of the garnets depart widely from parallelism in much of the area. Near the Chester and Athens Domes the early rotational axes (b_1) tend to average north-northwest (Fig. 14–4a), whereas the late axes (b_2) tend to plunge north-northeast to northeast (Fig. 14–4b). This is well illustrated east of Chester Dome. For the examples shown in Figures 14–2 and 14–3, the angles between the rotational axes within the schistosity, that is, between b'_1 and b_2 are 33° and 89° respectively. This non-parallelism of the two tectonic axes helps to explain the location of the detached lower hinge of the Star Hill Sigmoid. Thus, if the hinge was oriented in a north-northwest direction after rotation ceased about b_1, subsequent folding about the north-northeast axis associated with the development of the Townshend-Brownington Syncline would have bent it into a doubly plunging trough which plunges beneath the surface on the northwest and reappears on the southeast. Similarly the large open sigmoid fold (Chang et al., 1965, p. 54) on the east limb of the Green Mountain Anticlinorium northwest of Chester Dome (43°36'N 73°36'W) could represent the northerly extension of the

Star Hill Sigmoid. This interpretation implies a northward change in direction of the hinges of this structure toward an alignment more northeasterly than that of the later folding, in accord with the apparent early westward convexity of the hinges of the Ascutney sigmoid at the same latitude.

Early Paleozoic Metamorphism and Diastrophism

Some of the rotated garnets in the Early Cambrian (?) Pinney Hollow Formation show a more complicated history than any observed within the Siluro-Devonian strata. At localities U34a and S35j (Fig. 14–4) tectonometamorphic "angular unconformities" appear in many of the larger garnets of that formation (Fig. 14–7). In each of these garnets an inner zone has its enclosed relict schistosity truncated against a sharp boundary with an outer zone having different color, refractive index, included mineral assemblage, and orientation for its enclosed relict schistosity. The outer zone must therefore have grown under altered metamorphic conditions *after* a rotation of the inner zone, unrecorded in the garnet, and *before* the development of evidence for b_1 in the outer zone. A section of specimen S35j parallel to the schistosity shows that the inner core, containing chloritoid and staurolite, rotated about another axis during growth before event I. A Paleozoic diastrophism earlier than I, accompanied by metamorphism at least into the garnet zone, must have been present in the area. Zen (1967, p. 44–69) identifies a major Middle Ordovician diastrophism in the Taconic Mountains in which there was transport of material from the east, possibly off an earlier Green Mountain Anticlinorium expressed in Ordovician and older rocks. It seems plausible to equate the deformation indicated by the earliest rotation within such garnets with Zen's Taconic diastrophism. Further study of these unusual garnets may shed light on the regional distribution and nature of this diastrophism and its concomitant metamorphism.

RELATED PHENOMENA AND THEIR IMPLICATIONS

The points discussed above have further implications affecting interpretation of related phenomena associated with the late Paleozoic diastrophism. Some of the major implications and the presently available evidence relating to them follow. Others will be treated in a subsequent paper.

Divergence of Horizons *WRa* and *SPa* Toward the West

Clearly, with the flow toward the west centered in the Waits River Formation during event I, it would not be surprising for horizons *WRa* and *SPa* to diverge toward the west. Evidence of this divergence is prominent in eastern Vermont. Defining the separation between these horizons as apparent thickness, White and Jahns (1950, p. 190–191) give a figure of 24,000 feet (7,300 m) for the east limb of the Green Mountain Anticlinorium in central Vermont. Based on Thompson's work (1952, oral communication), a similarly derived figure not far south of Mt. Ascutney on the east limb of the Chester Dome is only 1,350 feet (410 m).

Comparison of attitudes of the two horizons on the east side of the Chester and Athens Domes provides a crude indication of westward thickening before II. Commonly, though not invariably, attitudes near horizon *SPa* are steeper than those near *WRa*. Because there has been considerable minor folding, however, attitudes measured in the field do not provide an accurate measurement of the divergence. A better measurement can be obtained by combining plunge data from the rotated garnets (b'_1) with regional divergence as shown on the geologic map (Fig. 14–4a). One

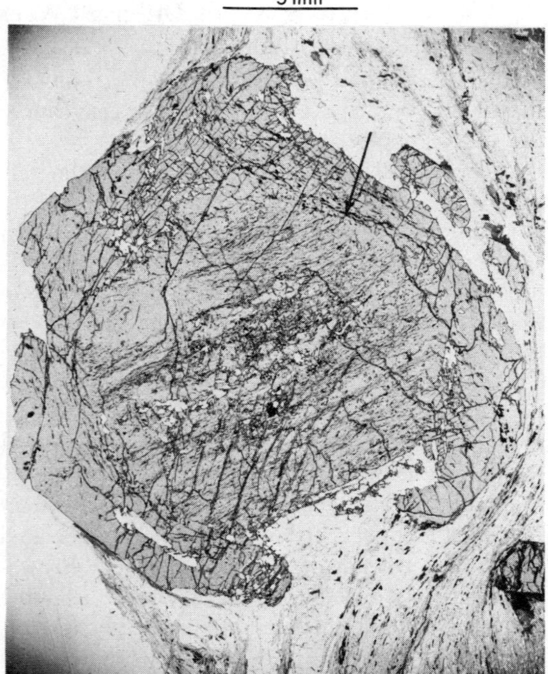

FIGURE 14–7. Garnet in staurolite-bearing paragonite schist of Pinney Hollow Formation from loc. S35j showing tectonometamorphic "angular unconformity" (arrow). View N39°E parallel to the strike. Garnet 20 mm across.

of the most conspicuous features of the geologic map on the east side of the two domes is the southward divergence of *WRa* and *SPa* in the plane of the map. This divergence cannot be explained by a decrease in average dip because the dips do not show any appreciable change. Thus, there is an *apparent* southward thickening of the material between *WRa* and *SPa* in this area. If it can be assumed that the divergence is in the direction of flow deduced from the rotated garnets (i.e. that it lies in the plane perpendicular to b'_1), then the gross attitudes of the boundaries of the section and, therefore, the angle between *WRa* and *SPa* can be determined. In the area near Saxtons River Village, the strikes of the two boundaries and the early rotation axis (b'_1) are plotted on the stereographic net. The assumption that the divergence is in the direction of flow means that b'_1 is parallel to both horizons *WRa* and *SPa*. Therefore the gross attitudes of the two surfaces are each specified by the great circle that contains b'_1 and the strike direction of the surface. Near Saxtons River, Vermont, horizons *WRa* and *SPa* strike N5°E and due N respectively. One large rotated garnet (P12a) at horizon *SPa* near Saxtons River Village has b'_1 oriented N3°W 27° up from the horizontal. Thus, the gross dips of horizons *WRa* and *SPa* are 74°E and 83°E respectively; the angle between them is therefore about 10°. Utilization of other garnets from the same locality gives values ranging from 8° to 13°.

Although additional more detailed work is needed, the gravity results of Bean (1953, p. 528–533, Pl. 3) for the Strafford and Pomfret Domes also suggest a westerly divergence of horizons *WRa* and *SPa*. Bean found residual negative gravity anomalies over these domes (after correction for a regional gradient) that indicate rocks of relatively low density at fairly shallow depths, supporting the belief that these structures are mantled gneiss domes similar to those to the south-southwest. Assuming that they are indeed mantled gneiss domes, any wedge-like thickening of the relatively dense mantling strata across one of these domes should displace the residual negative gravity anomaly due to the dome in a direction opposite to the direction of thickening. Bean's data (Bean, 1953, Pl. 3) do indeed indicate a south-southeasterly displacement of the anomaly over the Strafford Dome and a north-easterly displacement of it for the Pomfret dome, both of which are consistent with a general westerly thickening of the Waits River Formation.

Strain Phenomena Indicative of Westward Extending Flow During the First Tectonic Event

Quartz conglomerate (S21d, Fig. 14–8) and blastoporphyritic amphibolite in the Shaw Mountain Formation and crystal tuffs in the Standing Pond Volcanics contain pebbles and phenocrysts within Siluro-Devonian units that, by their resultant change in shape, reflect the superposition of deformations that have affected these units. Because of the much greater deformation associated with event I, one would expect its effects to dominate. Simple shear parallel to the bedding schistosity would lead to early maximum stretch axes canted at various angles to the west from perpendicular to the schistosity, dependent on local heterogeneities. The canting would be counterclockwise at *WRa* and clockwise at *SPa*; there would be no strain within the bedding planes.

What are the observed facts? Figure 14–9 shows the stretch directions of saussuritic plagioclase phenocrysts and quartzite pebbles (with Ω_s subtracted in order to give a comparison with Figure 14–4a). The direction of stretch agrees rather well with the direction of flow adduced from Figure 14–4a, especially considering that event II must have had an effect. An important complication appears near the hinges in tight folds, where the stretch tends toward parallelism with the fold axes. A second important complication is that the direction of stretch is invariably parallel to the bedding and schistosity of the rocks. If these were simple shearing planes, there would be no strain within them, and the direction of stretch should be inclined to them at an observable angle, in conflict with the strain and parallelism actually seen.

Deformed pebbles and rotated garnets in quartz conglomerate and quartzite of the Shaw Mountain Formation give additional evidence that the simple shear model is too restrictive. The oil-soaked serial slices of quartz conglomerate, illustrated in Figure 14–8, are cut to show the principal dimensions of the pebbles. Averaging of numerous careful measurements on these slices gives axial ratios, 8.76:0.779:0.147. At locality S22c, where pebbles show similar deformation, the rotation of garnets in the quartzite is only 15°, in contrast to 338° in calcareous schist a few feet away.

A better model for the deformation in the Waits River Formation accompanying event I would be that of extending flow, as in certain parts of a valley glacier (see Nye, 1952, p. 87,

89–90) or of a stratum of salt feeding into a salt anticline (Jones, 1959). In this approximation the deformation is the superposition of (1) a simple shearing motion with shearing planes parallel to the stratification and the direction of shear perpendicular to b_1 and (2) extension parallel to the bedding in the direction of flow and compression perpendicular to the bedding, with an intermediate extension or compression perpendicular to these two and parallel to b_1. The two kinds of deformation evidently differ in importance in the different strata. The first kind, which is recorded best by the rotated garnets, is important in the more easily deformed schists but is practically insignificant in the less easily deformed quartzites and amphibolites. The second kind is certainly important in the less easily deformed strata and may be of comparable importance in the more easily deformed strata, assuming no discontinuities at stratigraphic boundaries of strain components parallel to the stratification.

Models of deformation will be considered further in another paper (Rosenfeld, in preparation). An illuminating discussion has been given by Thompson (1950, p. 112–128).

Relationships of a Stratigraphic Uncertainty to Alternative Possible Mechanisms for Early Event

It is not as yet possible to give a unique diastrophic model of event I. This uncertainty results from the stubborn refusal of the Standing Pond Volcanics, in the face of repeated negotiations in the field, to yield any but ambiguous answers concerning the secret of this unit's stratigraphic facing. This seemingly easily obtainable geological fact is needed to decide between the two models that have been advocated.

In one model (Fig. 14–10) the Standing Pond Volcanics *underlie* the upper part of the Waits River Formation stratigraphically (for details and critical discussion see: Ern, 1963; Goodwin, 1963; Chang et al., 1965, p. 40–42, 56–62). The Waits River Formation along the east side of the Chester and Athens Domes lies in a synclinal trough; the Standing Pond Volcanics inside this trough thin to disappearance on the west limb. Both

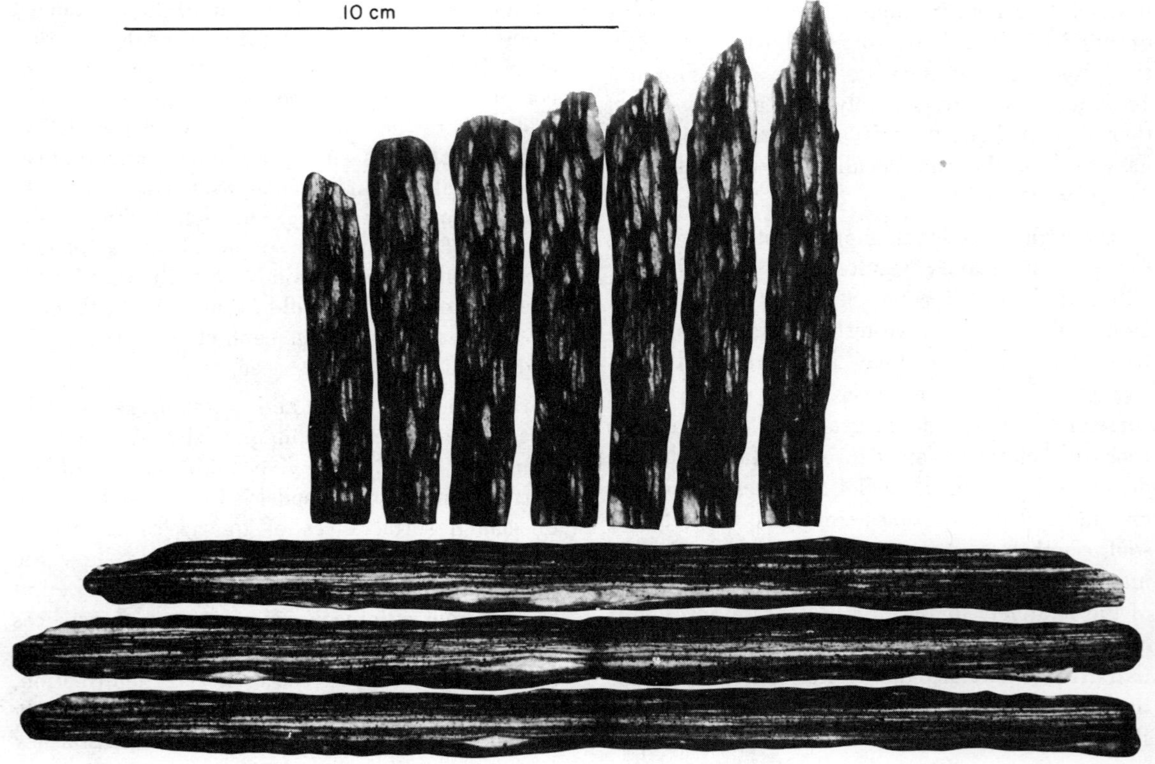

Figure 14–8. Oil-soaked serial sections of deformed conglomeratic quartzite of Shaw Mountain Formation from locality S21d viewed between crossed polars. The three slices at the bottom are parallel to the directions of greatest and least stretch, those above the direction of intermediate and least stretch. Attitude: N14°E 64°E. Plunge of long axes: N38°E 39°. Intermediate stretch less than 1 suggests convergent extending flow toward west-southwest during event I.

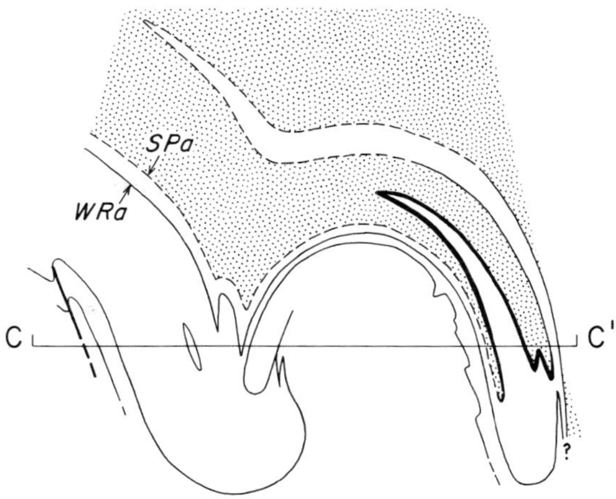

FIGURE 14–10. Schematic view of cross-section C–C′ if the Standing Pond Volcanics are older than the Waits River Formation.

the Ascutney and Star Hill Sigmoids are very large, but otherwise conventional, flexure slip drag folds in a westward-opening formerly recumbent syncline. This recumbent syncline, in turn, is the complementary lower half of an even larger formerly recumbent sigmoid fold whose western hinge at horizon SPa lay in or an unknown distance to the west of the axis of the Townshend–Brownington Syncline. The Gile Mountain Formation in the core of the westward-closing upper half of this giant sigmoid "roots" in the easternmost band of Gile Mountain Formation (Fig. 14–1). The kinematic information contained in this paper does not entirely preclude this model, although I find it difficult to envisage a sequence of emplacement of such a large nappe with a root in that particular area. The principal difficulty confronting this model is the fact that the axis of low-temperature metamorphism lies within the easternmost band of Gile Mountain Formation (Doll et al., 1961), implying that formerly more deeply buried masses of rock became colder than shallower rocks after tectonic processes raised them to the same level.

The other model has the Standing Pond Volcanics stratigraphically *overlying* the Waits River Formation, the relationship implied by most of the cross sections of Doll et al. (1961). This model has much to recommend it (Rosenfeld, 1960, 1965), but at present it is probably wise to leave the question open until better stratigraphic evidence is forthcoming. In this model the early westward flow in the Waits River Formation is intrastratal. The process is essentially like that demonstrated for the origin of salt anticlines

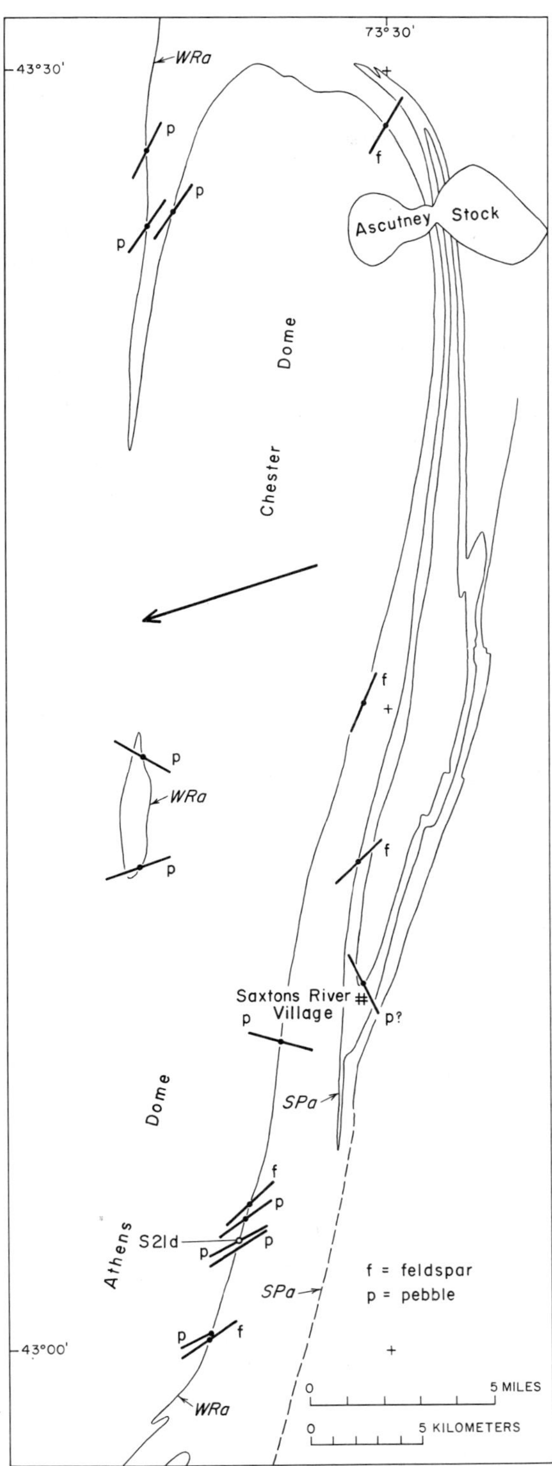

FIGURE 14–9. Map of axes of maximum stretch (heavy bars) of pebbles and phenocrysts with late rotation of strata, Ω_s, removed. Large arrow indicates direction of subhorizontal flow of event I.

(Jones, 1959; Seidl, 1927, Fig. 40), with the exception that the flowing schists were dense relative to the subjacent basement. In this model the higher density of the schists caused them to flow westward into a structural trough instead of an anticlinal crest, possibly as a result of being uplifted to the east. This model, illustrated in Figure 14–11, has one important advantage over the preceding one. It places a late synclinal trough containing Gile Mountain Formation, developed during event II, at the low-temperature metamorphic axis in what would have been a structurally high position at the cessation of event I. An additional item of evidence favoring a syncline in the same location occurs at locality R70a (Figs. 14–4 and 14–5), a few feet west of the garnet isograd. There the contained

schistosity in garnets of the Standing Pond Volcanics, on the east (long) limb of the Ascutney Sigmoid, is planar; the garnets show a small counterclockwise rotation about horizontal axes after growth. Other evidence cited earlier in this paper indicates that the Ascutney Sigmoid can be interpreted best as a giant flexure-slip drag fold. If these garnets formed *before* event I, they should therefore show a clockwise rotation of the planar bands relative to the schistosity. If they grew *during* event I, the included schistosity in cross-section should be in the form of a sigmoid spiral indicating clockwise rotation. They therefore must have grown *after* event I and *during* a quiescent interval before or during event II. This relatively late growth (probably very near the thermal maximum) is indicated further

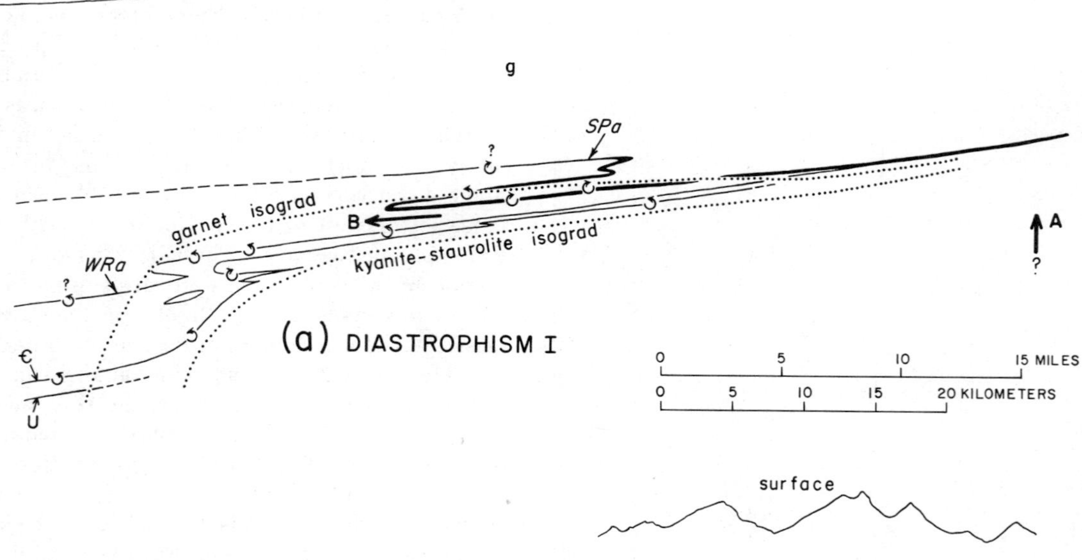

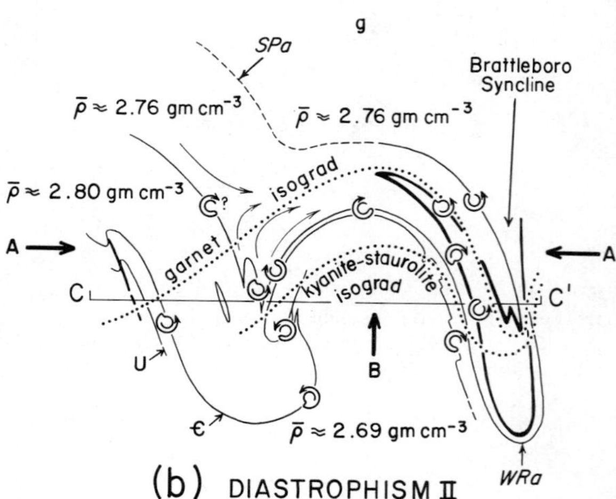

FIGURE 14–11. Schematic model for late Paleozoic diastrophism in southeast Vermont applied to evolution of section C–C′ if Standing Pond Volcanics are younger than Waits River Formation. (a) Event I consists of epeirogenic uplift to the east, IA, followed by west-southwest flow, IB, centered in the Waits River Formation. Small arcs with arrows give rotation during this stage. (b) Event II consists of east-west horizontal compression, IIA, and possibly concurrent buoyant (?) upthrust of mantled gneiss domes, IIB, with some possible easterly flow of material off Green Mountain Anticlinorium and over Chester Dome. Curved arrows in vicinity of Townshend-Brownington Syncline indicate inferred streamlines and directions of flow late in II. Rotation symbol shows sequence of rotations related to both I (inner) and II (outer). (a) and (b) are drawn to indicate the cross-sections at the close of I and II respectively. Depth below surface in (b) is based on petrologic criteria (Rosenfeld, 1968, p. 345).

by their proximity to the garnet isograd. They also grew east of the region affected by *clockwise* rotations associated with event II discussed in an earlier part of this paper. Their counterclockwise rotations are difficult to reconcile with the first model. The rotations are consistent with the late syncline to the east according to the second model in which the deformation was by flexure-slip, as in the case of deformation at horizon *WRa* in the Townshend–Brownington Syncline west of the domes.

Perhaps the chief objection to the second model is the relatively great rigidity required for the overlying rock units during event I while there was a flow toward the west underneath, centered in the Waits River Formation. Why weren't the overlying strata rafted along during the flow rather than remaining in place to provide the observed shear sense? Perhaps the analogy with salt structures is valid, and the calcareous schists and phyllites of the Waits River Formation were far more easily deformed during event I than the stratigraphically thick (as shown by the mineral assemblages) quartzo-feldspathic schists above them, which were also colder (and thus presumably even less easily deformed).

Because of the extensive extrapolations involved, experimental work, such as that of Heard (1963), cannot yet shed light on the reasonableness of this model. It is worthwhile, however, to anticipate the nature of an experimental confrontation. It would be difficult to allow a duration for event I of more than 50 million years (1.6×10^{15} sec). Shear strains of at least 20 deduced from rotated garnets in some schists therefore point toward an average strain rate $\dot{\gamma}$ greater than about 10^{-14} sec^{-1}. The temperatures during the growth of the garnets were probably within the range of 400 to 500°C; and the pressures were approximately 5 kb. The angle of tilt θ of the Waits River Formation could hardly have been greater than 10°. Given a thickness, h, of 3 kilometers, the shear stress, $\tau = \frac{1}{2}gh\Delta\rho \sin\theta$, would have been near 3 bars, or less, where g is the gravitational acceleration and $\Delta\rho$ is the density contrast. Thus the viscosity ($= \tau/\dot{\gamma}$) would have been less than 3×10^{20} cgs. It will be interesting to see what light future experiment will shed on this conjecture.

Possibility of Continuity Between Acadian and Appalachian Diastrophisms at Depth

The intriguing possibility presents itself that event I is Acadian and event II is Appalachian. The marine offlap that accompanied the development of the Catskill Delta to the west during the late Paleozoic suggests uplift to the east. The associated tilt toward the west might simultaneously have caused the westward flow associated with event I in the second diastrophic model above. The features of event II would seem to fit in well with the formation of the folds of the Valley and Ridge Province of the Appalachians, which are considered to have developed during the latest Paleozoic.

Two specific kinds of geochronological study would provide a check on this speculation. The relatively limited deformation of apophyses (as around the granite mass in the Guilford Dome) and other features associated with some syntectonic granites in eastern Vermont (see, for example: Dale, 1909, p. 25) indicates that they were intruded after event I but well before completion of event II. Geochronological study of these granites should provide a lower limit on the age of event I. Prompted by this observation, R. S. Naylor is performing such a study (1966, oral communication). Comparison of mineral assemblages encapsulated in garnets near horizon *SPa* at a number of localities indicates that fascicular hornblende outside of the garnet developed during event II. It may be possible to date event II by the method of Steiger (1964). This method might also be useful for dating event I, inasmuch as hornblendes in some units are essentially parallel to elongate pebbles and phenocrysts in adjacent rocks. This suggests that they grew during event I and were bent around folds during event II.

REFERENCES

Bean, R. J., 1953, Relation of gravity anomalies to the geology of central Vermont and New Hampshire: Geol. Soc. America Bull., v. 64, p. 509–538

Billings, M. P., Thompson, J. B., Jr., and Rodgers, J., 1952, Field Trip No. 1, Geology of the Appalachian Highlands of east-central New York, southern Vermont, and southern New Hampshire: Guidebook for Field Trips in New England, Geol. Soc. America, 65th Ann. Meeting, p. 1–71

Chang, P. H., Ern, E. H., Jr., and Thompson, J. B., Jr., 1965, Bedrock geology of the Woodstock Quadrangle, Vermont: Vermont Geol. Survey Bull., no. 29, 65 p.

Dale, T. N., 1909, The granites of Vermont: U. S. Geological Survey, Bull. 404, 138 p.

Doll, C. G., Cady, W. M., Thompson, J. B., Jr., and Billings, M. P., *compilers* and *editors*, 1961, Centennial geologic map of Vermont: Montpelier, Vermont. Vermont Geol. Survey, scale 1:250,000

Ern, E. H., Jr., 1963, Bedrock geology of the Randolph Quadrangle, Vermont: Vermont Geol. Survey Bull., no. 21, 96 p.

Gilluly, James, *in* Lovering, T. S., et al., 1938, Report of the committee on structural petrology: Division of Geol. and Geog., Nat. Research Council, Wash., D. C., p. 52–58

Goodwin, Bruce K., 1963, Geology of the Island Pond area, Vermont: Vermont Geol. Survey Bull., no. 20, 111 p.

Heard, Hugh C., 1963, Effect of large changes in strain rate in the experimental deformation of Yule marble: Jour. Geology, v. 71, p. 162–195

Jones, R. W., 1959, Origin of salt anticlines of Paradox Basin: Am. Assoc. Petroleum Geologists Bull., v. 43, p. 1869–1895

Nye, J. F., 1952, The mechanics of glacier flow: Jour. Glaciology, v. 2, p. 82–93

Paterson, M. S., and Weiss, L. E., 1961, Symmetry concepts in the structural analysis of deformed rocks: Geol. Soc. America Bull., v. 72, p. 841–882

Peach, B. N., Gunn, W., Clough, C. T., Hinxman, L. W., Crampton, C. B., Anderson, E. M., and Flett, J. S., 1912, The geology of Ben Wyvis, Carn Chuinneag, Inchbae and the surrounding country, including Garve, Evanton, Alness and Kincardine (explanation of sheet 93): Mem. Geol. Surv., Scotland, p. 1–189

Rosenfeld, J. L., 1954, Geology of the southern part of the Chester Dome, Vermont: Ph.D. thesis, Harvard University, 303 p.

———, 1960, Rotated garnets and the diastrophic-metamorphic sequence in southeastern Vermont [abs.]: Geol. Soc. America Bull., v. 71, p. 1960

———, 1961, The contamination-reaction rules: Amer. Jour. Sci., v. 259, p. 1–23

———, 1965, Further evidence of the nature of the major diastrophism in southeast Vermont [abs.]: Geol. Soc. America Spec. Paper 82, p. 167

———, 1968, Piezobirefringence around quartz inclusions in almandine and conditions of metamorphism [abs.]: Trans. Amer. Geophys. Union, v. 49, p. 345

———, (in preparation) The observation and interpretation of rotated garnets in metamorphic rocks

Segerstrom, Kenneth, 1956a, Geologic map of the Colrain Quadrangle, Massachusetts-Vermont, Bedrock Geology: U. S. Geol. Survey Geol. Quad. Map GQ–86

———, 1956b, Geologic map of the Shelburne Falls Quadrangle, Massachusetts, Bedrock Geology: U. S. Geol. Survey Geol. Quad. Map GQ–87

Seidl, E., 1927, Die Salzstöcke des deutschen (germanischen) und des Alpen-Permsalz-Gebietes; ein allgemeinwissenschaftliches Problem: Kali, Zeitschr. f. Gewinnung, Verarbeitung, und Verwertung der Kalisalze, v. 21, p. 34–38, 77–84, 109–111, 205–209, 240–244, 259–264, 295–312, 326–337, 346–360

Spry, Alan, 1963, The origin and significance of snowball structure in garnet: Jour. Petrology, v. 4, no. 2, p. 211–222

Steiger, R. H., 1964, Dating of orogenic phases in the central Alps by K–Ar age of hornblende: Jour. Geophys. Research, v. 69, p. 5407–5421

Thompson, J. B., 1950, A gneiss dome in southeastern Vermont: Ph.D. thesis, Massachusetts Inst. of Tech., 149 p.

White, W. S., and Jahns, R. H., 1950, Structure of central and east-central Vermont: Jour. Geology, v. 58, p. 179–220

Zen, E-an, 1967, Time and space relationships of the Taconic Allochthon and Autochthon: Geol. Soc. America Spec. Paper 97, 107 p.

Chapter 15

Nappes and Gneiss Domes in West-Central New England

JAMES B. THOMPSON, JR., PETER
ROBINSON, TOM N. CLIFFORD, AND
NEWELL J. TRASK, JR.

Upon the first day of field work in New Hampshire I saw the quartzite of Moose Mountain; and said to my companion, 'There is the key to the geology of this region.'—C. H. Hitchcock (1912, p. 138).

INTRODUCTION

A STRIKING FEATURE of the geology of western New England is the occurrence of two belts of mantled gneiss domes roughly paralleling the Connecticut River. The western belt extends south from east-central Vermont through the eastern Berkshire highlands of Massachusetts into western Connecticut. The eastern belt, comprising about twenty domes, extends from the Maine–New Hampshire boundary near Berlin, New Hampshire to Long Island Sound, east of New Haven, Connecticut. The eastern belt constitutes the Bronson Hill anticline of Billings (1956) which, in view of the complexities now apparent, we shall refer to as the Bronson Hill *anticlinorium.*

The domes of the Bronson Hill anticlinorium have a rough *en échelon* arrangement so that their long axes are oriented north–south, whereas the axis of the anticlinorium trends north-north-east (Pl. 15–1a). The rocks in the cores of the domes are mainly gneisses ranging from quartz diorite to granite in composition, and the mantling strata include a variety of gneisses, amphibolites, mica schists, quartzites and calc-silicate rocks of Paleozoic age. Our results, summarized below, show that the domes themselves are a late tectonic feature, preceded by the formation of at least three giant nappes. The tectonic style of the belt as a whole is, therefore, much like that of the Pennine Alps.

HISTORY OF INVESTIGATION

The anticlinal nature of the gneiss domes of the Bronson Hill anticlinorium was recognized by C. H. Hitchcock and his co-worker, J. H. Huntington (Hitchcock, 1877) who considered the distinctive gneisses in the cores to be "protogene" gneisses, older than the surrounding schists and quartzites. Somewhat later Hitchcock (1883, 1890) expressed the view that the core gneisses were largely metavolcanic in origin, and that the domes themselves marked the loci of volcanic islands, around and upon which the mantling strata were later deposited. Hitchcock's views at this time were, in fact, quite close to the interpretation now favored by us except that the core gneisses are now regarded as largely Ordovician metavolcanics and intrusives rather than as Laurentian ones. Emerson (1898) interpreted the cores as metamorphosed "conglomerate gneiss" of Cambrian age. After the turn of the century, however, both Hitchcock (1908, p. 165–166) and Emerson (1917, p. 241–243) changed their earlier views and regarded the core gneisses as igneous intrusives *younger* than the surrounding rocks. They explained the concordant marginal foliation as related to metamorphism and deformation of the partially solidified outer parts of the domes during the later stages of crystallization.

That certain of the metamorphosed strata of the Connecticut Valley region are Paleozoic in age became apparent with the discovery of fossils at Bernardston, Massachusetts (Hitchcock, 1835) and near Littleton, New Hampshire (Hitchcock,

1871, 1874). Hitchcock (1912) was aware of the need for detailed structural studies, but early attempts at Bernardston (Emerson, 1898, p. 253–299; Hitchcock, 1877, p. 428–457) and in the area near Skitchewaug Mountain (Pl. 15–1a, locality 6) (Hitchcock, 1912, p. 125–139) were unsuccessful owing to poor base maps and unrecognized complexities. Inverted sections, for example at Bernardston and on Skitchewaug Mountain, were not recognized as such, although Richardson (1931, p. 201) suggested that the quartzites of Skitchewaug Mountain might have been thrust to the west from New Hampshire.

Modern techniques of mapping and petrologic interpretation were first used in this belt in western New Hampshire in the 1930's by Billings (1937) and have been carried on more or less continuously since then by Billings and his students and associates (Hadley, 1942, 1949; Page, 1937; Lyons, 1955; Chapman, 1939, 1942, 1952; Kruger, 1946; Heald, 1950; Fowler–Billings, 1949; Moore, 1949). Related studies in Massachusetts were carried out more or less contemporaneously, by Balk (1941, 1942, 1956 a,b,c, 1957), Hadley (1949), and Willard (1951, 1952).

Billings, by detailed mapping and careful study of the Silurian and Devonian fossil localities near Littleton, New Hampshire, established a stratigraphic succession that he could again identify in the intensely folded and highly metamorphosed rocks of the Bronson Hill anticlinorium a few miles to the southeast. This succession can now be recognized as far south as east-central Connecticut (Rosenfeld and Eaton, 1956; Eaton and Rosenfeld, 1960), and has been verified by restudy of the long-known fossil locality at Bernardston, Massachusetts (Boucot et al., 1958) and by the discovery of some two dozen new localities between Moose Mountain, Hanover, New Hampshire and Springfield, Vermont (Boucot and Thompson, 1958, 1963; Thompson, unpublished data). The elaborate structures described here would never have been recognized without the tight stratigraphic control made possible by Billings' work farther north.

Another significant discovery was the recognition by Chapman (1939) and by Kruger (1946, see also Kruger and Linehan, 1941) that the bodies of granodiorite gneiss (Bethlehem Gneiss) southwest of Mascoma Lake, New Hampshire (Chapman) and east of Bellows Falls, Vermont (Kruger) are floored outliers, probably connected at one time to the Mount Clough pluton, a larger mass of the same type of rock that crops out over extensive areas east of the Bronson Hill anticlinorium.

The current series of investigations began in the spring of 1950 with the discovery by Thompson and A. J. Boucot of fossils of probable Silurian age in the quartzites at Skitchewaug Mountain (Billings et al., 1952). The occurrence was puzzling because the quartzites are surrounded and structurally underlain by rocks regarded as Devonian. During subsequent mapping by Thompson and by Ratté (1952), it soon became clear that a major recumbent fold, the Skitchewaug Nappe, was the reason for the inversion (Thompson, 1954, 1956). With this impetus, an extensive program of detailed re-mapping was begun by us and is still in progress. Our present interpretation is shown in generalized form on Plate 15–1 and discussed below.

ACKNOWLEDGMENTS

The present interpretation would not have been possible without the work of Billings and his earlier students. We are deeply grateful to Professor Billings for the many days he has spent in the field with each of us, for the many hours of discussion, and for the continuing, vigorous and witty criticism that has—we hope!—kept our interpretations within the bounds of logic and reason.

We are also grateful to the many geologists who have aided us at various stages of this work, but particularly: to A. J. Boucot for his careful study of our fossils; to C. A. Chapman for providing Thompson with a manuscript map of the Claremont area containing much useful unpublished data; to J. L. Rosenfeld for discussions on this and surrounding areas, and for permitting us to use unpublished data on the Vermont portion of the area in the Bellows Falls and Brattleboro quadrangles; to D. F. Eschmann for helpful discussion on the geology of the Athol quadrangle; and to G. J. F. MacDonald for access to the results of his preliminary studies near Bernardston, Massachusetts.

Thompson's field work was supported by funds of the Department of Geological Sciences, Harvard University. The work of Robinson began as part of a doctoral thesis at Harvard University and has been supported by a National Science Foundation Predoctoral Fellowship (1958–62), a Grant of the Penrose Bequest of the Geological Society of America (1959), a grant of the Reginald and Louise Daly Geological Fund of Harvard University (1959–61), the University of Massachusetts, and most recently by National Science Foundation Research Grant GA–390

(1966–68). The field work and part of the laboratory work by Clifford was carried out during tenure of fellowships from the Commonwealth Fund (now the Harkness Fund) of New York (1956–57) and the National Research Council of Canada (1962). The work of Trask was for a doctoral thesis at Harvard University and was supported by a Humble Oil Company Fellowship (1960–62), a National Science Foundation Predoctoral Fellowship (1962–64), and a grant of the Reginald and Louise Daly Geological Fund of Harvard University (1963). The generosity and kindness of these sponsoring organizations are gratefully acknowledged.

We also acknowledge the work of assistants in the field. Thompson was assisted by C. J. Venuto, E. L. Weinberg, G. J. F. MacDonald, A. Nicholson, E-an Zen, and P. J. Hart. Robinson was assisted by F. Graybeal, K. A. Pankiwskyj, J. Patrick, T. Loder, D. Halpin, A. Hine, P. Pinet, J. C. Boothroyd, and W. P. Freeborn. Trask was assisted by A. Broom. D. P. Drake assisted Robinson in compiling and lettering the illustrations.

We are also grateful to the members of the Blue Mountain Forest Association, Newport, New Hampshire, for permission to carry out geologic mapping in the Corbin Park game reserve, and to the Metropolitan District (Boston) Commission, Belchertown, Massachusetts, for providing access and assistance in the Quabbin Reservoir area.

STRATIGRAPHY AND LITHOLOGY

Littleton, N. H., Sept. 28, 1870: No longer call New Hampshire Azoic. Silurian fossils discovered today.—C. H. Hitchcock (1874, p. 468).

General Statement

The central Connecticut Valley is underlain by Paleozoic metamorphic and intrusive rocks unconformably overlain in Massachusetts by Triassic continental sedimentary rocks and basalts. Four distinctive groups of Paleozoic crystalline rocks are recognized: (1) a sequence of metamorphosed sedimentary and volcanic rocks (Fig. 15–1), (2) gneisses in the cores of domes, (3) large semi-concordant plutonic sheets, and (4) discordant plutons. Of these, the metamorphosed sedimentary and volcanic rocks are Middle Ordovician, Silurian, and Early Devonian in age, and comprise a sequence essentially identical to that established by Billings (1937) in the Littleton–Moosilauke area of New Hampshire, some 30 miles (45 km) north of the area of Plate 15–1a, except that his lowest unit, the Albee Formation is not exposed. Because there

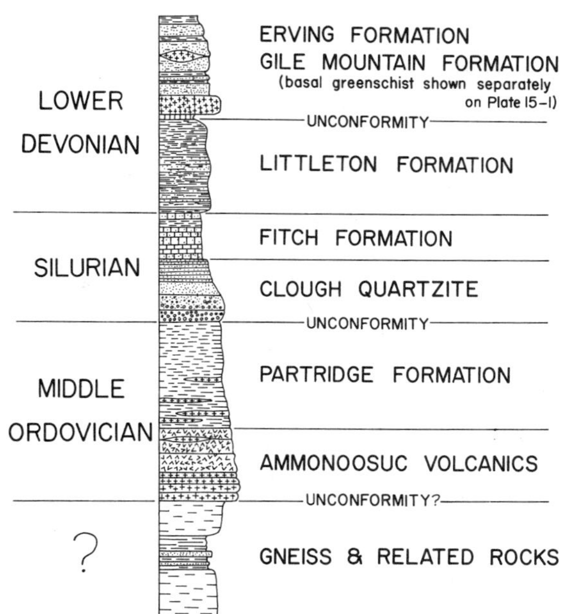

FIGURE 15–1. Stratigraphic column for the area shown on Plate 15–1a.

are several distinctive mapping units, this succession has been the key to the structural interpretation presented here.

The cores of the gneiss domes are occupied by massive gneisses that may be of intrusive derivation, and layered rocks of possible sedimentary and volcanic derivation. These core gneisses are overlain by metamorphosed volcanic rocks that are believed to be of Middle Ordovician age and that constitute the lowest unit of the sequence (1) above. Locally the gneisses are overlain directly by Silurian or younger strata. The relations at such contacts are consistent with the interpretation that the gneisses are pre-Silurian.

The semi-concordant plutonic sheets are Devonian or younger, and generally occur in metamorphosed Devonian rocks. They may be interpreted as either large sills or highly metamorphosed and partially mobilized Devonian felsic volcanics. Various types of discordant plutons cut the other rocks, but are themselves deformed and metamorphosed.

Ammonoosuc Volcanics and Partridge Formation

The Ammonoosuc Volcanics are metamorphosed lavas, pyroclastics, and volcanic sediments ranging in composition from basalt to rhyolite. Individual outcrops are massive to finely bedded with relict primary features including lava pillows and conglomerate beds. The mafic volcanics are metamorphosed to greenschists and amphibolites. Anthophyllite and cummingtonite amphibolites are characteristic, though far from dominant, in the southern part of the area. The felsic volcanics are now feldspar gneisses and schists. In many sections a crude internal stratigraphy is recognized with predominantly mafic volcanics at the base and predominantly felsic volcanics at the top. Unusual rock types include conglomerates with a marble matrix, and garnetiferous rocks interpreted as metamorphosed cherts related to volcanic activity. Sillimanite and kyanite-rich gneisses at the top of the felsic volcanics are metamorphosed products either of subaerial weathering or penecontemporaneous hydrothermal alteration.

The overlying Partridge Formation consists predominantly of mica schist containing black carbonaceous matter, commonly graphite, and an iron sulfide that is magnetic pyrrhotite in most occurrences. The sulfide causes the outcrops to weather rusty or dark brown with crusts of secondary oxides and sulfates, and led to such early names as "Alum Pond," "Brimstone Hill" and "Brimfield," Massachusetts. The schists are interpreted as metamorphosed marine shales and siltstones deposited in a strongly reducing environment. Less abundant rock types include calc-silicate and quartz granulites that are metamorphosed impure dolomitic and arenaceous sediments, and coticules of fine-grained garnet granulite (Clifford, 1960), presumably metamorphosed manganiferous cherts. Mafic and felsic volcanics similar to layers in the Ammonoosuc are also present, and locally make up more than 50% of the outcrops.

Rocks almost certainly correlative with the Ammonoosuc Volcanics and the Partridge Formation have been found in the Cupsuptic quadrangle of west-central Maine (Harwood, 1966; Green and Guidotti, this volume), where Middle Ordovician graptolites have recently been found in black slates like those of the Partridge Formation (Harwood and Berry, 1967). Near Somerset Junction, Maine, farther northeast, volcanic rocks resembling those of the Ammonoosuc carry shelly fossils (Boucot, 1961) indicating a somewhat earlier Middle Ordovician age (see also Neuman, this volume). This confirms earlier correlations of the Partridge and Ammonoosuc with the Cram Hill Formation and Barnard Volcanics of eastern Vermont (Billings, 1956; Doll et al., 1961). The Cram Hill can be traced northward to Middle Ordovician graptolite localities at Beauceville, Quebec (Cady, 1960; Berry, 1962).

Clough Quartzite and Fitch Formation

As anticipated by Hitchcock, the Clough Quartzite (Billings, 1937, 1956) has proved to be the most useful marker unit in deciphering the major structures of the Bronson Hill anticlinorium. It is mainly quartz conglomerate and orthoquartzite, and crops out in bold white ledges that form some of the higher summits in west-central New Hampshire and central Massachusetts. Locally the formation contains polymict conglomerates, micaceous quartzites, mica schists and calc-silicate rocks. The maximum proved thickness, free of structural duplication, is about 800 feet (250 m). The thickness varies abruptly along the strike and in places the Clough is absent. Some of this variation is due to tectonic thinning and large-scale boudinage, but Billings (1937) showed that it is partly a primary stratigraphic feature. Conglomerates, particularly the coarser ones, are most characteristic of the basal beds. In most occurrences the pebbles or cobbles are now highly stretched. The matrix of the conglomerates is generally orthoquartzite, or less commonly mica schist (locally containing fuchsite, Clifford, 1957).

The Clough generally rests on schists of the Partridge Formation, but locally lies on the Ammonoosuc Volcanics, and, at three localities, cited below, on core-gneisses of the domes.

Some of the topmost beds of quartzite contain lenticular pods of calcareous quartzite, marble, or calc-silicate rock. Where least deformed the calcareous quartzite contains abundant fossil remains. Edward Hitchcock (1835) first recorded the occurrence of these fossils at Bernardston, Massachusetts. More recently fossils have been found at several localities on Skitchewaug Mountain near Springfield, Vermont (Thompson, *in* Billings et al., 1952; Boucot et al., 1958), at numerous localities along the ridges of Croydon and Grantham mountains between Claremont and Mascoma Lake, N. H. (Boucot and Thompson, 1958, 1963), and on Moose Mountain, Hanover, N. H. More than twenty recognizable forms occur, even at localities in the sillimanite zone, and show that the Clough is late Early Silurian (Llandovery C3–C5).

The Fitch Formation (Billings, 1937) in the area of Plate 15–1a is mainly a mixture of calc-silicate granulite and biotite–plagioclase granulite, characterized at many localities by rows of conspicuous ellipsoidal pits formed by the removal of carbonates by weathering. In the northern part of the area these rock types are typically separated from the uppermost quartzites of the Clough by about 100 feet (30 m) of mica schist. The formation is not present everywhere, but has a maximum thickness of about 400 feet (120 m). Pelmatozoan columnals and plates were found in the Fitch on Croydon Mountain (Boucot and Thompson, 1963), and columnals were found 1.5 miles (2.4 km) northeast of Windsor, Vermont, but the formation is best dated from fossils near Littleton, New Hampshire (Billings and Cleaves, 1934), that are now regarded (Boucot and Thompson, 1963) as of Ludlow age.

Devonian Formations

The Littleton Formation (Billings, 1937) and its equivalents underlie extensive areas in New England and adjacent Canada. It is typically a gray mica schist or phyllite with a variable proportion of sandy or quartzitic beds from a fraction of an inch to 2 or 3 feet (2 cm to 1 m) thick. Minor rock types include rare calc–silicate rocks and some concordant biotite gneisses, some of them distinctly fragmental, that appear to be volcanic in origin. The schists of the Littleton Formation may be readily distinguished in most areas from those of the Partridge, even where the Clough and Fitch are absent. The Littleton is less carbonaceous, weathers gray rather than rusty, is more resistant to erosion, and typically underlies high ground. It also contains less biotite and generally greater abundance of staurolite and Al-silicates at appropriate metamorphic grades. The Littleton has yielded no fossils in the area of Plate 15–1a but contains Lower Devonian fossils farther north (Billings and Cleaves, 1934; Boucot and Arndt, 1960).

The Erving Formation, exposed in the village of Erving, Massachusetts, southeast of Crag Mountain, includes the Erving Hornblende Schist of Emerson (1917) and associated granulites and schists. In the Orange, Massachusetts, area it appears to rest unconformably on other rock units including the Littleton Formation. The most abundant rock-type consists of interbedded fine-grained plagioclase-biotite granulite, mica schist, and minor calc-silicate granulite. Layers of hornblende-epidote amphibolite, interpreted as basaltic flows and tuffs, are interbedded with the granulite, particularly near the base. Thin coticule beds are found in granulite and schist near amphibolite contacts, and marble and calc-silicate rock occur locally at the base. The Erving Formation is tentatively correlated on lithologic grounds with the Standing Pond Volcanics and adjacent parts of the Gile Mountain Formation in southern Vermont (Robinson, 1963, 1967a).

The rocks shown as Gile Mountain Formation on Plate 15–1a are coextensive with a broad area so designated in eastern Vermont by Doll et al. (1961). North of Springfield, Vermont, the Gile Mountain is separated from the Littleton Formation by pre-Silurian rocks, but from there south to Bernardston, Massachusetts the boundary is placed along a thin greenstone correlated by Doll et al. (1961) with the Standing Pond Volcanics. In places these greenstones have a thin conglomerate on their eastern side, toward the Littleton Formation. The Gile Mountain Formation resembles the Littleton in many ways but is less sandy and locally, particularly in the area west and northwest of Claremont, New Hampshire, contains beds from a few inches to six or more feet (10 cm to 2 m) thick, of siliceous limestone weathering to a brown porous crust.

Gneisses and Related Rocks in Cores of Domes

The gneisses and related rocks in the cores of domes consist of: massive feldspar-quartz gneiss of probable intrusive origin; layered felsic gneiss and amphibolite of probable volcanic derivation; quartzose gneiss, quartzite, calc-silicate rock, and minor mica schist of sedimentary derivation; and massive amphibolite in pre-metamorphic dikes.

The massive gneisses, ranging from granite to quartz diorite in composition, are dominant in most of the domes in New Hampshire and are extensive in the Warwick dome in Massachusetts. These have been assigned to the Oliverian Plutonic Series by Billings (1937, 1956), tentatively regarded by Billings and others as of Devonian age. Chapman (1939) reported that the Clough Quartzite on Moose Mountain (Pl. 15–1a) was feldspathized at its contact with the gneisses of the Mascoma dome, and Kruger (1946) showed the gneisses of the Alstead dome as invading the Littleton Formation. However, re-mapping of the Alstead dome has shown that the gneisses are not in contact with the Littleton Formation. The contact on Moose Mountain has been re-examined by Thompson (see also Naylor, this volume), and is here interpreted as an unconformity. None of the Clough appears to be missing, and the basal conglomerate is no more feldspathic than in some places where it rests on Ammonoosuc Volcanics.

In Massachusetts, conglomerates of the Clough Quartzite on Crag Mountain are in contact with gneisses of the Pelham dome, and Clough conglomerates southwest of Orange are in contact with massive gneisses of the Kempfield anticline. Emerson (1898, p. 213), Balk (1956b), and Robinson (1963) regard the conglomerates as resting unconformably on the gneisses. We therefore know of no place where the geologic relations demand a Silurian or younger age for the core gneisses. Furthermore, although early lead-alpha ages (Lyons et al., 1957) suggested a Devonian age, more complete isotopic studies by Naylor (this volume) have shown that these gneisses can be no younger than Ordovician, and thus are probably equivalent in age to the rocks assigned by Billings (1937, 1956) to the pre-Silurian Highlandcroft Plutonic Series near Littleton, New Hampshire.

Layered gneisses of volcanic derivation that make up much of the Keene dome, and the bodies of Monson Gneiss, are predominantly coarse-grained, biotite-plagioclase gneisses containing hornblende amphibolite layers and boudins. They are not easily separated from the Ammonoosuc Volcanics, but Robinson (1963) has given mineralogical and textural criteria by which they may be distinguished. Preliminary isotopic data on such rocks from the Mascoma dome by Naylor (this volume) suggest they also are Ordovician and represent the lower part of an Ammonoosuc volcanic pile.

Rocks of obviously sedimentary derivation occur only in the Pelham dome in Massachusetts. These show little resemblance to the known Cambro-Ordovician sequence of aluminous mica schists and minor mafic volcanics exposed on the east limb of the Berkshire anticlinorium twenty-five to thirty miles (40 to 50 km) to the west (Hatch et al., this volume). If the rocks of these two areas are of the same age, major facies changes must be postulated. On the other hand, it is possible that the rocks of the Pelham dome are Precambrian in age because they bear a close resemblance to some known Precambrian rocks in the Berkshire Highlands, in the Green Mountains, and in the Chester dome of Vermont (Doll et al., 1961).

Outcrops and thin layers of homogeneous foliated hornblende amphibolite are abundant in parts of some of the gneiss domes. In the southern part of the Keene dome, Moore (1949) recognized some of them as metamorphosed dikes. Intrusive contacts in the country rock of the dikes are cut by a metamorphic foliation. This foliation is in turn cut by dikes that are themselves folded, and the dikes and country rock are in turn cut by a yet younger foliation.

Bethlehem Gneiss and Kinsman Quartz Monzonite

These are foliated rocks ranging from quartz diorite to granite in composition. Different types are not shown separately on Plate 15–1a. The typical Bethlehem Gneiss is granodiorite that has a conspicuous foliation accentuated by a high content of biotite and muscovite. Potassic feldspar, where present, occurs as large augen or subhedral "*dents de cheval.*" With increasing abundance of potassic feldspar these rocks grade into Kinsman Quartz Monzonite containing less biotite, and typically less well foliated. Muscovite is rare in the Kinsman but a high alumina-content is indicated by the common presence of garnet, cordierite (Heald, 1950), or even sillimanite. The two types are chemically (Billings and Wilson, 1964) similar and their mineralogical differences may be explained in terms of reactions of the type: muscovite + biotite + quartz$\rightarrow$ garnet + potassic feldspar + H_2O, suggesting that the Kinsman-types reached their present mineralogy under somewhat hotter and possibly less hydrous conditions.

Both rock types, but particularly the Kinsman, locally crosscut adjacent units and contain disoriented xenoliths. On a regional scale, however, the contacts are concordant and these rocks form large sheetlike masses in the lower part of the Littleton Formation, only locally in contact with older rocks. They have been interpreted either as concordant plutons emplaced by forceful injection (Billings, 1937; Hadley, 1942; Chapman, 1939, 1942; Kruger, 1946; Moore, 1949; Heald, 1950) or as metasomatic replacements of the Littleton Formation (Chapman, 1952). We suggest that they might have been derived from ignimbrite masses in the Littleton Formation that have been mobilized by partial fusion. In north-central Maine the Lower Devonian rocks, in part surely correlative with the Littleton, contain bodies of ignimbrite (Rankin, this volume) on a scale comparable to that of the Bethlehem and Kinsman masses here. Chemical analyses of the Bethlehem and Kinsman (Billings and Wilson, 1964) are similar to those of the Devonian ignimbrites of Maine (Smith, 1933). Alumina is high relative to the alkalis in both sets of analyses, and mineralogic evidence of peraluminous types is common in both areas. In their westernmost areas of outcrop, the Bethlehem and Kinsman show clear evidence of having been significantly hotter than the rocks underlying them (Kruger, 1946; Chapman, 1953; Moore, 1949), but this fact does not necessarily contradict any of the above hypotheses. Numerous large complex pegmatites at or near *upper* contacts of the Bethlehem and Kinsman sheets have accounted for much of the feldspar, mica, and beryl production in west-central New Hampshire (Cameron et al.,

1954). In addition to containing typical Kinsman types, the Hardwick Pluton contains extensive dark biotite quartz diorites (Emerson, 1917; Fitzgerald, 1960; Mook, 1967), and may have different stratigraphic relationships than the plutons discussed above.

Cross-Cutting Intrusives

The Belchertown and Prescott Intrusive Complexes are extensive, regionally discordant plutons. The Belchertown Complex consists of biotite–hornblende quartz diorite and granodiorite, augite-bearing granodiorite, and lesser pyroxene hornblendite and mafic breccia (Guthrie and Robinson, 1967). It truncates the south end of the Pelham Dome, and cuts rocks as young as the Erving Formation. Near contacts, it shows strong lineation and foliation concordant with those in the country rock, indicating that it was present and largely solid during the last stages of regional deformation and metamorphism.

The Prescott Complex, consisting of biotite granodiorite and hornblende gabbro (Makower, 1964), cuts the Ammonoosuc, Partridge, Clough, and Littleton formations in the core of a major isoclinal syncline related to the nappes under discussion here. The rocks have a universally strong foliation and lineation as in the country rock, that must have been imprinted during the doming that followed formation of the nappes. A preliminary Rb/Sr whole rock isochron by R. S. Naylor (written communication, 1966) based on four specially collected specimens from the complex suggests an age of intrusion of about 380 million years (Middle Devonian), which, if correct, would indicate the nappes formed in the Lower or Middle Devonian.

Several cross-cutting bodies of binary granitic rock, some of them foliated, have been mapped farther north by Moore (1949), Fowler–Billings (1949), Heald (1950), and Kruger (1946).

STRUCTURE

General Statement

The structural features of the area shown on Plate 15–1a are the product of several major phases of deformation: an early phase of recumbent folding; one or more phases of gneiss dome formation and related folding; movements about steeply plunging axes in the western part of the area indicating an east-side-north couple; and a much later phase of high angle faulting. The early phase included both the nappes described in this paper, and other recumbent folds common in the Massachusetts part of Plate 15–1a (Robinson, 1963, 1967a). The phase of gneiss dome formation also included the formation of tight synclines and anticlines, and culminated in the development of the present Bronson Hill anticlinorium and the complementary synclinoria to the east and west. Both the nappes and the gneiss domes involve Lower Devonian rocks, hence are at least as young as Devonian. Several lines of evidence indicate that both the nappes and the domes were emplaced during the regional metamorphism. Some faults of the latest phase can be traced into the area of Triassic rocks in Massachusetts where their movements appear to have been more or less synchronous with late Triassic sedimentation. Local zones that contain minor folds associated with retrograde metamorphism appear to represent post-Devonian but pre-Triassic tectonic events.

By far the most useful technique in deciphering the structures shown on Plate 15–1a has been detailed stratigraphic mapping by careful tracing of distinctive lithologic units, notably the Clough Quartzite. Locally, as on Croydon and Grantham Mountains, there is nearly 2000 feet (600 m) of topographic relief, but over most of the area the relief is less than 500 feet (150 m). The three-dimensional analysis of the nappe structures has been greatly aided by the fact that the axial surfaces of the nappes have been strongly arched and in places involuted by the subsequent doming, as well as offset by major faults. Minor structural features related to the major fold phases have been used in interpreting the geometry but cannot be discussed adequately here.

Primary sedimentary structures have been used wherever possible to determine the direction of stratigraphic tops. Graded bedding, though rarely well preserved, may be found in some of the more massive sandy beds of the Littleton Formation and in some of the granulites of the Fitch Formation. Crossbedding is conspicuous in certain outcrops of the Clough Quartzite, particularly on the broad open ledges along the ridge of Grantham Mountain. The fossils in the Clough Quartzite, some of them in the sillimanite zone, occur in an easily recognized rock at the top of the formation and have now been found at three distinct tectonic levels. Coupled with the dating of the Fitch and Littleton Formations in areas to the north, this confirms most of the larger features of the present interpretation.

Three major nappes have been identified. In ascending order these are the Cornish Nappe, the Skitchewaug Nappe, and the Fall Mountain Nappe. Of these, the Skitchewaug Nappe is the most thoroughly known, and the only one for which the hinge lines can now be located with any certainty.

Cornish Nappe

The principal evidence for the Cornish Nappe is found in the extreme northwest part of Plate 15–1a where the elongate area of Gile Mountain Formation west of Claremont is overlain structurally by pre-Silurian rocks (Pl. 15–1b, sect. A–A′) that are continuous northward with those of the Coos (Boundary Mountain) anticlinorium of Billings (1956). From regional relations, the tectonic transport could only have been to the west or northwest and must have been at least 5 miles (8 km) though no hinges have yet been located.

The Cornish Nappe is possibly related to recumbent structures in eastern Vermont shown by Doll et al. (1961) and discussed by Rosenfeld (this volume). If the correlation outlined above, equating the hornblende schists of the Erving Formation with the Standing Pond Volcanics, is correct, then the Cornish Nappe in north-central Massachusetts could be interpreted (Robinson, 1963) as rooted in the Warwick dome (Pl. 15–1b, sect. E–E′). On this basis the Pelham dome would belong to the autochthon beneath the nappe. As already noted, the Pelham dome has features which distinguish it from other domes of the Bronson Hill anticlinorium and in which it resembles the domes of southeastern Vermont, and of Massachusetts and Connecticut west of the Triassic lowlands.

Skitchewaug Nappe

The first clear evidence for large-scale recumbent folding in this area emerged during mapping of the outlier of Silurian and pre-Silurian rocks of which Skitchewaug Mountain is a part (Pl. 15–1b, sect. B–B′). As will be seen from Plate 15–1a, the map pattern of the outlier is shaped like an arrowhead flying south. The outlier is within a doubly plunging syncline of Littleton having a north-south axis. The pre-Littleton rocks are everywhere overturned except in the flap extending south from the socket of the arrowhead where the rocks are a part of the upper limb of the Skitchewaug Nappe. Hinge-lines defined by corresponding points on the barbs of the arrow thus trend nearly east-west at the north end of the outlier. As no comparable rocks crop out to the north or west, a search for the root zone was carried out to the south and east (Thompson, 1954, 1956). It soon

became apparent that most of the quartzites mapped by Kruger (1946) as either Hardy Hill Quartzite, or as quartzites within the Littleton Formation, are in fact inverted Clough at the same tectonic level as the inverted Clough in the Skitchewaug Mountain outlier. East of Straw Hill (Pl. 15–1b, sect. B–B′, and Pl. 15–1a) the autochthonous Clough beneath the nappe and the inverted and upper limbs of the nappe can be seen in sequence, dipping eastward off the Unity dome and beneath the Mount Clough pluton.

Later mapping has shown that rocks belonging to the Skitchewaug Nappe extend at least into central Massachusetts. A large outlier farther north, on Croydon and Grantham mountains, shows the internal structure (Pl. 15–1b, sect. A–A′) and subsequent deformation of the nappe particularly clearly owing to the removal of vegetation and soil cover by a major forest fire in 1953. Still farther north, beyond the limits of Plate 15–1a, structures exist in the vicinity of Mount Cube (Hadley, 1942), on Hogback Mountain, and possibly also on Northey Hill near Littleton, New Hampshire (Billings, 1937), that appear to be similar. Preliminary re-examination of the Mount Cube area (Douglas Rumble, oral communication, 1966) appears to confirm the existence of the nappe in that area.

The work of Thompson and Clifford before 1960 had shown that all rocks mapped as quartzite within the Littleton Formation by previous workers must be considered as possible Clough Quartzite involved in previously undetected complex structure. The maps of the Keene–Brattleboro area (Moore, 1949) and Mount Grace quadrangle (Hadley, 1949) contained numerous such examples and pointed the way to further work. At Hyland Hill (Moore, 1949) an outlier of the Ammonoosuc Volcanics occurs in the western part of the Keene dome. Here a north-south trending graben through the outlier brings down rocks of a higher structural level. These rocks belong to the Partridge, Clough, and Littleton Formations. A continuous sub-horizontal quartzite layer is shown by Moore (1949) in the middle of the Littleton Formation, but investigation by Robinson in 1960 showed that the schist below the quartzite is typical Littleton Formation, whereas the rock above is rusty-weathering schist and amphibolite typical of the Partridge Formation. The quartzite is thus inverted Clough showing that part of Hyland Hill is an outlier of the Skitchewaug Nappe.

Clues to the presence of the Skitchewaug Nappe south of the Massachusetts State line are shown on the geologic maps of the Bernardston and Northfield quadrangles by Balk (1956a; 1956b) who mapped a persistent, south- and southeast-dipping quartzite layer in the area northeast of Bernardston. The quartzite contains fossils and interbeds of marble and calc-silicate granulite at the well-known fossil locality 1.0 mile (1.6 km) north of the village of Bernardston. The identification of Silurian fossils in the quartzite by Boucot et al. (1958) clearly indicates that the quartzite is older than the Littleton Formation dipping beneath it and probably represents the inverted limb of the Skitchewaug Nappe. The quartzite was shown by Doll et al. (1961) as Clough and this was confirmed by Trask (1964) who traced it north into New Hampshire, around the east side of Vernon dome. This belt of quartzite is structurally underlain by the Littleton Formation and overlain by the Partridge Formation. The sequence is therefore identified as the inverted limb of the Skitchewaug Nappe. Southwest of Spofford Lake, the Clough is missing but the Littleton and Partridge formations are separated by Fitch, and tops are shown by inverted graded bedding. Here the inverted limb has been strongly refolded about axes trending north-northwest. Two patches of Clough Quartzite immediately west of the Ashuelot pluton are in the normal upper limb of the Skitchewaug Nappe.

On the west flank of the Warwick and Pelham domes, the Erving Formation abuts directly against the Triassic border fault. Where Triassic cover is absent on the western (downthrown) side of the fault, the Erving Formation is juxtaposed by Partridge, Clough, and Littleton, belonging to or overlying the Skitchewaug Nappe. The area of poor exposure around Amherst is shown as Partridge on Plate 15–1a because Partridge types predominate. These rocks are more highly metamorphosed than those east of the fault and contain much pegmatite and granite, suggesting proximity to a mass such as the Ashuelot pluton.

The inverted Clough directly west of the main body of Monson Gneiss in Orange, is in the inverted limb of the Skitchewaug Nappe, as at Bernardston. The sequence that forms the upper limb of the nappe has been traced along the east sides of the Croydon, Unity, Alstead, and Keene domes, and of the main body of Monson Gneiss. It also encircles the Tully body. The narrow belts of Monson Gneiss, peripheral to the east side of the main body and wrapping around the south end of the Tully body, are interpreted as the core of a recumbent fold with the same movement sense as the nappe. Between the Keene dome and the main body of Monson Gneiss the upper limb

of the nappe passes through an isoclinal involution presumably formed during later extreme northward overturning of the Monson Gneiss. For about 5 miles (8 km) southwest of Bliss Hill, the pre-Littleton rocks in the core of the nappe are attenuated to as little as 120 feet (37 m). On Bliss Hill, and again to the southwest, the Partridge is missing and the Clough of the upper limb is considered to be in tectonic contact with the Littleton beneath the nappe. Both the attenuated core and the involution of the upper limb have been refolded about steep south-plunging axes.

Hinges, where the base of the Clough Quartzite on the inverted limb of the Skitchewaug Nappe is connected to the base of the Clough of the "autochthon" have now been located at seven points: (1) on the west side of the Mascoma dome near its south end; (2) on the east side of the Alstead dome near its south end; (3) and (4) on Surry Mountain between the Alstead and Keene domes; (5) west of Keene, New Hampshire; (6) on Bliss Hill, south of the Keene dome; and (7) within the Prescott Intrusive Complex. Hinges joining the inverted and upper limbs have been located at two localities on Skitchewaug Mountain, at two localities between Grantham Mountain and Mascoma Lake, and tentatively at two localities near Bellows Falls, Vermont.

The hinge-to-hinge amplitude of the Skitchewaug Nappe is about 3 miles (5 km) southwest of Mascoma Lake. It broadens southward to about 15 miles (24 km) in an area extending from Skitchewaug Mountain south into Massachusetts. In the central part of the map area (Pl. 15–1a) the inverted limb of the nappe is largely intact, but elsewhere, notably at localities in the Orange area and in the Croydon Mountain area, the Clough and Fitch are absent. Cliff exposures on the east slopes of Croydon and Grantham Mountains show that the Clough is there reduced to a string of boudins along the inverted Littleton–Partridge contact. In such places this contact might be regarded as a thrust flooring the nappe, but it is not yet possible to determine whether the surface is one of finite displacement or whether the Clough has simply been boudinaged in a zone of extreme attenuation.

Relationships near Mascoma Lake, near Surry Mountain, and near Bliss Hill show that in these areas the axial surface of the Skitchewaug Nappe and the axial surface separating it from the autochthon are close together (Fig. 15–2) in a thin layer of Partridge Formation.

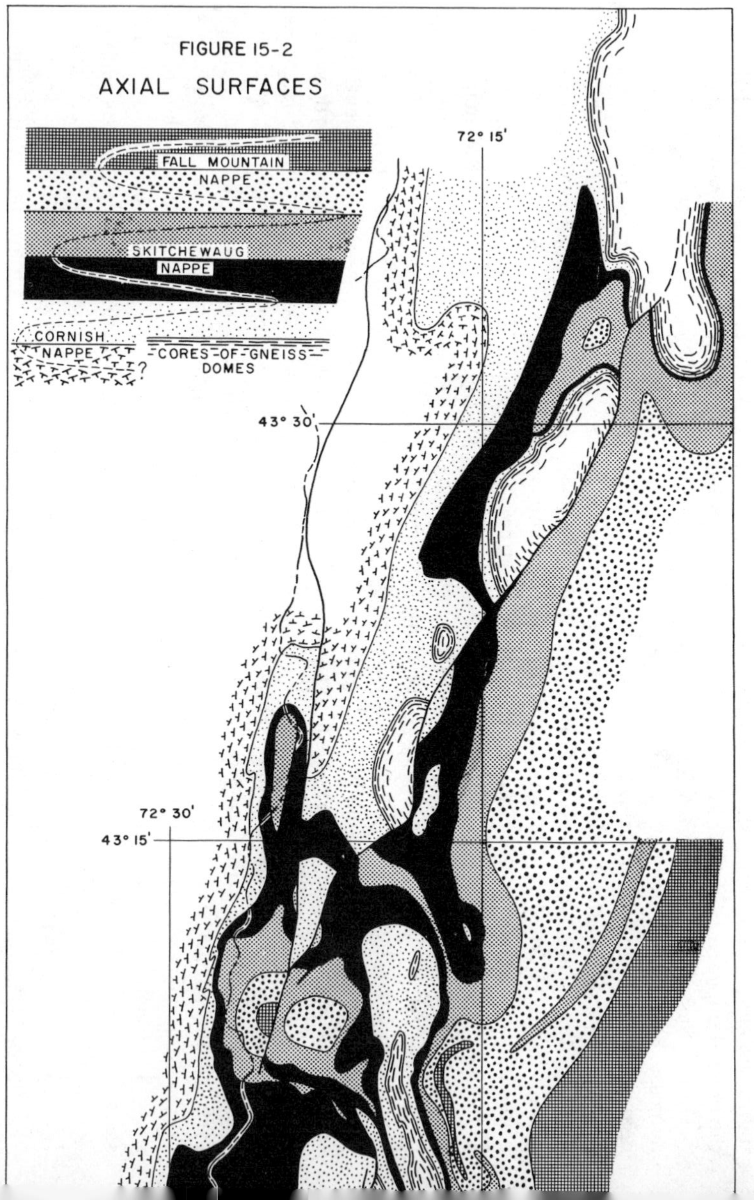

FIGURE 15-2
AXIAL SURFACES
FALL MOUNTAIN NAPPE
SKITCHEWAUG NAPPE
CORNISH NAPPE
CORES-OF-GNEISS DOMES
72° 15'
43° 30'
72° 30'
43° 15'

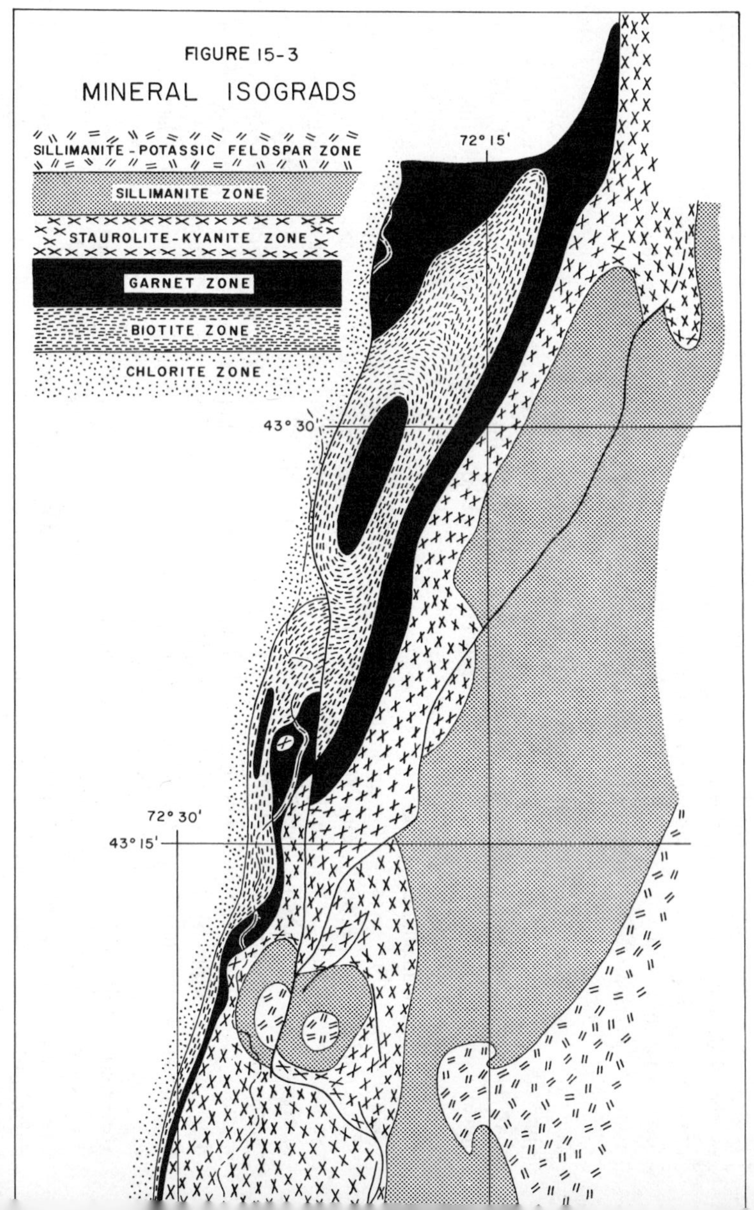

FIGURE 15-3
MINERAL ISOGRADS
SILLIMANITE-POTASSIC FELDSPAR ZONE
SILLIMANITE ZONE
STAUROLITE-KYANITE ZONE
GARNET ZONE
BIOTITE ZONE
CHLORITE ZONE
72° 15'
43° 30'
72° 30'
43° 15'

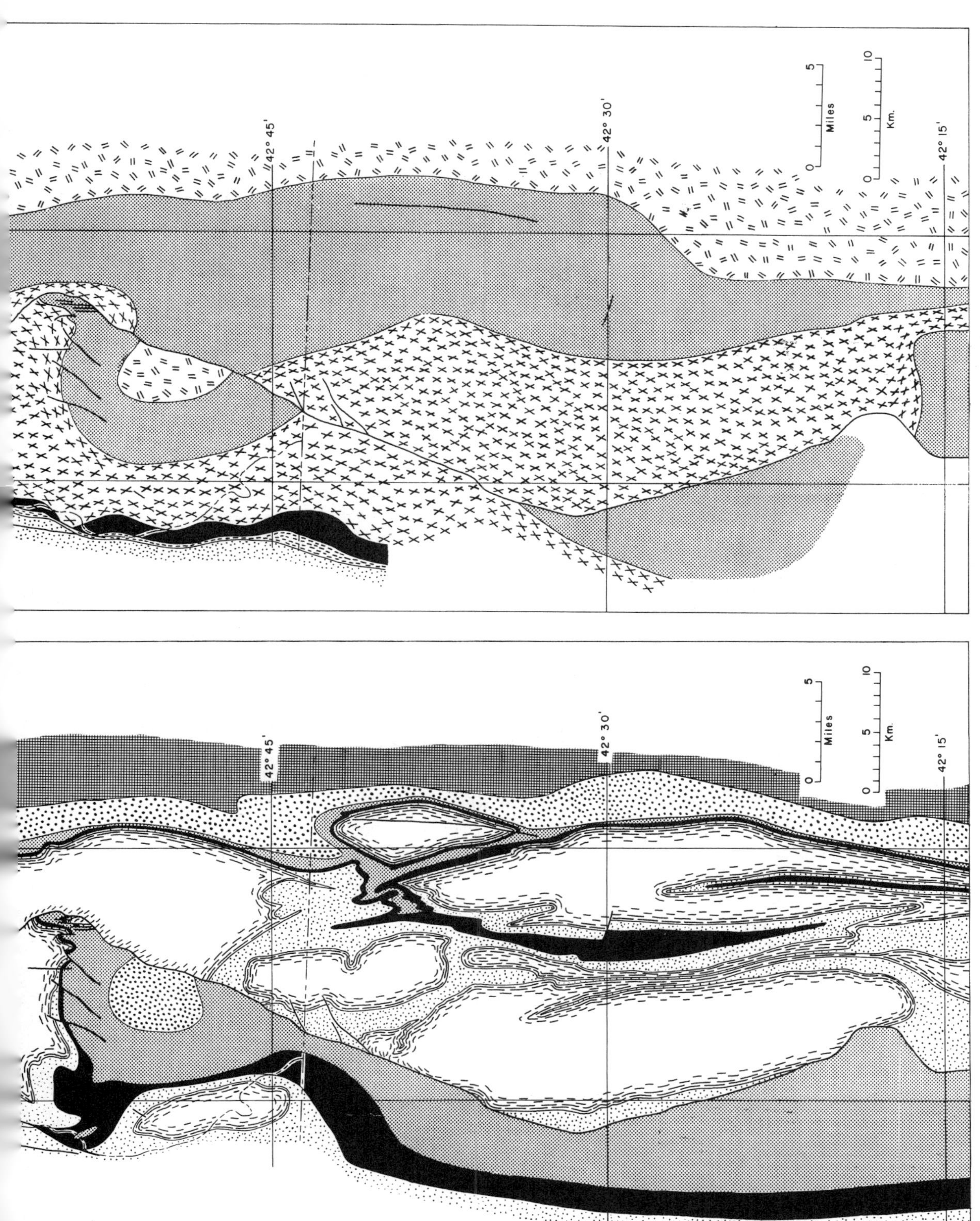

The gneisses of the Keene dome and of the domes north of it are clearly autochthonous relative to the nappe. At the north end of the main body of Monson Gneiss there is good evidence that these axial surfaces are strongly deformed by later movements. However, neither surface appears to enter rocks stratigraphically beneath the Partridge, except where narrow bands of Monson-like gneiss on the east side of the main body and around the south end of the Tully body are interpreted as lying within the nappe. It is thus evident that in the root zone these axial surfaces are no more than a few hundred feet apart. It is possible that they may meet, thus eliminating the inverted limb. If so, we may regard the Skitchewaug Nappe as rooted in a bedding thrust east of what is now the axis of the Bronson Hill anticlinorium.

Fall Mountain Nappe

In 1955, Clifford found a conglomerate on Fall Mountain, New Hampshire, east of Bellows Falls, Vermont, that was structurally underlain by Littleton, and overlain by highly metamorphosed schists of Partridge aspect. All of these rocks are above the Bellows Falls pluton of Bethlehem Gneiss, which is in turn above the Skitchewaug Nappe (Pl. 15–1b, sect. C–C′).

Similar occurrences of highly metamorphosed Partridge to the east near Lake Warren are surrounded by discontinuous conglomerate, quartzite, and calc-silicate granulite, and appear to overlie, in inverted sequence, the narrow band of Bethlehem Gneiss shown as connecting the Mount Clough pluton to the southwest spur of the Cardigan pluton. The schists bounded by the Ashuelot pluton and the Triassic border fault southwest of Keene, New Hampshire, appear to be Partridge above the Ashuelot pluton.

We interpret these three areas as outliers of a third major nappe, the Fall Mountain Nappe. Although the root zone has been investigated only by reconnaissance, it has been shown on Plate 15–1a as lying in rusty-weathering, Partridge-like schists and gneisses in a broad area occupying the western parts of the Lovewell Mountain and Monadnock quadrangles and in a narrow belt west of the Hardwick Pluton. No hinges have yet been located, but the hinge-to-hinge amplitude must be at least 15 miles (24 km). The septum of Littleton east of the main body of Monson Gneiss has been traced southward to the Connecticut state line (Peper, 1966, 1967) suggesting that the axial surface separating the Skitchewaug and Fall Mountain nappes can be traced at least this far south. If this axial surface extends farther south into Connecticut the interpretation presented here may be in conflict with that of Dixon and Lundgren (this volume).

Gneiss Domes

The gneiss domes of the Bronson Hill anticlinorium are complex, but a description of their general features is essential to an understanding of the regional pattern on Plate 15–1a. They have been cited by Eskola (1949, p. 470) as typical mantled gneiss domes, but their mode of origin, as interpreted here, differs significantly from that proposed by Eskola.

The more northerly domes, namely the Mascoma, Croydon, Unity, and Alstead domes are broad, doubly plunging arches separated by deep, isoclinal synclines. The axial surface of the Skitchewaug Nappe has been folded in several of these synclines. These northerly domes are less deeply eroded than those farther south and large portions of their roofs are still preserved, particularly in down-faulted areas. The Mascoma, Croydon, and Unity domes have nearly vertical, locally overturned dips on their western sides, and bedding and foliation attitudes show that the Mount Clough pluton overlies parts of them (Pl. 15–1b, sects. A–A′ and B–B′; see Griscom and Bromery, this volume), accounting for the unusual outcrop width of the pluton in that area. East of the Grantham fault, the Unity dome is largely buried, but its axis can be traced south in the mantling schists to a point about 3 miles (5 km) northeast of Lake Warren. The northwest lobe of the Keene dome extends similarly northward, beneath nearly flat mantling strata, to at least the south edge of the Bellows Falls pluton (Pl. 15–1b, sect. C–C′).

Farther south the domes are more deeply eroded and largely unroofed. There is strong overturning at the margins of the southerly domes where they have mushroomed and overridden one another. The northeast lobe of the Keene dome is overturned northward onto the south end of the Alstead dome and westward onto the main part of the Keene dome, resulting in a complex pattern near Surry Mountain (Trask and Thompson, 1967). The Vernon dome is apparently a tongue-like mass pointing upward from the east. The north end of the Warwick dome plunges east-southeast (W. P. Freeborn, unpublished map, 1967). The dome is divided into two lobes by a transverse syncline overturned to the south. The south lobe is overturned southwest onto the Pelham dome and south in a great mushroom-like overhang over the north-plunging end of the Kempfield anticline, itself overturned to the west.

Two of the domes contain early recumbent folds that involve both dome gneiss and mantle. These early folds are apparently unrelated to the larger nappes under consideration (Robinson, 1967a). The south end of the Keene dome is a highly compressed conical anticline superimposed on an earlier recumbent anticline and its related syncline involving gneiss, Ammonoosuc, and Partridge. Two hinge points on the recumbent syncline show that it trends northeast and that overfolding was from northwest toward southeast. A satellitic dome near the southern tip of the Keene dome has a core of Ammonoosuc volcanics and the form of a compressed cylinder plunging steeply southeast. The north end of the Pelham dome is a broad symmetrical arch plunging very gently north, superimposed on recumbent anticlines and synclines near Crag Mountain. The recumbent folds involve all units from Pelham Gneiss to Erving Formation and were apparently overfolded from west toward east. A narrow isoclinal syncline in the southern part of the dome may be related to similar early recumbent structures.

The main and Tully bodies of Monson Gneiss depart greatly in form from typical gneiss domes. The north end of the main body has the form of a syncline plunging 20 to 60 degrees south. The west limb is nearly vertical and the east limb dips 30 to 60 degrees west. The gneiss overlies an inverted sequence of younger units, and forms with them a series of folds in which younger rocks are exposed in anticlines and older rocks in synclines. The narrow band of Partridge Formation in the middle of the Monson Gneiss in the Quabbin Reservoir area is apparently a normal syncline on top of the main body. The Tully body has a bulbous, almost cylindrical form with both ends plunging southwest beneath the main body (T. M. Pike, unpublished map, 1967). The extreme, essentially recumbent, northward overturning of the north ends of the main and Tully bodies took place after the formation of the Skitchewaug Nappe and before the regional tight folding synchronous with dome formation. The Clough and Littleton formations on the upper limb of the nappe were folded into an isoclinal syncline during northward overturning of the main body, and the axial surface of this isoclinal syncline was then tightly folded during dome formation.

The main body of Monson Gneiss narrows southward but has been traced continuously into the Killingworth dome in southern Connecticut (Rodgers et al., 1959; Lundgren, 1962; Peper, 1966, 1967). South of Quabbin Reservoir it has been referred to as the "Monson anticline" and has been postulated as the core of a major nappe, the Colchester nappe, overfolded from west toward east (Dixon and Lundgren, this volume). Known relations seem to be compatible with formation of the Colchester nappe during the overfolding of the north end of the main body of Monson Gneiss, after the formation of the Skitchewaug and Fall Mountain nappes. In agreement with this conclusion, the outcrop pattern in the Monadnock quadrangle directly east of the Keene dome shows that major isoclinal folds here are overfolded from west toward east.

Late-Stage Folding and Faulting

Steeply plunging folds, typically showing an east-side-north movement, characterize the synclinorium flanking the Bronson Hill anticlinorium on the west and produced many of the folds shown along the western margin of Plate 15–1a. These folds have affected the Cornish Nappe and are believed to post-date the doming. Other late folds, described by Robinson (1963, 1967b) and Halpin (1965), are associated with retrograde metamorphic features in the Orange and Quabbin Hill areas.

The last major deformation recorded here resulted in the formation of the two master faults shown on Plate 15–1a, the Triassic border fault and the Grantham fault. Plate 15–1a shows details of these faults not previously published. Both are essentially normal faults, but the border fault probably has a slight east-side-south, strike-slip component and south of the Bellows Falls Pluton, where it dips steeply northeast, is locally a reverse fault. These faults and the faults affecting the western lobe of the Keene dome were probably active during Triassic sedimentation. Coarse conglomerates and angular breccias directly west of the border fault indicate that the fault scarp was a local source of coarse clastic sediments. The Grantham fault is downthrown on the east side so that the block between it and the Triassic border fault is a horst. All these faults are accompanied by intense silicification and alteration of the country rocks and are marked in many places by enormous bosses of quartz-rock showing an open drusy structure. Fresh slickensides are present at many localities.

PHYSICAL CONDITIONS DURING DEFORMATION

Overturning of Isogradic Surfaces

Metamorphic isograds delineating the boundaries between the chlorite, biotite, garnet, staurolite-kyanite, sillimanite, and sillimanite-potassic feldspar zones are shown in Figure 15–3 in order to emphasize their relations to the axial surfaces of the nappes. There is, as has long been known, an overall increase in metamorphic grade eastward across the Bronson Hill anticlinorium. However, there is also an obvious correlation between high metamorphic grade and high tectonic level, indicating a strong overturning of the isogradic surfaces. This overturning was apparent in earlier works (Chapman, 1939; Kruger, 1946; Moore, 1949) and has been discussed at some length by Chapman (1953). Now that the axial surfaces of the nappes are fairly well located, it is even more striking, particularly in the vicinity of the Bellows Falls and Ashuelot plutons and on Skitchewaug Mountain where higher grade rocks at higher tectonic levels are completely surrounded by lower grade rocks at lower tectonic levels.

It is thus evident that successive nappes carried successively hotter rocks westward over colder ones. Whether the Bethlehem Gneiss and Kinsman Quartz Monzonite are the cause, or the product, of the heating farther east that preceded the emplacement of the nappes must remain a question until the origin of these rocks can be definitely established. We can not prove whether these rocks are: (1) metasomatized Littleton schists as suggested by Chapman (1952); (2) of direct magmatic origin as has most generally been held; or (3) partially fused and mobilized Devonian volcanics. Clifford has found textural and mineralogic evidence that there is a contact aureole in the country rocks about the northwest base of the Bellows Falls pluton. This observation can probably be incorporated in any of the above hypotheses but is easily understood in terms of the last two whereby the material in the pluton would have had sufficient fluidity to move laterally into cooler regions.

Depth of Burial

The sequence of metamorphic zones in the area of Plate 15–1a is characterized by the direct transition of kyanite to sillimanite without the appearance of andalusite. Andalusite schists do occur, however, in areas about 25 miles (40 km) to the east of Plate 15–1a (Thompson and Norton, this volume). This suggests that pressures were near, but slightly higher than, the pressure of the kyanite-andalusite-sillimanite triple point at the time of final recrystallization. From the recent experimental studies of Newton (1966) the triple point pressure is about 4.2 kilobars corresponding to a depth of burial of about 15.5

kilometers or 50,000 feet. This is not unreasonable in view of the tectonic style of the area, and is probably minimal for the part in Massachusetts east of the Triassic border fault (Robinson, 1966).

In this connection it is interesting to note that Clifford found what appear to be mica pseudomorphs after chiastolite in the aureole at the northwest base of the Bellows Falls pluton. These pseudomorphs are, however, at a high tectonic level. Recently sillimanite pseudomorphs tentatively after andalusite have been identified in a small inclusion of Partridge schist in the middle of the Hardwick Pluton, and in the southern half of the long inclusion of Partridge in the Belchertown Complex, itself in the regional kyanite zone. In all these cases it is possible that the original porphyroblasts might have formed prior to the emplacement of higher nappes.

ORIGIN OF THE NAPPES AND DOMES

. . . the oldest group is disposed in no less than twenty-two areas of small size, scattered like the island of an archipelago.—C. H. Hitchcock (1883, p. 295). (Address, Vice-President, Section E., A.A.A.S., August, 1883)

The evidence just discussed makes it clear that the nappes were derived from an area of heated rocks farther east. Unfortunately the later doming has obscured the geometric relations at the time the nappes were emplaced and it is still not possible to say whether the nappes slid into place from an uplifted and heated mass to the east or whether they were ejected from a downfolded area to the east under the impetus of lateral compression. In this context it is interesting to note that similar nappes in southeastern Connecticut are believed to have moved toward the *east* (Dixon and Lundgren, this volume).

The geometric similarities between gneiss domes and salt domes are apparent and these have led to the suggestion (Thompson, 1950) that the mechanism of emplacement may also be similar. The rocks in cores of the domes consist dominantly of quartz and alkali feldspar, hence they are somewhat less dense than the average of the mantling strata. It is, therefore, possible that the domes are located either at the sites of thicker piles of Ordovician volcanic material, or where the Ordovician volcanics were locally thickened and invaded by intrusive material in pre-Silurian time. If so, the *en-échelon* arrangement of the domes would be a relict Ordovician feature, conceptually not far from the volcanic islands of the type suggested in 1883 by C. H. Hitchcock. On the other hand, the location and arrangement of the domes may be a late feature, related to a horizontal couple, as suggested by Hadley (1942). A feature possibly favoring the former hypothesis, however, is the relative thinness, or local absence, of Partridge in the domes as compared with the nappes and other areas nearby. This would suggest that the domes may correspond to topographic highs in the volcanic surface upon which the Partridge was deposited, and that the Ordovician volcanic achipelago invoked by stratigraphers (Kay, 1951, Pl. 1) may indeed have been here!

And lo! stratigraphy led all the rest.—M. P. Billings (1950, p. 445) (Address, Retiring Chairman, Section E, A.A.A.S., December, 1949)

REFERENCES

Balk, Robert, 1941, Devonian Bernardston Formation of Massachusetts restudied: Geol. Soc. Amer. Bull., v. 52, p. 2009–2010

————, 1942, The Pelham gneiss dome, Massachusetts [abs.]: Amer. Geophys. Union Trans., v. 23, p. 343–344

————, 1956a, Bedrock geology of the Massachusetts portion of the Bernardston quadrangle, Massachusett-Vermont: U. S. Geol. Survey Geol. Quad. Map GQ–90

————, 1956b, Bedrock geology of the Massachusetts portion of the Northfield quadrangle, Massachusetts, New Hampshire and Vermont: U. S. Geol. Survey Geol. Quad. Map GQ–92

————, 1956c, Bedrock geology of the Millers Falls quadrangle, Massachusetts: U. S. Geol. Survey Geol. Quad. Map GQ–93

————, 1957, Geology of Mount Holyoke quadrangle, Massachusetts: Geol. Soc. America Bull., v. 68, p. 481–504

Berry, W. B. N., 1962, On the Magog, Quebec, graptolites: Amer. Jour. Sci., v. 260, p. 142–148

Billings, M. P., 1937, Regional metamorphism of the Littleton-Moosilauke area, New Hampshire: Geol. Soc. Amer. Bull., v. 48, p. 463–566

————, 1950, Stratigraphy and the study of metamorphic rocks: Geol. Soc. Amer. Bull., v. 61, p. 435–448

————, 1956, The geology of New Hampshire, Part II—Bedrock geology: N. H. Plan. and Devel. Comm., 203 p.

———— and Cleaves, A. B., 1934, Paleontology of the Littleton area, New Hampshire: Amer. Jour. Sci., 5th ser., v. 28, 412–438

————, Rodgers, John, Thompson, J. B., Jr., 1952, Geology of the Appalachian Highlands of east-central New York, southern Vermont, and southern New Hampshire: Geol. Soc. America Guidebook for Field Trips in New England, Boston, Mass., p. 1–71

———— and Wilson, J. R., 1964, Chemical analyses of rocks and rock-minerals from New Hampshire: Part XIX, Mineral Resources Survey, New Hampshire Division of Economic Development, 104 p.

Boucot, A. J., 1961, Stratigraphy of the Moose River synclinorium, Maine: U. S. Geol. Survey Bull. 1111–E, 188 p.

———— and Arndt, R. H., 1960, Fossils of the Littleton Formation (Lower Devonian) of New Hampshire: U. S. Geol. Survey Prof. Paper 334–B, p. 41–53

————, MacDonald, G. J. F., Milton, Charles, and Thompson, J. B., Jr., 1958, Metamorphosed Middle Paleozoic Fossils from central Massachusetts, eastern Vermont, and western New Hampshire: Geol. Soc. Amer. Bull., v. 69, p. 855–870

———— and Thompson, J. B., Jr., 1958, Late Lower Silurian fossils from sillimanite zone near Claremont, New Hampshire: Science, v. 128, p. 362–363

———— and Thompson, J. B., Jr., 1963, Metamorphosed Silurian brachiopods from New Hampshire: Geol. Soc. Amer. Bull., v. 74, p. 1313–1334

Cady, W. M., 1960, Stratigraphic and geotectonic relationships in northern Vermont and southern Quebec: Geol. Soc. Amer. Bull., v. 71, p. 531–560

Cameron, E. N., Larrabee, D. M., McNair, A. H., Page, J. J., Stewart, G. W., and Shainin, V. E., 1954, Pegmatite investigations, 1942–1945, New England: U. S. Geol. Survey Prof. Paper 255, 352 p.

Chapman, C. A., 1939, Geology of the Mascoma quadrangle, New Hampshire: Geol. Soc. Amer. Bull., v. 50, p. 127–180

————, 1942, Intrusive domes of the Claremont-Newport area, New Hampshire: Geol. Soc. Amer. Bull., v. 53, p. 889–916

————, 1952, Structure and petrology of the Sunapee quadrangle, New Hampshire: Geol. Soc. Amer. Bull., v. 63, p. 381–425

————, 1953, Problem of inverted zones of metamorphism in western New Hampshire: Illinois State Acad. Sci. Trans., v. 46, p. 115–123

Clifford, T. N., 1957, Fuchsite from a Silurian(?) quartz-conglomerate, Acworth Township, New Hampshire: Am. Mineralogist, v. 42, p. 562–568

————, 1960, Spessartine and magnesium biotite in coticule-bearing rocks from Mill Hollow, Alstead Township, New Hampshire, U. S. A.: Neues Jahrb. Mineralogie Abh. (Stuttgart, Germany), Band 94, Festband Ramdohr, 2 Hälfte, p. 1369–1400

Doll, C. G., Cady, W. M., Thompson, J. B., Jr., and Billings, M. P., *compilers* and *editors*, 1961, Centennial geologic map of Vermont: Vermont Geol. Survey, Montpelier, Vt., Scale 1:250,000

Eaton, G. P. and Rosenfeld, J. L., 1960, Gravimetric and structural investigations in central Connecticut: Internat. Geol. Cong., 21st, Copenhagen, 1960, Rept., pt. 2, p. 168–178

Emerson, B. K., 1898, Geology of old Hampshire County, Massachusetts: U. S. Geol. Survey Monograph 29, 790 p.

————, 1917, Geology of Massachusetts and Rhode Island: U. S. Geol. Survey Bull. 597, 289 p.

Eskola, P. E., 1949, The problem of mantled gneiss domes: Geol. Soc. London, Quart. Jour., v. 104, p. 461–476

Fitzgerald, Kim, 1960, Geology of the western half of the Royalston quadrangle, Massachusetts: M. S. thesis, Univ. of Massachusetts, 120 p.

Fowler-Billings, Katharine, 1949, Geology of the Monadnock region of New Hampshire: Geol. Soc. Amer. Bull., v. 60, p. 1249–1280

Guthrie, J. O. and Robinson, Peter, 1967, Geology of the northern portion of the Belchertown Intrusive Complex: New England Intercollegiate Geol. Conf., 59th Ann. Mtg., guidebook, Amherst, Massachusetts, p. 143–153

Hadley, J. B., 1942, Stratigraphy, structure, and petrology of the Mt. Cube area, New Hampshire: Geol. Soc. Amer. Bull., v. 53, p. 113–176

————, 1949, Bedrock geology of the Mt. Grace quadrangle, Massachusetts: U. S. Geol. Survey Geol. Quad. Map GQ–3

Halpin, D. L., 1965, The geometry of folds and style of deformation in the Quabbin Hill area, Massachusetts: M. S., thesis, Univ. of Massachusetts, 93 p.

Harwood, D. S., 1966, Geology of the Cupsuptic quadrangle, Maine: U. S. Geol. Survey open file report 850, 259 p.

————— and Berry, W. B. N., 1967, Fossiliferous lower Paleozoic rocks in the Cupsuptic quadrangle, west-central Maine: U. S. Geol. Survey Prof. Paper 575–D, p. D16–D23

Heald, M. T., 1950, Structure and petrology of the Lovewell Mountain quadrangle, New Hampshire: Geol. Soc. Amer. Bull., v. 61, p. 43–89

Hitchcock, C. H., 1871, Helderberg corals in New Hampshire: Amer. Jour. Sci., 3rd Ser., v. 2, p. 148–149

—————, 1874, On Helderberg rocks in New Hampshire: Amer. Jour. Sci., 3rd Ser., v. 7, p. 468–476, 557–571

—————, 1877, The geology of New Hampshire, Part II—Stratigraphical geology: Concord, New Hampshire, 684 p.

—————, 1883, The early history of the North-American continent: Science, v. 2, p. 293–297

—————, 1890, Significance of oval granitoid areas in the lower Laurentian (abstract with discussion by G. H. Williams): Geol. Soc. Amer. Bull., v. 1, p. 557–558

—————, 1908, Geology of the Hanover, New Hampshire, quadrangle: Vermont State Geologist 6th Report, p. 139–186

—————, 1912, The Strafford quadrangle: Vermont State Geologist 8th Report, p. 100–145

Hitchcock, Edward, 1835, Report on the geology, mineralogy, botany, and zoology of Massachusetts, 2nd edition: Amherst, Massachusetts, 702 p.

Kay, Marshall, 1951, North American geosynclines: Geol. Soc. Amer. Mem. 48, 143 p.

Kruger, F. C., 1946, Structure and metamorphism of the Bellows Falls quadrangle of New Hampshire and Vermont: Geol. Soc. Amer. Bull., v. 57, p. 161–206

—————, and Linehan, Daniel, S. J., 1941, Seismic studies of floored intrusives in western New Hampshire: Geol. Soc. Amer. Bull., v. 52, p. 633–648

Lundgren, L. W., Jr., 1962, Deep River area, Connecticut—stratigraphy and structure: Amer. Jour. Sci., v. 260, p. 1–23

Lyons, J. B., 1955, Geology of the Hanover quadrangle, New Hampshire-Vermont: Geol. Soc. Amer. Bull., v. 66, p. 105–145

—————, Jaffe, H. W., Gottfried, David, and Waring, C. L., 1957, Lead-alpha ages of some New Hampshire granites: Amer. Jour. Sci., v. 255, p. 527–546

Makower, Jordan, 1964, Geology of the Prescott intrusive complex, Quabbin Reservoir quadrangle, Massachusetts: M. S. thesis, Univ. of Massachusetts, 91 p.

Mook, A. L., 1967, Petrography of the Athol quadrangle, Massachusetts: U. S. Geol. Survey open file report.

Moore, G. E., 1949, Structure and metamorphism of the Keene-Brattleboro area, New Hampshire-Vermont: Geol. Soc. Amer. Bull., v. 60, p. 1613–1670

Newton, R. C., 1966, Kyanite-andalusite equilibrium from 700° to 800°C: Science, v. 153, p. 170–172

Page, L. R., 1937, The geology of the Rumney quadrangle, New Hampshire: Ph.D. thesis, Univ. of Minnesota, 150 p.

Peper, J. D., 1966, Stratigraphy and structure of the Monson area, Massachusetts and Connecticut: Ph.D. thesis, Univ. of Rochester, 126 p.

—————, 1967, Stratigraphy and structure of the Monson area, Massachusetts and Connecticut: New England Intercollegiate Geol. Conf., 59th Ann. Mtg., guidebook, Amherst, Massachusetts, p. 105–113

Ratté, J. C., 1952, Bedrock geology of the Skitchewaug Mountain area, Claremont quadrangle, New Hampshire-Vermont: M.A. thesis, Dartmouth College, 56 p.

Richardson, C. H., 1931, The areal and structural geology of Springfield, Vermont: Vermont State Geologist 17th Rept., p. 193–212

Robinson, Peter, 1963, Gneiss domes of the Orange area, Massachusetts and New Hampshire: Ph.D. thesis, Harvard Univ., 253 p.

—————, 1966, Alumino-silicate polymorphs and Paleozoic erosion rates in central Massachusetts [abs.]: Amer. Geophys. Union Trans., v. 47, p. 182

—————, 1967a, Gneiss domes and recumbent folds of the Orange area, west-central Massachusetts: New England Intercollegiate Geol. Conf., 59th Ann. Mtg., guidebook, Amherst, Massachusetts, p. 17–47

—————, 1967b, Geology of the Quabbin Reservoir area, central Massachusetts: New England Intercollegiate Geol. Conf., 59th Ann. Mtg., guidebook, Amherst, Massachusetts, p. 114–127

Rodgers, John, Gates, R. M., and Rosenfeld, J. L., 1959, Explanatory text for preliminary geological map of Connecticut, 1956: Connecticut Geol. Nat. History Survey Bull., 84, 64 p.

Rosenfeld, J. L., and Eaton, G. P., 1956, Metamorphic geology of the Middle Haddam area, Connecticut—a progress report [abs.]: Geol. Soc. Amer. Bull., v. 67, p. 1823

Smith, E. S. C., 1933, Contributions to the geology of Maine V: Amer. Jour. Sci., 5th ser., v. 25, p. 225–228

Thompson, J. B., Jr., 1950, A gneiss dome in southeastern Vermont: Ph.D. thesis, Massachusetts Institute of Technology, 160 p.

—————, 1954, Structural geology of the Skitchewaug Mountain area, Claremont quadrangle, Vermont-New Hampshire: New England Intercollegiate Geol. Conf., 46th Ann. Mtg., guidebook, Hanover, N. H., p. 35–41

—————, 1956, Skitchewaug Nappe, a major recumbent fold in the area near Claremont, New Hampshire [abs.]: Geol. Soc. Amer. Bull., v. 67, p. 1826–1827

Trask, N. J., Jr., 1964, Stratigraphy and structure in the Vernon-Chesterfield area, Massachusetts, New Hampshire, Vermont: PhD. thesis, Harvard Univ., 99 p.

—————, and Thompson, J. B., Jr., 1967, Stratigraphy and structure of the Skitchewaug Nappe in the Bernardston area, Massachusetts and adjacent New Hampshire and Vermont: New England Intercollegiate Geol. Conf., 59th Ann. Mtg., guidebook, Amherst, Massachusetts, p. 128–142

Willard, M. E., 1951, Bedrock geology of the Mount Toby quadrangle, Massachusetts: U. S. Geol. Survey Geol. Quad. Map GQ-8

—————, 1952, Bedrock geology of the Greenfield quadrangle, Massachusetts: U. S. Geol. Survey Geol. Quad. Map GQ-20

Chapter 16

Structure of Eastern Connecticut[*]

H. ROBERTA DIXON AND LAWRENCE
W. LUNDGREN, JR.

INTRODUCTION

THE BRONSON HILL anticlinorium and Merrimack synclinorium, major structural units recognized in New Hampshire by Billings (1956), extend southward across Massachusetts and eastern Connecticut. The major structural features within the Connecticut parts of the anticlinorium and the synclinorium are shown in Figure 16–1. The Bronson Hill anticlinorium has been studied in detail in northern Massachusetts and New Hampshire and its extension in eastern Connecticut is generally similar to that farther north. In Connecticut it consists of the Monson and Glastonbury anticlines and the intervening Great Hill syncline, culminating in the Killingworth dome at the south. The Merrimack synclinorium, which occupies the area between the Monson anticline on the west and the Lake Char thrust fault on the east, has not been intensively studied except in eastern Connecticut. For this reason structural relationships and interpretations in the southern part of the Merrimack synclinorium are stressed in this paper. The fold pattern indicated by relationships in eastern Connecticut is at variance with the pattern suggested by Thompson et al. (this volume) for the synclinorium to the north. The two interpretations are in apparent conflict in the area of the Monson anticline.

Since 1955 most of eastern Connecticut has been mapped at a scale of 1:24,000 by the U. S. Geological Survey and the Connecticut Geological and Natural History Survey under a cooperative program between the two agencies. The geologic

map of eastern Connecticut (Fig. 16–2) is modified from the map by Goldsmith (1963) and is a compilation of the recent mapping. The geology of areas which have not yet been covered by modern mapping is taken from older maps and from reconnaissance by current workers.

The structural and stratigraphic interpretations discussed in this paper, and illustrated in the map and fence diagram of Figure 16–2 are largely the result of an evolution of ideas during the course of mapping in the area. We are grateful to numerous co-workers, assistants and advisors for the cooperative spirit of sharing information and ideas in many discussions and field trips. Specifically we would like to acknowledge the contributions of Mr. G. P. Eaton, Mr. Richard Goldsmith, and Mr. G. L. Snyder of the Geological Survey and Mr. Lawrence Ashmead as co-workers, and of Prof. M. P. Billings, Dr. L. R. Page, Prof. J. W. Peoples, Prof. John Rodgers and Prof. J. B. Thompson, Jr., for continued interest, advice and criticism during the work.

On the basis of the recent work, two major structural features have been recognized in the Connecticut portion of the Merrimack synclinorium.

1. A large recumbent syncline in the micaceous schists and calc-silicate gneisses which make up the synclinorium is overturned to the east; the fold was formed during the major phase of metamorphism in post-Early Devonian time and was subsequently refolded.

2. A large thrust fault borders the synclinorium on the south and east; rocks in the synclinorium were pushed east and southeast on the fault plane after the peak of thermal metamorphism and the first stage of major folding.

[*] Publication authorized by the Director, U. S. Geological Survey, and the Director, Connecticut Geological and Natural History Survey.

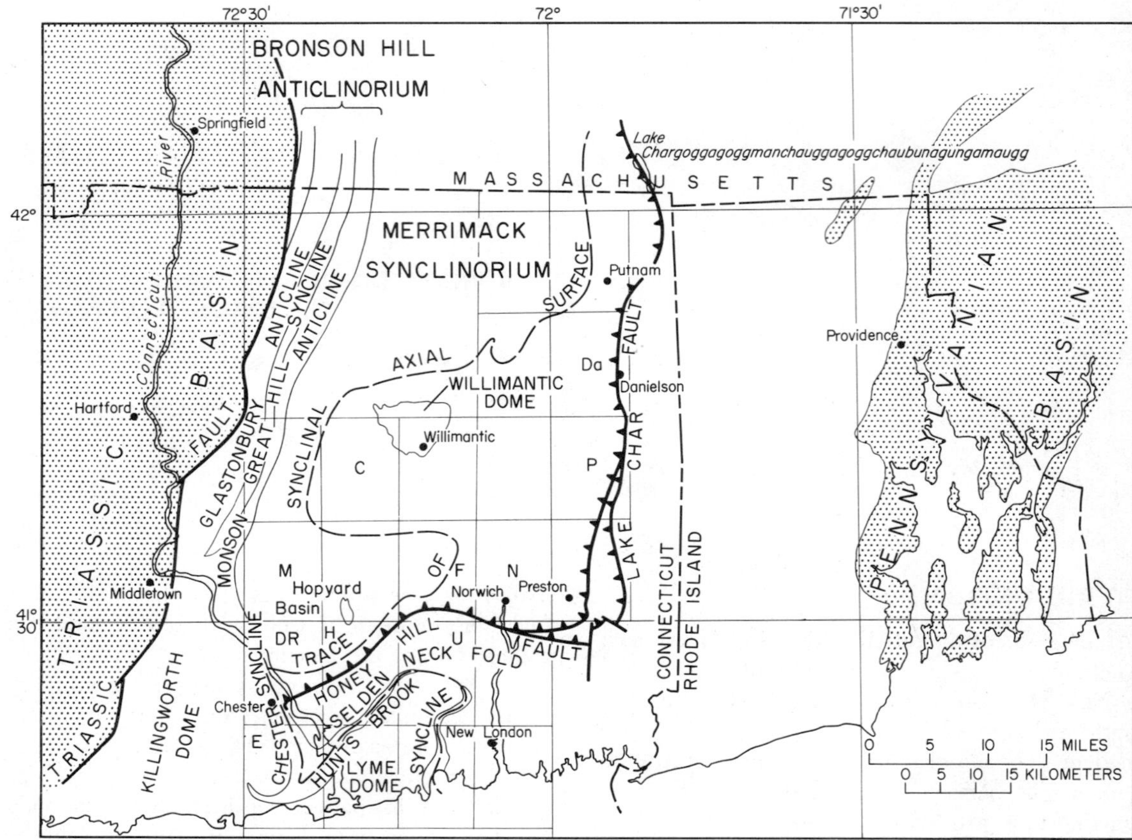

FIGURE 16–1. Index map of eastern Connecticut, showing major structural features. Quadrangle names: C, Columbia; Da, Danielson; DR, Deep River; E, Essex; F, Fitchville; H, Hamburg; M, Moodus; N, Norwich; P, Plainfield; U, Uncasville.

These structural features are illustrated in the fence diagram, Figure 16–2.

STRATIGRAPHY

All the rocks in the Merrimack synclinorium have been metamorphosed to staurolite grade or higher; no fossils have been found for direct dating of the rocks and few reliable sedimentary structures have been recognized. Relative ages of these rocks therefore are inferred primarily on the basis of correlations and structural interpretations. Table 16–1 gives the structural sequences and the inferred stratigraphic sequences of units in the west, central, and eastern parts of the area.

West of the Monson anticline, the sequence from Middletown Gneiss through Littleton Formation is well established as a normal stratigraphic sequence. The units have been traced across Massachusetts to connect with the Ammonoosuc (Middletown Gneiss), Partridge (Brimfield Schist), Clough, Fitch, and Littleton Formations of New Hampshire. The Middletown

is underlain by Monson Gneiss in the core of the Monson anticline and in the Killingworth dome area. Radiometric age analyses by Brookins and Hurley (1965) confirm field evidence that Monson is slightly older than Middletown. The New London, Mamacoke, and Plainfield Formations are, by superposition, older than the Monson (Goldsmith, 1966). These pre-Monson rocks are not exposed along the Monson anticline, but are exposed in complex folds of the coastal area, primarily the Selden Neck and Lyme domes.

On the eastern edge of the synclinorium the sequence from Quinebaug Formation through Scotland Schist has been internally established as a normal stratigraphic sequence by a few critical exposures of preserved sedimentary features. These indicate that the Tatnic Hill Formation is younger than Quinebaug and that the Scotland Schist is younger than the Hebron Formation. The rocks at the base of this sequence are thrust over the older Plainfield Formation. The eastern units have not yet been directly correlated with units of known age elsewhere. They extend

TABLE 16-1. Structural sequences across eastern Connecticut

Bronson Hill Anticlinorium		Merrimack	Synclinorium
West		**Central**	**East**
↑ Littleton Formation Fitch Formation Clough Quartzite UNCONFORMITY Brimfield Schist Middletown Gneiss Monson Gneiss	} Bolton Group, of Rodgers et al. (1959)	Monson Gneiss Middletown Gneiss ↓ Brimfield Schist Hebron Formation ↑ Tatnic Hill Formation Quinebaug and Monson Formations	↑ Scotland Schist Hebron Formation Tatnic Hill Formation Quinebaug Formation
Southeast costal units	{ New London Gneiss Mamacoke Formation Plainfield Formation	Willimantic dome units	—— Fault —— Plainfield Formation

Solid arrows indicate established stratigraphic sequence. Dashed arrows indicate inferred stratigraphic sequence.

northward into Massachusetts, but are cut off along a fault in southern Massachusetts and relationships across the fault have not yet been established.

Across the central part of the synclinorium, from Monson Gneiss on the west to Scotland Schist on the east, the sequence is inferred from structural interpretation to be inverted. The east limb of the Monson anticline is overturned so that east of the anticline Monson Gneiss overlies Middletown Gneiss (where present) which in turn overlies Brimfield Schist. On the east side of this belt and in the Hopyard basin Brimfield Schist structurally overlies Hebron Formation. Thus it appears that the rocks are progressively younger eastward from the Monson Gneiss to the middle of the Hebron Formation, and the Hebron Formation and Scotland Schist are the youngest units in the synclinorium. That this stratigraphic sequence is inverted has not yet been independently established. The Hebron and Scotland may be correlative with parts of the Bolton Group of Rodgers et al. (1959) west of the Monson anticline or the Hebron at least, and possibly also the Scotland may be older than the Bolton Group.

Plutonic rocks of varying composition and age occur primarily as sills in the metasedimentary and metavolcanic rocks. The unit labeled "g" in Figure 16-2 includes at least two separate rock types. East of the Monson anticline, the rocks are related granodiorite to quartz monzonite gneisses. They form sills, mostly in the Hebron, and have not been recognized in rocks older than the Tatnic Hill Formation. West of the Monson anticline the rocks are quartz monzonite to granite, and occur in anticlinal cores; the largest body is in the Glastonbury anticline. Radiometric work

by Brookins and Hurley (1965) and by Zartman et al. (1965) indicate that the gneisses east of the Monson anticline are older than those on the west side. The Sterling Plutonic Group ("sg" in Fig. 16-2) makes up a distinct group of gneisses which are associated with Monson and older rocks. They consist of sills of quartz monzonite, granite and alaskite gneiss. The Sterling gneisses occur primarily in the lower plate of the major thrust fault, and the only known occurrence of the rocks in the upper plate of the fault is in the core of the Willimantic dome (see Figs. 16-1 and 16-2).

RECUMBENT FOLD SYSTEM

The major recumbent fold system, as it is interpreted here and as shown in the fence diagram of Figure 16-2, is not a nappe in the classical sense, but is rather a system of folded folds. The major fold which has been mapped across the area is a large syncline, the Chester–Hunts Brook syncline (Fig. 16-1) which has been refolded about the axial plane of the Selden Neck fold (see Fig. 16-2). The Chester syncline was first mapped by Lundgren (1963, 1964) west of the Connecticut River. In the Essex quadrangle (Fig. 16-1), the axial plane of the Chester syncline bends east, and its continuation east of the river is called the Hunts Brook syncline (Goldsmith, 1961). The approximate trace of the axial surface of the recumbent syncline is indicated by a dashed line on Figure 16-1 and 16-2. Rocks west of this line are in the overturned limb of the syncline and east of this line they are in the normal limb, except in the area south of the Honey Hill fault. The basic argument for the recumbent syncline is that the Brimfield Schist, which structurally overlies the Hebron Formation in the Hopyard basin, is equivalent to the Tatnic Hill Formation, which structurally underlies the Hebron. This general interpretation was, in fact, first suggested by Percival (1842, p. 289) who compared rocks of the Tatnic Hill Formation (his Unit F) with the Brimfield Schist (his Unit D) and suggested ". . . the two might be considered as forming a whole, enclosing the range (E) on the East and West; the coarser grained rocks, partly with bucholzite [fibrolite], extending in a narrow band along the borders of the southwest extremity of the latter, apparently forming a connecting link between them." (Percival's Unit E is Hebron Formation.)

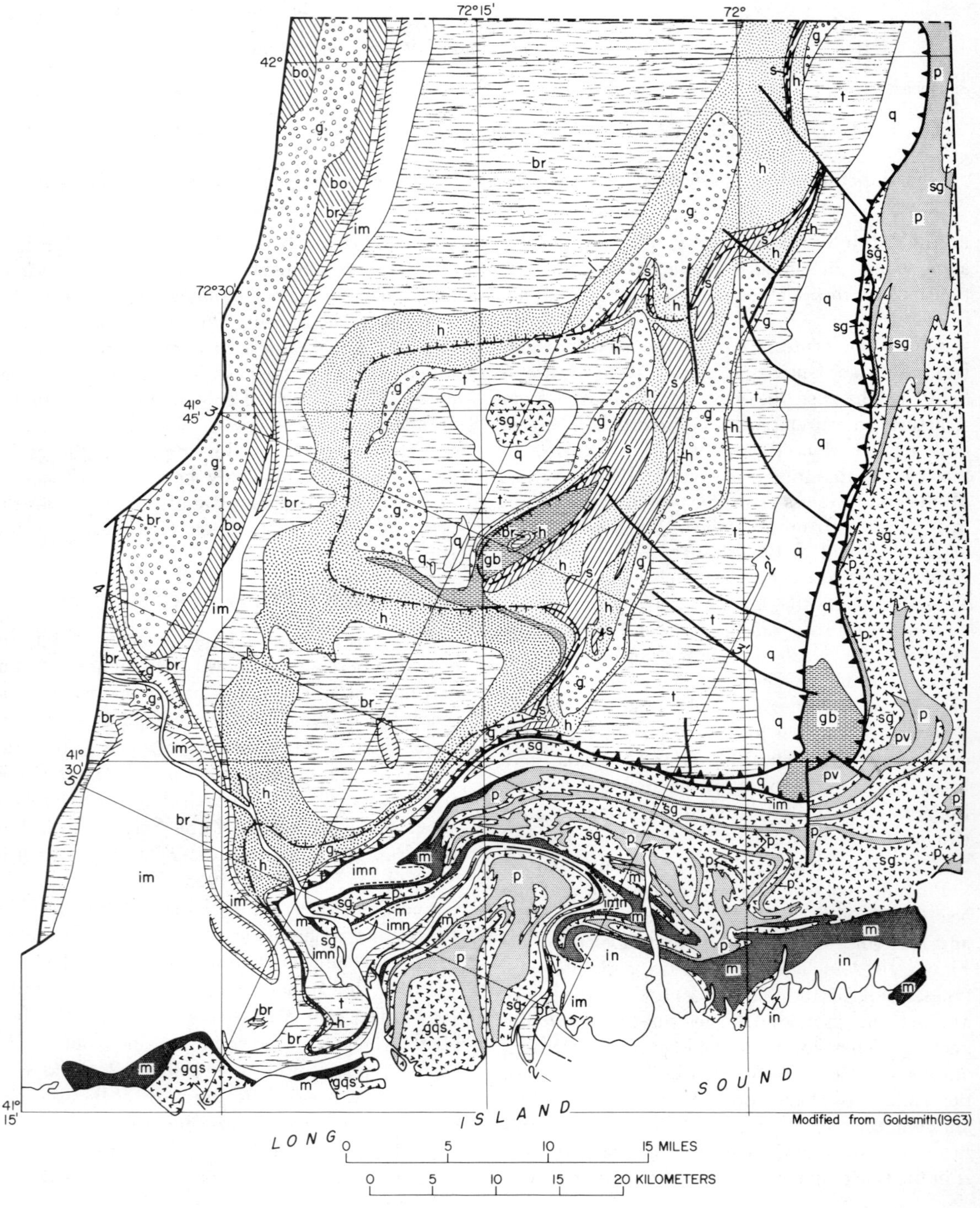

The critical evidence for the recumbent fold is displayed along the western edge of the synclinorium and south of the Honey Hill fault. Micaceous schist in the core of the Hunts Brook syncline south of the fault (Goldsmith, 1961; Lundgren, 1966) can be traced west into the Tatnic Hill Formation and Brimfield Schist in opposite limbs of the Chester syncline (Lundgren, 1963, 1964) west of the Connecticut River (Fig. 16–2). Monson Gneiss and older rocks are symmetrically arranged in opposite limbs of the Hunts Brook syncline, though some units are much thinned on the south and west limb (Goldsmith, 1961). West of the Connecticut River in the Chester syncline the micaceous schists are split into two belts separated by Hebron in the

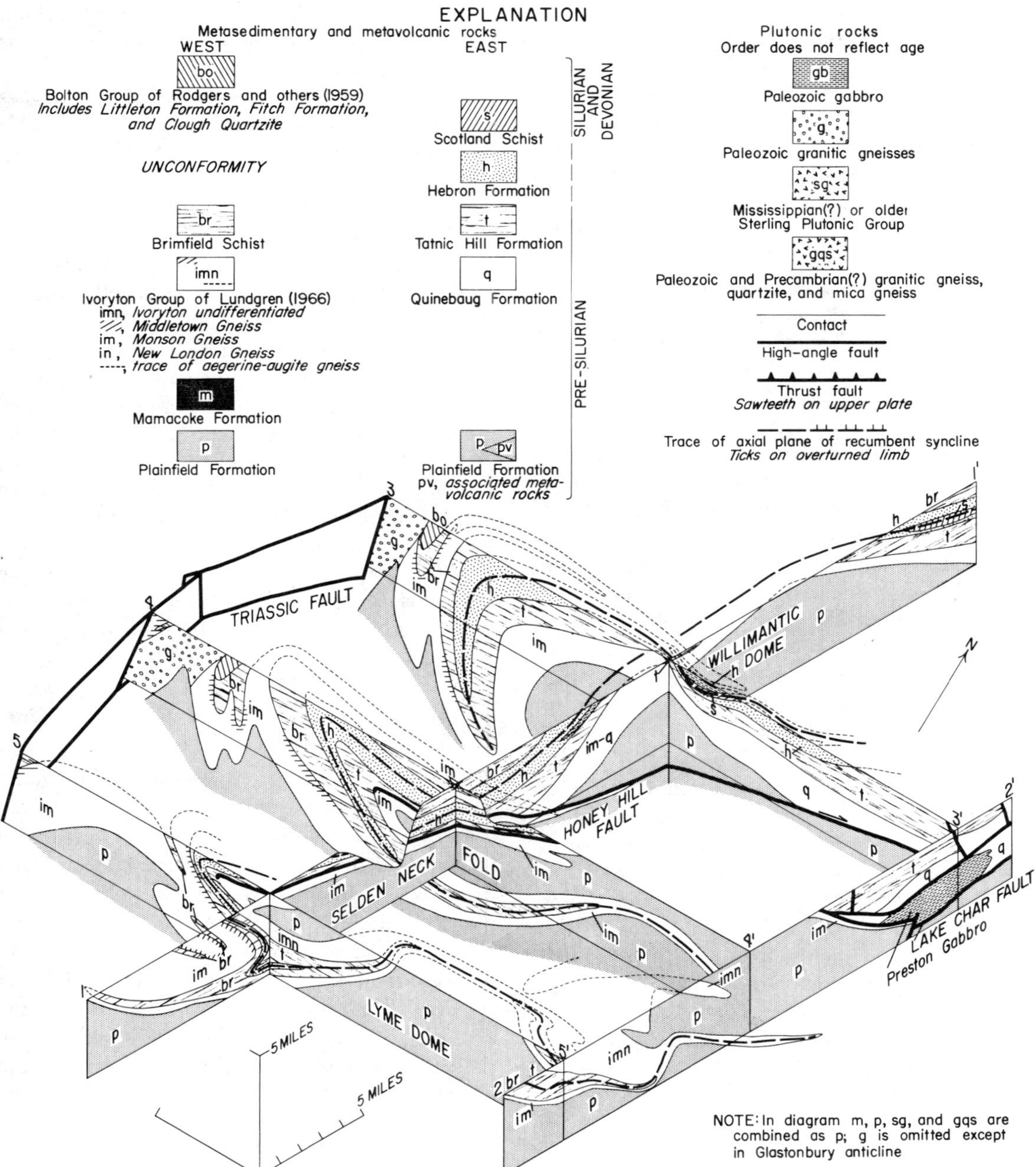

FIGURE 16–2. Generalized geologic map and fence diagram of eastern Connecticut.

synclinal core. The schist on the east limb is continuous with the Tatnic Hill Formation, and the schist on the west limb is continuous with the Brimfield Schist. Thus, the Brimfield and Tatnic Hill are equivalent units on opposite sides of the narrow appendix of Hebron Formation south of Chester. In the appendix the Chester syncline appears to be simply a tight overturned syncline with a steeply dipping axial surface. In the Moodus area, however, the steeply dipping Brimfield of the west limb is continuous with gently dipping Brimfield in the Hopyard basin (except where a stream cuts through the Brimfield), where Brimfield overlies Hebron. Within the Hopyard basin and structurally overlying the Brimfield Schist is an isolated synform of anthophyllitic gneiss and amphibolite which is indistinguishable from the anthophyllite gneiss of the Middletown Gneiss and in the basal Brimfield on the flanks of the Monson anticline. If this anthophyllite gneiss is Middletown or basal Brimfield, the sequence in the basin must be inverted (Fig. 16–2, section 4–4′). Similar evidence for older rocks overlying the Brimfield has not been found north of this area but should be looked for as mapping progresses in the area of the Brimfield of northern Connecticut and across Massachusetts.

The exact location of the trace of the axial surface of the recumbent syncline is not yet definitely established. Through much of the area it must lie within the Hebron Formation (see Fig. 16–2). Along the extreme eastern limit of the trace—that is, southeast and northeast of the Willimantic dome—Scotland Schist is isoclinally folded and is probably in the synclinal core. Vertical beds, indicating the axial zone of a recumbent fold, have so far been observed in only one exposure of Scotland. Because the center of the Willimantic dome is made up of a normal, right-side-up sequence of units, the axial surface must be warped up over the dome and its trace on the present ground surface would wrap around the south, west, and north sides of the dome. The axial surface of the syncline apparently is refolded around the axial plane of the Selden Neck fold and south of the Honey Hill fault it dips north under the Selden Neck fold (Fig. 16–2; see also Lundgren, 1964).

The core of the overlying anticline is exposed in the main belt of Monson Gneiss in the Monson anticline along the west edge of the synclinorium. Monson in the recumbent anticlinal core has been eroded away and is not exposed in Connecticut, so Monson Gneiss cannot be traced from the steep core into the recumbent part of the fold. The

Monson Gneiss belt is continuous from Massachusetts through Connecticut to Long Island Sound. North of the sound the gneiss of the Monson anticline merges with similar gneiss in the Killingworth dome, and traces through an area of intricate structure (Lundgren, 1964) around the south side of the Chester–Hunts Brook syncline into the Lyme dome.

In northern Connecticut, the east limb of the Monson anticline is overturned to the east, and toward the south it steepens to become almost vertical. Throughout this distance, anthophyllitic gneiss of the Middletown Gneiss occurs on both limbs of the Monson anticline; however, it is commonly thin and discontinuous so is not everywhere present. Similar anthophyllite gneiss has not, however, been observed at a comparable stratigraphic horizon either south of the Honey Hill fault, or in the Quinebaug Formation. Thus the presence of the apparently inverted anthophyllite gneiss in the Hopyard basin, 8 miles (13 km) east of all other known occurrences, is most easily explained as a result of eastward movement during recumbent folding.

If the relationships of the Chester and Hunts Brook synclines are as suggested in the fence diagram of Figure 16–2, the Monson anticline should be refolded essentially parallel to the refolding of the Chester syncline. It should thus be possible to follow the anticlinal axial surface around the bend of the Chester syncline and eastward into the Lyme dome, and progressively younger units should be exposed south of the axial surface; younger units have not, however, been found in the Lyme dome. The southern end of the Monson anticline has evidently been folded (Lundgren, 1964), but no satisfactory interpretation of that area can be made until the Killingworth dome is mapped in detail.

The general pattern of metamorphic zones across eastern Connecticut is consistent with the concept of recumbent folds which were overturned after the peak of thermal metamorphism. The lowest grade rocks, of kyanite-staurolite grade, occur only in the cores of the large synclines; the Great Hill syncline on the west and the easternmost part of the recumbent Chester syncline. The rocks in the fold limbs and in the core of the overlying anticline are of sillimanite-grade. Staurolite-grade Scotland Schist, where isoclinally folded in the synclinal core, is structurally overlain as well as underlain by sillimanite-grade rocks in the fold limbs. These relations indicate structural inversion of metamorphic zones as well as of stratigraphic units.

Numerous small-scale overturned folds in the Hebron and the Tatnic Hill Formations are also consistent with a general west-to-east direction of transport. Most of these folds have amplitudes of a few inches to a few feet, and though their patterns are consistent with the inferred direction of movement, they do not indicate the scale of movement. A fairly large recumbent fold, covering several miles, was mapped by Snyder (1967) on the west flank of the Willimantic dome. This fold also indicates eastward movement. Some small folds, however, indicate other directions of movement. Hansen et al. (1967), after a detailed study of minor folds in a small area of Hebron Formation in the Deep River area (Fig. 16–1), suggested south-to-north movement of the rocks at one stage of fold evolution. A detailed study of small-scale structures throughout the area is needed to interpret various possible stages of folding.

An alternative interpretation of the structure in the vicinity of the Merrimack synclinorium and Bronson Hill anticlinorium in Massachusetts and southern New Hampshire is suggested by Thompson et al. (this volume). These investigators have identified a series of recumbent folds overturned from east to west in that area and have traced them south to the Connecticut border. The suggested root zone of one anticline (Skitchewaug nappe) is the Monson anticline, and the root for the uppermost anticline (Fall Mountain nappe) is within or east of the synclinorium. These two recumbent anticlines are separated by a syncline which reaches the Connecticut border near the west edge of the belt of Brimfield Schist (Fig. 16–2). This interpretation is at present difficult to reconcile with geologic relationships observed in Connecticut, especially south and west of the Honey Hill fault; across much of eastern Connecticut, as well, several lines of evidence indicate a general west to east sense of movement.

THRUST FAULT

The large thrust fault across the southern and eastern parts of eastern Connecticut has been mapped in two segments, both of which are traced primarily by intense cataclasis of the rocks in and near the zone of the fault. The Honey Hill fault has been traced from west of the Connecticut River eastward to the Preston Gabbro area (Fig. 16–2), and the Lake Chargoggagoggmanchaug-gagoggchaubunagungamaugg fault (hereafter referred to as the Lake Char fault) has been traced from the lake of that name in southern Massachusetts south to the Preston Gabbro. The area of the gabbro itself has not yet been mapped

in detail, but reconnaissance in the area and the work by Sclar (1958) indicate that the fault splits north and west of the gabbro, one branch going beneath the gabbro and the other one above it. So far as can be determined within the limit of exposure, the Honey Hill–Lake Char fault is a single fault surface that is the contact between the Tatnic Hill, and Quinebaug Formations and the Preston Gabbro in the upper plate and the Monson and Plainfield Formations and interlayered Sterling gneisses in the lower plate. This surface lies within a zone of granulated rock, and shear displacements did occur throughout this zone. There are smaller thrust faults branching off the major fault surface in both the upper and lower plates, but only locally is there a narrow zone of imbricate faults in which rock types from above and below the fault are intermixed.

Recent aeromagnetic work in southeastern Connecticut has revealed a prominent linear east-west magnetic anomaly which extends from long. 72°8′W eastward across Rhode Island along lat. 41°28′N. (Bromery, 1967; Griscom and Bromery, this volume). Information is not yet available as to the trace of the anomaly west of long. 72°8′W. The anomaly is coincident with the Honey Hill fault only at the south end of the Preston Gabbro (Boynton and Smith, 1965; Sclar, 1958). West of the gabbro the anomaly diverges from the trace of the fault, and is located entirely within the lower plate of the north dipping fault. There are no indications of faulting on the geologic maps of either Goldsmith (1967) or Feininger (1965) along the trace of the anomaly west or east of the gabbro, nor is there any apparent relationship between the anomaly and the bedrock units exposed along it. The cause of the anomaly is not yet known, but present indications are that it is not related to the Honey Hill fault.

The plane of the Honey Hill–Lake Char fault is an irregular, warped surface and its trace follows a sinuous course with a generally shallow dip north and west. The dip of the Honey Hill fault steepens from about 20° north at Honey Hill in the Deep River quadrangle (Lundgren, 1963) to 55° north at the south end of the Preston Gabbro (Sclar, 1958). The Lake Char fault can be observed in two places in the Plainfield quadrangle (Dixon, 1965) dipping about 10° west; elsewhere the average dip of the foliation in the cataclastic rocks near the fault, and presumably of the fault itself, is about 25° west. Throughout much of the length of the fault as mapped so far, its trend remains approximately parallel to the structural trends of the rocks on either side, and the cataclastic rocks provide the main evidence for a fault. The zone of strong cataclasis consists of a belt 1,500 to 2,500 feet (450 to 750 m) thick. The rocks in this zone have been converted to mylonite and blastomylonite, and rocks in an even thicker zone show some granulation, which is locally intense. The part of the cataclastic zone above the fault contact is two or three times thicker than the part below it; subsidiary thrust faults also are more common in the upper plate than in the lower plate.

The cataclastic rocks are developed from a variety of rock types, and differ in their characteristics depending on the mineral composition of the original rock and the degree of granulation and recrystallization. Most cataclastic rocks are darker, better layered and more closely jointed and fractured than the normal non-cataclastic rock. They range from very fine-grained mylonite and aphanitic ultramylonite, in which all minerals are thoroughly granulated, to rocks in which the essential minerals are megascopically recognizable but which are seen microscopically to contain ragged and fractured grains in a fine-grained, granulated matrix. Within a few feet of the fault contact, most of the rock is mylonite or fine-grained blastomylonite. In the rest of the zone of strong cataclasis adjacent to both the Honey Hill and the Lake Char segments of the fault, layers of mylonite are interleaved with blastomylonite, both of which may be partially recrystallized. The blastomylonites are partially crushed rocks (augen gneisses) that contain porphyroclasts of zoned and fractured feldspar, and of garnet, muscovite, or hornblende where these minerals were present in the original rocks. The cataclastic rocks generally contain some recrystallized biotite, epidote and chlorite, but the principal mineral assemblages are those typical of sillimanite zone rocks.

Shearing, and the development of some mylonites and most blastomylonites apparently took place early in the development of the fault, while the rocks were still under metamorphic conditions such that middle amphibolite facies assemblages were stable. Pressure was, however, high enough that rocks in the fault zone were thoroughly granulated, partially recrystallized and locally thinned. Although the shear and the resultant granulation were most concentrated close to the fault, granulation was distributed through a layer of rock several thousand feet thick. These early mylonites and blastomylonites were themselves folded isoclinally, notably near Chester and in the Fitchville quadrangle (Fig. 16–1; Snyder, 1964), by later movements along the fault plane. More strongly cataclastic rocks—particularly aphanitic and retrogressed mylonite and ultramylonite dikes that locally branch from these layers—developed later than the blastomylonites. Small ultramylonite dikes are locally abundant along conjugate shears and along the axial surfaces of conjugate folds in the fault zone. Ultramylonites along the Honey Hill fault include clasts of blastomylonite and the dikes transect the foliation and limbs of small folds in the blastomylonite. On a somewhat coarser scale, a fault breccia near Danielson, either on the Lake Char fault or on a small fault subsidiary to it, contains clasts of mylonitized Plainfield quartzite and Sterling gneiss. Abundant closely spaced slickensided surfaces probably represent some of the latest movements in the fault zone. These surfaces most commonly are developed on planes of the earlier cataclastic foliation.

Major movement on the fault plane apparently was toward the southeast. In the Fitchville quadrangle (Snyder, 1964) the gneisses of the Tatnic Hill Formation in the upper plate are dragged into large west-plunging folds along the fault which indicate a southeastward movement of the upper plate. A pervasive lineation in the cataclasites along the Lake Char fault trends N. 60°–70° W. The lineation is probably an "a" lineation, produced during granulation by streaking out of mineral grains parallel to the direction of movement; it does not indicate whether movement was northwestward or southeastward, but the later is consistent with southeastward movement of the Honey Hill fault as well as with the regional geologic framework.

The amount of movement on the fault is difficult to estimate, and along much of it no actual displacement can be proven. West of the Connecticut River the trend of the cataclastic rocks along the Honey Hill fault bends sharply from southwest to southeast, parallel to the trend of the Chester syncline. For about 15 miles (24 km) east of this bend the fault trace remains parallel to the structural trends of the rocks both above and below it. Along the trace here, the gneisses of the Tatnic Hill Formation are in fault contact with the Monson Gneiss, but this contact is a normal stratigraphic contact south of the fault where the rocks are unfaulted. The first observable offset is from the drag folds in the Fitchville quadrangle east to the Preston Gabbro, where the rocks of the lower part of the Tatnic Hill Formation and the Quinebaug Formation are sharply truncated. This is the only place where a sharp discordance in structures was observed along the fault. The Lake Char fault is again subparallel to structural trends on either side of the fault zone, though in the upper plate it cuts across the Quinebaug Formation at a low angle. Across the Plainfield and Danielson quadrangles (Dixon, 1965, 1968), about 2,000 to 3,000 feet (600 to 900 m) of the Quinebaug is cut out from south to north. Reconnaissance just north of the Connecticut–Massachusetts State line indicates that the Quinebaug is very thin on the west side of the fault, and farther north it is probably cut out entirely. The Quinebaug Formation is thus progressively cut out northward. As the unit has not been observed where it is unfaulted, how much of the lower part of the Quinebaug may be missing at the south end of the fault cannot yet be determined.

Truncation of rock units beneath is even more difficult to demonstrate. The fault remains essentially parallel to the trend of these units, though at its eastern end the Monson Gneiss must eventually be cut out, or overridden. The Lake Char fault puts the Quinebaug Formation in fault contact with the Plainfield Formation and the Sterling gneisses but, as these units may be in normal stratigraphic sequence, this could mean omission of a few thousand feet of section or of none at all. The Quinebaug in the upper plate of the Lake Char fault is correlated with the Monson Gneiss in the lower plate of the Honey Hill fault. To the southwest the Monson is separated from the Plainfield by the New London Gneiss and the Mamacoke Formation, which together have a maximum estimated thickness of 2,500 feet (750 m; Goldsmith, 1966). Both the New London and the Mamacoke, however, lens out eastward along strike and in the Uncasville quadrangle (Goldsmith, 1967) the Monson is in direct contact with the Plainfield. The New London and Mamacoke may, therefore, never have been present in the eastern part of the area, and the succession of Quinebaug over Plainfield may be a normal one. Thus the low-angle truncation of the Quinebaug in the upper plate is the only demonstrable displacement along the Lake Char fault.

Although sharply truncated structures are not commonly observed along the exposed trace of the fault, there are indications that the fault must cut across structures beneath the synclinorium. The rocks in the center of the Willimantic dome include units (see Table 16–1) which to the south and east either are involved in the fault or occur only in the lower plate. There is, however, no indication of faulting or cataclasis of the rocks around the dome. The core of the dome is an alaskite gneiss similar to the alaskites of the Sterling Plutonic Group south and east of the fault. The alaskite is overlain successively by hornblende gneisses similar to Monson and Quinebaug and by micaceous schist and calc-silicate gneiss of the Tatnic Hill Formation. Below the fault the Sterling alaskite is associated with Monson and older rocks, and most likely the alaskite in the core of the dome is underlain by Monson or older rocks. We do not know how far north and west the fault extends under the synclinorium, but it must truncate the Monson and Sterling gneisses of the lower plate somewhere beneath the synclinorium in order for these rocks to appear, unfaulted, in the dome. We consider it probable that the refolded folds in the coastal area continue beneath and are cut off by the thrust fault (see fence diagram, Fig. 16–2).

The zone of intense cataclasis above the Lake Char fault is about 2,000 feet (600 m) thick, but almost all of the Quinebaug Formation and much of the lower part of the Tatnic Hill show some cataclastic textures which are locally intensely developed. The more strongly cataclastic rocks in the Tatnic Hill Formation can be related to repetitions of lithologic units within the formation, and probably indicate smaller thrusts within the upper plate. These faults (not shown on Fig. 16–2) are subparallel to the major thrust fault and probably branch from the Honey Hill fault in the south. One or more of these thrusts may have moderately large offset, as they apparently thrust higher grade (sillimanite-potassium feldspar) rocks of the lower part of the Tatnic Hill onto lower grade (sillimanite-muscovite) rocks of the Quinebaug Formation. The numerous small faults in the upper plate could account for the gross exaggeration in thickness of the Tatnic Hill Formation southward toward the Honey Hill fault. The rocks of the upper plate are also cut by numerous northwest-trending, steeply dipping faults. These faults apparently originate on or near the main fault, and probably formed fairly late.

There is also evidence for small- to moderate-scale thrusting in the Hebron and Brimfield of the overturned limb of the Chester syncline. Rocks near the contact between these units are locally granulated and crossed by ultramylonite dikes, and in the Brimfield small low-angle faults can be observed in outcrops and traced for at least thousands of feet (M. H. Pease and John Peper, oral communication, 1966). The apparent exaggerated thickness of the Brimfield in northern Connecticut could be explained, in part at least, by these thrust faults.

The Lake Char fault continues north into Massachusetts and a few miles north of the state line probably connects with one or more of the faults currently being mapped in northeastern Massachusetts (Skehan, this volume). Just where and how these faults come together has not yet been determined. Emerson (1917) showed an abrupt change in structural trends about 5 miles (8 km) north of the State line, where the north-trending units from Connecticut are abruptly truncated by essentially east-trending units. The faults apparently converge in this general area, but until the area is mapped in more detail it is difficult to say whether the two fault systems are continuous or whether one system cuts off the other.

TECTONIC EVOLUTION

Several lines of evidence from eastern Connecticut indicate a series of events that produced (1) large-scale recumbent folding, (2) progressive metamorphism to high amphibolite facies, (3) development of domes and basins, and (4) cataclasis and thrust faulting. There were at least two distinct stages of folding. Large-scale folding, metamorphism, and the inception of faulting were most likely overlapping aspects of one major tectonic event. Formation of broad domes and basins followed recumbent folding and the earlier stages of thrust faulting. Movement in the fault zone probably continued over a long time, and final movement was certainly later than any of the other events.

There is no direct evidence in eastern Connecticut for dating the major period of tectonic activity or determining the length of time that it encompassed. Folding and metamorphism of Silurian and Lower Devonian rocks in the Great Hill syncline indicate post-Early Devonian tectonic activity; by analogy with northern New England, the major period of fold evolution and of regional dynamothermal metamorphism to upper amphibolite facies was probably Middle to Late Devonian, that is, during the Acadian orogeny. It is possible that the pre-Silurian rocks may have been folded prior to deposition of the younger rocks, but this earlier folding has not been definitely established in this area. Zartman et al. (1965) suggest that radiometric age analyses of rocks and minerals from eastern Connecticut indicate an Acadian age for dynamothermal metamorphism, but their numbers do not support this statement. Whole rock Rb–Sr analyses of some of the gneisses shown in Figure 16–2 as "g" give an age of 430 ± 20 m.y. using a decay constant of $\lambda = 1.39 \times 10^{-11}$ year^{-1} which Zartman (personal communication, 1967) would now

prefer to use (the Rb–Sr numbers cited in the paper were based on a decay constant of $\lambda = 1.47 \times 10^{-11}$ year^{-1}). This number indicates a pre-Acadian age for the gneisses. The mineral ages cluster around 250 m.y., and reflect a Permian "event" which is well established in southeastern New England, but whose nature is not well understood. The radiometric work, thus, gives no firm evidence either for or against Acadian tectonic activity in the area.

Major isoclinal folding, which resulted in the Chester–Hunts Brook syncline, apparently accompanied dynamothermal metamorphism of the area. During the subsequent refolding the syncline was overturned to the east. The direction of movement during the refolding of the syncline is the same as the direction of movement on the Honey Hill–Lake Char fault, and it seems likely that refolding and faulting were synchronous. The major fault zone is located entirely within or below the normal limb of the recumbent syncline, and it is clear that neither the fault nor the recumbent fold system fits the picture of a typical alpine nappe in which a recumbent anticline is separated from its root zone and the underlying syncline by a thrust fault. If the fence diagram (Fig. 16–2) gives a fairly correct picture of the structural relationships of the area, the axial surface of the Chester syncline was tightly refolded around the north-northeast axis of the Selden Neck fold, and the upper limb of the refold was pushed eastward across the synclinorium. The Honey Hill fault is located on the upper limb of the Selden Neck fold, and probably movement in the fault zone began during the refolding. Subsidiary faulting apparently took place along other favorable planes higher in the fold structure. The ultramylonite dikes and cataclastic rocks in the axial zone of the Chester syncline (which would, under these circumstances, act like a bedding plane), near the Hebron–Brimfield contact, and within the Brimfield probably formed at a shallower level than the main thrust fault did. Inversion of metamorphic zones in the refolded structure suggests that the refolding was later than the peak of thermal metamorphism.

Evidence has been given earlier for several stages of movement in the fault zone. It is not clear whether these stages were discontinuous or were parts of a long continued event, although the apparent overlapping of events suggests more or less continuous movement. The earliest stages of the fault are indicated now only by the thick zone of cataclastic rocks, in which the granulation increases in intensity toward the fault. This granulation apparently started while the rocks of the area were still being metamorphosed and folded. Folding of the early cataclastic rocks, truncation of metamorphic isograds (Snyder, 1964; Goldsmith, 1967), ultramylonite dikes and breccias, and slickensided foliation surfaces all indicate that significant displacements continued in the fault zone after metamorphic conditions ceased. Most of the rocks in the fault zone are pre-Devonian; none of the post-Devonian granitic rocks of southern Connecticut and Rhode Island have been identified in the fault zone. Consequently age limits for the fault movement cannot be closely established. We believe that fault activity probably continued through the Pennsylvanian and possibly into Permian time. There is no evidence that the fault zone was involved in the Triassic tectonic activity that affected the Connecticut Valley; the final stages of movement on the fault probably represent the last pre-Triassic tectonic activity in the area.

The increasing structural discordance along the fault trace from Chester eastward to Preston and northward into Massachusetts may reflect differing depths of development of the fault zone. Thus the westernmost limit of the Honey Hill fault may have formed at the deepest level, and the fault plane may become progressively more shallow toward the east and the north. This is also suggested by the fact that many of the mylonites along the plane of the Lake Char fault are more strongly retrogressed to low-grade mineral assemblages than are those along the Honey Hill fault to the west. Fault breccia, such as that near Danielson, has not been observed near the Honey Hill fault. Both these characteristics are suggestive of lower pressures along the Lake Char fault, at least during the later stages of fault movement. Corollary to this suggestion is the assumption that the present trace of the Honey Hill fault was determined by doming and upwarping of the coastal area.

REFERENCES

Billings, M. P., 1956, The geology of New Hampshire, Part II, Bedrock geology: New Hampshire State Planning and Dev. Comm., 200 p.

Boynton, G. R., and Smith, C. W., 1965, Aeromagnetic map of the Old Mystic quadrangle and part of the Mystic quadrangle Connecticut: U. S. Geol. Survey Geophys. Inv. Map GP–544

Bromery, R. W., 1967, Geophysical evidence of a pronounced east-west lineament in southern New Eng-

land [abs.]: Geol. Soc. America, Northeastern Sect., Program, 1967 Ann. Mtg., Boston, p. 16

Brookins, D. G., and Hurley, P. M., 1965, Rb-Sr geochronological investigations in the Middle Haddam and Glastonbury quadrangles, eastern Connecticut: Am. Jour. Sci., v. 263, p. 1–16

Dixon, H. R., 1965, Bedrock geology of the Plainfield quadrangle, Connecticut: U. S. Geol. Survey Geol. Quad. Map GQ–481

————, 1968, Bedrock geology of the Danielson quadrangle, Connecticut: U. S. Geol. Survey Geol. Quad. Map GQ–696

Emerson, B. K., 1917, Geology of Massachusetts and Rhode Island: U. S. Geol. Survey Bull. 597, 289 p.

Feininger, Tomas, 1965, Bedrock geologic map of the Ashaway quadrangle Connecticut and Rhode Island: U. S. Geol. Survey Geol. Quad. Map GQ–403

Goldsmith, Richard, 1961, Axial plane folding in southeastern Connecticut: U. S. Geol. Survey Prof. Paper 424–C, p. C54–C57

————, 1963, Geologic sketch map of Eastern Connecticut: U. S. Geol. Survey open-file report

————, 1966, Stratigraphic names in the New London area, Connecticut: U. S. Geol. Survey Bull. 1224–J, 9 p.

————, 1967, Bedrock geology of the Uncasville quadrangle, Connecticut: U. S. Geol. Survey Geol. Quad. Map GQ–576

Hansen, Edward, Scott, W. H., and Stanley, R. S., 1967, Reconnaissance of slip-line orientations in parts of three mountain chains, in Carnegie Inst. Wash., Yearbook 65, p. 406–410

Lundgren L. W., Jr., 1963, The bedrock geology of the Deep River quadrangle, Connecticut: Connecticut Geol. and Nat. History Survey Quad. Rept. 13, 40 p.

————, 1964, The bedrock geology of the Essex quadrangle, Connecticut: Connecticut Geol. and Nat. History Survey Quad. Rept. 15, 36 p.

————, 1966, The bedrock geology of the Hamburg quadrangle, Connecticut: Connecticut Geol. and Nat. History Survey Quad. Rept. 19, 41 p.

Percival, J. G., 1842, Report on the geology of the state of Connecticut: New Haven, Conn.: Osborn & Baldwin, 495 p.

Rodgers, John, Gates, R. M., and Rosenfeld, J. L., 1959, Explanatory text for preliminary geological map of Connecticut, 1956: Connecticut Geol. and Nat. History Survey Bull. 84, 64 p.

Sclar, C. B., 1958, The Preston gabbro and the associated metamorphic gneisses, New London County, Connecticut: Connecticut Geol. and Nat. History Survey Bull. 88, 131 p.

Snyder, G. L., 1964, Petrochemistry and bedrock geology of the Fitchville quadrangle, Connecticut: U. S. Geol. Survey Bull. 1161–I, 64 p.

————, 1967, Bedrock geology of the Columbia quadrangle, Connecticut: U. S. Geol. Survey Geol. Quad. Map GQ–592

Zartman, Robert, Snyder, George, Stern, T. W., Marvin, R. F., and Bucknam, R. C., 1965, Implications of new radiometric ages in eastern Connecticut and Massachusetts: U. S. Geol. Survey Prof. Paper 525–D, p. D1–D10

Chapter 17

Origin and Regional Relationships of the Core-Rocks of the Oliverian Domes[*]

RICHARD S. NAYLOR

INTRODUCTION

RECENT FIELD and geochronologic study indicates that the core-rocks of the Oliverian Domes in the Connecticut Valley area are volcanic and plutonic rocks of probable Ordovician age (Naylor, 1967). These rocks are an integral part of a larger belt of Ordovician volcanic and plutonic rocks in the central part of the northern Appalachian geosyncline. This paper summarizes the conclusions of a recent study reinterpreting the geology of a representative Oliverian Dome, the Mascoma Dome east of Hanover, New Hampshire (Naylor, 1967), and discusses the regional relationships of the Oliverian core-rocks in light of the new findings. About one quarter of the Mascoma Dome is underlain by unstratified rocks with cross-cutting relationships. Field and geochronologic data indicate that these rocks were intruded prior to the deposition of late Lower Silurian mantle-units, and that none of the important crosscutting relationships are due to remobilization during later orogeny. These intrusive rocks, and similar rocks in the cores of some of the other Oliverian Domes, are probably related to intrusive rocks of the Highlandcroft Series. The remaining core-rocks of the Mascoma Dome and other Oliverian Domes are weakly stratified and are reinterpreted as metamorphosed volcanic rocks. These rocks are probably Ordovician and are not as old as Precambrian. They are an integral part of the Ordovician "belt of volcanoes and islands" discussed by Berry in another paper in this volume.

[*] California Institute of Technology Contribution No. 1466.

The author is deeply indebted to G. J. Wasserburg for support and critical encouragement of research on the Oliverian Domes which the author undertook for a Ph.D. thesis at the California Institute of Technology. The author is grateful to A. J. Boucot for nurturing his interest in the regional geology of the northern Appalachians, and to the following additional people for helpful discussions: A. L. Albee, M. P. Billings, W. M. Cady, Peter Robinson, John Rodgers, J. L. Rosenfeld, L. T. Silver, and J. B. Thompson, Jr. The author acknowledges fellowship and research support from the National Science Foundation and the California Institute of Technology.

THE OLIVERIAN PROBLEM

The Oliverian Domes lie in a belt which coincides approximately with the crest of the Bronson Hill Anticlinorium a few miles east of the Connecticut River. About twenty separate domes lie en-echelon along this belt from Long Island Sound to northeastern New Hampshire, a distance of about 250 miles (450 km). Each dome consists of a core of granite and gneiss encircled concordantly by a mantle of well-stratified metamorphic rocks. The core-rocks of these domes are igneous in composition and general appearance and have been reported to locally crosscut the mantling strata. Following Billings (1937, p. 502), most previous workers have interpreted all of the core-rocks as granitic rocks intruded during the Devonian (Acadian) orogeny. Although a variety of intrusive mechanisms have been proposed, it has always been difficult to explain why the Oliverian core-rocks are generally in contact with a single mantling unit of Ordovician metavolcanic rocks (Ammonoosuc Volcanics) along most of the 250-mile belt of domes. This is more the behavior to be expected of an older rock unit lying stratigraphically beneath the Ammonoosuc Volcanics and cropping out in a series of domes. This view was taken by Eskola (1949) in his classical paper on the origin of mantled gneiss domes, with the further suggestion that the rare crosscutting features are the result of remobilization of the older rocks during intense later metamorphism. Following the recognition of Precambrian rocks in the core of the nearby Chester Dome, a number of geologists (e.g. Robinson, 1963) have suggested that some of the Oliverian core-rocks may be Precambrian also. The difficulties of such an interpretation have been stressed by Billings (1956, p. 123–125; and personal communication, 1963) who maintained that evidence for the presence of remobilized rocks is lacking.

For the past several years the author has been engaged in a field and geochronologic study of the Oliverian problem. Some of the major conclusions from this study (Naylor, 1967) are summarized in this section, but there is insufficient space to present all of the data on which these conclusions are based. These data, together with a discussion of the more general problem of mantled gneiss dome will be published elsewhere. The chief problem in the study of gneiss domes is to establish which, if any, of the rocks formed at the time of metamorphism of the stratified rocks mantling the domes, and which, if any, are older. To search for traces of possible early history in some of the core-rocks, it is necessary to peer through the effects of later middle- and high-grade metamorphism and deformation. The large Mascoma Dome east of Hanover, New Hampshire, is well suited for study of the primary age relationships of the core-rocks because the later metamorphism was less intense there than in most of the domes farther south.

GEOLOGY OF THE MASCOMA DOME

The Mascoma Dome consists of a core of massive gneiss and granite encircled concordantly by a mantle of well-stratified, middle- and high-grade metamorphic rocks. In plan the core is an oval, elongate in a northerly direction, about 20 miles long and as much as 7 miles wide (32 by 11 km). The western contact of the core is nearly vertical and locally overturned, but on the other flanks the contact dips gently outward. Layering in the overlying strata wraps around the core in a surprisingly regular pattern indicating that the dome is structurally a doubly

plunging anticline. Minor folds are rare or lacking in the basal parts of the mantle, suggesting that the core-rocks behaved as a competent mass during the deformation which shaped the dome.

The Mascoma Dome was previously mapped at 1:62,500 scale by Chapman (1939) and Hadley (1942) who subdivided the core-rocks into mappable units of quartz diorite, granodiorite, quartz monzonite, and granite. The quartz diorite occurs as a border phase and the quartz monzontie and granite occur chiefly in a massive body in the northwest quadrant of the dome. Most of the core is underlain by gneissic granodiorite. Chapman (1939) and Hadley (1942) concluded that all of these rocks were intruded during the Devonian. The present author has re-mapped contacts in the northwestern part of the Mascoma Dome on a 1:10,000 scale airphoto base (Naylor, 1967), and has dated samples of the rocks by the Rb-Sr and U-Pb methods. The results of this study indicate that the core-rocks of the dome originated in two different ways. The quartz monzonite and granite (hereafter described as unstratified core-rocks) crosscut the other core-rocks. These massive, homogeneous, unstratified rocks are probably intrusive, but are demonstrably older than late Lower Silurian. Most of the quartz diorite and granodiorite (hereafter described as stratified core-rock) is faintly layered and is not crosscutting. The present author suggests these are mostly metamorphosed volcanic rocks lying stratigraphically beneath the Ammonoosuc Volcanics.

Mantle Units

The Ammonoosuc Volcanics are the lowest unit of the rocks that mantle the Mascoma Dome. This formation consists chiefly of epidote-amphibolite (plagioclase-hornblende-epidote gneiss) but contains about ten percent of layers of light-colored felsic rocks (fine-grained plagioclase-quartz-biotite granulite). The felsic layers can be traced long distances (as much as 5 miles in one case) along strike, and are not crosscutting (Naylor, 1967). They appear to be primary stratigraphic horizons within the volcanic rocks, and Billings (1937, p. 476–477) has argued convincingly that similar layers in less-metamorphosed exposures of the Ammonoosuc Volcanics originated as tuff and tuff-breccia. In the Mascoma area the Ammonoosuc Volcanics are about 400 feet (120 m) thick, but appear to have been tectonically thinned.

Quartzite, quartz-pebble conglomerate, and quartz-mica schist of the Clough Formation unconformably overlie the Ammonoosuc Volcanics and the core-rocks of the Mascoma Dome. In the southern part of the Mascoma area Boucot and Thompson (1963) have described late Early Silurian fossils from staurolite- and sillimanite-grade rocks from the Clough Formation. Overlying mantle-units include the Fitch Formation (Upper Silurian) and the Littleton Formation (Lower Devonian). The mantle-units are further discussed by Thompson et al. (this volume).

Stratified Core-Rocks

In the northern part of the Mascoma Dome the predominantly mafic Ammonoosuc Volcanics give way down-section to a great thickness of weakly stratified gneiss of intermediate igneous composition which constitutes most of the core-rock of the dome. The transition is not abrupt as there is a zone about a hundred feet thick (30 m) in which sporadic layers of Ammonoosuc-type amphibolite alternate with layers of the light-colored gneiss, and in which the Ammonoosuc Volcanics contain interlayers of light-colored felsic rocks. Chapman (1939) and Hadley (1942) interpreted the felsic layers and light-covered gneiss as lit-par-lit injections of Oliverian intrusive rock into the Ammonoosuc Volcanics, but as mentioned above, they are more probably stratigraphic interlayers of light-colored volcanic rock. The main mass of the intermediate gneiss is reinterpreted (Naylor, 1967) as a stratified unit, probably meta-volcanic, lying stratigraphically beneath the Ammonoosuc Volcanics. The contact between these two units is concordant both regionally and in detail, and nowhere appears crosscutting. A few small bodies of intrusive quartz diorite cross-cut layers of amphibolite near the top of the intermediate gneiss unit, but appear to constitute only a small fraction of the intermediate rocks. These bodies were nowhere observed to crosscut the Ammonoosuc Volcanics.

About three quarters of the core of the Mascoma Dome is underlain by this thick-layered, light-colored, weakly stratified gneiss of intermediate igneous composition. Faint lamination is visible within some layers, but strongly banded gneiss is absent. The layers average 5 to 15 feet (1.5 to 5 m) in thickness. The layers differ from one another mostly by minor variations in the concentration of dark minerals or in the degree of aggregation of biotite or quartz. A few layers contain abundant biotite-rich schlieren which are absent in adjacent layers. Sporadic amphibolite layers are conspicuous in the predominantly light-colored gneiss, and can be traced for a hundred feet or more in areas of good exposure. The layering is best displayed along the western margin of the dome where the gneiss, capped by resistant mantle-rocks, stands in bold cliffs. In the cliff exposures prominent, deeply-weathered cracks commonly separate the layers. The layers are quite regular in the competent gneiss and are not intricately deformed or contorted. About 2100 feet (640 m) of the gneiss is exposed in the northern part of the Mascoma Dome. The gneiss is intruded by granitic rocks in this area and its original stratigraphic thickness may have been considerably greater.

The gneiss consists chiefly of plagioclase, quartz, and biotite or hornblende. The plagioclase and most of the quartz occur in 1 to 2 mm equant grains. Some of the quartz is in aggregates. Biotite commonly occurs in centimeter-sized aggregates of flakes which impart a characteristic blochy appearance to exposed foliation surfaces. Gneiss in the uppermost layers mostly has the composition of quartz diorite, whereas gneiss in many of the lower layers contains small amounts of fine-grained potassium feldspar and has the composition of granodiorite.

Field and geochronologic research (Naylor, 1967) leads to the following conclusions regarding the stratified core-rocks: (1) The stratification was not created during the deformation which formed the dome, but is an older feature of the gneiss. (2) The gneiss lies stratigraphically beneath the Ammonoosuc Volcanics. This is indicated by the observation that the contact between the two units is concordant on both a local and a regional scale, together with the lack of positive evidence suggestive of an intrusive origin. (3) A volcanic origin for the protolith of the gneiss is plausible but not demonstrated conclusively. The gneiss is similar in composition to altered, marine, volcanic clastic rocks (Eocene Ohanapecosh Formation of Washington) described by Hopson (1964, p. 34). A Pb^{207}/Pb^{206} age of 450 ± 25 m.y. measured on zircon (Naylor, 1967, p. 57) probably dates the volcanic episode which produced the rock. (4) No rocks as old as Precambrian have been detected within the core of the Mascoma Dome.

Unstratified Core-Rocks

The remaining quarter of the core of the Mascoma Dome is underlain by rocks which occur in crosscutting bodies. The largest of these bodies is a sub-central pluton of granitic rocks (the Mascoma Pluton) occupying much of the northwest quadrant of the dome core. The granitic rocks are massive, homogeneous, and unstratified. Near the margins of the core the granite is moderately foliated, but in the central parts of the pluton foliation and lineation are only weakly developed. Thin, sharply-bounded aplite veins are common. Elongate biotite-rich schlieren (xenoliths?) are locally conspicuous, but nowhere constitute as much as one percent of the rock. Small shear zones occur throughout the pluton, but most of the rock is not predominantly cataclastic. Allowing for the effects of deformation at the time of dome formation, the massive, homogeneous granitic rocks closely resemble those of batholiths which are believed to have been intruded at intermediate depth. They lack the veined, streaky, or migmatitic appearance suggestive of remobilized granites.

The granitic rocks consist of plagioclase, quartz, microcline, and biotite with minor amounts of epidote and magnetite. The plagioclase and most of the quartz occur as a matrix of 1 to 2 mm equant grains. Some of the quartz occurs in elongate aggregates. Microcline occurs in irregular, centimeter-sized megacrysts which impart a characteristic appearance to the rock. The plutonic rock is mostly quartz monzonite, but microcline is sufficiently abundant locally for the rock to be called granite. The granitic rocks contain somewhat less quartz than does the stratified core-gneiss, which is much less potassic.

Field and geochronologic research (Naylor, 1967) leads to the following conclusions regarding the granitic rocks: (1) The granitic rocks crosscut the stratified core-rocks of the Mascoma Dome and may also crosscut the Ammonoosuc Volcanics. The granitic rocks, however, lie unconformably beneath the Clough Formation (Late Lower Silurian). Chapman (1939, Pl. 1) indicated that the Clough Formation was crosscut by the granite, but more detailed mapping by the present author using airphotos indicates that Chapman's contacts in the critical area were mislocated due to inaccuracies in the topography on the 1927 edition of the Mascoma quadrangle. Chapman's inference that the Clough Formation had been feldspathized by intrusion of the granite (Chapman, 1939, p. 168) does not appear supportable (see also Thompson et al., this volume). (2) The massive, homogeneous character of the rocks and the crosscutting contacts of the bodies in which they occur suggest that the granitic rocks are intrusive. (3) Pb^{207}/Pb^{206} ages of 450 ± 25 m.y. on zircon fractions, and Rb-Sr whole-rock ages of 440 ± 30 m.y. on samples from the granitic rocks probably date the time of intrusion. (4) The U-Pb data include that the granitic rocks do not contain significant amounts of zircons with Precambrain ages. Initial Sr^{87}/Sr^{86} ratios of 0.706 ± 0.003 suggest that the

granitic rocks did not form through simple fusion of Precambrian rocks rich in K and Rb. (5) Post-Lower Devonian remobilization or granitization do not appear to have been important in the formation of the granitic rocks.

Summary

Most of the core-rocks and the lowest unit of the mantle appear to be Ordovician (?) volcanic rocks. These rocks include at least 2100 feet (640 m) of predominantly felsic volcanic rocks (stratified core-gneiss) and at least 400 feet (120 m) of predominately mafic volcanic rocks (Ammonoosuc Volcanics). During or shortly following the volcanism, granitic rocks were intruded into at least the lower part of the volcanic sequence. It is not clear whether this occurred before or after emplacement of the Ammonoosuc Volcanics. Although the author favors the latter alternative, it is not impossible that the stratified core-gneiss was intruded by the granitic core-rocks and possibly regionally metamorphosed prior to deposition of the Ammonoosuc Volcanics.

Deposition of the upper Lower Silurian Clough Formation and younger mantle-units was preceded by a period of erosion in which the granitic rocks were at least locally unroofed. The major period of metamorphism occurred after deposition of the Lower Devonian Littleton Formation. Deformation accompanying middle-grade metamorphism shaped the core-rocks into a doubly plunging, asymmetric anticline (the Mascoma Dome) after formation of the recumbent folds discussed by Thompson et al. (this volume), but there was no significant granitization, remobilization, or intrusion associated with this event in the Mascoma area.

REGIONAL RELATIONSHIPS OF THE STRATIFIED CORE-GNEISS AND THE AMMONOOSUC VOLCANICS

Interpretation of the stratified core-gneiss of the Mascoma Dome as a stratigraphic unit lying beneath the Ammonoosuc Volcanics raises a stratigraphic problem which is discussed in this section. The Oliverian Domes lie to the southeast of the dashed line shown in Figure 17–1 in a belt between Locality 12 and Locality 22. With the possible exception of the northernmost dome, the other Oliverian Domes consist chiefly of rocks which are similar to the stratified core-gneiss of the Mascoma Dome and which probably originated in a similar manner. West of the dashed line in Figure 17–1 rocks beneath lateral equivalents of the Ammonoosuc Volcanics are quartz-rich arenites which bear no resemblance to the quartzo-feldspathic gneisses in the cores of the domes. The author suggests that the dashed line in Figure 17–1 is a facies boundary separating rock assemblages of grossly equivalent age. Such a relationship is consistent with the present age assignments of the metamorphic rocks and can be demonstrated along-strike in northern Maine, where lateral equivalents of these rocks have been dated by fossils.

Predominantly mafic volcanic rocks at the same stratigraphic level as the Ammonoosuc Volcanics are common in a belt along strike from the Mascoma area and in a belt a few miles to the west (Berry, this volume). In the least metamorphosed areas, these rocks consist chiefly of basalt, andesite, and spilite, with lesser amounts of trachyte and keratophyre, chert, and argillite. The volcanic rocks occur both as flows and as accumulations of pyroclastic material. The occurrence of pillow-lavas is fairly common. Fossils from these rocks are mostly of late Middle Ordovician age (graptolite zones 12 and 13; Berry, this volume). The predominantly mafic volcanic rocks include Unit Ovs (1) (numbers refer to references keyed in Figure 17–1); volcanic rocks of the Portage area (2); possibly the Dunn Brook Formation (3); some volcanic rocks of the Mount Chase area (4); greenstone and chert in the Shin Pond area (5); the Munsungun Lake and Bluffer Pond Formations (6); mafic volcanic units in the Beauceville Formation (7); possibly part of the Kennebec Formation (8); Arnold River Formation (9); part of the Dixville Formation (10); the Ammonoosuc Volcanics (11, 12, 15–20); volcanic rocks of the Orfordville Formation (13); Coburn Hill and Barnard volcanic members of the Missisquoi Formation (14); and the Middletown Formation (21). (The stratigraphic nomenclature of this paper follows that of the references cited and does not in all cases conform to the usage of the U. S. Geological Survey or the Stratigraphic Code.)

These predominantly mafic volcanic units are commonly overlain by and locally intercalated with dark, sulfide-rich and carbonaceous shale and equivalent metamorphic rocks including the following formations: upper member of Unit Ovs (1); rusty-weathering shale in the Portage area (2); slate in the upper part of the Beauceville Formation (7); Blind Brook Formation (6); part of the Dixville Formation (10); Partridge Formation (11, 12, 20); Cram Hill member of Missisquoi Formation (14); part of the Orfordville Formation (13); and the Collins Hill (Brimfield) Formation (21). These non-volcanic units above and intercalated with the volcanic rocks are

as much as several thousand feet thick at the western localities but are thinner and locally absent to the east.

In the rocks underlying the volcanic units there is a marked contrast in facies between quartz-rich arenite and argillite occurring northwest of the dashed line in Figure 17–1 and quartzo-feldspathic rocks occurring southeast of the line. The quartz-rich arenite and argillite facies includes the following units: Chase Lake Formation (6); lower part of the Beauceville Formation (7); Albee Formation (10–13); and the Moretown member of the Missisquoi Formation (14). In middle- and high-grade metamorphic terrane a characteristic feature of these units is the common occurrence of thin, schistose "pinstripe" partings in micaceous quartzite. The lower part of the Beauceville Formation (Berry, this volume) and the Chase Lake Formation (Hall, 1964, p. 37) have yielded late Middle Ordovician graptolites (zone 12). The quartzo-feldspathic rocks southeast of the line are represented by the following units: dacitic and andesitic tuff of the Shin Brook Formation (5; Neuman, 1964); stratified quartz diorite and granodiorite gneiss in the cores of the Oliverian Domes (12, 15–21); and probably the Monson and New London Gneiss (20–22). In the Shin Pond area (Locality 5) the Shin Brook Formation contains brachiopods dated as Early or early Middle Ordovician (Neuman, 1964).

In north-central Maine, the quartz-arenite facies rocks appear somewhat younger than the rocks of the quartzo-feldspathic facies. In the Shin Pond area (Locality 5), no representatives of the quartz-arenite facies are exposed between the rocks of the quartzo-feldspathic facies (Shin Brook Formation) and the overlying mafic volcanics. A similar relationship appears in the Oliverian Dome belt to the south, where rocks of the quartz-arenite facies (Albee Formation) are not exposed between the stratified core-gneiss of the domes and the Ammonoosuc Volcanics. It seems unlikely that Ordovician rocks of the quartz-arenite facies underlie the stratified core-gneiss of the domes. Where underlying rocks do appear beneath the stratified core-gneiss (Locality 22) they are well-stratified biotitic quartzite, sillimanite-biotite schist, marble, amphibolite and quartz - plagioclase - microcline - magnetite gneiss (Mamacoke and Plainfield Formations; Lundgren, 1962, p. 7; 1966, p. 21). The lithology of these rocks resembles that of the Grand Pitch Formation (which underlies the Shin Brook Fm. at Locality 5) more closely than that of the Albee Formation.

In northern Maine the late Middle Ordovician volcanic rocks give way abruptly eastward to non-volcanic slate, graywacke, and carbonate-bearing rocks of the same age (Localities 1 and 3; Chandler Ridge Formation and Carys Mills Formation; Pavlides, 1965; Pavlides and Berry, 1966; Berry, this volume). At several localities (1 and 3) rocks of these contrasting facies are exposed within a few miles of each other, but the facies boundary is covered by Silurian rocks. The southern extent of the eastern non-volcanic facies has not been established. Ordovician (?) rocks in the Dixfield quadrangle, Maine (locality 23), are non-volcanic, but the rocks are not fossiliferous and their correlation is uncertain (see also Osberg et al., this volume). These relationships

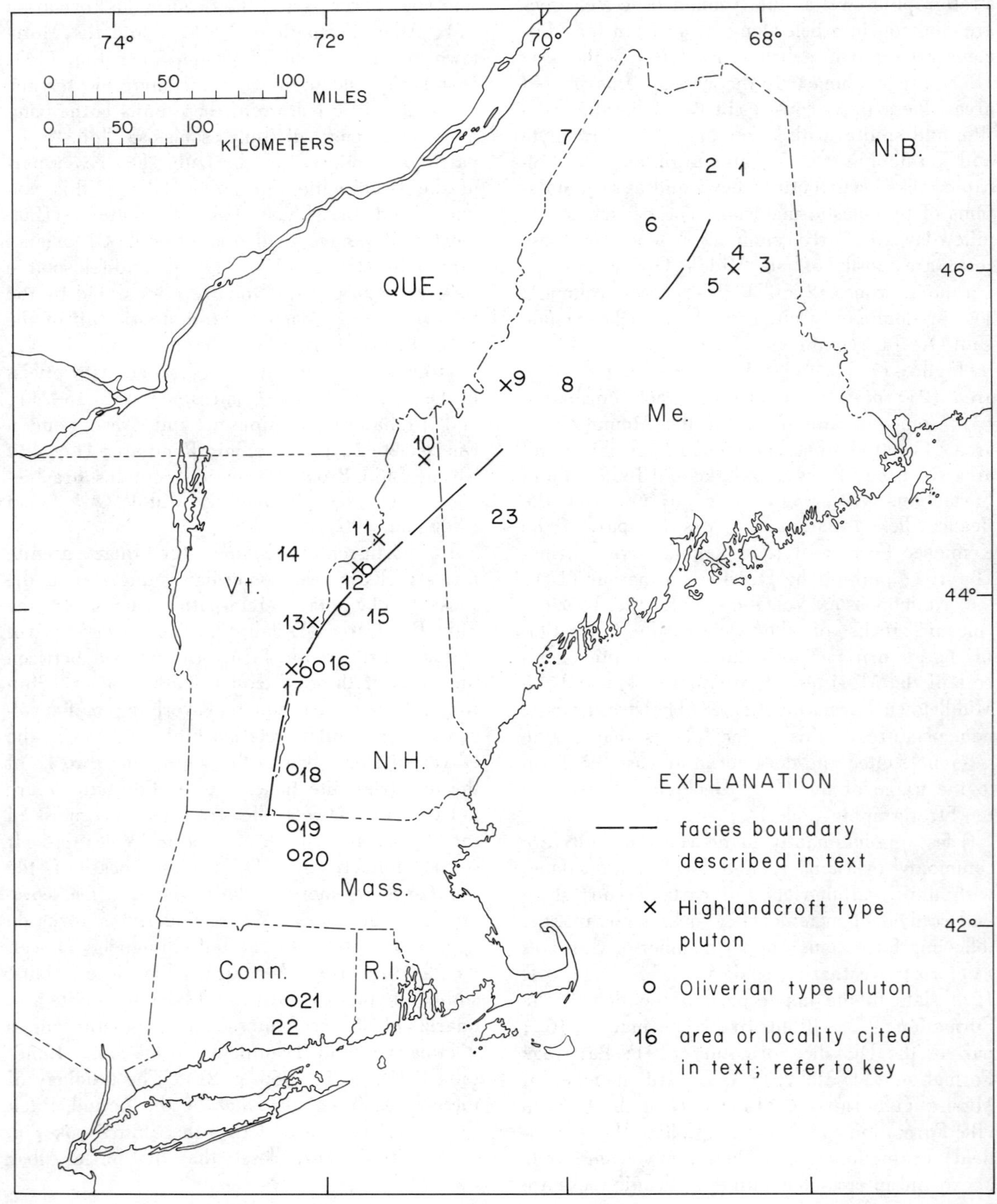

suggest that the Oliverian Dome belt lies near the eastern limit of the belt of Ordovician volcanic rocks.

In the northern part of the area the units occur in terrane characterized by low-grade regional metamorphism. Fossils are widespread, although only locally abundant, and provide satisfactory stratigraphic control. The sequences of units observed in the fossiliferous areas have been traced discontinuously along-strike into middle- and high-grade metamorphic terrane in southern New England. The constancy in lithology and stratigraphic sequence over long distances along strike is remarkable and contrasts sharply with the abruptness of facies changes across strike. In a broad sense the rock-stratigraphic correlation discussed in the previous paragraphs is probably also a time correlation, although it is not impossible that the sequence is time transgressive and becomes slightly older or younger southward. Late Early Silurian fossils have been identified in the overlying rocks in areas where the metamorphism is as intense as sillimanite-grade (Boucot and Thompson, 1963; Thompson et al., this volume), and indicate an upper limit to the age of the volcanic rocks even though fossils have not been found in the latter rocks south of about latitude 45°N.

In summary, the widespread upper Middle Ordovician sedimentary rocks and mafic volcanic rocks straddle the dashed line in Figure 17–1, but the underlying strata are different on the two sides of the line. West of the line the upper Middle Ordovician rocks are underlain by a sequence of Ordovician and Cambrian eugeosynclinal rocks whose thickness may locally exceed 50,000 feet (15 km) (Doll et al., 1961). Volcanic rocks and clastic rocks composed chiefly of volcanic materials make up about thirty percent of this section, but the rocks also contain an enormous volume of quartz and other clastic grains which were probably not derived from local volcanic sources. East of the dashed line in Figure 17–1 the thickness of volcanic rocks is about the same as to the west but the total section is thinner due to the absence of the thick sequence of Ordovician clastic rocks. Although the base of this eastern section is nowhere exposed, the underlying Cambrian clastic rocks do not appear as thick as to the west.

Precambrian gneiss basement is exposed in the Green Mountain Anticlinorium at the western margin of the eugeosyncline. That similar rocks underlie at least part of the eugeosynclinal belt is demonstrated by their reappearance in mantled gneiss domes of the Chester Dome group in southeastern Vermont. The correlations discussed in this paper indicate that most of the gneiss exposed in domes further east is of Ordovician (?) age. The oldest rocks known in the Oliverian belt

FIGURE 17–1. Map of New England and adjacent areas showing localities and features referred to in the text.

Key to Localities and References

No.	Locality	References
1	Presque Isle, Me.	Boucot et al. (1964)
2	Portage, Me.	Pavlides et al. (1964)
3	Houlton, Me.	Pavlides et al. (1964)
4	Island Falls, Me.	Ekren and Frischknecht (1967)
5	Shin Pond, Me.	Neuman and Rankin (*in* Caldwell, 1966)
6	Spider Lake, Me.	Hall (1964)
7	St. Magloire, Que.	Béland (1957)
8	Moosehead Lake, Me.	Boucot (1961)
9	Attean, Me.	Albee (1961); Albee and Boudette (in press)
10	Errol, N. H.	Green (1964)
11	Lancaster, N. H.	Billings (1956)
12	Littleton, N. H.	Billings (1937; 1956)
13	Fairlee, Vt.	Hadley (1942); Doll et al. (1961)
14	Central Vermont	Cady (1960); Doll et al. (1961)
15	Moosilauke, N. H.	Billings (1937; 1956)
16	Mascoma, N. H.	Chapman (1939); Hadley (1942); Naylor (1967)
17	Hanover, N. H.	Lyons (1955)
18	Keene, N. H.	Moore (1949)
19	Warwick, Mass.	Robinson (1963)
20	Pelham, Mass.	Balk (1956); Robinson (1963)
21	Middle Haddam, Conn.	Eaton and Rosenfeld (1960)
22	Deep River, Conn.	Lundgren (1962; 1966)
23	Dixfield, Maine	Pankiwskyj (1965)

and its extension along strike are Cambrian (?) slate and quartzite exposed in northeastern Maine, and schist and quartzite in southern Connecticut very tentatively assigned to the Cambrian (?). The author believes that the structure of the Oliverian Domes can be explained by deformation of massive gneiss and granite similar to and including the core-rocks exposed at the present erosion surface, and that it is not necessary to postulate the presence of a Precambrian gneissic basement at unseen depths to account for the structure of the domes.

Although the gross stratigraphy of the volcanic rocks is consistent over large areas, the individual volcanic units are irregularly distributed and vary abruptly in thickness. The volcanic rocks seem to be more resistant to folding than the sedimentary rocks and commonly occur as resistant masses in the cores of anticlinal structures. It is possible that these structures reflect sites where the deposits of volcanic rocks are thickest. Similarly it is possible that the distribution of the individual Oliverian Domes is partly controlled by variations in the original thickness of the core-rocks. A possible explanation of the en-echelon arrangement of the domes is that the axis along which the core-rocks are thickest trends more to the northeast than the trend of the folds that produced the domes. Along the flanks of the Green Mountain Anticlinorium and the Chester Dome, the regional structure seems to be influenced more by the presence of the Precambrian gneiss basement than by inhomogeneities in the Paleozoic strata. In these areas the volcanic rocks are turned on edge along with the sedimentary strata and do not occupy the cores of anticlines.

REGIONAL RELATIONSHIPS OF THE UNSTRATIFIED CORE-ROCKS

Pre-upper Lower Silurian granitic rocks (quartz diorite through granite) are common in the belt of volcanic rocks discussed in the previous section. These rocks occur chiefly as stocks of massive, porphyritic, somewhat sodic granodiorite and quartz monzonite. The texture and occurrence of these rocks are suggestive of granites intruded at shallow to intermediate depth. At many of the localities upper Lower or Upper Silurian strata overlie these granitic rocks nonconformably (Localities 5, 9, 12, 16, 19, and 21) or contain identifiable clasts of the granitic rocks (Localities 3, 5, and 9). Several localities in or near this belt contain Upper Ordovician conglomerates, but the author has found no mention

of clasts of the granitic rocks occurring in these conglomerates. Some of the granite bodies cut strata of late Middle Ordovician age, but this relationship cannot be demonstrated for all of the bodies, and some of the granitic rocks may be older. The close spatial and possibly temporal association of these granitic rocks with thick sequences of volcanic rocks suggests the possibility of a genetic relationship, but this has nowhere been demonstrated. Further field, petrologic, and isotopic study is needed to determine the relationships among these rocks, to establish the degree to which they constitute a related magma "series," and to determine the time interval over which the rocks were emplaced.

Billings (1937, p. 500; 1956, p. 46–48) assigned the name Highlandcroft Plutonic Series to the rocks of small granitic plutons occurring at Localities 11, 12, 13,. and 17. The type Highlandcroft Pluton is reported to crosscut the Albee Formation and the Ammonoosuc Volcanics and lies unconformably beneath the Upper Silurian Fitch Formation. Subsequent workers have recognized that granitic plutons at Locality 10 (Umbagog granodiorite) and Locality 9 (Attean quartz monzonite) are similar in age, lithology, and structural relationships to the Highlandcroft Series plutons. The Rockabema quartz diorite at Localities 4 and 5 also falls in this class. These plutons are indicated by the symbol "x" in Figure 17–1.

It is now apparent that the Mascoma Dome (Locality 16) contains in addition to the stratified core-gneiss a small pluton of quartz monzonite and granite which is older than upper Lower Silurian rocks (see description of unstratified Oliverian core-rocks above). Several of the other Oliverian Domes contain similar plutons which are denoted by the symbol "o" in Figure 17–1 (Localities 12, 15, and 17–21). The domes containing these plutons are broad compared to the more highly elongated domes (such as the Smarts Mountain and Croydon Domes north and south of Locality 16) whose cores consist almost entirely of the stratified core-gneiss. This broadening may be due to a buttressing effect of the massive plutons during later deformation. The large Pelham Dome in central Massachusetts (Locality 20) contains granitic rocks which differ in their mode of occurrence from those previously described. These granitic rocks are less massive, have a more migmatitic appearance, and occur more in sheet-like masses than in stock-like plutons. It is possible that these are granitic rocks, of pre-late Early Silurian age, which have been partially remobilized during high-grade post-Lower Devonian metamorphism.

The belt of Oliverian plutons is evidently a southward extension of the belt of Highlandcroft Series plutons. The plutons of both groups are characterized by similarities in the range of rock types and in the size and field relationships of the plutonic bodies. Most present differences in the appearance of the rocks are probably due to differences in the nature of later disturbances. Most of the Highlandcroft Series plutons lie in terrane affected by low-grade regional metamorphism. The rocks are commonly dark, intensely sheared and altered, and rich in chlorite and calcite. The plutons which occur sporadically in the cores of the Oliverian Domes lie in middle- and high-grade metamorphic terrane. These rocks are commonly lighter-colored, more massive and less sheared and altered, and generally lack chlorite and calcite. The plutons in the cores of the Oliverian Domes are intrusive into the quartzo-feldspathic stratified core-gneiss, whereas most of the Highlandcroft Series plutons occur on the quartz-arenite-facies side of the dashed line in Figure 17-1. A transitional case is afforded by the Whitefield pluton at Locality 12. This pluton lies in middle-grade terrane, and in appearance its rocks resemble those of the Olivernian plutons. The pluton resembles most of the Highlandcroft Series plutons in lying west of the facies boundary so that the plutonic rocks are in contact with the Albee Formation rather than the stratified Oliverian core-gneiss.

REFERENCES

Albee, A. L., 1961, Boundary Mountain Anticlinorium, west-central Maine and northern New Hampshire: U. S. Geol. Survey Prof. Paper 424–C, p. C51–C54

Albee, A. L. and Boudette, E. L., Geology of the Attean quadrangle, Somerset County, west-central Maine: U. S. Geol. Survey Bull., in press

Balk, Robert, 1956, Bedrock geologic map of the Millers Falls quadrangle, Massachusetts: U. S. Geol. Survey, Map GQ–93

Béland, Jacques, 1957, St. Magloire and Rosaire—St. Pamphile areas: Quebec Dept. Mines, Geological Report 76, 49 p.

Billings, M. P., 1937, Regional metamorphism of the Littleton-Moosilauke area, New Hampshire: Geol. Soc. America Bull., v. 48, p. 463–566

————————, 1956, The geology of New Hampshire, Part II—Bedrock geology: New Hampshire State Planning and Development Commission, 203 p.

Boucot, A. J., 1961, Stratigraphy of the Moose River Synclinorium, Maine: U. S. Geol. Survey Bull. 1111–E, p. 153–188

Boucot, A. J., and Thompson, J. B., 1963, Metamorphosed Silurian brachiopods from New Hampshire: Geol. Soc. America Bull., v. 74, p. 1313–1334

Boucot, A. J., Field, M. T., Fletcher, Raymond, Forbes, W. H., Naylor, R. S., and Pavlides, Louis, 1964, Reconnaissance bedrock geology of the Presque Isle quadrangle, Maine: Maine Geological Survey, Quadrangle Mapping Series, no. 2, 123 p.

Cady, W. M., 1960, Stratigraphic and geotectonic relationships in northern Vermont and southern Quebec: Geol. Soc. America Bull., v. 71, p. 531–576

Caldwell, D. W., ed., 1966, The Mount Katahdin region, Maine:New England Intercollegiate Geological Conference, Guidebook for 58th Ann. Mtg., Patten, Me., 61 p.

Chapman, C. A., 1939, Geology of the Mascoma quadrangle, New Hampshire: Geol. Soc. America Bull., v. 50, p. 127–180

Doll, C. G., Cady, W. M., Thompson, J. B., and Billings, M. P., compilers and editors, 1961, Centennial geologic map of Vermont: Montpelier, Vermont, Vermont Geol. Survey, scale 1:250,000

Eaton, G. P., and Rosenfeld, J. L., 1960, Gravimetric and structural investigations in central Connecticut: International Geological Congress, XXI Session, Part II, p. 168–178

Ekren, E. B., and Frischknecht, F. C., 1967, Geological-geophysical investigations of bedrock in the Island Falls Quadrangle, Aroostook and Penobscot Counties, Maine: U. S. Geol. Survey Prof. Paper 527, 36 p.

Eskola, P. E., 1949, The problem of mantled gneiss domes: Geol. Soc. Lond. Quar. Jour., v. 104, p. 461–476

Green, J. C., 1964, Stratigraphy and structure of the Boundary Mountain Anticlinorium in the Errol quadrangle, New Hampshire, Maine: Geol. Soc. America Special Paper 77, 78 p.

Hadley, J. B., 1942, Stratigraphy, structure and petrology of the Mt. Cube area, New Hampshire: Geol. Soc. America Bull., v. 53, p. 113–176

Hall, B. A., 1964, Stratigraphy and structure of the Spider Lake quadrangle, Maine: Ph.D. Thesis, Yale University, 153 p.

Hopson, C. A., 1964, The crystalline rocks of Howard and Montgomery Counties: The geology of Howard and Montgomery Counties, Maryland Geol. Survey, p. 27–215

Lundgren, L. W., Jr., 1962, Deep River area, Connecticut-stratigraphy and structure: Amer. Jour. Sci., v. 260, p. 1–23

————————, 1966, Bedrock geology of the Hamburg quadrangle Connecticut: Connecticut Geol. and Nat. Hist. Survey Quad. Rept. 19, 41 p.

Lyons, J. B., 1955, Geology of the Hanover quadrangle, New Hampshire—Vermont: Geol. Soc. America Bull., v. 66, p. 105–146

Moore, G. E., 1949, Structure and metamorphism of the Keene-Brattleboro area, New Hampshire, Vermont: Geol. Soc. America Bull., v. 60, p. 1613–1670

Naylor, R. S., 1967, A field and geochronologic study of mantled gneiss domes in central New England: Ph.D. Thesis, Calif. Inst. Technology, 123 p.

Neuman, R. B., 1964, Fossils in Ordovician tuffs, northeastern Maine: U. S. Geol. Survey Bull. 1181–E, 38 p.

Pankiwskyj, K. A., 1965, Geology of the Dixfield quadrangle, Maine: Ph.D. Thesis, Harvard University, 200 p.

Pavlides, Louis, 1965, Meduxnekeag group and Sprague-ville formation of Aroostook County, northeast Maine: U. S. Geol. Survey Bull. 1244–A, p. A52–A60

Pavlides, Louis, Mencher, Ely, Naylor, R. S. and Boucot, A. J., 1964, Outline of the stratigraphic and tectonic features of northeastern Maine: U. S. Geol. Survey Prof. Paper 501–C, p. C28–C38

Pavlides, Louis, and Berry, W. B. N., 1966, Graptolite-bearing Silurian rocks of the Houlton-Smyrna Mills area, Aroostook County, Maine: U. S. Geol. Survey Prof. Paper 550–B, p. B51–B61

Robinson, Peter, 1963, Gneiss domes of the Orange area, Massachusetts and New Hampshire: Ph.D. Thesis, Harvard University, 240 p.

Chapter 18

Stratigraphy of the Merrimack Synclinorium in West-Central Maine

P. H. OSBERG, R. H. MOENCH, AND
JEFFREY WARNER

INTRODUCTION

A MAJOR GEOLOGIC problem in eastern New England concerns the structural and stratigraphical relationships across the Merrimack synclinorium —a broad, northeast-trending synclinal tract that extends at least from eastern Connecticut (Billings, 1956; Dixon and Lundgren, this volume) to central Maine (Fig. 18–1). In west-central Maine, much of the central part of the synclinorium has been mapped in detail during the last 10 years or more. This work, representing the efforts of many geologists, has defined thick and lithologically varied stratigraphic sequences in different parts of the area, but how these sequences relate to one another is controversial. The focus of this paper is on the relationships of rock masses across the Merrimack synclinorium.

Figure 18–2 was compiled by Osberg, Moench, and Warner from several sources. Osberg is responsible for work in the Waterville, Vassalboro, and parts of the Norridgewock and Augusta quadrangles on the southeast; Moench for the Rangeley, Phillips and Rumford quadrangles in the northwest; and Warner for the Buckfield and part of the Augusta quadrangles. The Bryant Pond quadrangle was modified from a map by Guidotti (1965); the Dixfield and Farmington quadrangles were modified from Warner and Pankiwskyj (*in* Hussey, 1965); the Livermore quadrangle was modified from an unpublished map by Dabney Caldwell, and part of the Augusta quadrangle is from D. S. Barker (1961). Recent unpublished work by Allen Ludman in the Skowhegan quadrangle has contributed to our

interpretation in the northeast. We acknowledge our debt to all of these men. In addition we have benefited from field excursions and discussions with these and the many other geologists who have worked in west-central Maine. Moench is grateful to C. W. Wolfe of Boston University, and Warner is grateful to M. P. Billings and J. B. Thompson, Jr. of Harvard University for advice and encouragement during their work.

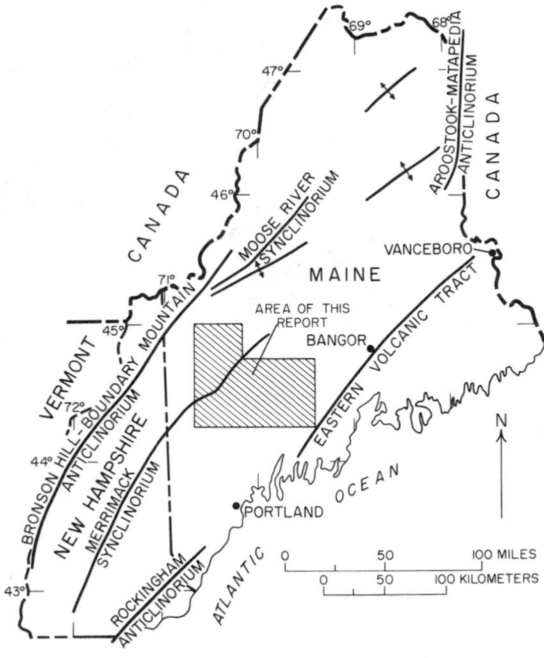

FIGURE 18–1. Index map of Maine and New Hampshire showing area of this report.

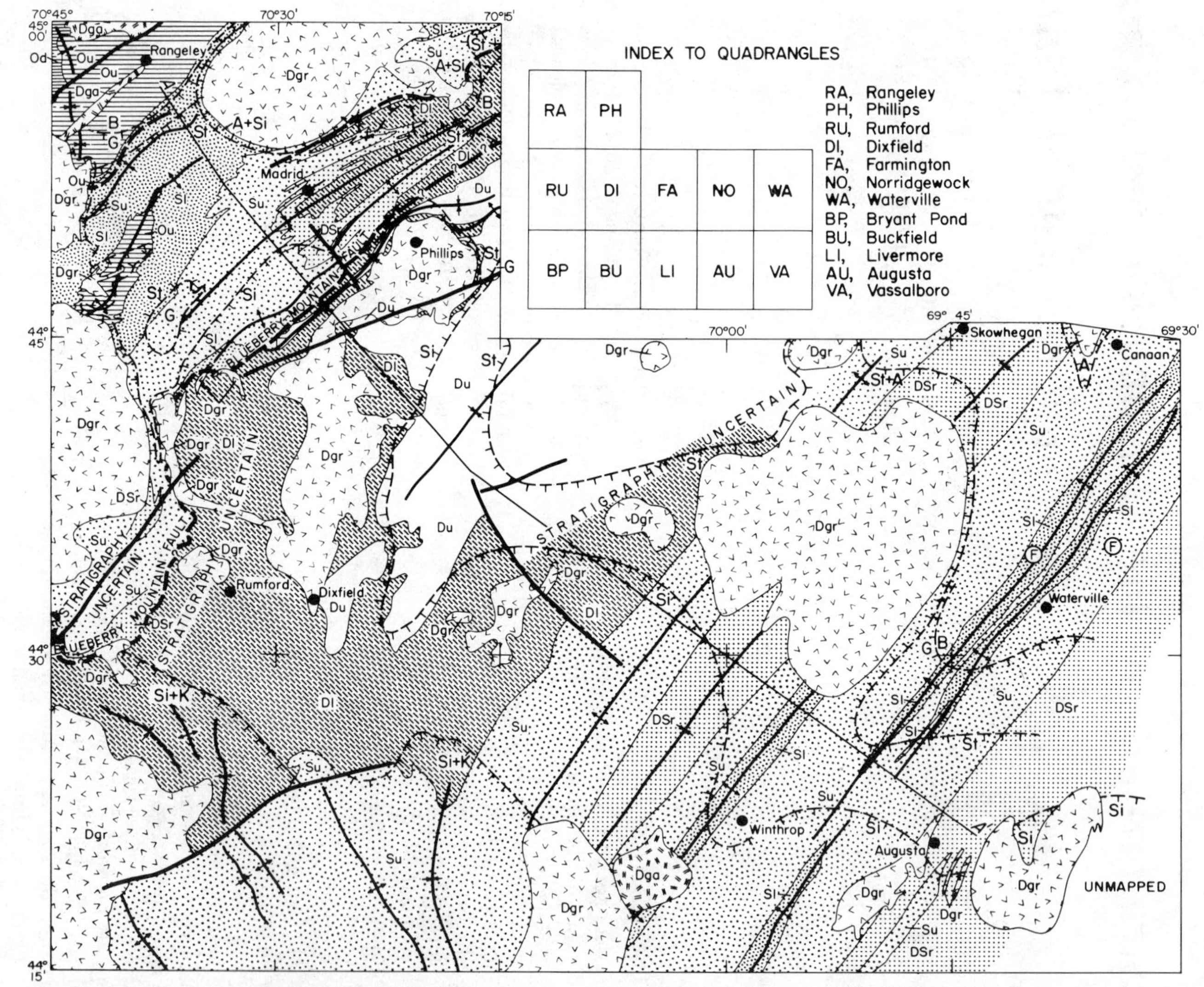

INDEX TO QUADRANGLES
RA, Rangeley
PH, Phillips
RU, Rumford
DI, Dixfield
FA, Farmington
NO, Norridgewock
WA, Waterville
BP, Bryant Pond
BU, Buckfield
LI, Livermore
AU, Augusta
VA, Vassalboro
RA | PH
RU | DI | FA | NO | WA
BP | BU | LI | AU | VA
Rangeley
Madrid
Phillips
Rumford
Dixfield
Skowhegan
Canaan
Waterville
Winthrop
Augusta
UNMAPPED
BLUEBERRY MOUNTAIN FAULT
STRATIGRAPHY UNCERTAIN
70°45'
70°30'
70°15'
70°00'
69°45'
69°30'
45°00'
44°45'
44°30'
44°15'

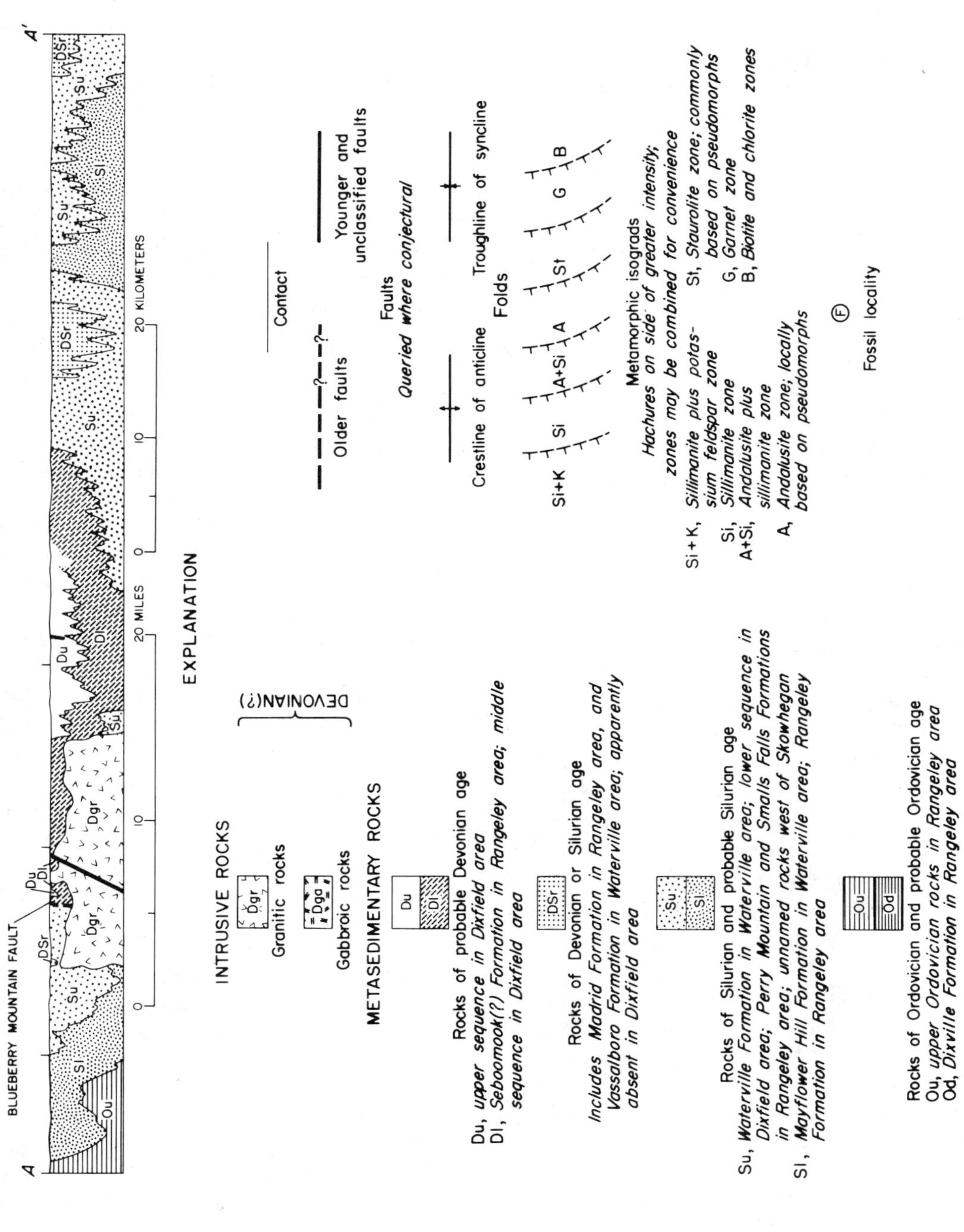

FIGURE 18–2. Tectonic map and sketch section of west-central Maine. Compiled by P. H. Osberg, R. H. Moench, and J. Warner, 1967.

GEOLOGIC SETTING

The Merrimack synclinorium is a broad complexly deformed synclinal tract that contains variably metamorphosed, dominantly eugeosynclinal clastic sedimentary rocks. In New Hampshire, Devonian rocks—the pelitic Littleton Formation—apparently form much of the synclinorium (Billings, 1956), whereas in west-central Maine, Silurian as well as Devonian(?) rocks are abundantly represented. The synclinorium is bounded on the northwest by the Bronson Hill—Boundary Mountain anticlinorium, and on the southeast by an incompletely understood tract of volcanic rocks that extends from Vanceboro to Portland (Fig. 18–1). Ordovician rocks that locally are unconformably overlain by Silurian or Devonian rocks are exposed in the Bronson Hill–Boundary Mountain anticlinorium (see Albee, 1961; Cady, 1960 and this volume; Boucot et al., 1964; Harwood and Berry, 1967; Green, 1964, and Green and Guidotti, this volume for discussions of the Bronson Hill–Boundary Mountain anticlinorium). The volcanic tract to the southeast has been mapped as an anticline in the Bangor–Vanceboro region (Larrabee and Spencer, 1963; Trefethen, 1940, written communication). Larrabee and Spencer (1963) reported Ordovician fossils from the volcanic rocks at Vanceboro. Southwest of Bangor, however, rocks of the volcanic tract may be Ordovician, Silurian, or Devonian; these alternatives are discussed by Hussey (this volume).

In Maine and northern New Hampshire the Bronson Hill–Boundary Mountain anticlinorium coincides approximately with an ancestral anticlinal tract that was locally emergent through much of the Silurian. Boucot et al. (1964) and Naylor and Boucot (1965) call the emergent parts the Somerset Island, and Cady (in press; also this volume) calls the ancestral anticlinal tract the Somerset geanticline. The Rangeley quadrangle (index map, Fig. 18–2) is immediately southeast of the emergent zone. An unconformity between Silurian rocks and Middle Ordovician and older rocks in the area immediately to the northwest (Harwood and Berry, 1967) defines the Taconic break. This unconformity has not been observed in the Rangeley quadrangle, where a thick and apparently conformable section of probable Upper Ordovician and Lower Silurian rock is exposed that is similar in many respects to the Ordovician–Silurian transition in northern Maine (Pavlides and Berry, 1966; Pavlides et al., this volume). Much of the area of the Merrimack synclinorium

may thus have existed as a trough of continuous sedimentation, perhaps from the Middle Ordovician through the Silurian and into the Devonian.

The metasedimentary rocks of the synclinorium were tightly folded along northeast-trending axes, intruded and contact metamorphosed by plutons of gabbroic to granitic composition, locally folded again about northwest-trending axes, and regionally metamorphosed. In contrast to the generally flat-lying recumbent structure that characterizes the southern part of the Merrimack synclinorium (Dixon and Lundgren, this volume; Thompson et al., this volume), bedding typically dips steeply in west-central Maine (Fig. 18–2, section A–A'). The major northeast-trending folds, however, are grossly open with low amplitude, for thin formations have broad areas of outcrop in the crests and troughs. Associated minor folds are commonly isoclinal, and have the characteristics of passive flow folds (Donath and Parker, 1964). Axial surfaces are typically steep and have subparallel slaty and phyllitic cleavage. Axes are typically subhorizontal in the central and eastern part of the area, but in the northwest they plunge at extremely variable angles, dominantly to the northeast, but locally to the southwest. Most if not all of the large intrusives of the area (Fig. 18–2) are probably Devonian. The mass of biotite granodiorite northeast of Skowhegan (Fig. 18–2) is one of many intrusives in Maine that have yielded Devonian radiometric ages (Faul et al., 1963). As the major northeast-trending folds deform Devonian rocks, but are cut by Devonian intrusives, this folding ended in Devonian time. Folding may have begun, however, while the Silurian sediments were accumulating (Moench, 1966).

Superposed on the major northeast-trending fold pattern are complex patterns of younger slip cleavages, small folds of various trends, and at places larger folds whose axial planes trend northwest (Fig. 18–2). On the scale of outcrops, evidence for this second folding consists of folded early folds and their associated slaty or phyllitic cleavage; on the scale of geologic maps, the evidence consists of deflections of the major northeast trend of stratigraphic units. Through much of the area and in northern Maine these superposed structures have been related genetically to the emplacement of plutons, and locally to post-metamorphic faulting (Guidotti, 1965; Moench, 1966, p. 1452–1453; Warner, 1967; and Pavlides et al., 1964, p. C34). However, from the geometry of conjugate folds in the Waterville quadrangle, Osberg (1968) suggests that the northwest-trending folds are related to regional northeast-southwest compression.

As shown by the isograds in Figure 18–2, metamorphic intensity ranges from the chlorite zone in the north and northeast to the sillimanite and the sillimanite-potassium feldspar zones (Guidotti, 1963; Evans and Guidotti, 1966) in the southwest. Throughout the area the isograds represent the combined effects of at least two metamorphic events. In regions of low- to middle-grade regional metamorphism, individual plutons have distinctive high-grade metamorphic halos. Both the plutonic and metamorphic rocks to the south appear to have been overprinted by a metamorphism that ended in Permian time (Faul et al., 1963).

Many faults, conveniently separable into older and younger faults, have been delineated in western Maine (Fig. 18–2). Three northeast-trending older faults, mapped by Moench in the northwest, were recognized primarily by major stratigraphic omissions. These faults do not have associated silification, brecciation, or cataclasis, nor do they offset metamorphic isograds.

One, the Blueberry Mountain fault, is named here, because it bears on the stratigraphic relationships described in this paper. In the Rumford quadrangle this fault crosses the core of a large northeast-trending anticline. Here, more than 10,000 feet (3 km) of section are believed to be missing along the fault. Although the Blueberry Mountain fault crosses a major anticline, this and other faults of the older generation are considered to be related to the major, early period of folding. The Blueberry Mountain fault has been deformed by younger folds, accounting for the sinuous trace of the fault in the Rumford quadrangle.

The younger faults are commonly characterized by zones of silicification, brecciation, cataclasis, and intense development of slip cleavage in the adjacent rocks. The younger east-northeast trending faults south of Phillips and in the Buckfield quadrangle offset metamorphic isograds (Fig. 18–2).

STRATIGRAPHY

For discussion of stratigraphy, the area of Figure 18–2 is conveniently divided into three subareas: (1) the Rangeley area in the northwest, comprising the Rangeley, Rumford and Phillips quadrangles northwest of the Blueberry Mountain fault; (2) the Dixfield area in the center and southwest, comprising the Dixfield, Farmington, Buckfield, and Bryant Pond quadrangles and the Phillips and Rumford quadrangles southeast of the Blueberry Mountain fault; (3) the Waterville area in the southeast, comprising parts of the Waterville, Norridgewock, Augusta, and Vassalboro quadrangles. The major stratigraphic units are briefly described in Tables 18–1 to 18–3. Many of the names of formations used in this paper are in general use by geologists working in west-central Maine, but they have not been formally proposed. In order to avoid using undefined names of formations, descriptions of new and revised formations are included as Appendix 1.

The section of the Rangeley area (Table 18–1) is probably more than 30,000 feet (9 km) thick, not including the Albee and Aziscohos Formations, which, to the west and northwest, directly underlie the Ordovician rocks of Table 18–1 (Green, 1964; Harwood and Berry, 1967; Green and Guidotti, this volume). Although no identifiable fossils have been found in rocks of the Rangeley area, Harwood and Berry (1967) correlate these rocks with the Dixville Formation of Green (1964) and with the Ammonoosuc and Partridge Formations (Billings, 1956) in New Hampshire. The Dixville Formation (Table 18–1) has been dated as upper Middle Ordovician (Harwood and Berry, 1967). The Rangeley Formation, which includes the polymictic Rangeley Conglomerate of Smith (1923) in its lower part, is similar to fossiliferous Lower Silurian conglomeratic rocks a few miles to the north (Boudette and Harwood, 1965). The fossils occur in calcareous quartz conglomerates that most closely resemble the uppermost part of the Rangeley Formation (Table 18–1). These quartz conglomerates, and those near the top of the Rangeley, probably correlate with the Clough Formation of Billings (1956) in New Hampshire. On the assumption that the overlying Perry Mountain and Smalls Falls Formations (Table 18–1) are absent at the classic section in New Hampshire, the next overlying units—the Madrid and Seboomook(?) Formations—are tentatively correlated respectively with the Middle or Upper Silurian Fitch Formation (Billings, 1956; Berry and Boucot, in press) and the Lower Devonian Littleton Formation (Billings, 1956) in New Hampshire, and with the Upper Silurian Hardwood Mountain and Lower Devonian Seboomook Formations (Boucot, 1961) in northwestern Maine. These correlations are inferred on the basis of similarities in lithology and sequence—prominent units of calcareous rocks overlain by thick gray metashales. It is recognized, however, that the Madrid is not as calcareous as the Fitch. Positive correlation must await the discovery of fossils in

TABLE 18-1. Stratigraphy of the Rangeley area, northwest of Blueberry Mountain fault (Moench, unpublished mapping)

Name and thickness	Probable age	Description
Seboomook(?) Formation >5000 ft (1700 m)	Devonian, possibly Late Silurian in part	Metashale; gray, variably massive, faintly cyclically bedded, and cyclically interbedded with white to light-gray locally calcareous metasandstone. Beds characteristically grade continuously from metasandstone, siltstone or semi-pelite into gray metashale; parallel lamination and cross lamination common
Madrid Formation 1000 ft (300 m)	Silurian or Devonian	Dominant metasandstone. Lower part is thin-bedded calcareous metasandstone, metasiltstone, metashale, and calc-silicate rock. Upper part at Madrid is thick-bedded light-gray slightly calcareous metasandstone with pods of calc-silicate rock, and subordinate gray metashale. Graded bedding and crossbedding common.
Smalls Falls Formation 2500 ft (900 m); pinches out to northeast	Silurian (?)	Sulfidic metashale and metasandstone; thick grit beds abundant in northeast. Lower part is cyclically bedded noncalcareous carbonaceous metashale and quartz-rich metasandstone. Upper part is calcareous sulfidic metasandstone and metasiltstone. Cross-lamination, convolute-lamination, graded bedding abundant.
Perry Mountain Formation 2000 ft (700 m)	Silurian (?)	Cyclically bedded quartz-rich metasandstone and muscovite-rich metashale; cross-lamination, convolute-lamination, and graded bedding abundant; tops of metasandstone beds sharp or abruptly gradational.
Rangeley Formation 1000 ft (300 m)	Silurian (?); upper part probably Llandoverian Stage (Boudette and Harwood, 1965)	Quartz-pebble metaconglomerate, conglomeratic metamudstone, arkosic metasandstone, and gray metashale. Abundance of coarse clastics diminishes southward. Graded bedding common.
4000 ft (1400 m) in north 2500 ft (900 m) in south		Gray metashale with subordinate arkosic metasandstone, quartz-rich polymictic metaconglomerate and conglomeratic metamudstone. Graded bedding common. Metaconglomerate abundant in north, absent in south.
4000 ft (1400 m; coarse clastics are Rangeley Conglomerate of Smith, 1923)		Polymictic metaconglomerate, arkosic metasandstone, and metashale. Metaconglomerate dominant in north, metasandstone dominant in center, and metashale dominant in south part of Rangeley area. Coarse clastics typically very thick bedded, massively graded.
Unnamed upper Ordovician rocks 4000–5500 ft (1400–1900 m)	Late and Middle Ordovician, possibly Early Silurian in part	Upper 1000 ft (300 m) is interlaminated light-gray non-sulfidic metashale, and locally calcareous metasiltstone, and metasandstone. Lower 3000–4500 ft (1000–1500 m) is metashale, metagraywacke, and acidic metavolcanics.
Dixville Formation 6000 ft (2000 m)	Middle Ordovician (Wilderness Stage; Harwood and Berry, 1967)	Massive to foliated greenstone overlain by black metashale and thin bedded metagraywacke.

the Madrid, which is shown as Silurian or Devonian on Figure 18–2. Assuming that these correlations are correct, great thicknesses of metamorphosed shale and local coarse clastic material are present in the Rangeley area that are not present in the classic sections of northern New Hampshire (Billings, 1956) and northwestern Maine (Boucot, 1961). These classic sections, however, are on the Somerset geanticline (Cady, this volume), where sedimentation was interrupted by uplift and erosion in Silurian time, and are thus expectably thinner than the Rangeley section, which represents continuous sedimentation in a subsiding basin throughout Silurian time.

The section in the Dixfield area (Table 18–2) may be more than 20,000 feet (6 km) thick. Guidotti (1965), Pankiwskyj (1964), Warner (1967), and Warner and Pankiwskyj (in Hussey, 1965) define many formations, of which only

TABLE 18-2. Stratigraphy of the Dixfield area. Simplified from Guidotti (1965), Pankiwskyj (1964), Warner (1967), Warner and Pankiwskyj (*in* Hussey, 1965), and Moench (unpublished mapping)

Name and thickness	Probable age	Description
Upper sequence 5800–7000 ft (2000–2400 m)	Devonian	Dominant metasandstone in upper part and metashale in lower part in west; metashale dominantly gray, locally sulfidic, variably massive to cyclically interbedded with subordinate white to gray metasandstone; graded bedding common; cross-lamination locally abundant, but not characteristic. Unit contains sparse calc-silicate rock and "ribbon" marble. Metasandstone of upper part commonly thick bedded, calcareous. Dominant metasandstone in east.
Middle sequence >5000 ft (1700 m)	Devonian, possibly late Silurian in part	Upper 250–750 ft (83–250 m) is Hildreths Formation; thick to thin bedded calcareous metasandstone, calc-silicate rock and local marble. Dominant metashale with sparse to abundant metasandstone interbeds and local thin bends of calc-silicate rock, separable into gray nonsulfidic and rusty-weathering sulfidic zones. Unit of fine- to medium-grained feldspathic metasandstone with pods and beds of calc-silicate rock present near base of middle sequence but extent unknown. Metashale typically poorly bedded, locally cyclically interbedded with subordinate metasandstone; graded bedding and parallel lamination common.
Lower sequence probably >4000 ft (1400 m)	Silurian	Metasandstone, calcareous metasandstone; subordinate metashale and calc-silicate rock; noncalcareous metashale of unknown thickness at base.

those of Guidotti (1965) have been formally named. Although stratigraphic relationships among these units are complex, the many units may be grouped into: (1) a lower sequence of thinly interbedded calc-silicate rocks and calcareous metasandstones, and noncalcareous metashales and metasandstones; (2) a middle sequence comprised largely of interstratified units of nonsulfidic and sulfidic metashales, but containing an apparently lenticular unit of metasandstone and subordinate calc-silicate rock and a thin, but widespread unit of metasandstone, calc-silicate rock and local marble—the Hildreths Formation of Warner (1967); (3) an upper sequence, composed of a lower zone of dominant nonsulfidic and locally sulfidic metashale, and an upper zone of metasandstone. The top of the middle sequence is placed at the top of the Hildreths Formation because this contact conveniently outlines the structure of a major part of the Dixfield area, and because the Hildreths Formation provides a convenient basis for comparison between the sections of the Dixfield and Rangeley areas.

The section of the Waterville area (Table 18–3) is in excess of 5,000 feet (1.5 km) thick. Three formations have been mapped and named, and two have been dated by fossils (see Osberg, 1968). The Mayflower Hill Formation, consisting of interbedded gray metashale and metasandstone, is probably late Llandovery. The overlying Waterville Formation, composed of metashale and subordinate quartz-rich metasandstones, and thin zones of limestone, is Wenlock or possibly Ludlow age. The Vassalboro Formation, on the basis of its stratigraphic position over dated Silurian rocks, is considered to be of Silurian or Devonian age.

CORRELATION ACROSS THE MERRIMACK SYNCLINORIUM

Precise correlation of map units across the Merrimack synclinorium is impossible because fossils have not been found in the section in the Rangeley area, and because units cannot be traced continuously from the Waterville into the Rangeley area. Tentative correlations may be drawn, however, on the basis of probable ages, lithology and gross lithologic sequence (Fig. 18–3). Although important differences exist between the units that are correlated from the Waterville to the Rangeley area, the gross sequence contains strikingly similar lithologies. Coarse feldspathic clastic sediments are abundant in the Rangeley Formation; finer feldspathic clastic sediments are present in the Mayflower Hill Formation. The Perry Mountain Formation and the Waterville Formation are thick units of dominantly thin

TABLE 18-3. Stratigraphy of the Waterville area (Osberg, 1968)

Name and thickness	Probable age	Description
Vassalboro Formation >1000 ft (300 m)	Devonian or Silurian	Dominantly slightly calcareous thick-bedded light-grayish-blue metasandstone and subordinate thin beds of gray metashale.
Waterville Formation 3000 ft (1000 m)	Silurian *Inocalus* from upper part suggests Wenlock or Ludlow age	Thinly laminated greenish-gray metashale and quartzite, and a prominent zone of thin-bedded limestone. To the west the unit contains at least two zones of limestone and thick beds of quartz-rich metasandstone.
Mayflower Hill Formation >1000 ft (300 m)	Silurian monograptids suggest a late Llandovery age	Light-bluish-gray to gray, locally coarse-grained metasandstone and metashale. Metasandstone beds commonly graded and locally display delicate cross-lamination.

bedded or laminated rocks. The Madrid Formation and the Vassalboro Formation are prominent units of calcareous clastic rocks. As discussed later we attribute differences between these units to facies changes.

The Mayflower Hill Formation on the southeast side of the synclinorium (Table 18–3) is equated approximately with part of the Rangeley Formation on the northwest side (Table 18–1) and is considered to be stratigraphically below all of the rocks of the Dixfield area (Figs. 18–2, 18–3). The Mayflower Hill and Rangeley are Early Silurian in age, and consist largely of interbedded gray metashales and feldspathic metasandstones that have many of the characteristics of turbidites, such as poor sorting, graded bedding, abundant inclusions of mud chips and clots. The presence of abundant metaconglomerates in the Rangeley Formation, together with their absence in the Mayflower Hill to the southeast, is consistent with facies changes that have been mapped in the Lower Silurian rocks in the Rangeley area. There the abundance and coarseness of conglomeratic rocks decrease to the southeast, indicating a source area to the northwest in Early Silurian time. The Waterville area was in a more distal part of the basin of sedimentation at that time, and received no conglomerate.

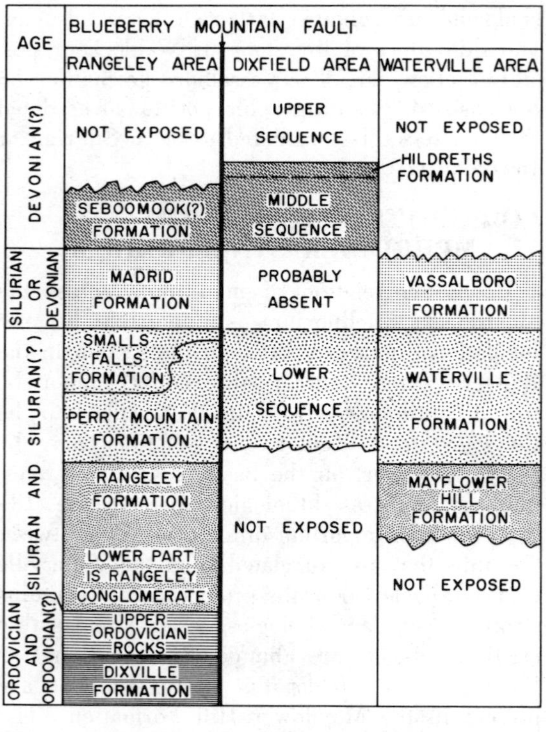

The Waterville Formation is equated with the Perry Mountain Formation of the Rangeley area, and with the lower sequence of the Dixfield area (Fig. 18–3). Both the Waterville and Perry Mountain are considered to be Wenlock or possibly Ludlow in age, and both are composed largely of cyclically bedded quartz-rich metasandstones and metashales. The strikingly cross-laminated and convolute-laminated beds of metasandstone that are abundant in the Perry Mountain are represented in the Waterville Formation by thin, commonly cross-bedded layers of metasandstone. At low metamorphic grades the metashales of both are pale greenish gray, although at higher grades the Perry Mountain metashales are muscovite-rich whereas those of the Waterville Formation tend to be biotite-rich. Major differences between the Perry Mountain Formation and Waterville Formation are the generally more thinly bedded to laminated character of the Waterville, and the presence in the Waterville of distinctive limestone units in contrast with the sparse thin beds of calc-silicate rock found in the Perry Mountain. These differences, like those between the Mayflower Hill Formation and the Rangeley Formation, suggest that the Waterville area was more distant from the source of clastics

FIGURE 18–3. Generalized correlation chart of the Rangeley, Dixfield, and Waterville areas. Patterns same as for Figure 18–2.

than the Rangeley area during Perry Mountain–Waterville sedimentation.

Correlation of the Waterville Formation (Table 18–3) with part or all of the lower sequence in the Dixfield area (Table 18–2) is established, at least within broad limits. The Waterville Formation extends continuously from Augusta to Winthrop across the intervening anticline (Fig. 18–2). To the northeast, from Winthrop to Canaan, the abundance of metasandstone increases in the Waterville. On the basis of similar sequence and lithology, the Waterville Formation at Canaan correlates across a syncline with the unnamed rocks, largely metasandstone and limestone, just west of Skowhegan, which in turn intertongue to the southwest with the more complexly interstratified metashales, metasandstones, and calc-silicate rocks of the lower sequence of the Dixfield area (compare with Warner and Pankiwskyj, *in* Hussey, 1965). In addition, the bedding characteristics of the Perry Mountain and Waterville Formations are present in the lower sequence of the Dixfield area.

The Vassalboro Formation is correlated with the Madrid Formation of the Rangeley area, but has not been recognized in the Dixfield area. The apparent absence of the Vassalboro Formation in the Dixfield area is a major problem, which is discussed later. Both the Madrid and Vassalboro are dominantly thick bedded, slightly calcareous, locally cross-laminated metasandstones, with some thinner-bedded calc-silicate rocks and gray metashale. The sequence within the Madrid Formation at Madrid—thin bedded calc-silicate rocks overlain by thick bedded less calcareous metasandstones—has not been recognized in the Waterville area. On the other hand, the sequence at Madrid is not as well developed in other parts of the Phillips quadrangle, and is thus not an important criterion for long range correlation.

Although the approximate correlation of the lower sequence of the Dixfield area with the Waterville Formation is established, the relationships of the middle and upper sequence with sections in the Rangeley and Waterville areas are not definite. We consider the middle sequence below the Hildreths Formation to correlate with the Devonian(?) Seboomook(?) Formation of the Rangeley area and that the upper sequence is stratigraphically well above any of the rocks exposed in the Rangeley or Waterville areas (Fig. 18–3). However, three alternatives are possible: (1) the lower part of the middle sequence might be a pelitic facies of the Madrid or Vassalboro; (2) the Madrid or Vassalboro was cut out by a fault, perhaps analogous to the Blueberry Mountain fault, along the boundary between the lower and middle sequences, and (3) the Madrid or Vassalboro was eroded prior to deposition of the probably Devonian middle sequence. An unconformity at the base of the Devonian has been recognized in the Moose River Synclinorium (Boucot, 1961).

Correlation of the middle sequence below the Hildreths Formation with the Seboomook(?) is made on the basis of lithologic similarity and position above the lower sequence, which we correlate with the Waterville Formation and Perry Mountain Formation. The dominant, non-sulfidic parts of the middle sequence below the Hildreths are gray metashales. Where sedimentary structures have not been obliterated by extreme metamorphism, these metashales are gray, and variably massive, faintly cyclically bedded, and distinctly cyclically interbedded with graded beds of metasiltstone or metasandstone. Bedding features, where preserved, are like those of the Seboomook(?) Formation. In addition, the middle sequence apparently lacks the coarse grained clastics of the Rangeley, and the Perry Mountain lacks the thick zones of massive gray metashale that are locally present in the middle sequence.

The correlation of the Hildreths Formation with the Madrid Formation is invalid because the lithologies and the sequences that contain them are dissimilar. Although both the Madrid and Hildreths Formations are dominantly calcareous metasandstones, the Madrid is much thicker than the Hildreths, lacks marble beds, and is commonly cross-laminated, unlike the Hildreths Formation. Further, the rocks below the Hildreths Formation are unlike the Smalls Falls and Perry Mountain Formations, below the Madrid.

The juxtaposition of the middle sequence with the Madrid, Smalls Falls and Perry Mountain Formations in the Phillips quadrangle, and with the Perry Mountain and Rangeley Formations in the Rumford quadrangle requires their separation by a major fault—the Blueberry Mountain fault. According to this interpretation, displacement would be minimal near the east side of the Phillips quadrangle, where Devonian(?) rocks are in contact across the fault. Displacement increases southwestward along the fault to the northeastern part of the Rumford quadrangle where Devonian(?) rocks are in contact with Lower Silurian rocks (Fig. 18–2). This fault has been mapped in the central part of the Rumford quadrangle, and is inferred to extend to the southwest approximately along the boundary between the Bryant Pond and Rumford quadrangles.

The relationships of the middle and upper sequences to the section of the Waterville area are uncertain. A prominent northwest-trending fault has been mapped in the Farmington and Dixfield quadrangles (Fig. 18–2). Southwest of this fault the stratigraphy of the lower, middle, and upper sequences is established. Northeast of the fault, however, the stratigraphy is uncertain. Warner and Pankiwskyj (*in* Hussey, 1965) equate calcareous metasandstone and black sulfidic slate northeast of the fault to the Hildreths Formation southwest of the fault; to the northeast the meta-sandstone and black slate are discontinuous. Alternatively, the limestone member in the unit mapped as Ludden Brook (local usage) by Warner and Pankiwskyj (*in* Hussey, 1965) north of the fault may be equivalent to the Hildreths. This unit has been traced into the northwest part of the Skowhegan quadrangle. Under either alternative, however, the rocks in doubt are Devonian(?). The presence of Devonian(?) rocks in the northwest part of the Skowhegan quadrangle requires the existence of an anticline between their outcrop and the syncline southeast of Skowhegan (Fig. 18–2).

SUMMARY

The Merrimack synclinorium is the dominant tectonic feature in west-central Maine. The trough of the synclinorium is defined by a thick mass of dominantly pelitic rocks of probable Devonian age—the Seboomook(?) Formation of the Rangeley area and the middle and upper sequences of the Dixfield area. The limbs of the synclinorium on either side of the Devonian(?) rocks are composed of Silurian and Silurian or Devonian rocks—the Rangeley, Perry Mountain, Smalls Falls and Madrid Formations on the northwest and the Mayflower Hill, Waterville and Vassalboro Formations on the southeast. Farther northwest the Middle Ordovician Dixville Formation is exposed, as well as a sequence of younger Ordovician rocks that grade upward and southeastward into the Rangeley Formation.

Observations from the northwest side of the area indicate that the synclinorium existed as a geosynclinal trough that received sediments continuously through much of Ordovician, Silurian and Devonian time. The Taconic orogeny, which is defined by a marked angular unconformity a short distance to the northwest of the area of Figure 18–2 (Harwood and Berry, 1967), was expressed in the Rangeley area by a great influx of coarse clastic sediments—the Rangeley Con-

glomerate, and the higher conglomerates of the Rangeley Formation—with no interruption of sedimentation. The Rangeley area at the time was close to the source of the coarse grained clastics, but the Waterville area, now 50 miles (80 km) to the southeast, was in a more distal part of the basin of sedimentation, and received only sparse coarse-grained, non-conglomeratic sediments. This pattern of sedimentation continued through later Silurian time, for the Waterville Formation is generally thinner bedded than the Perry Mountain Formation and contains prominent units of limestone, which are absent in the Perry Mountain. Later, the sulfide-rich Smalls Falls Formation may have accumulated in a strongly reducing basin of sedimentation. This basin probably was nearly restricted to the Rangeley area, for a possible equivalent of the Smalls Falls between the Waterville and Vassalboro Formations is thin or absent. The regional variations of the Silurian or Devonian Madrid Formation and Vassalboro Formation, and of the Devonian(?) rocks are too poorly understood to permit speculation on facies changes and provenance.

The foregoing conclusions are based on the assumption that we have correctly correlated the Mayflower Hill Formation with the Rangeley, the Waterville Formation with the lower sequence and with the Perry Mountain, the Vassalboro Formation with the Madrid, and the middle sequence with the Seboomook(?) Formation of the Rangeley area. Such correlation, over a distance of 50 miles (80 km) or more across a complex metamorphic terrane, is admittedly risky. It is consistent, however, with observed lithologies and sequences, considering facies changes, and with ages that have been established by means of fossils in the Waterville area, and inferred in the Rangeley area by correlation with dated rocks in nearby areas.

APPENDIX I

DESCRIPTIONS OF NEW AND REVISED FORMATIONS

Hildreths Formation (new name)

Type locality. The type locality of the Hildreths Formation is at Hildreths Mills, 0.7 mile (1.1 km) west of the height of land on the Wilton-Weld road (Maine Route 156), town of Perkins, north-central part of the Dixfield quadrangle. Typical outcrops are found along the road and in Bowley Brook. The Hildreths Formation was formerly called the Peru Formation by Pankiwskyj (1964).

Lithology. The Hildreths Formation consists of interbedded massive calc-silicate granulite, marble, and laminated metasandstone. Interbedding is commonly on a scale of ½ to 3 in. (1 to 8 cm).

Thickness. The Hildreths Formation has been estimated to be 200 to 800 ft. (60 to 240 m) thick.

Age. The age of the Hildreths Formation has been determined only approximately. It lies stratigraphically above rocks that are correlated with the Seboomook(?) Formation of Lower Devonian age. Thus, it is interpreted to be of Devonian(?) age.

Madrid Formation (revised name)

Type locality. The type locality for the Madrid Formation is along the Sandy River at the town of Madrid in the west-central part of the Phillips quadrangle.

Lithology. The upper part of the Madrid Formation consists of thick-bedded metasandstone and subordinate metashale. The lower part consists of thin bedded calcareous metasandstone, metashale, and subordinate calc-silicate granulite.

Thickness. The thickness of the Madrid Formation is estimated to be approximately 1000 feet (300 m).

Age. The Madrid Formation is assigned a Silurian (?) age. It lies conformably beneath rocks that are correlated with the Lower Devonian Seboomook Formation and lies above rocks that are interpreted to be of Silurian(?) age.

Mayflower Hill Formation (new name)

Type locality. The type locality of the Mayflower Hill Formation is on Mayflower Hill, just south of the campus of Colby College, in Waterville, in the south-central part of the Waterville quadrangle. Exposures may be studied at the quarry 1.1 miles (1.8 km) 0° from Colby College, on the campus of Colby College, and 1.3 miles (2.1 km), 97° from Oakland. These rocks in part were included originally in the Waterville Slate of Perkins and Smith (1925).

Lithology. The Mayflower Hill Formation consists of thick beds that grade from metasandstone at the base to metashale at the top. Some beds are slightly calcareous.

Thickness. The base of the Mayflower Hill is not exposed, but its outcrop is such as to suggest a thickness in excess of 1000 ft. (300 m).

Age. The Mayflower Hill Formation contains monograptids that are indicative of a Lower Silurian age.

Perry Mountain Formation (revised name)

Type locality. The type locality of the Perry Mountain Formation is on Perry Mountain, 9 miles (14.5 km) 134° from the town of Rangeley in the Rangeley quadrangle. Additional exposures of the Perry Mountain Formation may be seen in road cuts on Maine Route 4 south of Perry Mountain.

Lithology. The Perry Mountain Formation consists of cyclically interbedded white, quartz-rich metasandstone and light gray, mica-rich metashale. Cross-bedded and convolute features are characteristic of the metasandstone.

Thickness. The thickness of the Perry Mountain Formation is estimated to be approximately 2000 feet (600 m).

Age. The Perry Mountain Formation is conformably above the Rangeley Formation which is correlated with fossiliferous rocks of Lower Silurian age to the north, and it lies well below the Seboomook(?) Formation of Lower Devonian (?) age. On this basis the Perry Mountain Formation is interpreted to be of Silurian(?) age.

Rangeley Formation (revised name)

Type locality. The type locality for the Rangeley Formation is in Cascade Stream, 3.5 miles (5.6 km) 148° from the town of Rangeley in the Rangeley quadrangle. Other exposures of the Rangeley Formation may be observed on the hills south of Cascade Stream and along Maine Route 4, 2.5 to 4 miles (4 to 6.4 km) southeast of Rangeley.

Lithology. The Rangeley Formation is conveniently divided into three parts. The upper part consists of light-gray metashale, metasandstone, and subordinate quartz metaconglomerate. The middle part consists of metashale, metasandstone, and polymictic metaconglomerate. The lower part consists of polymictic metaconglomerate and metasandstone that wedge out into metashale to the south. The lower part is the Rangeley conglomerate of Smith (1923).

Thickness. The Rangeley Formation is approximately 9000 feet (2700 m) thick in the northern part of the Rangeley quadrangle but its thickness is less to the south.

Age. The Rangeley Formation is correlated on the basis of similarity of lithology and sequence with rocks containing Silurian fossils at Blanchards Pond in the Kennabago quadrangle. It is considered to be Silurian(?) in age.

Smalls Falls Formation (revised name)

Type locality. The type locality for the Smalls Falls Formation is at Smalls Falls 9.9 miles (16 km) 140° from the town of Rangeley in the Rangeley quadrangle. Numerous exposures occur in Chandler Mill Stream and in Sandy River in the vicinity of Smalls Falls.

Lithology. The Smalls Falls Formation is characteristically a black- and rusty-weathered, sulfide-rich metashale and metasandstone. Bedding is well displayed in water-washed outcrops in streams, but it is not obvious in weathered exposures. The upper part of the formation is calcareous. Beds of grit occur within the formation in the northeast part of its outcrop in the Phillips quadrangle.

Thickness. The Smalls Falls Formation is estimated to be 2500 feet (760 m) thick in the Rangeley quadrangle but it thins abruptly to the northeast.

Age. The age of the Smalls Falls Formation cannot be directly determined. Its stratigraphic position is between the Rangeley Formation (correlated with fossiliferous Silurian rock) and the Seboomook(?) Formation (Lower Devonian). A Silurian(?) age has been assigned to the Smalls Falls.

Waterville Formation (revised name)

Type locality. The type locality of the Waterville Formation is along the Kennebec River at the old campus of Colby College in the City of Waterville. Other outcrops just east of Maine Route 32, 0.9 mile (1.4 km) 110° from the city of Winslow in the south-central part of the Waterville quadrangle. Outcrops of limestone may be observed beneath the bridge connecting Waterville and Winslow.

Lithology. The Waterville Formation consists of cyclically and thinly laminated, light greenish-gray metashale and metasandstone. Interbedded gray limestone and metashale form a prominent member.

Thickness. The thickness of the Waterville Formation is estimated as 3000 feet (900 m).

Age. The Waterville Formation contains dendroid graptolites that suggest a Silurian age.

REFERENCES

Albee, A. L., 1961, Boundary Mountain anticlinorium, west-central Maine and northern New Hampshire: U. S. Geol. Survey Prof. Paper 424–C, p. C51–C54

Barker, D. S., 1961, Hallowell granite and associated rocks, south-central Maine: Ph.D. thesis, Princeton Univ., 240 p.

Berry, W. B. N., and Boucot, A. J., in press, Silurian of North America: Geol. Soc. America.

Billings, M. P., 1956, The geology of New Hampshire; Part II—Bedrock geology: New Hampshire State Planning and Development Commission, Concord, N. H., 203 p.

Boucot, A. J., 1961, Stratigraphy of the Moose River synclinorium, Maine: U. S. Geol. Survey Bull. 1111–E, p. 153–188

————, Field, M. T., Fletcher, Raymond, Forbes, W. H., Naylor, R. S., and Pavlides, Louis, 1964, Reconnaissance bedrock geology of the Presque Isle quadrangle, Maine: Maine Geol. Survey Quad. Mapping Series No. 2, 123 p.

Boudette, E. L., and Harwood, D. S., 1965, Fossils indicate margin onlap during Llandovery to Ludlow time: U. S. Geol. Survey Prof. Paper 525–A, p. A74

Cady, W. M., 1960, Stratigraphic and geotectonic relationships in northern Vermont and southern Quebec: Geol. Soc. America Bull., v. 71, p. 531–576

————, in press, Regional tectonic synthesis of northwestern New England and adjacent Quebec: Geol. Soc. America Spec. Paper

Donath, F. A., and Parker, R. B., 1964, Folds and folding: Geol. Soc. America Bull., v. 75, p. 45–62

Evans, B. W., and Guidotti, C. V., 1966, The sillimanite-potash feldspar isograd in western Maine, U. S. A.: Beitrage Mineral. und Petrol., v. 12, p. 25–62

Faul, Henry, Stern, T. W., Thomas, H. H., and Elmore, P. L. D., 1963, Ages of intrusion and metamorphism in the northern Appalachians: Am. Jour. Sci., v. 261, p. 1–19

Green, J. C., 1964, Stratigraphy and structure of the Boundary Mountain anticlinorium in the Errol quadrangle, New Hampshire-Maine: Geol. Soc. America Spec. Paper 77, 78 p.

Guidotti, C. V., 1963, Metamorphism of the pelitic schists in the Bryant Pond quadrangle, Maine: Am. Mineralogist, v. 48, p. 772–791

————, 1965, Geology of the Bryant Pond quadrangle, Maine: Maine Geol. Survey Quad. Mapping Series No. 3, 116 p.

Harwood, D. S., and Berry, W. B. N., 1967, Fossiliferous lower Paleozoic rocks in the Cupsuptic quadrangle, west-central Maine: U. S. Geol. Survey Prof. Paper 575–D, p. D16–D23

Hussey, A. M., II, ed., 1965, Field trips in southern Maine: New England Intercollegiate Geol. Conf. 57th Ann. Mtg., guidebook, Brunswick, Maine, 118 p.

Larrabee, D. M., and Spencer, C. W., 1963, Bedrock geology of the Danforth quadrangle, Maine: U. S. Geol. Survey Geol. Quad. Map GQ–221

Moench, R. H., 1966, Relation of S_2 schistosity to metamorphosed clastic dikes, Rangeley-Phillips area, Maine: Geol. Soc. America Bull., v. 77, p. 1449–1462

Naylor, R. S., and Boucot, A. J., 1965, Origin and distribution of rocks of Ludlow age (Late Silurian) in the northern Appalachians: Am. Jour. Sci., v. 263, p. 153–169

Osberg, P. H., 1968, Stratigraphy, structural geology, and metamorphism in the Waterville-Vassalboro area, Maine: Maine Geol. Survey Bull. No. 20, 64 p.

Pankiwskyj, K. A., 1964, Geology of the Dixfield quadrangle, Maine: Ph.D. thesis, Harvard Univ., 224 p.

Pavlides, Louis, Mencher, Ely, Naylor, R. S., and Boucot, A. J., 1964, Outline of the stratigraphic and tectonic features of northeastern Maine: U. S. Geol. Survey Prof. Paper 501–C, p. C28–C38

——————, and Berry, W. B. N., 1966, Graptolite-bearing Silurian rocks of the Houlton-Smyrna Mills area, Aroostook County, Maine: U. S. Geol. Survey Prof. Paper 550–B, p. B51–B61

Perkins, E. H., and Smith, E. S. C., 1925, Contributions to the geology of Maine, No. 1: A geologic section from the Kennebec River to Penobscot Bay: Am. Jour. Sci., v. 207, p. 204–228

Smith, E. S. C., 1923, The Rangeley conglomerate: Am. Jour. Sci., v. 5, p. 147–154

Warner, Jeffrey, 1967, Geology of the Buckfield quadrangle, Maine: Ph.D., thesis, Harvard Univ., 232 p.

Chapter 19

The Boundary Mountains Anticlinorium in Northern New Hampshire and Northwestern Maine

JOHN C. GREEN AND CHARLES
V. GUIDOTTI

INTRODUCTION

THE GEOLOGY of the Boundary Mountains, a highland area in northern New Hampshire, northwestern Maine, and adjacent Quebec, was almost unknown up to a few years ago except for regional reconnaissance during the nineteenth century (e.g. Logan, 1863; Hitchcock, 1877). In 1955, extensive and continuing study of the southwestern part of this area began, much of it as Ph.D. thesis problems under the guidance of M. P. Billings and J. B. Thompson, Jr., of Harvard University. This paper will discuss the state of knowledge as of 1967 of the area between long. 70°45′ W. and the western boundary of New Hampshire, and between lats. 44°45′ and 45°30′ N. (Fig. 19–1).

Sources of information for the various parts of this area are as follows: Woburn (lat. 45°23′ N., long. 70°52′ W.) area, Marleau (1957, 1958); Mt. Megantic area, McGerrigle (1935); Cupsuptic quadrangle (this and other quadrangles identified by letter symbols in Fig. 19–2), Harwood (USGS open file report, 1966); Arnold Pond quadrangle, western part, Green (unpublished notes); eastern half, Boucot, Griscom and Allingham (1964); Moose Bog, Second Lake and Indian Stream quadrangles, Green (in press and unpublished notes); Oquossoc quadrangle, Guidotti (unpublished notes); Errol quadrangle, Green (1964); Dixville and Averill quadrangles, Hatch (1963 and written communication 1962).

In this synthesis the writers have called upon much unpublished information from those who have recently mapped in the area. We are particularly indebted to David S. Harwood, U. S. Geological Survey, who has contributed much thought and many valuable comments during the preparation of this paper. Unfortunately Dr. Harwood was not able to join us as co-author. Nonetheless anyone familiar with the geology of the Boundary Mountains will recognize the great extent to which his work has influenced our thinking. Support to both authors from the Maine Geological Survey under the direction of Mr. Robert G. Doyle is also greatly appreciated. Much of Green's field work was supported by research grants from the Geological Society of America. We are also thankful to Mrs. Ruth Darden for drafting Figures 19–1 to 19–5.

STRATIGRAPHY

The area discussed (Fig. 19–1) includes the core and northwest flank of the broad Boundary Mountains anticlinorium (Cady, 1960) and, to the northwest, the first synclinal axis (Frontenac syncline) of the major, regional Connecticut Valley–Gaspé synclinorium (Cady, 1960). The core of the anticlinorium is composed of a complex series of Cambro-Ordovician rocks with local, unconformable outliers of Silurian metasediments. The rocks of the synclinorium are predominantly Devonian, resting unconformably (Taconic un-

255

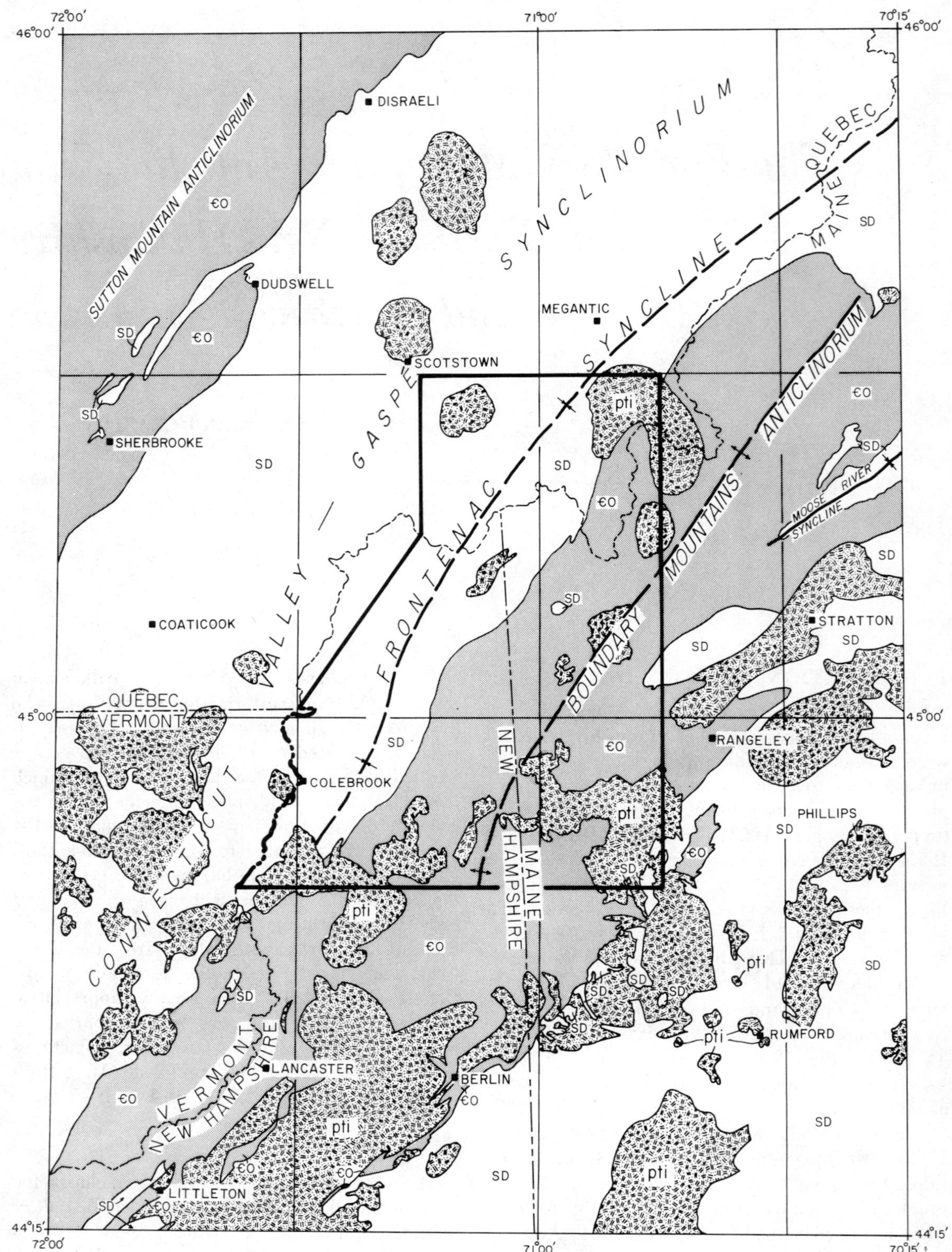

FIGURE 19–1. Index map. Heavily outlined area is area of geologic map, Figure 19–2. Coarse pattern (pti): Post-Ordovician intrusive rocks; Shading (€O): Cambro-Ordovician stratified and intrusive rocks; SD: Siluro-Devonian stratified rocks. Published sources of data: Billings, 1956; Boucot and others, 1964; Cady, 1960; Cooke, 1937, 1950, 1957; Doll and others, 1961; Green, 1964; Harwood, 1966; Hatch, 1963; Marleau, 1957; McGerrigle, 1935; Moench, 1966.

conformity; see Pavlides et al., this volume) on the Ordovician formations (Green, 1962). The stratified rocks are all regionally metamorphosed, with higher grades in the southeast part of the area. They are cut by intrusive rocks of several ages.

The stratified rocks beneath the Taconic unconformity have been divided into three formations (Fig. 19–3).

The Aziscohos Formation (Green, 1964) and the Albee Formation (Billings, 1937) are at the base of the exposed section. The Aziscohos apparently occupies the lower stratigraphic position; it differs from the Albee in its virtual lack of the "pinstripe" micaceous quartzite ("granulite") that characterizes the Albee Formation throughout western New Hampshire (Cady, 1960, p. 550) including the Errol and Second Lake quadrangles. Aside from the pinstripe quartzite, both formations are dominantly pelitic. The Aziscohos contains mostly carbonaceous phyllite and schist in its lower part and green phyllite in its upper part. In the upper part of the Albee, minor zones and belts of gray phyllite are present, as well as purple phyllite in the Cupsuptic quadrangle. Evidence from the Errol quadrangle (Green, 1964) suggests that the type Aziscohos underlies the Albee, but to the east and northeast Aziscohos-like rocks (Oad, Fig. 19–2) appear to overlie typical quartzites of the Albee Formation (Harwood, 1966). The solution to this apparent conflict may lie in lateral facies changes, involving eastward lensing out of the "pinstripe" quartzites, particularly in the upper part of the Albee Formation. This interpretation would preserve the relative stratigraphic positions of the Albee and Aziscohos, well established by graded beds elsewhere in the Errol and Oquossoc quadrangles. A lack of outcrop in some critical areas and need of more detailed mapping in others hampers solution of this problem. A thorough discussion of the alternatives is not possible here.

The above formations are intensely folded; estimates of thickness range from 17,000 to 4,500 feet (5200 to 1400 m), but the obviously rapid facies changes and lack of fossils or other stratigraphic markers make the estimates very crude. No identifiable fossils have been found, but as the Albee conformably underlies the fossiliferous Dixville Formation, it is probably Lower or Middle Ordovician. The Aziscohos cannot yet be dated more closely than Cambrian or Ordovician.

The Albee Formation is traceable into the type locality in west-central New Hampshire (Billings,

1937). It is correlated, by lithologic similarity and position in the stratigraphic sequence, with the Moretown Formation of north-central Vermont and the lower part of the Beauceville Formation of the Eastern Townships, Quebec. The Aziscohos is tentatively correlated, by the same criteria, with the Stowe and possibly Ottauquechee Formations of north-central Vermont, and with the Mansonville of the Eastern Townships. The Vermont and Quebec correlatives are on the east limb of the Green Mountain–Sutton Mountain anticlinorium. The upper, non-carbonaceous part of the Aziscohos may be lithically equivalent to rocks mapped as Albee Formation farther southwest in New Hampshire.

Overlying the Albee Formation is the Dixville Formation (Green, 1960, 1962, 1964; Hatch, 1963), consisting predominantly of carbonaceous slate, phyllite, schist and quartzite, as well as major units of metavolcanic greenstone and amphibolite. Throughout most of the area the metavolcanic rocks, which are locally pillowed, occupy an intermediate stratigraphic position within the formation (Green, 1964), but at about lat 45° N., on both sides of the major Albee–Aziscohos anticlinal axis, they appear at the base. Here they directly overlie non-carbonaceous phyllites and intertongue with the carbonaceous pelitic rocks of the Dixville Formation along strike. In the northeastern part of the area, the upper exposed part of the Dixville Formation, the Magalloway Member (Green, in press), consists primarily of coarse gray, green and tan metagraywackes and meta-arkoses with phyllite and minor metavolcanic rocks near 45°15′ N. Rocks similar to the Dixville Formation along strike in Quebec in the Woburn area have been called Arnold River Formation by Marleau (1958).

The Dixville Formation has been estimated to be from 5000 to 9000 feet (1500 to 2700 m) thick in this area, but the folding prevents any precise determination. A graptolite fauna collected by Harwood from carbonaceous slates in the southeastern part of the Cupsuptic quadrangle has been dated as late Middle Ordovician (Zone 12 of Berry, 1960; Harwood and Berry, 1967).

The distinctive combination of carbonaceous pelitic rocks and metabasalts that constitutes most of the Dixville Formation, along with its position in the pre-Silurian sequence, provides valuable bases for correlation. The Dixville is correlated with the Partridge and Ammonoosuc Formations of west-central New Hampshire, the Cram Hill and upper Moretown of north-central Vermont

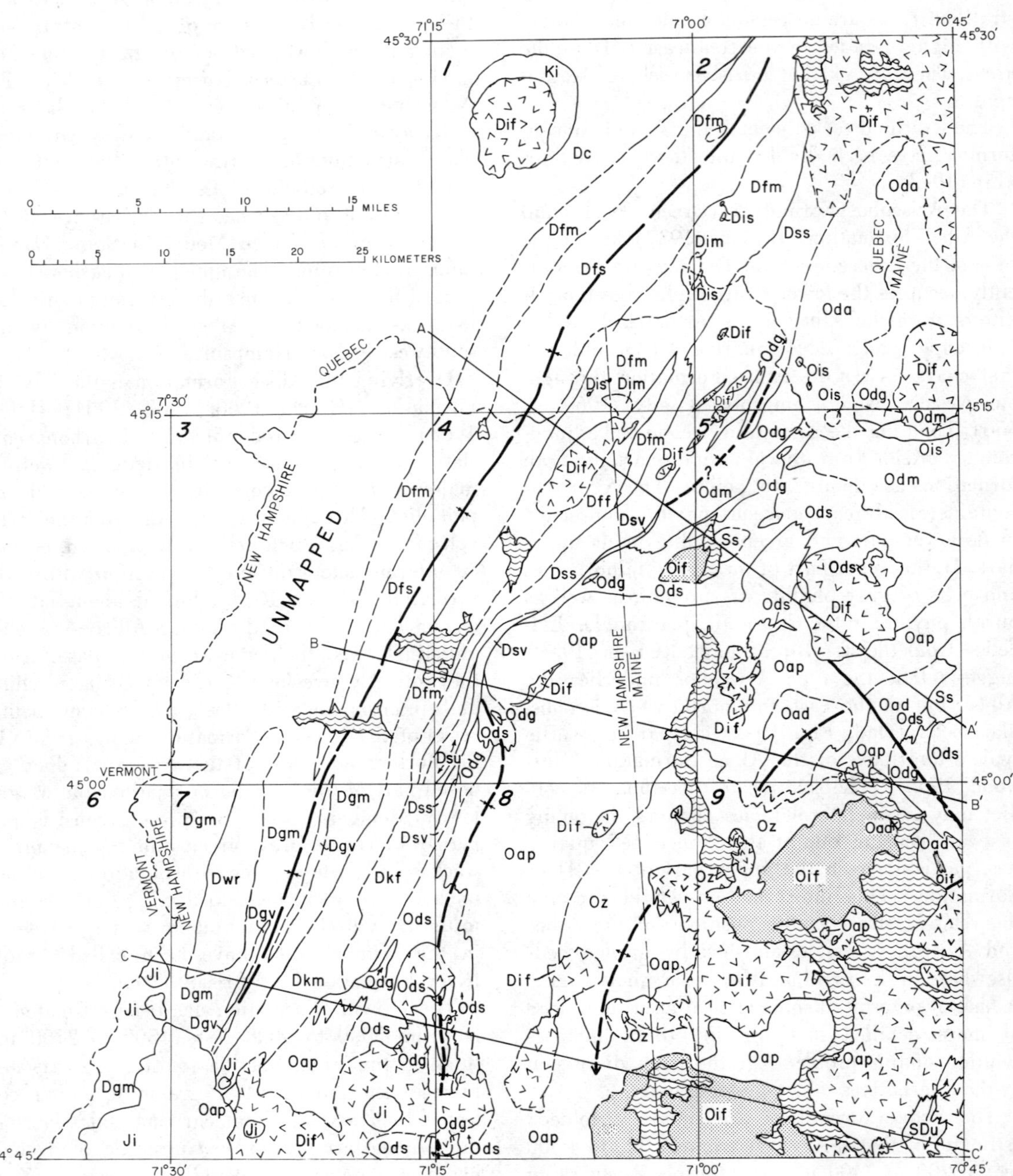

Figure 19–2.   Geologic map. Numbers at tops of quadrangles refer to their names:
1, Moose Bog (Mt. Megantic area in Canada); 2, Arnold Pond (Woburn area
in Canada); 3, Indian Stream; 4, Second Lake; 5, Cupsuptic; 6, Averill; 7,
Dixville; 8, Errol; 9, Oquossoc. A–A′, B–B′, C–C′ show traces of structure sections of Figure 19–4. Wavy pattern, lakes.

EXPLANATION

INTRUSIVE ROCKS

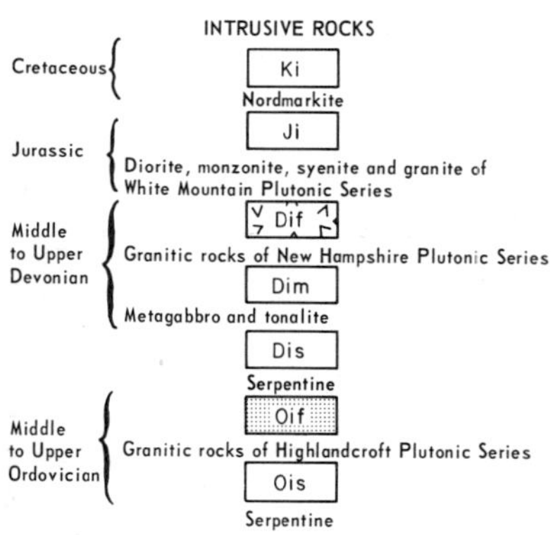

METAMORPHIC ROCKS

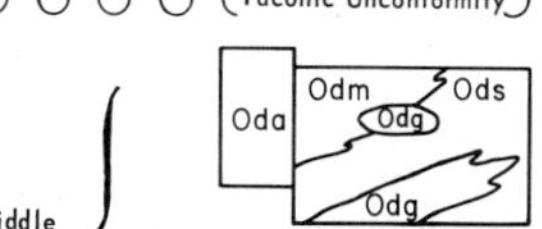

Lower Devonian (possibly includes some Upper Silurian)

Gile Mountain Fm.: Dgm, gray metagraywacke, phyllite, and schist; Dgv, mafic metavolcanic rocks
Waits River Fm.: Dwr, gray, calcareous and non-calcareous schist, quartzite and phyllite
Frontenac Fm.: Dfs, gray and green metagraywacke, phyllite and slate; Dfm, mafic metavolcanic rocks; Dff, felsic metavolcanic rocks
Kidderville Fm.: Dkm, mafic metavolcanic rocks with minor felsic metavolcanic rocks and carbon-aceous phyllites and quartzites; Dkf, felsic metavolcanic rocks and gray phyllites and quartz-mica schists
Seboomook Fm.: Dss, gray (minor green and purple) slate, gray phyllite and quartzite; Dsv, felsic metavolcanic rocks; Dsu, gray phyllites, mafic and felsic metavolcanic rocks
Compton Fm. Dc: Gray phyllite and micaceous quartzite

Siluro-Devonian — SDu — Pelitic to psammitic schists and gneisses

Upper and Lower Silurian — Ss — Limestone, conglomerate, argillite, graywacke, tuff

∿∿∿∿∿ Taconic Unconformity ∿∿∿∿∿

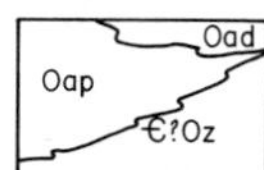

Middle Ordovician

Dixville Fm.: Ods, carbonaceous slate, phyllite, schist and quartzite; Odg, mafic metavolcanic rocks; Odm, silvery gray, green and tan coarse metagraywacke, meta-arkose, and phyllite, and minor metavolcanic rocks
Arnold River "Fm.": Oda, all of above rocks, north of 45°15, mostly Odm

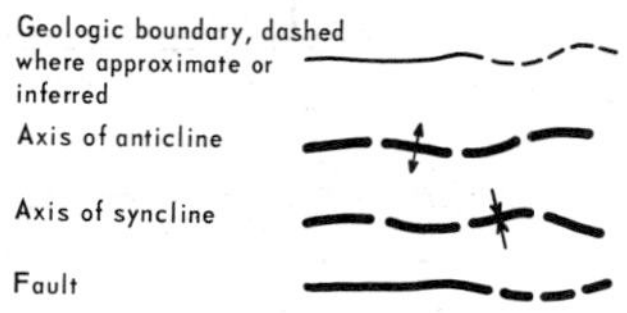

Upper Cambrian (?) to Middle Ordovician

Albee Fm.: Oap, greenish gray "pinstripe" mica-ceous quartzite, green phyllite and schist, green, purple and black slate, also some massive to pillowed greenstone; Oad, green slate and phyllite with locally abundant quartz stringers, purplish-gray slate and phyllite, carbonaceous phyllite, and thin greenstone lenses

Aziscohos Fm.€?Oz, green, pelitic phyllite and schist with abundant quartz stringers, also some dominantly carbonaceous phyllite and schist com-monly with quartz stringers.

STRUCTURE SYMBOLS

Geologic boundary, dashed where approximate or inferred

Axis of anticline

Axis of syncline

Fault

and the Bolton volcanics and metasediments of the upper Beauceville Formation in Quebec. The graptolites from the Cupsuptic quadrangle are correlated with the lower faunal assemblage in the slates (assigned to the Beauceville Formation) at Magog, Quebec. The Vermont and Quebec correlatives are on the northwest limb of the Connecticut Valley–Gaspé synclinorium, the New Hampshire correlatives on the southeast limb.

Above the angular Taconic unconformity in this area are several small remnants of Silurian rocks ranging in age, according to faunas thus far known, from late Early Silurian (upper Lland-overy; U. S. Geol. Survey, 1965, p. A74) to Late Silurian (lower Ludlow; Naylor and Boucot, 1965, p. 161). The Llandovery age was deter-mined from fossils collected by E. L. Boudette (U. S. Geol. Survey) from the Kennebago Lake quadrangle immediately east of the area, in quartzite associated with rocks similar to those at the eastern edge of the Cupsuptic quadrangle. The Silurian rocks include conglomerate, lime-stone, slate, argillite, quartzite, tuff and felsite. A very small group of limestone outcrops (not shown in Fig. 19–2) in the Second Lake quad-rangle near the Taconic unconformity has yielded corals and brachiopods that have been examined by W. A. Oliver, Jr., and A. J. Boucot, Jr., re-spectively; these fossils definitely indicate Silu-rian age, probably Ludlow (Oliver, written com-munication, 1963; Boucot, written communica-tion, 1963). Harwood (1966) estimates the greatest remaining thickness of the Silurian rocks to be about 3000 feet (900 m).

Northwest of the major Taconic unconformity that trends northeast-southwest across the area is a thick sequence of Silurian and Devonian rocks that have been assigned by various workers to the Seboomook, Frontenac, Compton, Kidder-ville, Gile Mountain and Waits River Formations (Fig. 19–3).

The lowest units have been mapped in the northeast part of the area as Seboomook Forma-tion (Perkins, 1925; Boucot, 1961) and in the southwest part as the Kidderville Formation (Hatch, 1963). Along much of the base of this sequence are felsic metavolcanic rocks, inter-bedded locally with carbonaceous phyllite and schist. In the vicinity of the New Hampshire–Maine State line, the felsite shows distinct pyro-clastic texture. The metavolcanic rocks at the base disappear to the northeast as the amount of gray slate, phyllite and micaceous quartzite in-creases. Northwest of and above these units is a

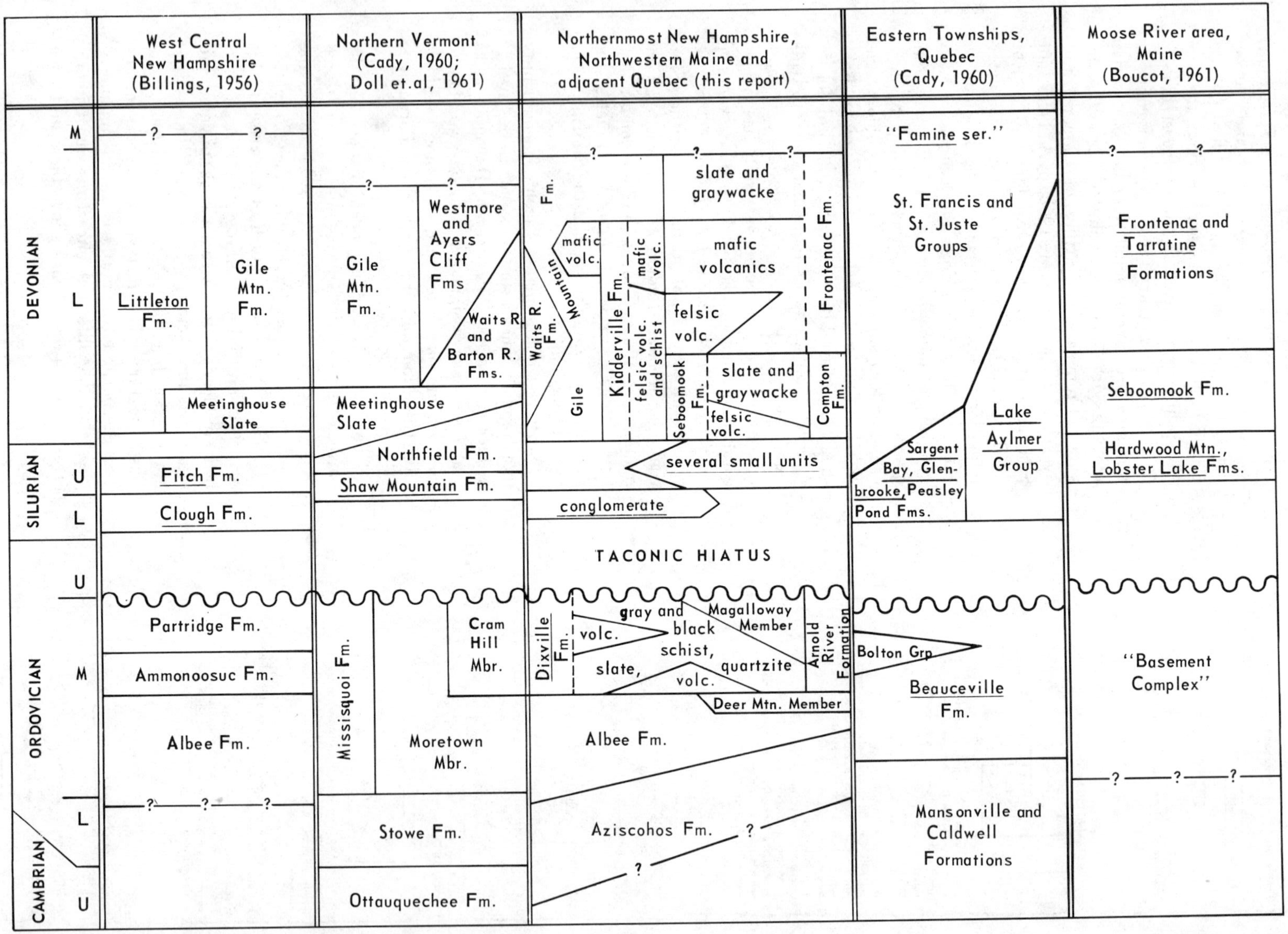

FIGURE 19-3. Generalized stratigraphic correlations for the Boundary Mountains anticlinorium. Queried boundaries indicate unknown limits due to erosion, non-exposure, lack of detailed mapping, or lack of stratigraphic control; blank areas are known hiatuses. Underlined formations are dated by fossils.

continuous band of predominantly mafic meta-volcanic rocks including volcanic-derived meta-sediments, which are assigned to the Frontenac Formation (McGerrigle, 1935; Marleau, 1958) in the northeast part of the area, and to the upper part of the Kidderville Formation in the Dixville quadrangle. Serpentinites and small metaperidotites that intrude these mafic meta-volcanic rocks in the vicinity of the Quebec–New Hampshire–Maine corner are exceptional in that they are evidently of Acadian, rather than Taconic, age; elsewhere in northern New England, serpentinites are not known to cut rocks younger than the Ordovician (Billings, 1956, p. 106). Perhaps they were Ordovician intrusions remobilized during the Acadian deformation, but they may also have been intruded for the first time during the Devonian volcanism or the Acadian orogeny. A thick sequence of felsic meta-volcanic rocks at the base of this unit extends for several miles southwest of the New Hampshire–Maine State line and intertongues with the metabasalts at the State line.

Overlying these volcanic rocks are gray phyllites, schists and micaceous quartzites of the Frontenac Formation (north of lat 45° N.) and the Gile Mountain Formation (south of lat 45° N.). From the distribution and shapes of pillows in greenstones in Quebec (Marleau, 1959) and gross stratigraphic repetition a major synclinal axis, the Frontenac syncline (Marleau, 1958; Albee, 1961), is thought to trend along this belt of metasediments. Thus the rocks to the north-west are assumed to be equivalent to units to the southeast. Many metavolcanic rocks, predom-inantly mafic, occur interbedded with gray meta-sediments northwest of this axis; these are tenta-tively correlated with the volcanic rocks of the Frontenac Formation. This volcanic-rich band is succeeded in turn to the northwest by more gray metasediments, assigned to the Compton Formation north of lat 45° N. (McGerrigle, 1935), and to the Gile Mountain Formation far-ther south. Not enough relict sedimentary fea-tures have yet been found to support or disprove the existence of this syncline in New Hampshire, and it is possible that this whole sequence is homoclinal, topping to the northwest. Another possible interpretation would place a major syn-clinal axis along the belt of Seboomook meta-sediments, so that the easternmost belt of metaba-salts of the Frontenac and Kidderville Formations (Dfm, Dkm, Fig. 19–2) would be Ordovician. However, metabasaltic pillows in this belt indicate

younger rocks to the northwest, throwing some doubt on this latter interpretation.

Surrounded by Gile Mountain pelitic and psammitic metasediments in the Dixville and Indian Stream quadrangles is a large area rich in calcareous rocks. It has been assigned by Hatch (1963) to the Waits River Formation, although he recognizes that the rocks may be simply a calcareous lens within the Gile Mountain. A metasedimentary unit with thick, graded gray-wacke beds (not shown in Figure 19–2) has been mapped by Hatch (1963) in the Gile Mountain Formation at the west edge of the map area at lat 45° N., but it has not yet been traced north-eastward into the Indian Stream quadrangle north of the Vermont line. Myers (1964) has traced this unit to the southwest in the Vermont portion of the Averill quadrangle. The graded beds show younging to the northwest.

In the southeast corner of the area, Guidotti found gneisses and schists of high metamorphic grade that continue into the Rangeley quad-rangle, where they are assigned to the Silurian or Devonian by R. H. Moench (1966) on the basis of lithologic similarity and sequence (Os-berg et al., this volume).

The post-Ludlow sequence in the hypothetical Frontenac syncline is estimated to total between 15,000 and 20,000 feet (4500 and 6000 m) in thickness, or even thicker if the sequence is homoclinal. It has so far produced no fossils, but does continue along strike (Albee, 1961) into the fossiliferous Lower Devonian (Boucot, 1961) Seboomook and Tarratine Formations of the Moose River syncline in Maine. The sequence unconformably overlies Upper Silurian (Ludlow) and older rocks and is cut by intrusive rocks cor-related with the Middle or Late Devonian New Hampshire Plutonic Series; it is therefore late Upper Silurian to Lower Devonian.

The Gile Mountain and Seboomook Formations are continuously traceable to the southwest and northeast, respectively, into their type localities in Vermont and Maine. The Compton and Frontenac Formations were first described from the Mount Megantic area (Fig. 19–2). The Waits River Formation in the Dixville quadrangle is corre-lated by lithic similarity with the widespread Waits River and Barton River Formations (Doll, 1951) in eastern Vermont, but as Hatch notes (1963, p. 31) various workers have used different carbonate/silicate ratios in distinguishing the Waits River from the Gile Mountain in separate areas; also, these calcareous units probably occur at more than one stratigraphic level and may not deserve formation status. The Gile Mountain–Frontenac metasediments are continuous to the northeast with the St. Francis–St. Juste Groups in Quebec (Cooke, 1937; Cady, 1960) and with the Lower Devonian Tarratine Formation in Maine (Albee, 1961); they are also correlated by lithic similarity and sequence with the Lower Devonian Littleton Formation of west-central New Hampshire. The mafic volcanic rocks assigned to the Kidderville, Frontenac and Gile Mountain formations may correlate broadly with the metabasaltic Standing Pond Member of the Waits River Formation in eastern Vermont (Rosenfeld, this volume).

It should be stressed that because the only diagnostic fossils so far found in the area have come from small and mostly isolated patches of Silurian rocks and from one locality in the Dixville Formation, correlation with other well-studied areas of the northern Appalachians is based primarily on lithic similarity, stratigraphic sequences, and continuity along strike. Because rapid lateral facies changes in this active, eugeosynclinal area may make all of these criteria unreliable for time correlation even within the limited area of this study, the regional correlations shown in Figure 19–3 are mostly only approximate and for the most part lithostratigraphic.

INTRUSIVE ROCKS

At least three major periods of intrusive activity are represented by plutonic rocks in the area. The earliest, believed to be of Late Ordovician (Taconic) age and to correlate with the Highlandcroft Plutonic Series of Billings (1956), includes a biotite-hornblende granodiorite (Umbagog Granodiorite) in the south-central part of the map area, a small hornblende granodiorite (Parmachenee Granodiorite) in the east-central part of the Second Lake quadrangle, and a foliated porphyritic adamellite (Adamstown Granite) in the north-central part of the Oquossoc quadrangle. These plutons intrude only pre-Silurian rocks, and are lithically similar to rocks of the Highlandcroft Plutonic Series. The Adamstown Granite also resembles some members of the Oliverian Plutonic Series (Billings, 1956, p. 48–50), now also considered to be Ordovician (see papers by Naylor and by Thompson et al., this volume). The Adamstown Granite has been strongly deformed and regionally metamorphosed, and the granodiorites have been locally deformed, especially at their contacts. Lead-alpha age determinations on zircons from the granodiorites, made by the U. S. Geological Survey (T. W.

Stern, written communication, 1962) range between 450 ± 50 m.y. (massive, central part of the Umbagog pluton) and 350 ± 50 m.y. (margin), implying crystallization in the Ordovician and with subsequent local recrystallization during the Acadian orogeny. The possibility remains, however, that these plutons are actually Acadian.

Stocks of syntectonic and post-tectonic biotite-muscovite adamellite and small lenses and stocks of gabbro and microgranite are widely distributed in all except the northwesternmost part of the area. They cut both the pre-Silurian and Siluro-Devonian stratified rocks (Fig. 19–5). They are thus post-Lower Devonian, and are correlated with the Acadian New Hampshire Plutonic Series on the bases of structural and compositional similarity. Their contacts range from sharply discordant to concordant to migmatitic, and many of the stocks have undergone local shearing. These structural relationships, as well as the width and nature of the mineral assemblages in their metamorphic aureoles, suggest successively deeper levels of intrusions from the northwest to the southeast part of the area.

High-level, post-tectonic stocks of diorite, syenite, and granite, characterized by lobate, cross-cutting contacts, and scattered basaltic and lamprophyric dikes cut the rocks in the southwestern part of the area. Because of their structure and composition they are assumed to belong to the Jurassic White Mountain Plutonic Series (Billings, 1956; also see Lyons and Faul, this volume) even though no radiometric age is available. A nordmarkitic partial ring dike of Cretaceous age (Fairbairn and others, 1961) forms part of the Mount Megantic massif in the northernmost part of the area, and is correlated with the other plugs and stocks of the Monteregian Hills to the west in Quebec.

STRUCTURE

When modern studies in this area began the entire area was tentatively considered to be part of a great anticlinal structure called the Coos anticlinorium (Billings, 1956, p. 111–112). Studies in the last ten years show the area to contain both a major anticlinorium of pre-Silurian rocks in the east (Boundary Mountains anticlinorium) and part of a synclinorium of Siluro-Devonian rocks in the west (Connecticut Valley–Gaspé synclinorium) as well as the regional Taconic unconformity lying along the shared limb. The anticlinorium trends southwesterly and corresponds generally with Billings' Gardner Mountain anticline, located sixty miles away in the Littleton–Moosilauke area of west-central

New Hampshire. Structural characteristics along the unconformity will be described first, followed by those of the pre-Silurian and then of the Siluro-Devonian rocks.

The major and minor structures in the metamorphic rocks are rather different on the two sides of the unconformity. The unconformity crosscuts major folds and stratigraphic units in the pre-Silurian rocks, but basal conglomerate has been seen only in the northeastern part of the area. The actual contact has not been recognized in outcrop. Abrupt wedging-out of units of the overlying Seboomook Formation suggest local faulting along the unconformity in the southwest part of the Second Lake quadrangle, and similar relationships are shown in the Dixville quadrangle at the base of the Kidderville Formation. The unconformity itself has been sharply folded near lat 45°15′ N., but in general appears to dip steeply northwestward. The several isolated patches of Silurian rocks indicate that the uncon-

formity was broadly to strongly folded over the Boundary Mountains anticlinorium but now has been mostly eroded away.

Several major folds are present in the pre-Silurian rocks. The dominant folds are the Rice Mountain–Hellgate Brook syncline, containing Dixville Formation in its core, and the Diamond Peaks anticline, with a core of Aziscohos Formation (Fig. 19–4). The appearance of the Magalloway Member of the Dixville Formation at about long 71° W. may represent the reemergence of the Rice Mountain–Hellgate Brook syncline from beneath the unconformity. If so, the syncline was further deformed during the Acadian folding to produce the synclinal tongue of Seboomook Formation immediately to the northeast. The Diamond Peaks anticline (Fig. 19–4) plunges south, as shown by the map pattern of the Albee Formation. The trends of both of these major folds swing from north at the southern edge of the area to northeast near the east edge

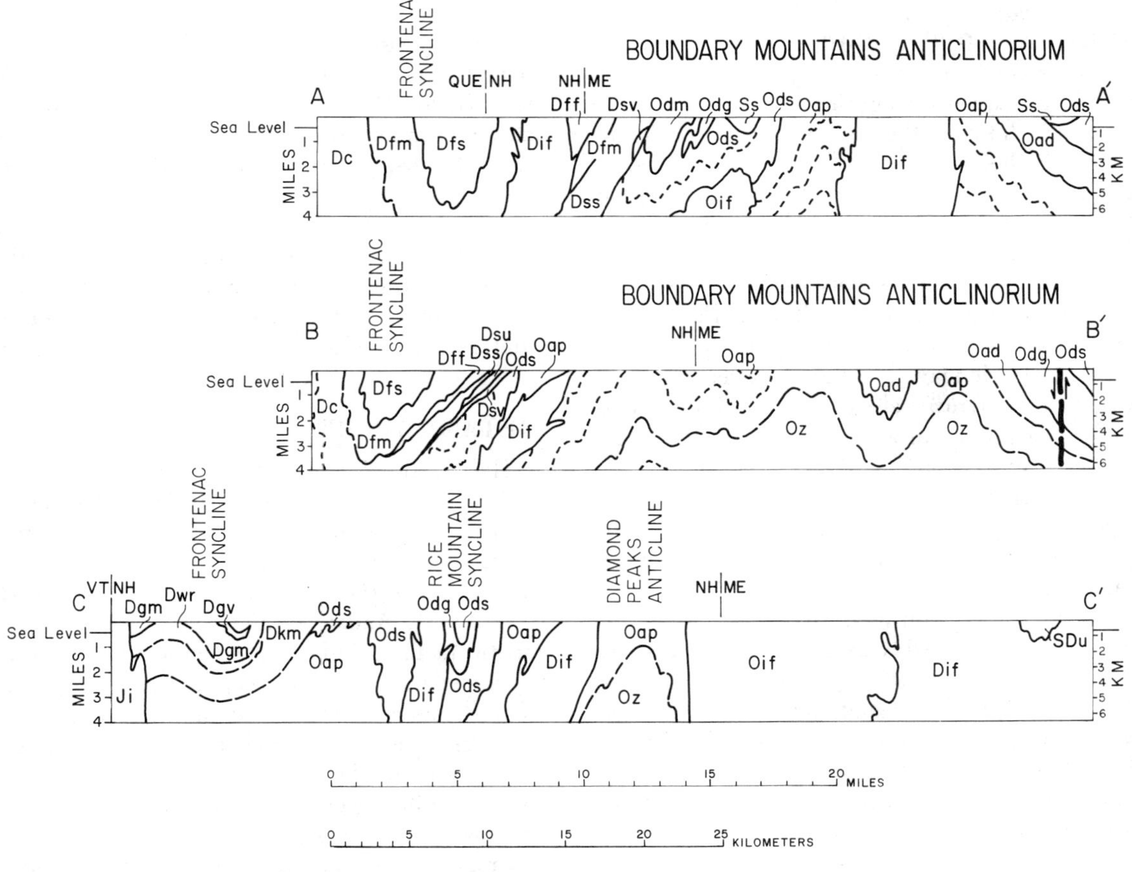

FIGURE 19–4. Interpretive geologic structure sections (for locations see geologic map, Fig. 19–2). Existence of Frontenac syncline assumed. Formation symbols as on geologic map. No vertical exaggeration; surface topography, negligible at this scale, is not shown.

of the area. A granitic pluton and its hornfelsic aureole southwest of the center of the Second Lake quadrangle appear to have acted as a buttress against which the syncline has been tightened during the Acadian folding.

The pre-Silurian rocks show evidence of two periods of deformation. Their minor folds and other structures plunge much more steeply than those in the Siluro-Devonian sequence, and in the Cupsuptic area their folding is tighter than that in the Silurian outliers. According to Harwood (1966), the intrusion of a large Acadian pluton in the central part of the Cupsuptic quadrangle has caused re-folding of earlier, isoclinal, Taconic folds along north to northeast axes. A secondary north-south fracture cleavage, approximately vertical, cuts the earlier folds and cleavage in the southern part of the Cupsuptic quadrangle and the northern part of the Oquossoc quadrangle, but its age has not been established. Possibly it is Acadian and resulted from the buttressing effect of the large Ordovician Adamstown Granite pluton in the Oquossoc quadrangle during the Acadian orogenic movement.

The only large-scale structural feature in the Siluro-Devonian rocks is the Frontenac syncline of Marleau (1958) and Albee (1961). As previously stated, this structure is hypothetical and the entire sequence from the unconformity to the northwest limit of the map area may be homoclinal. If the syncline does exist, as suggested by top directions based on greenstone pillows in Quebec and on a very few graded beds in New Hampshire, and by gross repetition of rock types, marked facies changes must occur between the east and west limbs. The position of the trace of the synclinal axis is uncertain in the Dixville quadrangle; it may lie either between or to the east of the narrow belts of mafic volcanic rocks in the Gile Mountain Formation. Following Hatch (1963), we assume that the axis passes between the greenstone lenses. The structure sections of Figure 19–4 are drawn on this assumption. Mapping in the Indian Stream quadrangle may resolve this uncertainty.

Hatch (1963) considers that the Waits River Formation in the Dixville quadrangle forms the core of an anticline, which is consistent with relations elsewhere in Vermont and with its relatively low eastward dips. However, this implies that the formation, at least 1000 feet (300 m) thick, must pinch out completely eastward within a distance of 2 to 5 miles (3 to 8 km) because it does not reappear as a mappable unit on the east limb of the Frontenac syncline. Hatch found here only two small outcrops of marble that might represent such a reappearance. If the Waits River does occupy this position between the Gile Mountain and the Kidderville, then the Kidderville (–Frontenac) volcanic rocks must underlie the Waits River in the anticline to the west. Alternatively, if the synclinal axis lies to the east of the thin greenstone bands in the Gile Mountain, these bands may be the greatly thinned equivalent of the Kidderville–Frontenac volcanic rocks. In this case the Waits River would be beneath all the volcanic units but would still have to pinch out somewhere west of the unconformity at the base of the Seboomook and Kidderville Formations.

In the Second Lake, Moose Bog, and Indian Stream quadrangles, minor fold axes in the Frontenac Formation plunge gently northeast or southwest, in contrast to the steep plunges that typify the pre-Silurian rocks. This contrast is taken to mean that the pre-Silurian rocks, having become deformed, consolidated, recrystallized, and strengthened during the Taconic orogeny, rose as the relatively rigid, coherent core of the Boundary Mountains anticlinorium whereas the weaker, perhaps poorly consolidated Devonian strata flowed passively off the mass rising to the southeast. Directions of movement of drag folds (Green, in press) support this interpretation. In the Dixville and Oquossoc quadrangles late-stage minor folds, unrelated to any visible lithic trends, deform the schistosity of the Siluro-Devonian rocks.

No faults of regional significance have been found in this area, and although the intense folding was certainly accompanied by widespread local shearing, large displacement along discrete fault surfaces does not appear to have been a major means of deformation.

METAMORPHISM

The stratified rocks of the entire area have undergone regional metamorphism; the metamorphic grade increases generally from the chlorite zone in the northwest to the upper sillimanite zone (Guidotti, 1966) in the southeast (Fig. 19–5). Andalusite and staurolite are characteristic in the middle-grade pelitic rocks, implying a moderate to shallow depth of metamorphism (Green, 1963). The Acadian granitic plutons are surrounded by narrow zones of sillimanite-grade rocks in the regional staurolite zone, and by andalusite-cordierite hornfelses and schists in the lower-grade regional zones. In the north-central part of the area, aphanitic hornfelses without recognizable porphyroblasts have been produced in the pelitic rocks. Garnet and stauro-

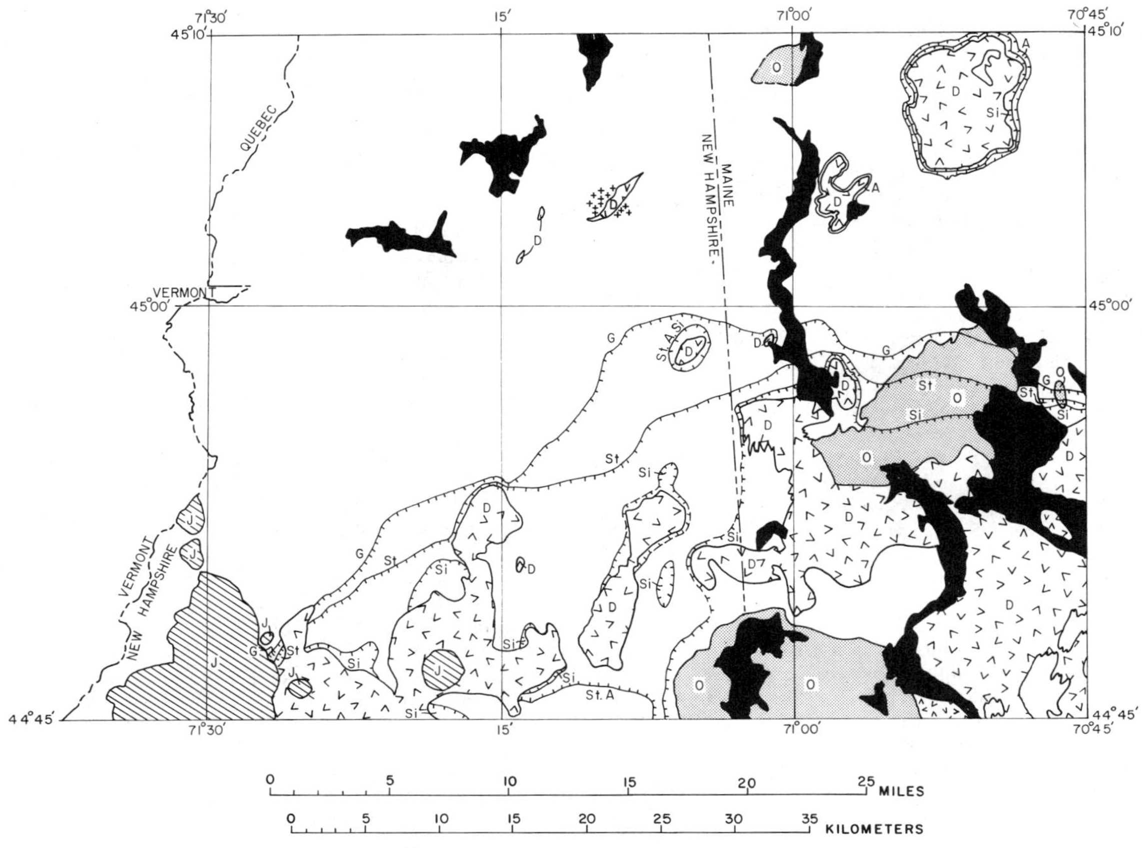

FIGURE 19–5. Metamorphic zones and intrusive plutons. Plutons: O = Ordovician (Highlandcroft magma series, Taconic); D = Devonian (New Hampshire magma series, Acadian); J. = Jurassic (White Mountain magma series). Isograds (hachures point toward higher zone): G = garnet; St = staurolite; A = andalusite; Si = sillimanite; + pattern = aphanitic hornfels. Biotite isograd is not clearly defined over most of the area, and is not shown. Black areas = lakes.

lite porphyroblasts have been rotated in some of the regionally metamorphosed schists. The regional metamorphism is thus interpreted as having occurred during the period of Acadian intrusion and folding, even though a few K/Ar radiometric studies of micas from schists of the Errol quadrangle have given a spread of Carboniferous ages (Hurley and others, 1958). Unfortunately such determinations are so fraught with hazards of interpretation, especially in an orogenic area such as this, as to greatly limit their value (see Lyons and Faul, this volume). Most of the contact-metamorphic aureoles and some of the regional metamorphic minerals have undergone some retrograde alteration.

PROBLEMS FOR STUDY

Several geologic problems remain to be solved in this area, most of which involve both structure and stratigraphy. Probably of greatest importance is the solution, by study of primary depositional features, of the question of the existence of the Frontenac syncline. The relationships between the various facies of the Albee–Aziscohos complex in the eastern third of the area need to be mapped in more detail in order to refine the structure in these older rocks. The structure and rock types of the "Arnold River Formation" in the northeast corner of the area are almost entirely unknown. Further detailed mapping might clarify the stratigraphic relations of the calcareous rocks assigned to the Waits River Formation, and of the various subunits of the Seboomook–Kidderville sequence. Radiometric study is needed before firm conclusions can be drawn on the ages of some of the intrusive rocks, of metamorphism, and of subsequent events that might have affected parent–daughter isotope ratios. Finally, the search for fossils should continue, particularly in the lower-grade rocks.

REFERENCES

Albee, A. L., 1961, Boundary Mountain Anticlinorium, west-central Maine and northern New Hampshire: U. S. Geol. Survey Prof. Paper 424–C, p. C51–54

Berry, W. B. N., 1960, Correlation of Ordovician graptolite-bearing sequences: Internat. Geol. Cong., 21st, Copenhagen, pt. 7, p. 97–109

Billings, M. P., 1937, Regional metamorphism of the Littleton-Moosilauke area, New Hampshire: Geol. Soc. America Bull., vol. 48, p. 463–566

————, 1956, The Geology of New Hampshire, part II—Bedrock geology: N. H. State Planning and Development Comm., Concord, 203 p. (includes map publ. 1955)

Boucot, A. J., Jr., 1961, Stratigraphy of the Moose River synclinorium, Maine: U. S. Geol. Survey Bull. 1111–E, p. 153–188

————, Griscom, Andrew, and Allingham, J. W., 1964, Geologic and aero-magnetic map of northern Maine: U. S. Geol. Survey Geophys. Inv. Map GP–312

Cady, W. M., 1960, Stratigraphic and geotectonic relationships in northern Vermont and southern Quebec: Geol. Soc. America Bull., vol. 71, p. 531–576

Cooke, H. C., 1937, Thetford, Disraeli, and eastern half of Warwick map-areas, Quebec: Canada Dept. Mines and Resources, Geol. Survey Mem. 211, 160 p.

————, 1950, Geology of a southwestern part of the Eastern Townships of Quebec: Canada Geol. Survey Mem. 257, 142 p.

————, 1957, Coaticook-Malvina area: Quebec Dept. Mines Geol. Rept. 69, 37 p.

Doll, C. G., 1951, Geology of the Memphremagog quadrangle and southeastern portion of the Irasburg quadrangle, Vermont: Vt. Geol. Survey Bull. no. 3, 113 p.

————, Cady, W. M., Thompson, J. B., Jr., and Billings, M. P., *compilers* and *editors*, 1961, Centennial geologic map of Vermont: Montpelier, Vermont, Vermont Geol. Survey, scale 1:250,000

Fairbairn, H. W., Faure, Gunter, Hurley, P. H., and Pinson, W. H., 1961, Evidence of the origin and time of separation of magmas of the Monteregian Hills, Quebec, from the development of radiogenic Sr^{87} [Abs.]: Geol. Soc. America Special Paper 68, p. 174

Green, J. C., 1960, Geology of the Errol Quadrangle, New Hampshire—Maine: Ph.D. thesis, Harvard University, 258 p.

————, 1962, Stratigraphy and correlation of the Boundary Mountain anticlinorium, northernmost New Hampshire [Abs.]: Geol. Soc. America Special Paper 73, p. 162

————, 1963, High-level metamorphism of pelitic rocks in northern New Hampshire: Amer. Mineralogist, vol. 48, p. 991–1023

————, 1964, Stratigraphy and structure of the Boundary Mountain anticlinorium in the Errol quadrangle, N. H.–Maine: Geol. Soc. America Special Paper 77, 78 p.

————, in press, Geology of the Connecticut Lakes-Parmachenee area, New Hampshire and Maine: Geol. Soc. America Bull.

Guidotti, C. V., 1966, Variations of the basal spacings of muscovite in sillimanite-bearing pelitic schists of northwestern Maine: Amer. Mineralogist, vol. 51, p. 1778–1786

Harwood, D. S., 1966, Geology of the Cupsuptic quadrangle, Maine: U. S. Geol. Survey Open File Report 850, 259 p.

————, and Berry, W. B. N., 1967, Fossiliferous lower Paleozoic rocks in the Cupsuptic quadrangle, west-central Maine: U. S. Geol. Survey Prof. Paper 575–D, p. D16–D23

Hatch, N. L., Jr., 1963, The geology of the Dixville quadrangle, New Hampshire: New Hampshire Dept. of Resources and Econ. Dev., Bull. 1, 81 p.

Hitchcock, C. H., 1877, Geology of New Hampshire, Pt. 2, Stratigraphical geology: Concord, N. H., 685 p.

Hurley, P. M., Fairbairn, H. W., Pinson, H. W., Winchester, J. W., and Gheith, M. A., 1958, Variations in isotopic abundances of strontium, calcium, and argon and related topics: 6th Ann. Prog. Rept. to U. S. Atomic Commission, Massachusetts Inst. Tech., Dept. Geology and Geophysics, 193 p.

Logan, W. E., 1863, Geology of Canada: Canada Geol. Survey Prog. Rept. to 1863: 983 p.

Marleau, R. A., 1957, Preliminary report on Woburn area, Electoral district of Frontenac: Quebec Dept. Mines, Geol. Surveys Br. Prelim. Rept. 336, 6 p., map

————, 1958, Geology of the Woburn, East Megantic, and Armstrong Areas, Beauce and Frontenac Counties, Quebec: D. Sc. thesis, Laval University, 184 p.

————, 1959, Age relations in the Lake Megantic Range, southern Quebec: Geol. Assoc. Canada Proc., vol. 11, p. 129–139

McGerrigle, H. W., 1935, Mount Megantic area, southeastern Quebec, and its placer-gold deposits: Quebec Bur. Mines Ann. Rept. (for 1934) pt. D, p. 63–104

Moench, R. H., 1966, Relation of S_2 schistosity to metamorphosed clastic dikes, Rangeley-Phillips area, Maine: Geol. Soc. America Bull., vol. 77, p. 1449–1462.

Myers, P. B., Jr., 1964, Geology of the Vermont portion of the Averill quadrangle, Vermont: Vermont Geol. Survey Bull. 27, 69 p.

Naylor, R. S. and Boucot, A. J., 1965, Origin and distribution of rocks of Ludlow age (Late Silurian) in the northern Appalachians: Amer. Jour. Sci., vol. 263, p. 153–169

Perkins, E. H., 1925, Contributions to the geology of Maine, No. 2, pt. 1, The Moose River sandstone and its associated formations: Amer. Jour. Sci., 5th Ser., vol. 10, p. 368–375

U. S. Geological Survey, 1965, Geological Survey Research 1965: U. S. Geol. Survey Prof. Paper 525A, 376 p.

STRATIGRAPHIC AND STRUCTURAL RELATIONSHIPS IN THE INNER, OR SOUTHEASTERN BELT

Chapter 20

Sedimentation, Tectonism, and Plutonism
of the Narragansett Basin Region

ALONZO W. QUINN AND GEORGE
E. MOORE, JR.

INTRODUCTION

THE WORK OF Billings and his associates, most of it in New Hampshire and Vermont, has demonstrated that the orogenic history of the northern Appalachians was exceedingly long and complex. Southern New England was involved in some of the orogenic events that affected the more northern areas, but in addition, it shows evidence of later orogenic events which cannot be demonstrated elsewhere in New England. The Narragansett Basin of Rhode Island and Massachusetts contains the only large mass of known Pennsylvania rocks in New England (Quinn and Oliver, 1962). As a consequence, the rocks of this basin provide the best evidence concerning late Paleozoic activity in the northern part of the Appalachian Mountain system. The Boston Basin contains probable Pennsylvanian rocks, but fossil evidence there is very scanty and not conclusively diagnostic of a Pennsylvanian age (Pollard, 1965). Our paper describes evidence for the earlier events as well as the later history in the Narragansett Basin. None of the evidence concerning earlier history is clearly different from that shown in other parts of New England, but specific correlations cannot yet be made.

Our paper is based mainly on recently published reports on the Rhode Island area, and partly on older reports concerning the Massachusetts area. Most of Rhode Island has been described in recent quadrangle maps and bulletins resulting from a cooperative program between the United States Geological Survey and the State of Rhode Island Development Council. Several reports on Massachusetts geology have been published under a similar program of cooperation between the U. S. Geological Survey and the Massachusetts Department of Public Works. Reports and maps used but not specifically cited elsewhere in this paper include those by Chute (1940, 1950), Hartshorn (1960), Koteff (1964), and Quinn (1963a). Parts of Massachusetts discussed here have not been mapped recently, however, and in Fig. 20–1 these areas are generalized and largely based on older mapping.

GROUPS OF ROCKS

The main groups of rocks in the Narragansett Basin region are: (1) older metamorphic rocks, including the Blackstone Series; (2) older plutonic rocks, which may belong to one or to several plutonic episodes; (3) plutonic and volcanic rocks of intermediate age; (4) Pennsylvanian rocks, mostly clastic sedimentary rocks, of the Narragansett Basin; and (5) young granitic rocks that are intrusive into the Pennsylvanian rocks. Figure 20–1 is a generalized geologic map of the area showing the distribution of these rocks. Modal analyses of the plutonic rocks are given in Table 20–1.

OLDER METAMORPHIC ROCKS

The Blackstone Series of the Blackstone River valley of Rhode Island includes 15,000 to 20,000 feet (4600 to 6100 meters) of quartz-mica schist, quartzite, greenstone, and marble (Emerson, 1917; Quinn, Ray, and Seymour, 1949; Quinn, 1959). These rocks must have resulted from deposition of a large thickness of clay, sand, silt,

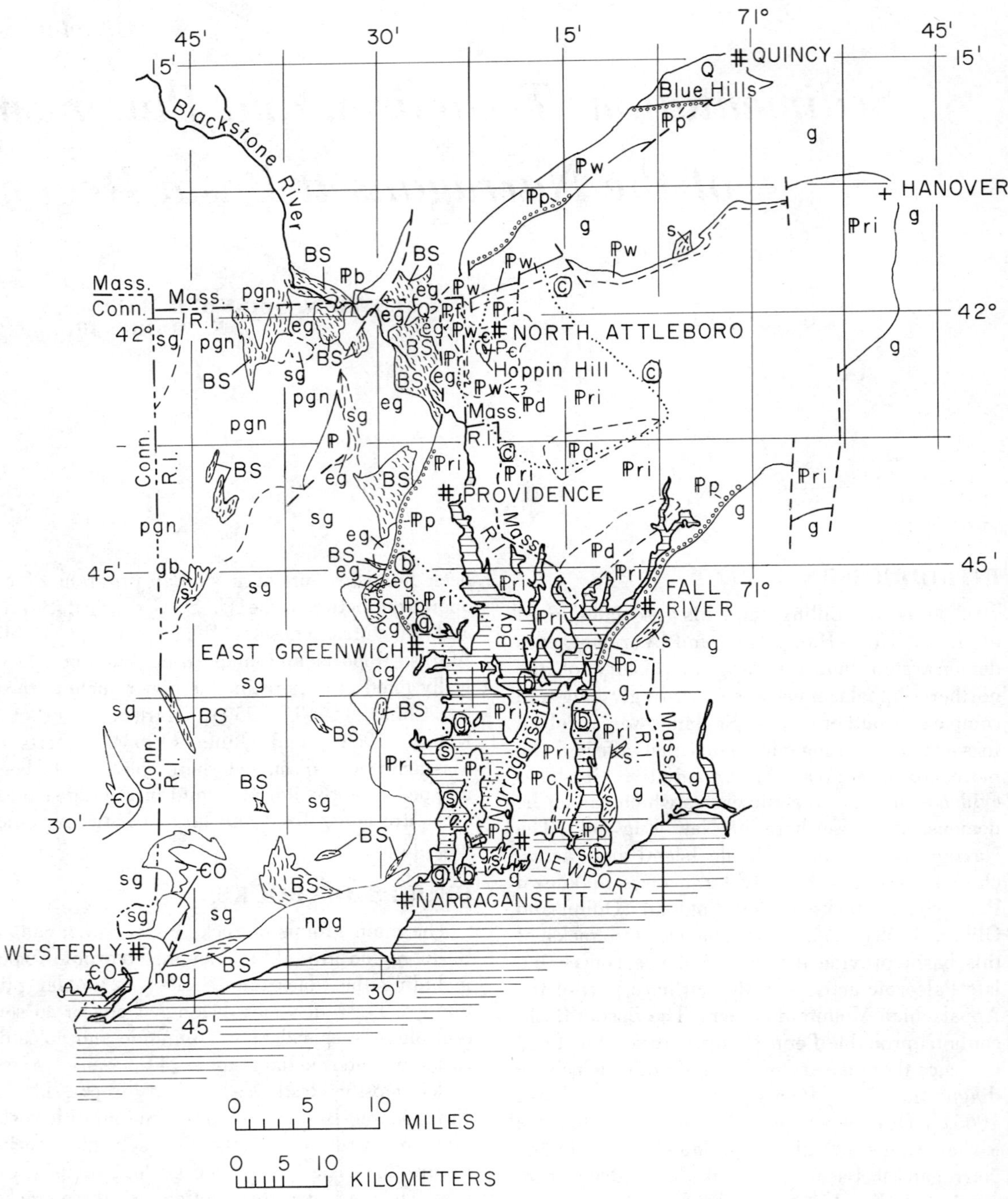

and lime mud. The rocks include layers of pyroclastic material, lava flows, and a few sills; most of these igneous rocks were of basaltic or andesitic composition. The sediments and volcanic rocks together were folded and subjected to regional metamorphism of moderate to high intensity. The Blackstone Series near and to the north of Providence strikes northwest and dips at moderate angles to the northeast. Some of the small bodies southwest of Providence have similar trends (Moore, 1958, 1959). These trends are sharply discordant to structural trends in nearby parts of New England. The several areas of Blackstone Series are separated from each other partly by intrusive granitic rocks and partly by overlying Pennsylvanian rocks. In the Blackstone valley the series includes the Mussey Brook Schist below, a thick quartzite in the middle (formerly correlated with the Westboro Quartzite), and the Sneech Pond Schist at the top

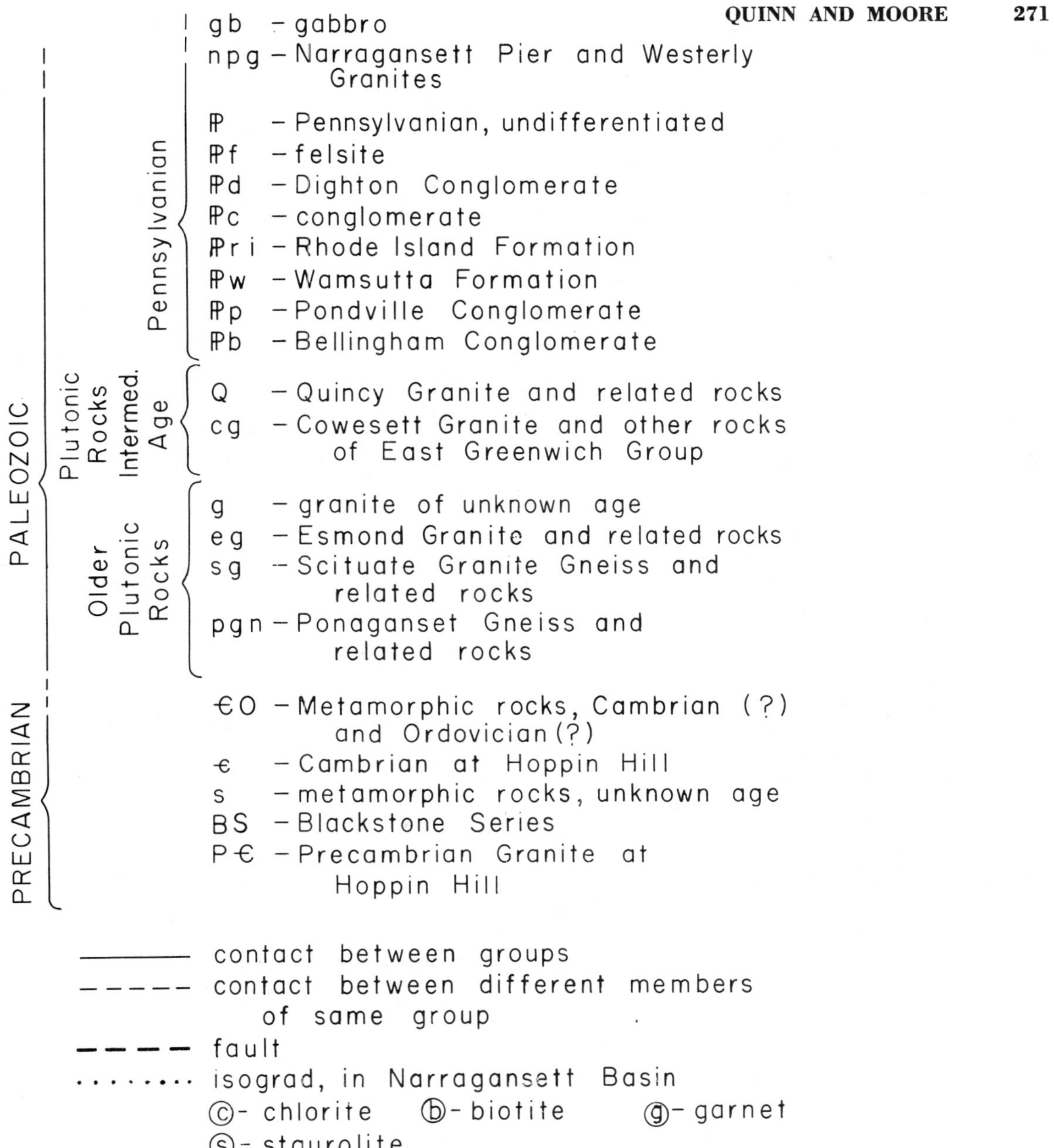

Figure 20–1. Generalized Geologic Map of Narragansett Basin Region; based on references cited in text, and also on Chute (1940, 1950), Hartshorn (1960), Koteff (1964), and Quinn (1963a).

(Quinn et al., 1949). The next area of Blackstone to the south, just west of Providence, includes similar rocks in a similar stratigraphic sequence. However, the smaller areas of metasedimentary rocks south and west of Providence show less similarity, and correlations are less certain.

Age of Blackstone Series

The Blackstone Series has been tentatively as-signed to the Precambrian, but the evidence is indirect and indefinite. Emerson (1917) correlated these rocks with the Marlboro Formation and the Westboro Quartzite of Massachusetts. We have abandoned these correlations because: (a) lithologic similarities are not close, (b) structural relations appear to be different, and (c) intervening granitic intrusives prevent tracing of formations from one area to another.

TABLE 20-1. Selected Modes of Plutonic Rocks (in volume percent)

WESTERLY GRANITE

Qtz	Kf	Plag	Biot	Musc	Acc	
27.5	35.4	31.4	3.2	1.3	1.2	16 sections, Fairbairn et al., 1951, p. 61.
24.3	27.8	41.3	5.5		1.1	6 sections, Quinn, 1943, at Bradford.

NARRAGANSETT PIER GRANITE

Qtz	Kf	Plag	Biot	Acc	
29.1	27.9	34.7	5.9	2.4	Moore, 1959, Watch Hill quad. (in GQ–117).
28.5	33.8	34.3	2.6	0.8	Moore, Carolina–Quonochontaug quad. (unpub.)

QUINCY GRANITE

Qtz	Kf	Rieb	Aeg	
39.7	51.4	6.5	2.4	Quinn, Pawtucket quad. (unpub.)

COWESETT GRANITE

Qtz	Kf	Plag	Biot	Acc	
42.6	52.6	0.7	3.2	0.4	Quinn, East Greenwich quad. (unpub.)

ESMOND GRANITE

Qtz	Kf	Plag	Biot	Acc	
31.7	43.1	18.5	1.2	5.5	Quinn, Pawtucket quad. (unpub.)
36.9	43.0	16.2	1.9	2.0	Quinn, Pawtucket quad. (unpub.)

QUARTZ DIORITE OF ESMOND GROUP

Qtz	Kf	Plag	Biot	Acc	
24.5		59.5	15.6	0.4	Quinn, Pawtucket quad. (unpub.)

TEN ROD GRANITE GNEISS

Qtz	Kf	Plag	Biot	Hb	Acc	
31.8	31.8	30.9	4.1		1.4	Moore, 1963, Coventry Center quad.
31.4	28.8	31.5	4.2	2.8	1.3	Moore, Hope Valley quad. (unpub.)
30.3	28.6	35.0	5.0		1.1	Moore, Hope Valley quad. (unpub.)

SCITUATE GRANITE GNEISS

Qtz	Kf	Plag	Biot	Hb	Musc	Acc	
24.8	45.4	27.2	1.7		0.2	0.7	Quinn, North Scituate quad. (unpub.)
31.7	31.6	30.2	3.7	2.0		0.8	Moore, 1963, Coventry Center quad.
35.3	42.1	15.7	5.5	0.4		1.0	Quinn, Crompton quad. (unpub.)

HOPE VALLEY ALASKITE GNEISS

Qtz	Kf	Plag	Biot	Acc	
36.1	28.3	32.7	1.8	1.0	Moore, Hope Valley quad. (unpub.)
43.0	34.2	19.6	1.4	1.5	Moore, Carolina–Quonochontaug quad. (unpub.)

PONAGANSET GNEISS

Qtz	Kf	Plag	Biot	Hb	Epid	Musc	Acc	
38.1	51.6	9.0	0.8			0.3	0.2	Quinn, 1967, Chepachet quad.
32.7	27.3	24.2	13.4	0.8	0.7		0.9	Quinn, 1967, Chepachet quad.
39.0	4.2	39.5	14.7		0.4		2.2	Moore, Coventry Center quad. (unpub.)

The Blackstone Series in northeastern Rhode Island is more metamorphosed than are the fossiliferous Lower Cambrian rocks at Hoppin Hill near North Attleboro (Dowse, 1950; Shaw, 1950). This suggests that the Blackstone Series is older than Lower Cambrian, but inasmuch as the nearest Blackstone Series rocks are 5 or 6 miles (8 to 10 km) from Hoppin Hill, the evidence is not strong.

A Precambrian or early Paleozoic age is suggested by the long sequence of events preceding the deposition of the Pennsylvanian rocks of the Narragansett Basin, even though these events cannot be assigned to definite geological periods nor can the time involved be stated in terms of years. In the East Greenwich area the Pennsylvanian rocks lie unconformably on the Spencer Hill volcanic rocks and on the Cowesett Granite, both of the East Greenwich Group (Quinn, 1952). The volcanic rocks contain boulders of Esmond Granite. The Esmond Granite is intrusive into the metamorphic rocks of the Blackstone Series. This record implies a sequence of events long enough to include: (1) deposition of 15,000 to 20,000 feet (4600 to 6100 m) of sediments and volcanic material to form the Blackstone Series; (2) deformation and metamorphism of these Blackstone rocks; (3) plutonic intrusion of the Esmond Granite, possibly also other separate intrusive events for other older plutonic rocks; (4) erosion long enough to expose the plutonic Esmond Granite; (5) volcanic activity to form the Spencer Hill volcanic rocks, which contain boulders of Esmond Granite; (6) intrusion of the Cowesett Granite and probably also the Quincy Granite, and (7) erosion long enough to expose the Cowesett Granite and the Quincy Granite. All these events happened before the deposition of the Pennsylvanian sediments of the Narragansett Basin. The length of time required for them probably was great enough to extend back to early Paleozoic or to Precambrian time.

Radiometric age determinations, which should ultimately help solve the question, but are not yet conclusive, do bear on the age of the Blackstone Series. The Northbridge Granite Gneiss of Massachusetts gave a Rb–Sr age of 535 ± 15 m.y. (Moorbath 1962) or 558 ±11 m.y. (Fairbairn et al., 1965). The Ponaganset Gneiss (Quinn, 1967) of Rhode Island is probably correlative with parts of the Northbridge Granite Gneiss, although the intervening area 5 or 6 miles (8 to 10 km) across has not been mapped in detail. The Ponaganset Gneiss is intrusive into the Blackstone Series, so a Cambrian age of the Ponaganset Gneiss would imply an early Cambrian or a Precambrian age of the Blackstone Series. Recent age determinations on the Quincy Granite have some significance for the considerably older Blackstone Series. The Quincy Granite has commonly been assigned to the Mississippian System (Lyons and Faul, this volume), and some radiometric ages agree with this. However, Zartman (in U.S. Geological Survey, 1967, p. A166–A167) obtained four K–Ar ages of 430–457 m.y. on hornblende of the Quincy Granite; he also obtained 9 Rb–Sr whole-rock ages of 340–400 m.y. on the Quincy Granite. The hornblende age would imply an Ordovician or greater age for the Blackstone Series.

These several lines of evidence leave the age of the Blackstone Series uncertain, but we believe that, altogether, the evidence favors a Precambrian age.

Other Older Metamorphic Rocks

Several other patches of metamorphic rocks differ from and are not assigned to the Blackstone Series, although most of these are also surrounded by plutonic rocks. Some of the metamorphic rocks in southwestern Rhode Island and southeastern Connecticut have been assigned to the Cambrian and Ordovician Systems (Feininger, 1965, a and b). There are other patches on the east side of the Narragansett Basin and also near the mouth of Narragansett Bay (Pollock, 1964). These are known only to be older than granitic rocks that are pre-Pennsylvanian.

Altogether, these several masses of older metamorphic rocks indicate a considerable amount of Precambrian or early Paleozoic sedimentation and vulcanism. Most of the rocks have the characteristics of eugeosynclinal deposits, but some especially in the Blackstone Series, are rather extensive clean quartzites. Their structural and metamorphic features resulted from moderate to severe deformation before the formation of the older plutonic rocks.

OLDER PLUTONIC ROCKS

Intrusive into the Blackstone Series and other older metamorphic rocks is a considerable variety of plutonic rocks, mostly in the range from quartz diorite to granite. Attention here will be given to those west of the Narragansett Basin; little is known about those on the east side except that they are different from those on the west side.

The older plutonic rocks west of the Narragansett Basin have been assigned to three main groups. Evidence is not clear as to how many episodes of plutonic activity these three groups represent.

The oldest and most extensive member of the three older groups is the Ponaganset Gneiss that crops out in northwestern Rhode Island (Quinn, 1967). It encloses several lenses of metadiorite and amphibolite that appear to be the earliest and most mafic parts of this plutonic group. Most of the Ponaganset Gneiss is very streaky and gneissic; individual layers are thin and range in composition from quartz diorite to granite. The rock probably was intruded syntectonically, although some of the gneissic structure may have been inherited from the metasomatism of an earlier schist. Some contacts with the Blackstone Series, however, indicate magmatic intrusion of the gneiss into the Blackstone rocks. At a few places, moderately large bodies of this gneiss have the composition of granite and are more massive; they seem to be the youngest parts of this formation. Table 20–1 shows the compositional range of this group as a whole.

The largest part of Rhode Island west of the Narragansett Basin is underlain by the group of gneisses that includes the Scituate Granite Gneiss, the Hope Valley Alaskite Gneiss, the Ten Rod Granite Gneiss, and other smaller units (Moore, 1958, 1959, 1963; Quinn, 1951). As a group they make up a complex body of batholithic proportions that extends westward into Connecticut. These rocks range from granodiorite to granite; probably most of the rock is quartz monzonite. The Scituate and the Hope Valley are generally even grained, though locally porphyritic, and are characterized by strong lineation. The lineation of the Scituate is shown by large oval splotches of biotite, and that of the Hope Valley by elongate rods of quartz. Much of the Scituate, and some of the Hope Valley, have well developed foliation. Most of the Ten Rod is porphyritic, though a non-porphyritic facies has been mapped locally. The Ten Rod is generally foliated and lineated; the lineation is shown by biotite streaks and by the phenocrysts, many of which have been partly granulated. Relationships between these rocks and the country rock of Blackstone and other metamorphic rocks, as seen in numerous places, indicate that the various units of the Scituate group were intruded as a liquid or partly liquid magma. The Scituate and the Hope Valley grade into each other and thus are of the same age. The Ten Rod is younger than the other two, for dikes of the Ten Rod cut the

Scituate in western Rhode Island. The generally strong foliation in this group, interpreted as a primary flow structure, suggests syntectonic intrusion. Commonly this flow structure is discordant to the schistosity of the older metamorphic rocks.

A second foliation, metamorphic in origin, is present in at least some of the rocks of this group. In much of the Scituate and the Hope Valley of the western part of Rhode Island the metamorphic foliation transects the primary flow structure at a high angle, though locally the metamorphic foliation parallels the primary flow structure. The relationships shown by the Hope Valley and by the Ten Rod in the southeastern part of their outcrop area were considered by Power (1959) and Nichols (1956) to indicate an origin by feldspathization of older schists. It might well be that these rocks resulted from granitization in the southeast, but from intrusion of magma elsewhere.

The third group within the older plutonic rocks is in northeastern Rhode Island, and it includes several bodies of quartz diorite, the Grant Mills Granodiorite, and the Esmond Granite (Quinn et al., 1949). Generally these rocks are less foliated than are the other older plutonic rocks, but some parts of this northeastern group do have a fairly well-developed foliation. A common characteristic of these rocks, especially of the Esmond Granite, is that the plagioclase, originally more calcic, has been altered to albite, with resultant formation of small crystals of epidote and muscovite within the plagioclase. Where contacts with the Blackstone Series are exposed, the structural relations commonly show evidence of magmatic intrusion.

A Cambrian age for the Ponaganset Gneiss is suggested by radiometric determinations (Moorbath et al., 1962). The other members of the older plutonic group may be correlative with the Oliverian or with the New Hampshire Plutonic Series of New Hampshire (Billings, 1956), but such a correlation cannot be established yet.

PLUTONIC AND VOLCANIC ROCKS OF INTERMEDIATE AGE

The older plutonic rocks resulted from rather large-scale, deep-seated, partly syntectonic igneous activity. After they were exposed by erosion, more igneous activity occurred. This activity differed from the earlier activity in several ways: (a) it involved some extrusive activity; (b) the intrusions were smaller; and (c) the compositions of the rocks were somewhat more alkalic. The rocks of intermediate age discussed in the present paper include the East Greenwich Group near East Greenwich (Quinn, 1952) and a few small intrusives of Quincy Granite and related granite porphyry in northeastern Rhode Island (Quinn et al., 1949).

East Greenwich Group

The East Greenwich Group includes the Spencer Hill volcanics (felsic flows, volcanic breccias, and volcanic conglomerates) and certain younger intrusive rocks (small masses of fine-grained granitic rocks, a granite porphyry, a larger mass of Cowesett Granite, and a related perthitic granite). The smaller intrusives are characterized by micropegmatite; they contain abundant microperthite and a little plagioclase in separate grains (hypersolvus rocks). Riebeckite is present in a very few places and purple fluorite is common as an accessory mineral in the igneous rocks and as a vein mineral. The layered rocks were tilted to steep angles and the whole mass was exposed by erosion before the deposition of the overlying Pennsylvanian rocks of the Narragansett Basin. Boulders of Esmond granite are contained in the volcanic conglomerate of the East Greenwich Group.

Quincy Granite

Small intrusives of Quincy Granite and a closely related granite porphyry crop out in northeastern Rhode Island. Larger bodies of these and related rocks (including the Blue Hills Granite Porphyry) are exposed in the Blue Hills and at Quincy, Massachusetts. The somewhat alkalic Quincy Granite is characterized by microperthite, minor separate plagioclase (hypersolvus granite), riebeckite, aegirite, and accessory fluorite. The rocks at East Greenwich are similar in composition although very little riebeckite or aegirite has been found there. The Quincy granite also appears to belong to a group that includes felsic volcanic rocks and some variety of small to moderately large intrusives.

The age of the Quincy Granite and related rocks, not yet established, has important regional implications. The "giant conglomerate" on the south slope of the Blue Hills of Massachusetts contains boulders of the Blue Hills Granite Porphyry and shows that this porphyry was weathered to boulders prior to the deposition of the Pennsylvanian conglomerate there (Chute, p. 91–114 in Skehan, 1964). This seems to establish a minimum age. The maximum age is not so firmly established. LaForge (1932, p. 35) wrote, "stocks and dikes of the Quincy Granite in Essex County cut the rocks of the Lynn volcanic complex, which is regarded as probably of early Devonian age." This early Devonian age was based on lithic similarity of the Lynn volcanic rocks to the Newbury volcanic rocks, which contain fossils of early Devonian (Toulmin, 1964, p. 17) or late Silurian age. Norman Cuppels (oral communication, 1967; also U. S. Geological Survey, 1966, p. A73) has discovered latest Silurian fossils in the Newbury volcanic rocks in the Georgetown quadrangle of northeastern Massachusetts and he confirms the similarity of the two sets of volcanic rocks. A key part of this age assignment is the lithic correlation of the fossiliferous with non-fossiliferous volcanic rocks. On the basis of these relations a Mississippian age has been given to the Quincy. A Rb–Sr whole-rock age of 325 ±15 m.y. (Bottino, 1963) agrees with that assignment. However, as previously mentioned, work by Zartman (written commun. and U. S. Geological Survey, 1967, p. A166–A167) indicates a minimum age of Ordovician for the intrusion of the Quincy. Recently Chute (1966, p. 28) concluded that the Quincy Granite and the Blue Hill Granite Porphyry are not so nearly contemporaneous as was formerly believed, but that the Quincy is distinctly older.

Part of the uncertainty here has to do with lithic correlation. Some of these doubtful correlations are between the Quincy Granite and granitic rocks north of Boston (Toulmin, 1961, 1964); others involve the volcanic rocks. At no place has the type Quincy Granite been seen to be intrusive into fossiliferous rocks.

PENNSYLVANIAN ROCKS OF NARRAGANSETT BASIN

The Narragansett Basin (Shaler et al. 1899; Quinn and Oliver, 1962) is a large synclinal mass extending northward from the mouth of Narragansett Bay to the northeast corner of Rhode Island, and thence northeastward to the vicinity of Hanover, Massachusetts. It is approximately 55 miles (90 km) long and, between Providence and Fall River (Fig. 20–1), about 18 miles (30 km) wide. A Pennsylvanian age of these rocks seems well established by abundant plant fossils that have been found in several places and apparently at different stratigraphic positions in the basin (Knox, 1944). Many of the rocks of the basin do not have fossils, so their ages are not definitely known. Similar rocks in two smaller basins about 5 miles (8 km) west of the main basin are correlated with those of the Narragansett Basin, although fossils have not been found in these smaller basins. The Pennsylvanian rocks of the basins lie with sharp discordance on the older metamorphic and plutonic rocks on both sides, so it is clear that the older rocks were folded and much eroded before the deposition of the Pennsylvanian rocks.

The rocks within the basin are predominantly clastic sedimentary rocks, chiefly coarse conglomerate, sandstone, lithic graywacke, siltstone, and shale. Several coal beds are also included. A rhyolite flow and two or three sheets of basalt near the northwest corner of the basin are exposed. Coarse clastic texture, irregular and discontinuous bedding, cross-bedding, fossils of land plants, and coal beds indicate that all or most of these rocks are non-marine. The coarse texture and the presence of unweathered feldspars and lithic fragments suggest that these sediments were washed into a lowland surrounded by rugged topography. Changes in texture, orientation of cross-bedding, and orientation of calamite stems, at different places in the basin, all suggest that the sediments were derived from the northeast (Towe, 1959). Only a few boulders in basal conglomerates can be identified as having come from nearby sources.

The rocks of the basin have been divided stratigraphically into the discontinuous Pondville Conglomerate at the base, the Wamsutta Formation, the very thick Rhode Island Formation, and the uppermost very coarse Dighton Conglomerate. The Wamsutta Formation, in the northwest corner of the basin, is mostly red. All the other units are gray to black. Total thickness has been estimated to be 12,000 feet (3700 m) (Shaler et al., 1899, p. 208–210) but because of sparse exposure and possible structural complications, the possibility of error is large. A thickness of 2200 feet (670 m) measured in a tunnel in Pawtucket, R. I. (Quinn, 1959) apparently represents only a small part of the whole.

The broadly synclinal structure of the Narragansett Basin is well established. Many smaller folds and some faults are well known, but the lack of traceable stratigraphic units makes it impossible to determine much of the structural detail within the basin.

The deformation of the Pennsylvanian rocks was accompanied by some metamorphism. The rocks in the northen part of the basin are essentially not metamorphosed, although they are firmly indurated. To the south the rocks are progressively more metamorphosed (Quinn and Glass, 1958), as is shown by the isograds on the map (Fig. 20–1). The pattern of isograds suggests that the rocks at the margins, which were more deeply buried, were most intensely metamorphosed. Slaty cleavage or schistosity is well-developed in the argillaceous rocks, and fracture cleavage in many of the coarser clastic rocks. Numerous outcrops show slaty cleavage parallel to the axial planes of minor folds. The deformation and metamorphism of these rocks also resulted in the elongation of pebbles and boulders in conglomerate. The stretched pebbles in the vicinity of Newport are elongated parallel to the axes of the folds. The ratio of dimensions of pebbles at Purgatory Chasm, a well-known locality east of Newport, is 1.0:2.4:5.7 and in the more northerly of the two subsidiary basins west of North Attleboro, the ratio is 1.0:2.5:13.5 (Agron, p. 22, and Hall, p. 54, *in* Quinn, 1963b). The rapid increase in degree of metamorphism in the southwest part of the basin, though accompanied by an increase in amount of deformation, may be at least in part related to the intrusion of the Narragansett Pier Granite.

Although this late Paleozoic deformation and metamorphism is clearly demonstrated only in the Narragansett Basin, nearby areas of older rocks probably were also affected. Lundgren (1966, a and b) suggests late Paleozoic metamorphism in Connecticut, and Brookins (1967) cites evidence for such late Paleozoic effects in Massachusetts. A recent age determination by B. J. Giletti (oral Communication, 1967) has a similar implication. Giletti and Quinn collected parts of veins that cut the Quincy Granite and that seem to be closely associated with the Quincy. These veins, which contain quartz, biotite, purple fluorite, and danalite are near the north border of Rhode Island and are about 1500 feet (460 m) west of the edge of the Narragansett Basin. A K–Ar age determination on the biotite gave 230 ± 8 m.y., implying late Paleozoic metamorphism of the vein.

YOUNG GRANITIC ROCKS

Near the mouth of Narragansett Bay and on the west margin of the Narragansett Basin are pegmatites and granitic rocks that intruded the Pennsylvanian rocks of the Narragansett Basin (Nichols, 1956). The granitic rocks are included in the Narragansett Pier Granite, which is abundantly exposed along the shore at Narragansett (formerly Narragansett Pier). A few small exposures are scattered along the south shore of Rhode Island (Moore, 1959) and more extensive exposures are present in the Westerly, R. I. area (Feininger, 1965a). An end moraine covers much of the south shore of Rhode Island, so exposures are not spaced closely enough to show whether the Narragansett Pier Granite forms a continuous mass, as shown on the map

(Fig. 20–1), or whether it is in separate small intrusives. The typical rock of this unit is a nearly massive reddish-pink medium-grained quartz monzonite. It is mostly even-grained, but locally porphyritic. Where locally contaminated by inclusions of schist, the rock is somewhat streaky and shows foliation and lineation. Irregular patches of granite having pegmatitic or aplitic texture are not uncommon. The presence of rotated schistose inclusions of the Rhode Island Formation, the general massiveness, and a lack of granulation indicate that the Narragansett Pier Granite is a post-tectonic intrusive, younger than the folding and metamorphism of the Pennsylvanian rocks.

In the Westerly area, the Narragansett Pier Granite and other rocks are cut by small intrusives of the Westerly Granite, mostly as south-dipping dikes. Several of these are shown on Figure 20–2. Similarity of composition and close geographic association suggest that the Narragansett Pier Granite and the Westerly Granite are genetically related to each other and that they were formed within a short time interval. The larger dikes of Westerly Granite are gray to light pink fine-grained quartz monzonite, so homogeneous that some of this rock from a quarry near Westerly was selected as one of the standards (G-1) used for analytic study of silicate rocks (Fairbairn et al., 1951). Though most of the rock is quartz monzonite, rocks from different bodies range from granite to granodiorite. Faint foliation parallel to the walls can be seen near the borders of some of the dikes, and the long dike southeast of Bradford shows a concentration of heavy minerals toward its base (Quinn, 1943; Hall and Eckelmann, 1961).

SUMMARY AND CONCLUSIONS

The development of the New England part of the Appalachians has been long and complex. The sequence of geosynclinal sedimentation, deformation and metamorphism, plutonic intrusion, and uplift has been partly or wholly repeated more than once. In this more southerly part of New England, the first recorded sequence occurred in Precambrian or early Paleozoic time, as shown by the Blackstone Series and the older plutonic rocks.

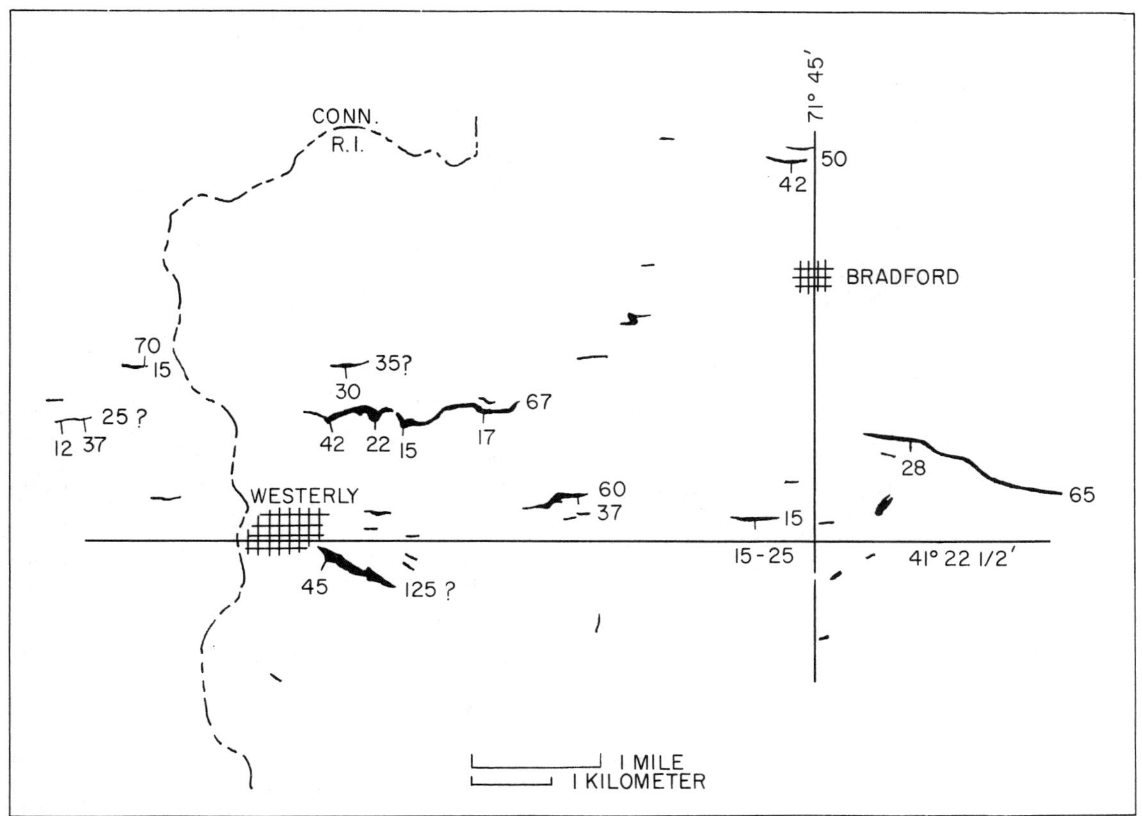

FIGURE 20–2. Bodies of Westerly Granite in the Westerly–Bradford area, near the southwest corner of Rhode Island. Thickness in feet and dip indicated where known. Modified from Feininger (*in* Quinn et al., 1963, p. 52).

The next part of the diastrophism here is recorded by the volcanic and plutonic rocks of the intermediate group. Apparently there was no large amount of geosynclinal sedimentation in this intermediate episode, but there was volcanism and plutonism. This was followed by another uplift and deep erosion.

While the topography was still rugged, a great thickness of clastic sediments was washed into a lowland area, now preserved as the Pennsylvanian rocks of the Narragansett Basin. A small amount of volcanic activity accompanied this sedimentation. The sedimentation was followed, probably in Pennsylvanian or Permian time, by deformation, metamorphism, and plutonic activity, the last resulting in the formation of the Narragansett Pier Granite and the Westerly Granite. Whereas late Paleozoic folding in much of the Appalachians was not closely associated with plutonism, southern Rhode Island does give clear evidence of such an association.

Altogether, the evidence in the Narragansett Basin region shows the same complexity, the repetitions of events, and the absence of fixed sequences that have become apparent in other mountain ranges (Gilluly, 1966).

REFERENCES

Billings, M. P., 1956, Bedrock geology, Pt. II of Geology of New Hampshire: N. H. State Plan. Devel. Comm., 203 p.

Bottino, M. L., 1963, Whole-rock Rb-Sr studies of volcanics and some related granites: 11th Ann. Prog. Rept. to U. S. Atomic Energy Commission, Massachusetts Inst. Tech., Dept. Geology and Geophysics, p. 65–84

Brookins, D. G., 1967, Rb-Sr age evidence for Permian metamorphism of the Monson Gneiss, west central Massachusetts: Geochim. et Cosmochim. Acta, v. 31, p. 281–283

Chute, N. E., 1940, Preliminary report on the geology of the Blue Hills quadrangle, Massachusetts: U. S. Geol. Survey and Massachusetts Dept. Public Works Co-op. Proj., Bull. 1, 52 p.

————, 1950, Bedrock geology of the Brockton quadrangle, Massachusetts: U. S. Geol. Survey Geol. Quad. Map GQ-5

————, 1966, Geology of the Norwood quadrangle Norfolk and Suffolk Counties, Massachusetts: U. S. Geol. Survey Bull. 1163–B, 78 p.

Dowse, A. M., 1950, New evidence on the Cambrian contact at Hoppin Hill, North Attleboro, Massachusetts: Am. Jour. Sci., v. 248, p. 95–99

Emerson, B. K., 1917, Geology of Massachusetts and Rhode Island: U. S. Geol. Survey Bull. 597, 289 p.

Fairbairn, H. W., Schlecht, W. G., Stevens, R. E., Dennen, W. H., Ahrens, L. H., and Chayes, Felix, 1951, A cooperative investigation of precision and accuracy in chemical, spectrochemical, and modal analysis of silicate rocks: U. S. Geol. Survey Bull. 980, 71 p.

————, Moorbath, S., Ramo, A. O., Pinson, W. H., Jr., and Hurley, P. M., 1965, Rb-Sr age of granitic rocks of southeastern Massachusetts and the age of the Lower Cambrian at Hoppin Hill: 13th Ann. Prog. Rept. to U. S. Atomic Energy Commission, Massachusetts Inst. Tech., Dept. Geology and Geophysics, p. 3–10

Feininger, Tomas, 1965a, Bedrock geologic map of the Ashaway quadrangle, Connecticut-Rhode Island: U. S. Geol. Survey Geol. Quad. Map GQ-403

————, 1965b, Bedrock geologic map of the Voluntown quadrangle, New London County, Connecticut, and Kent and Washington Counties, Rhode Island: U. S. Geol. Survey Geol. Quad. Map GQ-436

Gilluly, James, 1966, Orogeny and geochronology: Am. Jour. Sci. v. 264, p. 97–111

Hall, B. A. and Eckelmann, F. D., 1961, Significance of variations in abundance of zircon and statistical parameters of zircon populations in a granodiorite dike, Bradford, Rhode Island: Am. Jour. Sci., v. 259, p. 622–634

Hartshorn, J. H., 1960, Geology of the Bridgewater quadrangle, Massachusetts: U. S. Geol. Survey Geol. Quad. Map GQ-127

Knox, A. S., 1944, A Carboniferous flora from the Wamsutta formation of southeastern Massachusetts: Am. Jour. Sci., v. 242, p. 130–138

Koteff, Carl, 1964, Geology of the Assawompset Pond quadrangle, Massachusetts: U. S. Geol. Survey Geol. Quad. Map GQ-265

LaForge, Laurence, 1932, Geology of the Boston area, Massachusetts: U. S. Geol. Survey, Bull. 839, 105 p.

Lundgren, L. W., Jr., 1966a, Late Paleozoic metamorphism in southeastern Connecticut [abs.]: Geol. Soc. America, Northeastern Sect., Program, 1966 Ann. Mtg., Philadelphia, p. 30–31

————, 1966b, Muscovite reactions and partial melting in southeastern Connecticut: Jour. Petrology, v. 7, p. 421–453

Moorbath, S., 1962, Rb-Sr investigation of the Northbridge Granite Gneiss, Massachusetts: 10th Ann. Prog. Rept. to U. S. Atomic Energy Commission, Massachusetts Inst. Tech., Dept. Geology and Geophysics, p. 7–8.

Moore, G. E., Jr., 1958, Bedrock geology of the Hope Valley qudrangle, Rhode Island: U. S. Geol. Survey Geol. Quad Map GQ-105

————, 1959, Bedrock geology of the Carolina and Quonochontaug quadrangles, Rhode Island: U. S. Geol. Survey Geol. Quad Map GQ-117

————, 1963, Bedrock geology of the Coventry Center quadrangle, Rhode Island: U. S. Geol. Survey Bull. 1158A, 24 p.

Nichols, D. R., 1956, Bedrock geology of the Narragansett Pier quadrangle, Rhode Island: U. S. Geol. Survey Geol. Quad Map GQ-91

Pollard, Melvin, 1965, Age, origin, and structure of the post-Cambrian Boston strata, Massachusetts: Geol. Soc. America Bull., v. 76, p. 1065–1068

Pollock, S. J., 1964, Bedrock geology of the Tiverton quadrangle, Rhode Island-Massachusetts: U. S. Geol. Survey Bull. 1158–D, 16 p.

Power, W. R., Jr., 1959, Bedrock geology of the Slocum quadrangle, Rhode Island: U. S. Geol. Survey Geol. Quad. Map GQ-114

Quinn, A. W., 1943, Settling of heavy minerals in a granodiorite dike at Bradford, Rhode Island: Am. Mineralogist, v. 28, p. 272–281

————, 1951, Bedrock geology of the North Scituate quadrangle, Rhode Island: U. S. Geol. Survey Geol. Quad Map GQ-13

__________, 1952, Bedrock geology of the East Greenwich quadrangle, Rhode Island: U. S. Geol. Survey Geol. Quad Map GQ–17

__________, 1959, Bedrock geology of the Providence quadrangle, Rhode Island: U. S. Geol. Survey Geol. Quad Map GQ–118

__________, 1963a, Bedrock geology of the Crompton quadrangle, Rhode Island: U. S. Geol. Survey Bull. 1158B, 17 p.

__________, ed., 1963b, New England Intercollegiate Geol. Conf. Guidebook, 55th, Ann. Mtg., Providence, R. I., 56 p.

__________, 1967, Bedrock geology of the Chepachet quadrangle, Providence County, Rhode Island: U. S. Geol. Survey Bull. 1241G, 26 p.

__________ and Glass, H. D., 1958, Rank of coal and metamorphic grade of rock of the Narragansett Basin of Rhode Island: Econ. Geology, v. 53, p. 563–576

__________ and Oliver, W. A., Jr., 1962, Pennsylvanian rocks in New England: in Branson, C. C., ed., Pennsylvanian System in the United States, p. 60–73

__________, Ray, R. G., and Seymour, W. L., 1949, Bedrock geology of the Pawtucket quadrangle, Rhode Island-Massachusetts: U. S. Geol. Survey Geol. Quad. Map GQ–1

Shaler, N. S., Woodworth, J. B., and Foerste, A. F., 1899, Geology of the Narragansett Basin: U. S. Geol. Survey Mon. 33, 402 p.

Shaw, A. B., 1950, A revision of several early Cambrian trilobites from eastern Massachusetts: Jour. Paleontology, v. 24, p. 577–590

Skehan, J. W., S. J., ed., 1964. Boston area and vicinity: New England Intercollegiate Geol. Conf. Guidebook, 56th Ann. Mtg., Boston 120 p.

Toulmin, Priestly, 3rd, 1961, Geological significance of lead-alpha and isotopic age determinations of "alkalic" rocks of New England: Geol. Soc America Bull., v. 72, p. 775–779

__________, 1964, Bedrock geology of the Salem quadrangle and vicinity, Massachusetts: U. S. Geol. Survey Bull. 1163–A, 79 p.

Towe, K. M., 1959, Petrology and source of sediments in the Narragansett Basin of Rhode Island and Massachusetts: Jour. Sed. Petrology, v. 29, p. 503–512

U. S. Geological Survey, 1966, Geological Survey Research in 1965: U. S. Geol. Survey, Prof. Paper 525–A, 376 p.

U. S. Geological Survey, 1967, Geological Survey Research 1967: U. S. Geol. Survey, Prof. Paper 575–A, 377 p.

Chapter 21

Fracture Tectonics of Southeastern New England as Illustrated by the Wachusett-Marlborough Tunnel, East-Central Massachusetts

JAMES W. SKEHAN

GENERAL INTRODUCTION

IN 1960 AN unusual opportunity to study the fracture tectonics of southeastern New England was provided when the Metropolitan District Commission (M.D.C.) of the Commonwealth of Massachusetts drove a water tunnel, the Wachusett–Marlborough Tunnel, between two of the largest-scale fault zones recognized to date in this region, the Clinton–Newbury and the Bloody Bluff faults (Fig. 21–1). Crosby (1899) recognized both the fact of eastward thrust motions in the Clinton area and the intensely faulted nature of eastern Massachusetts as a whole. Later studies by Emerson (1917) and Hansen (1956) of the region traversed by the tunnel indicated that the area was of great structural complexity.

The Wachusett–Marlborough Tunnel, 42,013 feet (12,604 m) long, is 10 miles (16 km) northeast of Worcester (Fig. 21–1) and is driven from Wachusett Reservoir, Clinton, in a direction S. 42°36′48″ E. into Marlborough Township.

John Haller, Philip B. King, Randolph J. Martin III, George L. Snyder, and J. Tuzo Wilson provided stimulating written and oral discussions of the structures of this region. Linda S. Enggren and Nancy W. Hellier furnished valuable assistance in the preparation of this manuscript. The laboratory aspects of the research were funded by the National Science Foundation under Grant number GP–4969.

GEOLOGIC SETTING

The Wachusett–Marlborough Tunnel penetrates a highly folded and faulted eugeosynclinal sequence. The thrust block in which this tunnel is driven is bounded on the west by the Merrimack synclinorium of east-central Massachusetts and New Hampshire. The area to the east of the synclinorium consists of imbricate thrust sheets, the uppermost of the three recognized to date being the Wachusett–Marlborough and the lowermost the Boston Basin fault blocks. Between these lies an unnamed fault block.

Figure 21–3 is a geologic map compiled from Emerson (1917), Crosby (1899), Cuppels *in* Skehan (1964), and Bell (1967) and modified by the author as a result of preliminary surface mapping near the tunnel and Wachusett Reservoir. Emerson has mapped a conspicuous lithologic contact in the Newbury area between the Newburyport complex to the north and the Salem Gabbro-diorite to the south; in the Lowell area, between the Dracut Diorite and Merrimack Quartzite to the north and the Brimfield Schist and Andover Granite to the south; in the Ayer and Littleton area, between the Ayer Granodiorite to the north and the Brimfield Schist to the south; and in the Clinton area, between the Oakdale and Worcester Formations, intruded by the Ayer Granodiorite, and a second set of similar units. This line of Emerson's marks a boundary separating units of

281

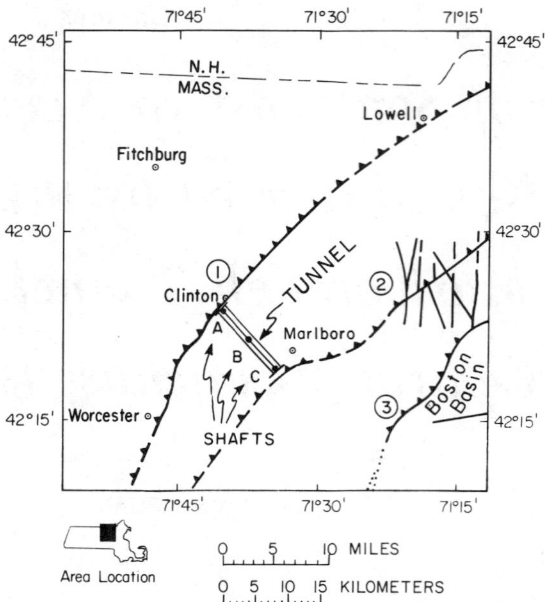

FIGURE 21–1. Index map showing the Wachusett–Marlborough Tunnel in relation to nearby cities and chief faults. Inset map shows the regional setting of the area. References: (1) Crosby, 1899; Skehan, 1967 a and b; (2) Cuppels, 1961, *in* Skehan, 1964; (3) Bell, 1948, 1967; Billings, 1929; LaForge, 1932.

strongly contrasting orientations. The formations to the west and north strike north to north-northeast; those southeast of this contact strike northeast. In the light of both subsurface and surface mapping, it seems clear that in the Clinton area the master fault of the region, the Clinton–Newbury fault, occupies the zone mapped in detail by Crosby (1899) and recognized by him as a zone of easterly thrust faulting.

Regional geologic considerations, especially those portrayed by Emerson (1917), as well as the development of the Clinton–Newbury fault in a prominent strike valley, suggest that this master fault zone is more important than, but closely related to, the three subsidiary fault zones described in detail in this paper (Fig. 21–4).

Pronounced contrasts in trends of simple Bouguer gravity anomalies (Bromery, 1967) are developed on either side of a line essentially coincident with the above lithologic boundary of Emerson. These contrasting geophysical trends thus reinforce the above interpretation based on geological data.

Figure 21–2 shows the relationship between the three subsidiary fault zones of this paper and the closely associated master fault of the region,

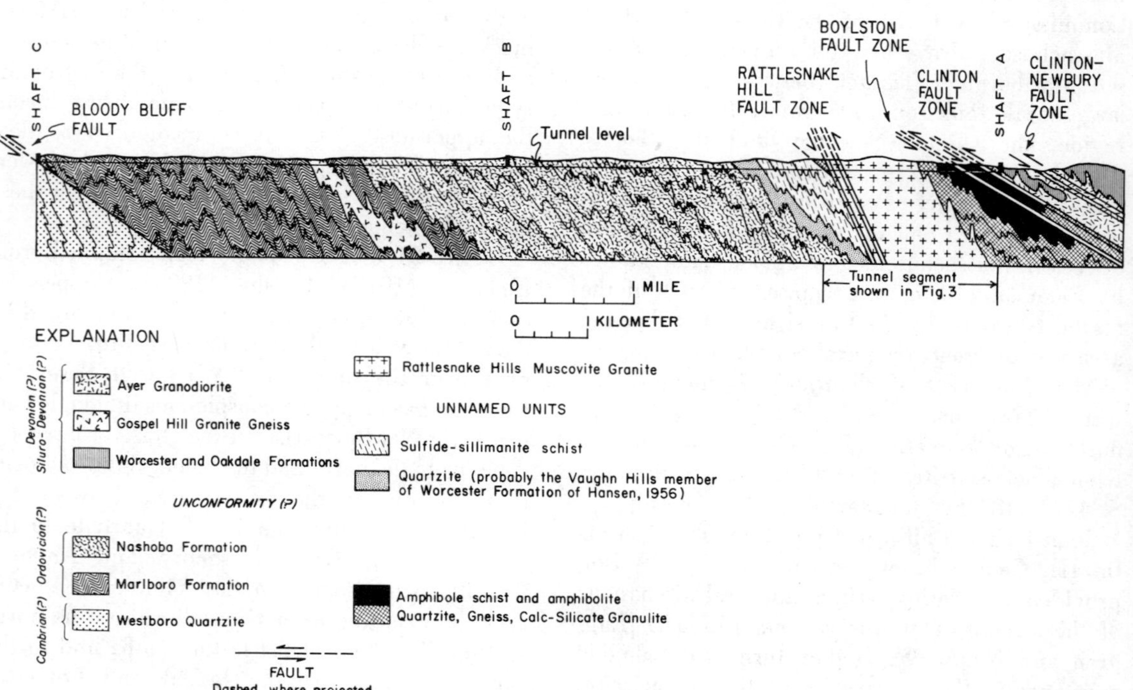

FIGURE 21–2. Regional cross section along the line of the Wachusett–Marlborough Tunnel showing major stratigraphic and plutonic units and the most clearly defined fault zones mapped or traced to date.

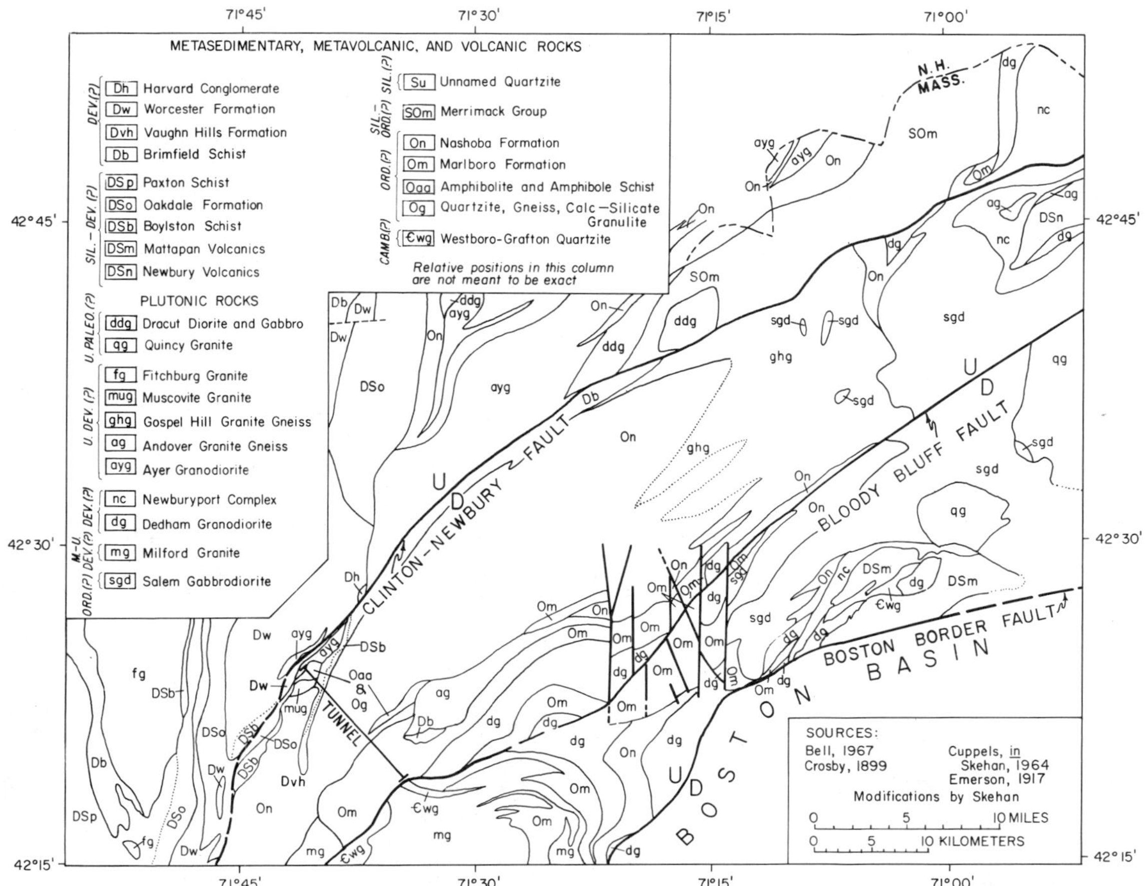

FIGURE 21–3. Generalized geologic map of northeastern Massachusetts showing lithologic units related to the interpretation of the major faults zones of the Wachusett–Marlborough Tunnel.

the Clinton–Newbury fault, lying approximately 1500 feet (450 m) northwest of Shaft A. The Wachusett–Marlborough fault block consists in large part of rocks in the sillimanite zone of metamorphism. The lower thrust blocks are composed of rocks of progressively lower metamorphic grade, the Boston Basin consisting of rocks in the chlorite zone. Preliminary compilation of information on the regional distribution of metamorphic zones by Goldsmith (1962) superimposed on a preliminary fault compilation by the author (Skehan, 1967a) indicates that metamorphic isograds in some instances coincide with faults, and in others are truncated by faults. In both cases the age of the dominant faulting is later than the metamorphism and is probably late Paleozoic to mid-Mesozoic.

This paper describes in detail the faults and joints for a segment of the tunnel from Sta. 0+00 to 76+00 (Fig. 21–2). Station numbers within the tunnel are referred to Shaft A (arbitrarily designated as Sta. 0+00). Therefore, Sta. 34+85

indicates a location 3,485 feet southeast along the tunnel line from Shaft A. This portion of the tunnel was selected for detailed structural analysis in order to determine the regional tectonic influence of the adjacent Clinton–Newbury Fault. Moreover, this segment is divided into two contrasting units, a heterogeneous stratigraphic section and a homogeneous unit, the latter consisting of the Rattlesnake Hill muscovite granite pluton. The analysis of structural features in rock units of contrasting characteristics was calculated to reveal valuable information on their mechanical response to large-scale deformative forces.

The close correlation between joints and faults suggests that detailed mapping of joints on the surface is potentially one of the most fruitful approaches to the recognition of attitude, location, and possibly also movement direction of regionally important covered faults. The abundance of information on faults available from the subsurface (Fig. 21–4) and the relative scarcity of such information along the surface line of the

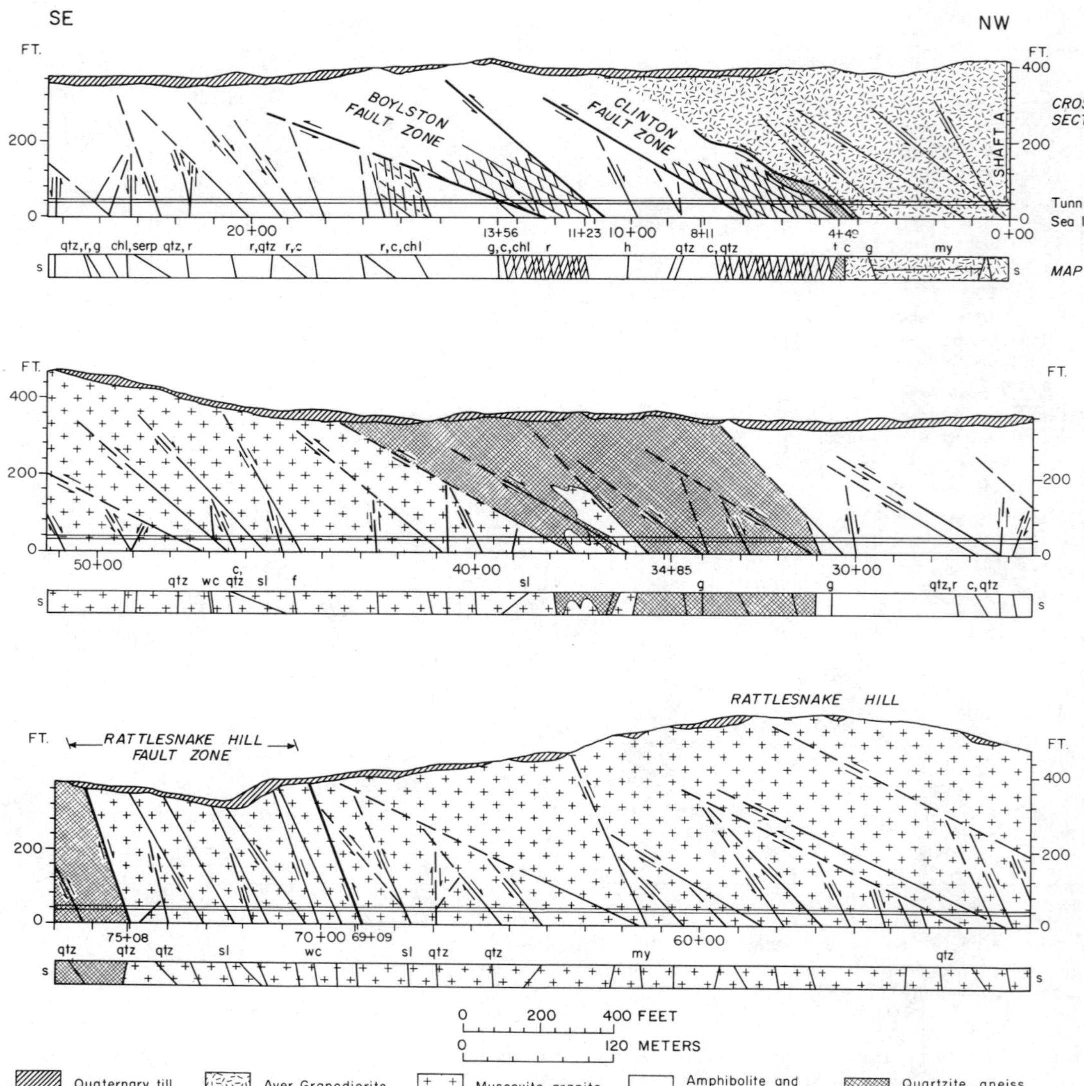

FIGURE 21–4. Generalized cross section and map of part of the Wachusett–Marlborough Tunnel (Sta. 0+00 to 76+00) looking southwest. No vertical exaggeration. Only those faults for which there are exact or generic surface indications are projected to the surface or to the base of the overburden. Patterns as follows: diagonal lines, Quaternary till; random dashes, Ayer Granodiorite; crosses, muscovite granite; blank (no pattern), amphibolite and amphibole schist; crosshatch, quartzite, gneiss, and calc-silicate granulite. Letter abbreviations as follows: c, calcite; chl, chlorite; f, feldspar; g, gouge; h, hematite; my, mylonite; qtz, quartz; r, rhodochrosite; s, line of section; serp, serpentine; sl, slickensides; t, talc; wc, white clay.

tunnel, due to widespread overburden, points to the need for improved indirect methods of recognizing at the surface the more important fault zones.

STRATIGRAPHY

The rocks exposed in the full length of the tunnel are interpreted as a sequence of intensely deformed metamorphic rocks of probable Ordo-

vician to Devonian or even Carboniferous age that dip generally northwest (Fig. 21–2; Skehan, 1967a and b). The lithology and sequence of these rock units are summarized in Table 21–1. Relations elsewhere in the region indicate that the two unnamed units in the upper part of the metamorphic sequence (Fig. 21–2) may be correlative with part of the Marlboro and Westboro Formations; also, that the unnamed units imme-

TABLE 21-1. Geologic units exposed in and near the Wachusett–Marlborough Tunnel with a brief description of each

Intrusive rocks (not necessarily in correct sequence)	
Rattlesnake Hill Granite	muscovite granite
Gospel Hill Granite Gneiss	porphyritic granite gneiss
Ayer Granodiorite	granodiorite gneiss
Metamorphic rocks	
Worcester and Oakdale Formations	phyllite, schist, quartzite, and calc-silicate granulite
– – – – – – – Clinton-Newbury fault, sequence broken – – – – – – –	
Unnamed	amphibole schist and amphibolite
Unnamed	quartzite, biotite gneiss, and calc-silicate granulite
– – – – – – – fault, sequence broken (?) – – – – – – –	
Unnamed	graphitic sulfide-sillimanite-andalusite schist, quartzite and calc-silicate unit
Unnamed	quartzite, non-graphitic, sillimanite-biotite schist, and interbedded quartzite and calc-silicate granulite unit (probably the Vaughn Hills Formation, Hansen, 1956)
– – – – – – – unconformity (?) – – – – – – –	
Nashoba Formation	sillimanite schist and migmatite in addition to same lithologies as Marlboro Formation
Marlboro Formation	amphibolite, granite gneiss, marble, calc-silicate granulite, cherty coticule and muscovite-biotite-chlorite schist
– – – – – – – Bloody Bluff fault, sequence broken – – – – – – –	
Westboro Formation	quartzite, and calc-silicate granulite

diately west of the Nashoba Formation may correlate with the lithologically similar Worcester and Oakdale Formations, faulted upon the Wachusett–Marlborough thrust block by movement on the Clinton–Newbury Fault (Fig. 21–2).

Metamorphic Rocks

The stratigraphic units exposed in the tunnel are an intensely folded and faulted, westerly-dipping, homoclinal sequence of metasediments and metavolcanics cut by intrusive rocks (Skehan, 1967a and 1967b). Figure 21–2 is a cross section through this sequence, which comprises the entire Wachusett–Marlborough thrust sheet, showing its relationship to the major bounding faults.

From oldest to youngest, the formations and the lithologic units comprising the sequence are described below. The Westboro Quartzite (Emerson, 1917) is a fine-grained, glassy quartzite and ranges from massive to thin-bedded in the upper and lower parts of the unit. The middle part of the formation is chiefly calc-silicate granulite.

The Marlboro Formation (Emerson, 1917) is characterized by massive to well-bedded amphibolite as well as biotite granite gneiss; thin beds of calcite and dolomite marble; epidote-calc-silicate granulite; well-bedded alternating amphibolite and cherty epidote-albite beds; well-banded pinstriped quartzite; finely bedded pink garnet-albite-quartz coticule and schist; buff to pink actinolitic cherty coticule; and calcareous muscovite-biotite-chlorite schist.

The Nashoba Formation (Hansen, 1956) contains lithologic units similar to the Marlboro, but these are present in different proportions. The Nashoba, moreover, also contains highly aluminous schists, characterized by the presence of sillimanite. Many units of this formation are distinctive in that pink to gray potassic feldspar and quartz are developed as lenses, layers and even as cross-cutting bodies, imparting to the rock features characteristic of migmatites.

The unnamed units immediately west of the Nashoba Formation (Fig. 21–2) are interpreted as stratigraphically above the latter formation. The lower unnamed unit consists of non-graphitic sillimanite-biotite schist, interbedded quartzite and calc-silicate granulite grading upward to thin-bedded quartzite in its upper part. This unit is probably the equivalent of Hansen's (1956) Vaughn Hills Member of the Worcester Formation. The upper unnamed unit consists of graphitic sulfide - sillimanite - andalusite - staurolite schist, quartzite, and calc-silicate granulite. On surface exposures this unit weathers to characteristic mottled rusty, black and yellowish green colors. The sulfide-sillimanite schist unit is probably the equivalent of the mica-schist facies of the Worcester Formation of Hansen (1956). The author leaves this unit unnamed until the question of whether the Worcester Formation of the Hudson–Maynard quadrangles can in fact be correlated with the Worcester Formation of the type locality. The lower quartzitic unit is distinguished from the sulfide schist unit by the development of a greater thickness of both the thin-bedded quartzite and calc-silicate beds in this basal unit. Moreover, the graphitic andalusite-sillimanite schist, characteristic of the upper unit, is not present in this part of the sequence. The lower quartzitic unit is, moreover, distinguished from the Nashoba Formation since the latter contains an abundance of metavolcanics, migmatites, and pegmatites. There is a marked decrease of such features in the unnamed units, suggesting the possibility that the boundary between them and the Nashoba marks an orogenic episode. The lower quartzitic unit may, therefore, be the stratigraphic equivalent, on the eastern shore, of the Clough Quartzite exposed on the western flank of the Merrimack synclinorium.

To the west of the quartzitic and sulfide-sillimanite schist units is a sequence consisting of gray to pink quartzite, biotite-plagioclase gneiss, and calc-silicate granulite. This unit is probably in fault contact with the sulfide-sillimanite schist unit and is lithologically more akin to the units west of the muscovite granite than to the sulfide-sillimanite unit (Fig. 21–2). It should be pointed out, however, that calc-silicate granulite and quartzite beds are present in both the sequences east and west of the Rattlesnake Hill pluton. If the author's preferred interpretation is correct (Fig. 21–2), the fault separating the sulfide-sillimanite schist unit from the unnamed quartzite-calc-silicate unit above it is probably a branch of the Rattlesnake Hill Fault Zone.

West of the Rattlesnake Hill pluton is the unnamed amphibole schist and amphibolite unit consisting of chlorite-epidote-amphibole schist and hornblende-biotite-epidote amphibolite respectively.

The Oakdale Formation (Emerson, 1917) consists of even-grained, glassy quartzite, biotite- and actinolite-quartzite, and calc-silicate granulite; the Worcester Formation (Perry and Emerson, 1903) consists of carbonaceous phyllite and graphitic sericite schist. The thin sequence (45.5 feet; 13.8 m) of calc-silicate granulite and gneiss stratigraphically above the unnamed mafic sequence (Fig. 21–4) may be a block of Oakdale lithology inserted into the tunnel sequence by faulting.

Intrusive Rocks

The Ayer Granodiorite (Jahns, 1952), a foliated and lineated, porphyritic to non-porphyritic, microcline-plagioclase-biotite granodiorite gneiss, is intrusive into both the Oakdale and the Worcester Formations.

The Gospel Hill Granite Gneiss (Fig. 21–2) is a gray to white, foliated, lineated, porphyritic,

biotite - muscovite - quartz - microcline - albite-garnet granite gneiss (Hansen, 1956). Hansen has interpreted it as having a metasomatic origin. Observations of contact relations with surrounding rocks in the tunnel sequence indicate that it is clearly intrusive. Structures developed in it indicate that it is probably syntectonic. The Gospel Hill granite in tunnel exposures contains a number of amphibolite and schist septa and is cut by pegmatites.

The Rattlesnake Hill muscovite granite is a massive, gray to white, medium-grained, quartz-albite-orthoclase-muscovite leuco-granite.

DESCRIPTION OF THE TUNNEL EXPOSURES

Fault Zones

Faults whose planar features or observed offsets are such that they are interpreted as having a displacement greater than the fault exposure in the tunnel are referred to as major or more important faults. Those fault zones, intercepted by this section of the tunnel, which have regionally significant displacements are (1) Clinton fault zone (Skehan, 1967a and 1967b) between Sta. 4+49 and 8+11; (2) the Boylston fault zone between Sta. 11+23 and 13+56; and (3) the Rattlesnake Hill fault zone between Sta. 69+07 and 75+18. The Clinton and Boylston fault zones are probably subsidiary faults or branches of the master fault of the region, the Clinton–Newbury fault zone.

The geologic cross section (Fig. 21–4) shows the interpreted relationship between the three most important fault zones and the lithologic units represented. The 3,485 foot (1045.5 m) sequence between Sta. 0+00 and 34+85 is composed of relatively massive mafic volcanic rocks, calc-silicate granulite, quartzite, feldspar-quartz gneiss and coarse microcline granodiorite gneiss. The volcanics make up about two thirds of the total breadth of outcrop in the tunnel. Of the 4,115 foot (1235 m) sequence between Sta. 34+85 and 76+00, consisting generally of massive muscovite granite, 162 feet (49 m) are calc-silicate granulite and biotite-quartz schist in the form of xenoliths and septa, and an intrusive sequence at the southeastern boundary of the granite. The Clinton, the Boylston and the Rattlesnake Hill fault zones correlate, in part at least, with conspicuous topographic lows at the surface. Rattlesnake Hill, a prominent topographic high, is developed over the main mass of muscovite granite.

The Clinton fault zone. The Clinton fault zone, 165 feet (49.5 m) thick, is developed largely in rocks ranging from massive amphibolite to moderately well-banded amphibole schist. Its northwestern boundary includes a body of calc-silicate granulite 45 feet (13.5 m) thick and is marked by a lens-shaped horse of banded porphyritic Ayer Granodiorite and a layer of talc 10 inches (25 cm) thick.

The tunnel section in amphibolite is characterized by numerous closely-spaced intersecting faults of several attitudes which produce diamond- to spindle-shaped blocks. The individual blocks commonly consist of an amphibolite core grading outward into a chlorite-amphibole schist, the latter interpreted as developed by hydrothermal alteration. Surfaces of the spindle-shaped blocks are intensely slickensided, some showing grooves throughout the 360° circumference of the block, indicating that it has been subjected to at least one complete rotation. Folds developed by faulting, slickensides and mineralized tension gashes in this zone indicate that easterly moving thrust and high-angle reverse faults are the rule.

The average attitude of the bounding faults and the dominant fault attitude within this zone is N.40°E., 30°NW. Prominent faults at and near the northwest boundary of this zone, between Sta. 4+49 and 4+66, are folded, locally developing dips of 87°NW. Calcite mineralized tension gashes (N. 60°E., 50°SE.) are developed essentially parallel to the axial plane of the folded faults.

The Boylston fault zone. The Boylston fault zone, 150 feet (45 m) thick, at Sta. 11+23 to 13+56, is bounded on the northwest by a fault having the attitude N.40°E., 40°NW. and on the southeast by a fault, N.45°E., 20°NW.; others, however, tend to develop conformably with the nearest bounding fault. Internal features of the Boylston fault zone are similar to those of the Clinton fault zone, except that rhodochrosite mineralization is characteristic of most joints in the center of the zone. Diamond- and spindle-shaped blocks of amphibole schist like those in the Clinton fault zone average 3–4 in. (7.5−10 cm) in long diameter with massive amphibolite cores. A 16.5-foot (5.1-m) thick block of graphite-chlorite-quartz-sericite schist appears to be faulted into this amphibolitic zone. The character of the tectonic movement in this fault zone is uniformly that of a thrust having an easterly tectonic movement direction. The attitude of the Boylston fault zone as a whole is estimated to be N.40E.; 30°NW.

The Rattlesnake Hill fault zone. The Rattlesnake Hill fault zone lies between Sta.

69+07 and 75+18. The tunnel penetrates the entire breadth of the Rattlesnake Hill muscovite granite pluton. Near both boundaries of the granite with the country rock, faulted calc-silicate granulite, quartzite, and schist blocks, interpreted as xenoliths, are encountered, as for example, at 35+84 to 38+07. From the latter location to 70+06, except for clay seams and other fault plane features, muscovite granite alone is present. The muscovite granite, whose intrusive contact is observed at 75+18, has been faulted at 75+19. Between 70+00 and 71+00 the granite contains several schistose fault blocks.

The character of the muscovite granite in the Rattlesnake Hill Fault Zone is greatly different from that northwest of Sta. 69+07. To the northwest, in spite of many faults (Fig. 21–4), the granite is generally medium- to coarse-grained, intermediate-gray to dark-brown in color. From 69+07 to 75+19, however, this unit is characterized as a light-gray, fine-grained, well-laminated, highly sheared muscovite-bearing rock interpreted as an intensely crushed and sheared part of the muscovite granite pluton. It is possible that this rock may be a highly deformed and faulted sequence of felsic metavolcanics having much the same mineral composition as the muscovite granite pluton. The presence, however, of a well-defined intrusive contact at 75+18 strongly suggests that the original rock, in which this fault zone was developed, was intrusive into the calc-silicate granulite unit as part of the main plutonic mass.

In the tunnel outcrop, the muscovite granite is recognized as being intensely sheared with conspicuous slip surfaces parallel to the banding, spaced at 6 inches to 2 feet (15 cm to 60 cm) apart. In hand specimen, this rock is so sheared as to take on the appearance of a finely banded crushrock with layers 1–2 mm thick generally showing well-defined drag folds. These folds, together with other movement direction indicators such as slickensides on fault planes, show that the tectonic movement has been in an easterly direction. The approximate attitude of the Rattlesnake Hill fault zone, as calculated from tunnel exposures, is N.50°E.; 65°NW.

Structural Analysis of Faults and Other Fracture Patterns

The faults represented by Figure 21–5A are developed in the heterogeneous sequence of metasediments and metavolcanics northwest of the Rattlesnake Hill muscovite granite pluton. Those represented by Figure 21–5B are in the pluton itself which is essentially a homogeneous unit.

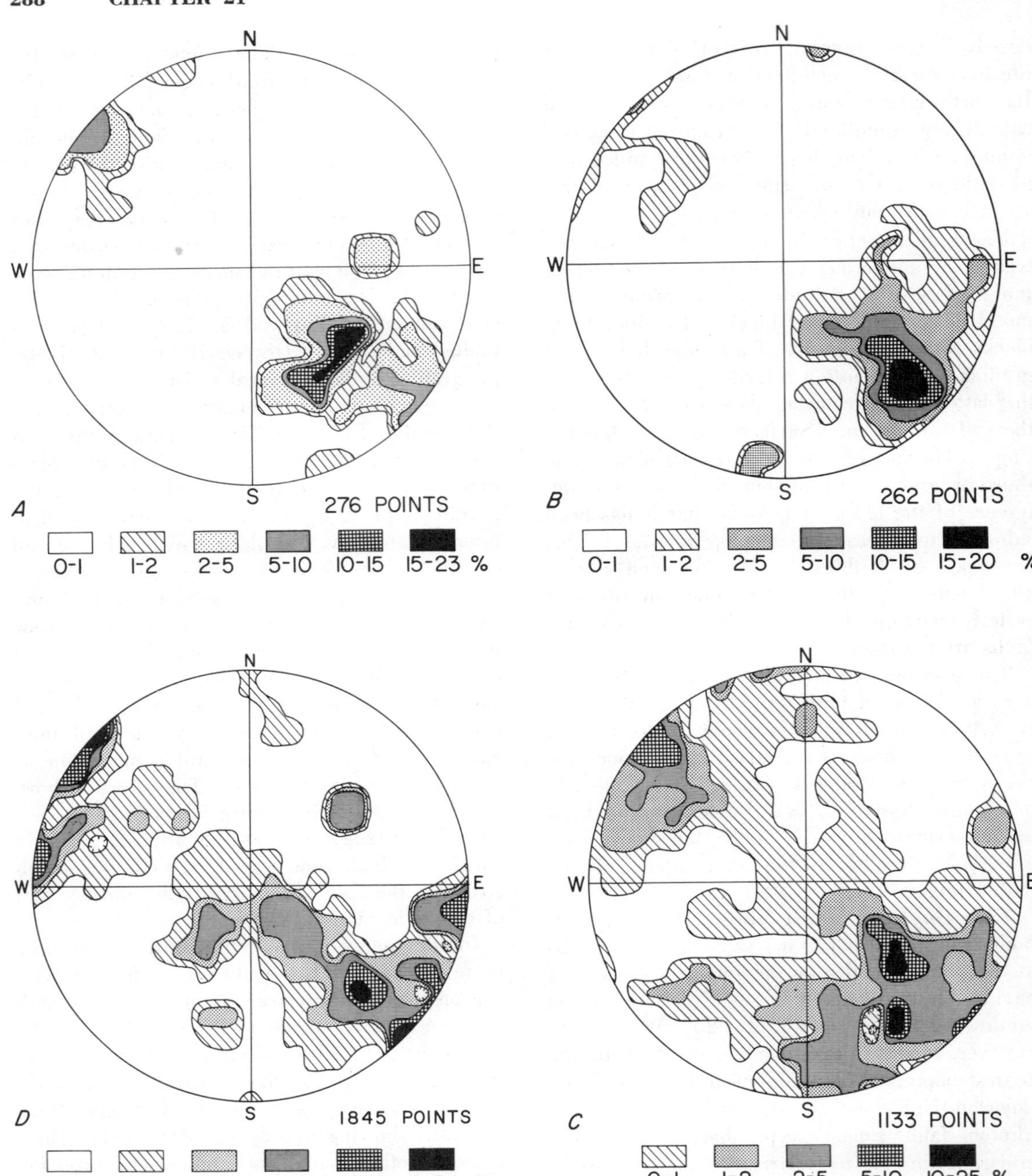

FIGURE 21–5. Orientation diagrams of normals to faults and joints between Sta. 0+00 and 76+00. Lower hemisphere projection on Schmidt equal area net. 5A Faults, Sta. 0+00 to 34+85. 5B Faults, Sta. 34+85 to 76+00. 5C Joints, Sta. 0+00 to 34+85. 5D Joints, Sta. 34+85 to 76+00.

These two divisions of this segment of the tunnel were selected in order to compare the attitudes of faults and joints in these contrasting lithologic units. In the interval 0+00 to 34+85, 276 major and minor faults were recorded, and in the interval 34+85 to 76+00, 262. In the respective intervals, 1133 and 1845 joints or joint sets were recorded.

An analysis of orientation diagrams (Figs. 21–5A and 21–5B) shows the following relationships: (1) faults striking N.40–50°E. and dipping 40–50°NW. are abundant in both types of rock; (2) the segment in heterogeneous rocks (0+00 to 34+85) shows a more pronounced development of faults of similar dip but more northerly strikes (N.25–40°E.) than that in the granite; (3) the segment in granite shows NE-trending faults with significantly steeper dips, 66–78°NW; (4) a small number of faults in granite strike N.80°W. and dip steeply NE. or SW., but faults in this orientation do not occur in the heterogeneous rocks.

The following is a detailed analysis of the data, as well as of diagrams constructed from the data, forming the basis for conclusions given above. Of the 276 major and minor faults in the segment of heterogeneous rocks (0+00 to 34+85) (Fig. 21–5A) 115 strike N.35–40°E. and dip 40–50°NW. A secondary maximum is at N.60°E. with 44 faults. In the section 34+85 to 76+00 (Fig. 21–5B), of the 262 major and minor faults, the attitude of 131 is N. 40–50°E.; 66–78°NW. A more detailed breakdown of faults in the granite shows that 63 strike N.40°E.; 41 strike N.45°E.; and 27, N.50°E.

Of the total number of faults in both segments of the tunnel, 170 are determined as reverse and 73 as normal faults. The dominant attitude of reverse faults is N.25–50°E.; 40–60°NW. A conspicuous number of reverse faults, however, show a wider range of dips from 35–80°NW. The strike of normal faults is dominantly N35–55°E., with dips from 65–75°NW. A high percentage of both reverse and normal faults, therefore, have essentially the same strike and dip. The range in dip for the higher concentrations of common orientations of reverse and normal faults lies between 35–80°NW., an angle of 45°. If this relationship is ultimately shown to mean that a significant number of such faults are conjugate sets, then one must explain why the sense of movement, portrayed in Figure 21–4, does not support the idea that these are conjugate sets. Such an explanation would demand that the deformation history of the region be of great complexity, possibly indicating that the principal stress directions have been rather variable.

In the segment in heterogeneous rocks (0+00 to 34+85), 103 are reverse faults as against 17 recognized as normal faults. In the muscovite granite, by contrast, more normal faults and comparatively fewer reverse faults were recognized than in the previous section, numbering 56 and 67 respectively.

In comparing reverse with normal faults, the following relations are evident: Faults in the heterogeneous segment are dominantly reverse faults striking from N.35–60°E. with low to intermediate dips. In the granite, on the other hand, reverse and normal faults are essentially comparable in numbers. They strike predom-inantly between N.40–50°E. with intermediate to steep dips.

JOINTS

Two orientation diagrams (Figs. 21–5C and 21–5D) show data on nearly 3000 joints or joint sets for the same intervals as Figs. 21–5A and 21–5B. The purpose of constructing these separate joint diagrams was to test for its value to surface investigations, whether a close correlation exists between faults and joints in this region, as well as to examine the nature of the response of these two units to joint-producing stresses.

Figure 21–5C, containing data on 1133 poles, shows the development of joints in the heterogeneous rock unit. The joints in this section are remarkably concentrated both as to strike and dip, with the dominant strike ranging from N.25–75°E. Dips range between 30–75°NW. This diagram shows two maximal concentrations in the southeast quadrant. A comparison of Figure 21–5A with 21–5C reveals a close correlation and partial coincidence between the more northerly maximum of Figure 21–5C and the maximum for faults. Thus faults striking N.25–60°E. have essentially the same dip as do the corresponding concentrations of joints. These joints are interpreted as being genetically related to such faults, and a significant number of the joints in this unit of heterogeneous rocks are therefore inferred to be shear joints.

On the other hand, joints striking from N.60–75°E. have no such corresponding development of faults although there is a maximum for faults of lower angles of dip having the same strike. Since the range in dip of the predominant joint and fault concentrations in this section is from about 40–75°W., this relationship may mean that a number of such joints are conjugate to faults.

Figure 21–5D, containing data on 1845 poles, shows the joint development in the Rattlesnake Hill pluton and reveals strong maxima for joints, one striking near N.45°E., dipping 50–70°NW.; and a second, striking N.45°E., dipping 80°NW. to 80°SE.

A comparison of Figures 21–5B and 21–5D reveals that both joints and faults have a partial coincidence of strike as well as dip in the higher concentrations. This suggests that a significant percentage of joints may be shear joints genetically related to the reverse faults. The nearly complete girdle of Figure 21–5D from southeast across the northwest quadrant suggests that the difference between distribution of minor concentrations observed in Figures 21–5B and 21–5D is due chiefly to the presence in the muscovite granite of post-tectonic sheeting or extension joints.

CONCLUSIONS

The strong correlation of attitude between reverse faults of the segment of the tunnel discussed in this paper, and the regionally important faults described by Bell (1967) suggests that structures of the Wachusett–Marlborough Tunnel are essentially typical of those developed in southeastern New England. Moreover, the relative abundance of recognized reverse and normal faults may be representative of the region as a whole.

Data from slickensides, from folds developed by faulting, and from related movement-direction indicators suggest that the average tectonic transport direction on westerly dipping reverse faults is easterly.

The analysis of fault and joint data (Fig. 21–5) leads the author to conclude tentatively that these fractures have been produced by forces which have had the same general direction throughout the history of movements on westerly-dipping reverse faults. The recorded deviations shown by the contours of lower value are not out of harmony with differences of response that could be expected from repeated adjustments by subsidiary fault blocks within a larger thrust unit.

The information recorded in Figure 21–5 is incomplete, however, if the movements on the Clinton–Newbury Fault have essentially obliterated the record of earlier fracturing in this segment adjacent to the fault. Other segments of the tunnel (Skehan, in press) at greater distances from the master fault retain the imprint of earlier tectonic activity in addition to the dominant overprint of low-angle westerly-dipping, as well as low-angle easterly dipping, reverse fault development.

On the basis of information from this study, and that of Bell (1967) and Lundgren (1963, 1966), the author concludes that in the segment of the tunnel of this study (0+00 to 76+00), easterly tectonic transport is characteristic of that phase of post-metamorphic orogenic activity dominated by westerly-dipping reverse faults.

REFERENCES

Bell, K. G., 1948, Geology of the Boston Metropolitan Area: Ph.D. thesis, Massachusetts Inst. of Tech., 390 p.

—————, 1967, Faults in eastern Massachusetts [abs.]: Program Geol. Soc. America, Northeastern Sect., Program, 1967 Ann. Mtg., Boston, p. 14

Billings, M. P., 1929, Structural geology of the eastern part of the Boston Basin: Am. Jour Sci. 5th Ser., v. 18, p. 97–137

Bromery, R. W., 1967, Simple Bouguer gravity map of Massachusetts: U. S. Geol. Survey, Geophys. Invest. Map GP–612

Crosby, W. O. 1899, Geology of the Wachusett Dam and Wachusett Aqueduct Tunnel of the Metropolitan Water Works in the vicinity of Clinton, Massachusetts: Technology Quarterly, v. 12, p. 68–96

Cuppels, N. P., 1961, Post-Carboniferous deformation of metamorphic and igneous rocks near the Northern Boundary Fault, Boston Basin, Massachusetts: U. S. Geol. Survey Prof. Paper 424–D, p. 46–48

Emerson, B. K., 1917, Geology of Massachusetts and Rhode Island: U. S. Geol. Survey Bull. 597, 289 p.

Goldsmith, Richard, 1962, Geologic map of New England, metamorphic zones: U. S. Geol. Survey open file map, scale 1:1,000,000

Hansen, W. R., 1956, Geology and mineral resources of the Hudson and Maynard quadrangles, Massachusetts: U. S. Geol. Survey Bull. 1038, 104 p.

Jahns, R. H., 1952, Stratigraphy and sequence of igneous rocks in the Lowell-Ayer region, Massachusetts: Geol. Soc. America, Guidebook for field trips in New England, 1952 Annual Mtg., Boston, p. 108–112

LaForge, Lawrence, 1932, The geology of the Boston area, Massachusetts: U. S. Geol. Survey Bull. 839, 105 p.

Lundgren, L. W., Jr., 1963, The bedrock geology of the Deep River quadrangle: Connecticut Geol. and Nat. History Survey Quad. Rept. 13, 40 p.

—————, 1966, The bedrock geology of the Hamburg quadrangle: Connecticut Geol. and Nat. History Survey Quad. Rept. 19, 41 p.

Perry, J. H. and Emerson, B. K., 1903, The geology of Worcester, Mass.: Worcester Nat. Hist. Soc., 166 p.

Skehan, J. W., S. J., ed., 1964, Boston area and vicinity: New England Intercollegiate Geol. Conf., Guidebook, 56th Ann. Mtg., Boston, 120 p.

—————, 1967a, Geology of the Wachusett-Marlborough Tunnel, east-central Massachusetts: a preliminary report: in O. C. Farquhar, ed., Economic Geology in Massachusetts, Univ., of Massachusetts Graduate School, p. 237–244

—————, 1967b, Evidence of easterly tectonic transport in the Appalachians of eastern Massachusetts [abs.]: Geol. Soc. America, Northeastern Sect., Program, 1967 Ann. Mtg., Boston, p. 56

—————, 1967c, Tectonic framework of southern New England and eastern New York: in Marshall Kay, ed., 1967, Symposium on the North Atlantic— geology and continental drift, Gander meeting, Am. Assoc. Petroleum Geologists, Memoir 12

Chapter 22

Stratigraphy and Structure of Southwestern Maine

ARTHUR M. HUSSEY II

INTRODUCTION

THE STRATIGRAPHY, structure, and age relations of the metamorphic rocks of southwestern Maine have been extensively revised as a result of mapping projects sponsored by the Maine Geological Survey over the past decade. Whereas Precambrian and Pennsylvanian ages were previously assigned to these rocks (Katz 1917; Keith, 1933) they are now considered to be of Ordovician to Early Devonian age. It is the purpose of this paper to discuss the stratigraphic and structural data that bear on the ages of the stratified rocks of southwestern Maine, and to mention some of the problems of interpretation of those data that make certain alternative interpretations necessary at present.

The information discussed here is drawn from work (both published and unpublished) of the writer throughout the area, and of Marc W. Bodine, Jr., in the Casco Bay quadrangle (lat 43°30′–43°45′N., long 70°–70°15′W.), Richard Gilman in the Kezar Falls and Newfield quadrangles (lat 43°30′–44°N., long, 70°45′–71°W.) and Richard Hatheway in the Wiscasset quadrangle (lat 44°–44°15′N., long 69°30′–69°45′W.). Because detailed studies are still in progress, many of the ideas and relations discussed here are preliminary and will undoubtedly be revised and refined in the future. The interpretations presented here are those of the writer and are not necessarily shared by those colleagues upon whose work they are based.

The writer acknowledges with gratitude the time spent by many colleagues in field visits to the area, particularly Marland Billings who has been an inspiration through his continuing interest in the progress of mapping and geological synthesis in southwestern Maine. Early drafts of this paper were reviewed by Robert G. Doyle, State Geologist of Maine; Professor Philip Osberg, University of Maine; and Dr. E-an Zen of the U. S. Geological Survey. Their comments and criticisms are sincerely appreciated.

GENERAL GEOLOGIC SETTING

Southwestern Maine is underlain by a diverse eugeosynclinal sequence of metasedimentary and metavolcanic rocks of Ordovician to Early Devonian (?) age, deformed during the Acadian orogeny in Middle or Late Devonian time (Fig. 22–1). These rocks range in metamorphic grade from lower greenschist to upper almandine amphibolite facies, and were metamorphosed during the Acadian orogeny and possibly again in Permian time (Faul et al., 1963). During a late stage of the Acadian orogeny, the stratified rocks were intruded extensively by calc-alkalic granitic and dioritic magmas of the New Hampshire Plutonic Series. Three suites of post-tectonic intrusive rocks (Fig. 22–2) were emplaced after the Acadian orogeny: (1) alkaline granite, syenite, and related alkalic rocks of the White Mountain Plutonic–Volcanic Series, intruded during Permian (?) to Jurassic (?) time; (2) basic and occasionally felsic dikes, intrusive into the metamorphic rocks and calc-alkalic and alkalic plutons, which may be related to Triassic igneous activity in the Bay of Fundy area and the Connecticut River Valley (this suite, because of the smallness of the units, is not shown on Figures

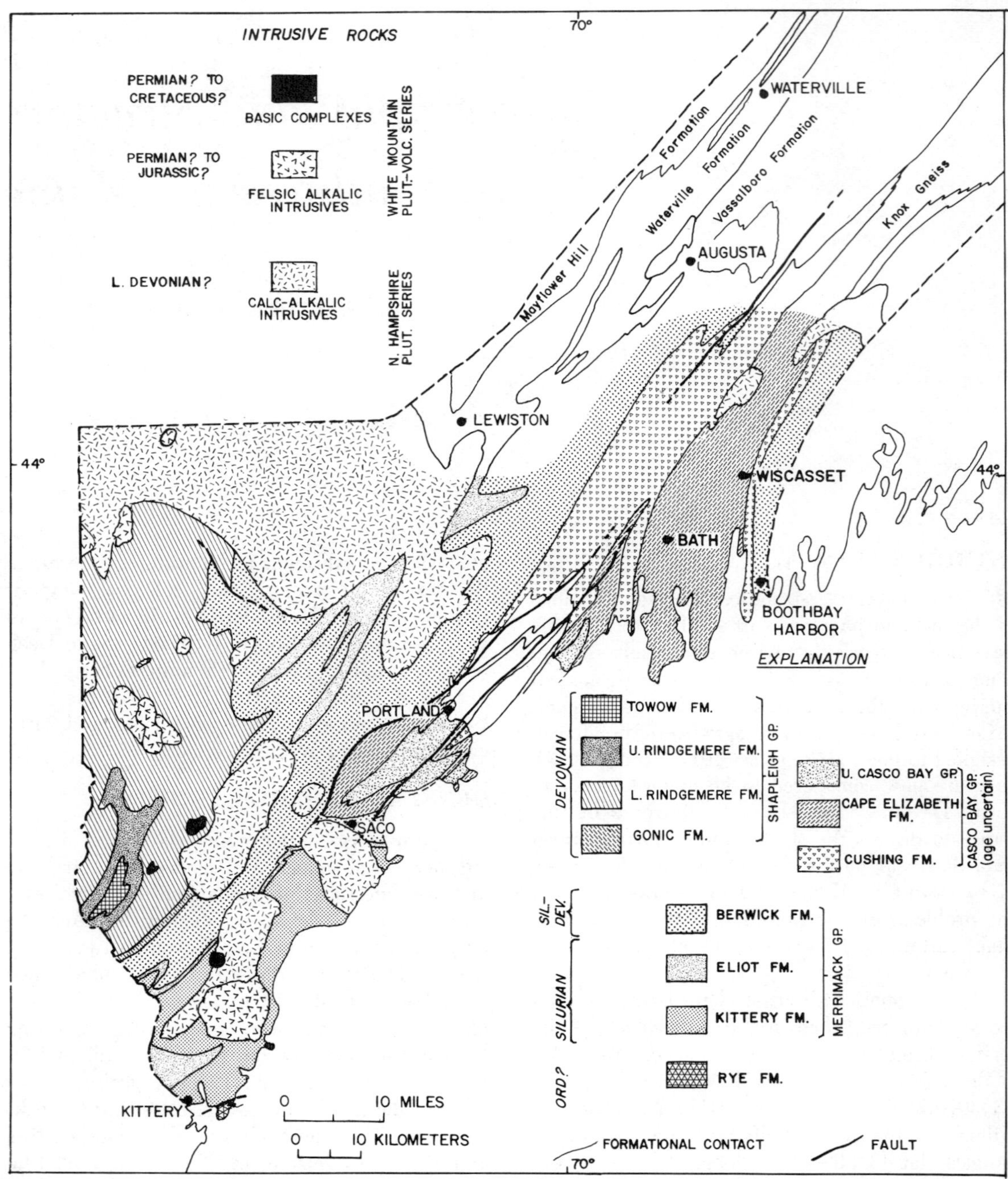

FIGURE 22–1. Generalized geologic map of southwestern Maine, modified from Preliminary Geologic Map of Maine (Doyle, 1967).

22–1 or 22–2); and (3) gabbroic complexes, correlatives of which in New Hampshire are included with the White Mountain Plutonic–Volcanic Series. The latter suite is younger than the dike suite and may have been intruded as late as the Cretaceous Period. A more detailed discussion of the igneous rocks is given by Hussey (1962); radiometric ages of these rocks are pending.

PREVIOUS WORK

Katz (1917) mapped and named the Kittery, Eliot, Berwick, Gonic, Rindgemere, and Towow Formations, and the individual formations of the Casco Bay Group which, unfortunately he could not delineate on his small-scale map. The unit to which he referred informally as the Algonkian complex in the very southwestern tip of the state

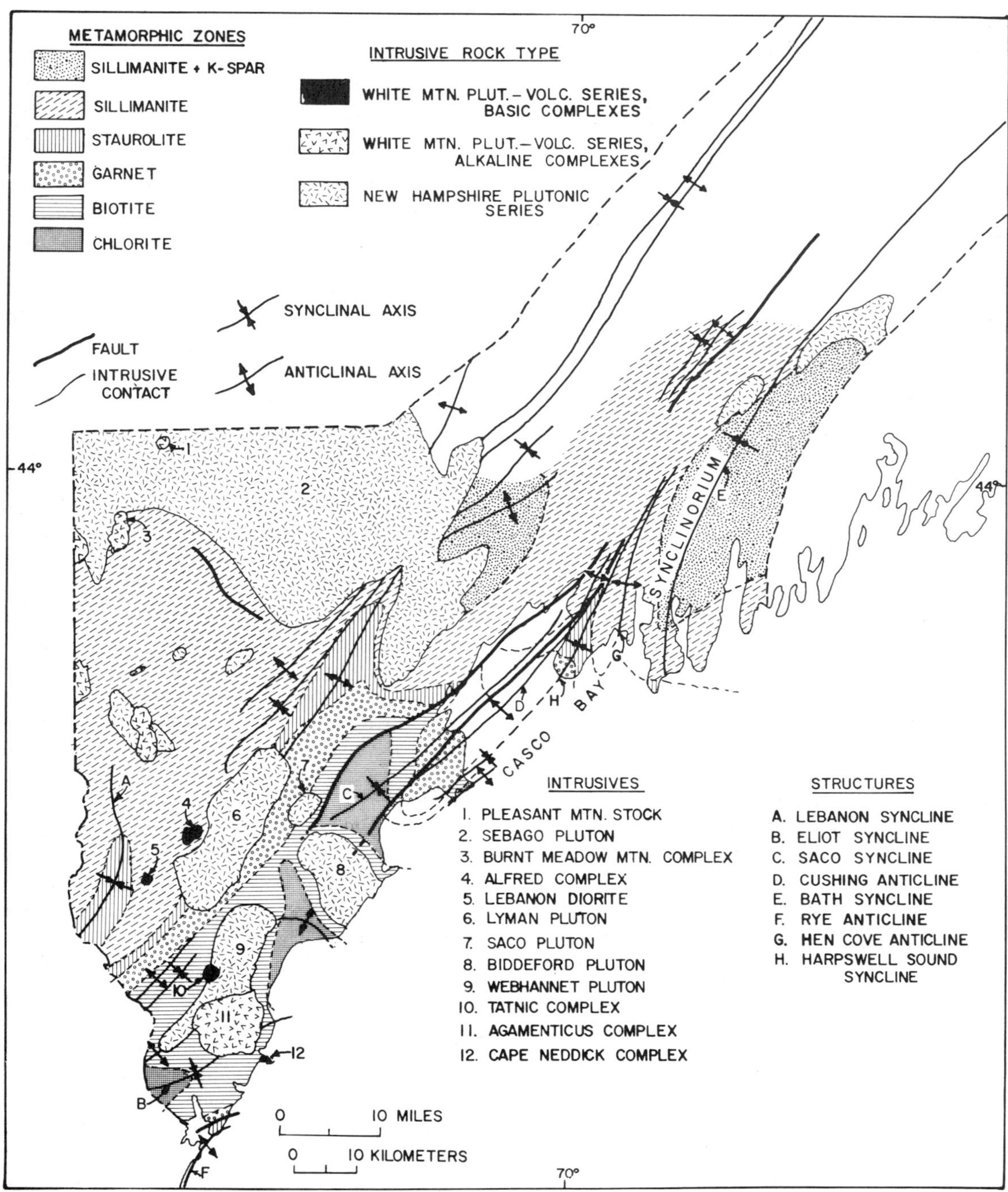

FIGURE 22-2. Generalized tectonic map of southwestern Maine showing major fold axes, faults, zones of regional metamorphism, and intrusive rock bodies.

and adjacent parts of coastal New Hampshire is now referred to as the Rye Formation (Billings, 1956). The unit here called the Cushing Formation was mapped by Katz as the Cushing Granodiorite on the assumption that it represents a deformed intrusive.

Fisher (1941) mapped the general area around Lewiston and provided the link between the low-grade fossiliferous metamorphics of the Water-ville area and the high-grade metamorphic rocks of southwestern Maine.

STRATIGRAPHY

The stratified rocks of southwestern Maine are divided into three groups: the Merrimack Group which includes the Kittery, Eliot, and Berwick Formations; the Shapleigh Group which includes the Gonic, Rindgemere, and Towow Formations;

and the Casco Bay Group which includes the Cushing Formation, Cape Elizabeth Formation, Spring Point Greenstone, Diamond Island Formation, Scarboro Formation, Spurwink Limestone, and Jewell Formation. The Rye Formation, which is primarily exposed in New Hampshire but which crops out in the southwestern tip of the map area, is not conveniently included in any of the groups and is discussed separately.

Stratigraphic sequence in the area of discussion has been established on the basis of (1) correlation of rock units in southwestern Maine with stratigraphically established units in adjacent parts of Maine and New Hampshire; (2) regional patterns of small-scale mesoscopic parasitic folds relative to macroscopic folds revealed through regional map patterns; and (3) contacts that have been observed in large-scale, nearly isoclinal mesoscopic folds. Primary indications of bedding tops, such as graded and cross bedding and channel scour-and-fill structure, have been observed frequently in rocks of the lower grades of metamorphism, particularly in the Kittery Formation, but are infrequently observed in the more highly metamorphosed rocks. Because of the virtual lack of unequivocal sedimentary tops evidence at observed contacts, the stratigraphic sequence discussed here is tentative and is contingent upon the postulate that the structural features used as tops evidence were formed early in the deformation and are not related to subsequent episodes of folding.

The general lithic character and the thickness of each of the formations are summarized in Tables 22–1 and 22–2, and the distribution of the formations is shown in the generalized geologic map of southwestern Maine (Fig. 22–1). Because of the small scale of the map, the individual formations of the Casco Bay Group above the Cape Elizabeth Formation are combined as one map unit.

Rye Formation

The lowest stratigraphic unit in the southwestern Maine area is the Rye Formation which Billings (1956) has divided into two members. The lower member, exposed almost exclusively in New Hampshire, consists of feldspathic and pelitic metasediments representing somewhat reworked volcanic detritus. The upper member consists predominantly of felsic metatuff, interbedded with subordinate metagraywacke and minor calcareous metagraywacke. Amphibolite occurs near the base of the upper member.

The Rye Formation is in contact with the Kittery Formation partly by faulting and partly by inferred disconformity. On the northwest side of Gerrish Island in Kittery (Lat. 43°, Long. 70° 41′) a major shear zone, in which both formations have been extensively crushed and mylonitized, coincides with the contact and would appear to be an extension of a major thrust fault shown by Goldsmith (1963) along part of the Rye–Kittery boundary in New Hampshire and continuing into Massachusetts. On the east side of Gerrish Island, on the other hand, distributions of outcrops of the two formations suggest a series of folds, perhaps close to the axis of the Rye anticline (Fig. 22–2). Lack of angular discordance between structures of the two formations rules against an angular unconformity between the two.

The metasedimentary and metavolcanic members of the Rye Formation are correlated with the Albee Formation and Ammonoosuc Volcanics of western New Hampshire respectively on the basis of similarity of lithology and position in a sequence (Billings, 1956). The Partridge Formation of western New Hampshire which lies conformably on top of the Ammonoosuc Volcanics is not present in the Kittery area of Maine and in southeastern New Hampshire; for this reason the Rye-Kittery contact on the east side of Gerrish Island is believed to be a disconformity.

Merrimack Group

The Kittery, Eliot and Berwick Formations constitute the Merrimack Group as defined by Billings (1956). These three formations, which form an apparently conformable sequence, have been separately mapped in southwestern Maine and New Hampshire southward approximately to the latitude 42° 50′; south of this latitude they have been mapped collectively as the Merrimack Group. The distribution of the three formations in southwestern Maine is shown in Figure 22–1.

The Kittery Formation consists of medium- to thin-bedded calcareous and non-calcareous quartzite and metagraywacke in which primary structures (graded bedding, cross lamination, small-scale scour and fill structure, and ovoid calcareous concretions) are common and well preserved. Coarse sand- to fine pebble-sized quartz and feldspar grains frequently form basal layers in the thicker graded beds of the formation. The perfection of graded bedding clearly suggests deposition by turbidity currents.

The Kittery Formation is conformable with the overlying Eliot Formation which crops out in three separate belts in southwestern Maine (Fig. 22–1), undergoing a progressive change in lithology from fine-grained calcareous metagray-

TABLE 22-1. General Lithology and Thickness of the Rye–Merrimack–Shapleigh Sequence

Formation			Lithology	Estimated thickness in feet (and meters)
SHAPLEIGH GROUP	Towow Formation		Rusty weathering sulfidic quartz-muscovite schist with minor interbedded graphitic phyllite	500+ (150+)
	Rindgemere Formation	Upper part	Moderately well-bedded, thin- to medium-bedded muscovite-biotite-quartz schist, quartz-mica schist, with or without sillimanite and /or staurolite	1000+ (300+)
		Lower part	Poorly-bedded gneissic quartz-biotite-muscovite-garnet schist, micaceous quartzite with sillimanite and/or staurolite. Includes some local calc-silicate gneiss (quartz-plagioclase-diopside-idocrase-calcite-grossularite assemblage common) and impure marble; some rusty-weathering quartz-mica schist, gneiss, and micaceous quartzite	5000–8000 (1500–2450)
	Gonic Formation		Silvery two-mica schist with staurolite and garnet; minor micaceous quartzite interbeds	1000 (300)
MERRIMACK GROUP	Berwick Formation		Thin to thick bedded calcareous quartzite and biotite quartzite, quartz-biotite-actinolite granulite, diopside- hornblende-plagioclase gneiss, and minor quartz-biotite-muscovite-garnet schist. Quartz-plagioclase-biotite gneiss and calc-silicate gneiss with salt and pepper texture in sillimanite zone	6000–8000 (1850–2450)
	Eliot Formation		Varying from calcareous phyllites to pelitic phyllites and schists—see discussion	1000 (300)
	Kittery Formation		Thin-bedded feldspathic, calcareous, and biotitic quartzites, biotite phyllite, quartz biotite schist and calc-silicate gneiss. Occasional beds of fine-granule quartzo-feldspathic conglomerate occur at base of thicker graded beds	6000–8000 (2850–2450)
	Rye Formation	Meta-volcanic member	Thin bedded porphyroblastic feldspar-quartz-biotite gneiss with thin interbeds of feldspathic metagraywacke. Includes some hornblende amphibolite, marble, and graphitic phyllite	1000–2000 (300–600)
		Meta-sedimentary member	Quartz-feldspar-biotite-muscovite-staurolite or sillimanite schist	1000+ (300+)

wacke and phyllite in the southwestern and eastern parts of the map area to metapelite with abundant staurolite or sillimanite in the northern part. The three belts are correlated by their stratigraphic position relative to the Berwick and Kittery Formations. In the south and east the Eliot is mineralogically similar to the Berwick and Kittery but differs in its lower percentage of quartz and the thinner and less competent character of its bedding. In the north, particularly between the Lyman and Sebago plutons (Fig. 22-2), the Eliot is distinctive in both its pelitic and its thin-bedded nature and includes a central member, 100 to 250 feet (30 to 75 m) thick, of fine-grained ribbon-bedded limestone with biotite phyllite or schist interbeds.

The Berwick Formation, which is lithically similar to the Kittery is a thick sequence of calcareous and non-calcareous metagraywacke and quartzite, with a tendency toward thicker and less well-developed bedding than that of the Kittery. Primary structures that are so well-developed in the Kittery Formation are less frequently observed in the Berwick, possibly because of the greater degree of metamorphism and recrystallization of the latter. Between Portland and Augusta (Fig. 22–1) the Berwick Formation has been extensively migmatized and re-

TABLE 22-2. General Lithology and Thickness of the Casco Bay Group

Formation	Lithology	Estimated thickness in feet (and meters)
Jewell Formation	Sulfidic and non-sulfidic two-mica schist with minor thin micaceous quartzite interbeds	200+ (60+)
Spurwink Limestone	Thin ribbon-bedded limestone with biotite phyllite or schist interbeds	100 (30)
Scarboro Formation	Sulfidic and non-sulfidic two-mica schist with minor thin micaceous quartzite interbeds. Identical to the Jewell Formation	500 (150)
Diamond Island Formation	Thin, evenly-laminated black sulfidic quartz-graphite schist and phyllite	0–150 (0–45)
Spring Point Greenstone	Thin bedded to massive chloritic schist, actinolitic schist or amphibolite depending on grade of metamorphism. Includes some feldspathic quartzite and metafelsite at the top	0–400 (0–120)
Cape Elizabeth Formation	Thin bedded quartz-plagioclase-biotite-muscovite schist, micaceous quartzite and some calc-silicate lenses	1000–5000 (300–1500)
Cushing Formation	Thin bedded to massive feldspar-quartz-biotite gneiss, amphibolite, quartz feldspar biotite schist and gneiss, calc-silicate gneiss, and occasional marble and sulfidic two-mica quartz schist	2000+ (600+)

(The whole table is labelled along the left margin: CASCO BAY GROUP)

crystallized to a salt-and-pepper textured quartzo-feldspathic biotite and calc-silicate gneiss, and includes a zone 500 to 1000 feet (150 to 300 m) thick of very sulfidic gneiss at the top. These gneisses were included by Fisher (1941) in his Pejepscot Formation but because he also included parts of the Cushing Formation in this unit, the writer recommends abandonment of the name "Pejepscot" and the use of "Berwick" instead.

The Berwick Formation has been traced without interruption from the Maine–New Hampshire border northeastward into the outcrop belt of the Vassalboro Formation of Osberg (in press), and southwestward by Sriramadas (1966) to the vicinity of Nashua, N. H. (Lat. 42° 45′, Long. 71° 30′), beyond which it loses identity in the Merrimack Group. The Berwick is believed to be conformable with the underlying Eliot Formation and with the overlying Gonic and Rindgemere Formations of the Shapleigh Group. Its relation to the Cushing and Cape Elizabeth Formations of the Casco Bay Group will be discussed in a later section.

Shapleigh Group

The name "Shapleigh Group" is proposed for the thick sequence of pelites consisting of the Gonic, Rindgemere and Towow Formations of Katz (1917) which lie conformably above the Berwick Formation in the western part of the map area (Fig. 22–1). The name is taken from the town of Shapleigh in northern York County which lies near the center of the outcrop belt of the group. Field work done since the compilation of the geology of southwestern Maine for the Preliminary Geologic Map of Maine (Doyle, 1967) has demonstrated the need for revision and redefinition of the subdivisions of the Shapleigh Group. Since detailed work is still in progress by Gilman in the Kezar Falls and Newfield quadrangles and by the writer in the Berwick quadrangle (lat 43°15′–43°30′N., long 70°45′–71° W.), however, the formational nomenclature used on the Preliminary Geologic Map of Maine is being maintained for this report.

The lowest unit of the Shapleigh Group is the Gonic Formation which is recognized only southwest of the Lyman Pluton (Fig. 22–2). The formation consists of rather poorly-bedded muscovite-rich pelite and micaceous quartzite. It lies conformably on the Berwick Formation as indicated by the interbedding of pelite with calc-silicate and biotite granulite and quartzite in the upper few feet of the Berwick Formation. The

Gonic grades imperceptibly upward into the lower part of the Rindgemere Formation.

The Rindgemere Formation, which is the most widespread unit of the Shapleigh Group, has been tentatively divided by the writer (Hussey, 1962) into two conformable units which are here referred to informally as the lower and upper parts of the Rindgemere pending redefinition at the completion of detailed field work. The lower part of the Rindgemere Formation consists primarily of poorly-bedded pelites and micaceous quartzites which have been extensively migmatized and injected with pegmatites. Sillimanite and pseudomorphs of muscovite after andalusite are common throughout. North of latitude 43°30′ the lower part of the Rindgemere includes thin units of calc-silicate granulite, marble and sulfidic schist which are too small to show at the scale of the map (Fig. 22-1). The upper part of the Rindgemere consists of essentially non-migmatized moderately well-bedded muscovite rich pelites and micaceous quartzites similar to the Gonic Formation except for better bedding.

The Towow Formation is the highest unit of the Shapleigh Group and crops out only in the center of the Lebanon syncline (Fig. 22-2). It is gradational to and conformable with the upper part of the Rindgemere from which it differs by the predominance of sulfide and presence of occasional interbeds of dark gray graphitic schist.

Casco Bay Group

The name "Casco Bay Group" was proposed by Katz (1917) for the sequence of pelitic metasediments and basic metavolcanics exposed in the area of Casco Bay. In the group, Katz included, in ascending stratigraphic order, the Cape Elizabeth Formation, Spring Point Greenstone, Diamond Island Formation, Scarboro Formation, Spurwink Limestone, Jewell Formation, and Macworth Slate. The Cushing Granodiorite of Katz which crops out within the belt of the Casco Bay Group has been clearly shown by Bodine (*in* Hussey, 1965, pp. 57–72) and the writer (*in* Hussey, 1965, pp. 4–24) to consist of the metamorphosed equivalents of volcanic rocks and feldspathic to calcareous sedimentary rocks; the writer here proposes that the Cushing be included as the basal unit of the Casco Bay Group. Based on reconnaissance work in the area the writer feels that the Macworth Slate is either part of the Cape Elizabeth Formation or a local member within the Cushing Formation. Distribution of the Cushing, Cape Elizabeth, and the upper formations of the Casco Bay Group are shown in Figure 22-1. Relations of the Casco Bay Group to the other groups will be discussed in a later section.

The Cushing Formation consists of a varied assemblage of metavolcanics and metasediments representing felsic and mafic volcanics, feldspathic sandstone, calcareous sandstone, limestone, siltstone and shale. Bedding is well developed in the metasedimentary units and weakly developed in the metavolcanics. The upper part of the formation is composed predominantly of blastoporphyritic metafelsite tuff and agglomerate. Where least recrystallized the fragmental character of these rocks is clearly indicated by such relict structures as stretched metafelsite clasts, and metacrysts of blue quartz and polysynthetically-twinned albite. The albite and quartz metacrysts are interpreted by the writer as relict crystal fragments of a crystal tuff. The lower part of the formation consists of a varied association of all the rock types noted above. The Cushing Formation is overlain conformably by the Cape Elizabeth Formation as indicated by interspersed Cushing-type metafelsite and amphibolite in the lower 100 feet (30 m) of the Cape Elizabeth. The relationship of the Cushing to the underlying sulfidic phase of the Berwick Formation is not known. The Cushing Formation pinches out abruptly south and west of its exposure in the Cushing anticline (Fig. 22-2) so that the Cape Elizabeth Formation overlies the Berwick at the southern end of the Casco Bay synclinorium.

The Cape Elizabeth Formation consists of metamorphosed thin beds of alternating gray sandstone and pelite with minor calc-silicate granulite lenses and pods. Local rusty-weathering metapelite members have been mapped at the base and in the lower parts of the formation and minor amphibolite lenses occur sporadically throughout the formation. The principal belt of the Cape Elizabeth Formation extending through Bath (Fig. 22-1) has been extensively migmatized and has been traced northward into the outcrop belt of the Knox Gneiss of Perkins and Smith (1925) and Trefethen (personal communication, 1950).

Conformably above the Cape Elizabeth Formation is a series of metamorphosed basaltic to andesitic lava flows and water-laid tuffs grading upward into metamorphosed feldspathic sandstone and felsite tuff. These rocks comprise the Spring Point Greenstone of Katz (1917). Above the Spring Point Greenstone is the Diamond Island Formation which, with its coal black color and strong iron-oxide and sulfate staining, is lithically the most distinctive unit in the Casco

Bay Group. Both the Spring Point Greenstone and the Diamond Island Formation pinch out to the southwest so that the Scarboro Formation directly overlies the Cape Elizabeth Formation around the southwestern nose of the Casco Bay synclinorium (Fig. 22–2). The Diamond Island Formation is succeeded upward by a conformable sequence of sulfidic and non-sulfidic metapelites which are strongly contorted and crumpled. These metapelites are divided into two stratigraphic units by the Spurwink Limestone, a very fine-grained ribbon-type limestone identical to the limestone unit of the Eliot Formation. The metapelites below the Spurwink were defined by Katz as the Scarboro Formation and those above as the Jewell Formation. The Jewell is preserved only in the troughs of doubly-plunging synclines in the Casco Bay area, and hence the nature of the top of the Formation and any overlying formations are unknown.

Rock Units East of the Casco Bay Group

Immediately east of the belt of the Cape Elizabeth Formation extending through Bath and Wiscasset (Fig. 22–1) a sequence of feldspathic metasediments with amphibolites and minor calc-silicate beds has been mapped and, because of lithic similarity, has been included with the Cushing Formation. No structural or primary sedimentary tops evidence indicates the stratigraphic position of these rocks relative to the Cape Elizabeth Formation. The contact, however, is conformable.

East of these Cushing-type rocks, in a belt extending from the ocean at Boothbay Harbor (Fig. 22–1) northeastward, is a thick sequence of thin-bedded quartzo-feldspathic biotite granulite and calc-silicate granulite quite similar in mineralogy to the Berwick Formation, with which it is included on Figure 22–1. A contact exposed in a highway cut near Wiscasset shows the conformability of these granulites with the Cushing-type unit on its western margin, but again no data bear on the relative age of the two units. These granulites have been traced northward into the Bucksport Formation of Trefethen (personal communication, 1950) and are probably equivalent to the Kellyland Formation of Larrabee (1964) and the Pale Argillite Division of the Charlotte Group of Alcock (1946) in New Brunswick and eastern Maine.

STRUCTURE

Folds

The major fold structures in southwestern Maine are the Rockingham anticlinorium, the crest of which may correspond to the Rye anticline in the very southern tip of the state (Fig. 22–2); the Casco Bay synclinorium; the Lebanon syncline which lies on the southeast flank of the Merrimack synclinorium (see Osberg et al., this volume); and the southern end of an anticlinorial belt defined by the outcrop bands of the Eliot Formation southwest of the Sebago Pluton (Figs. 22–1 and 22–2), and extending northeast of the Sebago Pluton through Lewiston into the outcrop belt of the Mayflower Hill Formation (Osberg, in press).

Because of its bearing on general stratigraphic interpretation the Casco Bay synclinorium merits more discussion. This is a doubly plunging structure made up of the Saco syncline, Cushing anticline, the Bath syncline and other lesser folds (Fig. 22–2). Of the individual folds the Bath syncline remains a questionable structure, defined by correlating the narrow belt of "volcanics" just east of Wiscasset with the much wider belt of the Cushing Formation west of Wiscasset. If this correlation is valid, a structure is created with the Cape Elizabeth Formation in the center, flanked first by Cushing- and then Berwick-type units. On the basis of relations to the west in the Hen Cove anticline (Fig. 22–2), where on a structural basis the Cape Elizabeth clearly overlies the Cushing Formation, the structural axis through Bath would be that of a syncline.

Much of the evidence for the stratigraphic order of the formations of the Casco Bay Group comes from the area of the Hen Cove anticline and the Harpswell Sound syncline (Fig. 22–2). In this area several localities have been found where contacts between formations or members of a formation are involved in large scale mesoscopic parasitic folds. The arch bend of the Hen Cove anticline has been observed at several shore localities along its strike and is the most firmly established structure in the area. Minor units within the Cushing Formation can be mapped around the nose of the fold or found in proper position on either limb. These larger scale mesoscopic folds are nearly isoclinal to somewhat open structures (difference of dip of limbs as much as 30° to 40°) with steep axial planes, gentle plunges and parabolic arch bends. Micas in the schists define a good axial-plane schistosity. Superimposed on these are smaller scale mesoscopic parasitic folds of a higher order, having the same general geometrical properties as the larger folds, except that they are more nearly isoclinal with tight parabolic arch bends. The writer interprets these folds to be a result of the first stage of folding in the area during the

Acadian orogeny. At many localities distinctly later folds can be seen superimposed on the earlier set, in which axial planes are variable but commonly rather gentle (30° to 50°) in dip, the plunges moderate, the arch bends very abrupt, the limbs quite open, and earlier schistosity crenulated and folded. These folds may be the result of late stages of Acadian deformation or a distinctly later orogeny (perhaps Appalachian).

Where major contacts have been involved in the primary-type folds, the sequence of the Casco Bay Group as previously indicated has always been upheld; no reversals have been noted. This is the basis for the stratigraphic sequence given.

Faults

Five major faults have been mapped in southwestern Maine (Fig. 22–2). The criteria, either singly or in combination, by which these faults have been recognized are (1) offset of formation contacts; (2) offset of metamorphic isograds; (3) juxtaposition of rocks of very different metamorphic grade; (4) juxtaposition of rocks of contrasting degree of migmatization; (5) thick breccia and mylonite zones; (6) repetition of strata; (7) topographic lineaments; and (8) juxtaposition of rocks of contrasting structural pattern and geometry. All are believed to be post metamorphic faults on the basis of offset or truncation of isograds, and in the case of one, strong trough-like topographic expression of the fault trace. The nature of movement on these faults has not yet been determined.

AGE RELATIONS OF THE METAMORPHIC ROCKS

General Statement

Katz (1917) suggested a Carboniferous (Pennsylvanian) age for the Kittery, Eliot, Gonic and Towow Formations, and those parts of the Casco Bay Group and the Rindgemere Formation that he mapped. The basis for the Pennsylvanian age assignment was a correlation of the Kittery and Eliot Formations with the Oakdale Quartzite and the Worcester Phyllite, respectively, in central Massachusetts which were considered to be of Pennsylvanian age. Billings (1956) has reviewed these correlations and pointed out the many uncertainties involved. For the Berwick Formation, Katz gave a pre-Carboniferous age and for the rocks of the Rye Formation he suggested a Precambrian age on the basis of lithic correlations to the Marlboro and Westboro Formations in eastern Massachusetts which were considered of Precambrian age.

A Precambrian age for all the high grade metamorphic rocks (other than those already assigned a Pennsylvanian age by Katz) was given by Keith (1933).

Fisher (1941) made the first advance toward a rational understanding of the age relations of these high-grade metamorphic rocks. He demonstrated, through his mapping, the equivalence of the high-grade metamorphic rocks of the Lewiston area with the fossiliferous low-grade metasediments of the Waterville area earlier mapped by Perkins and Smith (1925). Although many of his ideas concerning stratigraphic subdivisions and correlations have been revised, Fisher successfully dispelled the notion that had misled Keith, namely that high-grade metamorphism is a criterion of Precambrian age.

Age of the Rye Formation and the Merrimack and Shapleigh Groups

As a result of work sponsored by the Maine Geological Survey over the past decade, a correlation of the formations of the Merrimack Group in southwestern Maine with the fossilifeorus sequence in the Waterville area of central Maine has been reasonably established. On the basis of lithic similarity and position in a sequence, the Kittery, Eliot and Berwick Formations are correlated with the Mayflower Hill, Waterville and Vassalboro Formations respectively of Osberg (in press; also Osberg et al., this volume). Osberg reports that monograptid graptolites of late Llandovery age occur in the Mayflower Hill Formation, and dendroid graptolites of possible Wenlock to Ludlow age have been found in the Waterville Formation. Through the above lithic correlations, this indicates an Early Silurian age for the Kittery Formation, and a Middle to Late Silurian age for the Eliot Formation. The Berwick Formation, because of its position stratigraphically above the Eliot Formation, is then of Late Silurian to possibly Early Devonian age.

The Shapleigh Group is a continuation of the pelitic gneisses and schists of east-central New Hampshire mapped by Billings (1956) as the Littleton Formation of Early Devonian age. The writer correlates the Gonic and the lower, more poorly bedded part of the Rindgemere Formation with the formations of the Northern Sequence (Thompson Mountain, Shagg Pond, Billings Hill Formations, and Concord Pond, Wilbur Mountain, and Howard Pond Members of the Littleton Formation) of Guidotti (1965). These are essentially the same as the units comprising the "middle sequence" of Osberg et al. (this volume). The upper part of the Rindgemere

Formation and the Towow Formation are correlated with the cyclically bedded equivalents of
the upper Littleton Formation in west-central
New Hampshire and east-central Maine, and are
essentially equivalent to the cyclically bedded
pelites and rusty units of the "upper sequence" in
the Dixfield area (Osberg et al., this volume).

In view of the fact that the Shapleigh Group
is superimposed conformably on the Berwick Formation, it is likely that the entire Shapleigh
Group is of Early Devonian age.

The Rye Formation is considered to be of
Ordovician age on the basis of lithic correlation
of the upper metavolcanic member with the
Ammonoosuc Volcanics and the lower metasedimentary member with the Albee Formation of
western New Hampshire (Billings, 1956).

Stratigraphic Relationship and Age of the Casco Bay Group

A major question in southwestern Maine concerns the stratigraphic position of the Casco Bay
Group relative to the other formations and groups
of the area. The problem reduces to three alternatives, two of which are based on marked lithic
similarities of parts of the sequences involved,
and the third based primarily on internal structural simplicity and secondarily on lithic similarity.

*Alternative 1. The Casco Bay Group lies
stratigraphically beneath the Merrimack
Group and is of Ordovician age.* A marked
lithic similarity has been noted between the
Casco Bay Group, with its abundance of volcanic
rocks, and the sequence (oldest to youngest)
Albee–Ammonoosuc–Partridge in New Hampshire and the equivalent but stratigraphically
more complex Missisquoi Formation of Vermont.
Strong similarities exist between the Cape Elizabeth and Albee Formations, between the Cushing
Formation and the Ammonoosuc Volcanics, and
between the Scarboro–Jewell sequence and the
Partridge Formation. A major difficulty with this
lithically-based correlation is that the Albee-type
lithologies of the Casco Bay Group lie conformably above, not beneath, rocks of Ammonoosuc
affinity, although this may possibly be explained
by complex regional intertonguing of the different
lithologies.

If this correlation is valid, the Cushing Formation would equate with the metavolcanic member
of the Rye Formation; the equivalents of the remainder of the Casco Bay Group in the Kittery
area would be missing either because of erosion
of these units prior to deposition of the Kittery
Formation or because of faulting. The contact

of the Berwick Formation with the Casco Bay
Group would necessarily be a fault. Although a
fault has been mapped between the two along the
west side of the Casco Bay outcrop belt south of
Portland, no compelling evidence for one north
of Portland has been found. It is conceivable
that the rusty gneiss at the top of the Berwick
could represent the general location of such a
major fault. Apparent folds in the rusty gneiss
could either indicate a folded pre-metamorphic
fault or be resolved as a series of conjugate cross
faults. Furthermore, the calc-silicate and biotite
granulites on the east side of the Casco Bay
Group would not correlate with the lithically
similar Berwick Formation, but would be the
lowest stratigraphic unit in the area, assuming
that the Bath syncline is a valid structure. Equivalents of these granulites in New Hampshire and
Vermont are lacking.

*Alternative 2. The Casco Bay Group cor-
relates with the Eliot Formation and is
hence of Silurian age.* This correlation is suggested on the basis of lithic identity of the Spurwink Limestone and the ribbon-lime member of
the Eliot Formation, and on the general pelitic
character of the metasediments of the two sequences. Such a correlation also requires a fault
contact between the Casco Bay Group and the
Berwick Formation because of the synclinal character of the Casco Bay outcrop belt, and because
the Eliot Formation lies conformably below, not
above, the Berwick Formation. The volcanics
which form a significant part of the Casco Bay
Group and which are lacking in the Eliot Formation and other equivalent units in west-central
Maine (see Osberg et al., this volume) would
pinch out abruptly west of the outcrop belt of
the Casco Bay Group, as they do on strike to
the south within the outcrop belt of the Group.
Following this correlation, and assuming, as before, the validity of the Bath syncline, the granulites on the east side of the Casco Bay Group
would equate with the Kittery Formation which
is lithically similar.

*Alternative 3. The Casco Bay Group lies
conformably above the Berwick Formation
and is of Devonian age.* This alternative requires a correlation of the Casco Bay Group with
part of the Shapleigh Group, which is compatible
with the general pelitic character of the two
Groups. This offers the simplest and least tectonically involved correlation, and is the one that
was used in the compilation of the new state
geologic map.

The correlation is supported by the following
lines of reasoning:

1. Clear structural information in the area of the Hen Cove anticline (see Fig. 22–2) indicates that the Cushing Formation lies beneath the Cape Elizabeth Formation; hence on the west side of the Bath syncline, the units of the Casco Bay Group face toward the east. Similarly, the Eliot–Berwick sequence and the Mayflower Hill–Waterville–Vassalboro sequence of Osberg (in press) to the north (Fig. 22–1) also face east. Thus all units of the Merrimack Group (and correlatives) and the Casco Bay Group on the west side of the Bath axis face in the same direction.

2. Cushing- and Berwick-type lithologies are repeated in correct stratigraphic order on the east side of the Bath syncline, and here the contact between the two units is conformable.

3. Structural data such as bedding and foliation dips and strikes on either side of the contact do not indicate any discordancy between the Berwick and Cushing Formations on the west side of the Bath syncline north of Portland. South of Portland a discordancy is noted and is readily explained by the post-metamorphic fault mapped there.

If these relationships are valid, then the Casco Bay Group correlates with the Shapleigh Group in part. The Cape Elizabeth Formation would equate with the Gonic Formation and the non-rusty parts of the lower Rindgemere, the Scarboro and Jewell Formations with the local rusty units of the lower part of the Rindgemere, and the Spurwink Limestone with the calc-silicates of the lower part of the Rindgemere. The Cushing, Spring Point, and Diamond Island Formations would be missing from the Shapleigh Group by stratigraphic pinchout.

Until detailed mapping across the continuation of the Bath syncline to the north, and of areas to the east, has been completed, the problem of the stratigraphic position of the Casco Bay Group cannot be conclusively solved. Geophysical investigations of the contact zone between the Merrimack and Casco Bay Groups may be necessary tools in deciding between the alternatives of a faulted versus a conformable sedimentary contact. Until more conclusive information has come to light, it is proper to leave this problem in the form of alternatives.

REFERENCES

Alcock, F. J, 1946, Preliminary Map, St. Stephens, New Brunswick, Canada. Geological Survey Paper 46–2

Billings, M. P., 1956, Geology of New Hampshire: Part II—Bedrock geology: New Hampshire Planning and Development Commission, Concord, New Hampshire, 203 p.

Doyle, R. G., *ed.*, 1967, Preliminary geologic map of Maine: Maine Geol. Survey, 1:500,000

Faul, Henry, Stern, T. W., Thomas, H. H., and Elmore, P. L. D., 1963, Ages of intrusion and metamorphism in the northern Appalachians: Am. Jour. Sci., v. 261, pp. 1–19

Fisher, L. W., 1941, Structure and metamorphism of the Lewiston area: Geol. Soc. America Bull., v. 52, pp. 107–160

Goldsmith, Richard, 1963, Preliminary geologic map of New England, U. S. Geol. Survey open-file map

Guidotti, C. V., 1965, Geology of the Bryant Pond quadrangle, Maine: Maine Geol. Survey Quad. Mapping Series no. 3, 116 p.

Hussey, A. M., II, 1962, The geology of southern York County, Maine: Maine Geol. Survey Spec. Geologic Studies Series, no. 4, 67 p.

——————, *ed.*, 1965, Field trips in southern Maine: New England Intercollegiate Geol. Conf. Guidebook, 57th Ann. Mtg., Brunswick, Maine, 118 p.

Katz, F. J., 1917, Stratigraphy in southeastern New Hampshire and southwestern Maine: U. S. Geol. Survey Prof. Paper 108, pp. 11–29

Keith, Arthur, *compiler*, 1933, Preliminary geologic map of Maine: U. S. Geol. Survey, scale 1:1,000,000

Larrabee, D. M., 1964, Bedrock geologic map of the Big Lake quadrangle, Washington County, Maine: U. S. Geol. Survey Geol. Quad. Map GQ–358

Osberg, Philip H. (in press), Stratigraphy, structural geology and metamorphism in the Waterville-Vassalboro area, Maine, Maine Geol. Survey Bull.

Perkins, E. H., and Smith, E. S. C., 1925, A geological section from the Kennebec River to Penobscot Bay: Am. Jour. Sci., 5th Series, v. 9, pp. 204–228

Sriramadas, Aluru, 1966, Geology of the Manchester quadrangle, New Hampshire: New Hampshire Dept. of Resources and Econ. Dev. Bull. 2, 72 p.

IGNEOUS AND METAMORPHIC PETROLOGY AND PETROGENESIS

Chapter 23

Isotope Geochronology of the Northern Appalachians

J. B. LYONS AND H. FAUL

INTRODUCTION

WHEN RODGERS (1952) reviewed "absolute ages" in the Appalachians sixteen years ago, he could cite only two localities in Connecticut and one in New York where isotopic age data were available. In the intervening years, enormous strides have been made both in our understanding of regional geologic relations of the Northern Appalachians and in the development and application of isotopic dating techniques. Almost 500 isotopic ages have now been published for the area that extends from New York City to northeastern Newfoundland, and 120 lead-alpha dates are also available. Compilations of the radiometric data have been made by Goldsmith (1964) and Tilton (1965). Tectonic evolution has been summarized by Rodgers (1967). What the numbers now tell us about the geology, and what the geology and the numbers together now tell us about the fine structure of geosynclinal evolution, are the questions to which this review is addressed.

Drs. A. H. McNair, R. C. Reynolds, L. R. Page, R. S. Naylor, Priestley Toulmin, and R. E. Zartman have kindly given us the benefit of their comment and advice. Opinions expressed in this paper, however, are our own, as are any shortcomings and errors.

Most geologists are sufficiently sophisticated to appreciate why the word "absolute" has dropped from the literature of isotopic dating, and concise summaries of the problems involved are available in Faul (1966), Schaeffer and Zähringer (1966), Tilton and Hart (1963), and elsewhere. Decay constants have now been established with satisfactory accuracy for the commonly analyzed isotopes except Rb^{87}, for which λ's ranging from 1.39×10^{-11} y^{-1} to 1.51×10^{-11} y^{-1} have been used. Throughout this paper, we have calculated the Rb/Sr data with a λ of 1.47×10^{-11} y^{-1} because we take it to be the best independent value. Instrumental techniques have advanced to the stage where analyses for uranium, thorium, lead, rubidium, potassium, and argon can be performed routinely with a standard deviation of less than 2 percent, and isotopic ratios can be measured to one part in a thousand. Not all published isotopic data meet these high standards, but many do, and the geologist may properly ask, "Why, then, do we have all these problems in interpreting the data?"

In calculating an age from a set of analytical results, we necessarily assume a definite, and usually simple, model for the history of the system in question. It is a fundamental problem of geology that the models we erect for dating, as for other things, are too simple for the inherent complexity of natural events.

THE MEANING OF THE NUMBERS

Argon is quantitatively conserved in several minerals at room temperature, but a relatively small increase in temperature will cause it to leak out of the crystal. Increasing pressure counteracts the effect, but no experimental data exist that bear on this aspect of the problem, nor on the related one of K diffusion under varying *P-T* and chemical conditions. Inasmuch as pressure and temperature are not mutually independent in most geological situations, there is real indeterminacy in the interpretation of K–Ar data.

All we can say is that the argon clock starts when the rock cools to some relatively low temperature (perhaps 200–300°C for biotite), that it runs well as long as the rock remains cool, but that it is soon reset back to zero when the rock is reheated for any reason. Small crystals scattered in a fine-grained rock lose argon more quickly than do large crystals in a pegmatite, and mica loses argon at much lower temperatures than does hornblende. Many rocks in the Northern Appalachians lost all their argon in a gentle metamorphic event which produced no change visible in thin section; the loss of argon and the erasure of accumulated fission tracks may be the only tangible evidence that a rock has been thus affected.

The period of reheating may be gentle enough to cause only partial loss of argon. The clock then would be set back to some time later than crystallization and earlier than reheating, thus giving a *mixed age.*

Rb–Sr mineral ages are affected by reheating. The Rb–Sr system in biotites is very susceptible to hydrothermal alteration. Feldspars are fairly resistant, but tend to contain much original, non-radiogenic strontium, which limits their usefulness for dating. Pegmatite muscovite is probably the most resistant.

Strontium moving out of a rubidium-rich crystal during a period of reheating or alteration does not usually go very far before it is captured, mainly by calcium-rich feldspars and apatite. If the heating or alteration is sufficiently pervasive to cause equilibration of all strontium in a rock, and if the rock as a whole remains a closed system, one may analyze several mineral phases for Rb, Sr, and the Sr^{87}/Sr^{86} ratio, and plot a *mineral isochron* which will then give a reliable date for the time of that event.

Whole-rock Rb–Sr isochrons are a different matter; their interpretation involves the additional assumptions that all points on the isochron represent rocks isolated at the same time from some common environment, and that each of them had the same Sr^{87}/Sr^{86} ratio at that time. The whole-rock isochron dates the time of isolation and is not affected by later reheating as long as each rock remains a closed subsystem of this larger system.

Whether a rock is part of a larger system may be determinable from geologic considerations, but not always. Some published isochrons may give misleading ages because they include unrelated rocks, or because the rocks have formed open systems during metamorphism (Lanphere et al., 1964).

Lead-uranium data are notably sparse for the Northern Appalachians. Within the past 16 years, only 3 additional localities (Branchville, Connecticut, by Wasserburg et al., 1956; Redstone, New Hampshire, by Tilton et al., 1957, and Bear Mountain, New York, by Tilton et al., 1958) have been added to the list of places where concordant Pb^{206}/U^{238} and Pb^{207}/U^{235} determinations give what is generally conceded to be ironclad evidence of time of rock or mineral formation. Thus the ages 1145 and 975 (Bear Mountain, New York), 360 (Bedford, New York, and Branchville, Connecticut), 255 (Middletown, Connecticut) and 185 m.y. (Conway, New Hampshire) have become important "bench marks," both for calibrating K–Ar and Rb–Sr isotopic methods, and for comparison of ages. In all these localities, unfortunately, geologic relations fix only maximum limits on the geologic time scale, and the tectonic significance of the dates is based on indirect evidence.

Lead-alpha numbers should yield essentially the same evidence as Pb^{206}/U^{238} dates, but there are many built-in uncertainties, such as lead loss, variable Th/U ratios, etc., and the dates reported prior to 1960 are too low because of errors in the lead determinations (Rose and Stern, 1960). Consequently, the lead-alpha numbers serve, at best, as a rough guide to original time of formation, and are most meaningfully interpreted if K–Ar and Rb–Sr data are also available for the rock.

Figure 23–1 shows an age histogram of the Northern Appalachians and yields clues to significant patterns. The striking peaks at 240–250 m.y. and 360–370 m.y. on the K–Ar plot of 303 analyses, and the close correspondence of these numbers to our "bench mark" dates of 255 m.y. and 360 m.y. are indications of a non-random distribution. On the other hand, the 185 Rb–Sr determinations are not systematic, and we are unable to rationalize their spread with any consistent interpretation of geologic evolution.

GEOLOGIC DEVELOPMENT OF THE NORTHERN APPALACHIANS

Precambrian

It is now widely recognized (cf. Engel, 1963) that the Appalachians are located within an infrastructure whose radiometric dates range up to a maximum of about 1100 m.y. The tectonic–plutonic events immediately preceding the development of the Appalachian geosyncline proper have been consistently misnamed the "Grenville orogeny" (Gilluly, 1966); their effect was to re-

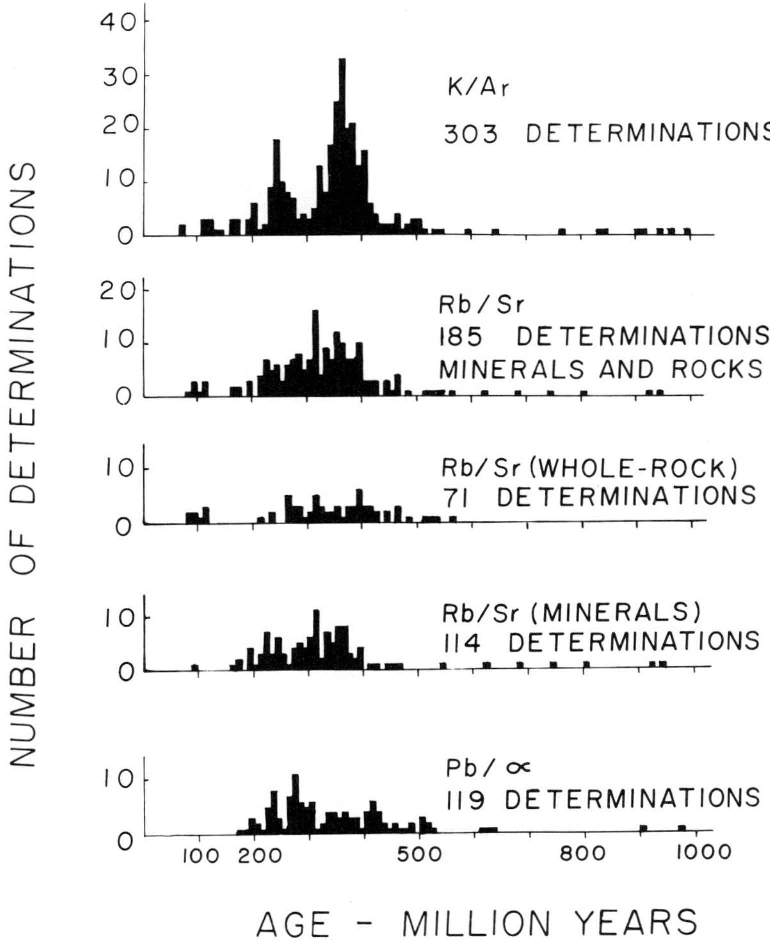

FIGURE 23-1. Histogram of K/Ar, Rb/Sr, and Pb/α ages for the Northern Appalachians.

set all nuclear clocks in a broad region from Eastern Canada through the Appalachians to Texas.

In eastern Newfoundland, in New Brunswick, and on Cape Breton Island, thick sequences of basaltic to rhyolitic volcanics intercalated with sediments lie unconformably below Lower Cambrian strata (Weeks, 1957, p. 131–137). Whole-rock Rb–Sr age determinations on these volcanics by Fairbairn et al. (1966) yield internally consistent dates ranging from 468 to 537 m.y., but these dates are at least 100 m.y. too young for latest Precambrian on any of the now accepted time scales.

Two Rb–Sr dates (Fairbairn, 1958; McCartney et al., 1966) on feldspar from the Holyrood granite of Newfoundland, which intrudes the Harbor Main (537 m.y.) volcanics, give an age of 565 m.y., whereas a whole-rock Rb–Sr determination on the granite (McCartney et al., 1966) gives an age of 580 m.y. The 468 m.y.

whole-rock Rb–Sr isochron for the Precambrian Coldbrook volcanics near St. John, New Brunswick (Fairbairn et al., 1966) is younger than the 500–640 m.y. K–Ar dates in the Precambrian terrane nearby (Lowden et al., 1963; Leech et al., 1963).

The Dedham, Northbridge, Milford and Hoppin Hill granites of southeastern Massachusetts are generally presumed to be consanguineous, and the Hoppin Hill granite lies unconformably below Lower Cambrian sediments. Rb–Sr whole-rock ages on all these granites (Fairbairn et al., 1965) lie near a 540 m.y. isochron. Considering the uncertainty, both stratigraphic and geochronologic, which is attached to the "base of the Cambrian," it is difficult to decide what to make of the consistently low measured Precambrian ages in the Northern Appalachians. Taken as a whole, the geochronologic evidence from the Maritime Provinces and New England points to the evolution of a post-Grenville Precambrian terrane in the region which is now largely the continental shelf; this picture is in agreement with geologic deduction (cf. Rodgers and Neale, 1963; Williams, 1964; Lilly, 1966). Lilly (1966) has proposed the name Avalonian to describe these Late Precambrian orogenic and plutonic events.

Cambrian

The lithologies, relative ages, and distribution of Cambrian rocks throughout the Northern Appalachians are reasonably well known (see Theokritoff, this volume) but meaningful geochronologic data are scarce. The few ages that fall in the 500–600 m.y. range (Fig. 23–1) are chiefly on Precambrian rocks, but one (518 m.y.; Fairbairn et al., 1960a) is on a Nova Scotia granite, probably of post-Cambrian age (Poole et al., 1964). No Cambrian volcanics have been reliably dated.

A Late Cambrian (?) orogeny has been described in southeastern Quebec (Cooke, 1955; Riordon, 1957), but the evidence has been questioned by Cady (1960, p. 564), and K–Ar dating (Wanless et al., 1966, p. 78) has not resolved this issue. There are no radiometric data that elucidate the recently described Late Cambrian (?) "Penobscot disturbance" of eastern Maine (Neuman and Rankin, *in* Caldwell, 1966).

Ordovician

Summaries of regional geologic relations in New Hampshire (Billings, 1956), Vermont and southeastern Quebec (Cady, 1960), Connecticut (Rodgers et al., 1959) and the Canadian Appalachians (Neale et al., 1961; Poole et al., 1964; and Weeks, 1957) emphasize not only the fundamental differences between Ordovician miogeosynclinal and eugeosynclinal sedimentation, but also two other important features— an apparent increase, relative to the Cambrian, of mafic and felsic volcanics, and evidence for non-synchronous and spatially localized tectonism in different parts of the geosyncline. Geologically, the two more significant tectonic events are the emplacement of plutons and the Taconic orogeny. The latter is generally identified as a profound unconformity separating Middle Ordovician from Middle Silurian strata (Zen, this volume; Neale et al., 1961), but it is absent along the 400 km. long Metapedia anticlinorium of Maine and the Gaspé (Pavlides et al., 1964; Boucot et al., 1964; Pavlides and Berry, 1966; Pavlides et al., this volume).

In Table 23–1, we have assembled some of the more interesting age determinations on rocks which are or may be of Ordovician age.

Billings (1937) was the first to identify Taconic-age intrusives (the Highlandcroft Series)

TABLE 23-1.   Some Ordovician Ages in the Northern Appalachians

Geochronologic age (m.y.)	Rock and Locality	Paleontologic-Geologic age	Method	Reference
405	Ammonoosuc volcanics (N.H.)	Pre-Silurian; Middle Ordovician (?)	Rb–Sr whole-rock	Brookins and Hurley, 1965
418	Middletown Formation (Conn.)	Correlated with Ammonoosuc Volcanics	Rb–Sr whole-rock	Brookins and Hurley, 1965
445	Monson Gneiss (Conn.)	Pre-Ammonoosuc Volcanics; age not known	Rb–Sr whole-rock	Brookins and Hurley, 1965
420	Mascoma (Oliverian) Granite (N.H.)	Pre-Silurian (?); or Middle Devonian (?)	Rb–Sr whole-rock	Naylor and Wasserburg, 1966
318	Glastonbury and Maromas Gneiss (Conn.)	Pre-Silurian (?); Correlated with Oliverian granites	Rb–Sr whole -rock	Brookins and Hurley, 1965
464	Gabbro; Ripogenus Dam (Me.)	Early Paleozoic (?)	K–Ar	Faul et al., 1963
479	Serpentinite; Thetford, P.Q.	Intrudes Middle Ordovician	K–Ar	Leech et al., 1963
495	Serpentinite; Mt. Albert, Gaspé, P.Q.	Middle Ordovician (?)	K–Ar	Lowden et al., 1963
438 (av.)	Granite, N.B.	Intrudes Middle Ordovician	K–Ar	Lowden et al., 1963
>500	Zircon in pegmatite (Conn.)	Early Paleozoic (?)	Pb–U	Zartman et al., 1965a
435	Andover granite (Mass.)	Post-Cambrian; post Upper Silurian (?)	Rb–Sr whole-rock	Fairbairn et al., 1967

in the Northern Appalachians on geologic grounds, but these rocks were not identified on structural-stratigraphic grounds in the Canadian Appalachians until 1960 (Poole et al., 1964, p. 68; and Naylor, this volume). The proposed assignment by Naylor and Wasserburg (1966) of the Oliverian plutons of New Hampshire to the Ordovician, as well as geologic or geochronologic evidence (Table 23–1; Poole et al., 1964) for Ordovician intrusives in Maine, Quebec, New Brunswick and Newfoundland, now make it clear that Taconic-age magmatism was widespread. The dates for the Canadian serpentinites (Table 23–1) and their close agreement with the 460 m.y. K–Ar age reported by Lapham and Bassett (1964, p. 665) for the ultramafics in Pennsylvania apparently confirm the inferred Ordovician age of the main ultramafic belt of the Northern Appalachians (Chidester, this volume).

Probably because of the effects of subsequent orogenies, many of the dates in Table 23–1 for southern New England are suspect. The age for the Glastonbury gneiss is clearly too low, and the Monson gneiss may actually be as old as Precambrian. The age of the Ammonoosuc and Middletown formations relative to the probably younger ultramafics of Quebec is anomalous. Field evidence (Castle, 1965) suggests that the Andover granite may be post-Silurian, in disagreement with radiometric determination. Similarly, it is interesting that a Newfoundland granite with a K–Ar age of 484 m.y. (Early Ordovician) intrudes Silurian formations (Lowden et al., 1963).

Hadley (1964, Fig. 1) has found in both the Central and Southern Appalachians a well-marked peak at 430–450 m.y. on an age histogram, and assigns this maximum to the emplacement of plutons during the Taconic orogeny. The smearing of the record in the Northern Appalachians apparently means: (a) a more complex and severe post-Taconic history; (b) the inability of isotopic dating to add significant detail to the Ordovician geologic record.

Silurian

A Late Llandovery (late Early Silurian) to Ludlovian (Late Silurian) northwestward transgression is represented over much of the Northern Appalachians by thin conglomerates succeeded by thin limestones. Naylor and Boucot (1965) have identified five Ludlovian lithologic belts, of which the eastern two, consisting of graywacke-slate and volcanic-shale, have eugeosynclinal affinities. Boucot (1962, p. 156) has identified a "Salinic Disturbance" separating Ludlovian

(Upper Silurian) and Gedinnian (Lower Devonian) rocks. There are no geologic data which prove that any of the plutons in the Northern Appalachians is of Silurian (or Salinic) age.

Bottino and Fullagar (1966) obtained whole-rock Rb–Sr ages of 390 m.y. for the Eastport Volcanics of northeastern Maine, which are latest Silurian or earliest Devonian, and for the nearby Hedgehog formation of Lower Devonian (Gedinnian, or New Scotland) age. These two dates agree with the commonly accepted 400 m.y. age (Holmes, 1959; Kulp, 1960; Faul, 1966) for the break between the Silurian and Devonian. However, the Lower Silurian (Llandovery) Arisaig Volcanics of Nova Scotia has a whole-rock Rb–Sr age of 384 m.y. (Bottino et al., 1963), and the Upper Silurian or Lower Devonian Newbury Volcanics of eastern Massachusetts has a whole-rock Rb–Sr age of 335 m.y. (Fairbairn et al., 1967). The two latter dates are mutually consistent, but inexplicably 60 m.y. too young for the accepted time scale. Zartman (written communication, 1967) suggests that tuffaceous acid volcanics may become closed systems for Rb–Sr only after recrystallization, which may be long after the initial volcanism; this may partly explain the anomalies.

Several granites, particularly in New Brunswick and Newfoundland, have Silurian K–Ar ages (430–400 m.y.), and intrude Silurian or older rocks. Therefore, as suggested by Wanless et al. (1966) Late Silurian plutonism may have occurred in the Northern Appalachians, but field geologic relations do not provide conclusive proof.

Devonian

Devonian sedimentation, volcanism, and tectonism have been ably summarized by Boucot (1962), Pavlides et al., (1964), Weeks (1957) and Neale et al., (1961). The onset of the Acadian orogeny is of major importance; its first major pulse occurred in the Late Emsian (the Schoharie–Onondaga break), although deformational pulses also can be identified in the Middle and Late Devonian (Boucot et al., 1964; Neale et al., 1961). Boucot (1962, p. 161) suggests that the Acadian orogeny may have begun first in northern Nova Scotia, possibly in Gedinnian (New Scotland) time, and worked southward, reaching the latitude of New York in Eifelian (Onondaga) time.

Abundant intrusion of granites throughout the entire Devonian is suggested by the radiometric data (Fig. 23–1), but this is not certain on geologic grounds (see Page, this volume). Most granites in New England appear to be Late Emsian (late Early Devonian) or younger on strati-graphic-structural evidence, but only two Appalachian granites in Canada are known to be post mid-Devonian (Poole et al., 1964). Ultramafics are rare; a few intrude Lower Devonian formations in western Maine and southeastern Quebec (Cady, 1960, p. 575; Green and Guidotti, this volume) and are undated.

The Northern Appalachians are one of the few regions where there are numerous age determinations for intrusives and volcanics of Devonian age that are closely bracketed on paleontologic-geologic grounds (Table 23–2). The beginning of the Devonian in the Holmes (1959) time scale (400 m.y.) was based on the Calais granites and is now confirmed by the Shiphead bentonite in the Gaspé Peninsula and the Hedgehog and Eastport Volcanics in Maine, within the uncertainty of the rubidium decay constant. The Kineo and Traveler rhyolite ages (340–350 m.y.) fix a time for the Emsian (late Early Devonian) and the Hog Island intrusion crudely confirms this date. The 374 m.y. age for the Fisset Brook volcanics (which is close to the Devonian–Carboniferous time break) is anomalous, and cannot be interpreted for the time being.

The coincidence of the 360 m.y. Pb/U dates on Bedford, New York, and Branchville, Connecticut, pegmatites with the 360 m.y. peak on the K–Ar histogram (Fig. 23–1) confirms our belief that the break represents an extremely important, widespread and accurately dated geologic event. The near-concordance of many 400–350 m.y. K–Ar and Rb–Sr mineral dates in Maine and Nova Scotia adds weight to the evidence. Some of the whole-rock Rb–Sr isochrons for Nova Scotia (Fairbairn et al., 1964) likewise lie in that range, but others are anomalously low, even on rocks whose individual K–Ar and Rb–Sr mineral ages are concordant at 350–400 m.y.

In New Brunswick, granites intrude Silurian rocks (Devonian sediments are not widespread), but reach the Carboniferous in only one locality. K–Ar dates for the granites are commonly in the 380–400 m.y. range (Tupper et al., 1959; Leech et al., 1963; Lowden et al., 1963; Wanless et al., 1966) and suggest an Early Devonian or possibly Late Silurian age. In New Brunswick, northern Maine, and northern Nova Scotia, a 360–400 m.y. K–Ar age is common, but the ages tend to decrease toward the south in Maine and Nova Scotia (Faul et al., 1963; Hurley et al., 1958; Fairbairn et al., 1960a; Poole et al., 1964) probably because of a Late Paleozoic overprint. Several

Canadian Survey K–Ar dates for Quebec and Newfoundland also suggest a Devonian isotopic age, but only a few of them yield stratigraphic proof that they are Early Devonian or younger.

With the exception of central and northern Maine, most of New England has a confused pattern of radiometric ages for rocks which are of inferred Acadian age; the problem is complicated by uncertainties in field relationships in several critical areas, and by smearing of much of the Acadian isotopic record by the late "Allegheny" disturbance.

Acadian age granites and pegmatites of Vermont and New Hampshire have K–Ar and Rb–Sr mineral dates which fall into the 290–340 m.y. range (Faul et al., 1963; Fairbairn et al., 1960b; Damon and Kulp, 1957). Lead-alpha dates in the same or synchronous granites in northern (Lyons et al., 1957), central (Webber et al., 1958), and southern (Quinn et al., 1957) New England fall into the same age bracket, but all these lead-alpha age determinations are erroneously low (Rose and Stern, 1960). More recent and presumably more accurate lead-alpha ages lie in the 310–450 m.y. range (Green, 1964; Faul et al., 1963).

In Massachusetts, concordant mineral and whole-rock Rb–Sr ages of 345–360 m.y. for the Chelmsford and Ayer granites (Zartman et al., 1965a; Fairbairn et al., 1967) infer Late Devonian plutonism; the Ayer granite, however, is reported to intrude fossiliferous Pennsylvanian (?) coal at Worcester, Massachusetts. Rb–Sr whole-rock ages (Fairbairn et al., 1967) of 435 m.y. for the Andover Granite, 390 m.y. for the Cape Ann Granite, 340 m.y. for the Salem Gabbro–Diorite, and 325 m.y. for the Newbury volcanics are not in accord with field observations which demonstrate that the Cape Ann intrudes the Salem (Toulmin, 1964), and suggest (Castle, 1965) that the Andover is younger than the Late Silurian or Early Devonian (Cuppels, 1967) Newbury Formation.

Whole-rock Rb–Sr ages of 405 m.y. in granites in eastern Connecticut (Zartman et al., 1965a), the 360 m.y. Pb–U pegmatite dates at Branchville, Connecticut, and Bedford, New York, (Nier et al., 1939; Wasserburg et al., 1956), as well as several concordant K–Ar and Rb–Sr ages on schists and gneisses in the 350–360 m.y. range in northeastern New Jersey and southeastern New York (Long, 1961, 1962; Long and Kulp, 1962) demonstrate widespread Devonian plutonism and metamorphism throughout this portion of the Appalachians.

TABLE 23-2. Radiometric–Paleontologic Ages of Some Devonian and Near-Devonian Rocks of the Northern Appalachians

Geochrono-logic age (m.y.)	Rock	Paleontologic age	Method	Reference
374 ± 32	Fisset Brook volcanics; basal Horton Group, Cape Breton Is.	Earliest Miss. (Kinderhook)	Rb–Sr whole-rock	Cormier and Kelly, 1964; McDougall et al., 1966
364 ± 10	Contact metamorphosed Seeboomook slate (Oriskany; sets minimum age for Lower Devonian) Maine	Oriskany or younger (Early Devonian)	K–Ar biotite	Hurley et al., 1959
365 ± 15	Contact aureole of Hog Island qtz. monz., Jackman, Me.	Oriskany or younger (Early Devonian)	K–Ar biotite	Hurley et al., 1959
340 ± 28	Hog Island qtz. monz.	Oriskany or younger (Early Devonian)	Rb–Sr biotite	Hurley et al., 1959
350 ± 10	Kineo rhyolite, Maine	Emsian (Oriskany-Onondaga, = late Early Devonian-Middle Devonian)	Rb–Sr whole-rock	Bottino et al., 1965
340 ± 10	Traveler rhyolite; unconformably overlain by lowest Middle Devonian; Maine	Emsian (late Early Devonian)	Rb–Sr whole-rock	Bottino et al., 1965
389 ± 20	Shiphead bentonite, Gaspé	Siegenian (Oriskany) (middle Early Devonian)	K–Ar sanidine	Smith et al., 1961
382 ± 26	Shiphead bentonite, Gaspé	Siegenian (Oriskany) (middle Early Devonian)	K–Ar biotite	Smith et al., 1961
391 ± 10	Hedgehog volcanics, Maine	Gedinnian (New Scotland; early Early Devonian)	Rb–Sr whole-rock	Bottino and Fullagar, 1966
390 ± 5	Eastport Formation, Maine (volcanics)	Gedinnian (Early Devonian) or Ludlovian (Late Silurian)	Rb–Sr whole-rock	Bottino and Fullagar, 1966
405 ± 8	Calais granite, Maine	Post-Silurian and pre-Late Devonian	K–Ar biotite	Faul et al., 1963
385 ± 8	Calais granite, Maine	Post-Silurian and pre-Late Devonian	Rb–Sr biotite	Faul et al., 1963

Carboniferous

Terrestrial or marine wedges of clastic Carboniferous rocks underlie portions of Newfoundland and northeastern Maine (Larrabee et al., 1965), but are best developed in the Nova Scotia–New Brunswick basin and in southeastern New England. In both latter areas, sediments are intercalated with basalts and rhyolites. Local unconformities and disconformities are common throughout the Carboniferous stratigraphic section (Poole et al., 1964, p. 64; Belt, this volume). Intrusives cut Carboniferous strata at St. John, New Brunswick, (Alcock, 1938) and at several places in southern Rhode Island and southeastern Connecticut (Quinn and Moore, this volume). Geologic relations and K–Ar dates (Wanless et al., 1967) demonstrate emplacement of magmatic rocks in Newfoundland in Latest Devonian or Earliest Mississippian time.

The ages of a series of alkaline granites in eastern Massachusetts remain uncertain (Table 23–3). The granites have generally been considered to be comagmatic, but radiometric dating casts doubt on this assumption. A post-Silurian age for the Cape Ann granite is possible (Castle, 1965) but unproven; a pre-Pennsylvanian and post-Cambrian age for the Quincy granite is certain (Chute, 1966; Quinn and Moore, this volume). Until the radiometric and geologic conflicts are clarified, the Mississippian (?) age assignment generally made for these rocks cannot be satisfactorily revised.

Several K–Ar determinations, which are Carboniferous, came from the Maritime Provinces. One of them (317 m.y.) (Wanless et al., 1966) is on a granite intruding Mississippian sediments near St. John, New Brunswick, and implies a Mississippian age for that intrusive; many of the others are apparently on older rocks which have lost argon. Carboniferous Rb–Sr whole-rock dates also turn up for several granites of probable Devonian age in southwestern Nova Scotia (Fair-

TABLE 23-3. Radiometric Dates on Alkaline Intrusions of Eastern Massachusetts

Age (m.y.)	Method	Formation	Comment	Reference
393 ± 10	Rb–Sr whole-rock	Cape Ann Granite		Bottino et al., 1963
345 ± 15	K–Ar Hornblende	Cape Ann Granite		Fairbairn et al., 1967
260 ± 30	Pb–α Zircon	Peabody and Cape Ann Granites	Minimum age	Webber et al., 1958
307 ± 25	Rb–Sr whole-rock	Quincy Granite		Bottino et al., 1963
234 ± 10	Rb–Sr whole-rock	Blue Hills Aporhyolite	Generally considered a phase of Quincy Granite	Bottino et al., 1963
232 ± 10	Rb–Sr whole-rock	Blue Hills Rhyolite	Same as above	Bottino et al., 1963

bairn et al., 1960a; 1964); these low ages may reflect Carboniferous deformation rather than Carboniferous intrusion as suggested by Fairbairn et al. (1964).

The only radiometrically dated extrusives which have been paleontologically dated are the basal Mississippian Fisset Brook volcanics of Nova Scotia (cf. Table 23–2; 374 m.y. whole-rock Rb–Sr age) and the Middle or Upper Pennsylvanian Wamsutta volcanics of Rhode Island (225 m.y. Rb–Sr whole-rock age; Bottino et al., 1963). K–Ar and Rb–Sr mineral ages of 260 ± 25 m.y. have been measured on granites (and related hornfels) intruding folded Pennsylvanian (Alleghenian) sediments of the Narragansett Basin of Rhode Island (Hurley et al., 1960). Lead-alpha (minimal) ages for the same granites (Quinn et al., 1957) average 235 m.y. The many Pb/U, K/Ar and Rb/Sr concordant ages near Middletown, Connecticut, are on pegmatites cutting rocks as young as the Devonian (Rodgers and Rosenfeld, 1959). Because there are no Permian sedimentary rocks in New England, the Rhode Island dates fix our 260 m.y. "bench mark" age between Late Pennsylvanian and Late Triassic. This is also the probable age of the misnamed "Alleghenian" orogeny (Woodward, 1957). This deformational episode, which was far less severe than the Acadian or Taconic, is recorded by the folding of the Carboniferous sedimentary basins, by magmatism, and by the pattern of 260 ± 30 m.y. isotopic ages in New England and Nova Scotia. The limits of the region over which the clock has been reset in New England extend from southwestern Connecticut into southwestern Maine, and do not correspond to the known limits of Carboniferous or younger sedimentation, magmatism, folding, or faulting. Explanations for the pattern of crustal reheating still remain speculative.

Triassic

No dates have been obtained in the Upper Triassic terrestrial sediments and volcanics of either New England or the Bay of Fundy area.

Erickson and Kulp (1961) report K–Ar ages of 190–200 m.y. for the Palisade Diabase and Watchung Basalt of the Newark Basin; these ages are probably valid for the Triassic of the Northern Appalachians.

Harakal (1966) has reported a mixed pattern of K–Ar mineral ages in western Rhode Island in which a 200 m.y. event is overprinted on an older 250 m.y. age. The younger event presumably reflects the Palisade Disturbance in an area 40 miles east of the Triassic basin in Connecticut.

Jurassic (?)

The 180 m.y. concordant Pb/U, K/Ar and Rb/Sr ages in the Conway granite (Tilton et al., 1957; Aldrich et al., 1958; Hurley et al., 1958) closely fix the time of emplacement of the White Mountain plutonic-volcanic series. Lead-alpha ages (Lyons et al., 1957) on several other rocks of the White Mountain series agree with the 180 m.y. date, but Faul et al. (1963) found a 130 m.y. K–Ar age for a hornfels of the Ascutney, Vermont, stock, which apparently belongs to the White Mountain series. Further determinations are in progress.

The closeness of the ages of the Triassic lavas (190 m.y.) and the White Mountain magma series (180 m.y.) suggests that the igneous events may be related. Petrographic correlations for the rocks, however, have not been established, and the mutual geologic relations are obscure. Diabase dikes are common throughout New England, are generally thought of as Triassic; however, they intrude ring-dike structures in southeastern Maine (Hussey, 1962), assigned to the White Mountain series. Funnel- or pipe-shaped gabbros in the same area are also younger than the ring-dike structures. De Boer (1967) has found that many of the diabase dikes of the Appalachians have a remanent magnetization corresponding to a Cretaceous age.

K–Ar ages of 115–144 m.y. on lamprophyre dikes in Newfoundland (Wanless et al., 1966; Wanless et al., 1967) represent the only other known Northern Appalachian rocks of possible Jurassic age.

Cretaceous

The main belt of White Mountain series plutons, trending north-northwest across New Hampshire, meets the eastern terminus of the east-trending Monteregian Hills plutons near the International Boundary. Despite their close proximity, however, the stocks of the Monteregian Hills are about 55 m.y. younger than those of the White Mountain series: K–Ar mineral and whole-rock ages (Lowden, 1960; 1961; Hurley, 1960) average 115 ± 10 m.y.; Rb–Sr whole-rock isochrons (Fairbairn et al., 1963) give an age of 124 m.y.; but Rb–Sr biotite ages (Fairbairn et al., 1963) average 99 m.y. The age pattern as a whole implies crystallization beginning 125 m.y. ago (Early Cretaceous) and subsequent redistribution of isotopes.

The Monteregian Hills are the youngest known intrusives of the Northern Appalachians. Some recent data by Zartman et al. (1965b) show a 120–150 m.y. pattern for alkalic intrusions along the west side of the Appalachians in Vermont, New York, and Virginia, indicating that the Monteregian intrusives are not a chronologically isolated event.

Volcanism, Tectonism, and Plutonism

The idea that there may be spatial and temporal interrelations among geosynclinal volcanics (particularly the felsic type), granitic plutons, and orogeny has been tested by Gilluly (1965) for the Cordillera of the United States, and found wanting. For the Northern Appalachians, the evidence which we have reviewed shows that correlations between major tectonism and major plutonism are more than mere coincidence, perhaps because here we see a deeper level of the earth's crust. Table 23–4 is a summary of the igneous-tectonic events in the Northern Appalachians as we now understand them. The times of volcanism shown in this table generally precede the times of intrusion, to which the radiometric dates apply.

Table 23–4 does not distinguish numerous lesser disturbances recorded by minor unconformities, partly because these are not synchronous over wide areas, and partly because both the geologic and chronologic records become increasingly obscure with time. While we endorse conclusions such as those by Eardley (1962, p. 202) that the Northern Appalachians show evidence for "a long succession of compressional impulses with accompanying intermittent volcanics and magmatic activity," we think that "superior" tectonism (to borrow his phrase) should be stressed. Such superior orogenies are marked by the development of granites, which cover roughly the same area as does intense tectonism; and include the granites related to Avalonian, Taconic, Acadian, and "Alleghenian" tectonism. Because the occurrence of Silurian granites is questionable, the only remaining intrusive rocks are the alkalic magmatic suites of Mississippian (?), Lower Jurassic and Lower Cretaceous ages, all of which lack evidence of associated areal tectonism, and all of which are of relatively minor areal extent (the White Mountain series constituting a possible exception). As association of alkaline complexes with atectonic intervals of geologic time is universal, it is not surprising to find it also in the Northern Appalachians. Outpouring of basalts

TABLE 23-4. Volcanism, Plutonism, and Tectonism in the Northern Appalachians

Radiometric age (m.y.)	Geologic age	Volcanism	Plutonism	Tectonism
115 ± 10	Early Cretaceous		Monteregian alkalic plugs	
180 ± 10	Early Jurassic	Felsic Volcanics	White Mountain Series	
195 ± 10	Late Triassic	Basalts		Palisade Disturbance
260 ± 20	Late Pennsylvanian or Early Permian		Granites	"Allegheny" Orogeny
320 ± 85	Mississippian (?)		Granites	
360 ± 30	Devonian	Felsic to Mafic Volcanics	Widespread Plutonism; Rare Ultramafics	Acadian Orogeny
400 ± 10	Late Silurian	Felsic to Mafic Volcanics	Granites (?)	Salinic Disturbance
450 ± 35	Middle Ordovician	Felsic to Mafic Volcanics	Widespread Plutonism; Major Ultramafics	Taconic Orogeny
500 (?)	Late Cambrian (?)	Felsic to Mafic Volcanics		Penobscot Disturbance
600 (?)	Late Precambrian (?)	Felsic to Mafic Volcanics	Granites	Avalonian Orogeny

in taphrogeosynclines, also to be anticipated, is represented by the Triassic effusives.

CRUSTAL AND CONTINENTAL ACCRETION

The idea that younger geosynclines lie progressively on the ocean side of cratonal areas, and that they represent additions of crustal material to the continental nucleus is an attractive hypothesis to many geologists. In the Northern Appalachians, the geologic-chronologic evidence which we have reviewed does not provide decisive evidence on this question.

The radiogenic Sr^{87} model of continent formation (Hurley et al., 1962) does provide another test of the accretion theory. The earth's mantle is impoverished in Rb relative to the crust, so that the rate of accumulation of Sr^{87} in the crust is much more rapid; Hurley et al. (1962) estimate that the differential crustal increase of Sr^{87}/Sr^{86} averages .010 per billion years. To take a probably oversimplified example, we might assume that the "Grenville" crust became isolated from the mantle a billion years ago. The Sr^{87}/Sr^{86} ratio in the mantle has probably not increased by more than .001 or .002 in the past billion years, and now lies in the range of .707 ± .002. A volcanic rock generated in the mantle 300 m.y. ago would show an initial Sr^{87}/Sr^{86} close to .707, but an anatectic granite created 300 m.y. ago from an average "Grenville" terrane should show an initial Sr^{87}/Sr^{86} of approximately .714, and thus betray its crustal origins. The pertinent data for the Northern Appalachians are assembled in Table 23–5.

The numbers in Table 23–5 tell no consistent story, and show that the demands of the accretion theory set standards which cannot be met with the data currently available. The fact that most of the initial ratios fall below .710 may indicate to some that the rocks have a mantle origin, that

TABLE 23-5. Initial Sr^{87}/Sr^{86} Ratios in Volcanic and Plutonic Rocks of the Northern Appalachians

Formation or Unit	Age	Initial Sr^{87}/Sr^{86}	Reference
Monteregian Hills	Early Cretaceous	.704	Fairbairn et al., 1963
Wamsutta Volcanics	Middle or Late Pennsylvanian		
Rhyolite		.721	Bottino et al., 1963
Basalt		.708	
Blue Hills Complex	Mississippian (?)		
Aporhyolite		.708	Bottino et al., 1963
Porphyry		.722	
Quincy Granite	Mississippian (?)	.728	Bottino et al., 1963
Cape Ann Granite	Mississippian (?)	.709	Bottino et al., 1963
Fisset Brook Volcanics	Early Mississippian (?)	.704 ± .002	Cormier and Kelly, 1964
Ayer Granite	Late Devonian (?)	.700 ± .010	Zartman et al., 1965a
Nova Scotia Granite	Middle (?) Devonian	.705 — .711	Fairbairn et al., 1964
Canterbury Gneiss	Devonian (?)	.705 ± .002	Zartman et al., 1965a
Andover Granite	Devonian (?)	.705	Fairbairn et al., 1967
Chelmsford Granite	Devonian (?)	.719	Fairbairn et al., 1967
Traveler Rhyolite	Early Devonian	.710	Bottino et al., 1965
Kineo Rhyolite	Early Devonian	.715	Bottino et al., 1963
Quartz Monzonite (Eastern Connecticut)	Devonian (?)	.700 ± .020	Zartman et al., 1965a
Eastport Volcanics	Early Devonian	.707	Bottino and Fullagar, 1966
Hedgehog Volcanics	Late Silurian or Early Devonian	.706	Bottino and Fullagar, 1966
Newbury Volcanics	Late Silurian	.709	Fairbairn et al., 1967
Arisaig Volcanics	Early Silurian	.709	Bottino et al., 1963
Ammonoosuc Volcanics	Middle (?) Ordovician	.707	Brookins and Hurley, 1965
Middletown Formation	Middle (?) Ordovician	.705	Brookins and Hurley, 1965
Glastonbury Gneiss	Ordovician (?)	.709	Brookins and Hurley, 1965
Maromas Gneiss	Ordovician (?)	.712	Brookins and Hurley, 1965
Monson Gneiss	Ordovician (?) or older	.707	Brookins and Hurley, 1965
Coldbrook Volcanics	Late Precambrian	.707	Fairbairn et al., 1966
Fourche Volcanics	Late Precambrian	.705	Fairbairn et al., 1966
Bull Arm Volcanics	Late Precambrian	.704	Fairbairn et al., 1966
Harbour Main Volcanics	Late Precambrian	.705	Fairbairn et al., 1966
Dedham Granodiorite	Late Precambrian	.705 ± .004	Fairbairn et al., 1965
Northbridge Granite Gneiss	Late Precambrian	.707 ± .004	Fairbairn et al., 1965
Hoppin Hill Granite	Late Precambrian	.705 ± .022	Fairbairn et al., 1965

there has been major continental accretion, and that most of the plutonic rocks did not originate by anatexis of geosynclinal roots. We conclude, however, that the data are ambiguous and that the Sr method is currently incapable of providing a decisive answer for the Appalachians.

CONCLUSIONS

In summarizing the applications of isotope dating to metamorphic rocks, Moorbath (1965, p. 264) has commented that, "these methods represent nothing more than a further stage in the automation of the time-honored geological hammer." How far has this particular hammer brought us in our understanding of the Northern Appalachians?

Despite our unusually high ratio of number-of-dates to outcrop-area in the Northern Appalachians, geochronology has been unable to unscramble all the details of tectonic history for all parts of the region. The disturbance of isotopic systems because of thermal, hydrothermal, and pressure effects during repeated cycles of tectonism necessitates the introduction of judgment and interpretation in evaluating the results. However, geochronology has shown us that the history of the Northern Appalachians stretches backward into the Precambrian, and that the overridingly important event was the Acadian orogeny. It has also demonstrataed how, as for the "Allegheny" orogeny, the probable limits of the deformed area may be picked out by the K-A method. It has unravelled a succession of post-Devonian igneous events for which the geological record is, at best, obscure, and has provided us with a series of paleontologic-geochronologic dates which are critical in constructing the geologic time scale. And, finally, even though the data are currently inadequate, it has furnished us with a theoretical test for two of the major problems in geology—the accretion of the continental crust, and the origin of magmas. What more can anyone ask of a hammer?

Note added in Proof

Thirty-five new K–Ar determinations by C. T. Harper (Can. Jour. Earth Sciences, 1967, v. 5, p. 49–59) show metamorphism of the slates of the Taconic Klippe 460–445 m.y. ago, a mixed age pattern (425–325 m.y.) along the Green Mountain anticlinorium thought to be due to Silurian–Early Devonian uplift, and a Late Devonian (345 ± 7 m.y.) pattern in eastern Vermont.

Fullagar and Bottino (Can. Jour. Earth Sciences, 1967, v. 5, p. 311–317) have redetermined the whole-rock Rb–Sr age of the Arisaig volcanics, which lie near the Ordovician–Silurian time boundary. The new determinations (407 ± 15 m.y. for $\lambda = 1.47 \times 10^{-11}$ y^{-1}, or 430 ± 15 m.y. for $\lambda = 1.39 \times 10^{-11}$ y^{-1}) is 23 m.y. older than their previous result (Bottino and Fullagar, 1966).

REFERENCES

Alcock, F. J., 1938, Geology of St. John region, New Brunswick: Canada Geol. Survey Mem. 216, 65 p.

Aldrich, L. T., Wetherill, G. W., Davis, G. L., and Tilton, G. R., 1958, Radoactive ages of micas from granitic rocks by Rb–Sr and K–Ar methods: Am. Geophys. Union Trans., v. 39, p. 1124–1134

Billings, M. P., 1937, Regional metamorphism of the Littleton-Moosilauke area, New Hampshire: Geol. Soc. America Bull., v. 48, p. 463–566

__________, 1956, The geology of New Hampshire, Part II, Bedrock geology: New Hampshire State Planning and Dev. Comm., 203 p.

Bottino, M. L., Pinson, W. H., Jr., Fairbairn, H. W., and Hurley, P. M., 1963, Whole-rock Rb–Sr ages of some Paleozoic volcanics and related granites in the northern Appalachians [abs.]: Am. Geophys. Union Trans., v. 44, p. 111

__________, Schnetzler, C. C., and Fullagar, P. D., 1965, Rb–Sr whole-rock age of the Traveler and Kineo rhyolites, Maine, and its bearing on the early Devonian [abs.]: Geol. Soc. America Spec. Paper 87, p. 15

__________, and Fullagar, P. D., 1966, Whole-rock Rb–Sr age of the Silurian-Devonian boundary in northeastern North America: Geol. Soc. America Bull., v. 77, p. 1167–1175

Boucot, A. J., 1962, Appalachian Siluro-Devonian, in Some aspects of the Variscan Fold Belt; 9th Inter-University Geol. Congress: Manchester Univ. Press, p. 155–163

__________, Field, M. T., Fletcher, R. C., Forbes, W. H., Naylor, R. S., and Pavlides, Louis, 1964, Reconnaissance bedrock geology of the Presque Isle quadrangle, Maine: Maine Geol. Survey Quad. Mapping Series, no. 2, 123 p.

Brookins, D. G., and Hurley, P. M., 1965, Rb–Sr geochronological investigations in the Middle Haddam and Glastonbury quadrangles, eastern Connecticut: Am. Jour. Sci., v. 263, p. 1–16

Cady, W. M., 1960, Stratigraphic and geotectonic relationships in northern Vermont and southern Quebec: Geol. Soc. America Bull., v. 71, p. 531–576

Caldwell, D. B., ed., 1966, The Mount Katahdin region, Maine: New England Intercollegiate Geol. Conf., 58th Ann. Mtg., guidebook, Shin Pond, Maine, 61 p.

Castle, R. O., 1965, A proposed revision of the subalkaline intrusive series of northeastern Massachusetts: U. S. Geol. Survey Prof. Paper 525-C, p. C74–C80

Chute, N. E., 1966, Geology of the Norwood Quadrangle, Norfolk and Suffolk Counties, Massachusetts: U. S. Geol. Survey Bull., 1163-B, 78 p.

Cooke, H. C., 1955, An early Paleozoic orogeny in the eastern townships of Quebec: Geol. Assn. Canada, Proc., v. 7, p. 113–121

Cormier, R. G., and Kelly, A. M., 1964, Absolute age of the Fisset Brook Formation and the Devonian-

Mississippian boundary, Cape Breton Island, Nova Scotia: Can. Jour. Earth Sci., v. 1, p. 159–166

Cuppels, Norman, 1967, Geologic implications of undeformed fossils in the Newbury Formation of pre-Acadian age in northeastern Massachusetts: Geol. Soc. America, Northeastern Section, Program, 1967 Annual Meeting Boston, p. 22

Damon, P. E., and Kulp, J. L., 1957, Argon in mica and the age of the Beryl Mt., N. H., pegmatite: Am. Jour. Sci., v. 255, p. 697–704

DeBoer, J., 1967, Paleomagnetic-tectonic study of Mesozoic dike swarms in the Appalachians: Jour. Geophys. Res., v. 72, p. 2237–2250

Eardley, A. J., 1962, Structural geology of North America (2d ed.): New York, Harper and Row, 743 p.

Engel, A. E. J., 1963, Geologic evolution of North America: Science, v. 140, p. 143–152

Erickson, G. P., and Kulp, J. L., 1961, Potassium-argon dates on basaltic rocks: N. Y. Acad. Sci. Ann., v. 91, p. 321–323

Fairbairn, H. W., 1958, Age data from Newfoundland: 5th Ann. Prog. Rep. to U. S. Atomic Energy Commission, Massachusetts Inst. Tech. Dept. Geology and Geophysics, (unpublished), p. 69

——————, Hurley, P. M., Pinson, W. H., and Cormier, R. F., 1960a, Age of the granitic rocks of Nova Scotia: Geol. Soc. America Bull., v. 71, p. 399–413

——————, Pinson, W. H., Hurley, P. M., and Cormier, R. F., 1960b, Comparison of the ages of coexisting biotite and muscovite in some Paleozoic granite rocks: Geochim. et Cosmachim. Acta., v. 19, p. 7–9

——————, Faure, G., Pinson, W. H., Hurley, P. M., and Powell, J. L., 1963, Initial ratio of Strontium 87 to Strontium 86, whole-rock age and discordant biotite in the Monteregian igneous province, Quebec: Jour. Geophys. Res., v. 68, p. 6515–6522

——————, Hurley, P. M., and Pinson, W. H., 1964, Preliminary age study and initial Sr^{87}/Sr^{86} of Nova Scotia granitic rocks by the Rb–Sr whole-rock method: Geol. Soc. America Bull., v. 75, p. 253–257

——————, Moorbath, S., Ramo, A., Pinson, H. W., and Hurley, P. M., 1965, Rb–Sr whole-rock isotopic analyses and the Cambrian-Precambrian problem in southeastern Massachusetts: Am. Geophys. Union Trans., v. 46, p. 173

——————, Bottino, M. L., Pinson, W. H., and Hurley, P. M., 1966, Whole-rock Rb–Sr and initial Sr^{87}/Sr^{86} of volcanic rocks underlying fossiliferous Lower Cambrian in the Atlantic Provinces of Canada: Can. Jour. Earth Sciences, v. 3, p. 509–521

——————, Bottino, M. L., Handford, L. S., Hurley, P. M., Heath, M. N., and Pinson, W. H., 1967, Radiometric ages of igneous rocks in northeastern Massachusetts: Geol. Soc. America, Northeastern Section, Program, 1967 Annual Meeting, Boston, p. 24–25

Faul, Henry, 1966, Ages of rocks, planets, and stars: New York, McGraw-Hill Book Co., 109 p.

——————, Stern, T. W., Thomas, H. H., and Elmore, P. L. D., 1963, Ages of intrusion and metamorphism in the northern Appalachians: Am. Jour. Sci., v. 261, p. 1–19

Gilluly, James, 1965, Volcanism, tectonism, and plutonism in the western United States: Geol. Soc. America Special Paper 80, 69 p.

——————, 1966, Orogeny and geochronology: Am. Jour. Sci., v. 264, p. 97–111

Goldsmith, Richard, 1964, Geologic map of New England: 1. General geology; 2. Metamorphic zones;

3. Bibliography of radiometric ages: U. S. Geological Survey open-file rept., 1:1,000,000

Green, J. C., 1964, Stratigraphy and structure of the Boundary Mountain anticlinorium in the Errol Quadrangle, New Hampshire-Maine: Geol. Soc. America Spec. Paper 77, 78 p.

Hadley, J. B., 1964, Correlation of isotopic ages, crustal heating and sedimentation in the Appalachian region: in W. D. Lowry, ed., Tectonics of the Southern Appalachians; Virginia Polytech. Inst., Dept. of Geol. Sci. Mem. 1, p. 33–44

Harakal, J. E., 1966, Potassium-argon ages of the Scituate Granite Gneiss, North-Central Rhode Island: M. S. thesis, Brown Univ., 32 p.

Holmes, Arthur, 1959, A revised geological time scale: Edin. Geol. Soc. Trans., v. 17, Part 3, p. 183–216

Hurley, P. M., 1960, Summary of K–Ar ages from the Monteregian Hills: 8th Ann. Prog. Rept. to U. S. Atomic Energy Commission, Massachusetts Inst. Tech. Dept. of Geology and Geophysics (unpublished), p. 283

——————, Pinson, W. H., and Fairbairn, H. W., 1958, Intrusive and metamorphic rock ages in Maine and surrounding areas [abs.]: Geol. Soc. America Bull., v. 69, p. 1591

——————, Boucot, A. J., Albee, A. L., Faul, H., Pinson, W. H., and Fairbairn, H. W., 1959, Minimum age of the Lower Devonian slate near Jackman, Maine: Geol. Soc. America Bull., v. 70, p. 947–949

——————, Fairbairn, H. W., Pinson, W. H., and Faure, G., 1960, K–Ar and Rb–Sr minimum ages for the Pennsylvanian section in the Narragansett Basin: Geochim. et Cosmochim. Acta., v. 18, p. 247–258

——————, Hughes, H., Faure, G., Fairbairn, H. W., and Pinson, W. H., 1962, Radiogenic strontium 87 model of continent formation: Jour. Geophys. Res., v. 67, p. 5315–5334

Hussey, A. M., II, 1962, The geology of southern York County, Maine: Maine Geol. Survey Spec. Geologic Studies Series, no. 4, 67 p.

Kulp, J. L., 1960, Geologic time scale: Science, v. 133, No. 3459, p. 1105–1114

Lanphere, M. A., Wasserburg, G. J., Albee, A. L., and Tilton, G. R., 1964, Redistribution of strontium and rubidium isotopes during metamorphism, World Beater Complex, Panamint Range, California, in Harmon Craig et al., eds.; Isotopic and cosmic chemistry; Amsterdam, Netherlands, North–Holland Publishing Co., p. 269–320

Lapham, D. M., and Bassett, W. A., 1964, K–Ar dating of rocks and tectonic events in the Piedmont of southeastern Pennsylvania: Geol. Soc America Bull., v. 75, p. 661–667

Larrabee, D. M., Spencer, C. W., and Swift, D. J. P., 1965, Bedrock geology of the Grand Lake area, Aroostook, Hancock, Penobscot, and Washington Counties, Maine: U. S. Geol. Survey Bull. 1201-E, 38 p.

Leech, G. B., Lowden, J. A., Stockwell, C. H., and Wanless, R. K., 1963, Age determinations and geological studies (Report 4): Canada Geol. Survey Paper 63–17, 140 p.

Lilly, H. D., 1966, Late Precambrian and Appalachian tectonics in the light of submarine explorations on the Great Bank of Newfoundland and in the Gulf of St. Lawrence—Preliminary views: Am. Jour. Sci., v. 264, p. 569–574

Long, L. E., 1961, Isotopic ages from northern New Jersey and southeastern New York, in Geochronology of rock systems: New York Acad. Sci. Ann., v. 91, p. 400–407

————, 1962, Isotopic age study, Dutchess County, New York: Geol. Soc. American Bull., v. 73, p. 997–1006

————, and Kulp, J. L., 1962, Isotopic age study of the metamorphic history of the Manhattan and Reading Prongs: Geol. Soc. America Bull., v. 73, p. 969–995

Lowden, J. A., 1960, Age determinations by the Geological Survey of Canada: Canada Geol. Survey Paper 60–17, 51 p.

————, 1961, Age determinations by the Geological Survey of Canada (Report 2): Canada Geol. Survey Paper 61–17, 127 p.

————, Stockwell, C. H., Tipper, H. W., and Wanless, R. K., 1963, Age determinations and geological studies (Report 3): Canada Geol. Survey Paper 62–17, 140 p.

Lyons, J. B., Jaffe, H. W., Gottfried, D., and Waring, C. L., 1957, Lead-alpha ages of some New Hampshire granites: Am. Jour. Sci., v. 255, p. 527–546

McCartney, W. D., Poole, W. H., Wanless, R. K., Williams, H., and Loveridge, W. D., 1966, Rb/Sr age and geological setting of the Holyrood granite, southeast Newfoundland: Can. Jour. Earth Sciences, v. 3, p. 947–957

McDougall, I., Compton, W., and Bofinger, V. M., 1966, Isotopic age determinations on Upper Devonian rocks from Victoria, Australia: a revised estimate of the Devonian–Carboniferous boundary: Geol. Soc. America Bull., v. 77, p. 1075–1088

Moorbath, S., 1965, Isotopic dating of metamorphic rocks, in W. S. Pitcher and G. W. Flynn, eds., Controls of Metamorphism: Oliver and Boyd, p. 235–267

Naylor, R. S., and Boucot, A. J., 1965, Origin and distribution of rocks of Ludlow age (Late Silurian) in the northern Appalachians: Am. Jour. Sci., v. 263, p. 153–169

————, and Wasserburg, G. J., 1966, Rb–Sr studies on mantled gneiss domes in central New England [abs.]: Am. Geophys. Union Trans., v. 47, p. 194–195

Neale, E. R. W., Beland, J., Potter, R. R., and Poole, W. H., 1961, A preliminary tectonic map of the Canadian Appalachian region based on age of folding: Can. Mining and Metallurgical Bull., v. 54, p. 687–694

Nier, A. O. C., Thompson, R. W., and Murphy, B. F., 1939, The isotopic composition of lead and the measurement of geological time; Part II: Phys. Rev., v. 55, p. 153–163

Pavlides, Louis, Mencher, Ely, Naylor, R. S., and Boucot, A. J., 1964, Outline of the stratigraphic and tectonic features of northeastern Maine: U. S. Geol. Survey Prof. Paper 501–C, p. C28–C38

————, and Berry, W. B. N., 1966, Graptolite–bearing Silurian rocks of the Houlton–Smyrna Mills area, Aroostook County, Maine: U. S. Geol. Survey Prof. Paper 550–B, p. B51–B61

Poole, W. H., Kelley, D. G., and Neale, E. R. W., 1964, Age and correlation problems in the Appalachian region of Canada, in F. F. Osborne, ed.; Geochronology in Canada: Royal Soc. Canada Spec. Pub. no. 8, p. 61–84

Quinn, A. W., Jaffe, H. W., Smith, W. C., and Waring, C. L., 1957, Lead-alpha ages of Rhode Island granitic rocks compared to their geologic ages: Am. Jour. Sci., v. 255, p. 547–560

Riordon, P. H., 1957, Evidence for a pre-Taconic orogeny in southeastern Quebec: Geol. Soc. America Bull., v. 68, p. 389–394

Rodgers, John, 1952, Absolute ages of radioactive minerals from the Appalachian region: Am. Jour. Sci., v. 250, p. 411–427

————, 1967, Chronology of tectonic movements in the Appalachian region of eastern North America: Am. Jour. Sci., v. 265, p. 408–427

————, Gates, R. M., and Rosenfeld, J. L., 1959, Explanatory text for Preliminary geological map of Connecticut, 1956: Connecticut Geol. and Nat. History Survey Bull. 84, 64 p.

————, and Neale, E. R. W., 1963, Possible Taconic klippen in western Newfoundland: Am. Jour. Sci., v. 261, p. 713–731

Rose, Harry, Jr., and Stern, T. W., 1960, Spectrochemical determination of lead in zircon for lead-alpha age measurements: Am. Mineralogist, v. 45, p. 1243–1256

Schaeffer, O. A., and Zähringer, J., eds., 1966, Potassium argon dating: New York, Springer-Verlag, 234 p.

Smith, D. G. W., Baadsgaard, H. Folinsbee, R. E., and Lipson, J., 1961, K–Ar age of Lower Devonian bentonites of Gaspé, Quebec, Canada: Geol. Soc. America Bull., v. 72, p. 171–173

Tilton, G. R., 1965, Compilation of Phanerozoic data for eastern North America, in G. W. Wetherill, ed., Geochronology of North America: Nat. Acad. Sci.–Nat. Res. Council, Pub. 1276, p. 181–220

————, Davis, G. L., Wetherill, G. W., and Aldrich, L. T., 1957, Isotopic ages of zircon from granites and pegmatites: Am. Geophys. Union Trans., v. 38, p. 360–371

————, Wetherill, G. W., Davis, G. L., and Hopson, C. A., 1958, Ages of minerals from the Baltimore gneiss, near Baltimore, Maryland: Geol. Soc. America Bull., v. 69, p. 1469–1474

————, and Hart, S. R., 1963, Geochronology: Science, v. 140, p. 357–366

Toulmin, Priestley, III, 1964, Bedrock geology of the Salem quadrangle and vicinity, Massachusetts: U. S. Geol. Survey Bull. 1163–A, 79 p.

Tupper, W. M., et al., 1959, Intrusive granites in New Brunswick: 7th Ann. Prog. Rept. to U. S. Atomic Energy Commission, Massachusetts Inst. Tech. Dept. of Geology and Geophysics (unpublished), p. 187–194

Wanless, R. K., Stevens, R. D., Lachance, G. R., and Rimsaite, J. Y. H., 1966, Age determinations and geological studies—K–Ar isotopic ages (Report 6): Canada Geol. Survey Paper 65–17, 101 p.

————, Stevens, R. D., Lachance, G. R., and Edmonds, C. M., 1967, Age determinations and geological studies—K–Ar isotopic ages (Report 7): Canada Geol. Survey Paper 66–17, 120 p.

Wasserburg, G. J., Hayden, R. J., and Jensen, K. J., 1956, A^{40}–K^{40} dating of igneous rocks and sediments: Geochim. et Cosmochim. Acta, v. 10, p. 153–165

Webber, G. K., Hurley, P. M., and Fairbairn, H. W., 1958, Relative ages of eastern Massachusetts granites by total Pb ratio in zircon: Am. Jour. Sci., v. 254, p. 574–583

Weeks, L. J., 1957, The Appalachian region, in Geology and economic minerals of Canada: Canada Geol. Surv. Econ. Geol. Series, no. 1, p. 123–205

Williams, H., 1964, The Appalachians in northeastern Newfoundland—a two-side symmetrical system: Am. Jour. Sci., v. 262, p. 1137–1158

Woodward, H. P., 1957, Chronology of Appalachian folding: Am. Assoc. Petroleum Geologist Bull., v. 41, p. 2312–2327

Zartman, R., Snyder, G., Stern, T. W., Marvin, R. F., and Bucknam, R. C., 1965a, Implications of new radiometric ages in eastern Connecticut and Massachusetts: U. S. Geol. Survey Prof. Paper 525-D, p. D1–D10

——————, Brock, M., Heyl, A. V., and Thomas, H. H., 1965b, K–Ar and Rb–Sr ages of some alkalic intrusive rocks from central and eastern United States: Geol. Soc. Am. Special Paper 87, p. 187–188

Chapter 24

Paleozoic Regional Metamorphism in New England and Adjacent Areas

JAMES B. THOMPSON, JR. AND
STEPHEN A. NORTON

INTRODUCTION

PALEOZOIC ROCKS have been regionally metamorphosed over extensive areas in the Northern and Maritime Appalachians. The area with which we are directly concerned here is in central and western New England and adjacent parts of New York and Quebec (Fig. 24–1). Paleozoic rocks having mineral assemblages typical of the biotite zone or of higher grade zones are also known from areas to the northeast of that of Figure 24–1, principally in Nova Scotia and Newfoundland, but information from these areas is not yet sufficient for a meaningful regional compilation.

The existence of an unusual display of metamorphic rocks in New England and adjacent areas was recognized in the early state surveys. Important contributions to the petrography of these rocks were made by Hitchcock (1841), Hawes (1878), and Emerson (1898, 1917). Names such as "cummingtonite," "fibrolite," and "sillimanite" all derive from New England localities or occurrences. Systematic mapping of mineral zones in the Northern Appalachians, however, was not begun until the 1930's with the work of Billings (1937) in northern New Hampshire and of Balk (1936) and Barth (1936) in southeastern New York and adjacent Connecticut.

The sources of information for the compilation of Figure 24–1 are many. Recent state compilations for New York (Broughton et al., 1962), Vermont (Doll et al., 1961), New Hampshire (Billings, 1955), and Maine (Doyle and Hussey, 1967), all provide information on the distribution of metamorphic zones in their respective

areas. The mineral zones shown in Figure 24–1 have been somewhat modified from those shown on the above maps by using information presented elsewhere in this volume and in other published reports listed in the references at the end of this paper. For Massachusetts, Connecticut, and Rhode Island we have relied mainly on recent published areal reports (see references), and also on a map compiled by Goldsmith (1962).

It became clear in the course of compilation that some of the lines in Figure 24–1 are closely controlled, whereas others are but tentative extrapolations. Enough detailed studies exist, however, to show that the general pattern of Figure 24–1 is nowhere grossly in error.

ACKNOWLEDGMENTS

We wish to thank our many colleagues for constructive discussions and for the many scattered pieces of information that have aided us in the preparation of Figure 24–1. We also wish to make more specific acknowledgment to those investigators who have contributed unpublished data in critical areas, notably Charles E. Bickel (western Penobscot Bay area, Maine), Paul C. Hess (south-central Massachusetts), John L. Rosenfeld (central Connecticut), and Robert W. Schnabel (west-central Massachusetts).

CHOICE OF ISOGRADS

The boundaries drawn in Figure 24–1 have been dictated mainly by the nature of the available data. Most investigators have based their schemes of zonal mapping on a sequence of index minerals (or, less commonly, index assemblages) that follows or closely parallels the sequence used by Barrow (1893, 1912) in the Scottish Highlands. The isograds in the original reports are generally described as based on "first appearances" of key minerals or mineral pairs in metamorphosed clastic sedimentary rocks, with smoothing of the lines appropriate to the scale of the map. Such first appearances are dependent not only on externally imposed conditions of temperature, pressure, and the activities of components having comparatively free mobility, such as H₂O, but also on the original bulk composition of the rock.

The mineral zones shown on Figure 24–1 are mapped on assemblages found in pelitic rocks, but these zones may be correlated approximately with the mineral facies developed in metamorphosed mafic igneous rocks, and with index minerals found in calc-silicate rocks, as shown in Table 24–1. The correlations are based on interstratified occurrences of the different rock-types, and show the most commonly observed relationships. There is considerable variation, however, from area to area, particularly in the assemblages in calc-silicate rocks as compared to the others. Tremolite and actinolite, for example, occur in carbonate rocks associated with biotite-

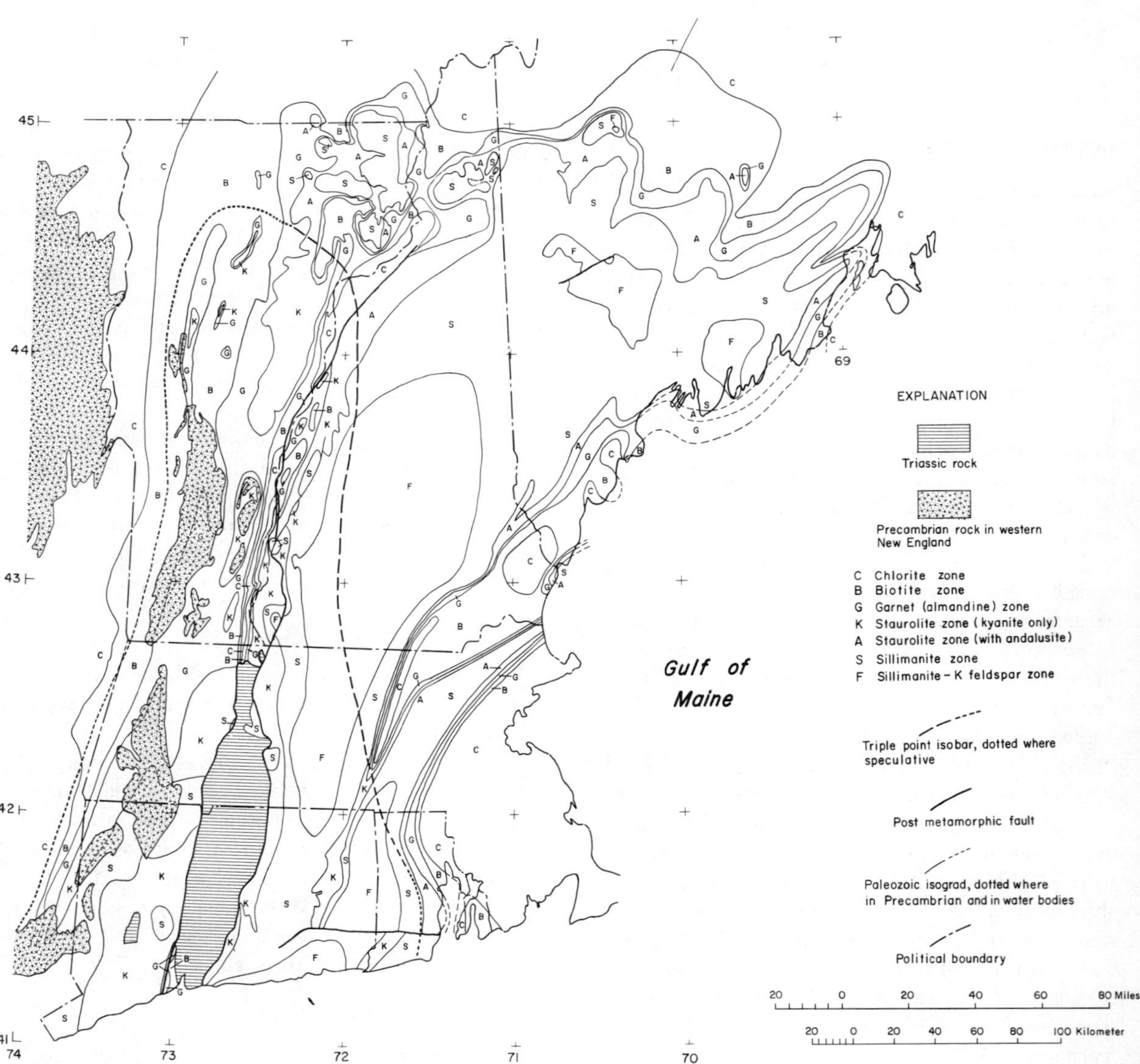

FIGURE 24–1. Paleozoic regional metamorphism in New England and adjacent areas. Precambrian rocks in eastern New England and of the New York City area are not shown.

TABLE 24-1. Approximate correlation of mineral zones in pelitic rocks, as shown in Figure 24-1, with characteristic mineral facies developed in associated mafic rocks and with index minerals found in calc-silicate rocks.

Pelitic Rocks	Mafic Rocks	Calc-silicate Rocks
Biotite zone	Greenschist facies	Talc, phlogopite
Garnet zone	Epidote-amphibolite facies	Tremolite, actinolite, epidote, zoisite
Staurolite zone	Amphibolite facies	Diopside
Sillimanite zone		Grossularite, scapolite
Sillimanite–K feldspar zone	Hornblende–Pyroxene Granulite facies	Forsterite

zone pelites in western Vermont, but do not appear until the staurolite zone in parts of northeastern Vermont (White and Billings, 1951). Many of these variations may be explainable in terms of limited mobility of CO_2 (Billings and White, 1950).

For the interpretation of isograds in terms of conditions of metamorphism, the isograds should ideally be drawn so as to minimize the effects of variation in initial bulk composition. Procedures have been devised using various continuous and discontinuous variations in mineral facies (Thompson, 1957) to achieve this, but such procedures are laborious and have been carried out in only a few selected areas (see, for example, Green, 1963; Guidotti, 1963; Albee, 1965, and this volume; Evans and Guidotti, 1966). First appearances, however, may very nearly fulfill the above requirement in areas where rocks having a considerable variation in bulk composition are intimately interstratified. In such areas the mapped isograds probably approximate fairly closely the first possible appearances of the index minerals, or mineral pairs, in assemblages containing quartz and muscovite, consistent with the local pressure, temperature, and activity of H_2O. The success of the method thus depends both on the range of bulk compositions available and on the scale of the interstratification of the various rock types relative to the scale of the map. For metamorphosed clastic sedimentary rocks the index minerals that have been used by most investigators can be regarded, to a first approximation at least, as phases in the system SiO_2–Al_2O_3–FeO–MgO–K_2O–H_2O (see Thompson, 1957, for more complete discussion), and many isogradic reactions in such rocks can be described fairly accurately as reactions in that system, even in the presence of minerals containing other components.

The isograds of Figure 24–1 may now be considered briefly in terms of generalized reactions for the first appearances of the principal index minerals or assemblages. The reactions that follow will be classed as *simple* or *complex*. The *simple* reactions are balanced using numerical coefficients consistent with the simplified formulas listed in Table 24–2. Each reaction is, in fact, the resultant of two effects, one involving the Fe-components, and the other the Mg-components. These have been combined for conciseness in Table 24–2 as the stoichiometric form is otherwise identical. These reactions correspond, in general, to displacements of specific three-phase fields in an A-F-M representation of the mineral assemblages (Thompson, 1957; Albee, this volume, Fig. 25–2). As such they represent continuous reactions or variations in facies, except where the minerals involved are colinear in A-F-M projection. This exception, however, appears unlikely for ferromagnesian minerals except at the end-member limits. The *complex* reactions, on the other hand, are all discontinuous reactions, and cannot be uniquely balanced even with the simplified formulas of Table 24–2 unless the ratio Fe:Mg for each phase is specified. These ratios will in general differ from point to point along such a reaction boundary. Each *complex* reaction, however, may be regarded as the sum of two *simple* ones. The arrows in each reaction, as written, indicate a prograde displacement.

If the range of rock compositions in a given metamorphic terrane were such that all possible compositions occur that can appear in an A-F-M projection, then the first appearance reactions as written below would be essentially the reactions for the isograds as mapped and would represent the first possible appearance of the indicator. If, on the other hand, the range of compositions is restricted in a given terrane to a part of the

TABLE 24-2. Simplified formulas used in balancing simple reaction equations.

Mineral Name	Abbreviations	Simplified Formulas
Kyanite	Kya	Al_2SiO_5
Andalusite	And	Al_2SiO_5
Sillimanite	Sil	Al_2SiO_5
Kaolinite	Kao	$Al_2Si_2O_5(OH)_4$
Pyrophyllite	Pyp	$Al_2Si_4O_{10}(OH)_2$
Staurolite	Stt	$(Fe,Mg)_2Al_9Si_4O_{23}(OH)$
Chloritoid	Ctd	$(Fe,Mg)Al_2SiO_5(OH)_2$
Almandine	Alm	$(Fe,Mg)_3Al_2Si_3O_{12}$
Chlorite	Chl	$(Fe,Mg)_7Al_4Si_4O_{15}(OH)_{12}$
Biotite	Bio	$K(Fe,Mg)_3AlSi_3O_{10}(OH)_2$
Muscovite	Mus	$KAl_3Si_3O_{10}(OH)_2$
Paragonite	Par	$NaAl_3Si_3O_{10}(OH)_2$
K-feldspar	Ksp	$KAlSi_3O_8$
Albite	Alb	$NaAlSi_3O_8$
Calcite	Cal	$CaCO_3$
Dolomite (ankerite)	Dol	$Ca(Mg,Fe)(CO_3)_2$
Quartz	Qtz	SiO_2
Hematite	Hem	Fe_2O_3
Magnetite	Mag	Fe_3O_4
Rutile	Rut	TiO_2
Ilmenite	Ilm	$FeTiO_3$

A-F-M projection, then the isograd mapped may represent some other reaction corresponding to a first appearance only in the restricted range of compositions found. Several of the more likely among these somewhat specialized reactions are given below, but no attempt has been made to be exhaustive. It is further possible, with a restricted range of composition, that the reactions for the isograd as mapped may involve phases for which oxides such as Na_2O, CaO, Fe_2O_3, and TiO_2 are essential. In this regard we may write stoichiometric relations such as the following to show how the effective bulk composition in A-F-M projection may be affected:

$$Par = Alb + (Al_2O_3) + H_2O \qquad (1)$$

$$Mag = Hem + (FeO) \qquad (2)$$

$$Ilm = Rut + (FeO) \qquad (3)$$

These may be combined with the stoichiometric relations among pairs of A-F-M phases to write simple reactions such as:

$$Par + 3Chl + 9Qtz \rightarrow Alb + 7Alm + 19H_2O \qquad (4)$$

or

$$5Hem + Chl \rightarrow 5Mag + 2Ctd + 2Qtz + 4H_2O \qquad (5)$$

or complex reactions involving (1), (2), or (3) with sets of three A-F-M phases. Reaction (4) above might correspond to a garnet isograd in rocks of restricted composition and (5) to a chloritoid isograd in rocks of restricted composition. Reaction (5) is also of interest in that it is basically a dehydration reaction although easily mistaken for a deoxidation reaction on casual examination of the rocks. It is probably one of the key reactions involved in the disappearance of purple or red color from many phyllites in the lower biotite zone.

I. The boundary between the chlorite and biotite zones in Figure 24–1 has been drawn mainly on the first recorded appearance of either biotite or chloritoid. These two events are probably not, in general, coincident, but we cannot separate them on data now available. The rocks in which these minerals first appear are also very fine-grained and many occurrences may have been missed. Stilpnomelane appears to form at a somewhat lower grade than either biotite or chloritoid, and can be misidentified as biotite.

First appearances of biotite are generally recorded in rocks having the composition of graywackes, arkoses, siltstones, or siliceous volcanic rocks. In these rocks biotite probably forms by a simple reaction of the type:

$$3Chl + 13Ksp \rightarrow 7Bio + 6Mus + 12Qtz + 5H_2O \qquad (6)$$

Data on coexisting biotite and chlorite indicate that the first appearance of biotite by this reaction should be in rocks having a high value of $MgO:FeO$. In rocks containing carbonates, however, biotite may appear by a reaction of the type:

$$3Dol \text{ (or Ankerite)} + Ksp + H_2O \rightarrow Bio + 3Cal + 3CO_2 \qquad (7)$$

First appearances of chloritoid occur in rocks having a high ratio of $Al_2O_3:(FeO + MgO)$ and probably arise through reactions such as:

$$Chl + 5Kao \rightarrow 7Ctd + 7Qtz + 2H_2O \quad (8)$$

or:

$$Chl + 5Pyp \rightarrow 7Ctd + 17Qtz + 4H_2O \quad (9)$$

Chloritoid is strongly siderophile relative to chlorite, hence its early appearance should be favored by a high value of FeO:MgO. For a more extensive discussion of rocks near the biotite isograd, see Zen (1960).

II. The boundary between the biotite and garnet zones is based on the first appearance of almandine. In aluminous rocks almandine may appear through simple reactions of the type:

$$Ctd + 2Chl + 6Qtz \rightarrow 5Alm + 13H_2O \quad (10)$$

In more typical pelites, however, the reaction of first appearance is probably:

$$6Chl + Mus + 15Qtz \rightarrow 13Alm + Bio + 36H_2O \quad (11)$$

Garnet is siderophile relative to chlorite, biotite, and even chloritoid, hence is favored in either of the above reactions by a high value of FeO:MgO. Garnets are also strongly manganophile but most authors have not extended their garnet isograds so as to include low-grade garnets known to be notably manganiferous. Simple reactions of the type:

$$3Ctd + Bio + 3Qtz \rightarrow 2Alm + Mus + 3H_2O \quad (12)$$

may produce almandine in specific rocks but the assemblage Qtz-Mus-Bio-Ctd is not reported in rocks from the biotite zone, hence it is unlikely that (12) is the reaction for any first appearance isograd yet drawn.

III. The boundary between the garnet and "staurolite" zones in Figure 24–1 has been drawn on the first appearance of either staurolite, kyanite, or andalusite. In highly aluminous rocks, kyanite (or andalusite) appears to precede staurolite, probably by way of a reaction such as:

$$Pyp \rightarrow Kya \text{ (or And)} + 3Qtz + H_2O \quad (13)$$

Reported occurrences of pyrophyllite, however, are rare and this reaction is best regarded as tentative. In more normal pelitic rocks the appearance of kyanite or andalusite is probably related to a complex reaction of the type:

$$Ctd + Chl + Mus + Qtz \rightarrow Kya \text{ (or And)} + Bio + H_2O \quad (14)$$

or, following the appearance of staurolite:

$$Stt + Chl + Mus + Qtz \rightarrow Kya \text{ (or And)} + Bio + H_2O \quad (15)$$

These complex reactions are closely related to the following simple reactions, each of which may be responsible in a given rock for the first appearance of kyanite or andalusite:

$$3Chl + 7Mus + Qtz \rightarrow 13Kya \text{ (or And)} + 7Bio + 18H_2O \quad (16)$$

$$3Ctd + Mus + Qtz \rightarrow 4Kya \text{ (or And)} + Bio + 3H_2O \quad (17)$$

$$6Stt + 4Mus + 7Qtz \rightarrow 31Kya \text{ (or And)} + 4Bio + 3H_2O \quad (18)$$

First appearances of staurolite may arise through complex reactions of the type:

$$Ctd + Chl + Kya \text{ (or And)} \rightarrow Stt + Qtz + H_2O \quad (19)$$

$$Ctd + Kya \text{ (or And)} + Bio \rightarrow Stt + Mus + Qtz + H_2O \quad (20)$$

$$Ctd + Chl + Mus \rightarrow Stt + Bio + Qtz + H_2O \quad (21)$$

$$Alm + Chl + Mus \rightarrow Stt + Bio + Qtz + H_2O \quad (22)$$

or through simple reactions such as:

$$4Ctd + 5Kya \text{ (or And)} \rightarrow 2Stt + Qtz + 3H_2O \quad (23)$$

$$31Ctd + 5Mus + Qtz \rightarrow 8Stt + 5Bio + 27H_2O \quad (24)$$

$$4Chl + 55Kya \text{ (or And)} \rightarrow 14Stt + 15Qtz + 17H_2O \quad (25)$$

$$31Chl + 55Mus \rightarrow 26Stt + 55Bio + 20Qtz + 173H_2O \quad (26)$$

Reactions (19), (23), and (25) are observable only in highly aluminous rocks. Chlorites at these metamorphic grades are highly magnesian and simple reactions such as (25), (26), and particularly (16) will be observable only in rocks with relatively high Mg:Fe ratio.

IV. The boundary between the staurolite and sillimanite zones can be related in many areas to the conversion of kyanite or andalusite to sillimanite. Much sillimanite, however, particularly in the less aluminous rocks seems to be produced by simple reactions of the type:

$$6Stt + 4Mus + 7Qtz \rightarrow 31Sil + 4Bio + 3H_2O \quad (27)$$

(Billings, 1937, p. 552) or:

$$Alm + Mus \rightarrow 2Sil + Bio + Qtz \quad (28)$$

and by the complex reaction:

$$Stt + Mus + Qtz \rightarrow Sil + Alm + Bio + H_2O \quad (29)$$

It is probable that these last are responsible for much of the common association of fibrolitic sillimanite with biotite and muscovite.

V. The boundary between the sillimanite and sillimanite-K feldspar zones has been drawn on the first reported appearance of the *pair* sillimanite-K feldspar. This may be represented by the complex reaction:

$$\text{Muscovite + Na plagioclase + Quartz} \rightarrow$$
$$\text{Sillimanite + K-Na feldspar + H}_2\text{O} \quad (30)$$

following which further sillimanite and K feldspar may be produced by the simple reaction:

$$\text{Muscovite + Quartz} \rightarrow \text{Sillimanite}$$
$$+ \text{K-Na feldspar + H}_2\text{O} \quad (31)$$

It should be emphasized (see Thompson, 1961) that (30) represents neither the first appearance of sillimanite, nor the complete disappearance of muscovite. Reactions related to the above have been studied in some detail by Guidotti (1963); Evans and Guidotti (1966); and Lundgren (1966). Cordierite-bearing and kinzigitic gneisses have been reported (Sclar, 1958; Barker, 1962; Snyder, 1964) from pelitic rocks in the sillimanite–K feldspar zone.

A FOSSIL ISOBARIC SURFACE

In the northern and eastern areas of Figure 24–1, the Al-silicate preceding sillimanite is andalusite, and in the areas to the west and south it is kyanite. A dashed line has been drawn on Figure 24–1 such that all occurrences of andalusite believed to have formed during regional metamorphism lie to the east or northeast of it (see Doll et al., 1961; Woodland, 1963, 1965; Wilson, 1965; Albee, this volume, Fig. 25–1). This line is therefore regarded as the approximate trace, on the present erosion surface, of a fossil isobaric surface corresponding to the pressure of the kyanite-andalusite-sillimanite triple point. Estimates of the pressure of this triple point have been subject to repeated experimental revision in recent years. A recent value of 4 kilobars has been proposed by Newton (1966); this pressure would correspond to that developed by a rock overburden of some 15 kilometers (50,000 feet). The style of the deformation accompanying much of the metamorphism in southwestern New England (see papers in this volume by Rosenfeld, by Dixon and Lundgren, and by Thompson et al.) suggests plastic, almost fluid-like behavior, and it does not seem likely that significant "overpressures" in excess of the lithostatic pressure could have been operative here. With complex overfolding, moreover, the depth of burial during metamorphism may well exceed the thickness of the original stratigraphic cover by several fold. The implications as to the volume of rock removed from New England since the late Paleozoic are surprising, but by no means impossible.

The mineralogic control for locating the fossil isobar cannot be maintained beyond northeastern Vermont. Its possible continuation, however, has been deduced from regional geologic relations and is shown as a dotted line extending northwestward, westward, then southward through western Vermont and eastern New York. The location is uncertain in detail, but must be within the zone bounded on the west by the Appalachian front, and on the east by the westernmost sillimanite isograd in Massachusetts, Connecticut, and southeastern New York. We may reasonably infer also that this surface is anticlinal in form and that it plunges gently northerly. Its dip, on the west, is probably steep.

PALEOZOIC METAMORPHISM OF PRECAMBRIAN ROCKS

The Precambrian rocks of the eastern Adirondacks were metamorphosed in Precambrian time and are referred by Broughton et al. (1962) to a "hypersthene zone," corresponding, in its mineral assemblages, to the higher grade part of the sillimanite-K feldspar zone as developed in the Paleozoic rocks of New England. The rocks occupying the core of the Green Mountains in Vermont, the Berkshire Highlands in Massachusetts and Connecticut, and the Housatonic and Hudson Highlands in Connecticut and New York, were probably similar to those in the eastern Adirondacks when first buried by Paleozoic sedimentary rocks. The present nature of these Precambrian rocks, however, has been much altered by the subsequent Paleozoic metamorphism.

Polymetamorphic effects are most conspicuous in the Precambrian rocks in areas where the later Paleozoic metamorphism is of fairly low grade. Where the Paleozoic rocks are in the staurolite zone, or at a higher grade, the reconstitution of the Precambrian rocks is much more thorough, and relict features of an earlier metamorphic event can be recognized only with difficulty. In the Green Mountains and Berkshire Highlands, where the Paleozoic rocks are in the biotite or garnet zone, relict features have been noted by Brace (1953), Skehan (1961), Chang et al. (1965), and Norton (1967).

TIMING OF THE PALEOZOIC METAMORPHISM

From information summarized elsewhere in this volume (Rosenfeld; Dixon and Lundgren; Thompson et al.) it is evident that, at least in southwestern New England, the main period of regional metamorphism overlapped two or more major stages of deformation. These events can be dated geologically only as later than Early Devonian (the age of the youngest rocks metamorphosed) and earlier than Triassic. Rocks as young as Pennsylvanian, however, have been regionally metamorphosed in southern Rhode Island (Quinn and Moore, this volume) and possibly in east-central Massachusetts (Perry and Emerson, 1903).

The distribution of the higher grade rocks in several distinct belts or "highs" separated by narrower "troughs" of lower grade rocks suggests the possibility that these may have developed at different times. The available radiometric data summarized in this volume (Lyons and Faul; Naylor) are not definitive on this matter, but there is evidence supporting both mid-Devonian (Acadian) and late Paleozoic periods of recrystallization. The possibility of pre-Silurian regional metamorphism in northern Vermont and southern Quebec is discussed by Albee (this volume). The isograds of Figure 24–1 may thus represent the combined effects of more than one metamorphic event, at least locally.

FOLDING AND FAULTING OF ISOGRADIC SURFACES

In western New Hampshire and central Massachusetts, there is evidence (Thompson et al., this volume) that isogradic surfaces are strongly overturned. This appears to be related to the emplacement of nappes derived from an already heated source area to the east.

The isograds in the Connecticut Valley region have also been offset by faults related to the development of the Triassic Lowlands. The larger faults in this area have been shown on Figure 24–1. The Honey Hill fault (Dixon and Lundgren, this volume) in southeastern Connecticut also offsets the isograds in that area. The existence of later, probably post-metamorphic, faulting in eastern Massachusetts (Skehan, this volume) may also measurably offset the isograds, but this has not yet been demonstrated. Certain areas of closely spaced isograds, as shown on Figure 24–1, may actually have discontinuities in grade related to later faulting.

METAMORPHISM AND PLUTONIC ACTIVITY

The plutonic rocks assigned to the New Hampshire Plutonic Series (Billings, 1937, 1956) appear in most areas to be roughly contemporaneous with, or slightly younger than, the main regional metamorphism. Their distribution relative to the higher grade metamorphic zones shows a striking correlation as seen in a comparison of Figure 24–1 (this paper) with the geologic map accompanying this volume. A genetic relationship is thus highly likely, but it is by no means clear which is the primary and which the secondary feature (see Page, this volume). Contact aureoles around the plutons assigned to the New Hampshire Plutonic Series are obscure or absent in much of southwestern New England. These aureoles are, however, well-defined in northeastern Vermont, northern New Hampshire, and in northwestern Maine (Green and Guidotti, this volume); it is thus clear that in these areas the magmas must have originated elsewhere, presumably at greater depth. The presence of distinct contact aureoles should in general be most characteristic of plutonic bodies emplaced at relatively shallow depths. This is also indicated by the close correlation between the distribution of such contact aureoles relative to the triple-point isobar for the Al-silicates.

Traditionally the plutonic masses have been assigned a primary role in the development of any spatially associated regional metamorphism (Barrell, 1921; Page, this volume). It is at least equally plausible that the plutons are, in fact, end products of the metamorphic process rather than its primary cause. Certain observations are at least consistent with an anatectic origin for the New Hampshire Plutonic Series:

(1) The rock types are, on the whole, highly siliceous and also characterized by an excess of alumina over the sum of the alkalies and lime. This is commonly expressed mineralogically by the presence of muscovite or garnet. These are compositional characteristics that would be expected if the magmas were formed by partial fusion of eugeosynclinal sediments.

(2) Migmatites and abundant pegmatites are characteristic of the sillimanite and sillimanite-K feldspar zones. Recent studies of phase equilibria pertinent to reactions at this metamorphic grade suggest temperatures at or near the melting intervals for many common rocks, particularly if the activity of H_2O is high.

(3) The near absence of mafic rocks that can be assigned unambiguously to the New Hampshire Plutonic Series is difficult to explain if the magmas originated in the mantle or in the deepest part of the crust.

If the above arguments are correct, the problem of a primary heat source is set one step back (or one level deeper in the earth), but this problem would arise eventually, anyway, even if the plutonic masses were assigned a primary role.

REFERENCES

References consulted in the course of compilation of Figure 24-1 but not cited in the text are listed separately at the end of references.

Albee, A. L., 1965, Phase equilibria in three assemblages of kyanite-zone pelitic schists, Lincoln Mountain quadrangle, central Vermont: Jour. Petrol., v. 6, p. 246–301

Balk, Robert, 1936, Structural and petrologic studies in Dutchess County, New York. Part I. Geologic structure of sedimentary rocks: Geol. Soc. America Bull., v. 47, p. 685–774

Barker, Fred, 1962, Cordierite-garnet gneiss and associated microcline-rich pegmatite at Sturbridge, Massachusetts and Union, Connecticut: Am. Mineralogist, v. 47, p. 907–918

Barrell, Joseph, 1921, Relations of subjacent igneous invasion to regional metamorphism: Am. Jour. Sci., 5th series, v. 1, p. 1–19, 174–186, 245–267

Barrow, George, 1893, On an intrusion of muscovite-biotite gneiss in the south-east Highlands of Scotland: Geol. Soc. London, Quart. Jour., v. 49, p. 330–358

————, 1912, On the geology of lower Dee-side and the southern Highland border: Geol. Assoc., Proc., v. 23, p. 268–284

Barth, T. F. W., 1936, Structural and petrologic studies in Dutchess County, New York. Part II. Petrology and metamorphism of the Paleozoic rocks: Geol. Soc. America Bull., v. 47, p. 775–850

Billings, M. P., 1937, Regional metamorphism of the Littleton–Moosilauke area, New Hampshire: Geol. Soc. America Bull., v. 48, p. 463–566

————, 1955, Geologic map of New Hampshire: U. S. Geological Survey. Scale 1:250,000

————, 1956, The geology of New Hampshire, Part II, Bedrock geology: Concord, New Hampshire, State Planning and Dev. Comm., 203 p.

————, and White, W. S., 1950, Metamorphosed mafic dikes of the Woodsville quadrangle, Vermont and New Hampshire: Am. Mineralogist, v. 35, p. 629–643

Brace, W. F., 1953, The geology of the Rutland area, Vermont: Vermont Geol. Survey Bull. 6, 124 p.

Broughton, J. G., Fisher, D. W., Isachsen, Y. W., and Rickard, L. V., *compilers* and *editors*, 1962, Geologic map of New York—1961, scale 1:250,000: New York State Mus. and Sci. Service Geol. Survey Map and Chart Ser., no. 5 (text, 42 p.)

Chang, P. H., Ern, E. H., and Thompson, J. B., Jr., 1965, Bedrock geology of the Woodstock quadrangle, Vermont: Vermont Geol. Survey Bull. 29, 65 p.

Doll, C. G., Cady, W. M., Thompson, J. B., Jr., and Billings, M. P., *compilers* and *editors*, 1961, Centennial geologic map of Vermont: Vermont Geol. Survey, scale 1:250,000

Doyle, R. G., and Hussey, A. M., II, *compilers* and *editors*, 1967, Preliminary geologic map of Maine: Maine Geol. Survey, scale 1:500,000.

Emerson, B. K., 1898, Geology of old Hampshire County, Massachusetts, comprising Franklin, Hampshire, and Hampden Counties: U. S. Geol. Survey Mon. 29, 790 p.

————, 1917, Geology of Massachusetts and Rhode Island: U. S. Geol. Survey Bull. 597, 289 p.

Evans, B. W., and Guidotti, C. V., 1966, The sillimanite-potash feldspar isograd in western Maine, U. S. A.: Contr. Mineral. and Petrol., v. 12, p. 25–62

Goldsmith, Richard, 1962, Geologic map of New England—Metamorphic zones: U. S. Geol. Survey open file map.

Green, J. C., 1963, High-level metamorphism of pelitic rocks in northern New Hampshire: Am. Mineralogist, v. 48, p. 991–1023

Guidotti, C. V., 1963, Metamorphism of the pelitic schists in the Bryant Pond quadrangle, Maine: Am. Mineralogist, v. 48, p. 772–791

Hawes, G. W., 1878, The mineralogy and lithology of New Hampshire: *in* Hitchcock, C. H., *The Geology of New Hampshire, III,* part IV, p. 1–251

Hitchcock, Edward, 1841, Final report on the geology of Massachusetts: I, J. S. and L. Adams, Amherst; II, J. H. Butler, Northhampton, 831 p.

Lundgren, L. W., Jr., 1966, Muscovite reactions and partial melting in southeastern Connecticut: Jour. Petrol., v. 7, p. 421–453

Newton, R. C., 1966, Kyanite-andalusite equilibrium from 700° to 800°C: Science, v. 153, p. 170–172

Norton, S. A., 1967, Geology of the Windsor quadrangle, Massachusetts: U. S. Geol. Survey open file rept. 896, 210 p.

Perry, J. H., and Emerson, B. K., 1903, The geology of Worcester, Massachusetts: Worcester Nat. History Soc., 166 p.

Sclar, C. B., 1958, The Preston gabbro and the associated metamorphic gneisses, New London County, Connecticut: Conn. Geol. Nat. History Survey Bull. 88, 136 p.

Skehan, J. W., S. J., 1961, The Green Mountain anticlinorium in the vicinity of Wilmington and Woodford, Vermont: Vermont Geol. Survey Bull. 17, 159 p.

Snyder, G. L., 1964, Petrochemistry and bedrock geology of the Fitchville quadrangle, Connecticut: U. S. Geol. Survey Bull. 1161–I, p. 1–63

Thompson, J. B., Jr., 1957, The graphical analysis of mineral assemblages in pelitic schists: Am. Mineralogist, v. 42, p. 842–858

————, 1961, Mineral facies in pelitic schists, *in* G. A. Sokolov, *ed.,* Physico-chemical problems of the formation of rocks and ores (in Russian): Akad. Nauk. S. S. S. R., Inst. Geol. Ore Deposits, Petrog., Mineral., Geochem., p. 313–325; English summary

White, W. S., and Billings, M. P., 1951, Geology of the Woodsville quadrangle, Vermont—New Hampshire: Geol. Soc. America Bull., v. 62, p. 647–696

Wilson, J. R., 1965, Bedrock Geology of the Ossipee Lake Area, New Hampshire: Ph.D. Thesis, Harvard Univ., 177 p.

Woodland, B. G., 1963, A petrographic study of thermally metamorphosed pelitic rocks in the Burke area, northeastern Vermont: Am. Jour. Sci., v. 261, p. 354–375

————, 1965, The geology of the Burke quadrangle, Vermont: Vermont Geol. Survey Bull. 29, 65 p.

Zen, E-an, 1960, Metamorphism of Lower Paleozoic rocks in the vicinity of the Taconic Range in west-central Vermont: Am. Mineralogist, v. 45, p. 129–175

For New Hampshire, Connecticut, Massachusetts, and Rhode Island, the following additional sources have been used:

Connecticut Geologic and Natural History Survey Bulletins 40, 51, 55, 56, 76, 78, 86, 88, 89

Connecticut Geologic and Natural History Survey Miscellaneous Series 1, 3, 5, 8

Connecticut Geologic and Natural History Survey Quadrangle Reports 3, 4, 5, 6, 7, 8, 9, 11, 13, 14, 15, 16, 17, 19, 21, 22

New Hampshire Department of Resources and Economic Development Bulletins 1 and 2

U. S. Geol. Survey Bulletins 1038, 1158 A–E, 1161 I, 1163 A, B, D, 1194–C

U. S. Geol. Survey Geol. Quad. Maps, GQ Series 1, 3, 5, 8, 13, 16, 17, 20, 42, 85, 86, 87, 90, 91, 92, 93, 105, 108, 114, 117, 118, 121, 127, 139, 144, 199, 200, 265, 335, 370, 392, 403, 426, 427, 436, 466, 467, 468, 481, 494, 507, 537, 574, 575, 576, 592, 609, 642, 655

Chapter 25

Metamorphic Zones in Northern Vermont

ARDEN L. ALBEE

INTRODUCTION

THE METAMORPHOSED pelitic rocks in north-central and northwestern Vermont are strikingly different from those in northeastern Vermont. North-central Vermont is a schistose, regionally-metamorphosed terrane, typical of the kyanite-sillimanite facies series of Miyashiro (1961), whereas northeastern Vermont is a hornfelsic, "regional-contact" metamorphosed terrane more typical of the low-pressure intermediate, or Buchan, facies series of Miyashiro (1961). These pronounced differences in the type of metamorphism may be reflections of different times as well as different physical conditions of metamorphism.

Higher-grade assemblages in the eastern part of the western metamorphic belt are extensively altered to lower-grade mineral assemblages, although similar higher-grade assemblages somewhat farther west have not undergone such alteration. This observation suggests the possibility that a younger, eastern metamorphic belt has been partially superimposed on an older, more westerly metamorphic belt and that the low-grade adjustment occurred synchronously with higher-grade metamorphism in the eastern belt. The available isotopic dates can be interpreted to indicate that the two belts were not formed simply as stages of a single period of metamorphism, but have substantially different ages.

Regardless of their relative ages, distinct differences between the two belts imply that the physical conditions of metamorphism were different. Published assemblages have indicated a sequence of metamorphic reactions in eastern Vermont quite unlike those in western Vermont, leading one to expect that the difference in con-ditions between the two metamorphic belts might result in different chemical fractionation of Mn, Fe, and Mg between the phases. As will be shown, no differences in the sequence of metamorphic reactions can be demonstrated and the apparent differences between the two belts are due mainly to gross differences in rock composition. Moreover, the temperature dependency of the partition coefficients is so small that it is obscured by other effects.

GENERAL RELATIONSHIPS OF THE EASTERN AND WESTERN METAMORPHIC BELTS

Figure 25–1 shows the distribution of metamorphic zones and of rock units important to a discussion of the metamorphism of the pelitic rocks in northern Vermont. This figure is based primarily on the state geologic map (Doll et al., 1961), which should be consulted for sources of more detailed descriptions of the rock units and structural features.

In this paper the discussion of metamorphic zones is confined to pelitic schist, the most common rock type in northern Vermont. The isograds of Figure 25–1 are defined by the presence of biotite or chloritoid, pyralspite garnet, kyanite-chloritoid, staurolite, and sillimanite in quartz-muscovite bearing rocks. Most of the western half of the area of Figure 25–1 lies outside the garnet isograd in the biotite-chloritoid zone. Elongate areas of garnet and kyanite-chloritoid zone rocks are associated with the crests of several major anticlines—the Green Mountain anticlinorium, the Northfield Mountain anticlinorium, and the Worcester Mountain anticlinorium. These individual areas will be re-

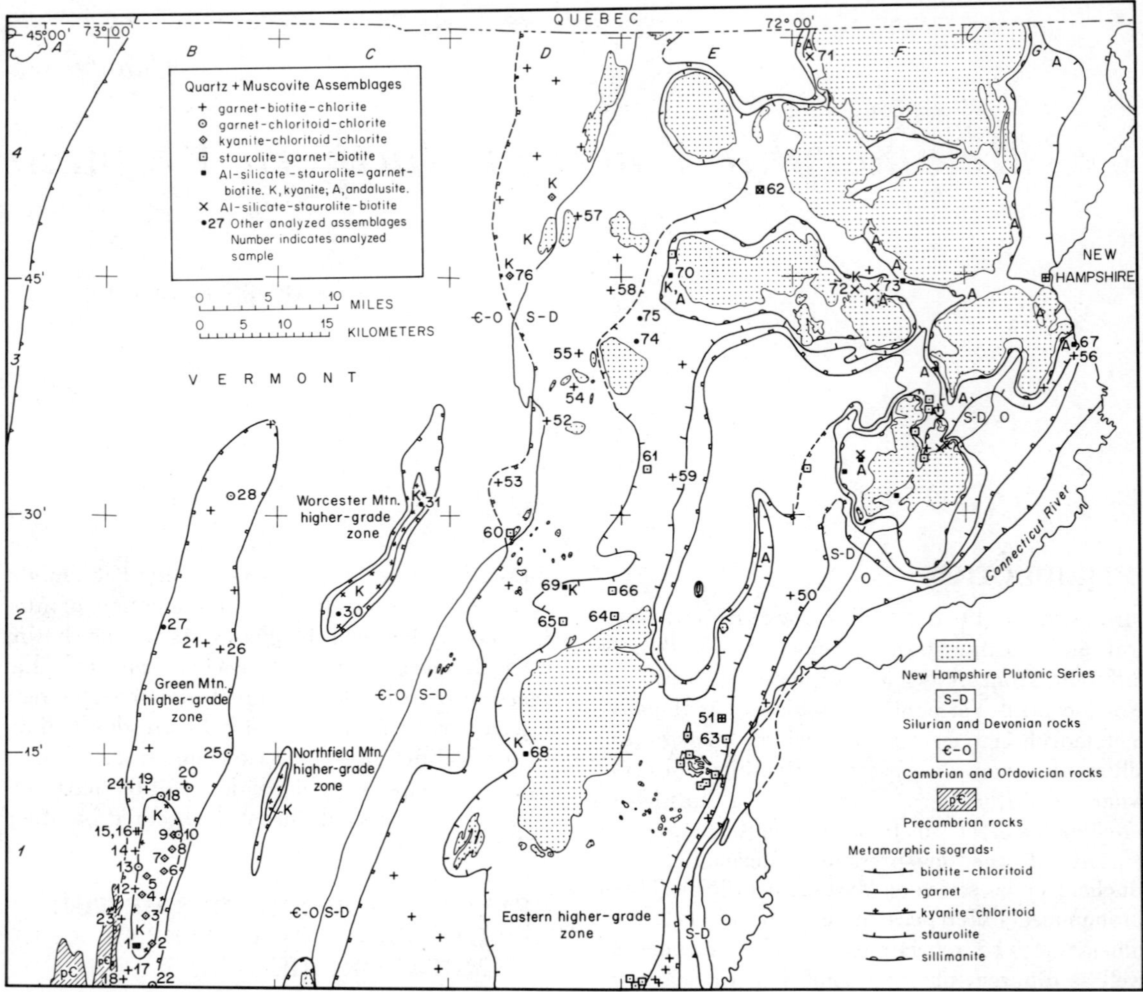

FIGURE 25–1. Metamorphic map of northern Vermont. Grid letters along top and grid numbers along left side are used to locate quadrangles mentioned in text.

ferred to as the Green Mountain higher-grade area, the Green Mountain garnet zone, etc. (see Fig. 25–1). The kyanite-chloritoid and garnet-zone rocks associated with the Worcester and Northfield Mountain anticlines have been partially altered to assemblages of the biotite-chloritoid zone, although similar rocks of the kyanite-chloritoid and garnet zones in the axial region of the Green Mountain anticlinorium show no such retrograde features.

Rocks of the garnet, staurolite, and sillimanite zones occur in the eastern half of the map area. The grade of metamorphism rises abruptly eastward and then drops to biotite and chlorite zone in the extreme eastern part of the area. These higher grade rocks, which will be referred to as the eastern higher-grade area, are spatially related to the intrusive rocks of the New Hampshire Plutonic Series. Isotopic dates (Hurley et al.,

1958, p. 31; Faul et al., 1963) on biotite from two such intrusives in the south-central part of Figure 25–1 suggest that the age of the eastern higher-grade metamorphism is about 350 m.y.

Textures of the rocks in the western part of the map area are distinct from textures in most of the eastern portion. Rocks in the western metamorphic belt are characterized by well-developed schistosity, either parallel or transverse to bedding. Within rocks of the garnet- and kyanite-chloritoid zones, schistosity, metamorphic segregations, and porphyroblastic minerals which grew during the deformation, are abundant. Even within the biotite-chloritoid zone a change in metamorphic intensity eastward and westward from the axial region of the Green Mountain anticlinorium is suggested by the character and abundance of these features. In contrast the eastern metamorphism is characterized by porphyroblastic minerals which grew distinctly later than the deformation and which have grown in and replaced a fine-grained matrix, characteristically slaty rather than schistose; metamorphic segregations are rare.

These differences between the eastern and western areas, in conjunction with the retrogradation of the higher grade rocks of the Worcester and Northfield Mountains, support the hypothesis that rocks of the western part of Figure 25–1 show, predominantly, the effects of an earlier metamorphism associated with the formation of the Green Mountain anticlinorium and that the rocks of the eastern part of the area show, predominantly, the effects of a later metamorphism associated with the New Hampshire Plutonic Series. In terms of this hypothesis the effects of the two metamorphic stages cannot be clearly distinguished in the intervening biotite-chloritoid-zone rocks, but are marked by partial retrogradation of the higher grade rocks in the Worcester and Northfield Mountains.

On the basis of geologic relationships it has been generally assumed that both the eastern and western metamorphism occurred in Devonian time, even though the western metamorphism at the present level of exposure is taken to be somewhat older. However, isotopic age data (Lowden et al., 1963, p. 104; Rickard, 1965; Cady, written communication, 1963; Harper, 1968) strongly suggest that the western metamorphism and deformation was pre-Silurian. The most suggestive measurement in this regard is a K–Ar date of 430 m.y. on coarse-grained muscovite from kyanite-garnet schist of the Worcester Mountain higher-grade area (near Locality 30) (Cady, written communication, 1963; cited by Rickard, 1965, p. 530). The growth of very coarse-grained muscovite, kyanite, and garnet in the Worcester Mountains was associated with the formation of a steep crenulation schistosity, which is typically highly transverse to bedding and approximately parallel to the axial planes of the major folds. Kyanite and garnet porphyroblasts have, subsequently, undergone extensive retrogradation to pseudomorphic aggregates consisting of assemblages of the biotite-chloritoid zone. Hence, the 430 m.y. age is probably a minimum age for the metamorphic growth of kyanite, garnet, and coarse-grained muscovite and for the formation of the schistosity. Furthermore, one may speculate that K–Ar ages on the fine-grained muscovite pseudomorphs after kyanite would yield the approximately 350 m.y. age of the eastern metamorphism.

The existence of a significant period of metamorphism and deformation during Ordovician time has also been inferred by Rosenfeld (this volume) on the basis of a detailed study of the rotational history of garnets in probable Cambrian rocks in southeastern Vermont.

This author's interpretation, not shared by most investigators in Vermont, is that most of the deformational and metamorphic features in the rocks in northwestern Vermont along the Green Mountain anticlinorium are pre-Silurian, probably middle Ordovician. Older isotopic ages have been preserved in the extreme northwestern part of Figure 25–1, but loss of daughter products caused the isotopic dates to decrease to 350 m.y. to the east and southeast. The west edge of the Devonian metamorphic overlap is believed to trend more easterly than the axis of the Green Mountain anticlinorium so that the west edge crosses the axis of the Green Mountain anticlinorium just south of Figure 25–1. Verification of this interpretation must, however, await further isotopic dating.

METHOD OF STUDY AND ACKNOWLEDGMENTS

This paper presents results from the study of about 2000 thin sections of pelitic rocks from the area of Figure 25–1. Most of these are from the central part of the area, that mapped by the author and his former colleagues of the U. S. Geological Survey, W. M. Cady and A. H. Chidester. The observations in the eastern part of the area are based on samples collected at localities suggested by the published quadrangle reports and on samples provided by many of those responsible for the mapping, as well as upon assemblages cited in the published reports on each quadrangle.

The sequence of mineral assemblages in this area will be illustrated by Thompson projections (Thompson, 1957) of the $K_2O–Al_2O_3–FeO–MgO$ tetrahedron (Fig. 25–2). A three-phase assemblage on this projection, such as garnet-biotite-chlorite + quartz + muscovite, contains the six essential components SiO_2, Al_2O_3, K_2O, MgO, FeO, and H_2O; it is trivariant in P, T, and activity of H_2O (a_{H_2O}). The same "three-phase" assemblage in an actual rock will also be trivariant in P, T, and a_{H_2O} if any additional phase introduces an additional component and any additional component introduces an additional phase. The compositions of the phases in such an assemblage may vary continuously due to changes in P, T, a_{H_2O}, and activity of additional components, which do not introduce an additional phase. Hence, by choosing samples with the maximum number of co-existent phases for analytical study, the composition of the phases, as well as the chemical fractionation between phases, will provide information on variation in the external variables. In these rocks MnO acts as such an additional component and the compositions of phases in the "three-phase" assemblages vary continuously with a_{MnO} (or MnO content) as well as with P, T, and a_{H_2O}. The locations of all such "three-phase" samples, as well as those analyzed, are indicated on Figure 25–1.

Mn–Fe–Mg or Mn–Ca–Al were simultaneously analyzed on an Applied Research Laboratory (EMX) microprobe analyser using the correction procedures of Bence and Albee (1968). Two different microscope fields of view in which all coexisting ferromagnesian minerals were present were selected from polished thin sections of each sample. In each field 3–5 grains of biotite, chloritoid, chlorite, and staurolite were analyzed for Mn, Fe, and Mg by ten 10-second analyses with a one micron beam to establish homogeneity, and by two 100-second analyses with a 20 micron beam in larger homogeneous areas of the grain. The Ca content in these minerals is very low and was not analyzed. Biotite, chlorite, and staurolite are typically more homogeneous than chloritoid, which in some samples may have varied by as much as five percent from the value quoted. Almost all the garnet porphyroblasts are zoned with high-Mn cores and low-Mn rims; they were sampled in traverses or randomly to locate points with the maximum-Mn and minimum-Mn values, and those points were then analysed for Mn, Ca, and Al. In general, the garnet values quoted are for for the most nearly euhedral garnet with the maximum range of Mn. Fe^{2+} and Fe^{3+} were calculated by normalization of $Ca + Mg + Fe + Mn + Al$ to the garnet formula: $(Ca, Mg, Mn, Fe^{2+})_3$ $(Fe^{3+}, Al)_2 (SiO_4)_3$. This procedure is subject to the cumulative analytical error, but the Fe^{3+}/Al ratio (≈ 0.02) is reasonably consistent from rim to core in most samples and from sample to sample.

The analytical data is presented on log-plots (Figs. 25–3 and 25–5) showing atom-fraction Mg/Fe and Mn/Fe of biotite, chlorite, and chloritoid and Mg/Fe^{2+}, Mn/Fe^{2+}, and Ca/Fe^{2+} for the rim and core of a single garnet in the sample. The samples are ordered in each assemblage by increasing Mg/Fe of chlorite or biotite. Sample locations are given in Figure 25–1.

The log-plot has the advantage over a tabular presentation of data that equal partition coeffi-

cients (K_D) are represented by equal line segments between sample points because:

$$\log K_D \overset{\text{Mg-Fe}}{\underset{\alpha - \beta}{}} = \log \frac{(Mg/Fe)_\alpha}{(Mg/Fe)_\beta}$$
$$= \log (Mg/Fe)_\alpha - \log (Mg/Fe)_\beta$$

The use of a small card on which to mark the length of line segments is recommended for following the discussion in this paper. The log-plot illustrates the relationships between total composition, assemblage, and location more readily and on fewer diagrams than more conventional plots (i.e., log Mg/Fe_α *vs.* log Mg/Fe_β), which, however, illustrate more readily the variation of K_D for a given mineral pair.

The author is indebted to W. M. Cady, A. H. Chidester, R. A. Christman, J. G. Dennis, L. M. Hall, L. S. Hollister, and R. H. Konig for providing some samples used in this study, and to R. H. Jahns, W. S. White, and R. A. Woodland for providing some unpublished modes, whose assemblages are shown on Figure 25–1. The microprobe was maintained and operated by A. A. Chodos; Lily Ray performed many of the computations; and A. E. Bence wrote CITRAN programs for the corrections. A. E. Bence and L. S. Hollister, as well as the editors, have reviewed the paper. Financial support has been provided by National Science Foundation Grants G-18911 and GP-3570 and the California Institue of Technology (Contribution Number 1465).

PETROLOGY OF PELITIC SCHISTS OF THE WESTERN METAMORPHIC BELT

Biotite–Chloritoid Zone

About half the area of Figure 25–1 lies in the biotite-chloritoid zone. The grain size of the major constituents tends to increase with the grade of metamorphism. Many of the rocks in the extreme west side still preserve their sedimentary textures and structures with only the matrix recrystallized. Rocks in the axial region of the Green Mountain anticlinorium are now porphyroblastic albite gneiss, but preserve layering parallel to bedding. Rocks in the eastern part of the western belt are commonly phyllites and preserve little evidence of bedding even though they are much finer grained than the gneiss.

Textural features indicate that the deformation related to the formation of the Green Mountain anticlinorium continued after the porphyroblastic minerals had formed. For example, chloritoid plates have been rotated by crenulation related to slip cleavage and have been cut by the slip cleavage. The patterns of inclusion trains in albite porphyroblasts indicate a complicated and varied relationship between albite growth and deformation. Sigmoid patterns commonly indicate growth during rotation, but evidence of rotation after the end of growth is also common. Locally, in the axial region of the Green Mountain anticlinorium, albite porphyroblasts contain tiny euhedral garnets, although the ground mass contains only chlorite. Garnet apparently formed before the albite porphyroblasts had attained their full size and was retrograded except where armoured by albite. Post-growth rotation of these porphyroblasts is also observed.

Biotite occurs in less-aluminous rocks and chloritoid in rocks with a high, although not anomalously high, Al content, but both minerals are absent in most rocks. The compositional range of chlorite and muscovite from pelitic schists permits the muscovite-chlorite assemblage to encompass the composition of the most common argillaceous and many arenaceous sedimentary rocks (Albee, 1952).

Important mineral assemblages observed in the quartz-muscovite bearing rocks of the biotite-chloritoid zone are:

chloritoid-chlorite-paragonite
chlorite-paragonite-albite
chlorite-albite ± garnet armoured in albite
biotite-chlorite-albite ± perthitic feldspar

Paragonite occurs with chloritoid and chlorite, but not with biotite; albite does not occur with chloritoid. K-feldspar-bearing assemblages are rare and the K-feldspar occurs as relict detrital grains. As K-feldspar in association with biotite and chlorite represents a possible stable assemblage in this grade of metamorphism, it is assumed that the observed assemblage is stable.

The mineral assemblages observed for this grade of metamorphism are indicated in Figure 25–2a. The mineral compositions are based on optical properties because detailed analytical work has not yet been undertaken on these rocks. An assemblage containing chloritoid + chlorite + an Al-silicate such as pyrophyllite is also possible in this grade of metamorphism, but would occur only in relatively high Al and high Mg/Fe rocks. Although a number of chloritoid + chlorite rocks have been x-rayed, no pyrophyllite has yet been found.

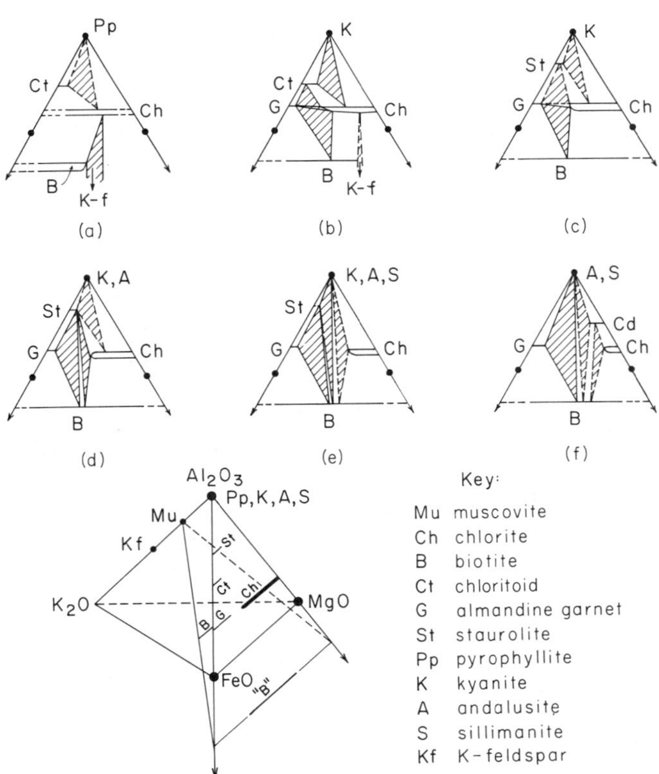

FIGURE 25–2. Thompson projections of the K_2O–Al_2O_3–FeO–MgO tetrahedron illustrating the observed sequence of quartz-muscovite assemblages with increasing metamorphic grade and the compositions of coexistent phases. The western metamorphic belt is represented by (a) and (b); the eastern belt by all six. Three-phase assemblages are ruled; solid lines indicate observed three-phase assemblages and the compositions of their coexistent phases.

Garnet and Kyanite-Chloritoid Zones in the Green Mountain Anticlinorium

Garnet-zone rocks occur in a north-trending band 3–6 miles (5–10 km) wide along the crest of the Green Mountain anticlinorium. Rocks of kyanite-chloritoid zone occur in the southern part of this band within an Al-rich unit, the Mt. Abraham Schist. Physical conditions appropriate to this zone may actually extend somewhat farther north than shown on Figure 25–1, but it cannot be recognized because the appropriate bulk-rock composition is absent.

Important mineral assemblages observed in the quartz-muscovite bearing rocks of the garnet zone are:

> garnet-chloritoid-chlorite-paragonite
> garnet-chlorite-paragonite-albite
> garnet-biotite-chlorite-albite

The kyanite-chloritoid zone is characterized by the additional assemblage:

> kyanite-chloritoid-chlorite-paragonite

Paragonite occurs with chloritoid, chlorite, garnet, and kyanite, but not with biotite; albite does not occur with chloritoid or kyanite.

Rocks of the kyanite-chloritoid zone from a small area west of Mt. Grant in the Lincoln Mountain quadrangle (Figure 25–1, locality number 1) have been studied in great detail. Major element, minor element, and oxygen isotopic analyses have been reported on the coexisting minerals for each of the major assemblages (Albee, 1965; Taylor, Albee, and Epstein, 1963; Garlick and Epstein, 1967). The assemblages and compositions of coexistent phases for these kyanite-chloritoid-zone rocks are shown on a Thompson projection (Fig. 25–2b), indicating that the three major assemblages can coexist under the same P, T, and a_{H_2O} in rocks of somewhat different composition (see also Greenwood, 1967). The constancy of the partition of Fe–Mg–Mn as well as of the minor elements and oxygen isotopes provides convincing evidence that equilibrium has been attained and that the equilibrium temperature was the same for all samples within this small area.

The Mn, Fe, and Mg contents of Fe–Mg silicates have been determined by microprobe analysis for samples from the three major assemblages selected from throughout the Green Mountain higher grade zone. The garnet-biotite-chlorite assemblage occurs throughout the higher grade area, but the occurrence of the kyanite-chloritoid-chlorite and the garnet-chloritoid-chlorite assemblages is restricted, with a few exceptions, to the immediate vicinity of the highly aluminous

Mt. Abraham Schist. Hence, it is probable that the analysed samples of the garnet-biotite-chlorite assemblage represent a broader temperature range than the samples of the other two assemblages.

Figure 25–3 shows atom fraction of Mg/Fe and Mn/Fe of biotite, chlorite, and chloritoid and Mg/Fe^{2+}, Mn/Fe^{2+}, and Ca/Fe^{2+} of garnet. Typically the garnet is systematically zoned with Mn/Fe^{2+} decreasing from core to rim and Mg/Fe^{2+} increasing or remaining constant. The single exception (LA-351) has higher Mn/Fe^{2+} and lower Mg/Fe^{2+} from core to rim. In most samples Ca-zoning is much smaller than that of Mn. Typical zoning is illustrated by the following mineral formulas:

LA-10q, core: $(Ca_{.56}Mn_{.87}Mg_{.10}Fe^{2+}_{1.47})_{3.00}$-$Fe^{3+}_{.05}Al_{1.95})_{2.00}(SiO_4)_3$

LA-10q, rim: $(Ca_{.57}Mn_{.35}Mg_{.19}Fe^{2+}_{1.89})_{3.00}$-$Fe^{3+}_{.05}Al_{1.95})_{2.00}(SiO_4)_3$.

The systematic zoning corresponds to that predicted by a Rayleigh depletion model (Hollister, 1966). This model predicts a bell-shaped profile for Mn since its large partition coefficient results in progressive depletion of Mn from other minerals in the rock surrounding an individual garnet as it grows. The coexistent minerals continuously re-equilibrate so that, at any particular time, they are in equilibrium with an infinitesimally-thin outer layer of the garnet. Accurate analysis of the extreme edges of garnet is difficult and in this reconnaissance analytical study, continuous profiles were analyzed on only a few garnets. In most samples, a point near the rim with the minimum Mn-content was taken as the rim composition. Despite the fact that this value may not represent the exact rim composition, it is noteworthy that the partition between garnet rim and other phases does not exhibit gross discrepanices (i.e., the distance between the corresponding points on Figure 25–3 is fairly constant). In the generalizations and conclusions on chemical fractionation which follow, the composition of the garnet rim, not the garnet core, is considered, and greater deviation is anticipated than for other minerals.

The data of Figure 25–3 indicates a consistent fractionation between coexistent phases:

Mg/Fe: chlorite > biotite $\gg$ staurolite $\cong$ chloritoid $\gg$ garnet rim

Mn/Fe: garnet rim $\gg$ chloritoid $\cong$ staurolite $\gg$ chlorite > biotite

Moreover, since the samples shown on Figure 25–3 are ordered by increasing Mg/Fe of chlorite in each assemblage, it is readily apparent that both the Mg/Fe and the Mn/Fe of all minerals in each of the three assemblages increase quite systematically with increasing Mg/Fe of chlorite. These systematic variations, even though there is considerable deviation, indicate the existence of several relationships:

1. The distance between points, which indicate the Mg/Fe (or the Mn/Fe) for any two phases in a given assemblage, is nearly the same in all samples; hence the partition coefficients (K_D) are nearly the same in all samples regardless of differences in Mg/Fe, Mn/Fe, T, or other variables;

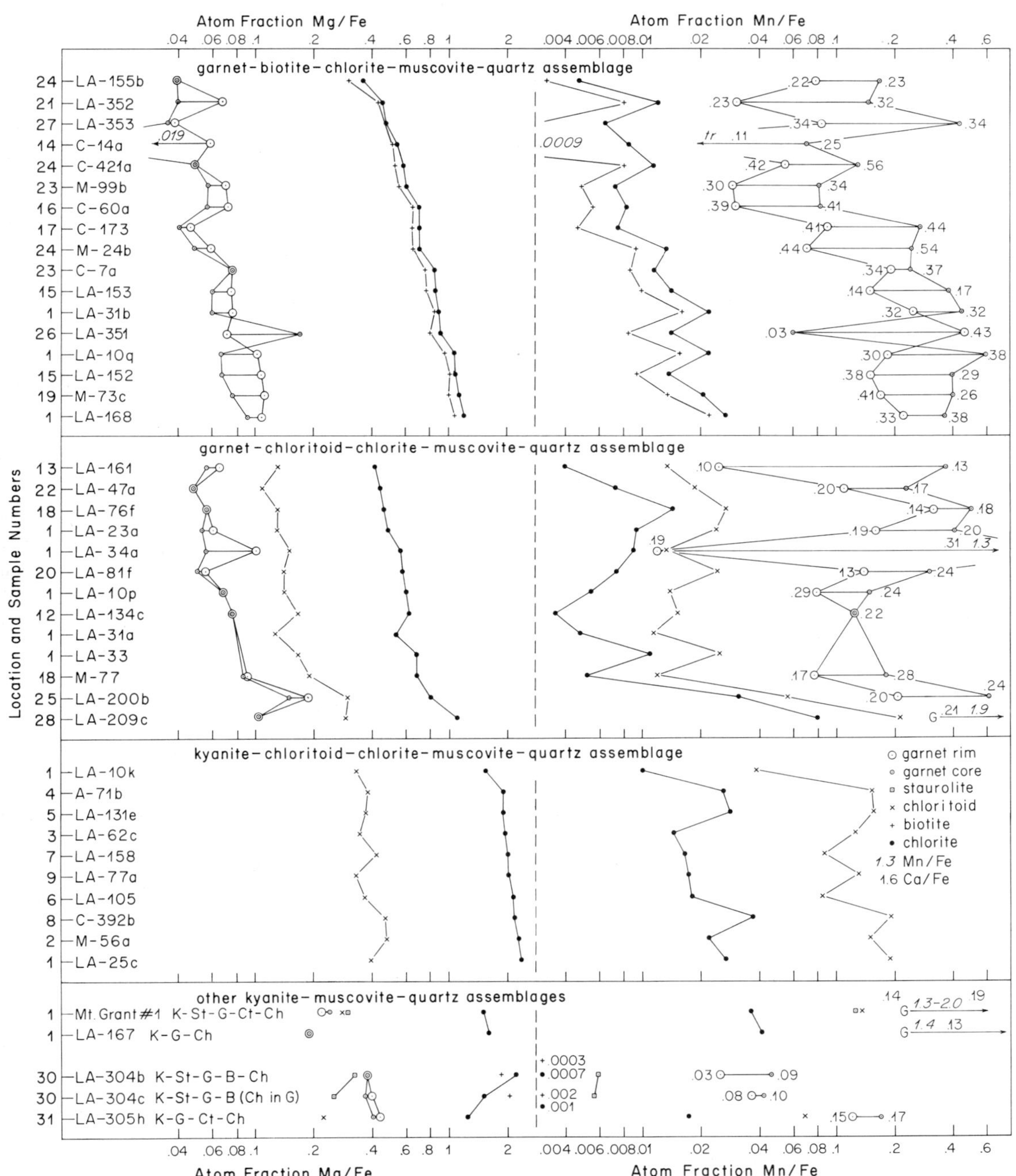

FIGURE 25–3. Atom fraction Mg/Fe and Mn/Fe in coexistent phases in samples from the western metamorphic belt. Atom fraction Mg/Fe²⁺, Mn/Fe²⁺, and Ca/Fe²⁺ are shown for garnet. Equal distances between points for coexistent phases represent equal partition coefficients.

2. The Mg/Fe varies systematically with the Mn/Fe in each phase of a given assemblage;

3. Knowing these relationships the Mg/Fe and Mn/Fe of all phases for a given assemblage can be predicted from the Mg/Fe or the Mn/Fe of a single phase.

Figure 25–4 is a ternary Mn–Fe–Mg diagram showing schematic tie lines which fit the observed relationship between Mg/Fe and Mn/Fe in garnet and chlorite with constant $K_D{}^{Mg/Fe}_{G/Ch}$ and $K_D{}^{Mn/Fe}_{G/Ch}$. This illustrates the manner in which the bulk Mn-content of the rock controls the Mg/Fe, as well as the Mn/Fe, of all the phases.

The data do deviate from the exact relationships indicated above, and it is necessary to determine whether these deviations can be correlated with the metamorphic grade of the individual samples by means of the spatial relationships shown on Figure 25–1. The data for the garnet-biotite-chlorite assemblage include three samples from a single locality (number 24) near the low grade margin of the garnet zone and three samples from a single locality (number 1) in the kyanite-chloritoid zone. The samples from each locality should represent nearly identical temperatures of metamorphism, but the temperatures of the two localities should be distinctly different because of the difference in metamorphic grade. However, the total variation in the partition coefficients at each locality overlaps the differences between the two localities. As the partition coefficients do not differ significantly for the two sets of samples, we cannot distinguish a significant temperature difference.

This finding might have been predicted from studies on the partition of Mg and Fe between calcic pyroxenes (cpx) and orthopyroxenes (opx) (Bartholomé, 1962; Kretz, 1963). $K_D{}^{Mg/Fe}_{opx/cpx}$ is about 0.54 for an estimated temperature of 670°C and about 0.73 for 1150°C (Kretz, 1963). If we assume a similar temperature dependency—i.e. that near 500°C a 100°C change in temperature changes K_D by 10 percent of its value—the observed variation in $K_D{}^{Mg/Fe}_{Ct/Ch}$ at locality 1 is equivalent to the effects of a temperature variation of about 100°C.

Although the Mg/Fe and Mn/Fe for phases from locality 24 are somewhat lower than for the corresponding phases from locality 1, this difference cannot be assigned to a continuous variation with P, T, or activity of H_2O as suggested by Thompson (1957, p. 855) because it may represent a difference in bulk composition, as illustrated by Figure 25–4. In the other two assemblages the samples from locality number 1 encompass most of the total variation of all the samples, both in their compositions and in their partition coefficients.

As noted previously (Albee, 1965, p. 281) the partition coefficients for garnet-chlorite and chloritoid-chlorite differ in the two assemblages containing these pairs, even in samples from a single locality (number 1). This emphasizes the dependency of the partition coefficients upon compositional variation of the phases as well as upon temperature variation. It also points out the difficulty of comparing partition coefficients for mineral pairs which are chosen from different assemblages in order to sample a large difference in metamorphic grade.

Two apparently anomalous assemblages from the Mt. Grant area (locality number 1) have also been studied. Sample LA-167 has the assemblage kyanite-garnet-chlorite, which would appear to be incompatible with the more common assemblages. However, the Mn/Fe^{2+} of its garnet is about 10 times greater than that of garnet from other samples although the Mg/Fe and Mn/Fe of the coexisting chlorite is similar to that of chlorite from the kyanite-chloritoid-chlorite assemblage. Sample Mt. Grant-1 has the assemblage kyanite-staurolite-garnet-chloritoid-chlorite. The Mg/Fe and Mn/Fe of chlorite and chloritoid are intermediate to values from the kyanite-chloritoid-chlorite and garnet-chloritoid-chlorite assemblages, although its garnet has Mn/Fe^{2+} about 10 times greater. The staurolite has Mn/Fe and Mg/Fe similar to that of chloritoid, but the total Fe content is relatively low for staurolite. A wave-length scan indicated a high Zn content, $Zn/Fe \cong 2(Mn/Fe)$ corresponding to nearly two percent by weight of ZnO. In each of the

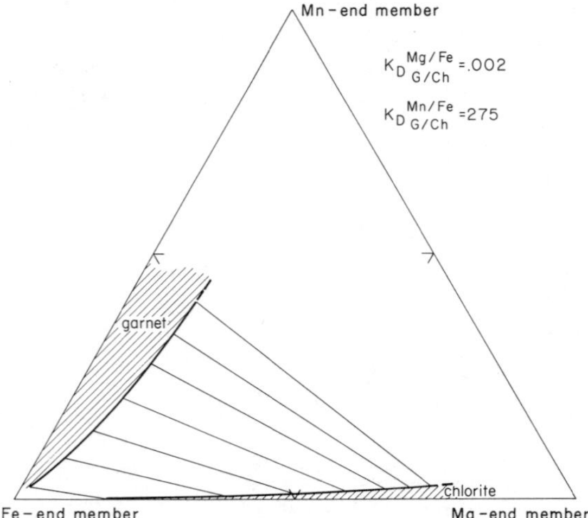

FIGURE 25–4. Ternary Mn–Fe–Mg diagram with schematic tie lines illustrating the generalized relationships between the compositions of coexistent garnet and chlorite.

samples garnet is probably an additional phase stabilized by the high bulk Mn content and in the second sample staurolite is probably stabilized by the high Zn content.

Quite clearly neither the systematic compositional variation nor the rather small variation in the partition coefficients can be used as valid indicators of metamorphic grade in the Green Mountain higher grade area. It is probable that the temperature effect on the partition coefficients is so small that it is masked by sampling and analytical errors, by the complications introduced by the garnet zoning, and by the effect of other compositional variables. Nevertheless, the strikingly systematic relations in the partition of Mn, Mg, and Fe between phases suggest that equilibrium has been closely approached.

Higher-Grade Rocks of the Worcester and Northfield Mountains

Garnet- and kyanite-zone rocks occur within the Stowe Formation in an elongate belt nearly 20 miles (32 km) long in the Worcester Mountains and in a belt about 10 miles (16 km) long in the Northfield Mountains (Fig. 25–1). These higher grade areas are along the crests of major anticlines, similar in position to the higher grade areas along the crest of the Green Mountain anticlinorium.

The generally fine-grained schist and phyllite of the Stowe Formation grade into coarse-grained rocks, which are poorly schistose due to the coarseness of grain; bedding is rarely preserved. Lenticles and irregularly shaped masses of coarsely crystalline quartz are abundant. In the kyanite zone partly altered porphyroblasts of garnet and kyanite form as much as 35 percent and 50 percent, respectively, of the rock. Pod-shaped bodies, commonly as much as 1 by 5 feet (0.3 by 1.5 m), consist almost entirely of kyanite and ilmenite, and some layers consist predominantly of kyanite and ilmenite.

The mineral relations are complicated by the partial alteration of kyanite, garnet, and biotite to mineral assemblages characteristic of a lower grade. However, the important assemblages of the higher-grade quartz-muscovite rocks prior to retrogradation seem to have been:

Garnet zone
garnet-chloritoid-chlorite
garnet-biotite-chlorite
Kyanite zone
kyanite-chloritoid-chlorite
kyanite-chloritoid-garnet-chlorite
kyanite-garnet-chlorite
kyanite-garnet-biotite
kyanite-staurolite-garnet-biotite-chlorite

Within the Hyde Park quadrangle (C-3, Fig. 25–1), paragonite is common in chloritoid-chlorite assemblages and occurs rarely in garnet assemblages, but neither paragonite nor albite has been found with kyanite.

Kyanite is partially to completely altered to an aggregate of fine-grained muscovite, commonly with chloritoid; garnet is altered to pseudomorphic aggregates of chlorite and magnetite. Much of the chloritoid is entirely within the kyanite pseudomorphs and formed during the alteration, but chloritoid also occurs in large plates interstitial to kyanite blades. This interstitial chloritoid is inferred to be part of the earlier metamorphic assemblage. Similarly both primary and secondary chlorite and muscovite are present. Staurolite has been found at a single locality (number 30) in the assemblage kyanite-staurolite-garnet-biotite-chlorite in a rock with very little retrograde alteration. The common assemblage prior to retrogradation seems to have been kyanite-chloritoid-chlorite, the same one found in the kyanite-chloritoid zone of the Green Mountain anticlinorium.

Interpretation of retrograde textures also suggests that kyanite-garnet-chlorite-muscovite-quartz was a common assemblage prior to the retrograde alteration. In the Green Mountain higher-grade area this assemblage is stabilized in several instances by a high Mn content. An even higher-grade reaction may have occurred marking the transitions from garnet-chlorite compatibility to kyanite-biotite compatibility and from chloritoid to staurolite, because the assemblages kyanite-garnet-biotite-muscovite-quartz and kyanite-staurolite-garnet-biotite-chlorite-muscovite-quartz have been noted in several samples in the Montpelier quadrangle (C-2, Fig. 25–1). X-ray studies show that the kyanite has altered to muscovite, rather than pyrophyllite during the retrograde alteration. This is not attributed to addition of K during alteration; it is suggested, instead, that the assemblage kyanite-garnet-biotite-muscovite-quartz was common and that it has altered to chlorite-chloritoid-muscovite-quartz.

These inferences suggest that, at its peak, the metamorphic grade for the western metamorphism was somewhat higher in the Worcester and Northfield Mountain higher-grade areas than in the Green Mountain higher-grade area.

Comparison of analytical data on three samples —LA-304b, LA-304c, LA-305h—with samples from the Green Mountain higher-grade area indicates several anomalies in the distribution co-efficients. The anomalous distribution coefficient for Mg–Fe between biotite and chlorite in LA-304c is not unexpected. The chlorite occurs entirely as inclusions in the garnet, but garnet with Mg/Fe^{2+} greater than Mg/Fe in coexisting chloritoid or staurolite (see Fig. 25–5) is unique to these three samples. Moreover, the Mg/Fe^{2+} of these garnets is 2–10 times that of the garnet in all the other samples. The data indicate that the phases are not all in mutual chemical equilibrium, which is to be expected in view of the petrographic evidence.

Petrology of Pelitic Schists of the Eastern Metamorphic Belt

The eastern metamorphic belt includes rocks ranging in grade from the chlorite zone to the sillimanite zone; biotite, garnet, staurolite, and sillimanite isograds are delineated in Figure 25–1. The distribution of index minerals is rather erratic due, in part, to the lower proportion of pelitic rocks and, in part, to the irregular distribution of intrusive masses of the New Hampshire Plutonic Series. The higher grade rocks have a distinct spatial relationship to the intrusions. On the other hand, some compositional control of index minerals is suggested by the large area of biotite-zone rocks within the Lyndonville (E-3, Fig. 25–1) and St. Johnsbury quadrangles (E-2) which corresponds closely to the distribution of a highly calcareous unit, the Waits River Formation (see State geologic map, Doll et al., 1961).

The texture of these rocks, except in the extreme southeastern portion of Figure 25–1, is distinct from that of rocks in the western high-grade areas. The quartz-mica matrix is finer grained and they are more apt to be called phyllite, slate, or hornfels than schist, even in high-grade areas. Porphyroblasts of garnet, staurolite, andalusite, kyanite, biotite, chlorite, and ilmenite crosscut all primary and secondary foliation surfaces and, except in the extreme southeastern portion of Figure 25–1, formed after all rotation along the foliation surfaces had ceased. This is in marked contrast to the mode of occurrence of these minerals in the western metamorphic area. Most of the porphyroblastic minerals are inclusion filled and commonly contain so much graphite as to nearly obscure their form. A distinctive feature is the common occurrence of rayed-garnet porphyroblasts due to the inclusion of graphite flakes at the junctions of growing dodecahedral faces.

The western limit of the garnet zone of the eastern metamorphic belt has been modified (dashed portion) from that shown on the State geologic map (Doll et al., 1961) so as to include all known examples of such post-deformational garnet porphyroblasts. Lower grade zones to the west are detectable only where they have caused partial retrogradation of older higher-grade rocks as in the Worcester and Northfield Mountains. Lower-grade rocks to the east along the Connecticut River have not been studied in detail, but may include rocks that are retrograde from a pre-existing higher grade of metamorphism.

The assemblage garnet-biotite-chlorite-muscovite - quartz - plagioclase occurs throughout the garnet zone, and there is a general absence of assemblages more aluminous than the garnet-chlorite-muscovite join of the Al_2O_3–K_2O–FeO–MgO tetrahedron. High-Al assemblages are known only in the extreme northwest part of the garnet zone (locality 76 and vicinity) where the Umbrella Hill Conglomerate has the critical assemblages:

kyanite-chloritoid-chlorite-muscovite-quartz
chloritoid-chlorite-muscovite-quartz

Garnet is absent from these rocks. Paragonite occurs in some samples and pyrophyllite occurs in association with kyanite in several samples.

The assemblage chloritoid-biotite-muscovite-quartz $\pm$ garnet has been reported in a number of localities (Dennis, 1956, p. 80; Eric and Dennis, 1958, p. 50; Hall, 1959, p. 33). Sample 11-16, provided by Dennis, contained garnet-hornblende-biotite and sample LA-357, provided by Hall, contained garnet-biotite-chlorite. Despite special efforts, examples of chloritoid-biotite assemblages have not been verified. It appears that the garnet-chlorite-muscovite join was stable throughout the garnet zone; chloritoid does not appear in these rocks because their bulk compositions lie below the garnet-chlorite join in Figure 25–2c, and they are generally much less aluminous than those in the western metamorphic belt.

Analytical data for 10 samples of the garnet-biotite-chlorite assemblage are given in Figure 25–5; included are two samples (LA-264b and LA-260) from within the staurolite zone in which the chlorite is probably retrograde in origin. The other samples show a consistent partition of elements, but, as in the western belt, neither the compositional variation nor the variation in the partition coefficients can be simply related to metamorphic grade by reference to the spatial relationships. The garnets are more homogeneous and Mn/Fe^{2+} of garnet does not show as sys-

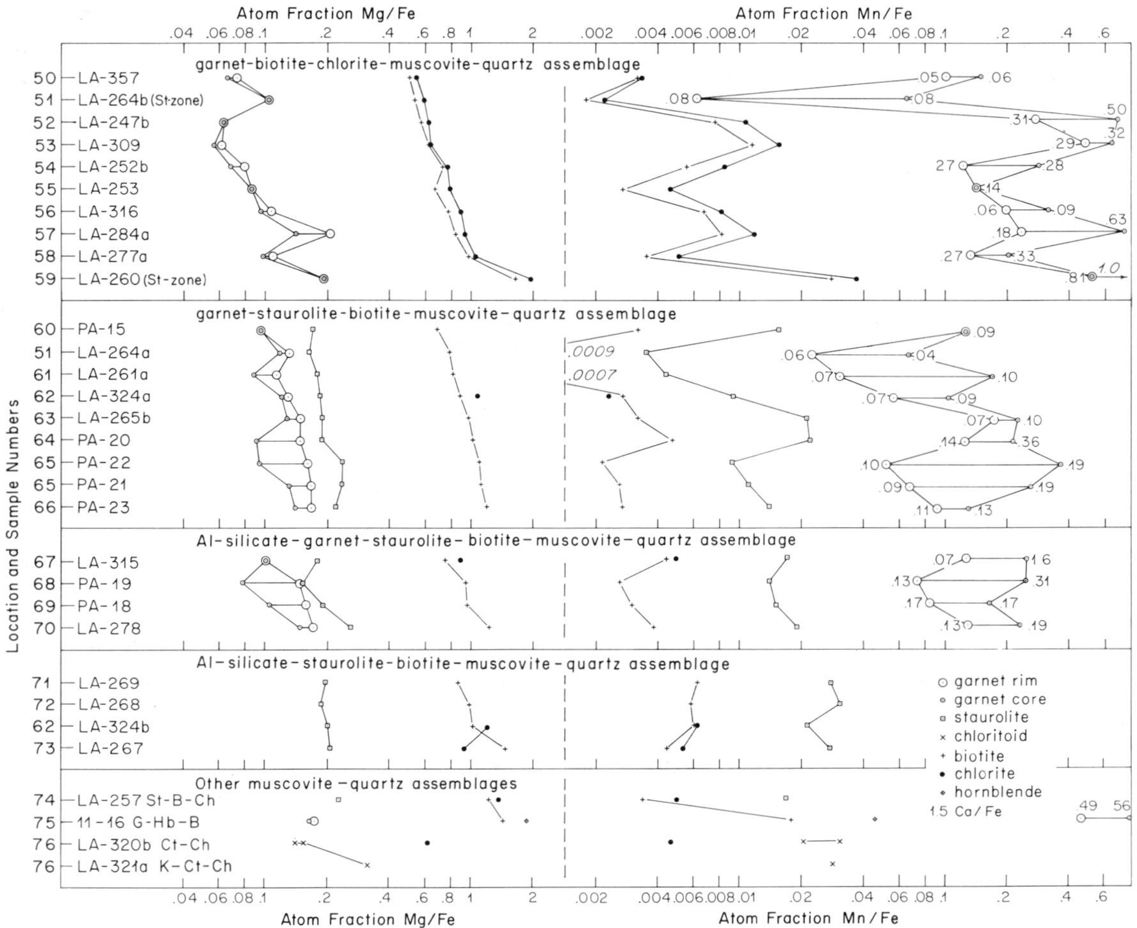

FIGURE 25–5. Atom fraction Mg/Fe and Mn/Fe in coexistent phases in samples from the eastern metamorphic belt. Atom fraction Mg/Fe²⁺, Mn/Fe²⁺, and Ca/Fe²⁺ is shown for garnet. Equal distances between points for coexistent phases represent equal partition coefficients.

tematic a variation with Mg/Fe of the phases. Comparison of the data for the garnet-biotite-chlorite assemblage from Figures 25–3 and 25–5 indicates that the individual minerals have a similar range of composition in the eastern and western metamorphic belts. However, the garnet rims in the western belt have a somewhat lower Mn/Fe relative to chlorite and biotite than those in the eastern belt, and, as a result, $K_D{}^{\mathrm{Mn/Fe}}_{\mathrm{G/Ch}}$ and $K_D{}^{\mathrm{Mn/Fe}}_{\mathrm{G/B}}$ are somewhat larger for the eastern rocks.

The staurolite isograd is marked by the appearance of staurolite in the characteristic assemblage staurolite - garnet - biotite - muscovite - quartz-plagioclase as shown in Figure 25–2d. Hence, staurolite appears in these rocks through the reaction:

garnet + chlorite + muscovite =

biotite + staurolite

This reaction occurs at a higher grade than the transformation of chloritoid to staurolite and of pyrophyllite to kyanite. The western limit of the staurolite zone has been modified (dashed portion) from that shown on the State geologic map (Doll et al., 1961) so as to exclude a group of samples with the assemblage garnet-biotite-chlorite. Analytical data are given for 9 samples of the staurolite-garnet-biotite assemblage in Figure 25–5. There is a consistent partition of elements, very little variation in partition coefficients, and a fairly systematic variation of Mg/Fe with Mn/Fe in all phases. The partition coefficients for garnet-biotite do not differ markedly from those in the garnet-biotite-chlorite assemblage, and no temperature dependency can be discerned.

Within the staurolite zone, andalusite or kyanite also occurs in the assemblages Al-silicate-garnet-staurolite-biotite and Al-silicate-staurolite-biotite (Fig. 25–2e). As indicated on Figure 25–1, kyanite occurs to the southwest and andalusite to the northeast; in the intermediate zone andalusite and kyanite occur together and occur with sillimanite adjacent to larger intrusive bodies. These higher-grade assemblages, the coexistence of the three Al-silicate polymorphs, and the paragenetic relationships of the Al-silicates have been described in detail by Woodland (1963, 1965), who infers that temperature and pressure conditions must have closely approached those of the triple point of the Al-silicate polymorphs.

The analytical data in figure 5 indicate partition coefficients and Mg/Fe and Mn/Fe of the phases essentially the same in the Al-silicate-staurolite-biotite $\pm$ garnet assemblages as in the staurolite-garnet-biotite assemblages. Reference to Figure 25–2 suggests that the biotite of these assemblages should show a greater difference in composition than actually exists. The four-phase assemblage Al - silicate - staurolite - garnet - biotite contains an additional phase which is not related to an excessive Mn content nor an excessive Ca content. A more general explanation for the common appearance of this assemblage is that the garnet is relatively non-reactive, as implied by its pronounced zoning, and persists as a relic even after changing conditions or composition have placed the rock out of the stability field of garnet. The chlorite in these assemblages is typically retrograde in origin and in several samples has distinctly anomalous partition coefficients.

Woodland (1963, p. 365) indicates that staurolite breaks down in the highest grade rocks and is typically absent in andalusite and sillimanite rocks, but no samples of the highest grade assemblage Al-silicate-garnet-biotite have yet been analysed. Cordierite has been observed at the margins of several intrusives and the assemblage cordierite-Al-silicate-biotite-muscovite can coexist stably with the other Al-silicate assemblages in rocks with higher Mg/Fe. Both of these assemblages are shown on Figure 25–2f.

Petrologic Relationships of the Western and Eastern Metamorphic Belts

When the reconnaissance analytical study was begun, it was believed that the eastern and western metamorphic belts formed at different times under metamorphic conditions different enough to result in a different sequence of meta-

morphic reactions. The supposed occurrence of the assemblage garnet-biotite-chloritoid in eastern Vermont had suggested a sequence of metamorphic reactions (Hall, 1959, p. 95) quite different from that in western Vermont and it was expected that several questions could be answered:

1. Whether the close approach to chemical equilibrium, which had been demonstrated in the Mt. Grant area (locality number 1), was typical of both metamorphic belts.

2. How the detailed sequence of assemblages differed in the two belts.

3. Whether changes in partition coefficients and compositions of phases might be used to measure smaller differences in metamorphic grade than those which result in the appearance or disappearance of a phase as suggested by Thompson (1957, p. 855).

4. Whether metamorphic conditions differed sufficiently as to result in differences in Mg/Fe and Mn/Fe partition in the two belts.

The systematic fractionation suggests that a close approach to chemical equilibrium has been attained in most samples from both belts. The exceptions are mainly those in which nonequilibrium was clearly suggested by the textural relationships. The constancy of partition coefficients involving the rims of zoned garnet porphyroblasts is consistent with the assumptions involved in the Rayleigh depletion model of garnet zoning.

No evidence has been found to suggest a different sequence of metamorphic reactions in the two belts, except for the presence of andalusite rather than kyanite in northeastern Vermont, and the major apparent differences in sequence are simply due to differences in bulk composition. Figures 25–2a and 25–2b represent the sequence in western Vermont and Figures 25–2c, 25–2d, 25–2e, and 25–2f represent the higher grades in the eastern belt. In northwestern Vermont there is a sufficient variety of bulk rock compositions to verify the assemblages lying above the garnet-chlorite join, whereas in northeastern Vermont the bulk rock compositions lie below the garnet-chlorite join and the more aluminous assemblages for the lower grades are not present. It is probable that Figures 25–2a and 25–2b represent the lower-grade sequence of reactions for the eastern metamorphic belt as well as for the western belt.

The results of this study suggest that in general, as well as in northern Vermont, the partition of Mn, Mg, and Fe between ferromagnesian silicates will not provide a simple measure of the variation in metamorphic grade nor of the differences in metamorphic conditions in two metamorphic belts. Neither the partition coefficients nor the variation in the composition of the phases can be simply related to metamorphic grade, either over differences in metamorphic grade smaller than those which result in the appearance or disappearance of a phase or over the entire range of metamorphic grade.

Partition coefficients for a given mineral pair differ significantly in different assemblages at the same locality and the same, metamorphic grade, but show no significant variation in a single assemblage representing a broad range of metamorphic grade despite the differences in T, Mn/Fe, and other variables. It must be concluded that the temperature dependency of these partition coefficients is so small that it is masked by sampling and analytical errors, complications introduced by garnet zoning, and the effect of other variables.

The variation in composition of the individual phases over the entire range of metamorphic grade studied does not differ significantly from that found in the garnet zone. In particular, the Mg/Fe variation in the individual phases is so strongly dependent on the Mn content of the rock that no dependency on metamorphic grade can be recognized.

Nevertheless, the fact remains that the rocks in the eastern and western metamorphic belts are distinctly different, although the differences are now seen to be almost entirely textural rather than a different sequence of assemblages. It seems most likely that the cause of the differences is primarily kinetic, which may be attributed in part to different heating rates, different amounts of pore fluid, and different extent of pervasive deformation at the time of high temperature.

REFERENCES

Albee, A. L., 1952, Comparison of the chemical analyses of sedimentary and metamorphic rocks [abs.]: Geol. Soc. America Bull., v. 63, p. 1229

————, 1965, Phase equilibria in three assemblages of kyanite-zone pelitic schists, Lincoln Mountain quadrangle, central Vermont: Jour. Petrology, v. 6, p. 246–301

Bartholomé, P., 1962, Iron-magnesium ratio in associated pyroxenes and olivines, *in* Petrologic Studies—A volume to honor A. F. Buddington: New York, Geol. Soc. America, p. 1–20

Bence, A. E., and Albee, A. L., 1968, Empirical correction factors for the electron microanalysis of silicates and oxides: Jour. Geology, v. 76, p. 382–403

Dennis, J. G., 1956, The geology of the Lyndonville area, Vermont: Vermont Geol. Survey Bull., v. 8, 98 p.

Doll, G. G., Cady, W. M., Thompson, J. B., Jr., and Billings, M. P., *compilers* and *editors*, 1961, Centennial geologic map of Vermont: Vermont Geol. Survey, Montpelier, scale 1:250,000

Eric, J. H., and Dennis, J. B., 1958, Geology of the Concord–Waterford area, Vermont: Vermont Geol. Survey Bull., v. 11, 66 p.

Faul, Henry, Stern, T. W., Thomas, H. H., and Elmore P. L. D., 1963, Ages of intrusion and metamorphism in the northern Appalachians: Am. Jour. Sci., v. 261, p. 1–19

Garlick, G. D., and Epstein, Samuel, 1967, Oxygen isotope ratios in coexisting minerals of regionally metamorphosed rocks: Geochim. et Cosmochim. Acta, v. 31, p. 181–214

Greenwood, H. J., 1967, The N-dimensional tie-line problem: Geochim. et Cosmochim. Acta, v. 31, p. 465–490

Hall, L. M., 1959, The geology of the St. Johnsbury quadrangle, Vermont and New Hampshire: Vermont Geol. Survey Bull., v. 13, 105 p.

Harper, C. T., 1968, Isotopic ages from the Appalachians and their significance: Can. Jour. Earth Sciences, v. 5, p. 49–59

Hollister, L. S., 1966, Garnet zoning: An interpretation based on the Rayleigh fractionation model: Science, v. 154, p. 1647–1651

Hurley, P. M., et al., 1958, 6th Ann. Prog. Rept. to U. S. Atomic Energy Commission, Massachusetts Inst. Tech., Dept. Geology and Geophysics, 193 p.

Kretz, R., 1963, Distribution of magnesium and iron between orthopyroxene and calcic pyroxene in natural mineral assemblages: Jour. Geology, v. 71, p. 773–785

Lowden, J. A., Stockwell, C. H., Tipper, H. W, and Wanless, R. K., 1963, Age determinations and geological studies: Canada Geol. Survey Paper 62-17, 140 p.

Miyashiro, A., 1961, Evolution of metamorphic belts: Jour. Petrology, v. 2, p. 277–311

Rickard, M. J., 1965, Taconic orogeny in the western Appalachians: Experimental application of microtextural studies to isotopic dating: Geol. Soc. America Bull., v. 76, p. 523–536

Taylor, H. P., Albee, A. L., and Epstein, Samuel, 1963, O^{18}/O^{16} ratios of coexisting minerals in three assemblages of kyanite-zone pelitic schist: Jour. Geology, v. 71, p. 513–522

Thompson, J. B., Jr., 1957, The graphical analysis of mineral assemblages in pelitic schists: Am. Mineralogist, v. 42, p. 842–858

Woodland, B. G., 1963, A petrographic study of thermally metamorphosed pelitic rocks in the Burke area, northeastern Vermont: Am. Jour. Sci., v. 261, p. 354–375

————, 1965, The geology of the Burke quadrangle, Vermont: Vermont Geol. Survey Bull., v. 28, 151 p.

Chapter 26

Evolution of the Ultramafic Complexes
of Northwestern New England*

ALFRED H. CHIDESTER

INTRODUCTION

THIS PAPER IS BASED on detailed studies of parts of the belt of ultramafic rocks in Vermont and northwestern Massachusetts (see Fig. 26–1) carried out by the U.S. Geological Survey. In their geotectonic setting, stratigraphic and structural relations, range of variation, and metamorphic features, the ultramafic rock complexes are representative of most of the ultramafic rocks of New England and of many in southeastern Quebec. The aim of the paper is to relate the varied petrologic and structural features of the ultramafic complexes to the tectonic setting, sedimentary environment, and structural and metamorphic history of the belt; and on the basis of these relations to present an interpretation of the evolutionary history of the complexes that accounts for their striking variations and that is generally valid for the ultramafic complexes of New England and southeastern Quebec.

It is gratifying to express here my indebtedness to Professor Marland Billings, who directed and carried out the early part of the U.S. Geological Survey's program of talc investigations in Vermont, in 1944–1945, and who has subsequently maintained a lively interest in these studies that has been very stimulating and helpful. I take pleasure, also, in acknowledging the considerable contributions to this paper of my colleagues in the Vermont studies, Dr. Wallace M. Cady and Prof. Arden L. Albee.

* Publication authorized by the director, U. S. Geological Survey.

REGIONAL GEOLOGY

The country rocks of the belt of ultramafics are stratigraphically entirely below the unconformity at the base of the Silurian section (see Cady, in press). They comprise phyllite and schist, both of which are locally graphitic and albitic, greenstone or amphibolite, and minor quartzite and impure marble. The lithologic units commonly intergrade rather abruptly by small-scale intertonguing and by change in mineral composition, both parallel and normal to the trends of the units. They represent metamorphosed sedimentary and volcanic rocks that range in probable age from earliest Cambrian to Ordovician. The protoliths of the predominant metamorphosed sedimentary rocks are graywacke and shale. The thickness of the restored section of rocks above the older Precambrian and below the Silurian ranges from 50,000 to 60,000 feet (15,000 to 18,000 m) in central Vermont to less than 35,000 feet (10,000 m) in northern and southern Vermont (see Cady, in press). These features are characteristic of a deeply subsided trough that was rapidly filled with clastic and volcanic rocks —the eugeosynclinal zone of the early to middle Paleozoic orthogeosyncline of the northern Appalachians (see Cady, 1967).

The country rocks of the belt range in metamorphic grade from the greenschist facies to the epidote-amphibolite facies. The rocks are characterized by chlorite, biotite, almandine garnet, epidote, actinolitic hornblende, hornblende, and kyanite, in various combinations appropriate to the metamorphic grade and rock composition.

The metamorphic grade of these rocks is generally related to the depth of erosion: areas of higher grade are associated with structural highs; of lower grade, with structural lows.

The dominant structural feature of the belt of ultramafic rocks is the Green Mountain anticlinorium (see regional geologic map, this volume; see also Doll et al., 1961). Most of the belt is east of the anticlinorial axis, where the rocks dip chiefly eastward in a steep homoclinal succession, interrupted near the eastern side of the belt by the Worcester and Northfield Mountain

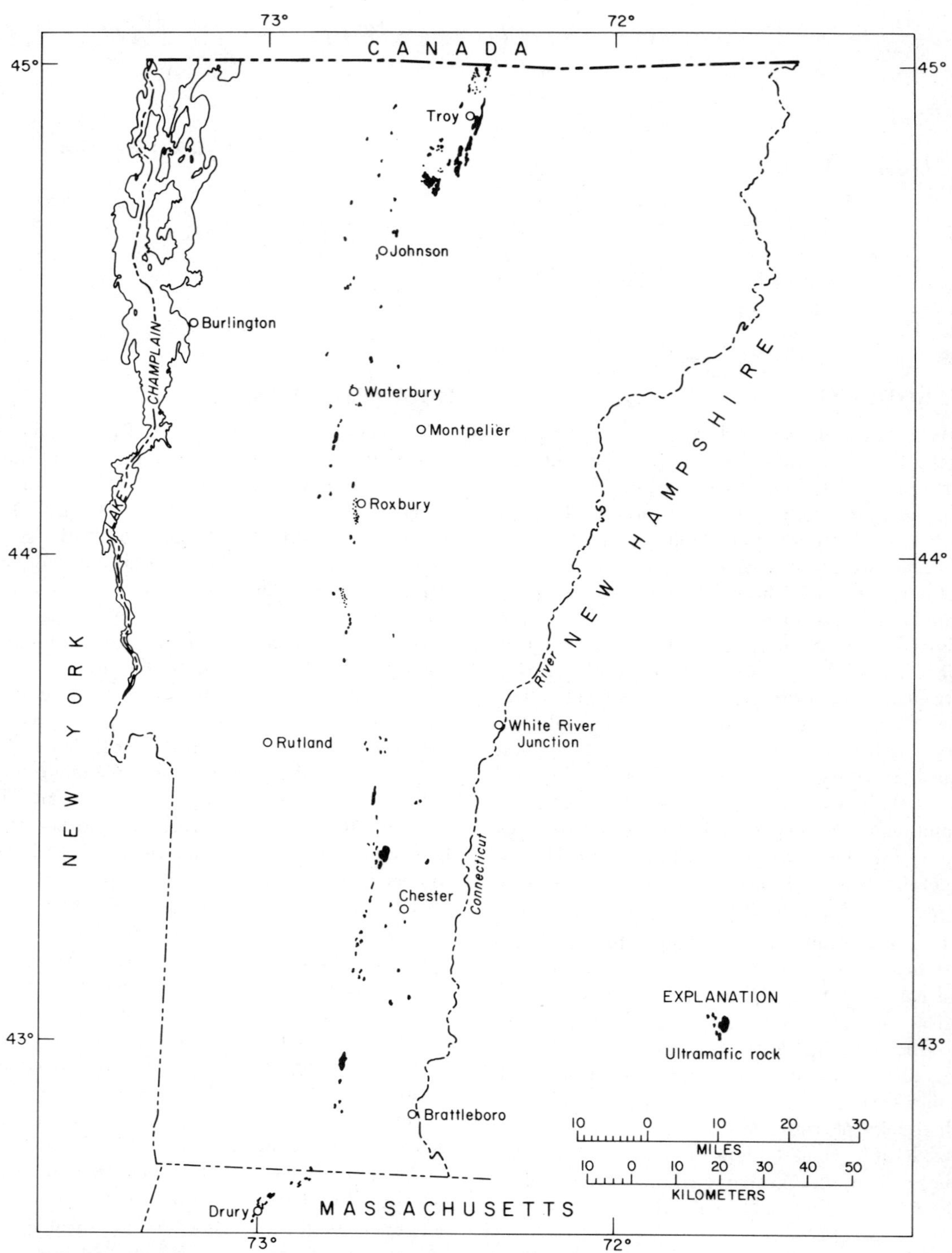

FIGURE 26–1. Map showing the distribution of ultramafic rocks in Vermont and northwestern Massachusetts.

anticlines in northern and central Vermont, and by several domes in southern Vermont and northwestern Massachusetts. Where the western edge of the belt crosses the anticlinorial axis in northern Vermont, the rocks range through gentle dips to steep westward dips in the west limb.

Two styles of folds, which correspond to two age groups distinguished on the basis of intersecting relations, occur in the belt. The older folds are isoclinal and plastic; their form surfaces are bedding planes. The younger folds are open and nearly parallel (concentric); their form surfaces are schistosity that in most places virtually parallels bedding, but locally transects it markedly. In places, the older (plastic, isoclinal) folds are deformed by the younger (open, parallel) folds, and the axial planes and axial plane schistosity of the older folds are the form surfaces of the younger folds. Slip cleavage parallel to the axial planes of the younger folds transects both bedding schistosity and the "transecting schistosity" that is parallel to the axial planes of the older folds.

Throughout most of the belt, the older folds are only a few inches or a few feet in amplitude and are greatly subordinate to the younger folds. In parts of the belt, however, particularly in northern Vermont, older folds attain amplitudes of tens or hundreds of feet, and locally they have a marked effect on the map pattern. The younger folds are ubiquitous and are related in distribution, pattern, style, and age to the Green Mountain anticlinorium and the Worcester and Northfield Mountain anticlines, whose main stages of development (in Devonian time) essentially coincided with the formation of the domes in southern Vermont and northwestern Massachusetts (see Rosenfeld, this volume).

Faults are generally small and subordinate features of the belt. Most are exposed by mining operations in the immediate vicinity of ultramafic rock complexes, and have only a few feet to a few tens of feet of displacement. Most have a strong strike-slip component, but a few are thrust faults. It is possible that better exposures elsewhere would reveal that such minor faults are common throughout the belt.

GENERAL GEOLOGY OF THE COMPLEXES

The distribution of known ultramafic rocks in Vermont and northwestern Massachusetts is somewhat spotty, but the overall distribution from north to south in the belt is about uniform. The association of greenstone and amphibolite with the belt of ultramafic rocks is striking, but the ultramafics occur in all rock types in the belt. Many occur within greenstone and amphibolite, others are at the upper or lower contact, many others are in schist and phyllite, and some of the larger ultramafic complexes crosscut several rock types.

Other intrusive igneous rocks are rarely associated with the ultramafic complexes. In a few places, thin dikes of camptonite or spessartite (Chidester, 1962, p. 84), and less commonly granite, cut the ultramafic rocks.

The ultramafic complexes are typically podlike or lenticular in form. Their principal planes parallel the regionally dominant transecting schistosity. The bodies range in length from as little as 30 feet (10 m) to as much as 6 miles (10 km), and in thickness from a few tens of feet to as much as 5000 feet (1600 m). Their vertical or down-dip dimension is generally not determined, but in a few, that dimension is known to approach or exceed the strike length exposed at the surface.

The majority of ultramafic complexes are of intermediate size—a few hundred to a thousand feet long and a few tens to a few hundreds of feet thick. They consist typically of a central core of serpentinite, massive at the center and schistose at the margins, surrounded successively outward by a shell of talc-magnesite rock a few inches to as much as 80 feet (25 m) thick and a shell of steatite (talc rock) somewhat variable but generally about 3 feet in thickness. The steatite is adjoined by a "blackwall" alteration zone, consisting in most places of chlorite or successive zones of chlorite and biotite. The smaller ultramafic bodies contain only a small core of schistose serpentinite or consist entirely of talc-magnesite rock and steatite. The very large ultramafic complexes contain, within the serpentinite core, masses of only slightly to moderately serpentinized dunite and minor peridotite, and locally contain appreciable slip- and cross-fiber chrysotile asbestos, in a few places in economic quantities (Cady et al., 1963). Such large complexes locally contain talc-magnesite rock and steatite at the margins and are bordered by blackwall; but in many places such talcose rocks are absent and serpentinite is adjoined by serpentine-chlorite rock and rodingite (calc-silicate rock) rather than blackwall. The two kinds of contact relations are illustrated in Figure 26–2. In addition to the peripheral zones of talcose rocks, many bodies in the intermediate and large size ranges contain tabular masses, as much as 30 feet (10 m) thick, of talc-magnesite rock which locally grades into magnesite rock.

The very large ultramafic complexes in which appreciable dunite and peridotite occur are confined to northern and southern Vermont, where both the accumulation of pre-Silurian eugeosynclinal sediments was thinner than in other parts of the belt and erosion has exposed deeper parts of the section. Smaller, completely serpentinized and extensively steatitized complexes also occur in these parts of the belt. On the other hand, in central Vermont where the sedimentary section is thicker and erosion has not reached so far toward the eugeosynclinal axis, all the complexes are of small to intermediate size and contain no dunite or peridotite.

PETROGRAPHY OF THE COMPLEXES

The minerals and textural features of most of the complexes are predominantly metamorphic in origin, but relict primary textures and minerals are preserved in varying degree. Relict igneous features characterize the serpentinite, talc-magnesite rock, and most of the steatite; poly-metamorphic and relict sedimentary features characterize the contact alteration zones and the outer

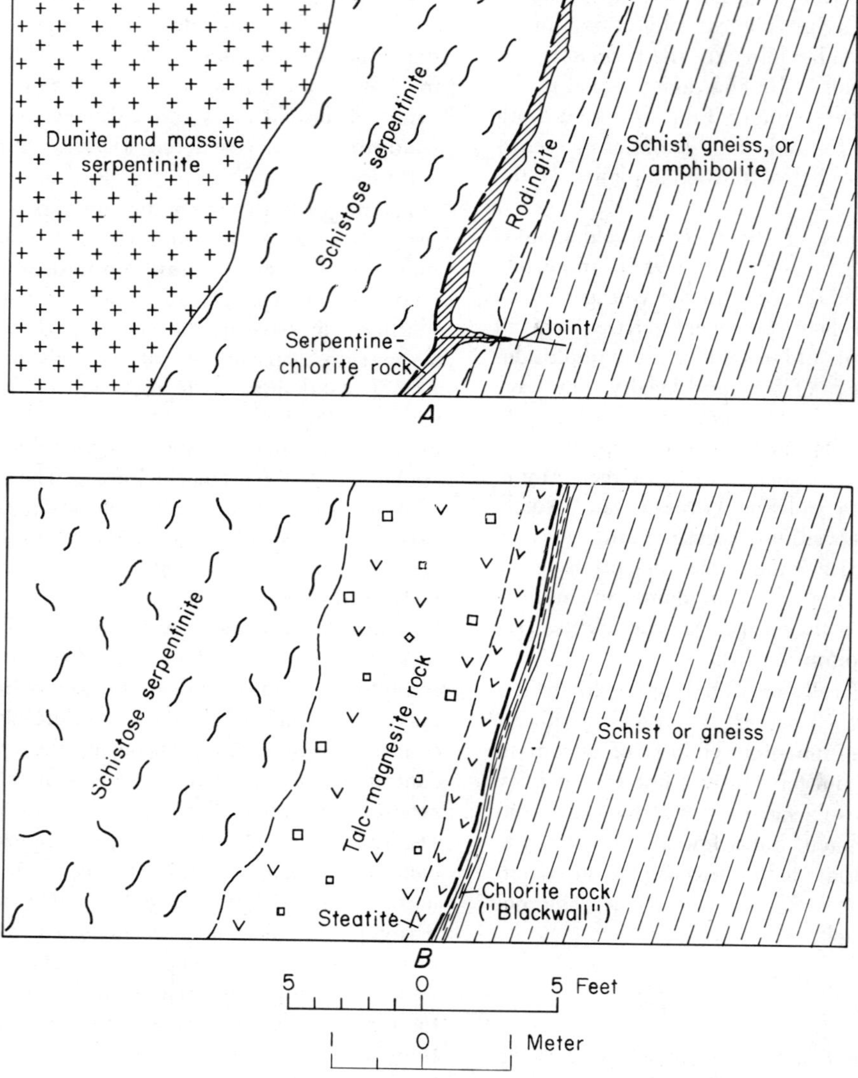

FIGURE 26–2. Diagrammatic sketch maps illustrating the two principal types of zonal relations at the contacts of ultramafic complexes. A, zonation of rodingite (calc-silicate rock) and serpentine-chlorite rock; B, zonation of chlorite rock ("blackwall") and steatite. Heavy broken line marks inferred position of original contact of the intrusive rock.

part of the steatite. The more significant petrographic features are summarized below.

Ultramafic Rocks

Olivine is the principal mineral of the dunite and peridotite, pyroxene is sparse in dunite and minimal in peridotite, and chromite is rare in both. Both dunite and peridotite display layering, marked by differences in mineralogy, color, and texture, from an inch to as much as 3 feet (1 m) thick. The transition from dunite to peridotite is abrupt, but in places is gradational over a few inches. The olivine in both the dunite and peridotite has a sugary granular texture; the pyroxene occurs as small subhedral to anhedral grains. In a few places crystals of olivine as much as 2 inches (5 cm) long display an incipient mosaic pattern like the predominant sugary granular texture, indicating that the sugary granular texture is largely cataclastic in origin. Chromitite occurs sparsely in sharply bounded layers rarely as much as a foot (30 cm) thick.

Massive serpentinite is chiefly antigorite but contains appreciable wispy shreds of chrysotile; schistose serpentinite is almost entirely antigorite. Massive serpentinite displays relict layering and relict textures of the parent igneous rock; schistose serpentinite consists of microscopic to macroscopic chips of randomly oriented platy serpentine (antigorite) separated by thin schistose zones in which the serpentine particles are aligned. The serpentinite displays no discernible evidence of volume change, even where serpentine invades dunite and peridotite in an intricate pattern.

Serpentine veins are of two kinds—asbestiform (chrysotile) and massive or columnar (picrolite); in places the two are intergrown. Each kind occurs both as cross-fiber (or cross-column) and slip-fiber (or slip-column) veins. Veins of cross-fiber habit are virtually confined to dunite, peridotite, and massive serpentinite; of slip-fiber habit, to schistose serpentinite and shear zones. Cross-fiber veins are tabular, wedgelike, or lenticular. They generally have matching walls, and matching irregularities are at opposite ends of individual fibers or columns. Seams of magnetite are common at the vein walls or along partings in the vein. Marginal zones of alteration, composed of chrysotile, antigorite, brucite, and magnetite, are distinctive of cross-fiber veins and are particularly conspicuous in dunite. Slip-fiber veins are small and are distributed along innumerable tiny shear surfaces in schistose serpen-

tinite, except for occasional conspicuous veins in shear zones.

Talc-magnesite displays chiefly metamorphic textures but contains ubiquitous though rare relict chromite. Vestiges of schistosity in the parent serpentinite are common, and in a few places relict layering from parent dunite and massive serpentinite is preserved. Magnesite-quartz rock, which is transitional with talc-magnesite rock, displays much the same textural relations and relict features. Steatite is generally schistose but locally massive. In most steatite, the only relict features are sparse grains of chromite; but adjacent to the blackwall zone, relict textures and minerals of the country rock are common in the steatite.

Rocks of the Contact Alteration Zones

The two types of contact alteration zones that are characteristic of the ultramafic complexes display distinctive associations that appear to be mutually exclusive: the blackwall zone is invariably associated with steatite, and the serpentine-chlorite rock and rodingite zone appears always to adjoin serpentinite without an intervening zone of steatite. These relations hold for contacts of septa and inclusions within the ultramafic bodies as well as for the main contacts of the bodies. In all the contact rocks, where the contact alteration zones cut across layering in the country rock, the undistorted pattern of relict layering in the zones indicates no discernible change in volume.

Rocks of the *blackwall and steatite* association manifest several zonal patterns that primarily reflect grade of metamorphism and composition of the country rock. In the lower greenschist facies, the blackwall in siliceous schist consists essentially of chlorite—highly magnesian next to the steatite, more ferriferous next to the schist; where the schist is albitic, a narrow zone at the outer margin of the blackwall is enriched in albite porphyroblasts, and pseudomorphs after albite are discernible in the pattern of chlorite in the blackwall. Adjacent to greenstone, amphibolite, and impure marble, the blackwall is thin, irregular, and consists of chlorite and tremolite or actinolite; sprays of tremolite also occur in the outer part of the steatite.

With increasing grade of metamorphism, biotite becomes increasingly abundant at the outer margin of the blackwall zone adjacent to siliceous schist, and tremolite increases in the outer part of the steatite zone adjacent to amphibolite. In the upper greenschist facies and the albite-epidote-amphibolite facies, the blackwall in siliceous schist consists of a chlorite zone next to the steatite and a biotite zone next to the schist, each as much as 3–4 inches (10 cm) thick. Next to amphibolite, tremolite is strikingly developed in the outer part of the steatite zone to thicknesses of as much as a foot (25 cm).

Serpentine-chlorite rock and *rodingite* (calc-silicate rock) have a sequential concentric relation to the ultramafic body; locally, however, narrow fingers of serpentine-chlorite rock project beyond the rodingite zone along joints. The serpentine-chlorite rock zone is rather uniformly 1–3 in. (3–8 cm) thick. The thickness of the rodingite zone varies with the composition of the country rock and with structural relations: next to siliceous schist, the rodingite is about 3 feet (1 m) thick; next to amphibolite, 2–12 inches (5–30 cm); at strongly crosscutting contacts where the country rock is sheared and faulted, as much as 10 feet (3 m).

Serpentine-chlorite is uniformly fine-grained and massive. Serpentine predominates over chlorite at the inner margin of the zone, next to serpentinite, and chlorite predominates in the outer part, next to rodingite. In the outer part, also, relict minerals and textures of the rodingite increase toward the rodingite in a transition zone a few millimeters thick. Rodingite varies widely in grain size (from fine to coarse) and in mineralogy. Two to five of the following minerals constitute distinctive assemblages: diopside, clinozoisite, zoisite, grossularite, vesuvianite, prehnite, chlorite, calcite, and magnetite. Diopside and clinozoisite predominate at the inner margin of the rodingite, chlorite and clinozoisite at the outer margin; garnet and vesuvianite are most abundant at the inner margin. Other variations reflect chiefly original differences in composition of relict beds.

STRUCTURAL FEATURES OF THE COMPLEXES

Structural elements of the ultramafic complexes include primary igneous features, features inferred to be related to emplacement, and features related to the post-emplacement history of the complexes. The relations between these features, and between them and the regional structural features, are critical to interpreting the history of the complexes.

The contacts of the ultramafic complexes are chiefly concordant with the regionally dominant transecting schistosity, which bears an axial-plane relation to the older folds. In most places the contacts are also virtually parallel to bedding schistosity; but locally, near older folds, they

diverge markedly from bedding, as does the transecting schistosity. The contacts crosscut the transecting schistosity in a step-wise manner, to result in cumulative discordancies up to several tens of feet. At many places, such crosscutting starts at small older folds, where bedding and transecting schistosity diverge. In a few localities where older folds as much as a few hundred feet in amplitude are conspicuous, the contacts of the complexes conform to the gross pattern of the folds but crosscut the bedding schistosity in detail.

Tabular septa and irregular inclusions of country rock in the ultramafic rocks are minor but locally conspicuous features of the complexes. All the septa have the same kind of schistosity as the country rock, and most are tabular parallel to the transecting schistosity. Many of the septa are either attached at one end to the wallrock, or their place of detachment can be inferred with fair confidence. However, many of the xenoliths, particularly the smaller ones, cannot be matched with the nearby wallrock.

Layering in dunite, peridotite, and massive serpentinite is generally subparallel to nearby contacts of the ultramafic complexes or to the inferred attitude of hanging walls that have been eroded away. Marked departures from this relation are associated with zones of pronounced shearing and faulting.

Schistose serpentinite, which occurs at the margins of a central core of massive serpentinite, or which in the smaller bodies constitutes the entire serpentinite mass, contains distinctive features that have been termed shear polyhedra (see Chidester and others, 1951, p. 7; Chidester, 1962, p. 24, 71). These consist of irregular, close-packed polyhedral units of massive serpentinite surrounded by thin, irregular shells of schistose serpentinite. The size of the polyhedra is irregular, but tends to decrease toward the outer part of the serpentinite mass. In the outermost parts, the massive units are commonly reduced to small chips in a schistose matrix, and the serpentinite has a pervading irregular schistosity generally parallel to the walls of the complex.

In contrast to the concentric zones of schistose serpentinite, the tabular zones of sheared serpentinite commonly transect boundaries between dunite, massive serpentinite, and sheared serpentinite. The serpentinite in the shear zones generally has a crude and irregular schistosity approximately parallel to the trend of the shear zones. In a few well-defined and regular shear zones, two dominant conjugate shear surfaces intersect in a rhombic pattern whose long axis is about parallel to the trend of the shear zone. The tabular zones of sheared serpentinite transect the schistosity of the marginally distributed sheared serpentinite.

The contacts of most ultramafic complexes are not noticeably affected by open folds of the younger set, because throughout most of the belt younger folds of more than a few inches amplitude are sparse. However, in favorable structural situations, such as near the crest of the Green Mountain anticlinorium and in the vicinity of subsidiary anticlines and synclines, contacts of the complexes are strongly folded in conformity with the younger folds in the country rock. In a few strongly folded bodies that have been almost completely altered to talc-carbonate rock and steatite, folds in the hanging wall (the footwall is commonly not well exposed by mining operations) display a characteristic peculiar form: The anticlinal crests are sharp, almost knife-edge, whereas the synclinal troughs are broadly rounded; the resultant profile is like that of oscillation ripple marks. Not uncommonly, steatite is injected a few inches into the country rock along the sharply angled anticlinal crest.

Where the contacts of ultramafic complexes are affected by the younger folds, the layering in dunite, peridotite, and massive serpentinite, and the schistosity in sheared serpentinite, are also folded. Though similar in pattern to the folds in the country rock, these folds in the ultramafic rocks are more diversely oriented.

Schistosity and cleavage related to the younger folds are rare and poorly developed in the ultramafic rocks. In a few places, the schistose serpentinite adjacent to highly folded wallrock displays a pronounced, though widely spaced, slip cleavage parallel to the axial planes of the folds. Steatite has a very pronounced schistosity parallel to the prevailing transecting schistosity in the adjacent wallrock, but only rarely displays a cleavage parallel to the axial planes of the younger folds. Where such cleavage is present it is a crude, widely spaced fracture cleavage that fans 10° or more from parallelism with the axial planes.

Faults in the ultramafic complexes are confined mainly to the vicinity of the contacts (see above, under "Regional Geology"), and generally die out into shear zones. In a few places, sharply bounded faults offset contacts of the complexes by several tens to a few hundreds of feet.

Joints are generally inconspicuous features of the ultramafic rocks. Sets of conjugate joints locally control the distribution of some cross-fiber asbestos veins, but they show no systematic relation to regional structures. The rare but striking talc-carbonate veins are controlled by sets of nearly horizontal and nearly vertical joints.

ORIGIN OF THE ULTRAMAFIC IGNEOUS ROCKS

Although most of the ultramafic complexes contain little or no unmetamorphosed igneous rock, large masses of relatively fresh igneous rock are present in some of the complexes. This fact, together with the distribution of relict igneous minerals in the serpentinite and talcose rocks, indicates an igneous parentage for most of the ultramafic rocks; relict textures and minerals indicate that only small amounts of the rocks in the complexes are derived from the metamorphic country rocks. Consequently, an account of the evolution of the ultramafic complexes depends on the mode of origin postulated for the igneous rocks, which remains highly speculative.

The intergradational layered relations of the dunite, peridotite, and chromitite suggest that they were formed by gravity accumulation during rhythmically varied crystallization of a somewhat more complex magma. Some of the thinner, discontinuous, compositionally different layers may have formed by shearing of the solidified mass during transport under orogenic stress. Similar relict layering in the massive serpentinite, in many of the larger shear polyhedra, and in some of the talc-carbonate rock, suggests that the protolith of all the derivative ultramafic rocks was similarly formed.

The near coincidence of the belt of ultramafic rocks with that of abundant metamorphosed mafic volcanic rocks and its location in the eugeosynclinal zone of the early Paleozoic orthogeosyncline suggests a genetic relation between the ultramafic plutonic rocks and the mafic volcanic rocks. Two kinds of genetic association appear to be almost equally plausible. One model postulates that the ultramafic igneous rocks were formed by accumulation from a complex basaltic magma of unspecified origin, that the liquid fraction was extruded as volcanic materials and intercalated with sediments in the geosynclinal section, and that the ultramafic cumulates were emplaced tectonically during subsequent orogeny. The other model postulates that the mafic volcanic rocks were formed during early stages of partial fusion of the upper mantle during downbuckling of the early Paleozoic geosyncline, and that the ultramafic rocks were formed during the later stages of fusion.

Each model allows for many variations on the central theme, but all versions are similar when modified to take into account the geologic relations of the complexes and the tectonic history of the belt. Both basic models must account for the rhythmically varied layering of the dunite, peridotite, and chromitite, and both must be compatible with the active orogenic setting. The two models differ chiefly in the postulated composition of the magma from which the ultramafic rocks were immediately derived. The postulated composition depends on the role assigned to partial fusion of the deep crustal or upper mantle rocks.

EMPLACEMENT AND STRUCTURAL HISTORY OF THE ULTRAMAFIC ROCKS

The locally cross-cutting relations of the contacts of the ultramafic complexes and the prevalence of septa and inclusions of country rock in the complexes indicate that the ultramafic rocks were emplaced by forceful intrusion. The general concordance of the contacts with the dominant transecting schistosity that is parallel to the axial planes of the early folds, the cross-cutting of early folds by the contacts, and the control exercised by transecting schistosity on the shape of septa and inclusions of country rock, all demonstrate that the ultramafic rocks were emplaced after the early stage of folding. The general absence of high-temperature contact-metamorphic zones at the margins of the bodies, the peripheral distribution of sheared serpentinite, the shear polyhedra in the schistose serpentinite, and the cataclastic texture of the dunite, further indicate that the ultramafic rocks were emplaced in the solid state at temperatures not much different from those of the enclosing wallrocks; however, variations in contact relations suggest small but significant differences in the temperature of different ultramafic rock bodies at the time of emplacement. Layering in dunite and peridotite suggests that such layered units were carried along essentially intact in the central parts of the larger masses of cumulates, mechanically buffered by yielding in the marginal zones.

The mechanism by which the ultramafic rocks were emplaced in the solid state at their present sites is believed to be intimately related to the process of serpentinization, and is discussed below in more detail under "Origin of the Metamorphic Rocks and Veins."

The conformity of the contacts of the complexes with the pattern of younger folds, and the presence in schistose serpentinite of slip cleavage parallel to the axial planes of these folds, demonstrate that the more recent period of folding was later than the emplacement of the ultramafic rocks. The local presence in the steatite of a weak fracture cleavage parallel to the axial planes of younger folds suggests that the last stages of the later folding may have extended into the period of steatitization of the complexes. Minor faulting, shearing, and jointing took place chiefly during the late stages of the later period of folding, but locally occurred after folding had virtually ceased. During this stage, joints and fractures were formed extensively in dunite and massive serpentinite, and transecting shear zones were formed in the ultramafic complexes.

ORIGIN OF THE METAMORPHIC ROCKS AND VEINS

The structural and petrologic relations of the serpentinite, serpentine veins, and talcose rocks of the ultramafic complexes indicate that they were formed by several distinct metamorphic processes, which have consistent sequential relations and consistent relations to the regional tectonic history. The contact-rock associations (steatite and blackwall; rodingite and serpentine-chorite rock) display similar consistent relations.

Serpentinite

Pervasive serpentinization of dunite and peridotite to form both massive and schistose serpentinite took place early in the tectonic history of the complexes. The peripheral distribution of the sheared and schistose serpentinite, the general conformity of the schistosity of the serpentinite to the contacts of the complexes, and the distribution of shear polyhedra in the schistose serpentinite all suggest that the complexes were extensively serpentinized prior to emplacement. This conclusion is supported by the general lack of high-temperature contact-metamorphic effects in the country rock adjacent to the complexes. A logical and appealing interpretation of these relations is that serpentinization of the dunite and peridotite proceeded concomitantly with transport of the ultramafic rock by kneading of the solid mass through the country rock under tectonic stress, and that water for serpentinization was derived from the adjacent country rocks. Transport is considered to have taken place spasmodically, with intermittent periods of movement separated by periods of immobility. Shearing of the mass during periods of movement facilitated the entry of water and consequent serpentinization during intermittent periods of immobility, and serpentinization increased the mobility of the mass during subsequent periods of movement, primarily by dehydration of the serpentine under shear stress at high pressure and temperature, in the manner described by Raleigh and Paterson (1965, p. 3965) and by Reicker and Rooney (1966, p. 196). Serpentinization was not confined to the highly sheared marginal zones, but pervasively altered the adjacent unsheared dunite and peridotite partly or entirely to massive serpentinite.

Serpentine Veins and Other Veins

Serpentine veins, including cross- and slip-fiber chrysotile asbestos and picrolite and their marginal alteration zones, were formed after emplacement of the ultramafic rocks, when tectonic transport and pervasive serpentinization of the rocks had ceased. In places, particularly in areas of complex younger folding near the Green Mountain anticlinorial axis, minor shearing after emplacement generated tension fractures in which chrysotile asbestos formed. Dunite, peridotite, and massive serpentinite yielded locally by fracture along local joint systems, but more commonly along diversely oriented and somewhat irregular cracks; these became the loci of cross-fiber veins of chrysotile asbestos and picrolite. Sheared and schistose serpentinite yielded by slippage along shear surfaces, which generated innumerable tiny fracture openings; these became the loci of slip-fiber veins of chrysotile asbestos and picrolite.

In cross-fiber veins, including those in which the fiber axes are markedly oblique to the walls, irregularities along opposite walls match down to the finest details, demonstrating that the veins were formed by fracture filling with almost no replacement of the walls. Such evidence is lacking in the tiny slip-fiber veins, but presumably they formed under similar conditions.

The favored interpretation for the origin of the chrysotile asbestos and picrolite veins, based on the observed relations in northern Vermont, is as follows: In response to local stresses generated by minor regional shearing movements, tensional fractures were opened incrementally in the ultramafic rocks. In the massive serpentinite and dunite, the chief component of relative movement was normal to the fracture walls; in the sheared serpentinite, parallel to the walls. As the openings formed, diffusion of material from the walls and water from the fractures led to deposition of serpentine in the fractures and formation of the marginal alteration zones adjacent to the fractures. Where the rate of deposition in the veins equalled the rate of opening of the fractures, chrysotile asbestos was formed, with the fiber axes parallel to the direction of relative movement of the opposite walls. Where the rate of

opening appreciably exceeded the rate of deposition, picrolite was formed, with the coarse fibers and columns poorly oriented.

Magnetite, calcite, and brucite formed essentially contemporaneously with the serpentine veins and under the same conditions. They record the concentration and local movement of material that could not be accommodated in the serpentine minerals.

Talcose Rocks and the Steatite–Blackwall Association

A sequential association is clearly demonstrated by the invariable replacement relation and the peripheral distribution of talcose rocks with respect to serpentinite, along with the pseudomorphic relation of talc to cross-fiber chrysotile asbestos. The temporal separateness of serpentinization and steatitization is less evident. The inference of separateness rests upon the fact that the extent of serpentinization is unrelated to the width of the talc-carbonate rock zone and that serpentinite is strongly affected by the younger folds, whereas steatite and talc-carbonate are affected only weakly or not at all. On these bases, steatitization is inferred to be the latest metamorphic episode in the geologic history of the ultramafic rocks. Though of essentially simultaneous origin, talc-carbonate rock (and carbonate-quartz rock) and steatite were formed by very different and essentially unrelated metamorphic reactions. The consistent relation between the width of the talc-carbonate rock zone and that of the steatite zone suggests that the talc-carbonate rock reaction triggered or facilitated the steatite-blackwall reaction by the release of water during the alteration of serpentinite to talc-carbonate rock.

Available evidence indicates essentially constant-volume replacement relations. Comparison of rock compositions then indicates that the alteration of serpentinite to talc-carbonate rock was primarily due to carbon dioxide metasomatism. The fundamental processes were addition of CO_2 and loss of H_2O; additions and losses of other components were negligible. Loss of water to the adjacent country rock may have triggered or facilitated the steatite-blackwall reaction (see Chidester, 1962, p. 89ff. for supporting analytical data).

The distribution of carbonate-talc-quartz rock and carbonate-quartz rock appears to be controlled by zones of faulting or intense shearing, which would furnish ready access for CO_2. These rocks appear to represent intermediate and end stages in the process of carbon dioxide metasomatism.

The replacement of albite porphyroblasts and other metamorphic minerals of the country rock by the chlorite and biotite of the blackwall demonstrates that the blackwall formed during or after the peak of regional metamorphism. The age equivalence of the blackwall and steatite are indicated by the fact that neither is cut by the other and by the generally sympathetic relation between the width of the two zones. Their genetic relation is further indicated by the following petrologic considerations.

As in the case of the talc-carbonate rock reaction, evidence for constant-volume replacement is the basis for determining the steatite-blackwall reaction. On this basis, comparison of the compositions of serpentinite, steatite, blackwall, and country rock demonstrates that the reaction was essentially one of interchange of constituents between serpentinite and schist, leading to a zoned distribution of steatite and blackwall between serpentinite (or talc-magnesite rock) and country rock. The reaction chiefly involved loss of Mg and H_2O from the serpentinite to the blackwall zone and loss of Si from the country rock to the steatite zone. Where the adjacent country rock was amphibolite or carbonate rock, and hence low in Si, the reaction was repressed and the steatite and blackwall zones were poorly developed.

As the steatite-blackwall reaction progressed, both the steatite and blackwall zones widened, chiefly at the expense of serpentinite (or talc-magnesite rock) and country rock, respectively. The inner margin of the blackwall zone also altered slowly to talc, so that the outer part of the steatite zone is derived from the country rock.

Other reactions in areas of low-grade metamorphism were minor, but in albitic schists Na was displaced outward at the advancing blackwall-schist boundary to form striking concentrations of albite porphyroblasts; where the schist was graphitic, graphite also was displaced outward. In areas of higher grade metamorphism, the activities of mobile components critical to the formation of chlorite and biotite attained such values that chlorite and biotite were concentrated in successive zones outward from the steatite zone. Where the wallrock was amphibolite, Ca migrated into the steatite zone to form a marginal shell of tremolite.

Serpentine-Chlorite Rock and Rodingite

The absence of contact rocks of the serpentine-chlorite rock and rodingite association in most ultramafic rock complexes suggests that the rocks of this association were formed under unusual conditions. The presence, in identical country rock, of serpentine-chlorite rock and rodingite within a few feet of blackwall and steatite indicates that the two associations must have been formed during different metamorphic episodes, and therefore presumably under different physical conditions. The fact that serpentine-chlorite rock and rodingite occur only adjacent to serpentinite, at contacts where talcose rocks are absent, suggests that such contact rocks are older than steatite and blackwall, and probably are related to the emplacement of the ultramafic rocks. The association of serpentine-chlorite rock and rodingite exclusively with large ultramafic complexes that contain masses of relatively fresh dunite and peridotite suggests that such bodies of rock were appreciably hotter during emplacement than the other bodies of ultramafic rock. The narrowness of the rodingite zone indicates, however, that the ultramafic rocks were not much hotter than the surrounding country rock. The cumulative evidence is that the serpentine-chlorite rock and rodingite were formed under rather special conditions of contact metamorphism.

The replacement relations of rodingite to country rock and of serpentine-chlorite rock to rodingite are sequential. These relations permit, but do not require, that the replacements represent a single complex reaction. The local projection of tongues of serpentine-chlorite rock along fractures beyond the rodingite zone suggests the possibility that the serpentine-chlorite reaction was separated from the rodingite reaction by a discrete time interval, and thus took place at a lower temperature.

Rodingite was formed by alteration and replacement of the country-rock minerals chiefly by diopside, garnet, vesuvianite, and clinozoisite. Two stages are distinguishable: (1) alteration to diopside, clinozoisite, and garnet, and (2) replacement of these minerals by a more calcic garnet and vesuvianite. These relations, together with the differences in mineralogy that reflect differences in protoliths and composition of adjacent beds, indicate that the rodingite reaction involved only relatively minor migration of components, and that the successive mineral reactions record the effects of changes in temperature and, probably, the activities of volatile constituents.

The uniform character of the serpentine-chlorite rock and the zonal distribution of serpentine and chlorite, grading from predominantly serpentine at the contact with serpentinite to predominantly chlorite at the contact with rodingite, demonstrate that the serpentine-chlorite rock reaction was metasomatic, involving wholesale conversion of rodingite progressively to chlorite and serpentine, chiefly by addition of H_2O and of Mg from the ultramafic body.

The foregoing relations and conclusions suggest the following interpretation for the origin of the rodingite and serpentine-chlorite rock assemblage:

A large body of ultramafic rock containing considerable amounts of dunite and peridotite reached its present site at temperatures significantly, but not greatly, above those of the surrounding country rock. The emplacement temperature of a body was a function of its size and its distance of transport from the site of accumulation. Because the mass of ultramafic rock was slightly hotter than the surrounding terrane, the country rocks were contact metamorphosed in a thin zone adjacent to the ultramafic rocks.

The mineral assemblages characteristic of the rodingite are seemingly of abnormally high grade partly because dunite and peridotite in the mass behaved essentially as sinks for water, thereby lowering the activity of water in the country rock and lowering the temperature of formation of the mineral assemblages characteristic of the rodingite. The dehydration of serpentine under shearing stress at high pressures and moderate temperatures, advanced above (see p. 350–351) as a major factor in facilitating tectonic transport of the mass, suggests a second factor that may have been effective, perhaps even dominant, in lowering the activity of water in the country rock immediately adjacent to the ultramafic rocks: At sufficiently high temperatures, sheared serpentinite in such a tectonically transported mass would be markedly water-deficient when the mass came to rest and would be a very effective sink for water from the adjacent country rock.

As hydration of the ultramafic rocks proceeded, the activity of water in the ultramafic rock and the immediately adjacent country rock slowly rose and the temperature decreased. When the rising water pressure and falling temperature reached critical values, rodingite became unstable in contact with serpentinite. The two rocks then reacted to form a narrow zone of serpentine-chlorite rock, which enlarged by encroachment upon the bordering rodingite. The advancing wave of alteration, dependent upon the addition of Mg derived from the serpentinite, was characterized by formation of chlorite next to the rodingite. As the wave of alteration advanced

farther, early-formed chlorite was altered progressively to serpentine.

The interpretation outlined above suggests that rodingite and serpentine-chlorite rock were probably formed sequentially in a narrow time interval, that this interval coincided with that in which the asbestos and other serpentine veins were formed, and that the physical conditions conducive to the formation of one were also conducive to formation of the other.

ROLE OF THE TECTONIC AND SEDIMENTOLOGICAL ENVIRONMENT

The relation of size and character of the ultramafic complexes to their tectonic and metamorphic setting, considered in the context of genetic concepts outlined in the foregoing sections, illuminates the role that the tectonic history and sedimentological environment played in the development of the complexes.

The structural relations of the contacts and septa of the complexes demonstrate that the ultramafic rocks were emplaced by forceful intrusion. The structural history and the concepts advanced for derivation and emplacement of the ultramafic rocks lead to the conclusion that the stage of igneous derivation occurred during or near the end of the eugeosynclinal deposition. Eugeosynclinal deposition in this belt ended with the disturbance that produced the early folds and that culminated in the unconformity at the base of the Silurian section. Tectonic stresses generated during this disturbance mobilized the ultramafic igneous rocks and kneaded them through the overlying water-saturated sediments. At successive positions the transported ultramafic rocks successively encountered more extensively folded country rock, so that emplacement of the ultramafic rocks at their present positions essentially post-dated the early folds. Stratigraphic relations indicate that exposed complexes were emplaced at considerable depths. Under such conditions, in a thick water-saturated geosynclinal pile, small or far-travelled bodies would be extensively serpentinized and would be finally emplaced at temperatures near those of the enclosing rocks; large or short-travelled bodies would be less extensively serpentinized, and would be emplaced at higher temperatures. Thus the family of characteristics displayed by the complexes appears to depend significantly on the tectonic and sedimentological environment, and is related to the size of the mass of ultramafic rock, the distance that the mass has travelled through wet sediments, and the local erosional pattern.

REFERENCES

Cady, W. M., 1967, Geosynclinal setting of the Appalachian Mountains in southeastern Quebec and northwestern New England, *in* T. H. Clark, *ed.*, Appalachian tectonics: Royal Soc. Canada Spec. Pub. no. 10, p. 57–68

Cady, W. M., in press, Regional tectonic synthesis of northwestern New England and adjacent Quebec: Geol. Soc. America Spec. Paper

Cady, W. M., Albee, A. L., and Chidester, A. H., 1963, Bedrock geology and asbestos deposits of the upper Missisquoi Valley and vicinity, Vermont: U. S. Geol. Survey Bull. 1122–B, 78 p.

Chidester, A. H., 1962, Petrology and geochemistry of selected talcbearing ultramafic rocks and adjacent country rocks in north-central Vermont: U. S. Geol. Survey Prof. Paper 345, 207 p.

Chidester, A. H., Billings, M. P., and Cady, W. M., 1951, Talc investigations in Vermont, preliminary report: U. S. Geol. Survey Circ. 95, 33 p.

Doll, C. G., Cady, W. M., Thompson, J. B., Jr., and Billings, M. P., *compilers* and *editors*, 1961, Centennial geologic map of Vermont: Vermont Geol. Survey, Scale, 1:250,000

Raleigh, C. B., and Paterson, M. S., 1965, Experimental deformation of serpentinite and its tectonic implications: Jour. Geophys. Research, v. 70, p. 3965–3985

Reicker, R. E., and Rooney, T. P., 1966, Weakening of dunite by serpentine dehydration: Science, v. 152, no. 3719, p. 196–198

Volcanism Related to Tectonism in the Piscataquis Volcanic Belt, an Island Arc of Early Devonian Age in North-Central Maine*

DOUGLAS W. RANKIN

INTRODUCTION

It struck me that the outline of this mountain [Horse Mountain] on the southwest of the lake, and of another beyond it, was . . . in the main like Kineo, on Moose-head Lake, having a similar but less abrupt precipice at the southeast end; . . .—H. D. Thoreau (1864).

A BELT OF NEARLY contemporaneous and litholog-ically similar volcanic and shallow intrusive rocks extends about 100 miles (160 km) across north-central Maine from Heald Mountain on the southwest to Snowshoe Mountain on the north-east (Fig. 27–1). Igneous rocks of this belt are predominantly felsic and are mostly extrusive rhyolites. Shallow intrusive bodies also occur. The slightly younger and very much larger Katahdin batholith is probably genetically related to the volcanism but is not discussed in this paper. The volcanic rocks are in a geosynclinal environment and are of Early Devonian (Becraft to Schoharie) age. Felsic volcanic rocks of this age have not been recognized elsewhere in Maine. The name Piscataquis volcanic belt is here pro-posed after Piscataquis County, the location of the bulk of the volcanic rocks. The belt is considered to be part of an island-arc system and parallels a series of en echelon synclinoria, part of the fold pattern produced during the Acadian orogeny. There is a close genetic relationship between the loci of volcanism and the development of the syn-

clinoria. The purpose of this paper is to syn-thesize recent findings by several workers on various parts of the belt and to outline the de-velopment of the belt. The large mass of rhyolite at Traveler Mountain at the northeast end of the belt (Fig. 27–1) will be described in more detail than are the other volcanic centers because it is the largest of the bodies of volcanic rocks and the one with which I am most familiar.

PREVIOUS WORK AND ACKNOWLEDGMENTS

Passing by Mt. Kenio, my attention being attracted by its singular appearance, I stopped to ascend and examine the mountain. It is the termination of a pen-insula, which extends for some distance into the lake on the eastern side, opposite the mouth of Moose river. As we approached the mountain from the south-west it had the appearance of a huge, artificial wall of stone, rising directly out of the water. Hodge (1838, p. 55).

Much of the rhyolite in north-central Maine occurs as isolated patches of mountain-forming rock scattered along the belt. These steep-sided mountains are among the more prominent topo-graphic features of the area and as such have attracted the attention of most visiting naturalists, past and present. C. T. Jackson's assistant, J. T. Hodge, climbed Mt. Kineo in 1837 and made the first recorded geological observations on the Piscataquis volcanic belt. Hodge (1838, p. 56) stated that Mt. Kineo is ". . . composed en-tirely of a bluish hornstone, like flint, exceedingly hard and compact." Jackson (1838, p. 125)

* Publication authorized by the Director, U. S. Geo-logical Survey.

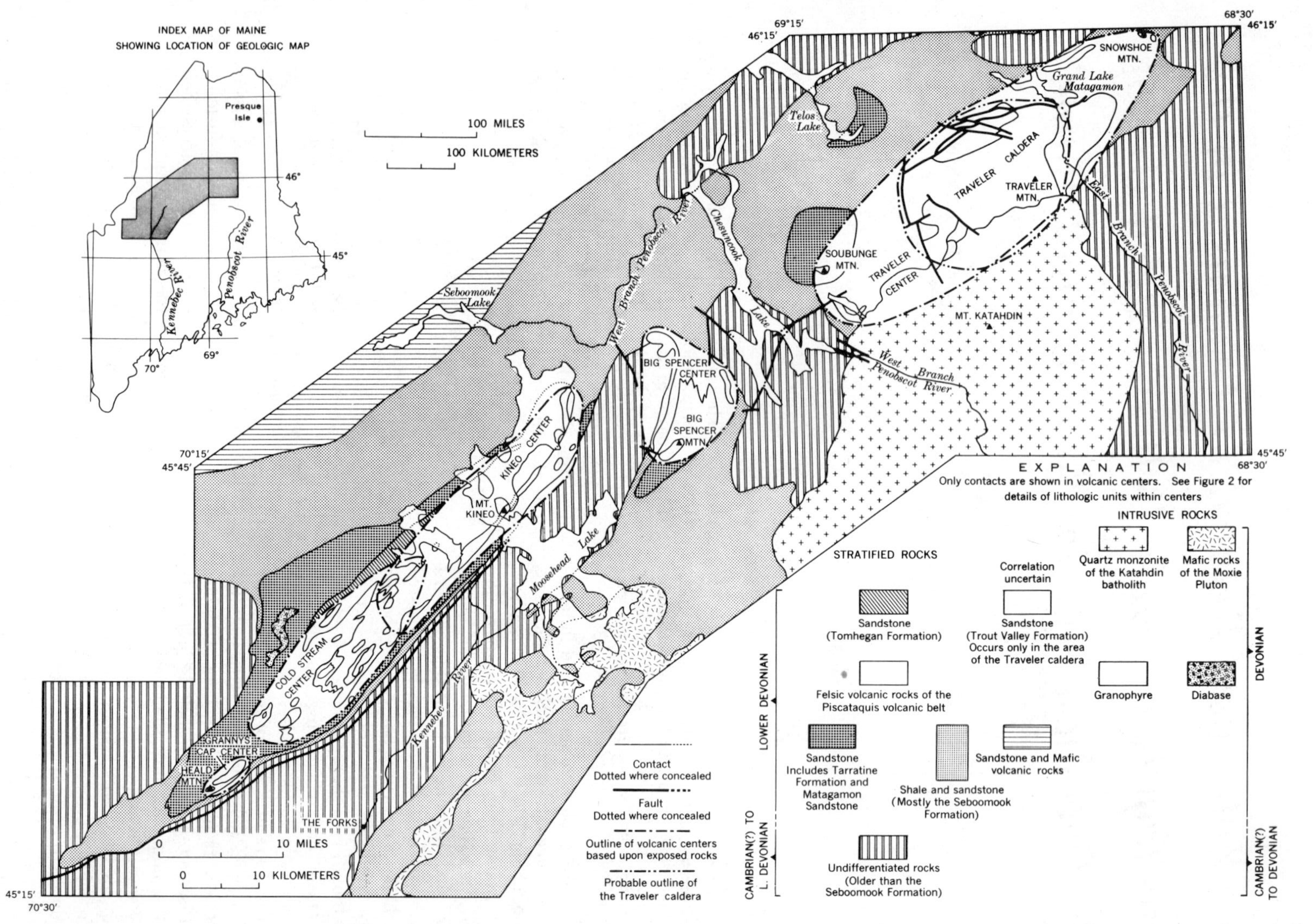

FIGURE 27–1. Generalized geologic map showing volcanic centers and geologic setting of the Piscataquis volcanic belt. Modified from Boucot (1961; written communication, 1966), Boucot and others (1964a), Doyle (1967), and Rankin (unpublished data).

apparently thought this rock was a hornfels produced where ". . . trap rocks have acted upon siliceous slate."

C. H. Hitchcock (1861, p. 404; 1862) who visited both Mt. Kineo and Traveler Mountain called the rhyolite siliceous slate, hornstone, or flint. By 1874, however, he recognized the volcanic nature of the rhyolite on Mt. Kineo, and presumably also of that in the Traveler Mountain area, although he did not specifically refer to the latter (Hitchcock and Huntington, 1874, p. 206). E. S. C. Smith (1925, 1930, 1933), in a series of papers, described the occurrence of rhyolites from a number of localities within the Piscataquis volcanic belt and noted that the rhyolites did occur in a belt.

Within the last 20 years, all the known igneous bodies of the Piscataquis volcanic belt have been at least briefly studied by one of five workers and the geologic setting is now reasonably well understood. This paper is essentially a synthesis of their work. Much of it is unpublished, and I am most grateful to these men for making their hard-won information available to me. A. J. Boucot of the University of Pennsylvania mapped the southern half of the belt for the U.S. Geological Survey. I would particularly like to thank Professor Boucot for access to and use of extensive unpublished data and for lending me his collection of rhyolite samples and thin sections. Without his perceptive mapping of lithologic types within the Kineo Volcanic Member, albeit with resulting interpretations different from mine, this synthesis would not be possible.

E. W. Heath mapped the area around Big Spencer Mountain for a M.S. dissertation at the Massachusetts Institute of Technology, and Andrew Griscom of the U.S. Geological Survey mapped the area around Soubunge Mountain. Snowshoe Mountain at the northeast end of the belt is in the quadrangle mapped recently by R. B. Neuman (1967), also of the U.S. Geological Survey.

My own interest in the Piscataquis volcanic belt began in 1955 while studying the Traveler Mountain area for a Ph.D. dissertation at Harvard University under the guidance of Professors M. P. Billings and J. B. Thompson, Jr. Subsequently, in one week in the fall of 1966, I visited Long Pond Mountain, Blue Ridge, Mt. Kineo, Big Spencer Mountain, and Soubunge Mountain.

Finally, I would like to acknowledge many helpful and stimulating discussions on volcanism with R. L. Smith and R. A. Bailey of the U.S. Geological Survey.

VOLCANIC CENTERS

Most of the volcanic rocks of the belt are grossly similar in appearance. These are flinty aphanitic rocks that break with a conchoidal fracture and contain 5–20% small phenocrysts (1–3 mm in length). They range in color from light gray through various shades of greenish gray and bluish gray to nearly black. In general, the darkest rocks contain the least altered phenocrysts and the best preserved primary textures.

In spite of the overall similarity, felsites in one part of the belt differ from those in another in phenocryst content, bulk chemistry, or structure. This suggests that there were several volcanic centers along the belt, each with its own chemical and eruptive characteristics. The shallow intrusive rocks, in general, reflect the mineralogy of the volcanic center to which they are closest. Whether or not any of these shallow intrusive bodies were actually conduits for the extrusive rocks is not known. Only at the largest center is there evidence for the location of the vent area.

At least five major volcanic centers have been recognized; two are overlapping and some may be multiple. The centers are, from northeast to southwest: Traveler, Big Spencer, Kineo, Cold Stream, and Grannys Cap (Fig. 27-1). In the section that follows, the volcanic rocks and eruptive character of each center will be discussed. The principal features of the centers are given in Table 27-1.

About 18 useful chemical analyses exist for the rhyolites of the Piscataquis volcanic belt; 6 are published (Smith, 1925, 1930, 1933), and the other 12 were made during my own study and that of Boucot. The weight percent silica of these analyses (constituents excluding H_2O and CO_2 normalized to 100 percent) ranges from 70.35 to 76.45. Nearly all the analyses, including those of rocks from the Traveler center that contain phenocrysts of calcic andesine (Table 27-1), plot in the rhyolite field of the classification of Rittmann (1952). The present composition of ancient volcanic rocks bears an uncertain relationship to the original composition of the volcanic product. For the purposes of this paper, however, the analyses will be accepted at face value, as will the Rittmann classification of rhyolite.

TABLE 27-1. Characteristics of volcanic centers of the Piscataquis volcanic belt

| | | Cold Stream | | | | Traveler | |
Volcanic Center	Grannys Cap	General	Chase Stream Mtn. area	Kineo	Big Spencer	Pogy Member	Black Cat Member
Formation Name	Kineo Volcanic Member of the Tomhegan Formation				Big Spencer Rhyolite	Traveler Rhyolite	
Composition	Probably rhyolite (no analyses)	Rhyolite (peraluminous)	Rhyolite (peraluminous)	Rhyolite (peraluminous)	Rhyolite (peraluminous)	Rhyolite	
Dominant extrusive product	Lava	Ash-flow tuff	Lava	Lava	Ash-flow tuff	Ash-flow tuff	Highly compacted ash-flow tuff
Volume of erupted material	1 cu mi	several cu mi		$<$ 10 cu mi	several cu mi	$>$ 100 cu mi	
Prominent phenocrysts	5%	15–20%	5–10%	8%	10%	15%	10%
Quartz	x	x	x (rare)	x	x	x	x (present in some samples, but atypical of member)
Sanidine		x (sparse)			x	x (rare)	
Plagioclase*	x ($\sim$An$_{10}$)	x ($\sim$An$_{10}$)	x ($\sim$An$_{10}$)	x ($\sim$An$_{10}$)	x ($\sim$An$_{10}$)	x ($\sim$An$_{47}$)	x ($\sim$An$_{47}$)
Mafic silicate				x?		x (altered)	x
Garnet		x	x (large)		x	x (rare)	x (rare)
Other			Apatite prominent				
Overlying rocks	None	Tomhegan Formation			None	Trout Valley Formation	
Underlying rocks	Tarratine Formation					Matagamon Sandstone**	

* Plagioclase from the Traveler Rhyolite is strongly zoned (An$_{40}$ to An$_{54}$). Plagioclase from the other centers is weakly zoned or nonzoned.
** The sandstone body at Soubunge Mtn. is included here in the Matagamon Sandstone.

TRAVELER CENTER AND TRAVELER RHYOLITE

Passing northerly, we find a steep high ledge or mountain, known by an inelegant name, which is a little back from the lake, and proves to be siliceous slate, being a continuation of this rock from the Traveller.
Hitchcock (1861, p. 404).*

General

The northeasternmost and by far the largest of the centers is near Traveler Mountain (Fig. 27–2) where about 80 cubic miles (330 cu km) of rhyolite occur within a structural depression thought to represent an ancient caldera. These rocks have generally been called the "Traveller Rhyolite," a name used by F. W. Toppan in 1932 in an unpublished Master's thesis at Union College, Schenectady, New York. Because this thesis cannot be considered a publication, the rock has never received a formal name. It is here called the Traveler Rhyolite after extensive exposures on Traveler Mountain, Maine. The spelling of Traveler Mountain is that which appears on the U. S. Geological Survey quadrangle map by that name.

Welded ash-flow tuff, commonly with well-developed columnar jointing makes up most of the unit. Lava, once thought to be a major constituent of the upper part of the Traveler Rhyolite (Rankin, 1961), is now known to be a very minor constituent. For this I am indebted to R. L. Smith of the U. S. Geological Survey, who visited the Traveler Mountain area with me in 1966 and pointed out evidence hitherto unobserved. Minor air-fall tuff and breccia occur locally. Little sedimentary rock is interbedded, and no fossils have been recognized.

The Traveler Rhyolite has been divided into the basal Pogy Member and the overlying Black Cat Member. Their aggregate thickness is at least 11,000 feet (3400 m). Rocks of the upper member differ from those of the lower in having few or no quartz phenocrysts and in possessing more obvious planar structure. The contact between the two members is gradational in terms of the percentage of quartz phenocrysts present.

Several granophyric intrusive bodies are associated spatially with the Traveler center. They are scattered over a distance of 34 miles (55 km) from southwest of Harrington Lake northeast to Snowshoe Mountain (Fig. 27–2). The largest body, north of Grand Lake Matagamon, is probably a sill about 5 miles (8 km) long which has a maximum thickness of about 1,000 feet (300

* Horse Mountain on the Traveler Mountain 15-minute quadrangle map of the U. S. Geological Survey.

meters). The granophyres are probably intrusive equivalents of the rhyolite, but it is unlikely that any were actual feeders for the volcanic rocks because the compositions of plagioclase phenocrysts in the intrusive and extrusive rocks are different.

Pogy Member

The lower member is here called the Pogy Member for exposures on North Pogy Mountain (Fig. 27–2). It is composed largely of welded ash-flow tuff containing 15% phenocrysts, of which one-third are quartz and most of the rest are plagioclase. Altered mafic phenocrysts are sparingly present. Lithic fragments, including felsite, diabase, sandstone, and shale are present in most samples. On the east side of Traveler Mountain the Pogy Member is about 3,000 feet (900 m) thick.

The contact between the Pogy Member and the underlying Matagamon Sandstone is well exposed at the base of the cliffs on Horse Mountain and Bald Mountain (Fig. 27–2). At both localities, a sharp, grossly conformable contact separates massive rhyolite containing no interbedded sedimentary rocks and the underlying obviously stratified rocks (Rankin, 1965, Fig. 4). This contact is defined as the base of the Pogy Member and may be traced discontinuously at both localities for distances of several hundred feet (a few hundred meters). The uppermost 20 feet (6 m) or so of the Matagamon Sandstone contains scattered pebbles of felsite and beds of tuffaceous sandstone as well as normal sandstone and shale of the formation, indicating that some volcanic activity preceded the emplacement of the massive rhyolite at these localities.

At Horse Mountain the basal 2–4 feet (0.6–1.2 m) of the Pogy Member consists of non-welded tuff in which undeformed, devitrified shards are clearly visible. This grades up into welded ash-flow tuff that makes up the rhyolite of the Horse Mountain cliffs. Collapsed pumice fragments occur a few feet (a meter or so) above the base, and deformed and flattened shards are visible in a thin section of a sample collected 5 feet (1.5 m) above the base. Columnar jointing becomes obvious in the rhyolite a few feet above this and is prominent on the 600-foot (200-m) cliff face. No stratification was observed on accessible parts of the cliff. Thus, at least on Horse Mountain, ash-flow tuffs form the base of the Traveler Rhyolite. The ash-flow sheet was hot enough so that upon cooling columnar jointing developed down to within a few feet of its base.

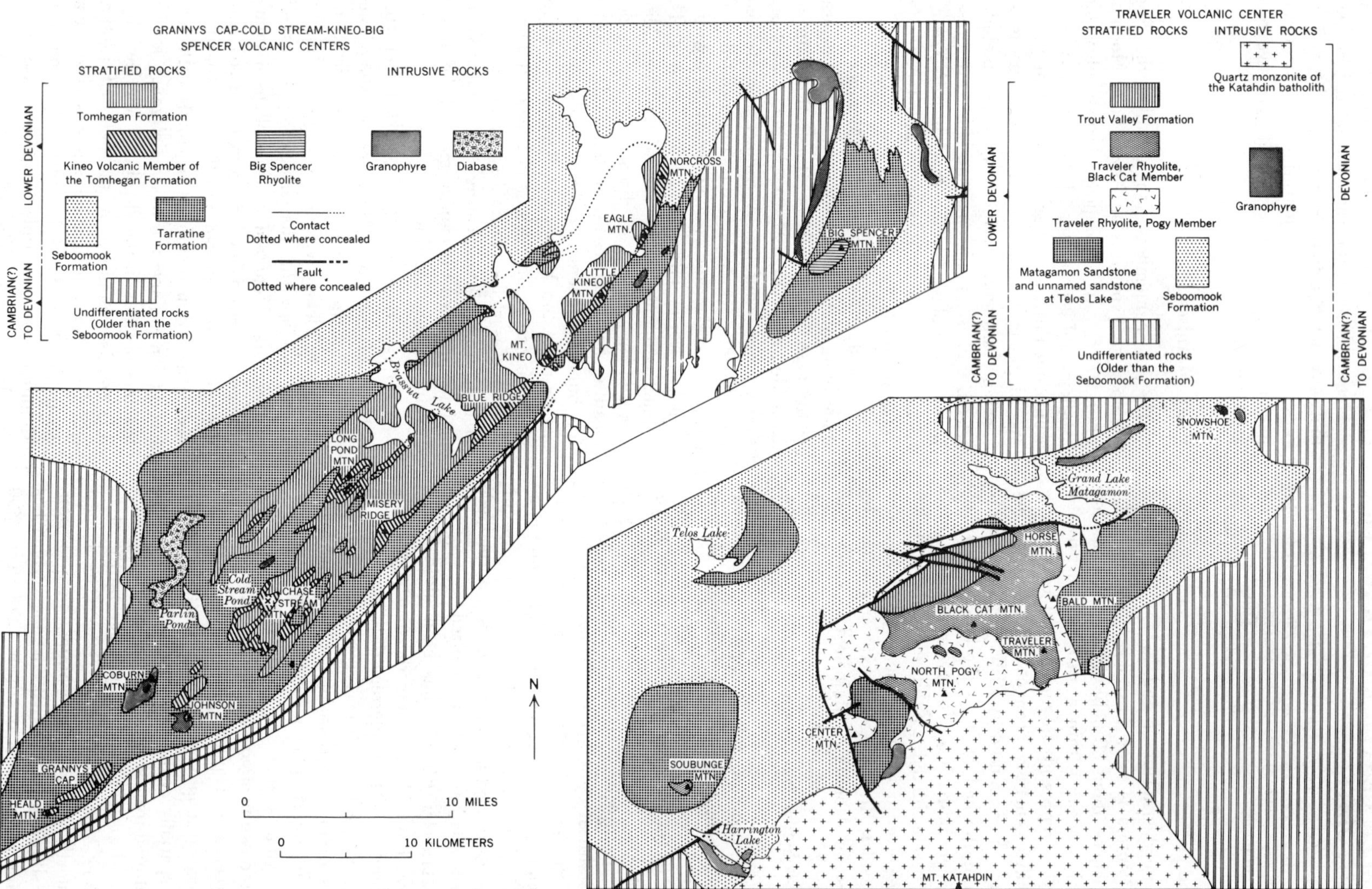

FIGURE 27–2. Generalized geologic maps of volcanic centers in the Piscataquis volcanic belt. See Figure 27–1 for regional setting.

Compaction features such as flattened shards were probably produced by plastic deformation at elevated temperatures rather than collapse accompanying devitrification, because devitrified shards at the base of the ash-flow sheet are undeformed. The welded ash-flows, therefore, are indicative of subaerial eruption and deposition (Rankin, 1960) rather than subaqueous eruption and deposition of the type described by Fiske (1963).

Several lines of evidence indicate that the welded ash-flow sheets were deposited on unconsolidated sediments. First, the considerable volcanic material in the upper 20 feet (6 m) or so of the Matagamon indicates that volcanism and sedimentation overlapped. Next, irregularities occur along the contact at the base of the massive rhyolite in small scale, with relief of a few inches to a foot (several cm) or so, and in large scale, with relief of 15–20 feet (5–6 m). These irregularities may be due to scouring of unconsolidated sediments by ash-flows. Finally, there are numerous clastic dikes in the base of the rhyolite at Bald Mountain and at Horse Mountain. The filling of the dikes may be either sandstone or tuffaceous sandstones, and the dikes vary from small anastomosing dikes, segments of which may be less than an inch thick, to larger ones as much as 20 feet (6 m) thick and at least several tens of feet long. These dikes suggest that unconsolidated sediments were forced up into fractures in the bottom of welded ash-flow tuffs.

Black Cat Member

The upper member of the Traveler Rhyolite is here called the Black Cat Member after excellent exposures on the slopes of the north peak of Black Cat Mountain (Fig. 27-2). This member is at least 8,000 feet (2,400 m) thick on Traveler Mountain. The petrographically similar rhyolite of Soubunge Mountain is thought to be an outlier of the Black Cat Member. Andrew Griscom (written communication, 1966) estimates that the thickness of rhyolite on Soubunge is 800 feet (250 m).

The Black Cat Member is composed largely of highly welded and compacted ash-flow tuff. Typically the welded tuff contains 10% phenocrysts, of which about 75% are plagioclase, 20% are augite, and 5% are magnetite. Biotite and hornblende phenocrysts may also be present, and in some samples from Soubunge Mountain, hornblende phenocrysts are more abundant than augite. Quartz phenocrysts, where they occur, rarely constitute as much as 5% of the total phenocrysts.

The ash-flow sheets are highly compacted. Length-to-thickness ratios of 20:1 are common, if not typical in collapsed pumice fragments. Rotated phenocrysts are also common, and small-scale flow folds occur locally. These features suggest that the Black Cat Member was emplaced as a succession of very hot ash flows, some of which continued to move downslope after or during welding and compaction.

Volcanism

The temperature of the magma that produced the Black Cat Member was apparently higher than that of the magma that produced the Pogy Member. At the time of eruption, liquids of the Pogy Member were crystallizing both quartz and feldspar (present as phenocrysts in the welded tuff), under conditions analogous to the thermal trough in the synthetic granite system of Tuttle and Bowen (1958), whereas liquids of the Black Cat were crystallizing only feldspar. The slightly higher total phenocryst content of the Pogy also indicates that it was somewhat more crystallized at the time of eruption. These observations are consistent with the idea that the basal Pogy Member came from the cooler upper part of a magma chamber and the overlying Black Cat came subsequently from deeper in the magma chamber. A further observation supporting this is that exotic lithic fragments are relatively common in the Pogy, but not in the Black Cat.

The following discussion intends to show that features of the Traveler Rhyolite are consistent with a model, rather than to demonstrate the validity of the model. Whether or not an ash flow will weld involves the interaction of several complex variables, a major one of which is emplacement temperature (Smith, 1960, p. 824–825; Boyd, 1961, p. 414–423). Pressure exerted by the mass of overlying tuff will do more welding in a thicker ash flow than in a thinner one of the same composition and emplacement temperature (see Smith 1960; Boyd, 1961). The influence of load upon welding in the Traveler Rhyolite is difficult to evaluate because the thickness of cooling units is unknown. At the risk of oversimplification, let us assume that cooling units were, in general, no thicker in the Black Cat Member than in the Pogy Member.

Boyd (1961) calculated that for a rhyolite magma, the greater the initial H_2O content, the greater the heat loss during emplacement of the ash flow. If we assume that cooling units are roughly the same thickness and the same composition, the minimum welding temperature should have a limited range depending largely upon the final H₂O content. Because the liquidus temperature of a rhyolite magma decreases with increasing H_2O content (steam saturation pressure), the difference between the liquidus temperature and the welding temperature, that is, the available cooling interval, decreases with increasing H_2O content. (Boyd, 1961, Fig. 11). The combined effect of these factors, using Boyd's model, is that a liquid with a lower initial H_2O content will have a higher emplacement temperature than one with a higher initial H_2O content. Other factors being equal, therefore, one would expect a drier magma to form a more densely welded tuff than a wetter one.

Kennedy (1955) suggested that at equilibrium the H_2O content of a magma should decrease downward in the magma chamber. Thus, assuming that the Pogy Member came from higher in the magma chamber than the Black Cat, it may have been wetter as well as cooler. It is certainly true that unaltered plagioclase phenocrysts are noticeably more common in the Black Cat than in the Pogy, and that mafic phenocrysts in the Pogy are invariably so altered that they cannot even be identified. Perhaps the alteration is deuteric, caused by abundant volatiles rising through the Pogy ash flows.

If it is correct that the Black Cat Member liquids were both hotter and drier than those of the Pogy Member, and if there are no additional complicating factors, we would predict that the emplacement temperature of the ash flows of the Black Cat should have been higher than that of the Pogy. The greater degree of compaction of the Black Cat, as well as presence of flowage features in the Black Cat are consistent with this prediction.

Much remains unknown about the Traveler Rhyolite. Neither flow units nor cooling units have been mapped. The chemistry is poorly known. Future work may reveal bulk compositional differences between the Pogy and Black Cat Members. The time interval involved in the eruptions may always be unknown. Nevertheless, the existing evidence is consistent with the Pogy Member having had a lower magma temperature and emplacement temperature and a higher initial H_2O content than the Black Cat Member. These suggestions, in turn, can be related to gradients in temperature and H_2O content in the magma chamber.

The main mass of the Traveler Rhyolite occupies a structurally depressed, roughly quadrilateral area about 8 by 12 miles (12 by 19 km) in extent (Fig. 27–2). It is bounded by high-angle faults on its north and west sides and is in-

truded by the quartz monzonite of Mt. Katahdin on its south side. In the center of this area the thickness of the volcanic rocks is at least 11,000 feet (3,400 m). Using an average thickness of 5,000 feet (1,500 m) as a conservative estimate, the volume of rhyolite within the structural depression is on the order of 80 cubic miles (330 km³); it may originally have been much greater than this. Smith (1960, Fig. 3) plotted volumes of pyroclastic-flow fields in cubic kilometers on a logarithmic scale and found that they range over seven orders of magnitude from 0.001 km³ to nearly 10,000 km³). The Traveler Rhyolite falls in the sixth order of magnitude on this scale (100 to 1,000 km³) and thus is among the larger pyroclastic-flow fields of the world. Smith (1960) notes that ash-flow deposits of orders 5, 6, and 7 are associated with subsidence structures in all examples where the source area is known. Probably the structural depression occupied by the Traveler Rhyolite originated as a caldera, and the name Traveler caldera is here given to the depression.

Some structural evidence indicates that caldera collapse began before the eruption of the bulk of the Black Cat Member. On Traveler Mountain, the rhyolite is in a north-northwest-trending syncline with the Black Cat Member occupying the trough. The Black Cat Member on the west side of the synclinal axis (toward the caldera center) is many times thicker than it is on the east side of the axis. Further, the compaction foliation in the two small patches of the Black Cat Member southwest of Black Cat Mountain is structurally discordant with that of the underlying Pogy Member.

Except for the isolated body in Soubunge Mountain, nothing is preserved today of the Traveler Rhyolite outside the main caldera. By analogy with modern calderas where the bulk of the material in pyroclastic-flow fields is preserved outside the caldera, we may surmise that the original volume of material was several times the estimated 80 cubic miles (330 km³). A blanket of welded tuff 800 feet (250 m) thick surrounding the Traveler caldera to a distance from the center as far as Soubunge Mountain would more than double the volume of material.

BIG SPENCER CENTER AND BIG SPENCER RHYOLITE

The only known extrusive rocks of the Big Spencer Center are on Big Spencer Mountain, 18 miles (29 km) southwest of Soubunge Mountain (Fig. 27–2). These are mineralogically distinct from the Traveler Rhyolite and from the rhyolites near Moosehead Lake to the west (Table 27–1) and are here called the Big Spencer Rhyolite. Heath (1959) estimated the preserved thickness of rhyolite to be 2,400 feet (730 m), but this seems excessive by about 1,000 feet (300 m). The Big Spencer Rhyolite consists largely of welded tuff. About 200 feet (60 m) of bedded tuff and tuffaceous sandstone is exposed at the base of the volcanic pile on the southwest end of the mountain.

Big Spencer Mountain is a steep-sided northeast-trending ridge that rises nearly 2,000 feet (600 m) above the surrounding country. The Big Spencer Rhyolite forms the upper 800–1,000 feet (250–300 m) of the mountain, and the lowest rhyolite outcrops on the northeast and southwest ends of the ridge are about 2.3 miles (3.7 km) apart. A traverse up each end of the ridge suggests that the stratigraphy within the Big Spencer Rhoylite may be fairly simple, although any correlation between the two ends of the mountain is clearly speculative. The preserved rocks indicate at least two cycles of ash-flow eruptions. In each cycle, moderately compacted garnet-bearing welded tuff is overlain by highly compacted garnet-free welded tuff. The more highly compacted welded tuff here as well as in the Traveler caldera may represent hotter liquids erupted from deeper in the magma chamber after higher, cooler liquids were exhausted. The garnets are now being studied; they are clearly phenocrysts. They are commonly euhedral and much fractured or present as crystal fragments. Eutaxitic structure bends around the crystals, and some crystals form growth aggregates with quartz phenocrysts. In a general way, if pressure and temperature have opposite effects upon the stability fields of minerals, one might expect garnet to crystallize from cooler liquids rather than hotter ones of the same composition. This assumes that in a magma chamber, the temperature gradient is more significant than the pressure gradient relative to the stability of garnet at the liquidus. Such a mechanism might explain the presence of garnet phenocrysts in moderately compacted welded tuffs stratigraphically beneath garnet-free highly compacted welded tuff.

On the southwest end of Big Spencer Mountain, the lowest exposed volcanic rocks are bedded tuffs and tuffaceous sandstones. The bedded tuffs consist of aphanitic vitric tuff, vitric-crystal tuff, and crystal-vitric tuff. The tuffaceous sandstone contains a much higher ratio of quartz to feldspar than is present in the phenocrysts of the welded tuff, indicating an admixture of epiclastic material. Some graded bedding was observed. These rocks may represent subaqueous parts of ash flows that crossed a Devonian shoreline between the northeast and southwest end of the mountain.

A sample of tuff collected by Heath on the south side of Big Spencer contains a large number of spherical inclusions as much as 7 mm in diameter. He noted that the mineralogy of the spheres appeared to be the same as that of the matrix and suggested that the inclusions were spherulites on the basis of megascopic examination only. In thin section, these are clearly accretionary lapilli. They possess many of the features described by Moore and Peck (1962), such as a decrease in grain size from core to rim, multiple shells (only two shells have been observed here), and lapilli fragments. In the present study, fragments of accretionary lapilli were observed in the bedded tuff at the southwest end of the mountain, but no unbroken lapilli. Moore and Peck point out that the occurrence of accretionary lapilli indicates that the rock enclosing the lapilli was deposited on land or possibly in shallow water. This is in complete agreement with other evidence of Big Spencer Mountain. Moore and Peck also point out that most occurrences of lapilli of known source are within a few miles of the vent and that the vent must have been above water or in shallow water.

It is pointless to try to estimate the volume of material erupted from the Big Spencer Center. An isolated 1,000-plus-foot (300-plus-meter) pile of welded tuff, however, can only be a remnant of a much larger pyroclastic field, whose volume was measured in cubic miles rather than fractions thereof.

KINEO VOLCANIC MEMBER IN THE KINEO CENTER

All the felsic volcanic rocks of Becraft to Schoharie age in the Piscataquis volcanic belt west and southwest of Big Spencer Mountain were included in the Kineo Volcanic Member by Boucot (1961). He placed the member in the overlying Tomhegan Formation, although no fossils are present in the volcanic rocks (see Boucot, 1961, p. 164) and other evidence is questionable. It seems more likely that the volcanic rocks are either a late phase of the underlying Tarratine Formation, or intermediate in age between the Tomhegan and Tarratine. We now know, further, that the Kineo as Boucot defines it includes a variety of rhyolites from several volcanic centers. A revision of stratigraphic nomenclature is clearly in order, but it is not an essential part of the following discussion. Use of the name Kineo Volcanic Member will continue here for rhyolites of the remaining centers in the Piscataquis volcanic belt.*

At the type locality on Mt. Kineo (Fig. 27–2), the Kineo Volcanic Member consists of massive, flow-banded rhyolite containing about 8 percent phenocrysts of quartz and plagioclase. Some rocks also contain minor amounts of chlorite-magnetite aggregates that are probably pseudomorphous after pyribole. Rhyolite of this type is found in a number of lenticular bodies for a distance of about 24 miles (38 km) from Norcross Mountain on the northeast to Misery Ridge and Long Pond Mountain on the southwest. Together these bodies form the Kineo Center.

Flow banding is commonly visible in both outcrop and thin sections. Chaotic flow banding similar to that found in younger rhyolite lava flows is present in many outcrops. Columnar jointing is not common, but it does occur in the body of rhyolite lava on Long Pond Mountain, and pieces of columnar joints are found along the shore of Moosehead Lake at the foot of Mt. Kineo. On Eagle Mountain, massive rhyolite is overlain by a conglomerate of poorly rounded cobbles and boulders of massive felsite in a matrix of gray siltstone (Boucot, 1961).

Thus the major product of the Kineo Center was rhyolite lava. Mapping by Boucot (1961) suggests that these lavas accumulated in local thick lenses or domes. The larger lenses such as that on Blue Ridge have thicknesses of as much as 4,000 feet (1200 m). Boucot (written communication, 1966) found that massive rhyolite on Little Kineo Mountain grades northeast into flow breccia and that massive rhyolite on Eagle Mountain grades southwest into flow breccia. Flow breccia and sedimentary rock containing a high percentage of volcanic detritus also occur at the southwest end of Blue Ridge along the shore of Brassua Lake.

Each lens must have had its own vent, although it is uncertain, for example, how many vents existed in the rather continuous belt of rhyolite from Blue Ridge to Little Kineo Mountain. Probably there were several source areas here. Based upon a sampling of flow-banding attitudes on the northeast and southwest ends of Mt. Kineo, the rhyolite dips away from the center of the penin-

sula, suggesting that one source area was near Mt. Kineo.

Boucot (1961; written communication, 1966) found only one thin lens of the Kineo Volcanic Member on the northwest limb of the Moose River synclinorium—located on the west shore of Moosehead Lake. A sample collected by him from this locality is a devitrified vitric tuff containing the same assemblage of phenocrysts present in the lavas of the Kineo Center. The vitric tuff indicates that there was also minor explosive activity from the Kineo Center.

Except for the lava flows on Long Pond Mountain and on the southwest shore of Brassua Lake, the lavas of the Kineo Center are exposed only along the southeast flank of the Moose River synclinorium. The Kineo Center thus consists of a series of lithologically similar rhyolite domes along a fairly well defined northeast linear trend that may represent a fracture zone. The original northeastern extent of the Kineo-type domes is, of course, unknown. The only clue is that felsites of the Piscataquis volcanic belt were erupted in a sedimentary environment that was depositing sand rather than mud. The underlying Tarratine Formation grades laterally into the more shaly Seboomook Formation at about the northeastern-most rhyolite dome of the Kineo Center. This change may indicate that this was approximately the northeastern extent of the Kineo Center. If the foregoing analysis is correct, the total volume of material erupted along the 24-mile (38-km) extent of the center is probably several cubic miles (10–20 km³) rather than several tens of cubic miles.

KINEO VOLCANIC MEMBER IN THE COLD STREAM CENTER

The Cold Stream Center is a large, poorly understood center southwest of, but overlapping the Kineo Center. Rocks of this center are scattered along 19 miles (27 km) of the Moose River synclinorium and are characterized by the presence of conspicuous garnet phenocrysts. Extrusive rocks of this center include probable lava flows around Chase Stream Mountain and ash-flow tuffs scattered over the whole center but best preserved near Cold Stream Pond. Grit and conglomerate containing greater than 50 percent volcanic material (including felsite fragments, embayed quartz grains of obvious volcanic origin, plagioclase grains of indeterminate origin, and, in some, fragments of garnet) are present near Long Pond Mountain, the southwestern arm of Brassua Lake, and 4 miles (6.4 km) northeast of Parlin Pond. Boucot (1961) included these in

---

* Boucot (this volume) now uses the name "Kineo Rhyolite," a name which also appears on the preliminary Geologic Map of Maine (Doyle, 1967). Further, Boucot (this volume) gives the age of the "Kineo Rhyolite" as Esopus, that is, intermediate between the ages of the Tarratine and Tomhegan Formations.

the Kineo Volcanic Member, although future work may indicate that they are more properly part of the overlying Tomhegan Formation. On Figure 27–1 these patches of conglomerate are included in the Cold Stream Center. Sizable bodies of intrusive garnet rhyolite and granophyre on Coburn and Johnson Mountains (Fig. 27–2) are the largest intrusive bodies associated with this center.

Ash-flow tuff is probably the most abundant extrusive product of the Cold Stream Center. The tuff is not bedded and has a maximum thickness of about 1,000 feet (300 m) (Boucot, written communication, 1966). It is more probably the product of ash flows than ash falls. Devitrified shards and pumice fragments are clearly visible in thin section. Commonly the shards are flattened and alined, but because resulting foliation is also the metamorphic foliation, it is not possible to determine whether the tuff was welded or not. Randomly oriented welded tuff fragments included in the ash-flow tuff, however, suggest that there was some welding.

Rhyolite of the Cold Stream Center on Chase Stream Mountain is texturally and mineralogically different from the ash-flow tuff. According to Boucot, outcrops are massive. In thin section neither flow banding nor fragmental texture is visible. The same undiagnostic texture and the same phenocryst assemblage occur in nearby small felsic intrusives. The rhyolite on Chase Stream Mountain is probably lava, although the evidence is not compelling. Garnet phenocrysts are very conspicuous and are as large as 5 mm in diameter. In thin section, apatite crystals are larger (many are about 0.5 × 0.3 mm) and thus more conspicuous than in other felsites of the Piscataquis volcanic belt.

Relations between the ash-flow tuff and lava are unknown. Boucot's mapping shows that is some ash-flow tuff associated with the lava on Chase Stream Mountain and some lava associated with ash-flow tuff on Cold Stream Pond. Whether these garnet-bearing rhyolites represent one center or more cannot be determined at present. Ash-flow tuff of the Cold Stream Center did extend into the area of the Kineo Center, however. One of Boucot's samples collected from the bottom of the rhyolite pile on Misery Ridge is a garnet-bearing "vitric-crystal" tuff, presumably originating from Cold Stream Center. Garnet-bearing tuff is also found as far northeast as the southwest arm of Brassua Lake.

The ash-flow sheets from the Cold Stream Center were probably originally thinner than those from the Big Spencer Center and certainly a great deal thinner than those from the Traveler caldera. Nevertheless, the scattered patches of tuff preserved today suggest that these sheets were extensive and that the volume of material erupted from the Cold Stream Center was at least several cubic miles.

KINEO VOLCANIC MEMBER IN THE GRANNYS CAP CENTER

The southwesternmost center of the Piscataquis volcanic belt is represented by felsite exposed on Heald Mountain and Grannys Cap. The principal extrusive product was probably lava very similar to that of the Kineo Center. The mode of occurrence of the volcanic rocks, as mountains of massive felsite, is also suggestive of the rhyolite domes of the Kineo Center. One sample collected by Boucot shows excellent flow banding, but in general the groundmass is more recrystallized than that from any other center. The greater degree of recrystallization is reasonable in view of the regional increase in metamorphic grade southwest across Maine. Allanite and biotite occur in the recrystallized groundmass. There are no chemical analyses of felsite from the Grannys Cap Center, but the close petrographic similarity to the rhyolite of Mt. Kineo suggest that it is probably rhyolite.

Boucot (written communication, 1966) gives the maximum preserved thickness of the felsites as 1,000–2,000 feet (300–600 m). He points out that the felsites could be slightly older than other felsites of the Piscataquis volcanic belt. From Big Spencer southwestward, the Misery Quartzite Member of the Tarratine Formation is a mappable unit about 500 feet (150 m) below the base of the Kineo Volcanic Member (Boucot, 1961; Heath, 1959). The Misery Quartzite is not present beneath the lavas of the Grannys Cap Center. This may be either because of a wedge-out of the quartzite or because the felsite is older. The age difference, if any, is not thought to be significant.

MAFIC ROCKS RELATED TO THE PISCATAQUIS VOLCANIC BELT

Consanguineous mafic rocks are, in general, conspicuously absent from the Piscataquis volcanic belt; no known extrusive mafic rocks are associated with the felsites. Boucot (1961; written communication, 1966) mapped a few small gabbro sills and dikes that intrude the Tarratine, Seboomook, and older formations in the southwestern part of the belt in the vicinity of the Cold Stream and Grannys Cap Centers. They are all probably older than the Acadian orogeny and if all are the same age, they would be Early Devonian but younger than the Tarratine Formation. A larger body of diabase crops out northwest of Parlin Pond within the terrane of the Tarratine Formation. Boucot (written communication, 1966) is uncertain whether the diabase is a sill or a flow. Both the map pattern and the grain size are more suggestive of an intrusive than an extrusive mafic rock. Extensive alteration suggests that the diabase is pre-Acadian orogeny. The diabase and gabbro intrusion are, thus, roughly the same age as the felsic volcanism of the Piscataquis volcanic belt.

The much larger mafic Moxie pluton southeast of the Moose River synclinorium (Fig. 27–1) appears to be younger than the felsic volcanism. Espenshade (1967) suggests that it is postorogenic. Although it, like the Katahdin batholith, is undoubtedly part of the larger magmatic episode, further consideration of the Moxie is beyond the scope of this paper.

STRATIGRAPHIC RELATIONS OF THE PISCATAQUIS VOLCANIC BELT

The extrusive felsites of the belt rest upon marine sandstones of Becraft–Oriskany age. The only exception is the rhyolite of Norcross Mountain at the northeast end of Moosehead Lake which overlies more pelitic rocks. Even here, sandstone is close by. These sandstones are facies of the upper part of the more pelitic Seboomook Formation that underlies much of northern Maine (see Boucot et al., 1964a). The upper part of the Seboomook contains marine fossils, also of Becraft–Oriskany age.

There is almost a one-to-one correspondence between the location of these lenses of sandstone and penecontemporaneous felsic volcanism in Maine. At present, the only known sandstone containing fossils of Becraft–Oriskany age that does not underlie felsite is the unnamed sandstone on Telos Lake. This area has not been studied in detail, and future work may reveal felsite here also. One may reasonably ask if there is not a genetic relationship between the sandstone and the Piscataquis volcanic belt.

Boucot (oral communication, 1959) characterizes the fauna of the sandstone lenses as being a shallow-water one compared with that of the Seboomook Formation. Welded tuff at the base of the Traveler Rhyolite and Big Spencer Rhyolite further suggest very shallow water for the immediately underlying sandstone. Ash-flow tuff that may or may not be welded is probably present at the base of the Kineo Volcanic Member in

the Cold Stream Center. Thus the volcanic centers are along a high-standing welt relative to the deeper water basins on either side.

Sedimentary rocks that overlie the felsites are probably mutually correlative, although their correlation is less certain than that of the underlying rocks. The Tomhegan Formation, which occupies the trough of the Moose River synclinorium, contains marine fossils of Schoharie (Oliver's Zone B) age (Boucot et al., 1964b). The Tomhegan also contains much felsite detritus near the contact with the underlying Kineo Volcanic Member.

The main body of the Traveler Rhyolite is overlain by the Trout Valley Formation of Dorf and Rankin (1962), a name here adopted by the U.S. Geological Survey. The Trout Valley contains much felsite detritus near its base, including a massive felsite conglomerate lentil. Minor amounts of sideritic sandstone and black sideritic ironstone, abundant impressions of terrestrial plants (psilophytes), and less common fossils including ostracodes, estherids(?), and eurypterid scales are present in the formation. It was probably deposited in brackish water. Unfortunately, the age of this unit is less well defined than that of the Tomhegan and age assignments range from late Early Devonian (Dorf and Rankin, 1962) to Middle Devonian (Boucot et al., 1964b, p. 93–95). The uncertainty depends in part upon whether the Onondaga Limestone of New York, which paleontologists commonly use as a reference, is considered to be of Early or Middle Devonian age and in part upon the ambiguity of cross correlations of Devonian flora and fauna (see Dorf and Rankin, 1962, p. 1003; Schopf, 1964, p. D49; Boucot et al., 1964b, p. 95).

The Trout Valley Formation is less deformed than the underlying rocks and clearly truncates major structures in the Traveler Rhyolite. The relatively undeformed condition of the Trout Valley may be due to its posttectonic age if the formation proves to be a correlative of the post-Acadian Mapleton Sandstone near Presque Isle, Maine, some 50 miles (80 km) northeast of Traveler Mountain. The Mapleton Sandstone also contains spiny psilophyte remains, and the dating of these plants as Middle Devonian rather than Late Devonian (Schopf, 1964) has brought a correlation between the Trout Valley and Mapleton within the realm of possibility. Boucot et al. (1964b, p. 94–96) suggest this correlation and then use the Middle Devonian age for the Trout Valley Formation, thus derived, to bracket the end of a first phase of the Acadian orogeny in the Traveler Mountain area. On the other hand, if the Trout Valley is a correlative of the Lower Devonian Tomhegan Formation, the relatively undeformed condition of the Trout Valley may be due to its shielded tectonic position above the thick competent Traveler Rhyolite.

Professor Henry N. Andrews, who is making a continuing study of the flora of the Trout Valley Formation, considers the flora to be of late Early Devonian age and more primitive than that of the Mapleton Sandstone (H. N. Andrews, written communication, Nov. 1967). Boucot (this volume) has recently found evidence in the Eastern Townships of Quebec that the Acadian orogeny occurred there in post-Onondaga time rather than in an interval between the Early and Middle Devonian. To the extent that the Acadian orogeny affected central Maine and the Eastern Townships at the same time, and that the Andrews age assignment is correct, the Trout Valley Formation is roughly correlative with the Tomhegan Formation and was shielded from the forces of the Acadian orogeny by the underlying Traveler Rhyolite.

AGE OF VOLCANISM

The Becraft–Oriskany (Early Devonian) age of the underlying sandstones establishes the maximum paleontologic age for the volcanic episode. The overall similarity of the felsites and the setting suggests that they were roughly coeval. Thus the Schoharie (Early Devonian) age for the Tomhegan Formation places a minimum age on the volcanism.

One set of published absolute ages also suggests that volcanism within the belt was roughly the same age. Bottino et al. (1966) obtained Rb–Sr whole rock ages of 360 ± 10 m.y. and 360 ± 15 m.y. for the Traveler Rhyolite and the Kineo Volcanic Member respectively. These absolute ages would appear too young, however, in comparison with the age obtained on a quartz monzonite stock that intrudes fossiliferous slate of the Seboomook about 10 miles (16 km) northwest of Parlin Pond. Hurley et al. (1959) by K–Ar and Rb–Sr measurements on quartz monzonite and hornfelsed slate determined the age of this stock (probably post-Acadian) as about 360 m.y.

CONCLUSIONS: VOLCANISM AND TECTONISM

The Traveler Rhyolite is not peraluminous, but most of the rhyolites from the other centers are. The peraluminous character presumably accounts for the presence of garnet phenocrysts and suggests that liquids may have been generated by partial melting of sediments in the deeper parts of the Appalachian geosyncline. This concept is further supported by the great bulk of felsites relative to mafic rocks of the same age.

Volcanic centers of the Piscataquis volcanic belt are located along a positive welt of Lower Devonian sandstones. The existence of the welt or ridge is indicated by a shallow-water fauna of the standstones and by subaerial volcanism in at least three of the volcanic centers. Deeper water coeval pelitic sediments flank both sides of the sandstone ridge.

Several factors probably contributed to the formation of the welt. The welt may have been caused in part by the swelling of the crust along the zone of magma formation. The sandstone ridge may also have been in part inherited from an earlier episode of volcanism that more or less paralleled the Piscataquis volcanic belt and that may have been much more extensive. Boucot et al., (1964b, p. 75–88) summarized the occurrence of strata of New Scotland (early Early Devonian) age in northern Maine, eastern Quebec, and northern New Brunswick. They point out that where strata of New Scotland age are thickest (thicker than 2,000 feet, or 600 m) they commonly contain more than 50 percent volcanic rocks. Further, these thicker strata with the contained volcanic rocks are restricted to a relatively narrow belt about 280 miles (450 km) long, extending from near The Forks, Maine (Fig. 27–1), to the north shore of Chaleur Bay, Quebec. The southwestern part of this belt more or less coincides with the Piscataquis volcanic belt. New Scotland age volcanism appears to have been more varied than the later episode and on the average more mafic. The summary by Boucot et al. (1964b) would suggest that andesite is the most common volcanic rock but that basalt and rhyolite also occur. The presence of abundant older volcanic rocks would easily explain the abundance of sand-sized felsite clasts in the Matagamon Sandstone well below (hundreds of feet) the base of the Traveler Rhyolite. Thus, the Piscataquis volcanic belt may have been part of a real island-arc system of Early Devonian age. Earlier volcanism (New Scotland age) was probably centered to the northeast and was dominantly of intermediate composition. Later, the center of volcanic activity shifted southwest to the Piscataquis volcanic belt, and the products were chiefly rhyolites. Thompson and others (this volume) have suggested that the Bethlehem Gneiss and Kinsman Quartz Monzonite of New Hampshire may be metamorphosed ash-flow tuffs correlative with the rhyolites of the Piscataquis volcanic belt. Thus the Early Devonian island arc may also have extended as far southwest as east-central Massachusetts.

In the Traveler Mountain area, the eruption of great volumes of ash flows was accompanied and followed by collapse and caldera formation. This collapse could easily account for the unconformable occurrence of the brackish-water sediments of the Trout Valley Formation above several thousands of feet (meters) of subaerially erupted volcanic rocks. There is no need to appeal to compressive orogenic movements of the Acadian orogeny. Further, all observed lithic clasts in the conglomerates of the Trout Valley are felsite, suggesting a local source area (a volcanic island) rather than a larger landmass created by orogenic movements (Rankin, 1958).

Two other centers, Big Spencer and Cold Stream, produced ash flows, the original volume of which was probably many times the volume of the preserved rocks. The ash-flow sheets of the Cold Stream Center are overlain by marine sediments, indicating some environmental change. The eruption of large volumes of ash flows and, to a lesser extent, lava flows may have initiated a general downwarping of the region by the withdrawal of support from below. Finally, during the Acadian orogeny, the already complex series of troughs along the Piscataquis volcanic belt was the locus of further downwarping to give rise to the en echelon synclinoria present today.

REFERENCES

Bottino, M. L., Schnetzler, C. C., and Fullagar, P. D., 1966, Rb–Sr whole-rock age of the Traveler and Kineo Rhyolites, Maine, and its bearing on the duration of the Early Devonian [abs.]: Geol. Soc. America Spec. Paper 87, p. 15

Boucot, A. J., 1961, Stratigraphy of the Moose River synclinorium, Maine: U. S. Geol. Survey Bull. 1111-E, p. 153–183

Boucot, A. J., Griscom, Andrew, and Allingham, J. W., 1964a, Geologic and aeromagnetic map of northern Maine: U. S. Geol. Survey Geophys. Inv. Map GP–312

Boucot, A. J., Field, M. T., Fletcher, Raymond, Forbes, W. H., Naylor, R. S., and Pavlides, Louis, 1964b, Reconnaissance bedrock geology of the Presque Isle quadrangle, Maine: Maine Geol. Survey Quad. Map. Ser., no. 2, 123 p.

Boyd, F. R., 1961, Welded tuffs and flows in the rhyolite plateau of Yellowstone Park, Wyoming: Geol. Soc. America Bull., v. 72, p. 387–426

Dorf, Erling, and Rankin, D. W., 1962, Early Devonian plants from the Traveler Mountain area, Maine: Jour. Paleontology, v. 36, p. 999–1004

Doyle, R. G., *ed.*, 1967, Preliminary geologic map of Maine: Augusta, Maine, Maine Geol. Survey, scale 1:500,000

Espenshade, G. H., 1967, Petrology and structure of the northeastern part of the Moxie pluton, Maine [abs.]: Geol. Soc. America, Northeastern Section Meeting, Boston, Mass., Program, p. 24

Fiske, R. S., 1963, Subaqueous pyroclastic flows in the Ohanapecosh Formation, Washington: Geol. Soc. America Bull., v. 74, p. 391–406

Heath, E. W., 1959, Stratigraphy and structure of the Roach River syncline, Piscataquis County, Maine: M. S. Thesis, Massachusetts Inst. Technology, 125 p.

Hitchcock, C. H., 1861, Geology of the Wild Lands, *in* Holmes, Ezekiel and Hitchcock, C. H., Preliminary report upon the natural history and geology of the State of Maine: Maine Board Agriculture, 6th Ann. Rept., p. 377–442

————, 1862, Geology of Maine: Maine Board Agriculture, 7th Ann. Rept., p. 835–842

————, and Huntington, J. H., 1874, Geology of the northwest part of Maine: Am. Assoc. Adv. Sci. Proc., v. 22, pt. 2, p. 205–214

Hodge, J. T., 1838, Report on the Allagash section, from the Penobscot to the St. Lawrence River, *in* Jackson, C. T., Second annual report on the geology of the public lands belonging to the two states of Massachusetts and Maine: Boston, Dutton and Wentworth, State Printers, p. 46–68

Hurley, P. M., Boucot, A. J., Albee, A. L., Faul, Henry, Pinson, W. H., and Fairbairn, W. H., 1959, Minimum age of the Lower Devonian slate near Jackman, Maine: Geol. Soc. America Bull., v. 70, p. 947–950

Jackson, C. T., 1838, Second report on the geology of the state of Maine: Augusta, Maine, Luther Severance, 168 p.

Kennedy, G. C., 1955, Some aspects of the role of water in rock melts, *in* Poldervaart, Arie, *ed.*, Crust of the earth: Geol. Soc. America Spec. Paper 62, p. 489–504

Moore, J. G., and Peck, D. L., 1962, Accretionary lapilli in volcanic rocks of the western continental United States: Jour. Geology, v. 70, p. 182–193

Neuman, R. B., 1967, Bedrock geology of the Shin Pond and Stacyville quadrangles, Penobscot County, Maine: U. S. Geol. Survey Prof. Paper 524-I, p. 1–37

Rankin, D. W., 1958, Lower Devonian nonmarine sediments in the vicinity of Traveler Mountain, north-central Maine [abs.]: Geol. Soc. America Bull., v. 69, p. 1632

————, 1960, Paleogeographic implications of deposits of hot ash flows: Internat. Geol. Cong., 21st, Copenhagen 1960, Rept., pt. 12, p. 19–34

————, 1961, Volcanic history of the Early Devonian Traveler felsite, north-central Maine [abs.]: Geol. Soc. America Spec. Paper 68, p. 250

————, 1965, The Matagamon Sandstone, a new Devonian formation in north-central Maine: U. S. Geol. Survey Bull. 1194-F, 9 p.

Rittmann, A., 1952, Nomenclature of volcanic rocks proposed for use in the catalogue of volcanoes and key-tables for the determination of volcanic rocks: Bull. Volcanol., ser. 2, t. 12, p. 75–102

Schopf, J. M., 1964, Middle Devonian plant fossils from northern Maine: U. S. Geol. Survey Prof. Paper 501-D, p. 43–49

Smith, E. S. C., 1925, Contributions to the geology of Maine, Number 2, Part II. The igneous rock of Mt. Kineo and vicinity: Am. Jour. Sci., v. 10, p. 437–444

__________, 1930, Contributions to the geology of Maine, Number 4. The geology of the Katahdin area. Part I, a new rhyolite from the state of Maine: Am. Jour. Sci., v. 19, p. 6–8

__________, 1933, Contributions to the geology of Maine, Number 5. An occurrence of garnet in rhyolite: Am. Jour. Sci., v. 25, p. 225–228

Smith, R. L., 1960, Ash flows: Geol. Soc. America Bull., v. 71, p. 795–842

Thoreau, H. D., 1864, The Maine Woods: Ticknor and Fields, Boston, Mass.

Tuttle, O. F., and Bowen, N. L., 1958, The origin of granite in the light of experimental studies in the system $NaAlSi_3O_8$–$KAlSi_3O_8$–SiO_2–H_2O: Geol. Soc. America Mem. 74, 153 p.

Chapter 28

Devonian Plutonic Rocks in New England*

LINCOLN R. PAGE

INTRODUCTION

THE INTRUSIVE ROCKS of Devonian age in New England have been referred by most recent investigators to either the Oliverian or the New Hampshire Plutonic Series† of Billings (1935). Younger, post-Devonian, igneous rocks have been assigned to his White Mountain Plutonic-Volcanic Series or to the Pennsylvanian and Triassic; the pre-Devonian plutons have been assigned to his Highlandcroft Plutonic Series or to the Precambrian. In 1935, Billings (p. 28) recognized that some of the isolated igneous bodies, such as the French Pond Granite and granite of Hedge Hog Hill, which he included in the New Hampshire Plutonic Series in the Littleton and Moosilauke quadrangles, New Hampshire, (44°00′–44°45′ lat., 72°00′–72°15′ long.), differed in habit and mineralogy from other members of the series, but he concluded that they were definitely of Devonian age. In 1946, Billings and Keevil (p. 815) concluded that the French Pond and Lebanon Granites (43°30′–43°15′ lat., 72°15′ long.) are Oliverian in age (see also Chapman, 1939; Kaiser, 1938; Lyons, 1955). Chapman, Billings, and Chapman (1944) described two unusual types of Oliverian rocks in the Mt. Washington quadrangle (44°15′–44°30′ lat., 71°15′–71°30′ long.); these are referred to in Billings and Keevil (1946, p. 806, 815), who indicated that these rocks do not fit the normal evolutionary pattern of the Oliverian Series. In 1956, Billings (p. 65–69) described the Hillsboro Plutonic Series in southern New Hampshire but did not show this separately on the State map because it included various igneous rocks that

did not seem to fit with the other plutonic groups. Billings and Keevil (1946, p. 815) also recognized some unusual hornblende-rich units in the Oliverian Series. Later, Cady (in press) recognized the need for additional subdivision of the plutonic rocks of Devonian age.

The purpose of this paper is to suggest various criteria that can be used to refer rocks to one of three plutonic series, (1) the Oliverian,‡ (2) the New Hampshire, and (3) a new Late Devonian series, rather than to designate all the individual plutons that might be part of each group. The new Late Devonian Plutonic Series (Fig. 28–1) includes a group of late posttectonic plutons, differentiated at depth, that range in composition from ultramafic to granitic and that were emplaced in part by shouldering aside their walls and in part by stoping. They have distinctive mineralogy, textures, and structures and are surrounded by metamorphic aureoles that are superimposed on the foliation and other structures of regional metamorphism. This Late Devonian series is not strictly equivalent to the Hillsboro Plutonic Series of Billings, nor to the New England Plutonic Series of Cady (in press).

These three Devonian plutonic groups have specific structural, mineralogical, and textural differences (see Table 28–1) that stem from differences in (1) the original or parent magma from which they were derived, (2) the trend and type of magmatic differentiation, (3) the temperature of the wall rock at the time of intrusion, and (4) tectonic history. Of prime importance is the distinction of these three groups as pretec-

---

* Publication authorized by the Director, U. S. Geological Survey
† Originally designated as "magma series."

‡ The Oliverian Plutonic Series is considered to be of Devonian age in this paper because of its geologic relationship to Lower Devonian rocks; however, others, including fellow contributors to this volume, subscribe to an Ordovician age.

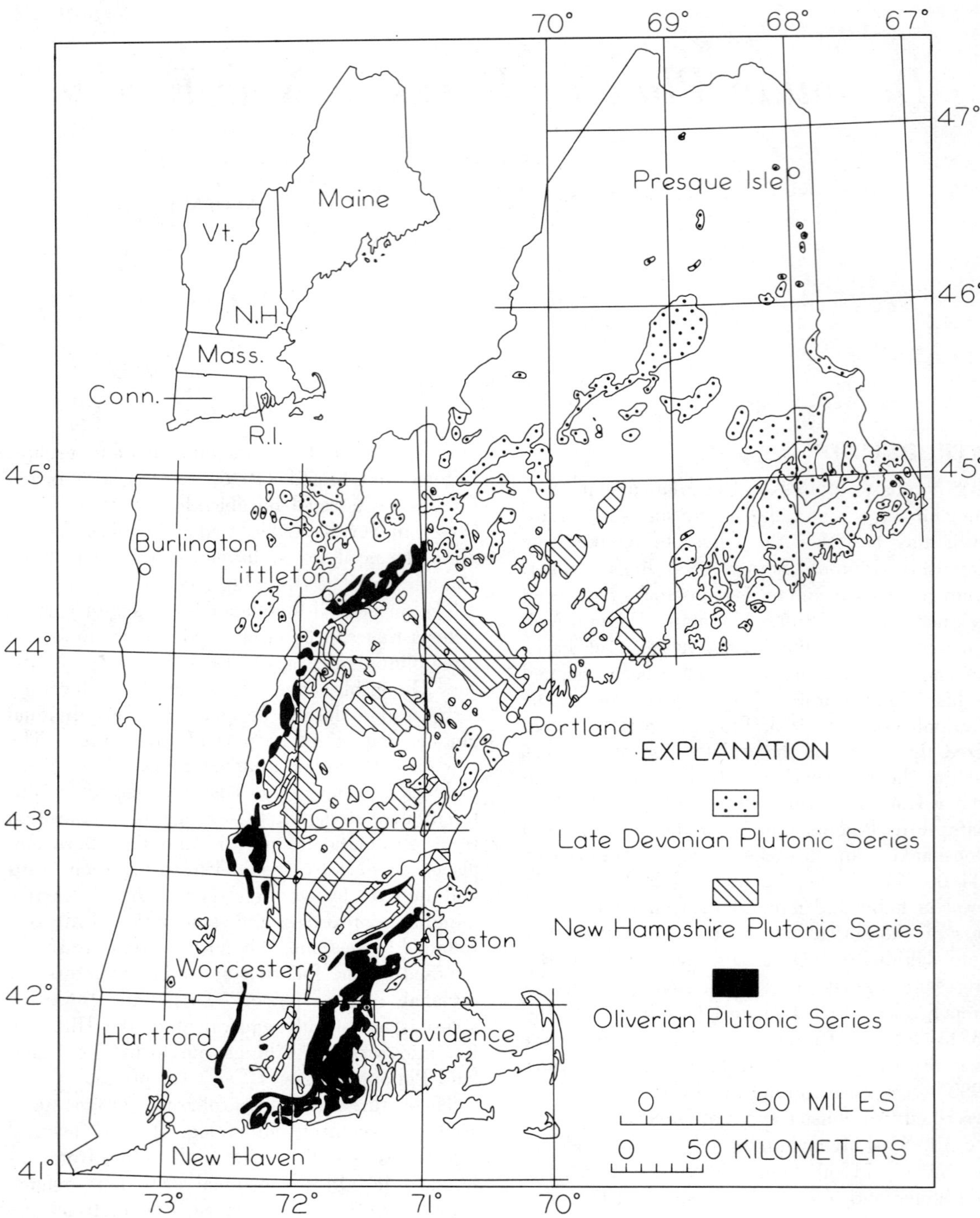

FIGURE 28–1. Distribution of Devonian plutonic rocks in New England. Outline of plutons based on compilations by Goldsmith (1964) and Doyle (1967).

tonic (Oliverian), syntectonic and early post-tectonic (New Hampshire), and late posttectonic (Late Devonian), as well as their distinction from pre-Silurian (Highlandcroft and Precambrian) and post-Devonian (White Mountain, Pennsylvanian, and Triassic) groups.

Newly discovered ore bodies in Maine and New Brunswick, which are associated with Upper Devonian igneous rocks, have indicated the economic importance of recognizing these subdivisions correctly. Massive sulfide deposits spatially associated with members of the Oliverian plutonic rocks have been known for years; many of the commercial pegmatites of New England, but no metallic sulfide deposits, are adjacent to rocks of the New Hampshire plutons; and base metals, precious metals, tin, molybdenum, tungsten, nickel, and cobalt deposits are related spatially to members of the Late Devonian Plutonic Series. Copper, silver, uranium, barium, and base metals are associated with rocks of Triassic age.

The writer has not made detailed investigations and systematic personal examination of all the plutonic bodies in New England and has had to depend in large part on the work of other (Billings, 1956; Doll et al., 1961; Goldsmith, 1964; Doyle, 1967) in preparing Figure 28–1. The writer, however, must accept responsibility for all revisions of their published data. More detailed study in the field will undoubtedly result in the separation of some of the larger bodies into units belonging to two or more plutonic series and the reassignment of others to different groups.

ACKNOWLEDGMENTS

The opinions expressed in this paper are based on detailed geologic mapping by the writer in the Rumney quadrangle, New Hampshire, from 1933 to 1937 (Page, 1937, 1939a and b, 1940a and b; Fowler-Billings and Page, 1942), mapping and study of pegmatite deposits in Maine, New Hampshire, and Connecticut in 1942 (Cameron et al., 1954), and the writer's observations during supervision of the regional mapping program of the U. S. Geological Survey in New England from 1960 to 1967. During this time field observations, including radiometric studies, have been made on representative parts of most of the intrusive bodies shown on Figure 28–1 in Connecticut, Massachusetts, Rhode Island, central and southeastern New Hampshire, and western, northern, and eastern Maine; detailed investigations of many of these bodies have been made under his supervision. In addition, numerous other plutonic bodies have been visited in Vermont, Quebec, New Brunswick, and Nova Scotia.

The writer is deeply grateful to his colleagues in the U. S. Geological Survey; to members of the Maine, Vermont, and Connecticut Geological Surveys; to the numerous graduate students at Harvard University working in New Hampshire, Vermont, and Maine under Professors M. P. Billings and J. B. Thompson; to those at Yale University working in Connecticut under Professor John Rodgers; to Professor James W. Skehan, S.J., of Boston College; to students from the Massachusetts Institute of Technology working in Maine under Professors A. J. Boucot and Ely Mencher; to geologists of the Quebec and New Brunswick Departments of Natural Resources; and to members of the Geological Survey of Canada who have contributed greatly through discussion and through the thoroughness of their detailed mapping studies.

VOLCANISM ASSOCIATED WITH DEVONIAN IGNEOUS ROCKS

The oldest known Devonian igneous rocks, pre-Oliverian in age, are dikes, sills, and flows of mafic to felsic composition that seem to be essentially contemporaneous with the enclosing Lower Devonian rocks dated by fossils. Such units were folded and metamorphosed during the Acadian orogeny. In the early work in New Hampshire (Billings and Cleaves, 1934; Billings, 1934, 1935; Page, 1937), the Acadian orogeny was considered to have taken place in the Middle and/or Late Devonian. Boucot et al. (1964), however, working in northern Maine, have shown that the metamorphism that took place during this period of orogeny ended in late Early or early Middle Devonian time,[*] because the unmetamorphosed Mapleton Sandstone of Middle Devonian age rests with marked angular unconformity on Lower Devonian slates. Similar relationships are shown by the Trout Valley Formation of late Early Devonian age and the older quartz latite of the Traveler Mountain quadrangle, Maine (46°00′–46°15′ lat., 68°45′–69°00′ long.) (Dorf and Rankin, 1962; Rankin, *in* Caldwell, 1966; Neuman and Rankin, *in* Caldwell, 1966; Rankin, this volume), of Early Devonian age. Volcanic rocks are present in the unmetamorphosed Perry Formation of Late Devonian age of southeastern Maine (45°00′ lat., 67°07½′ long.), (Doyle, 1967). Thus, the Acadian was both preceded and followed by volcanism.

---

[*] An oral communication from A. J. Boucot after this manuscript was prepared indicates Middle Devonian time as the end of the Acadian orogeny. See also Boucot, this volume.

The oldest Lower Devonian effusive rocks and the related dikes and sills are most abundant in terranes within a few miles of the Oliverian rocks in west-central New Hampshire and were domed by the Oliverian rocks. The younger Lower Devonian latite effusives in Maine seem to have a close spatial relationship to rocks of the Late Devonian Plutonic Series and form a zone along the northern side of the plutonic rocks (see Rankin, this volume). However, the quartz monzonite of Mt. Katahdin, part of the Late Devonian Plutonic Series, cuts these latites in the Traveler Mountain area, and the general lack of mafic units and the dominant latitic composition of these volcanic rocks suggest a closer relationship in source to the New Hampshire Plutonic Series. Basalt flows and other volcanic rocks, interlayered with Upper Devonian sedimentary rocks of the Perry Formation in southeastern Maine, represent the youngest dated Devonian volcanism and are probably related to the Late Devonian Plutonic Series. The ring-dike structures described by Chapman (this volume) along the Maine coast are probably calderas of this generation of volcanism. Thus, each period of plutonism was preceded or accompanied by volcanism.

AGE OF LATE DEVONIAN PLUTONIC SERIES

Stratigraphic Age

The age of the Late Devonian Plutonic Series is difficult to determine precisely on the basis of geological data. There is ample evidence that it is post-Acadian and post-regional metamorphism because the contact-metamorphic aureoles around the intrusive bodies are superimposed on, and in places obliterate, regional foliation. Inclusions of foliated metamorphic rocks in part have remained essentially unchanged and in part are changed to hornfels, and folds in the wall rock have been truncated or refolded. In eastern Maine and New Brunswick Mississippian rocks lie unconformably on these plutonic rocks. Fossiliferous conglomerate of Late Devonian age in southeastern Newfoundland is cut by a granite (Poole et al., 1964) considered to be a member of this new plutonic series. Thus, the geologic evidence suggests that the age of these intrusive rocks is Late Devonian.

Radiometric Age

Radiometric ages of granites and other igneous rocks in New England show a wide range in values for rocks considered Devonian on the basis of field evidence (Lyons and Faul, this volume; Naylor, this volume), and much more work is needed in order to understand the distribution of values reported to date. Many of the values appear inconsistent with the time scale and the ages determined by geological relationships. Nevertheless, isotopic dates have aided greatly in unraveling the plutonic history of New England. As these are reviewed at length elsewhere in this volume, only two selected ages will be mentioned, one near the north and one near the south end of the Upper Devonian plutonic terrane. The radiometric age of the quartz monzonite of Mt. Katahdin that cuts Lower Devonian rocks in the Traveler Mountain area, Maine, is given as 358 m.y. by Griscom (*in* Caldwell, 1966) (See also Faul et al., 1963, p. 8, who give 356 and 361 m.y.; Bottino et al., 1965; Hurley et al., 1959.) A 345 ± 15 m.y. age is given for Rb^{87}/Sr^{86} whole rock, to the so-called "Ayer" granite[*] by Zartman et al. (1965), in Massachusetts at the southeast end of the area in which these rocks are recognized. Thus, it would seem that the Late Devonian Plutonic Series is at least in part Late Devonian in age.

OLIVERIAN PLUTONIC SERIES

Textural and Structural Characteristics

Field relations, internal structure, and mineralogy indicate that the Oliverian Plutonic Series was intruded into unmetamorphosed Ordovician or Cambrian rocks, under a different set of conditions than any of the earlier or later igneous rocks. These differences (Table 28–1) can be used to determine its mode of intrusion, type of magmatic differentiation, tectonic history, and relative age.

In the type locality (Oliverian Brook, Moosilauke quadrangle, New Hampshire) and adjoining areas, the Oliverian plutonic rocks have features observable in the field that should serve as criteria for recognizing the same rocks in other areas. These features are: (1) internal primary igneous foliation, lineation, and structures indicating a domal character of the pluton; (2) bedding of the wall rocks paralleling the internal primary igneous foliation; (3) pervasive secondary cataclastic foliation and lineation paralleling regional foliation; it lies at angles to the primary igneous foliation and lineation at the

---

[*] This sample from the Worcester, Massachusetts area is called "Ayer" granite but comes from a body 0.25 mi. (0.4 km) south of the so-called Worcester coal mines and is quite distinct from bodies called Ayer (New Hampshire Plutonic Series) by most investigators.

north and south ends of the bodies, but is parallel in strike and, in places, in dip at the sides of the bodies; (4) mafic outer rock units that grade inward into more felsic units with dikes and apophyses of felsic units cutting more mafic units; (5) sills of various compositions interlayered with the mafic and felsic gneisses and schists of the wall rocks; (6) small pegmatites, aplites, and quartz veins in the intrusive rock and in the wall rocks, characterized by cataclastic textures and structures; and (7) distinctive buff to pink microcline and yellowish to gray plagioclase in the most felsic rocks and gray to greenish feldspar in the more mafic units.

TABLE 28-1. Significant differences in the three Devonian Plutonic Series

	Oliverian Plutonic Series	New Hampshire Plutonic Series		Late Devonian Plutonic Series
	Pretectonic	Syntectonic	Early posttectonic	Late posttectonic
Mode of emplacement	Forceful injection—as domes	Forceful injection—as sill-like bodies	Stoping—as irregular stocks	Forceful injection and stoping—as sill-like and cylindrical stocks
Emplaced in	Sedimentary and volcanic rocks	→– – – – – – – Metamorphic and igneous rocks – – – – – – –→		
Differentiated	In situ	At depth	At depth	At depth
Relation to wall rocks	Concordant with bedding	Concordant with foliation and in part with bedding	Discordant with foliation and bedding	In part concordant with walls, in part discordant
Metamorphic effect on wall rocks	None observed	Progressive regional upgrading	Local retrogressive aureoles	Local progressive upgrading contact-metamorphic aureoles
Primary internal structure	Gneissoid	Strikingly gneissoid	Locally gneissoid	Banded or locally gneissoid
Primary lineation	Outward plunging	Parallel to fold axes	None	None or local
Regional metamorphic structure	Gneissic	None*	None	None
Regional lineation	Parallel to fold axes	Parallel to axes of folds in both bedding and foliation	None or locally parallel to axes of folded foliation	Locally parallel to axes of folded foliation
Local metamorphic structure	Mylonite streaks	Mylonite streaks	Mylonite streaks	?
Texture	Sugary, blastoporphyritic	Foliated, porphyritic, rarely equigranular	Equigranular and subporphyritic	Equigranular, prophyritic
Color of potassium feldspars	Buff-pink	Gray-white	Gray-white	Flesh-pink
Color of plagioclase feldspars	Greenish-yellowish-white	Gray-white	Gray-white	Chalky white-gray
Characteristic mafic minerals	Biotite, minor hornblende	Biotite	Biotite**	Pyroxenes, hornblende, biotite
Type of associated pegmatites	Gneissic	Crossfoliate, segregation, layered, gneissoid, homogeneous, zoned	Segregation, homogeneous, zoned	Segregation, homogeneous, zoned
Type of associated aplites	Gneissic	Crossfoliate, gneissoid, nonfoliate	Gneissoid and nonfoliate	Gneissoid and nonfoliate
Abundance and size of associated pegmatites	Rare and small	→– – – – – Numerous and large – – – – – –→		Rather rare and small
Associated veins, replacement bodies, magmatic segregation deposits	Granulated quartz veins, massive sulfide veins, and replacement deposits	→– – – Nongranulated quarts veins – – – –→		Nongranulated quartz veins, massive sulfide veins, replacement, and magmatic segregation deposits
Relative radioactivity of comparable rock types	Low	Low	Medium	High

* Primary gneissoid structure developed during crystallization of magma under regional stress.
** Accompanied by muscovite.

These features suggest that the Oliverian magmas were (1) forcefully injected, (2) differentiated in situ, and (3) metamorphosed subsequent to intrusion. The primary foliation parallels the walls of the Oliverian intrusive rocks, reflecting the domal arrangement of the wall rock. The foliation is crossed by the pervasive cataclastic secondary foliation that is parallel to the regional trends of the Acadian folds in the country rock. The early foliation is marked by large grains and aggregates of biotite, feldspar, and quartz, the later one by fine-grained aggregates or crystals that give the rock a sugary appearance. The secondary foliation, though dominantly cataclastic, is also in part the result of recrystallization and is marked by the development of fine-grained oriented biotite. In places, this secondary structure is the most obvious foliation and is accompanied by pronounced linear structures, such as distinctive large blotches of fine-grained biotite flakes. The secondary foliation crosses aplites, pegmatites, and quartz veins with no systematic relation to their contacts. This foliation is difficult to recognize in the sill-like bodies in the country rock. The sills, however, commonly contain the two sizes of biotite, feldspar, and quartz that distinguish them from the enclosing light-colored, commonly layered, finer grained feldspathic gneisses into which they are intruded, and which have only one foliation. The texture of the Oliverian rocks is a composite of relict igneous fabric and the superimposed metamorphic fabrics of both cataclastic and recrystallization origin. Primary feldspars and quartz are granulated and have well-developed mortar structure, and primary biotite is bent and shredded. The plagioclase at the type locality is oligoclase in all rock units, in keeping with the grade of metamorphism of the area. In lower grade country rocks, such as the terrane of the Highlandcroft Plutonic Series, the plagioclase would be albite in contrast to the more varied plagioclase compositions noted in the various rock units of the postmetamorphic Late Devonian Plutonic Series.

The recrystallized secondary minerals are normally much finer grained than the primary minerals. Pyroxene is altered to hornblende, hornblende to biotite, and plagioclase to epidote, carbonate, and less calcic plagioclase. Tourmaline in pegmatites is crushed and streaked out in patches; it is only partially recrystallized. The felsic gneisses of the wall rocks in many places are volcanoclastic rocks of various original grain sizes and compositions and do not show the cataclastic structures because the metamorphism resulted in almost complete recrystallization of the clastic grains. These wall-rock gneisses are layered and may have a continuous foliation, whereas the igneous rocks have a discontinuous foliation. Some original large clasts in the wall-rock gneisses, particularly feldspar and quartz, however, have been crushed and flattened during regional metamorphism. Such grains are in a matrix of recrystallized minerals of different origin than the larger crushed grains. Thus, these wall-rock gneisses can be separated from those of intrusive origin.

In many places in New England near large faults of Permian-Triassic age, the Oliverian rocks may have a third foliation. This foliation is restricted to narrow zones, in places associated with mylonite and pseudotachylite, and is developed across the primary igneous and the Acadian secondary foliation. Where this late cataclastic structure cuts the wall-rock gneisses, it may be difficult to separate them from Oliverian rocks except by the presence of layering, slight differences in composition, the local nature of the last foliation, and crosscutting relationships at the contact.

In pre-Devonian intrusive rocks, the secondary structures and minerals are also present, but these rocks lack the pronounced primary structures of the Oliverian plutonic rocks and are intruded across the strike of the wall-rock formations.

The major Oliverian plutons are composite bodies. The older units are of dioritic composition, and grade from the walls inward into granodiorite, monzonite, and granite. Associated sills of various compositions are found in the wall rocks, and apophyses of younger units cut older ones. The Oliverian rocks were both porphyritic and equigranular, but as a result of metamorphism they have become blastoporphyritic or sugary in texture. Few bodies of gabbro seem to belong to this group, although the amphibolitic border zone of the Lebanon, New Hampshire, pluton may represent a gabbroic border zone (Lyons, 1955).

Regional Relations

Rocks of the Oliverian Plutonic Series form the core of the Bronson Hill anticline along the Connecticut River in western New Hampshire. They extend northeastward into Maine and southward into Massachusetts and Connecticut (Fig. 28–1). Other plutons of this type occur in eastern Massachusetts and Rhode Island, but much confusion exists as to which rocks belong to this group. A few workers have recognized and mapped the two foliations in these igneous rocks; others tend to equate any light-colored gneiss in an anticline with an Oliverian pluton. Many of these light-colored gneisses were originally felsic volcanics.

The radiometric ages determined for these rocks (Naylor, this volume; Lyons and Faul, this volume) suggest that this group of rocks may be pre-Silurian in age. Older Ordovician plutonic rocks (Highlandcroft), however, characteristically contain blue quartz (probably the result of higher titanium content), in most places lack the relict igneous foliation, and crosscut the regional trend of other Ordovician rocks. The doming of Lower Devonian rocks and other geologic considerations very strongly suggest that Oliverian plutons are post-Lower Devonian and pre-Acadian, despite the radiometric ages.

NEW HAMPSHIRE PLUTONIC SERIES

Textural and Structural Characteristics

The New Hampshire Plutonic Series for the most part was intruded and crystallized under Acadian orogenic stresses; a few plutons crystallized soon after these stresses were released. The older members of this series form elongate, gneissoid* bodies parallel to Acadian folds or faults, wrap around domal masses of the Oliverian Series, and were injected along regional foliation in many places; these are syntectonic. The younger members of this series grade from foliated to nonfoliated rock and in part are in elongate masses and in part in irregular bodies that crosscut the Acadian structures; these younger members are clearly posttectonic. Unlike the Oliverian plutons that differentiated in situ, these were differentiated at depth and injected as separate bodies that crystallized with a relatively uniform composition. They are gray or white rocks, as distinct from the Oliverian and Late Devonian Plutonic Series. On the map (Fig. 28–1), the various elongate bodies of the New Hampshire plutons form a flattened oval, elongated to the northeast, with ends in Maine and Connecticut. Two large, roughly circular masses in east-central New Hampshire and in west-central Maine, which may mark the center of igneous activity, are in the center of the oval.

The syntectonic members of this group are strikingly foliated and as a result have been considered metamorphic rocks by some investigators. However, the criteria listed below indicate that

the members of this series are igneous rocks crystallized under regional stress and are gneissoid rather than gneissic in structure; the outer edges of individual bodies and the older units of this series have the most pronounced foliation. This primary foliation parallels in detail the irregularities of the contact inward for a few feet before assuming a regional trend parallel to the Acadian folds and regional foliation.

The criteria that allow recognition of the syntectonic members of this group (Table 28–1) are: (1) occurrence in elongate bodies that reflect the shapes of the main Acadian folds and Oliverian plutons; (2) presence of well-developed gneissoid structure that in detail parallels the walls of the body but in general parallels the regional trend of Acadian foliation; (3) typical igneous minerals and textures, grain size increasing inward from the contacts; (4) bent or fractured biotite and potash feldspar sealed by groundmass minerals having igneous textures; (5) apophyses and sill-like offshoots of younger units into older units; (6) absence of systematic variation in composition of bodies from the walls inward; (7) presence in the plutons of aplite and pegmatite of two ages, the earliest being segregation and crossfoliate types that have a distinctive primary foliation parallel that of the host, the younger ones having only normal gneissoid flow banding or zoning; (8) mineral lineation and crinkling or minor folds in the gneissoid structure paralleling similar structures in the wall rocks; (9) the wall rocks, in most places, being regional metamorphic rocks of at least staurolite grade; (10) general absence of metamorphic minerals and foliations in the plutons, except in local zones of wall-rock assimilation, younger shearing, faulting, or intrusion; (11) presence of potassium feldspar of two generations: lenticular phenocrysts in early more foliate units and parts of units, and euhedral, but elongate, crystals in later units, embayed by myrmekite and groundmass minerals, including late potassium feldspar (Page, 1939b), and (12) presence of discoidal and lenticular inclusions, in which foliation and lineation parallel that of primary gneissoid structure.

The above criteria clearly show that the foliated rocks of the New Hampshire Plutonic Series have a distinctive mode of intrusion, mode of differentiation, and mineralogy, texture, and structure that allow the units of this series to be separated from other pre- and posttectonic rocks. Many of these features indicate that these rocks were forcefully injected after development of the Oliverian domes, the main Acadian folds, and the

---

* The term gneissoid is used to distinguish primary foliation of rocks with igneous textures, formed by crystallization of magma under stress, from the secondary foliation or gneissic structure of rocks with metamorphic textures, formed by recrystallization of rock under stress.

early regional metamorphic structures, and yet the rocks crystallized under the same general regional stresses. The gneissoid structure in the plutons and in the early crosscutting pegmatites parallels the regional metamorphic structures and indicates that the early units of this series crystallized under Acadian stresses and therefore are syntectonic.

The New Hampshire Plutonic Series was differentiated at depth into separate magma types, which were intruded and cooled without appreciable differentiation in situ. There are separate intrusions of gabbroic, dioritic, quartz dioritic, granodioritic, monzonitic, quartz monzonitic, or granitic composition, dikes and sills of the more felsic units cutting the more mafic units. Pegmatites are very abundant as small stocks, dikes, and sills of a variety of compositions and internal structure. In the major units there was minor differentiation in place, because younger felsic rocks are found as irregular segregations and dikes in the more mafic rocks. The variations in the plutons are rarely as systematic as are similar variations in the Oliverian rocks.

Detailed examination of the contact of the Bethlehem Gneiss and the Kinsman Quartz Monzonite in New Hampshire and equivalent rocks in Massachusetts and Connecticut shows that grain size decreases near the contact and the gneissoid structure becomes coarsely schistose in places. The foliation closely parallels all irregularities of the contact, but within inches, or at most a few feet, it becomes parallel to the long axis of the body itself and the main fold system. This foliation must have been developed while the Acadian folding stresses were still active, or it would not have formed across the early aplites and pegmatites. Flowage of partly crystalline magma under considerable stress is indicated by the fracturing and offsetting of the earlier feldspar and the bending of biotite without crushing the later feldspar and quartz of the groundmass or offsetting adjacent contacts. Lenticular potassium feldspar grains, not aggregates, and crystals of feldspar wrapped in biotite also are evidence of crystallization under stress. The phenocrysts may contain inclusions of groundmass minerals that outline crystal boundaries rather than earlier foliation, as would be the case if they were metacrysts (Page, 1939b). The gneissoid foliation is so well developed and persistent that it has been considered a secondary regional metamorphic structure by some investigators, and as a result, in places, these rocks have been correlated with felsic gneisses of volcanic origin.

The New Hampshire Plutonic Series rocks are various shades of black, gray, or white; pink members, as in the Oliverian Plutonic Series, are unknown, although in parts of what has been mapped as the Fitchburg pluton and in some associated pegmatites, the potash feldspar may have a buff to pink color. Hornblende is rare; the main mafic mineral is biotite. Large crystals or lenses of potash feldspar are commonly the most distinctive mineralogical feature. These phenocrysts are early, the crystals were aligned and fractured by flowage, and the lenses owe their shape to growth under stress, not cataclasis. Sphene and garnet are abundant locally.

Gray to white biotite and biotite-muscovite granite dikes, stocks, and elongate bodies cut the strongly foliated, syntectonic members of this series. These are syntectonic to posttectonic granites identified by: (1) close spatial relationship to syntectonic bodies; (2) obscure foliation parallel to their walls; (3) emplacement by stoping and intrusion along fractures, or along zones of weakness paralleling Acadian folds; (4) retrograde metamorphism in the wall rocks at their contacts; (5) uniform fine- to medium-grained texture; and (6) the abundance of pegmatites both in the granite and in the surrounding wall rocks.

These late granitic rocks are similar to the gray granites of the Late Devonian Plutonic Series but differ in that they show little evidence of forceful injection, they lower rather than raise the metamorphic grade of the adjacent country rock, and they are associated with abundant pegmatites.

Pegmatites

Both variety and abundance of pegmatites characterize terranes of the New Hampshire Plutonic Series. The major pegmatite districts of New England are, for the most part, related to this series, and the composition, internal structure, and spatial relationships of the pegmatites are significant in understanding the magmatic differentiation history of these rocks. The pegmatites in this series are of many different ages, as determined by crosscutting relationship of minor bodies; each has specific characteristics of structure, texture, and mineralogy, but all are white, gray, or buff. In order of decreasing age in any one pluton, these are (1) cross-foliated pegmatites; (2) segregation pegmatites; (3) layered pegmatites; (4) gneissoid pegmatites; (5) homogeneous pegmatites; and (6) zoned pegmatites (Page et al., 1953; Cameron et al., 1949).

The cross-foliated pegmatites and aplites are rare and are mostly within the gneissoid members of the plutonic rocks. A few occur in schistose or gneissic country rock, and both varieties have a foliation identical to that of the host. The foliation, like the gneissoid structure of the host, was formed by crystallization of magma under stress, not by reorientation of crystallized grains as in the case of Oliverian pegmatites. In a few places, flow banding and zoning can also be recognized in the same body.

Segregation pegmatites in the early units of the New Hampshire Plutonic Series may be foliated like the cross-foliated ones, but the segregation pegmatites of the younger, equigranular granites lack a foliation and are commonly homogeneous or zoned. These are biotite, muscovite, potassium feldspar, plagioclase, and quartz-rich rocks that occasionally contain tourmaline and garnet.

The layered pegmatites have a very heterogeneous composition. Many contain units of sugary grained, banded, feldspar-garnet rock that is cut by dikes, sills, and irregular masses of gneissoid, homogeneous, or zoned pegmatite. Others are composed of alternate bands or zones of pegmatitic material that are repeated in an unsystematic manner. These composite bodies are commonly large enough to be mappable, and if the feldspathic layers are large enough, they may be an economic source of feldspar. They are similar in composition to the gneissoid pegmatites and rarely contain beryl, sheet muscovite, and lithium minerals.

The gneissoid pegmatites both crosscut and parallel structures in the host rock but are most common in the wall rocks near gneissoid diorite, quartz diorite, and quartz monzonite. They form abundant stringers, dikes, and sill-like bodies that follow the foliation of the host and are commonly confused with coarse-grained felsic layers of gneisses of other origins. The gneissoid structure formed parallel to the contacts during flowage and crystallization of magma under stress, so the mineral grains tend to be lenticular, but all combinations of lenses and crystals occur. In some bodies, part of the micas, feldspars, or quartz grew at right angles to the walls and later were enclosed in the same minerals that formed parallel to the walls during flowage of the magma. In other pegmatites, segregations of unoriented minerals occur at random or in central locations in the gneissoid body; thus decreased stress conditions are recorded. In others, gneissoid pegmatite forms the wall zone of zoned pegmatites. The gneissoid pegmatites may be of mappable size, but commonly they are small, numerous, and closely spaced stringers that produce a coarse-grained lit-par-lit gneiss. The lack of obvious chilling at the margins of many of these pegmatites, as in the segregation and cross-foliate varieties, suggests these were intruded during metamorphism. These pegmatites are primarily composed of biotite, muscovite, plagioclase, graphic granite, microcline-perthite, and quartz. Garnet, tourmaline, or magnetite are the only common accessory minerals.

The homogeneous pegmatites in general have a uniform composition and texture, but in places they may grade into gneissoid or zoned pegmatites. Commonly, a thin border zone in the homogeneous pegmatites is a fine-grained unit of muscovite, biotite, feldspar, and quartz, in which the crystals are oriented at right angles to the walls; the center of the pegmatite shows haphazard orientation of coarser grains of these same minerals. Some of these pegmatites have gneissoid outer units that crystallized under intrusive or other stress; the non-gneissoid central parts formed under static conditions.

Zoned pegmatites grade into the gneissoid and homogenous varieties and are by far the least numerous of pegmatites. They have been described in detail by many authors (Cameron et al., 1949) because they are the sources of most economically important pegmatite minerals and are the most complex mineralogically. Their zonal structure is the result of magmatic differentiation in situ under static conditions. Some, however, have outer gneissoid zones formed by flowage stresses. Zoned pegmatites, like the other varieties, were crystallized from magmas injected under intrusive stress, as indicated by the development of induced foliation at their edges and other structures caused by flowage of viscous liquids (Cameron et al., 1949).

All the above types of pegmatites, except the segregation pegmatites, can be concordant bodies in both the gneissoid rocks of the New Hampshire Plutonic Series and their foliated wall rocks; less commonly, they are crosscutting. In the late equigranular granites, however, only the segregation, homogeneous, and zoned varieties occur, and crosscutting bodies are more numerous than concordant ones in their wall rocks.

The close spatial relationship of large zoned bodies and the gneissoid granodiorites strongly suggests that these potassium-rich pegmatites represent a fractionation of the more potassic parts of the magma by filter pressing potassium-rich material into the wall rocks; thus by loss of material, an original quartz-monzonitic magma would be converted to granodiorite.

Regional Relations

The more elongate bodies of the New Hampshire Plutonic Series are at the outer edges of the oval-shaped area of distribution; in the type area in New Hampshire, in order of decreasing age, they are the Bethlehem Gneiss, the Kinsman Quartz Monzonite, and the Concord Granite. Less elongate bodies occur in the central part of the oval in central New Hampshire and west-central Maine. Most of the rocks of the New Hampshire Plutonic Series in New England are in or close to sillimanite-grade rocks; a few are in staurolite-grade rocks (See also Thompson and Norton, in this volume). Many of the oldest elongate bodies were intruded at or near the base of the Devonian rocks; the younger plutons are higher stratigraphically. The wall rocks may also be Silurian and Ordovician rocks. This is distinct from the Oliverian plutons that most commonly are intruded into Ordovician felsic gneisses; in Connecticut and Rhode Island the latter also intrude Cambrian(?) rocks.

The distribution of isograds of regional metamorphism, suggests that the central part of the batholith of the New Hampshire plutons was in the central New Hampshire and western Maine, in about the center of the Merrimack synclinorium. This is so, because the elongate, strongly gneissoid bodies are at the edge of the area of high metamorphic grade, whereas the less gneissoid, more circular masses are in the center of this area. Increase in volume, caused by successive intrusions of magma from depth and by increased temperature, produced an elongate dome or blisterlike central batholith area. Satellitic sill-like bodies of New Hampshire plutons extend outward to the zone of Oliverian domes along the Bronson Hill anticline on the west and to the Oliverian domes of Rhode Island and eastern Massachusetts on the east and south. This blisterlike batholithic area within the Merrimack synclinorium had the effect of vertically uplifting the center of the synclinorium and was the locus of the highest grade metamorphic rocks. Subsequently, at the end of Paleozoic time and during post-Triassic time, major thrust faults cut across the southern part of the Merrimack synclinorium (K. G. Bell, 1967, written communication) and cut the west limb of the Bronson Hill anticline, disrupting the original zonal pattern of the regional metamorphic isograds in these areas, producing what seem to be very steep gradients. Thus, the original relationship of the New Hampshire Plutonic Series and grade of metamorphism is obscure in places.

The minimum age of the New Hampshire Plutonic Series cannot be obtained by direct geologic relationships because rocks of the Series do not extend into areas of Middle or Upper Devonian rocks. Therefore the age is based on the fact that the earliest rocks of the New Hampshire Plutonic Series were intruded along Acadian folds and metamorphic foliation, and that the intrusive rocks also contain crosscutting aplites and cross-foliated pegmatites in which the primary igneous foliation is parallel to the igneous foliation in the wall rocks. Also, in places, the sedimentary wall rocks have been metamorphosed to sillimanite schists in which the sillimanite is oriented parallel to the mineral lineation in the plutons. This indicates that such schists formed under the same stresses as the plutons after the start of the main period of folding; thus the plutons are truly syntectonic. The younger members of the series form a group varying from syntectonic to posttectonic rocks that downgrade sillimanite to muscovite and that are intruded by stoping into relatively brittle metamorphic rocks.

Radiometric ages on rocks of the New Hampshire Plutonic Series are conflicting and need more careful analysis before the correct ages can be designated (See Lyons and Faul, this volume.) Late Paleozoic faulting and other events have put a 250 m.y. overprint on many of these rocks and their associated pegmatites as well as on many other Devonian rocks.

LATE DEVONIAN PLUTONIC SERIES

Textural and Structural Characteristics

Many of the plutonic rocks in New England cut Lower Devonian rocks but do not fit the criteria for the Oliverian and New Hampshire Plutonic Series. Similar Devonian rocks in New Brunswick (Potter, 1967, p. 6–7) are overlain unconformably by Mississippian rocks. These intrusives which are placed in a new series, herein referred to as the Late Devonian Plutonic Series, form a complete sequence of ultramafic rocks, gabbros, norites, granodiorites, monzonites, quartz monzonites, and granites. They all are similar in that: (1) they occur in separate intrusive bodies that show little magmatic differentiation in situ; (2) they are equigranular rocks having layering in the mafic varieties but having little igneous foliation in the felsic units; (3) they all are enclosed in hornfels or other metamorphic aureoles that have been superimposed on regionally metamorphosed wall rocks; (4)

they shoulder aside and refold the walls in part, but elsewhere contain abundant inclusions of regionally metamorphic and older igneous rocks from their walls; (5) they rarely contain large or abundant pegmatites or include pegmatite intruded into the wall rocks; where present, the pegmatites commonly contain flesh-pink potassium feldspar; (6) the granodiorites, monzonites, and granites commonly contain hornblende, flesh-pink potassium feldspar, and chalk-white plagioclase, in part as mantled grains or rapakivi structures; in a few places the potassium feldspar is gray or white; (7) they are in general more radioactive than are Oliverian or New Hampshire plutonic rocks; and (8) they commonly are spatially associated with metallic mineral deposits.

The lack of metamorphic features readily distinguishes both the mafic and felsic members of this series from those in the Oliverian and older plutons; the absence of well-developed igneous foliation separates this group from the syntectonic rocks of the New Hampshire Plutonic Series. The presence of flesh-pink potassium feldspar, hornblende, and chalk-white plagioclase in the felsic units readily distinguishes most of these felsic rocks from the early posttectonic units of the New Hampshire Plutonic Series. The gray felsic rocks of the Late Devonian Plutonic Series present a much more difficult problem, however, for in hand specimen, they resemble the Concord Granite and related rocks of the New Hampshire Plutonic Series. Here, the field relations are important; individual bodies that contain both gray and pink rocks belong to the new series. Gray rocks that raise, or maintain, the metamorphic grade of the enclosing rocks belong to the new series; those that downgrade their wall rocks belong to the New Hampshire Plutonic Series.

More difficulty will be found in separating the rock of the new Series from those related to the White Mountain Plutonic–Volcanic Series of Early Jurassic(?) or Late Triassic age. However, it seems possible to identify members of the White Mountain Plutonic–Volcanic Series (1) on the basis of the alkalic composition; (2) by the lack of significant contact-metamorphic effects in the wall rocks; (3) by differences in texture and other igneous structures; and (4) by their much higher radioactivity and thorium content.

The distinctive features of this group of rocks indicate that the Late Devonian Plutonic Series: (1) was emplaced in previously regionally metamorphosed rocks that were deformed both by flow and fracture; (2) was differentiated at depth and intruded as separate bodies in a complete series from ultramafic to felsic; and (3) is a source of metallic ores.

The hornfels around an individual body may be miles wide. The hornfels grades from high-grade rocks, which contain sillimanite, andalusite, and cordierite near the intrusive body, outward to merge with the regional metamorphic rocks. Acadian foliation is destroyed in the inner parts of the contact zone. Where the pluton was intruded into rocks of garnet and staurolite grade, the high-temperature inner part of the hornfels may be quite narrow and not mappable as such; intrusions into the sillimanite zone would leave the rocks unchanged at the contact.

Where rocks of this series have been emplaced by shouldering aside parts of their walls, previous fold axes have been folded, earlier linear structures distorted, and new ones developed; slip or fracture cleavage cuts the earlier Acadian flow cleavage, and fractures and faults were formed and in places were filled by the intruding rock. All these structures indicate expansion upward and outward. Commonly, only one side of the body shows good evidence of plastic deformation of the wall rocks; the other is crosscutting, and along this part the inclusions are so numerous that they form an injection breccia in places. The inclusions are recognizable metamorphic wall rocks, which have haphazard orientation of the foliation as distinct from the rare inclusions in the Oliverian and more numerous ones in the syntectonic New Hampshire plutonic rocks in which they are well oriented parallel to the primary foliation in the igneous rock.

Pegmatites

The scarcity of pegmatites of large size associated with rocks of the Late Devonian Plutonic Series distinguishes this group from rocks of the New Hampshire Plutonic Series. The flesh-pink feldspar of the few pegmatites present is another distinguishing feature and contrasts with the granulated buff-pink feldspars of the Oliverian pegmatites.

Radioactivity and Mineral Deposits

The overall radioactivity of Devonian intrusive rocks varies widely, primarily with a variation in potassium content, in feldspars and micas, and less sharply with uranium and thorium content (see Birch et al., this volume). Rocks of the Late Devonian Plutonic Series seem to have a somewhat higher overall radioactivity than do rocks of the other series, though this radioactivity is much lower than that of the White Mountain Plutonic–Volcanic Series. With care, all these rocks can be separated on this basis (see Billings and Keevil, 1946).

Metallic mineral deposits have been noted near the Upper Devonian plutonic rocks in Canada as well as in New England. These sulfide deposits replace regionally metamorphosed rocks and themselves are cut by faults of both late Paleozoic and Triassic age. In general, the sulfide deposits have an extremely low radioactivity, as in the Bathurst district, New Brunswick, whereas many mineral deposits of post-Pennsylvania or post-Triassic age in Connecticut and Massachusetts contain more than normal uranium. Sulfides within the ultramafic and mafic rocks show no signs of being affected by regional metamorphism and are clearly related to the enclosing rock. The sulfide ores related to the Upper Devonian plutonic rocks contain Ni, Co, Cu, Zn, Pb, Sn, Mo, W, and precious metals in contrast to those of the Oliverian group which are mainly Cu, Zn, Pb, and W.

Regional Relations

The geologic features of the Late Devonian Plutonic Series that fix its age are the contact-metamorphic effects superimposed on rocks that were regionally metamorphosed during the Acadian orogeny, the unconformable Mississippian rocks, and the presence of unmetamorphosed Triassic(?) dikes. Thus, this plutonic series could be Middle or Late Devonian; however, the Belleoram Granite Pluton, correlated with this series, in the Fortune Bay area, southeastern Newfoundland, cuts dated Upper Devonian conglomerate (Poole et al., 1964). Spatial relations and composition of related Upper Devonian volcanic rocks also favors a Late Devonian age. These geologic data are supported by radiometric ages in part (see Lyons and Faul, this volume), but the data are not clearly interpretable at present.

SUMMARY

In summary, the members of the three Devonian plutonic series in New England are pre-, syn-, and posttectonic (Acadian) in age and can be separated on the basis of internal and external structure, mode and place of magmatic differentiation, and various mineralogical and textural characteristics. Isotopic age determinations give conflicting results but in general substantiate the relative order of age of the three Devonian plutonic series.

REFERENCES

Billings, M. P., 1934, Paleozoic age of rocks of central New Hampshire: Science, n.s., v. 79, no. 2038, p. 55–56

————, 1935, Geology of the Littleton and Moosilauke quadrangle, N. H.: Concord, New Hampshire State Plan. and Devel. Comm., 51 p.

————, 1956, The geology of New Hampshire, Part II, Bedrock geology: Concord, New Hampshire State Planning and Dev. Comm., 203 p.

————, and Cleaves, A. B., 1934, Paleontology of the Littleton area, New Hampshire: Am. Jour. Sci., 5th ser., v. 28, no. 168, p. 412–438

————, and Keevil, N. B., 1964, Petrography and radioactivity of four Paleozoic magma series in New Hampshire: Geol. Soc. America Bull., v. 57, no. 9, p. 797–828

Bottino, M. L., Schnetzler, C. C., and Fullagar, P. D., 1965, Rb–Sr whole rock age of the Traveler and Kineo Rhyolites, Maine, and its bearing on the duration of the Early Devonian [abs.]: Geol. Soc. America Special Paper 87, p. 15

Boucot, A. J., Field, M. T., Fletcher, Raymond, Forbes, W. H., Naylor, R. S., and Pavlides, Louis, 1964, Reconnaissance bedrock geology of the Presque Isle quadrangle, Maine: Maine Geol. Survey Quad. Mapping Ser., no. 2, 123 p.

Cady, W. M., in press, Regional tectonic synthesis of northwestern New England and adjacent Quebec: Geol. Soc. America Spec. Paper.

Caldwell, D. B., ed., 1966, The Mount Katahdin region, Maine: New England Intercollegiate Geol. Conf., 58th Ann. Mtg., guidebook, Shin Pond, Me., 61 p.

Cameron, E. N., Jahns, R. H., McNair, A. H., and Page, L. R., 1949, Internal structure of granitic pegmatites: Econ. Geology, Mon. 2, 115 p.

Cameron, E. N., Larrabee, D. M., McNair, A. H., Page, J. J., Stewart, G. W., and Shainin, V. E., 1954, Pegmatite investigations, 1942–45, in New England: U. S. Geol. Survey Prof. Paper 255, 352 p.

Chapman, C. A., 1939, Geology of the Mascoma quadrangle, New Hampshire: Geol. Soc. America Bull., v. 50, no. 1, p. 127–180

Chapman, C. A., Billings, M. P., and Chapman, R. W., 1944, Petrology and structure of the Oliverian magma series in the Mount Washington quadrangle, New Hampshire: Geol. Soc. America Bull., v. 55, no. 4, p. 497–516

Doll, C. G., Cady, W. M., Thompson, J. B., Jr., and Billings, M. P., compilers and editors, 1961, Centennial geologic map of Vermont: Vermont Geol. Survey, Montpelier, scale 1:250,000

Dorf, Erling, and Rankin, D. W., 1962, Early Devonian plant from the Traveler Mountain area, Maine: Jour. Paleontology, v. 36, no. 5, p. 999–1004

Doyle, R. G., ed., 1967, Preliminary geologic map of Maine: Maine Geol. Survey, Augusta, 1:500,000

Faul, Henry, Stern, T. W., Thomas, H. H., and Elmore, P. L. D., 1963, Ages of intrusion and metamorphism in the northern Appalachians: Am. Jour. Sci., v. 261, p. 1–19

Fowler-Billings, Katharine, and Page, L. R., 1942, The geology of the Cardigan and Rumney quadrangles, New Hampshire: Concord, New Hampshire State Plan. and Devel. Comm., 31 p.

Goldsmith, Richard, 1964, Geologic map of New England. 1. General geology; 2. Metamorphic zones; and 3. Radiometric ages: U. S. Geol. Survey open-file report, 3 sheets, scale 1:1,000,000

Hurley, P. M., Boucot, A. J., Albee, A. L., Faul, H., Pinson, W. H., and Fairbairn, W. H., 1959, Minimum age of the Lower Devonian slate near Jackman, Maine: Geol. Soc. America Bull., v. 70, no. 7, p. 947–950

Kaiser, E. P., 1938, Geology of the Lebanon granite, Hanover, N. H.: Am. Jour. Sci., 5th ser., v. 36, no. 212, p. 107–136

Lyons, J. B., 1955, Geology of the Hanover quadrangle, New Hampshire–Vermont: Geol. Soc. America Bull., v. 66, no. 1, p. 105–146

Page, L. R., 1937, The geology of the Rumney quadrangle, New Hampshire: Ph.D. thesis, Univ. Minnesota, 150 p.

————, 1939a, The geology of the Rumney quadrangle, New Hampshire: Summaries Ph.D. theses, Univ. Minnestoa, v. 1, 5 p.

————, 1939b, Introduction of feldspar into inclusions, Ellsworth, New Hampshire [abs.]: Am. Mineralogist, v. 24, no. 3, p. 190

————, 1940a, Geologic map and structure sections of the Rumney, New Hampshire, quadrangle: Concord, N. H., New Hampshire State Highway Comm.

————, 1940b, Igneous and metamorphic rocks of the Rumney quadrangle, New Hampshire [abs.]: Geol. Soc. America Bull., v. 51, no. 12, p. 1936–1937

Page, L. R., and others, 1953, Pegmatite investigations 1942–45, Black Hills, South Dakota: U. S. Geol. Survey Prof. Paper 247, 229 p.

Poole, W. H., Kelley, D. G., and Neale, E. R. W., 1964, Age and correlation problems in the Appalachian region of Canada, in Geochronology in Canada: Royal Soc. Canada Spec. Pubs. no. 8, p. 61–84

Potter, R. R., 1967, Geologic investigations in New Brunswick: New Brunswick Mines Branch, Inf. Circ. 67–1, p. 37

Zartman, Robert, Snyder, G. L., Stern, T. W., Marvin, R. F., and Bucknam, R. C., 1965, Implications of new radiometric ages in eastern Connecticut and Massachusetts: U. S. Geol. Survey Prof. Paper 525–D, p. D1–D10

A Comparison of the Maine Coastal Plutons and the Magmatic Central Complexes of New Hampshire

CARLETON A. CHAPMAN

INTRODUCTION

A GROUP OF ESSENTIALLY post-orogenic plutons in southeastern Maine shows both similarities with and differences from the well-known Mesozoic plutons of the White Mountain magma series. The disposition, apparent shape, composition, and internal and external structural relations of the plutons in both areas suggest that they formed from discrete, roughly equant bodies of magma. Evidence based on orientation and distribution of feldspar crystals and inclusions of country rock indicates that the felsic as well as the mafic plutons represent largely crystal cumulates which built up by sedimentation above the floors of individual magma chambers. The co-existence of two magmas (granitic and syenitic in New Hampshire; granitic and gabbroic in Maine) in the late stages of development is suggested by the complicated intrusion sequence, the presence of composite dikes, and the heterogeneity of many felsic bodies. These conclusions may explain why intrusion centers in individual central complexes appear to migrate with time, why ring-dikes are preponderantly felsic, why the subsiding block within a ring-fracture is finally arrested, and why the subsidence may be repeated. Major differences in the two areas may be attributed largely to difference in depth of formation and nature and rate of rise of the magma.

I am very much indebted to Marland Billings for introducing me to the concepts of ring-dikes and cauldron subsidence and for stimulating me in the larger study of magmatic central complexes. In my study of the Maine area I have been greatly assisted by the work of many student associates, and to these men I express my sincere gratitude. Although this paper draws upon their observations, it is not necessarily in agreement with their interpretations and conclusions. In addition, I wish to acknowledge with thanks the many helpful comments and suggestions offered by Paul Bateman, Donald Henderson, Arthur Hussey, Walter White, and E-an Zen during the critical reading of the manuscript.

In such a brief comparative account of the Maine and New Hampshire plutons as this, restrictions must be made to apparently outstanding features, relations, and implications. Emphasis will be placed on field data; petrographic details will be covered elsewhere. Descriptive treatment of the New Hampshire magmatic central complexes, which constitute the White Mountain magma (plutonic) series, is intentionally brief to permit a more detailed account of the Maine coastal plutons, much less voluminously represented in the literature. For additional information on the White Mountain magma (or plutonic) series the reader is referred to the regional treatise by Billings (1956), which presents a very fine summary and extensive bibliography on the subject.

MAJOR FEATURES OF THE MAINE COASTAL PLUTONS

Nature and Distribution

Well exposed along the eastern half of the Maine coastal region are numerous granitic intrusions of roughly circular, stock-like form known as the Maine coastal plutons (stippled in Fig. 29–1). Most of these bodies are between 5 and 10 miles (8–16 km) in diameter and appear as discrete masses. Contiguous and nearly contiguous bodies locally form small groups. The plutons cut rocks of the Bays-of-Maine igneous complex as well as older stratified volcanic and sedimentary units (Fig. 29–1). They have been dated as early Late Devonian (Faul et al., 1963).

The Bays-of-Maine igneous complex (random-dash patterns in Fig. 29–1; Chapman, 1962a), which extends for at least 175 miles (280 km) along the Maine and New Brunswick coast, is a mafic-felsic assemblage emplaced largely within and beneath a thick pile of volcanic and sedimentary rocks of Silurian and Devonian age. Along its northwestern flank the complex invades the Ellsworth Schist of low-grade metamorphic rank and pre-Middle Silurian age. The complex was probably emplaced over a long period of time, but the plutonic phases observed appear to have formed during the late stages of the Acadian orogeny. Emplacement probably occurred in a series of pulses, controlled by a steeply inclined and mechanically unstable zone roughly parallel to the tectonic strike.

The complex is divided into two compositional phases: an older gabbroic phase and a younger granitic-granophyric phase. Volumetrically the mafic phase may be many times more abundant

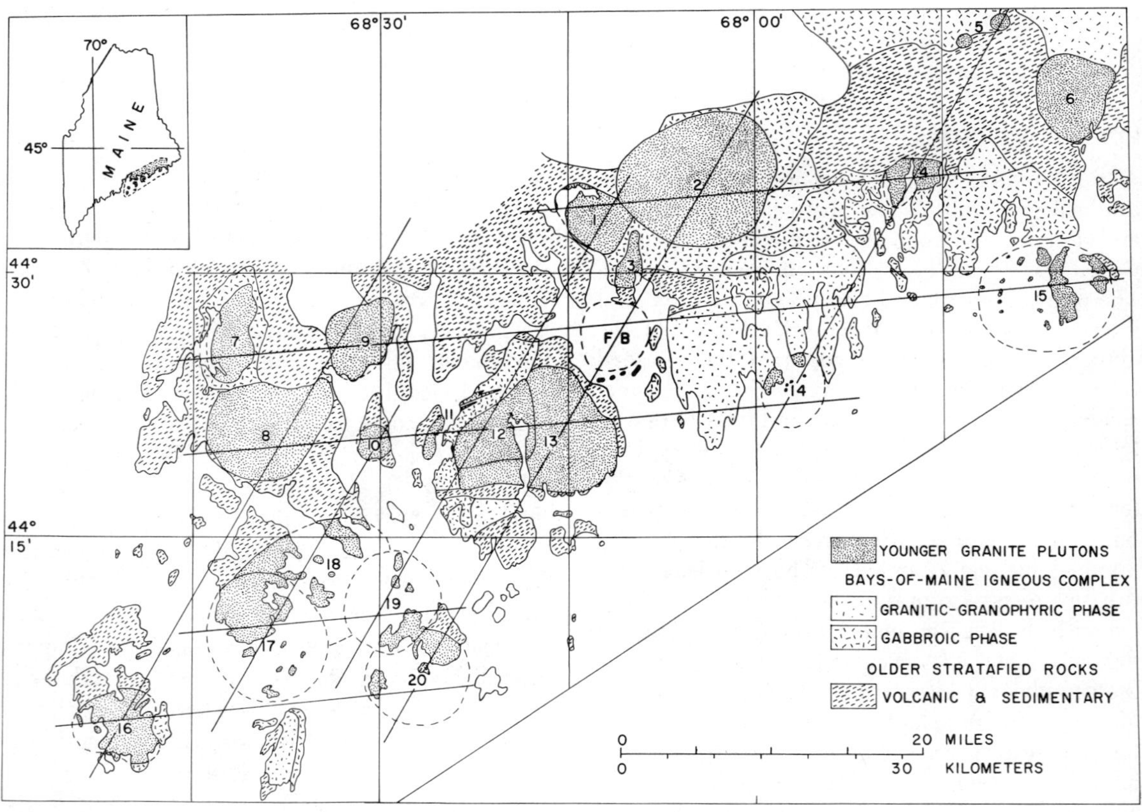

FIGURE 29–1. Distribution of the Maine coastal plutons. 1, North Sullivan pluton; 2, Tunk Lake pluton; 3, Sorrento pluton; 4, Harrington pluton; 5, Centerville intrusions (shape poorly known); 6, Jonesboro pluton; 7, South Penobscot pluton; 8, Sedgwick pluton; 9, East Blue Hill Pluton; 10, Long Island pluton; 11, Bartlett Island ring-dike; 12, Somesville pluton; 13, Cadillac Mountain pluton; 14, Corea pluton; 15, Great Wass pluton, 16, Vinalhaven pluton; 17, Stonington pluton; 18, Oak Point pluton; 19, Swans Island pluton; 20, Minturn pluton; FB, Frenchman Bay structural basin, probably formed by subsidence above an unexhumed granitic pluton. Patternless islands—geology not known. Plutons 4 and/or 6 may not belong to the younger granites. Heavy straight lines—pluton lattice pattern.

than the felsic one. The main part of the complex is visualized as an elongate body of layered rocks with numerous sub-horizontal, sheet-like or tongue-like projections, which formed largely beneath and within the older volcanic and associated sedimentary rocks. Capping most of the main mafic portion of the complex and some of its satellites are sheets composed of felsic rock, which represents the granitic-granophyric phase.

Not only are the Maine coastal plutons younger than the Bays-of-Maine igneous complex, but they appear to have formed in a somewhat different tectonic environment. For this reason the coastal plutons are considered to be separate from the great complex rather than merely a late phase.

As has been pointed out (Chapman, 1963) for other areas of the world, the Maine coastal plutons show reticulate arrangement. This is brought out on the geologic map (Fig. 29–1) where the lattice directions N30°E and N85°E have been superposed. For convenience these two directions may be taken to represent the disposition pattern of these plutons. Precision is not to be implied by the two values chosen, but there is a strong tendency for the bodies to be aligned roughly NNE and E-W. The fit is reasonably good and especially so when we consider that great distances are involved, that many plutons have irregular outline, that many plutons are probably composite, and that large portions of many plutons (particularly the southern ones) are covered by water. Two additional linear directions, roughly N60°E and N32°W, may be noted, but these appear to be less representative. In this net considerably fewer plutons fall on lattice nodes. These directions, however, roughly bisect the lattice angles shown in Figure 29–1 and further substantiate an orderly disposition for the plutons.

Many plutons appear elongate parallel to one or another of the alignments. In the Mount Desert complex (plutons 11, 12, and 13 of Figure 29–1) there appears to have been a migration of intrusion centers with time along the east-west lattice line. The acute bisectrix of the lattice lines (Fig. 29–1) parallels the Appalachian trend and that of the Bays-of-Maine igneous complex; the obtuse bisectrix parallels a prominent group of mafic dikes. These relations are considered in more detail elsewhere (Chapman, 1967).

Contact Relations With Country Rock

The plutons show sharp, smooth contacts and discordant relations against the country rock. In most cases a well-defined metamorphic aureole has developed, up to roughly a mile from the contact, in which a wide variety of hornfelses may be found. Biotite, cordierite, anthophyllite, andalusite, sillimanite, microcline, hornblende, grossularite, stilpnomelane and diopside are some of the more diagnostic minerals.

Near many contacts the country rock is commonly jointed, shattered or brecciated. The Cadillac Mountain pluton, for example is bordered by a severely disturbed zone as much as half a mile in width. The intensity of fracturing increases inward toward the granite and culminates in a mass of thoroughly jumbled breccia. The breccia consists of angular to rounded blocks of hornfels, generally up to a few feet across. A few blocks over 300 feet (100 m) long, however, have been seen. The breccia matrix is formed largely of pulverized and recrystallized country rock, and occasional small dikes of granite cut through the breccia matrix and breccia blocks. One striking feature of the contact zones is the remarkably small number of granitic apophyses. Those observed do not extend far from the general contact.

A second striking feature is the apparent change in dip of the country rock near the granite contacts. In general the country rock appears to be dragged downward so that gently dipping layers have been steepened or have undergone a reversal in dip. This tendency for layers to incline inward toward the pluton's center gives rise to what is here called an external centripetal dip pattern. In steeply dipping or massive country rock such rotational movement is not readily detected.

The most striking example of an external centripetal dip pattern has developed around the Cadillac Mountain pluton on Mount Desert Island (Chapman, 1953, 1962b). Here the pattern extends completely around the pluton. Along the northeast, east, southeast, and south portions, roughly horizontal beds of the Bar Harbor Series dip toward the contact at 20°–70°. Vertical dips have also been observed locally. Along the north side, foliation in the Ellsworth Schist has been shifted so as to dip more directly toward the contact. Along the northwest, west, and southwest sides the layering in the gabbroic phase of the Bays-of-Maine igneous complex is inclined at a low to moderate angle toward the contact.

Far to the southwest on Vinalhaven Island, layering in the gabbroic country rock shows a well-developed external centripetal pattern along the east, southeast, and south sides of the Vinalhaven pluton. Inward dips have been observed locally (Stewart, 1956, and Wingard, 1961) around the Oak Point, East Blue Hill, Sedgwick, Long Island, and South Penobscot plutons.

Composition and Texture

The Maine coastal plutons are composed almost entirely of true granite referred to earlier (Chapman, 1962a) as the "younger granite." Both two-feldspar and one-feldspar granites are represented but two-feldspar granites are far more common. Generally the two-feldspar granites are medium- to coarse-grained, light gray to pink, and allotriomorphic granular. A few are hypidiomorphic granular. These granites are composed essentially of quartz, perthitic microcline, and plagioclase (An_{10-30}). Small amounts of biotite and less commonly hornblende are present. Much of the granite is subporphyritic to porphyritic with subhedral to rounded feldspar, commonly showing rapakivi texture. Fine examples are found in the Great Wass (Terzaghi, 1940), Oak Point (Stewart, 1956; Wingard, 1961), Tunk Lake (Karner, 1963) and Corea plutons.

One-feldspar granite constitutes most of the Cadillac Mountain pluton, the outer part of the Tunk Lake pluton (Karner, 1963), and much of the Sorrento pluton (Karner, 1963). This rock is distinguished from the other granites by its composition, texture, and color. It is medium- to coarse-grained and medium-gray to greenish-gray where fresh. Upon slight weathering it becomes pinkish to deep salmon pink due to finely disseminated hematite dust. A striking hypidiomorphic texture is typical and is produced by closely packed subhedral to euhedral alkali feldspar grains (perthitic) and interstitial quartz and small amounts of mafics. Hornblende is generally present but biotite occurs locally; and in contact phases, sodium pyroxene and iron-rich olivine and magnetite may be present.

Superficially a particular pluton may appear quite uniform. Detailed model studies, however, show significant variations in composition and/ or texture. In the Tunk Lake pluton, for example, Karner (1963) shows that the granite ranges widely but very systematically from margin to center to give a concentric or annular pattern of transitional types. An outer ring of pyroxene granite passes gradually through hornblende granite to biotite granite to biotitic quartz monzonite at the center.

Contact Zones

Within one or two hundred yards (100–200 m) of the contact the granite may differ considerably from that of the cores. Commonly the texture is finer-grained, the mafic content higher, and the amount of K-feldspar lower. Plagioclase may be more abundant, and its An-content higher. In the Tunk Lake pluton (Karner, 1963) the quartz content is highest near the contact.

Near the contacts of some plutons inclusions of the country rock are more numerous, larger, more angular, and less altered. In the Cadillac Mountain pluton, inclusions are extremely numerous where the granite comes in contact with the surrounding zone of breccia; but they diminish in size and number and become somewhat more rounded and altered away from the contact. A noticeable foliation is found near some contacts. This structure, due to preferred orientation of mafic minerals and feldspar grains, dips steeply and strikes parallel to the trend of the contact. Elongate or slabby inclusions here exhibit the same preferred orientation.

Inclusion and Feldspar Crystal Orientation

Most of the plutons possess systematic structural patterns due to the planar orientation of slabby inclusions and/or tabular feldspar crystals. The internal structure is readily determined throughout the plutons although inclusions are generally more numerous near contacts. It should be noted that the inclusions about which we are here concerned generally lie more than one or two hundred yards (100–200 m) from the wall rock and are smaller and more rounded than those closer to the contact. The latter, by contrast, show diverse orientation, appear less altered, and are undoubtedly more locally derived.

The internal structure of some plutons is saucer-shaped with very low to horizontal dips over most of the body and centripetal inclinations at low angles in the marginal portion. This structural pattern is most clearly demonstrated by the inclusions of the Cadillac Mountain pluton (Chapman, unpublished data). Some bodies show a marked departure from this simple basin-like structural pattern as if complicated by folds and faults. Several plutons exhibit more of a bowl-shaped pattern with steep marginal dips and intermediate to low dips near the center. Such a structure is well shown in the North Sullivan (D. B. Allen, oral communication 1967), the Great Wass (J. G. Ward, oral communication 1967), and probably the Oak Point (Stewart, 1956) plutons.

MAJOR FEATURES OF THE WHITE MOUNTAIN MAGMA SERIES

The rocks of the White Mountain magma series of New Hampshire are of Early Jurassic age (Lyons and Faul, this volume) and lie within a belt roughly 40 miles (65 km) wide, 150 miles (240 km) long, and trending N15°W (Fig. 29–2). These mildly alkaline plutons cut older igneous and metamorphic rocks formed in Late

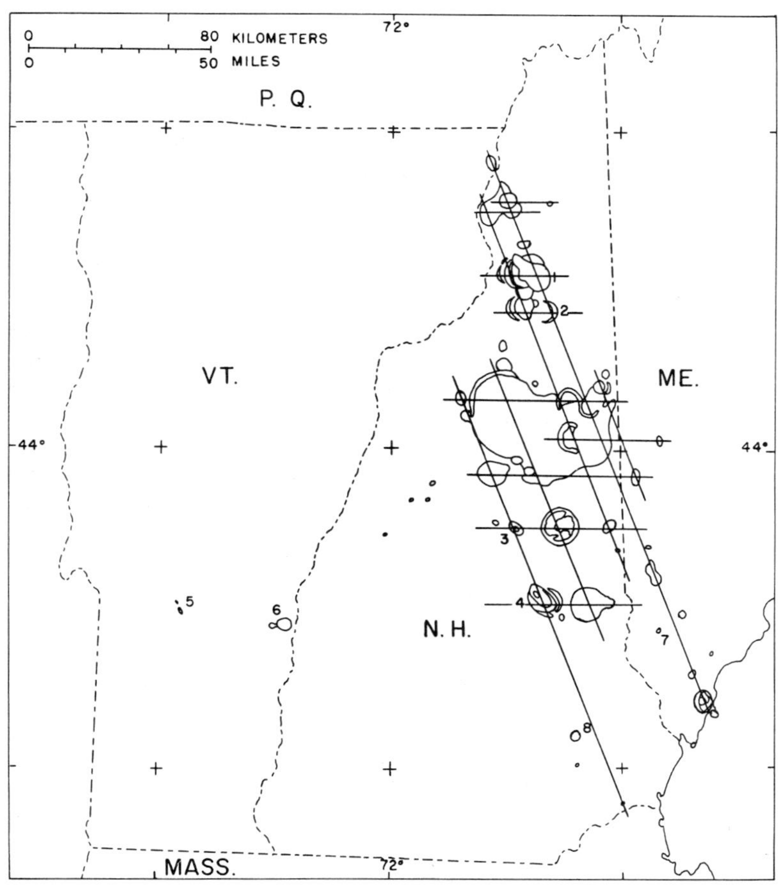

Figure 29–2. Distribution of the White Mountain magma series. 1, Percy complex; 2, Pliny complex; 3, Red Hill complex; 4, Belknap Mountain complex; 5, Cuttingsville complex; 6, Ascutney Mountain complex; 7, Lebanon pluton; 8, Pawtuckaway complex. Heavy straight lines—pluton lattice pattern.

Devonian time. The metamorphic rocks are of low to high grade; are composed largely of mica schist, gneiss, quartzite, and amphibolite; and represent stratified units of Middle Ordovician to Early Devonian age. It seems reasonable to extend this belt northwestward for an additional 50 miles (80 km) or more to include the alkaline intrusions of the Monteregian Hills of southern Quebec. The possible southerly extension of this belt for at least another 50 miles to include certain plutons in the general vicinity of Boston is treated elsewhere (Chapman, 1968).

More important perhaps than the belt-like distribution of plutons is their reticulate arrangement (Chapman, 1963, 1967). This is made strikingly apparent in Figure 29–2 where the lattice directions N22°W and E–W have been superposed. The tendency for the plutons and central complexes to lie at nodal points in the chosen lattice is more pronounced than in the case of the Maine coastal plutons. These two directions may be taken to represent the disposition pattern within the belt. Again, precision is not to be implied by these chosen values, and the more general bearings NNW and E–W may be preferred. Elongate complexes and plutons are common in the New Hampshire area and generally trend parallel to either lattice line. Elongate complexes are presumably due to a series of successive contiguous intrusions whose centers are roughly linearly disposed. The phenomenon known as

migration of intrusion centers, therefore, is probably more apparent than real. Examples of E–W elongation are found in the Percy, Pliny, and Ascutney Mountain complexes; elongation in the NNW direction is represented by the Belknap, Cuttingsville, and Red Hill complexes. In contrast to the Maine coastal lattice pattern, the obtuse bisectrix of the New Hampshire lattice pattern parallels the tectonic strike.

Contact relations with country rock are, for the most part, very similar in both the Maine coastal and New Hampshire belts of plutons. External centripetal dip patterns, so well developed around the Maine coastal plutons, however, are not detected around the plutons of the White Mountain magma series (Chapman and Chapman, 1940). Locally, apparent in-tilting of country-rock is suggested, but these relations may be merely coincidental. A feature present in the New Hampshire area, but not observed in association with the Maine coastal plutons, is a well-defined centripetal dip pattern of bedded rocks enclosed within ring-dikes. It is found in the Moat Volcanics (Billings, 1928, Kingsley, 1931), where it is restricted to the concave side of certain ring-dikes. The volcanic rocks are not found on the convex or external side of the ring structure.

Certain compositional and textural differences between the rocks of the Maine and New Hampshire areas should be noted. Whereas the Maine coastal plutons are almost entirely granite, the rocks of the White Mountain magma series consist roughly of 80% granite, 10% quartz syenite, and 10% syenite. Mafic and intermediate rocks are relatively rare. One-feldspar granites are relatively more common and the varieties of alkali feldspar and mafic minerals are greater in the White Mountain magma series.

Contact zones may be slightly finer grained than the cores of the large plutons, but the nature and distribution of angular inclusions are not strikingly different from those of the Maine coastal plutons. Small, rounded, and highly altered inclusions are widespread and may be abundant locally, particularly in some syenitic bodies.

Crystal- and inclusion-orientation patterns in the felsic plutons have not been reported, and even reference to mineral- and inclusion-orientation is rare. The mafic bodies, however, have well-defined, mappable structural patterns due to mineral layering (some rhythmic and some graded), mineral parallelism, igneous lamination, and inclusion parallelism. The planar structures tend to dip moderately to steeply and form bowl-shaped or funnel-shaped patterns. Good examples include the Ascutney Mountain complex (Chapman and Chapman, 1940), the Pawtuckaway Mountain complex (Roy and Freedman, 1944), and the gabbroic complexes of extreme southwestern Maine (Hussey, 1962).

THEORY OF STRUCTURAL CONTROL

A theory of grand-scale structural control on the formation of various magmatic central complexes and plutons has already been suggested (Chapman, 1963, 1967). In essence it is postulated that two sets of steep fracture zones formed deep within the earth. Selective melting along these zones gave rise to mafic magma, which concentrated at fracture-zone intersections to form rounded chambers and moved upward through the crust. In addition to the lattice arrangement, both the shape of the plutons and the apparent migration of intrusion centers within a complex may be accounted for.

The apparent absence of igneous activity at certain nodes in the lattice pattern, as determined by surface mapping, may be due to any of several reasons. (1) Melting along one or both of the controlling zones at the particular node may have been inadequate to develop a magma chamber. (2) Local conditions may have prevented a magma chamber from ascending to the present surface level. (3) Small bodies of igneous rocks at lattice nodes may have been overlooked in field mapping or may be concealed by surface mantle.

Numerous small pipes and breccia-filled vents have been found in the New Hampshire belt. Some of these lie on well-established lattice lines and undoubtedly represent conduits between buried magma chambers and the surface. Bodies of this dimension could readily escape detection in field mapping. Others may have been misinterpreted as dikes. In southwestern Maine the possible existence of one small igneous body was first suggested to Hussey (1962) by a magnetic high on a geophysical map. Subsequent field search revealed the presence of the poorly exposed body of Lebanon Diorite, which is here considered a nodal intrusion. Through careful geological and geophysical field search, a number of new centers of igneous activity may be disclosed. The vicinity of nodes on the better established lattice lines should be the most fruitful areas for search.

EMPLACEMENT OF THE MAINE COASTAL PLUTONS

In considering the origin of the Maine coastal plutons it seems certain that we are dealing with rocks derived from intrusive granitic magma. Evidence for such a conclusion includes: (1) the metamorphic aureoles of hornfelsic rock surrounding the plutons containing numerous higher temperature minerals than are found in the more remote country rock; (2) a marked reduction in grain size of the granitic rock against its contact; (3) the abrupt manner in which the plutons cross-cut structural features of the country rock; (4) apophyses of fine-grained granite filling dilated fractures in the country rock; (5) angular inclusions of country rock, with sharp contacts and diverse orientations, commonly re-broken, dilated, and veined by the granite; (6) perthitic feldspar inverted from high-temperature form (Stewart, 1956; Karner, 1963); and (7) planar structures due to oriented inclusions and feldspar crystals striking parallel to contacts and discordant with planar structures of adjacent country rock. This last feature indicates a primary or relict primary structure in a magma; it does not necessarily indicate magma flow.

The mode of emplacement for many of the plutons has long been considered (Chapman, 1953, 1962b) to be by ring-fracturing and subsidence. This conclusion is based on the following: (1) sub-circular outline of plutons; (2) smoothly-curved sharp boundaries; (3) paucity of granitic apophyses; (4) presence of ring-dikes and possible cone sheets; (5) ring-dikes passing into shatter-zones and breccias; (6) contact zones of brecciation around the plutons; (7) web joint patterns; (8) web vein patterns; (9) web breccia patterns; (10) external centripetal dip patterns; (11) rapid magma ascent indicated by chilled contacts.

Items 1–4 are frequently cited evidence but items 5–7 are less frequently encountered in the literature. The contact breccia zone and arcuate shatter zones of the Mount Desert complex are beautifully developed and probably represent the best exposed anywhere (Chapman, 1953, 1962b). Around the other plutons, breccia appears only locally. Web joint patterns (composed of radial and tangential joint sets) in the country rock are not readily disclosed because they are camouflaged by earlier and/or later joint sets. Items 8–10 appear to be characteristic of the Maine coastal plutons. Web vein and web breccia patterns in the country rock appear to express a deformation similar to and perhaps contemporaneous with that which created the web joint pattern. The web vein pattern is best developed in the contact breccia zone where the country rock was originally well-bedded. It is conspicuous, however, in gabbroic country rock. The long narrow zones of intensely brecciated rock that form the web breccia pattern appear confined to gabbroic country rock. The question of origin of this type of breccia is moot, but the shape and disposition of these particular breccia zones leave little doubt as to their tectonic significance. A centripetal dip pattern in country rock marginal to a pluton is strong evidence for subsidence. The rare reference in the literature to such patterns is undoubtedly due in part to the difficulty with which such patterns may be recognized. The excellent exposures and highly favorable conditions for development explain why external centripetal dip patterns are so readily reognized in the Maine coastal area.

I have shown elsewhere (Chapman 1966a, 1966b, 1967) that postulating a floored magma chamber helps to answer many problems relative to central complexes. The detailed reasoning behind the conclusions that such magma chambers do have floors is quite involved and cannot be given here. It is based, however, on several lines of evidence only one of which is pertinent to this study.

The saucer-shaped structural pattern of inclusions already described may be explained most logically as a sedimentational feature. It is believed that the elongate and tabular inclusions, which for some time were maintained in suspension in the magma, came to rest under the force of gravity on a rather level floor in the magma chamber. In the Maine area this floor was the top of a subsided block, as described below. As growing crystals also settled toward this floor, the inclusions came to rest most stably with their longer dimensions in a nearly horizontal position and became embedded in a continually thickening layer or crystal aggregate. More rapid accumulation at the outer portion of the aggrading floor may be indicated by the centripetal dips observed here for the inclusions. As the deposit of crystals and inclusions continued to build up, the "consolidation front" in the magma chamber advanced principally upward from the floor toward the roof. Advancement downward from the roof and inward from the walls of the chamber must have been comparatively insignificant. The consolidation pattern for these granitic plutons, therefore, is very similar to that proposed for the Skaergaard intrusion (Wager and Deer, 1939). The granites, furthermore, can in large part be considered cumulates in the sense of Wager et al. (1960). This interpretation helps to explain many textural features of the plutons, provided we make allowances for certain postcumulation changes. It seems reasonable that the cumulates were deposited upon the subsided block which came to rest on a floor of the magma chamber (Chapman, 1966a).

The recent investigation by Helgesen and Karner (1965) on the distribution of zircon crystals in the Tunk Lake pluton is particularly pertinent here. I feel that the zircon data may be considered analogous to those collected on elongate inclusions in the Cadillac Mountain pluton (Chapman, unpublished data) and, therefore, support my interpretation that the rocks of the Tunk Lake pluton are also cumulates. I would also interpret the concentric lithologic pattern of the Tunk Lake pluton (Karner, 1963) as due to erosion of a body composed of saucer-shaped layers of different petrographic types stacked one above another.

Convection currents within the granitic magma chamber seem required. Such currents would help to explain the gradual reduction both in size and number of inclusions through the marginal portion of the pluton. Again convection currents could have kept inclusions in suspension for a sufficiently long period of time to permit them to react thoroughly with the magma and to assume nearly universally a rounded form. The limited size, rounded nature, and degree of alteration of inclusions indicate further that relatively little xenolithic material was added to the magma chamber through overhead piecemeal stoping subsequent to the onset of consolidation there.

This conclusion in turn suggests that each pluton occupies a pocket in the crust which was rapidly formed and simultaneously filled with magma (cauldron subsidence). Subsidence along a ring-fracture created a large chamber filled coevally with magma which chilled against the cold country rock to form an effective barrier against piecemeal stoping. Had the melt been displaced upward largely by piecemeal stoping, a much slower process, it is unlikely that a chill zone would have formed.

Perhaps in some plutons the subsided block disintegrated into several large segments which settled differentially. This would lead to a highly irregular depositional surface for the cumulate and complicated structural pattern in the granite. Marked departure from a simple saucer-shaped internal structure in some plutons may be due to post-accumulation deformation brought about perhaps by renewed subsidence or adjustment of the underlying block. Both structural and petrographic evidence indicate that subsidence was repeated at least several times for some plutons. Post-cumulation phenomena are now being intensively investigated in several of the plutons; and, although it is premature to announce any very positive conclusions, a few points seem quite clear. Post-cumulation changes are extensive and commonly controlled by deformation. They include recrystallization marked by textural and compositional changes. Faulting and diking of the incompletely consolidated rock seem evident, and shearing of a crystal mush appears to have been an extensive phenomenon. The bowl-shaped structure patterns of some plutons may have resulted from pronounced settling and slumping in the pre-solidification stage.

EMPLACEMENT OF THE WHITE MOUNTAIN MAGMA SERIES

Evidence that the rocks of the White Mountain magma series were derived from magmas is similar to that indicated for the Maine coastal plutons. Virtually all geologists who have studied the series agree that the major controlling mechanism of emplacement in the observable portion of the crust involved ring-fracture stoping. Again the evidence is similar to that indicated for the Maine coastal plutons. A few differences, however, might be noted. Ring-dikes are much more numerous in the New Hampshire area. Although web vein and web breccia patterns have not been reported from the country rock around the White Mountain magma series, radial and tangential dike and joint patterns are found (Chapman, R. W., 1954). External centripetal dip patterns in the country rock indicating subsidence are not detected in the New Hampshire area. On the other hand, the well developed internal centripetal dip patterns in bedded rocks, not found in the Maine coastal area, are associated with some New Hampshire plutons. Not only do such patterns indicate subsidence, but the preservation of layered volcanic rocks (Moat Volcanics) only on the concave side of certain ring-dikes is a strong point in favor of subsidence within a ring-fracture.

Saucer-shaped structural patterns of inclusions and feldspars have not been reported for the felsic stocks of the White Mountain magma series, and planar structures within these bodies have been observed apparently only rarely. The presence of mineral layering and igneous lamination in some of the small plutons composed of mafic rock, however, implies some relatively firm support or floor upon which the crystals could be deposited. The bowl-shaped structural patterns here suggest post-depositional slump or settling. At least the mafic bodies appear to have consolidated in a manner similar to that of the Skaergaard intrusion. In addition, many felsic rocks of the White Mountain magma series, like certain rocks of the Maine coastal plutons, show textures characteristic of cumulates. It is suggested, furthermore, that the concentric lithologic pattern shown on the map of the Red Hill complex (Quinn, 1937) represents layers of syenitic rocks, nearly horizontal originally, analogous to those already noted for the Tunk Lake pluton.

The limited size and high degree of rounding and reaction exhibited by the inclusions in many stocks again suggest a rather active role for convection in the magma chamber. If the stock-like plutons are composed largely of cumulates, then we might conclude, as we did for the Maine coastal plutons, that overhead piecemeal stoping, within the magma chambers was quite ineffective subsequent to the onset of crystallization. There is evidence, however, that rather limited xenolithic showers did occur. Some plutons contain arcuate blocks a mile (1.6 km) long, apparently not associated with abundant smaller inclusions (Chapman and Chapman, 1940; Chapman, R. W., 1954). Such falls probably consist of roof rock, isolated largely by arcuate fractures and buried in an aggrading layer of crystals.

Some rock textures resemble recrystallization textures in rocks of the Maine coastal plutons and suggest post-cumulation changes due to deformation. Without the assistance of planar structures it is extremely difficult to estimate the relative importance of post-cumulation slump.

EVOLUTION OF THE MAGMA AND MAGMA CHAMBER

To account for many structural and petrological relations observed in magmatic central complexes from various parts of the world, I proposed (Chapman, 1967) a model of the evolution of the magma and its chamber. This model, though provisional, serves as a point of departure and will be outlined briefly. It is in general agreement with the recent proposal of Hamilton and Myers (1967).

Mafic magma, generated along steep fracture zones or zones of low pressure (probably in the mantle), moves up rapidly and into the crust where it possesses considerable superheat. Original lenticular pockets of melt are quickly transformed to more cylindrical shapes, particularly along fracture-zone intersections, and then to more equant forms by rock spalling and plastic flowage (Fig. 29–3). Ascent through the crust is accomplished by melting, stoping, and plastic. flow. Melting is enhanced in the more felsic portion of the crust, and granitic magma forms and segregates on top of the mafic melt under the dome-like roof of the magma chamber.* Heat from the mafic melt, supplied now largely by

---

* It should be emphasized that the coexistence of two implicated liquid phases may be due to two entirely different reasons: (1) liquid immiscibility and (2) lack of an opportunity to homogenize. The latter case is considered to hold here. Factors such as limited time, insufficient agitation, density differences, high viscosities, and low diffusion rates all tend to impede homogenization.

crystallization, serves to activate and augment the granitic magma. This mechanism of magma rise is similar to that proposed by Holmes (1931) and Daly (1933) and may involve zone melting (Dickson, 1958). As the mafic melt crystallizes, the floor of the magma chamber builds up gradually with cumulates and xenolithic material; and the mafic melt, in part contaminated by assimilation, eventually becomes differentiated not to a granitic one but to a syenitic one (Holmes, 1931).

At this stage the magma chamber normally reaches the upper part of the crust where stoping is more active. Xenoliths now engulfed by the overlying granitic melt may settle through the syenitic melt below to the chamber floor. It is calculated, on the basis of this model, that by the time the magma chamber comes to within roughly its diametral distance from the earth's surface the syenitic melt will be formed, if indeed it is to form at all. At this stage the roof rocks are susceptible to disintegration by fracturing, cauldron subsidence, piecemeal stoping, and volcanism; and the felsic magma will consolidate quickly.

Applying this model of a floored polymagmatic chamber to the White Mountain magma series, we may account for the abundance of granitic and syenitic rock, the paucity of mafic types, certain apparent irregularities in the intrusive and extrusive sequences, the heterogeneous character of certain felsic bodies, and the granitic margins on some syenitic stocks. See Figure 29–4.

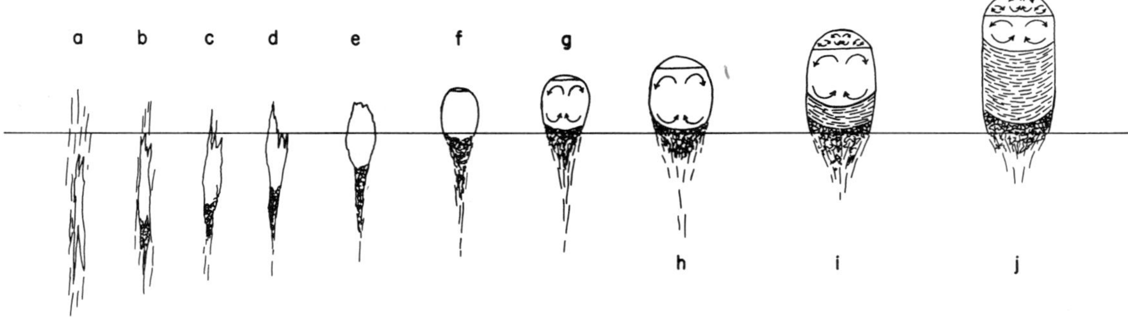

FIGURE 29–3. Evolution of a magma chamber. Horizontal line roughly represents crust-mantle boundary. (a) Mafic magma generated in fracture zone in mantle. (b) Widening and upward enlargement aided by slabbing and stoping. (c) Base closes in plastically. (d) Top enters base of crust. (e) Chamber widens in cross section. (f) Chamber well within crust and nearly equant in vertical cross section. Capping of granitic melt may develop. (g) Convection pattern relatively simple. (h) Chamber nearly equant in three dimensions, granite melt steadily increasing. (i) Crystallization contributes to ultramafic cumulate, good convection in both melts. (j) Granitic melt and mafic cumulate increasing, intermediate melt diminishing.

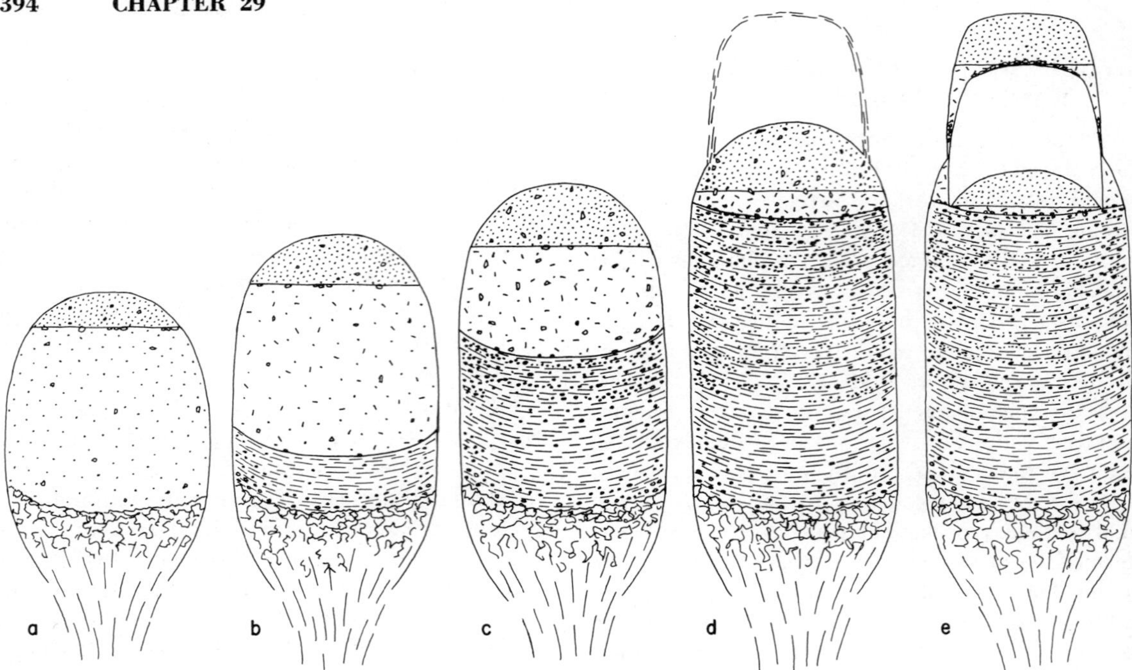

FIGURE 29–4. One style of emplacement for the White Mountain magma series. (a) Ellipsoidal chamber of basaltic melt capped by granitic melt near base of crust. (b) By crystallization and assimilation mafic melt becomes more dioritic. Granitic magma augmented by melting. (c) Further change leads toward monzonitic magma overlain by increased amount of granitic melt. (d) Top of chamber about diametral distance from earth's surface, floor built up largely by mafic cumulate. Granitic melt overlies syenitic one. Ring-fracture and subsidence initiated. (e) Paraboloidal block settles to floor. Granitic magma forced up ring-fracture first, followed by syenitic melt. Granitic melt trapped beneath block is available for later intrusion as ring-dike or central stock.

Turning to the Maine coastal plutons where genetically related syenite and gabbro are not recognized, this concept of a polymagmatic chamber may appear superfluous. If syenite did form here, it must lie buried beneath the granite cover. The presence of composite dikes (mafic-felsic), some of which cut the granite, and the repeated intrusion of mafic dikes alternating with intrusions of granite suggest that granitic and mafic melts coexisted in the Maine coastal area and that a syenitic melt never formed. Wager et al. (1965) adopt a similar model for certain Tertiary igneous rocks on the Isle of Skye. The absence of syenite in the Maine area may be due to the rapid rise of mafic melt high into the crust so that solidification took place through rapid dissipation of heat without differentiation to a syenitic melt. Furthermore, extensive stoping may have so chilled the mafic melt at its interface with granitic melt that a solid floor formed at this level and prevented any large amount of subsequent lower melt from escaping to higher levels. See Figure 29–5.

SUMMARY AND CONCLUSIONS

1. The reticulate patterns of magmatic activity in the two areas suggest that the plutons developed from discrete bodies of magma formed deep within the earth at the nodes of two sets of steep fracture zones of low pressure.

2. In Maine the lattice angle is roughly 55°, the acute bisectrix parallels Appalachian trend, and the obtuse bisectrix parallels a set of mafic dikes. In New Hampshire the lattice angle is roughly 68°, and the acute bisectrix is normal to the Appalachian trend.

3. The magmatic plutons of the two areas have developed from closed polymagmatic chambers of roughly equant form.

4. Ring-fracturing, generally accompanied by subsidence, in potentially weak roof rocks led to the final emplacement of the now exposed portions of the igneous bodies. Centripetal dip patterns in adjacent country rock and spider web patterns of felsic veins and breccia zones constitute new and additional evidence in the Maine area.

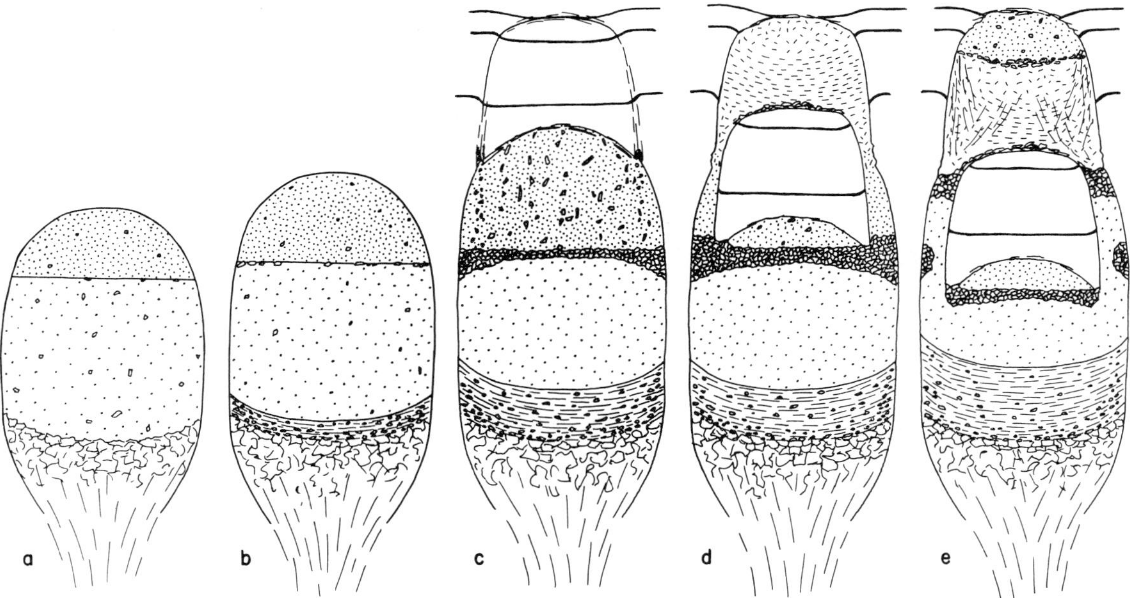

FIGURE 29–5. One style of emplacement for the Maine coastal plutons. (a) Chamber of basaltic melt rapidly emplaced well up in crust, develops much granitic magma. (b) Granitic melt rapidly augmented, crystallization of mafic melt begins. (c) Intense stoping leads to rapid ascent of chamber and construction of a xenolithic floor at the felsic-mafic magma interface. Sagging and ring-fracture initiated. (d) Subsided block rests on xenolithic floor. Granitic melt crystallizes above block to form saucer-shaped structure pattern. (e) Xenolithic floor collapses in part due to fusion by mafic melt. Block settles to equilibrium position. Granitic cumulate sags and slumps to form bowl-shaped structure pattern. Fresh granitic melt reaches roof zone. Mafic melt becomes more dioritic.

5. In the New Hampshire area a granitic melt (syntectic) overlying a syenitic one (partial differentiate of basaltic magma) filled the magma chamber just prior to the final emplacement. Consequently syenitic rocks show a preference for ring-dikes whereas granitic ones show a preference for stock-like plutons. Here the subsiding block came to rest on the chamber floor at the base of the syenitic melt.

6. In the Maine area a syenitic magma probably never formed beneath the granitic melt. Exposed portions of the plutons, therefore, are composed entirely of granitic rocks. The subsiding block appears to have been stopped, at least temporarily, at the felsic-mafic magma interface by a constructed floor, composed of a thick layer of xenolithic material.

7. Saucer-shaped structural patterns, formed by oriented inclusions and feldspar crystals, are well formed as cumulates on a chamber floor or, perhaps more precisely, on the top of the subsided block, which rested on the chamber floor. In New Hampshire such structural patterns are not reported except in the case of mafic bodies where a similar interpretation seems justified.

8. Marked departures from a simple saucer-shaped structural pattern and the presence of bowl-shaped patterns suggest post-accumulation slumping and deformation brought about perhaps by renewed subsidence or differential movement of the underlying block.

9. Quantitative study of inclusions, relative to size, shape, frequency, distribution, and orientation, shows: (1) convection was continually operative during consolidation and (2) overhead piecemeal stoping was ineffective subsequent to the onset of consolidation.

10. Very briefly expressed, the differences between the plutons and complexes in the two areas may be attributed largely to differences in depth of formation and nature and rate of ascent of the magma.

REFERENCES

Billings, M. P., 1928, The petrology of the North Conway quadrangle in the White Mountains of New Hampshire: Am. Acad. Sci., v. 63, no. 3, p. 67–137

——————, 1956, The geology of New Hampshire, Part II, Bedrock geology: New Hampshire State Plan. Devel. Comm., Concord, 203 p.

Chapman, C. A., 1953, Cauldron subsidence at Mount Desert Island, Maine: [abs.]: Geol. Soc. America Bull., v. 64, p. 1406–1407

————, 1962a, Bays-of-Maine igneous complex: Geol. Soc. America Bull., v. 73, p. 883–888

————, 1962b, The geology of Mount Desert Island, Maine, with explanation and descriptive field guide: Urbana, Illinois, 52 p.

————, 1963, Structural control on magmatic central complexes in central New England: [abs.]: Geol. Soc. America Spec. Paper 73, p. 128

————, 1966a, Paucity of mafic ring-dikes—evidence for floored polymagmatic chambers: Am. Jour. Sci., v. 264, p. 66–77

————, 1966b, Origin of igneous central complexes and formation of ring-dikes: [abs.] Geol. Soc. America Spec. Paper 87, p. 31

————, 1967, Magmatic central complexes and tectonic evolution of certain orogenic belts: in Etages Tectoniques, Colloque de Neuchâtel 18–21 avril 1966, Baconniere, Neuchâtel, p. 41–52

————, 1968, Intersecting belts of post-tectonic "alkaline" intrusions in New England: Illinois State Acad. Sci. Trans., v. 61, p. 46–52

Chapman, R. W., 1954, Criteria for the mode of emplacement of the alkaline stock at Mount Monadnock, Vermont: Geol. Soc. America Bull., v. 65, p. 97–114

————, and Chapman, C. A., 1940, Cauldron subsidence at Ascutney Mountain, Vermont: Geol. Soc. America Bull., v. 51, p. 191–212

Daly, R. A., 1933, Igneous rocks and the depths of the earth: McGraw–Hill Book Co., New York, 508 p.

Dickson, F. W., 1958, Zone melting as a mechanism of intrusion—a possible solution to the room and superheat problem [abs]: Am. Geophys. Union Trans., v. 39, no. 3, p. 513

Faul, Henry, Stern, T. W., Thomas, H. H., and Elmore P. L. D., 1963, Ages of intrusion and metamorphism in the Northern Appalachians: Am. Jour. Sci., v. 261, p. 1–19

Hamilton, Warren and Myers, W. B., 1967, The nature of batholiths: U. S. Geol. Survey Prof. Paper 554–C, 30 p.

Helgesen, J. D. and Karner, F. R., 1965, Zircon variation in the Tunk Lake Granite, southeastern Maine: North Dakota Acad. Sci., Proc., v. 19, p. 204–212

Holmes, Arthur, 1931, The problem of the association of acid and basic rocks in central complexes: Geol. Mag., v. 68, p. 241–255

Hussey, A. M., II, 1962, The geology of southern York County, Maine: Dept. Econ. Devel. Augusta, Maine, Spec. Geol. Studies Series, no. 4, 67 p.

Karner, F. R., 1963, Petrology of the Tunk Lake granite pluton, south-eastern Maine: Ph.D. Thesis, Univ. of Illinois, 94 p.

Kingsley, Louise, 1931, Cauldron-subsidence of the Ossipee Mountain: Am. Jour. Sci., v. 222, p. 139–168

Quinn, Alonzo, 1937, Petrology of the alkaline rocks at Red Hill, New Hampshire: Geol. Soc. America Bull., v. 48, p. 373–402

Roy, C. J., and Freedman, Jacob, 1944, Petrology of the Pawtuckaway Mountains, New Hampshire: Geol. Soc. America Bull., v. 55, p. 905–920

Stewart, D. B., 1956, Rapakivi granite in the Deer Isle region, Maine: Ph.D. Thesis, Harvard University, 150 p.

Terzaghi, R. D., 1940, The Rapakivi of Head Harbor Island, Maine: Am. Minerologist v. 25, p. 111–122

Wager, L. R., and Deer, W. A., 1939, The petrology of the Skaergaard intrusion of Kangerklugssaug, East Greenland: Meddel. om Grønland, bd. 105, no. 4, 352 p.

————, Brown, G. M., and Wadsworth, W. J., 1960, The types of igneous cumulates: Jour. Petrol., v. 1, p. 73–85

————, Vincent, E. A., Brown, G. M., and Bell, J. D., 1965, Marscoite and related rocks of the Western Red Hills complex, Isle of Skye: Roy, Soc. London, Phil. Trans., Series A., v. 257, p. 273–307

Wingard, P. S., 1961, Geology of the Castine—Blue Hill area, Maine: Ph.D. Thesis, Univ. of Illinois, 138 p.

SECTION VI

GEOPHYSICS

Chapter 30

Gravity Anomalies in Northwestern New England*

W. H. DIMENT

INTRODUCTION

THIS STUDY IS PRIMARILY concerned with the geologic significance of gravity anomalies in Vermont and adjacent parts of Massachusetts, New Hampshire, and New York. Similar patterns of anomaly are found for the length of the Appalachian system and in other linear mountain systems such as the Alps, Pyrenees, and Carpathians. If the anomalies are caused by similar processes, the solutions postulated for one region should not be inconsistent with facts in another.

With the wealth of data available for comparison, it is perhaps surprising that the geologic significance of the major gravity anomalies is not better understood. This reflects, in part, the difficulty of untangling the effects of different anomalous masses and, in part, our lack of knowledge concerning the lower crust and uppermost mantle where the masses responsible for the major anomalies are located. The situation could be much improved by detailed seismic refraction traverses and a few judiciously chosen deep drill holes.

History

Gravity exploration in New England proceeded in several phases. The earliest pendulum observations (1909–1914) were too widely spaced to correlate gravity with geology, but later measurements (1934–1937) were more closely and appro-

priately spaced (Duerksen, 1949). On the basis of these, Longwell (1938, 1943) made some of the first interpretations. About the same time, Woollard began his series of investigations, the New England part of which he summarized in 1948. The first detailed survey of large extent was conducted by Bean (1953) in a strip across central Vermont and New Hampshire and a part of New York. His survey proved rewarding, and other surveys soon followed (Bromery, 1967; Diment, 1953; Fitzpatrick, 1959; Joyner, 1963; Kane and Bromery, 1966 and this volume; Simmons, 1964; Thompson and Miller, 1958), so that most of New England and adjacent regions are now covered by stations spaced closer than 6–10 km. Most of these data were included in the Bouguer anomaly map of the United States (Woollard and Joesting, 1964).

Data

All gravity data reported here are relative to a value of 980,622 mgal at the Dominion Observatory, Ottawa, Canada. The link between Ottawa and northern Vermont was established by Fitzpatrick (1959). Subsequent surveys of Joyner (1963) and Simmons (1964) are based on the same datum, and rechecking of base stations by the various investigators indicates no significant errors. That part of Bean's (1953) data included in this report was readjusted to the Ottawa datum (see Joyner, 1963, for details). Bouguer anomalies were obtained by reduction to sea level using a density of 2.67 g/cm^3. Terrain corrections

---

* Publication authorized by the Director, U. S. Geological Survey.

were applied where it appeared that they would exceed 2 mgal.

The locations of the gravity stations used to contour the regional gravity map are shown in Figure 30–1. Station locations in a detailed survey in the northern part of the area are shown on the gravity map of that area (Fig. 30–10). For details of the gravity maps and sections on the periphery of the area of Fig. 30–1, it is best to refer to the results of the adjacent surveys (Bromery, 1967, Joyner, 1963, and Simmons, 1964) which overlap the area of Figure 30–1 and provide improved coverage.

ACKNOWLEDGMENTS

Part of the data summarized here was reported in a thesis (Diment, 1953) written under the direction of Francis Birch, M. P. Billings, and J. B. Thompson, and supported by funds from the Committee on Experimental Geology and Geophysics of Harvard University. More detailed gravity work was done in the northern part of the area during 1961 and 1962 for the U. S. Geological Survey. I am indebted to J. L. Meuschke and P. W. Philbin for conducting the aeromagnetic survey in the vicinity of Plattsburgh, and to A. L. Baldwin, Rudolph Raspet, and E. D. Saunders for substantial assistance in gathering the gravity data. John C. Roller of the U. S. Geological Survey kindly permitted me to view his preliminary interpretations of seismic refraction data from the southern Appalachians.

MAIN GRAVITY HIGH AND FLANKING GRAVITY LOWS

General Features

The nomenclature applied to principal gravity features of western New England is illustrated in Figure 30–2, and the relation of these features to major geologic units is shown in Figure 30–3. Among the general observations that can be drawn from these figures and from comparison of the Tectonic Map of the United States (Cohee et al., 1962) with the Bouguer Gravity Anomaly Map of the United States (Woollard and Joesting, 1964) are:

1. The axis of the main gravity high is nearly colinear with the axis of the Green Mountain anticlinorium in northern and central Vermont. To the south, the gravity axis is east of the apparent axis of the uplift.

2. The axis of the western gravity low is nearly colinear with the axes of the Middlebury–Hinesburg–St. Albans synclinoria.

3. The near coincidence of these gravity and geologic features strongly suggests that the configurations of the masses responsible for the gravity anomalies are related to geologic structure of the region.

4. Although there is a general coincidence of the main high with the anticlinorium and of the western low with the synclinoria, these relations are imperfect in detail. For example, the steep gradient between the high and low cuts across the Precambrian of the Green Mountain anticlinorium in quadrangles H–5 and I–5 (Fig. 30–3). Farther south (J–5 and K–5) the axis of the high is distinctly east of the Precambrian of the Berkshire Hills. Clearly, the high and the western low cannot be ascribed entirely to a density contrast between exposed Precambrian rocks of the anticlinorium and sedimentary rocks of the synclinorium.

5. The Bouguer gravity relief between the main high and the western low is large. Anom-

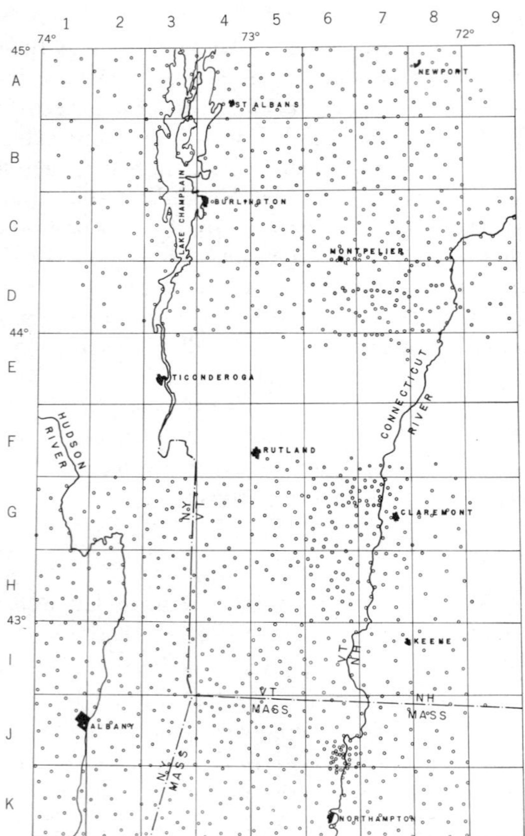

FIGURE 30–1. Distribution of gravity observations. The central part of the area, where no observations are shown, was surveyed in greater detail by Bean (1953). Locations of observations for a more detailed survey in the northern part of the area are shown on the gravity map in Figure 30–10.

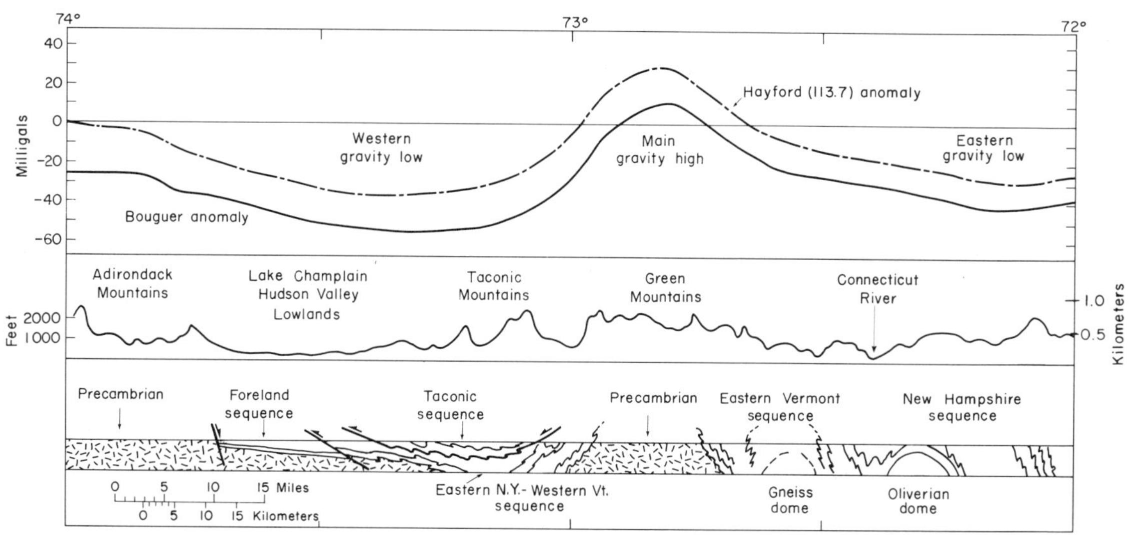

FIGURE 30–2. Gravity and topography profiles and generalized geologic section along line AA′ of Figure 30–3. Geology generalized from Billings et al. (1952). There is no vertical exaggeration in the geologic section.

alies range from nearly +15 mgal (G–5) to nearly −75 mgal (I–3). Application of isostatic corrections increases the gravity relief by 5–10 mgal (Heiskanen, 1948; Joyner, 1963). The large relief is another strong argument against the predominance of the effect of a density contrast between the sediment and Precambrian rocks, as it will be shown in a later section that a contrast of over 0.15 g/cm³ is unlikely unless large amounts of mafic rocks are involved.

6. The gravity field near the western edge of the Lake Champlain lowland and in the Adirondacks is irregular. Simmons (1964) has shown that these anomalies are at least in part associated with the Precambrian anorthosites of the Adirondack Mountains. To what extent the "anorthosite anomalies" contribute to the gravity field in the Champlain valley is not known.

7. The Taconic allochthon is in the region where the western low is best developed, thus suggesting another relation between structure and gravity anomalies. The slates and phyllites of the Taconic sequence are relatively dense rocks (Bean, 1953) and are unlikely to contribute to the low.

8. There is a suggestion that several gravity features cut across both the western low and main high. Both the Burlington cross-anticline (Cady, 1945) and the Monkton cross-anticline (Shaw, 1958) appear to be related to gravity axes that extend across the Lake Champlain valley (Fig. 30–4).

9. In a more regional sense, the axis of gravity high, or perhaps more appropriately the base of the steep gravity gradient between the western low and main high, marks a line of fundamental change in the crust for the length of the Appalachian mountain system. Miogeosynclinal rocks generally lie to the west or northwest of this line; metasedimentary, metavolcanic, and intrusive rocks of the eugeosyncline generally lie to the east or southeast. The gravity high is always within the eugeosynclinal province. The axis of the western low is usually near the eastern edge of the miogeosyncline, but in the southern Appalachians it lies to the east, in regions of exposed Precambrian rocks, a fact which could be explained by extensive overthrusting from the east (see King's excellent summary, 1964).

10. The eugeosynclinal region has been greatly uplifted with respect to the miogeosynclinal region. Apparently the region of the gravity high has been an axis of uplift at one time or another from the Ordovician until the Mesozoic. The structural relief of the Green Mountain anticlinorium is on the order of 10 km. The axis of the gravity high is roughly in the same position as the structural axis of the Mesozoic uplift which caused the Triassic basins to be disposed as they are (Longwell, 1943).

11. Felsic intrusive rocks are mainly found east of the main high. They increase in abundance from the region of the high into the region of the eastern low. These rocks are less dense

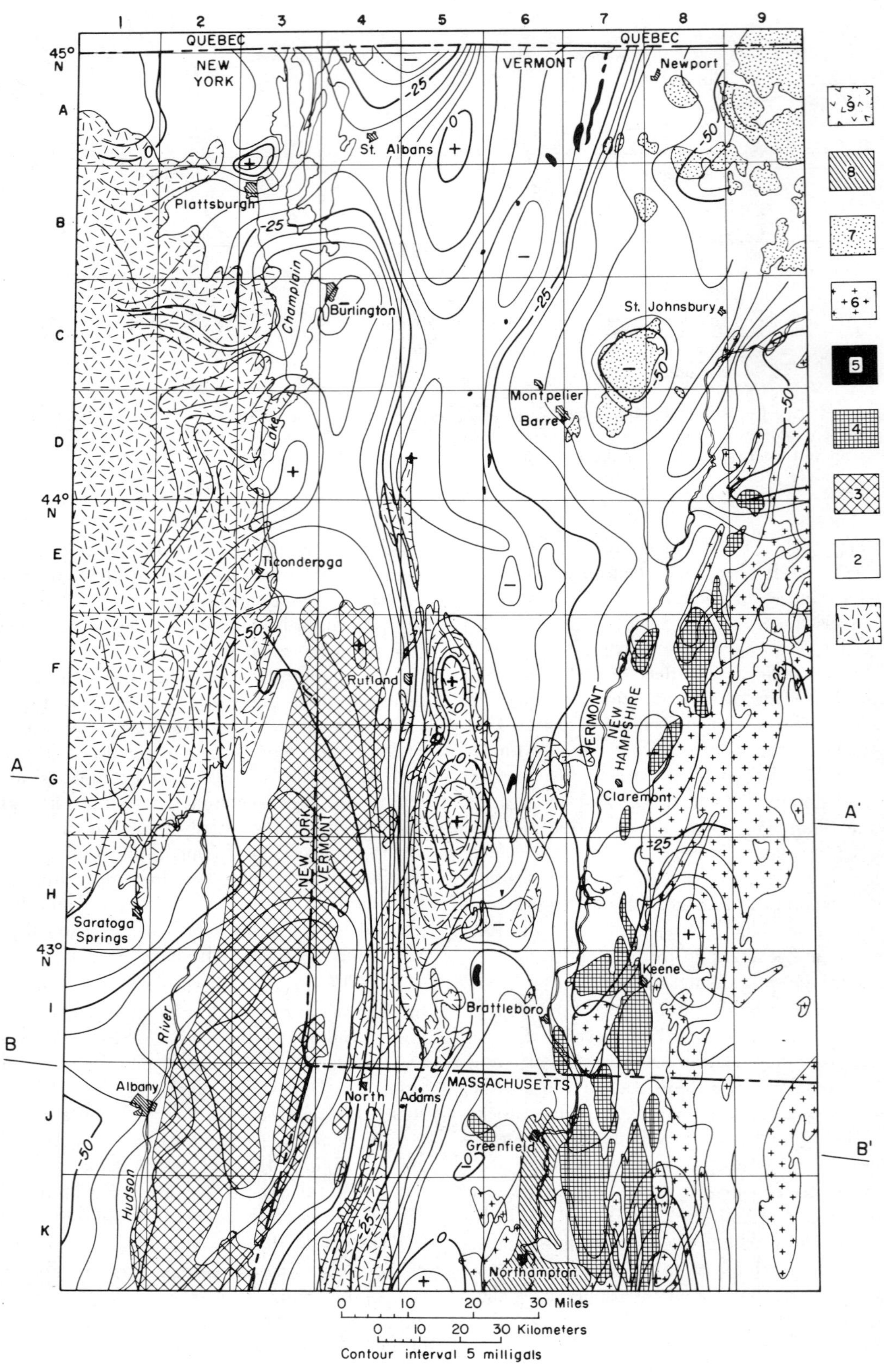

1
2
3
4
5
6
7
8
9
QUEBEC
QUEBEC
NEW YORK
VERMONT
Newport
45° N
A
St. Albans
Plattsburgh
B
Burlington
St. Johnsbury
C
Montpelier
Barre
44° N
D
Ticonderoga
E
Rutland
F
A
G
A'
Claremont
NEW YORK
VERMONT
VERMONT
NEW HAMPSHIRE
H
Saratoga Springs
43° N
Keene
B
I
Brattleboro
Albany
North Adams
MASSACHUSETTS
J
Greenfield
B'
Hudson River
K
Northampton
Lake Champlain
0   10   20   30 Miles
0   10   20   30 Kilometers
Contour interval 5 milligals

than their metasedimentary hosts and undoubtedly contribute significantly to the eastern low.

12. In New England, as well as in the rest of the Appalachians, there is a suggestion that the eastern low might be divided into two parts. The western part, which flanks the main high to the east, is the region of mantled gneiss domes and of some syntectonic felsic plutons. The eastern part is the region of abundant post-orogenic intrusive rocks such as those of the White Mountain Plutonic–Volcanic Series in northern New Hampshire.

Anomalies Caused by Miogeosynclinal Sedimentary Rocks

How large is the anomaly caused by the low density sedimentary rocks of the synclinoria of the Lake Champlain–Hudson valley region? For an answer we require estimates of the thickness of these sedimentary rocks and of the density contrast between them and the Precambrian basement rocks and/or the eugeosynclinal metasedimentary rocks to the east.

The maximum thickness of the miogeosynclinal sequence in western New England is probably less than 5 km, although it may be twice this in the deeper parts of the geosyncline in the central and southern Appalachians (Flawn, 1967). The density data (Bean, 1953, Diment, 1953, Joyner, 1963) indicate that the density of the miogeosynclinal sequence is probably not less than 2.7 g/cm^3. Important to this estimate is the fact that the rocks of the sequence, particularly near its eastern edge, are well indurated, so that porosity has negligible effect on density and that dolomites form a sufficient fraction of the section to cancel the effect of the low-density sandstones. Except for the metabasalts and ultramafic rocks, the average density of the schist and gneiss of the eugeosynclinal region and of the Precambrian basement can hardly exceed 2.85 g/cm. Adopting 0.15 g/cm^3 as the maximum density contrast, we get a maximum anomaly of about 30 mgal for the effect of a 5 km thickness of miogeosynclinal sedimentary rocks. Inasmuch as the gravity relief in western New England is about 90 mgal,

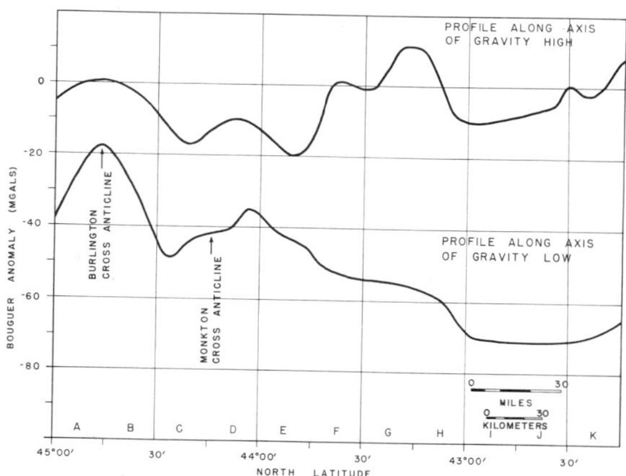

FIGURE 30–4. Bouguer anomaly profiles along the axes of the western low and the main high. Anomalies along axes have been projected to a meridianal line.

additional causes for the major anomalies must be sought. This does not mean that the effect of the sedimentary rocks can be ignored, for in places the gravity pattern is dominated by the contrast between them and denser rocks to the east. A good example is near latitude 45°N where the dense sequence containing metabasalts is thrust over the miogeosynclinal rocks (see also Griscom and Bromery, this volume; Clark and Eakins, this volume) Here, a near-surface density contrast of about 0.2 suffices to produce the major part of the gravity relief observed in the east-west direction.

Density Variations at Depth

The preceding arguments indicate that the masses responsible for the main high and possibly the western low must be sought at depth. At least three origins might be considered: (1) The mass configurations existed before the onset of Paleozoic orogenies; these configurations controlled the development of tectonic and gravity patterns during later orogenies; (2) The configurations originated as the result of structural deformation of a normal continental region or of a transitional region at the continental margin; and (3) The configurations originated primarily from intrusion of igneous material.

FIGURE 30–3. Bouguer anomaly map. Symbols for geologic units are: (1) Precambrian rock of Adirondack massif and Green Mountain–Berkshire Hills anticlinorium, and gneisses of Chester, Athens, Sadawga, and other domes. (2) Paleozoic sedimentary, metasedimentary, and volcanic rocks (except Taconic allochthon). (3) Taconic allochthon. (4) Oliverian Plutonic Series. (5) Ultramafic bodies (largely serpentinite). (6) New Hampshire Plutonic Series, including some undifferentiated Highlandcraft Plutonic Series. (7) Felsic intrusive rocks of eastern Vermont. (8) Triassic sequence. (9) White Mountain Plutonic–Volcanic Series. Geology generalized from Billings (1955), Doll et al. (1961), Emerson (1917), Fisher et al. (1962), and compiled map of this volume.

In its simple form, the first origin seems unlikely, but it is listed here to emphasize the point that the crust was certainly not a mass of uniform density at the onset of orogenesis. The anomalies of the Adirondack massif have been mentioned in this regard. Moreover, the existence of the anomalies of the foreland and midcontinent, particularly such striking features as the midcontinent and Michigan basin highs with their flanking lows (see Woollard and Joesting, 1964), indicates that the Precambrian fabric cannot be ignored even though we do not have the information to incorporate it in our interpretation.

Structural deformation of the crust and upper mantle provides an easy explanation for the major anomalies. Deformation of the Mohorovicic and/or Conrad "discontinuities" could produce the observed anomalies. The idea is attractive because, at least in New England, the region of the main high is generally one of uplift and the region of the western low is one of depression. Furthermore, the amplitudes of the required deformations are of the same order as the structural relief deduced from geological mapping. Lack of exact coincidence of gravity features and observable structure could be explained away on the basis of the asymmetry of the deformations, thrusting, and the like. The sharpness and complexity of structures required to explain the main high are not necessarily objectionable in an environment where the rocks were sufficiently plastic to have permitted the development of such mobile features as mantled gneiss domes.

An equally plausible interpretation for the gravity high is that it is the result of dense masses intruded into the crust; the flanking gravity lows could then owe their origin to structural depression or to depletion of dense materials within the upper mantle through intrusion into the crust or extrusion onto the surface.

No doubt, both structural deformation and large-scale plutonic activity were involved. Each phenomenon has several implications, which we now explore in greater detail.

Structural Deformation

Glennie (1933) suggested crustal warping as an explanation for certain gravity anomalies in India. Longwell (1943), Bean (1953), and Diment (1953, 1956) considered its application to the gravity anomalies of the Appalachians. Using Leet's (1941) seismic model for the New England crust, the idea was tested for profiles exhibiting the maximum gravity relief in western New England (Figs. 30–5 and 30–6). These configurations suggest the following: (1) the

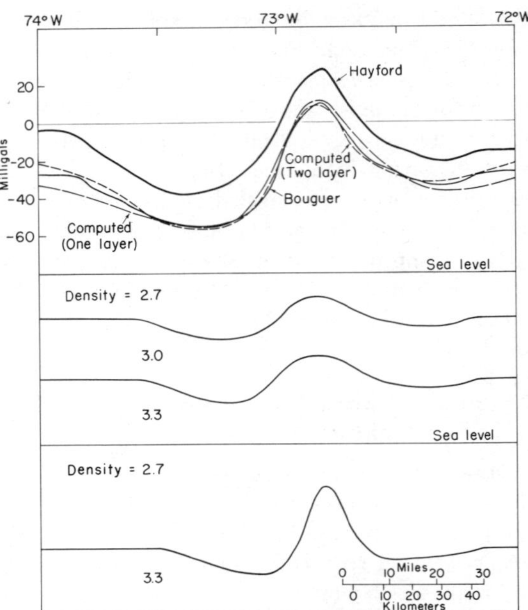

FIGURE 30–5. Two mass distributions, caused by crustal warping, that will produce anomalies along profile AA' of Figure 30–3. No vertical exaggeration.

amplitude of warps required of a one-layer crust is larger than those required of a two-layer crust. (2) The amplitude of the warp required is considerably less where the eastern low is poorly developed (Fig. 30–6), or if the eastern low can be regarded as the result of felsic intrusive rocks close to the surface. (3) If the gravity features are the result of relief on the Conrad discontinuity alone, the required deformation is large (Fig.

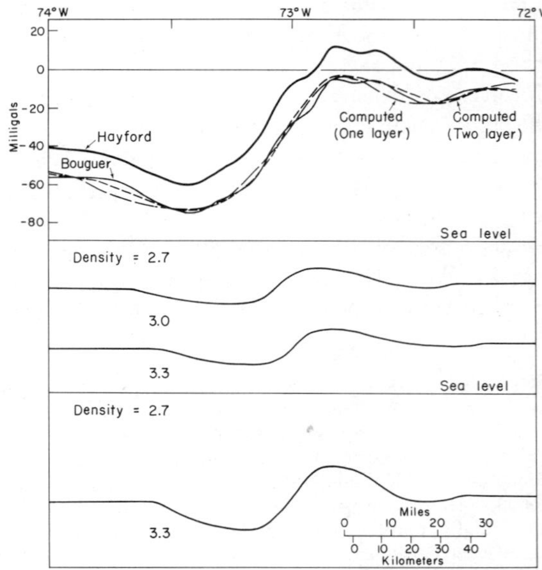

FIGURE 30–6. Mass distributions for profile BB'; see caption for Figure 30–5 for further data.

30–7), especially where the eastern low is well developed. (4) If we insist that both the western and eastern lows are caused by anomalous masses deep within the crust, the dense mass responsible for the gravity high must come close to the surface (say 5–10 km) in order to offset the effect of the deep low-density masses. However, if either of the anomalous masses responsible for the lows is shallow the dense mass need not be so large. (5) Although the configurations of the anomalous bodies are drawn with smooth boundaries, considerable modification is permissible; for example, the bodies could be bounded by faults, or could result from combinations of folds and intrusions of dense material.

Another variant on the crustal warping scheme involves buckling a crust whose density increases continuously with depth. The situation is illustrated in the western part of the section in Figure 30–8. Here, the density increase with depth in each column is assumed to be everywhere the same; hence, the density distribution can be approximated by vertical columns. The columns are arbitrarily cut off near the Mohorovicic discontinuity; had they not been, the density contrasts required would be less.

Figure 30–8 schematically illustrates another mechanism for producing the gravity high within the crust. The crust east of the steep gradient between the western low and the main high is regarded as more dense because of the eugeosynclinal metasedimentary rocks and abundant mafic materials. The eastern low is regarded as a consequence of light felsic plutonic rocks which are more widely distributed at depth than their surface expression would indicate. The concept of

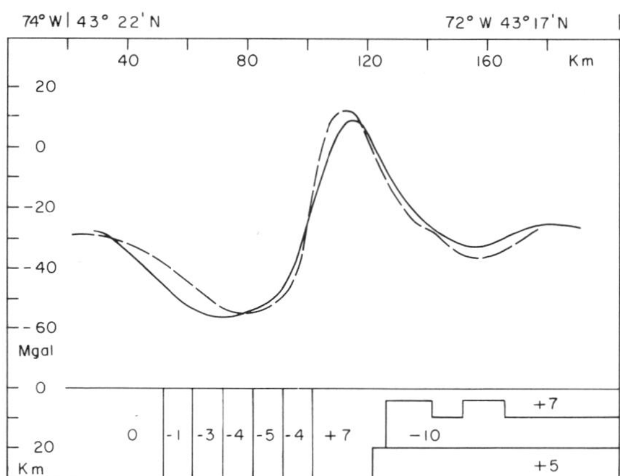

FIGURE 30–8. Mass distributions for profile AA′ assuming that density increases continuously with depth. Numbers indicate density contrast (g/cm³) times 100. Solid line, observed values; dashed line, computed values.

this type of denser upper crust beneath the gravity high may also have application to other parts of the Appalachians; however, it fails to explain local details such as the coincidence of the main gravity high with the light gneisses of the central part of the Green Mountain anticlinorium.

Preliminary interpretation of seismic refraction studies in the southern Appalachians (J. C. Roller, 1967, written communication) suggests that thickening of an intermediate layer of velocity 6.7 km/sec is mainly responsible for the gravity high. Such velocities are compatible with rocks of gabbroic composition. If these refraction data are applicable to the Appalachians as a whole, we would be restricted to a density contrast of about 0.3 g/cm³, and some variant on the model of figure 30–7 would then seem preferable.

Deformation of a normal continental crust is implied in the preceding illustrations. Undoubtedly, the situation was much different when the orogeny began. It is possible that the continental margin was close to the western edge of the eugeosyncline (see Rodgers, this volume) and that the sediments of the eugeosyncline were deposited on a subsiding oceanic or transitional crust. Subsequent uplift of such a region could bring the dense lower part of the former transitional crust close to the surface in the vicinity of the gravity high.

In the preceding discussion, we have also assumed that there is no significant lateral variation in density within the uppermost mantle. The gravitational effect of such a variation is probably small (Clark and Ringwood, 1964, p. 81), but it may be present as a consequence of the transition from a continental to oceanic region and/or as a consequence of warping density

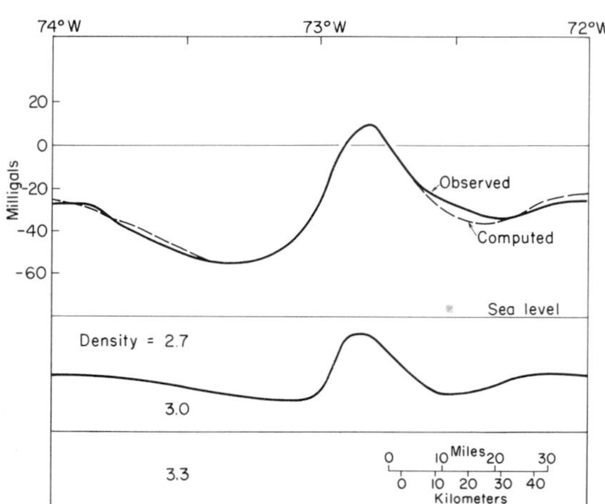

FIGURE 30–7. Mass distributions for profile AA′, assuming that the gravity data results from relief on the Conrad discontinuity alone.

stratification in the uppermost mantle. As the origin of such variations is linked to production of basalts, its discussion is deferred until the next section.

Large Mafic and Ultramafic Intrusions

A likely interpretation of the gravity high is that it represents a large mass of dense rock intruded into the crust. The idea is not new; de-Cizancourt (1948) suggested this to explain a similar anomaly in the Pyrenees. Similarly the gravity high over the Ivrea zone of the Alps has long been linked to the mafic and ultramafic rocks of this zone (Niggli, 1946). Recent summaries of the geophysical and tectonic data for the Alps (Laubscher, 1966, Berckhemer and Hersey, 1966) indicate that a mass of anomalously high seismic velocity occupies the greater part of the crust in the vicinity of the Ivrea zone. Fitzpatrick (1959) regards the main gravity high in Quebec as a consequence of intrusion of ultramafic material, and suggests that partial serpentinization of this material may have been responsible for uplift. Griscom (1963) has suggested a belt of mafic intrusions emplaced along fracture zones as an explanation for the Appalachian high.

The idea is attractive because small masses of mafic or ultramafic rocks are exposed in the vicinity (but not necessarily along the axes) of each of the gravity highs referred to above (see Chidester, this volume). Moreover, intrusive masses seem more compatible in shape with the mass configurations required to explain the main gravity high than are those resulting from simple crustal deformations.

It is not clear whether we are dealing with mafic or ultramafic rocks or a mixture containing significant quantities of both. The preliminary seismic-refraction results from the southern Appalachians (J. C. Roller, 1967, written communication) differ from those in the Ivrea zone of the Alps. In the Ivrea zone, Berckhemer and Hersey (1966) indicate velocities of 7.2–7.4 km/sec at depths less than 10 km below the surface and Laubscher (1966) shows a depression of the Mohorovicic discontinuity. Velocities of 6.7 km/sec generally represent the intermediate layer, and presumably indicate materials of gabbroic composition (Birch, 1961) or serpentinized peridotite, but velocities of 7.2–7.4 km/sec are rather high for such materials and could represent slightly altered mantle material.

If the mass is gabbroic, the density contrast between it and the surrounding crustal rocks can hardly exceed 0.3 g/cm^3. Even with this density contrast, the mass must occupy at least half the crust in regions where the gravity high is best developed. The size required of the dense mass becomes larger if we exclude the possibility of a density contrast in the lower crust (sub-Conrad), or if we insist that the anomalous masses responsible for the flanking lows are deep. If the anomalous mass is ultra-mafic, the density contrast would be larger, and smaller sizes are permissible. However, regardless of whether the mass is mafic or ultramafic, it must occupy a large fraction of the crust and, at least in places, come close to the surface. Furthermore, the size of anomaly requires that the mass, especially if it is gabbroic, be relatively uncontaminated; it could not be a fracture zone containing subordinate amounts of mafic material as is implied in some models.

It is important to know whether material is mafic (basalt-eclogite) or ultramafic (dunite-peridotite). If one accepts intrusive bodies of gabbroic composition as the cause of the gravity high and requires that this material be derived from the mantle not far below the crust, then several consequences are possible, depending on the petrological model adopted for the upper mantle (see Clark and Ringwood, 1964, for discussion) and on how far into the mantle the rupture containing the mafic material extends.

If the uppermost mantle is regarded as a mixture of peridotite and eclogite, extraction of the eclogite through partial fusion and intrusion into or onto the crust would leave a less dense and less radioactive peridotitic residuum. We might think of this low-density zone in the mantle as more widespread than the zone into which the mafic rocks were intruded, extending out under the region of the western gravity low. Furthermore, the zone might be considered as a cause for the lows flanking the main high, for the compensatory uplift of the region of the main high, and, to a lesser extent, for uplift near the eastern edge of the western low. In order for this low-density zone to be significant in these respects, either the density contrast must be substantial or its vertical extent must be large. The density contrast between eclogite and peridotite is about 0.25 g/cm^3 (Clark and Ringwood, 1964); clearly we must start with a mantle rich in eclogite if this mechanism is to be important.

Another aspect of this model, which introduces further complexity, is that such large mafic intrusions would probably also occupy a part of the uppermost mantle. If the gabbroic material is chemically different from the material it intrudes, then the transition from crust to mantle might take place at a different depth interval below the anomaly than below either side. If the basalt-eclogite transition takes place at a shallower depth than the normal crustal thickness, as Pakiser (1965) suggests for the eastern United States, a heavy root would be present in the crust. On the contrary, if the transition zone extends into the mantle, a light root would underlie the region of the gravity high. Preliminary analysis of the seismic refraction results in the southern Appalachians (J. C. Roller, 1967, written communication) provides no evidence for such roots. Under the Ivrea zone of the Alps, the Mohorovicic discontinuity is anomalously deep (Laubscher, 1966, Fig. 2); this may reflect higher temperatures in this younger region.

Preferred Model

Selection of a preferred interpretation involves a choice between an intrusive and a structural model and a choice as to whether separation of basaltic magmas from the mantle left a significant density variation within the mantle. The information available does not permit clear decisions. Moreover, both intrusive and structural mechanisms may be important and may in some ways be indistinguishable, because both involve large scale lateral and vertical transport of mantle material.

The highly schematic model of Figure 30–9 is intended to illustrate a possible crustal configuration in western New England. The principal features of this model are:

1. The preorogenic crust is regarded as a transitional one between continental and oceanic crusts with the granitic layer thinning to the east.

The Paleozoic eugeosynclinal sedimentary rocks are supposed to have been deposited on the eastern edge of this transitional crust and upon an oceanic crust. The felsic intrusive rocks are regarded mainly as a product of the partial fusion of these sedimentary rocks. It seems necessary that the main part of the eugeosynclinal sedimentary rocks was deposited directly on an oceanic crust, for otherwise we would face the problem of imbalance in density and radioactivity (Birch Roy, and Decker, this volume) that would result from a thick granitic part of the crust below the present metasedimentary rocks of the eugeosynclinal region. It is possible, of course, that excess granitic material could have been removed by refusion and subsequent intrusion into and extrusion onto the eugeosynclinal sedimentary rocks, but the volumes removed would have to have been very large.

2. The main gravity high is regarded as the result of uplift of dense sub-Conrad rocks along a narrow zone of deformation at the western edge of that high. The zone of deformation may not have been a single zone and probably exhibited an en echelon pattern as evidenced by offsets in the axis of the main gravity high. The uplift presumably occurred near the end of the middle Ordovician (Zen, this volume) and was probably coupled with (a) the subsidence of the eugeosyncline in response to the load of accumulating sediments and volcanic material and (b) the subsidence of the miogeosyncline and emplacement of the Taconic klippen. The uplift and denudation which exposed the eugeosynclinal rocks took place much later. The positioning of the breaks along which the Ordovician and postorogenic uplifts took place may account for some of the

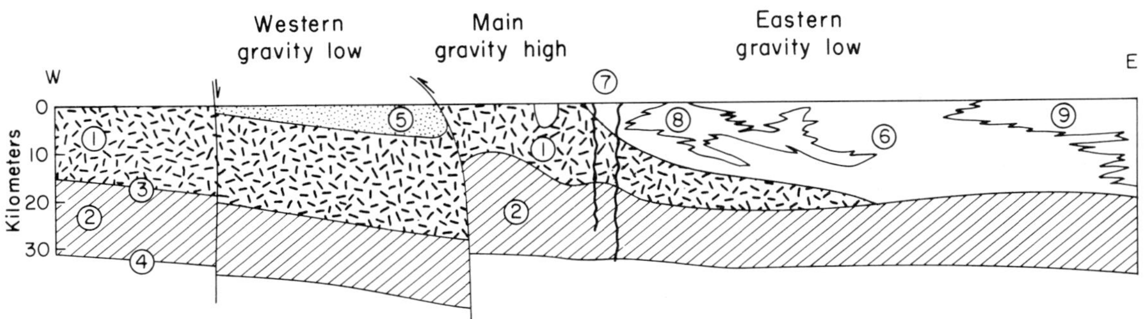

FIGURE 30–9. Hypothetical schematic E–W cross-section illustrating distribution of principal density units. (1) Precambrian granitic layer of the continental crust. (2) Intermediate layer of continental crust or oceanic crust and part of oceanic mantle transformed to gabbro and/or serpentinite. (3) Conrad discontinuity. (4) Mohorovicic discontinuity. (5) Sedimentary rocks of the miogeosyncline. (6) Metasedimentary and metavolcanic rocks of the eugeosyncline. (7) Ultramafic rock (largely serpentinite). (8) Calc-alkalic syntectonic intrusive rocks. (9) Alkaline postorogenic intrusive rocks.

difficulties in comparing the relation between gravity and geologic patterns in New England and in the central and southern Appalachians. It seems likely that the exact positions and the number of breaks might be controlled by local conditions and preexisting fabric.

3. The western low is regarded as the result of depression of lighter crustal material into denser realms.

4. The eastern low is thought to be the result of the relative paucity of dense sub-Conrad material at shallow depths and the relative abundance of light felsic intrusive rocks. Variations in the gravity pattern may be the result of mafic intrusive rocks and displacement at and below the Conrad that accompanied the uplift of the eugeosynclinal metasedimetary rocks.

5. Mafic volcanic material forms a significant part of the eugeosynclinal sequence. Perhaps mafic intrusive masses (not shown in Figure 30–9) also contribute significantly to the gravity field; in this model, it is supposed that they do not.

6. The ultramafic masses are regarded as minor features generated by alteration of the sub-Conrad materials at the time of the uplift causing the main high.

ANOMALIES OVER FELSIC INTRUSIVE ROCKS

Felsic intrusive rocks increase in abundance from the region just east of the main gravity high into central New Hampshire. The felsic rocks are generally some 0.1 g/cm^3 less dense than their metasedimentary hosts, and negative gravity anomalies result (see also Kane and Bromery, this volume). Bean (1953) and Joyner (1963) have analyzed several of these anomalies and concluded that the density contrasts evident at the surface do not persist to depths of more than about 5 km. Similar conclusions have been reached by Bott (1956) for other regions, but he estitmated some depths up to about 12 km.

The Barre pluton (C–7 in Fig. 30–3) is an example. The 20 mgal anomaly over this pluton can be explained by a mass the shape of a truncated cone having a density contrast of 0.1 g/cm^3 and extending to a depth of 7 km (Diment, 1953). Gravity evidence suggests no deeper extent for other plutons in the area surveyed. We are left with the possibilities that these intrusions are floored, and that either the density of the felsic masses increases or the density of the enclosing rocks decreases with depth.

The increasing abundance of light felsic rocks to the east has an important bearing on the interpretation of the main gravity high. If the eastern low is caused by these light intrusive rocks, then the high can be explained by an abrupt increase in density at the edge of eugeosynclinal belt, modified in the eastern part by the light intrusive rock.

The over-generalization in the preceding statements becomes apparent when the different magma series are considered in more detail. Although most of the intrusive rocks are felsic, some are mafic or intermediate in composition. For an example, gravity highs over several postorogenic plutons of the White Mountain Plutonic–Volcanic Series clearly indicate mafic or intermediate rocks beneath surface exposures of granitic rocks (Joyner, 1963; Griscom and Bromery, this volume).

CROSS-TRENDING ANOMALIES

Several anomalies seem to be aligned in a direction roughly perpendicular to the main tectonic trends. Some of these, such as those associated with the intrusive rocks of the Monteregian Hills just to the north of the area surveyed, are very pronounced and suggest that similar igneous rocks should be sought to the south where similar anomalies are known.

Comparison of gravity profiles along the axes of the main high and western low (Fig. 30–4) suggest that some features affect both. The correlation is particularly evident in the northern part of the area, where gravity culminations correspond to transverse structural axes in the Lake Champlain lowland. The correlation between the axes of the high and low is not as good to the south; but by and large suggests that anomalous masses do extend across the main tectonic trends and effect both the western low and main high.

The Monteregian Hills

Eight small plutons are exposed in a zone extending east-southeast from Montreal into the region of the main gravity high. The rocks are alkaline but range from felsic (nordmarkite) to mafic (olivine essexite) (Dresser and Denis, 1944). The plutons are definitely postorogenic; K/Ar dates suggest late Mesozoic ages (Fairbairn et al., 1962).

Fitzpatrick's (1959) analysis of the gravity anomalies indicates that : (1) Gravity highs (as large as 19 mgal) are associated with the individual plutons, but the size of the anomaly depends on the proportion of mafic material and the size of the pluton (average diameter 4 km). (2) The zonation of rock types within an individual pluton is not horizontal, but more nearly vertical, as indicated by anomaly offsets over zoned plutons. (3) The vertical dimensions of the plutons are roughly the same as the horizontal. (4) The regional gravity high which parallels the trend of the hills decreases in amplitude to the east, suggesting that the mass causing it becomes deeper or more felsic.

The Plattsburgh Alignment

About 70 km south of the Monteregian Hills, a series of gravity highs are aligned in a direction approximately parallel to the Monteregian trend (Fig. 30–10). Several of the anomalies were detected in earlier surveys (Diment, 1953; Simmons, 1964); Simmons (1964) and Thompson and Miller (1958) show their possible extension to the west and north.

The rocks responsible for these gravity highs are not exposed at the surface. The anomalies are located just north of the Adirondack massif in a region covered by less than a kilometer of Cambrian and Ordovician sedimentary rocks. Analysis of gravity and magnetic data (Diment, 1964) indicates that the top of the anomalous rocks under the sharpest anomaly (Plattsburgh) is no more than a kilometer from the surface, about the same as the thickness of the sedimentary rocks in this locality. Therefore, the available geophysical evidence does not tell us whether these could be Precambrian features, or must be younger, perhaps Mesozoic like the intrusives of the Monteregian Hills.

The magnetic anomaly patterns near Plattsburgh (Fig. 30–11) are remarkably similar to the gravity patterns. This indicates that the same body is responsible for both types of anomaly. The contrasts in density and intensity of magnetization must be at least 0.3 g/cm^3 and 0.005 emu, respectively; thus these large anomalies must be the result of mafic or ultramafic rock.

Several investigators (for example, Fitzpatrick, 1959; King, 1951) have related the Monteregian intrusive bodies to the broad zone of east-southeast-trending fracture in the vicinity of Ottawa. Perhaps the Plattsburgh alignment also owes its origins to intrusive bodies along faults related to this zone. The idea is geometrically attractive because both the Plattsburgh and Monteregian alignments seem to extrapolate into the Ottawa graben. The possibility of a disparity in age between the Plattsburgh and Monteregian intrusions is not necessarily objectionable, because the change of main tectonic trends, from north to northeast, no doubt existed early in the orogenic history of the Appalachians and anomalous geologic activity near this change in trend might be expected throughout the tectonic evolution of the

region. Indeed, the east-northeast alignment of epicenters (Smith, 1962) in the region suggests that the Plattsburgh–Monteregian zone is still active.

The regional gravity high, upon which the Plattsburgh anomaly is superimposed, extends across the Lake Champlain lowland where it roughly coincides with a structural feature known as the Burlington cross-anticline (A–4 and B–4 of Fig. 30–10 and Fig. 30–4). The gravity and magnetic relief across this structure, is large (Fig. 30–12) and cannot be explained by contrasts between the Precambrian rocks and the Paleozoic sedimentary rocks. As the density and susceptibility contrasts probably extend deep within the Precambrian, we are left with the notion of a dense and magnetic mass below this cross-trending feature (see also Griscom and Bromery, this volume).

Near Milton (B–4), on the south flank of the gravity feature, there is a change in facies in the post-Dunham Cambrian rocks, such that carbonate rocks occur to the south and slates to the north. The change has been variously interpreted as the intertonguing of shale and carbonate rock (Cady, this volume), as a positive axis separating a carbonate-filled and a clastic-filled basin (Shaw, 1958), and as a transition from a shallow-water to deep-water depositional environment (Rodgers, this volume; Zen, this volume).

Adopting the idea of a transition from a shallow-water to deep-water environment, we might regard the Plattsburgh alignment as the southern edge of an east-southeast trending rift or basin, along which mafic intrusions were localized. However, the Plattsburgh alignment may be simply the result of intrusions related to the cross-trending structural uplift.

Monkton Cross-Anticline

As in the Plattsburgh region, a cross-trending structural axis (Cady, 1945) nearly coincides with a gravity high which appears to extend across the Lake Champlain valley into the region of the main gravity high. Here again, anomalies caused by Precambrian anorthosite may be important; for example, at least at its west end, the low north of the axis probably results from a body of anorthosite.

Cuttingsville and Ascutney Stocks

There are two stocks of alkaline rocks in Vermont on the flanks of the main high; one near Ascutney (G–7) and the other near Cuttingsville (G–5). The Ascutney stock (Daly, 1903, Chapman and Chapman, 1940) is about 8 km across, and the Cuttingsville stock (Eggleton, 1918) is

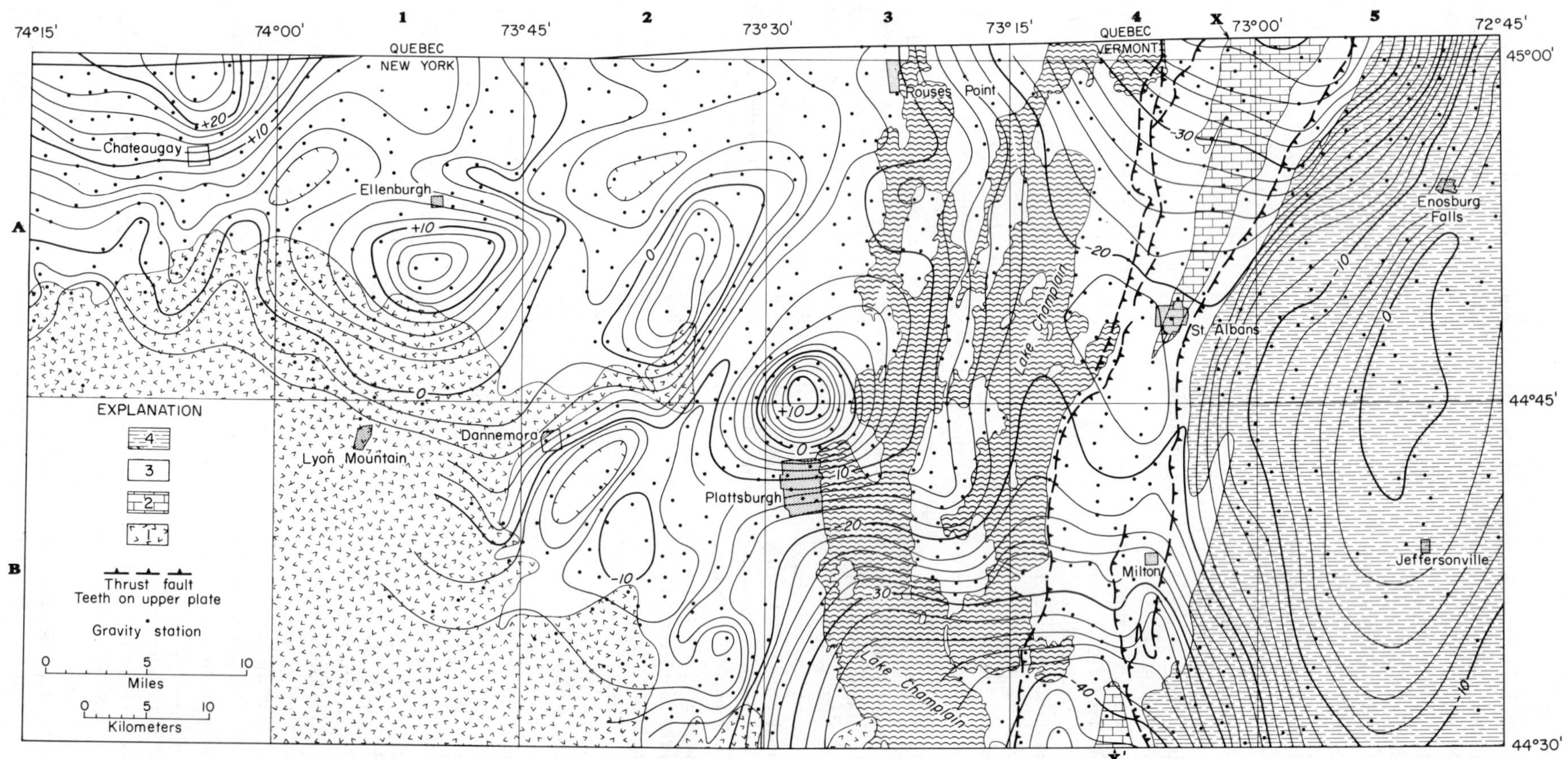

Figure 30–10. Bouguer anomaly map of northeastern New York and northwestern Vermont. Geologic symbols as follows: (1) Precambrian rocks of the Adirondack massif. (2) Ordovician rocks in the cores of the St. Albans and Hinesburg syn- clinoria. (3) Undivided Cambrian and Ordovician rocks. (4) Cambrian rocks of the Hinesburg thrust slice, belonging to the eugeosynclinal sequence. Contacts from Doll et al. (1961) and Fisher et al. (1962). Wavy pattern is Lake Champlain.

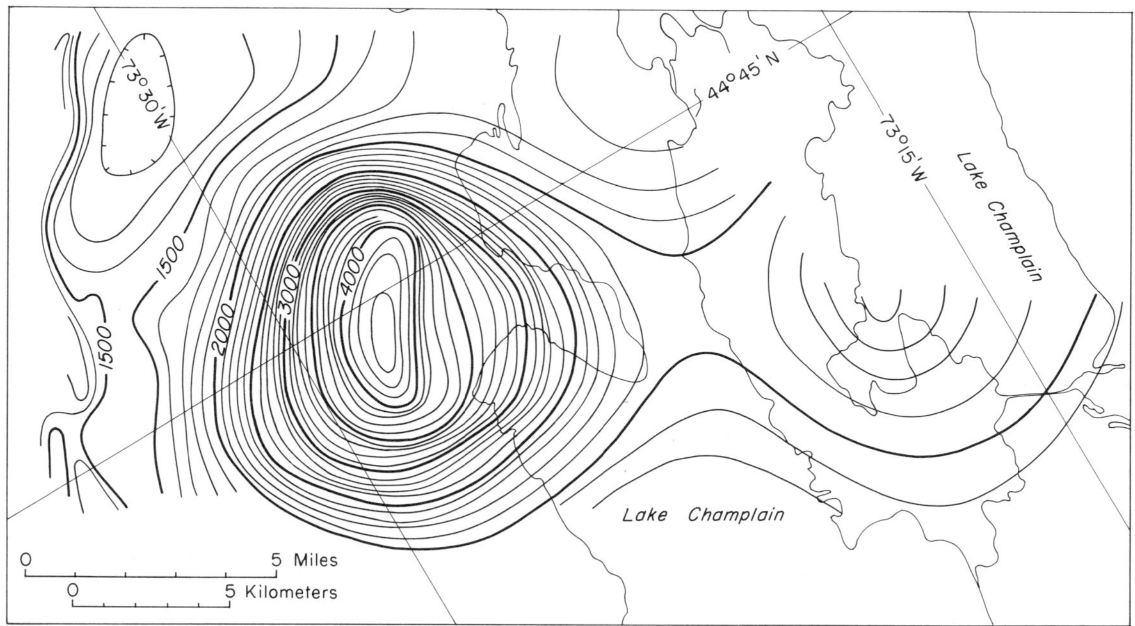

FIGURE 30–11. Aeromagnetic map in the vicinity of Plattsburgh, N. Y. Contour interval 200 gammas. Datum is arbitrary. Flight elevation 300 meters barometric.

less than 3 km across. The rocks of these two stocks are largely syenite with some granite, diorite and gabbro. These postorogenic rocks are similar to the rocks of the White Mountain Plutonic–Volcanic Series. Faul et al. (1963) assign a Mesozoic age to the Ascutney stock.

Although these stocks are small, they are of special interest because they are the only exposed postorogenic alkaline masses (with the exception of some of the dikes) in the area surveyed. They are alkaline, like the rocks of the Monteregian Hills, but are predominantly felsic rather than mafic.

Each stock is associated with a local gravity low, although the gravity pattern near the Ascutney stock is complicated by the presence of the low density gneisses of the Chester dome (see also Griscom and Bromery, this volume). In a more regional sense, the gravity along the axis of the main high is lower at the latitude of the stocks. Perhaps this is evidence for an extensive mass of low density material that was intruded into the anomalous mass that causes the main gravity high. Gravity provides no evidence for the extension of this zone to the east or west of the main high, but this is not surprising if one assumes that the density contrast is between the light intrusive rocks and the dense mass responsible for the main high.

TACONIC ALLOCHTHON

The Taconic allochthon is one of the more prominent geologic features of western New England. It extends from lat 43°45'N. (Fig. 30–3) southward out of the area surveyed. The allochthon is now regarded as a series of six discrete tectonic slices that have been thrust into their present position from the region of the Green Mountain–Berkshire Hills anticlinorium (Zen, 1967, this volume).

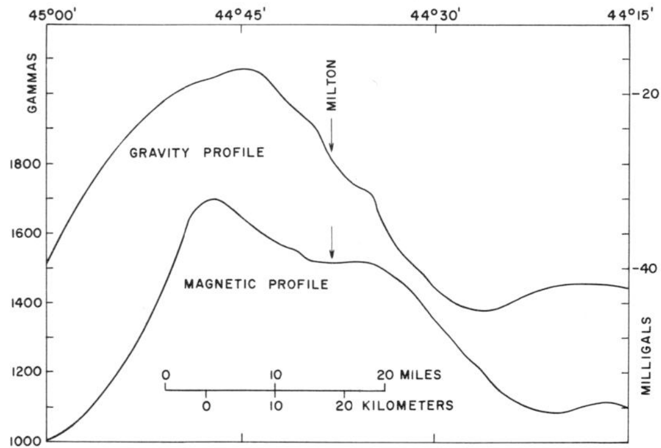

FIGURE 30–12. Aeromagnetic and gravity profiles along line X-X' (Fig. 30–10) across Plattsburgh alignment along eastern shore of Lake Champlain. Flight elevation 1000 meters barometric.

The gravity anomalies provide little information concerning the configuration of the sequence because there is little density contrast between the sequence and the surrounding rocks; however, the anomalies indicate anomalous crustal conditions in its vicinity. The allochthon is located in the region where the western low is best developed and where the gravity relief is largest (Figs. 30–3 and 30–4). It is also in the region where the structural relief seems largest. It is particularly notable that the northern end of the Taconic allochthon closely coincides with an abrupt southward increase in gravity along the axis of the gravity high. These observations seem to suggest that the gravity relief may be an indicator of structural relief at least in a limited locality, even though such a relation clearly cannot be widely applied. However, if we accept this premise and note the offset in the axis of the main gravity high in southern Vermont, we might expect that thrust sheets emplaced by gravitational sliding would be complex in form because the axis of uplife is offset.

Zen (this volume) has shown that the region in which the Taconic allochthon was emplaced was quite unstable during the Middle Ordovician and that the allochthon was emplaced shortly thereafter by gravity sliding of the sequence from the region of the rising Green Mountain–Berkshire Hills anticlinorium. It is reasonable to suppose that the subsidence and uplift were coupled in the sense that load impressed on the crust by the emplacement of the allochthonous sequence caused additional lateral transfer of dense mantle materials into the region of the main gravity high and depression of the light sialic crust into denser realms. Thus, we might expect to find greater gravity relief in the vicinity of large klippen than elsewhere. This is certainly so in northwestern New England, where the gravity relief north of the Taconic allochthon is small; indeed it is smaller here than in the rest of the Appalachians, a fact that may be related to the presence of the Adirondack massif. The notion that the development of gravity relief is abetted by emplacement of large thrust masses is compatible with the idea that the Precambrian of the Blue Ridge, which is located partly in the western gravity low, is largely a series of overthrust masses. Subsequent erosion of these masses would tend to produce uplift which may have been important in the evolution of the Valley and Ridge Province.

REFERENCES

Bean, R. J., 1953, Relation of gravity anomalies to the geology of central Vermont and New Hampshire: Geol. Soc. America Bull., v. 64, p. 509–538

Berckhemer, H. and Hersey, J. B., 1966, Some features of the Alpine–Mediterranean orogenesis, *in* W. H. Poole, *ed.*, Continental margins and island arcs: Geol. Survey of Canada, paper 66–15, p. 114–123

Billings, M. P., 1955, Geologic map of New Hampshire: U. S. Geol. Survey, scale 1:250,000

———, Rodgers, John, and Thompson, J. B., Jr., 1952, Geology of the Appalachian Highlands of east-central New York, southern Vermont, and southern New Hampshire: Geol. Soc. America Guidebook for Field Trips in New England, Boston, Mass., p. 1–71

Birch, Francis, 1961, The velocity of the compressional waves in rocks to 10 kilobars, 2: Jour. Geophys. Res., v. 66, p. 2199–2224

Bott, M. H. P., 1956, A geophysical study of the granite problem: Geol. Soc. London, Quart. Jour., v. 112, p. 45–67

Bromery, R. W., 1967, Simple Bouguer gravity map of Massachusetts: U. S. Geol. Survey, Geophys. Inv., GP–612

Cady, W. M., 1945, Stratigraphy and structure of west-central Vermont: Geol. Soc. America Bull., v. 56, p. 515–586

Chapman, R. W., and Chapman, C. A., 1940, Cauldron subsidence of Ascutney Mountain, Vermont: Geol. Soc. America Bull., v. 51, p. 191–212

Clark, S. P., Jr., and Ringwood, A. E., 1964, Density distribution and constitution of the mantle: Rev. Geophys., v. 2, p. 35–88

Cohee, G. V., *ed.*, 1962, Tectonic map of the United States: U. S. Geological Survey, scale 1:2,500,000

Daly, R. A., 1903, Geology of Ascutney Mountain, Vermont: U. S. Geol. Survey Bull. 209, 122 p.

deCizancourt, H., 1948, Essai d'interprétation de la tectonique profonde des Pyrénées: Soc. Geol. France Bull., v. 18, p. 271–284

Diment, W. H., 1953, A regional gravity survey in Vermont, western New Hampshire, and eastern New York: Ph.D. thesis, Harvard Univ., 176 p.

———, 1956, A regional gravity survey in Vermont, western New Hampshire, and eastern New York [abs.]: Geol. Soc. Amer. Bull., v. 67, p. 1688

———, 1964, Gravity and magnetic anomalies in northeastern New York [abs.]: Geological Society of America Abstracts for 1963: Geol. Soc. Amer. Special Paper no. 76, p. 45

Doll, C. G., Cady, W. M., Thompson, J. B. Jr., and Billings, M. P., *compilers* and *editors*, 1961, Centennial geologic map of Vermont: Montpelier, Vermont, Vermont Geol. Survey, scale 1:250,000

Dresser, J. A., and Denis, T. C., 1944, Geology of Quebec, Descriptive geology: Quebec Dept. of Mines, Geol. Rept. 20, 544 p.

Duerkson, J. A., 1949, Pendulum gravity data in the United States: U. S. Coast and Geod. Survey, Spec. Publ. 244, 218 p.

Eggleton, J. W., 1918, Eruptive rocks at Cuttingsville, Vermont: Am. Jour. Sci., v. 45, p. 377–410

Emerson, B. K., 1917, Geology of Massachusetts and Rhode Island: U. S. Geol. Survey Bull. 597, 289 p.

Fairbairn, H. W., Faure, Gunter, Hurley, P. M., and Pinson, W. H., 1962, Evidence of the origin and time of separation of magmas of the Monteregian Hills, Quebec, from development of radiogenic Sr^{87} [abs.]: Geol. Soc. Amer. Special Paper no. 68, p. 174

Faul, Henry, Stern, T. W., Thomas, H. H. and Elmore, P. L. D., 1963, Ages of intrusion and metamorphism in the northern Appalachians: Am. Jour. Sci, v. 261, p. 1–19

Fisher, D. W., Isachsen, Y. W. Rickard, L. V., and Offield, T. W., *compilers* and *editors*, 1962, Geologic map of New York 1961: New York State Mus. and Sci. Service Geol. Survey Map and Chart Ser. no. 5, scale 1:250,000

Fitzpatrick, M. M., 1959, Gravity in the Eastern Townships of Quebec: Ph.D. Thesis, Harvard Univ. 133 p.

Flawn, P. T., *chairman*, 1967, Basement Map of North America: Am. Assoc. Petroleum Geologists and U. S. Geol. Survey, scale 1:7,500,000

Glennie, E. A., 1933, Crustal warpings: Royal Astron. Soc. Monthly Notice Geophys. Suppl., v. 3, p. 170–176

Griscom, Andrew, 1963, Tectonic significance of the Bouguer gravity field of the Appalachians [abs.]: Geol. Soc. Amer. Special Paper no. 73, p. 163

Heiskanen, W., 1948, On the geoid study of the International Isostatic Institute: Bull. Geodesique, new ser., no. 7, p. 50–54

Joyner, W. B., 1963, Gravity in north-central New England: Geol. Soc. America Bull., v. 74, p. 831–858

Kane, M. F., and Bromery, R. W., 1966, Simple Bouguer gravity map of Maine: U. S. Geol. Survey, Geophys. Inv. Map GP–580

King, P. B., 1951, The tectonics of middle North America—Middle North America east of the Cordilleran system: Princeton, Princeton Univ. Press, 203 p.

———, 1964, Further thoughts on tectonic framework of the south eastern United States: *in* W. D. Lowry, *ed.*, Tectonics of the southern Appalachians, Virginia Polytech. Inst., Dept. Geol. Sciences, Mem. 1, p. 5–32

Laubscher, H. P., 1966, The project for deep drilling in the Swiss Alps, *in* D. C. Findlay and C. H. Smith, *eds.*, Drilling for scientific purposes: Geol. Survey of Canada, Paper 66–13, p. 14–24

Leet, L. D., 1941, Trial travel times for northeastern America: Seismol. Soc. America Bull., v. 31, p. 324–334

Longwell, C. R., 1938, Geologic interpretation of the gravity anomalies in northeastern United States [abs.]: Am. Geophys. Union Trans., 19th Ann. Mtg. pt. 1, p. 84.

———, 1943, Geologic interpretations of gravity anomalies in the southern New England–Hudson valley region: Geol. Soc. America Bull., v. 54, p. 555–590

Niggli, Ernst, 1946, Über den Zusammenhang zwischen der positiven Schwereanomalie am Südfuss der West und der Gesteinzone von Ivrea: Eclogae Geol. Helvetiae, v. 39, p. 211–228

Pakiser, L. C., 1965, The basalt-eclogite transformation and crustal structure in the western United States: U. S. Geol. Survey Prof. Paper 525–B, p. B1–B8

Shaw, A. B., 1958, Stratigraphy and structure of the St. Albans area, north-western Vermont: Geol. Soc. America Bull., v. 69, p. 519–567

Simmons, Gene, 1964, Gravity survey and geological interpretation, northern New York: Geol. Soc. America Bull., v. 75, p. 81–98

Smith, W. E. T., 1962, Earthquakes of eastern Canada and adjacent areas, 1534–1927: Dominion Observatory Ottawa, Publ., v. 22, p. 271–301

Thompson, L. G. D., and Miller, A. H., 1958, Gravity measurements in southern Ontario: Dominion Observatory Ottawa Publ., v. 19, p. 321–378

Woollard, G. P., 1948, Gravity and magnetic investigations in New England: Am. Geophys. Union Trans., v. 29, p. 306–317

———, and Joesting, H. R., 1964, Bouguer gravity anomaly map of the United States: U. S. Geol. Survey, scale 1:2,500,000

Zen, E-an, 1967, Time and space relationships of the Taconic allochthon and autochthon: Geol. Soc. America Special Paper no. 97, 107 p.

Chapter 31

Gravity Anomalies in Maine*

M. F. KANE AND R.W. BROMERY

INTRODUCTION

THIS REPORT is an analysis of a Bouguer gravity map of Maine (Kane and Bromery, 1966). The gravity map was compiled from measurements at over 8,000 stations spaced about 1 mile (1.6 km) apart on traverses most of which are separated by 5 miles (8 km) but a few by as much as 10 miles (16 km). With this amount of station separation, gravity anomaly sources less than 5 miles (8 km) in horizontal extent will be reflected incompletely, except where the location of stations is such that critical anomaly characteristics are mapped fortuitously. Gravity anomalies whose sources are of greater areal extent are well mapped.

The gravity map of Maine (Fig. 31–1) is characterized by a large group of distinctive anomalies superimposed on a rather regular background field. Comparison with the geology of Maine (Doyle, 1967) shows that gravity anomalies clearly correlate with known geologic features. The major conclusions of this report are as follows: (1) Regional gravity variations in Maine have their source, at least in part, in the upper part of the crust; (2) large local gravity anomalies are due to igneous plutons; and (3) at depth, the lateral dimensions of some felsic plutons are very great compared with their outcrop area and vertical dimensions.

GEOLOGY

The most comprehensive publication on the geology of Maine is the recently published geologic map of the State (Doyle, 1967). Reports

* Publication authorized by the Director, U. S. Geological Survey.

on the particular aspects of the geology of the area are given in several places in this volume.

In general, the bedrock of Maine consists of tightly folded, well-indurated, stratified rocks of Paleozoic age. They are intruded in many places by a wide variety of igneous rocks, dominantly felsic in composition and mostly Devonian in age. The igneous plutonic rocks range from isolated outcrops to large exposures tens of miles in breadth. In some areas, there are substantial volumes of volcanic rock. The prevailing trend of the geology is northeasterly; it is exhibited both by structural and outcrop trends in rocks of sedimentary and volcanic origin, and in elongation and alignment of igneous plutonic exposures.

The northern third of the State appears relatively free of intrusive rocks whereas the southern two thirds appear dominated by them. The most widespread rock in the northern area, occupying most of the area northwest of line A–B (Fig. 31–1) is cyclically bedded, dark-gray slate and metasandstone of Devonian age. South of this line, metamorphosed rocks ranging in age from Cambrian to Devonian are exposed. Their lithology encompasses a wide range including siltstones, shales, sandstones, graywackes, and conglomerates. Volcanic rocks are commonly present as part of each lithic unit. Calcareous rocks occur in scattered areas throughout the state and appear to be more prevalent in the south.

The stratified rocks have all been regionally metamorphosed. The degree of metamorphism ranges from high in the southwestern third of the State to low in much of the northern two-thirds. Part of the change in metamorphic grade takes place in a narrow but irregular zone joining these two areas (Doyle, 1967). Many of the plu-

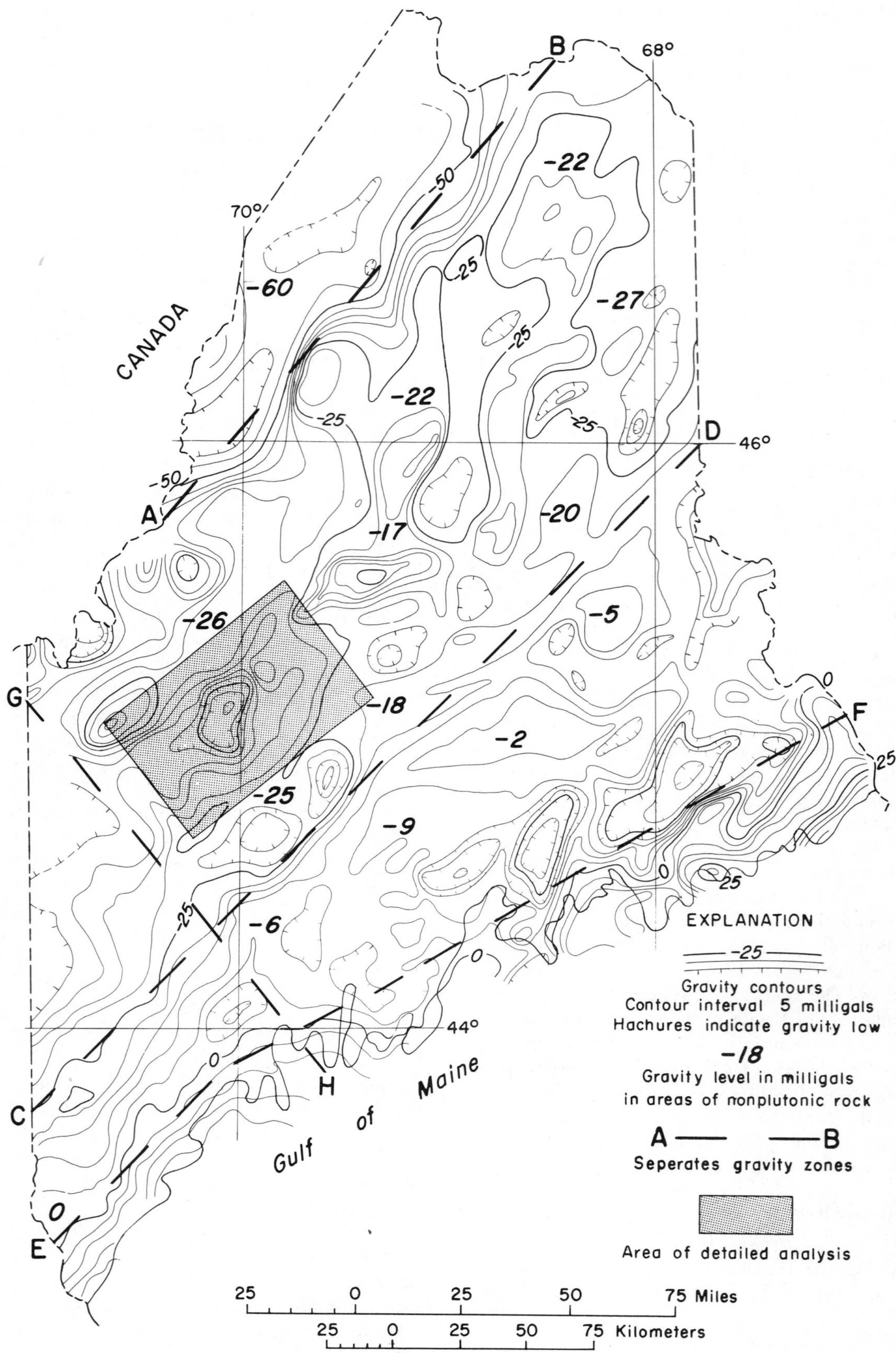

FIGURE 31–1. Bouguer gravity map of Maine. From Kane and Bromery, 1966.

tonic exposures in the north are surrounded by conspicuous rims of contact-metamorphosed stratified rocks.

The rocks southwest of, and in the vicinity of, line G–H (Fig. 31–1) are by far the most structurally complex. Many of the exposed rocks are gneisses, schists, and phyllites, reflecting the high degree of metamorphism pointed out above. Similar rocks are also common in areas that intervene among the plutonic rocks in the coastal region east of long. 69° W. The rocks northwest of line A–B (Fig. 31–1) are conspicuously lacking in structural elements. The remainder of the state appears intermediate in structural complexity. The foregoing summary reflects in part the progress of mapping in various parts of the State.

In the area enclosed by lines A–B and G–H, there is a crude symmetry in the distribution of gross rock units about a northeast-trending direction. The central part, comprising slightly over half the area, is dominated by rocks of sedimentary origin, widespread sections of which show little or no structure. This central zone is flanked by two wide zones of dominantly felsic intrusive rock. Most of the mafic plutonic rocks occur on the outer margins of these felsic zones; mafic plutonic rock, for example, is common along much of the coastal area east of long. 69° W. This crude symmetry does not hold, however, for the area to the southwest, which is characterized by high metamorphic grade and structural complexity (see Hussey, this volume).

GRAVITY

Density

Because of generally low porosity, the density of the rocks of Maine is largely a reflection of mineral density; therefore, gravity highs should occur over large coherent mafic masses (relative to host rock of sedimentary origin). Previous investigations (Bott and Smithson, 1967) in similar terrain show that gravity lows should occur over masses of felsic composition. As a matter of fact, both types of anomalies are present, but the gravity-low felsic-intrusion occurrence is far more common and probably of greater importance. The densities of the major constituent minerals can be used to derive normative rock densities according to their volume percentages. On this basis we have adopted the following for igneous plutonic rocks:

Normative felsic rock density 2.65 g/cm³
Normative mafic rock density 3.0 g/cm³

Although the high variability of the mineralogy of sedimentary rocks makes it difficult to assess an average density for them, the existence of such a regional average is suggested by the persistent background-gravity field levels over broad areas of the sedimentary rock. The density of a relatively pure quartzite (2.65 g/cm³) establishes a critical lower density limit for nonporous sedimentary rock masses. Although potassic and some sodic feldspars are less dense than quartz, their presence in pure form as a major rock mass of sedimentary origin is not considered likely. More mafic minerals, calcic feldspars, or carbonate minerals cause an increase in density as do minerals such as biotite, which develop during the early stages of metamorphism. This analysis shows that the highly indurated rocks of sedimentary origin are denser than felsic rock units, but the magnitude of the difference is uncertain. The problem of a plausible magnitude for the density of indurated sedimentary rocks is a complex one, and while a complete analysis is not made here, some evidence is given below to suggest what the magnitude might be.

If a gravity anomaly is of large amplitude and has steep gradients, a quantitative analysis will sometimes yield a significant limiting density contrast. In essence, the procedure consists of matching the steep gradients of an anomaly while holding the amplitude constant; either the slopes of the sides of the theoretical models or the density contrast are varied but with somewhat different results. At some point in the analysis, it becomes clear that there is a small range of density contrast, regardless of slope, that will give a satisfactory match between the computed and measured anomalies. Simmons (1964, p. 81) and Garland (1965, p. 112), for example, have used this method to derive densities for Precambrian metamorphic host rock in areas in North America.

A similar method was used by one of the authors in analyzing anomalies in two areas in northern Maine. The first analysis (Kane, 1960) was made of the *positive* gravity anomaly over a mafic pluton in central Maine; the derived density contrast between the mafic pluton and the host rock was +0.20 g/cm³. The second analysis was made of the *negative* anomaly over a felsic pluton in east-central Maine (Kane et al., 1966, p. 99); the result was −0.15 g/cm³. Two properties of the results of these analyses should be noted. First, the gravity field over the host rock for both areas is similar and is representative of the gravity field level over a very wide region; the results therefore have implications beyond the immediate environment of the plutons. Second, the two solutions give a consistent result of 2.8 g/cm³ for the host rock density.

Specific gravity measurements were made on a large group of rock samples from Maine collected by Professor Arthur J. Boucot, California Institute of Technology. Measurements were also made on samples taken from collections made by R. B. Neuman and Louis Pavlides of the Geological Survey and by the authors. The average values for the rocks are felsic plutonic rocks, 2.60 g/cm³ (94 samples); mafic plutonic rocks, 2.91 g/cm³ (151 samples); rocks of sedimentary origin, 2.65 g/cm³ (601 samples). The sampling was not ideal because most specimens were obtained from the northwestern area and eastern coastal area; also, as much as 20 percent of many of the rock specimens was visibly weathered. Nonetheless the results indicate the general order of densities.

For felsic and mafic plutonic rocks, the agreement between the mineral density computation (2.65 and 3.0 g/cm³) and specific gravity measurements (2.60 and 2.91 g/cm³) is good when one considers that weathering lowers the overall rock density. For rocks of sedimentary origin, the agreement between the curve-matching technique (2.8 g/cm³) and specific gravity measurements (2.65 g/cm³) is poor. The authors concluded that the curve-matching technique is the more valid of the two because it avoids such problems as representative sampling, variations in density with depth, and differential weathering. Consequently, we have assumed the result of curve-matching for the interpretation.

Interpretative Technique

The normal method of gravity interpretation is identification of anomaly, assumption as to its cause, and computation of the gravity field of a suitable theoretical model. Analysis is assumed to be complete when computed and observed fields are in reasonable agreement. Although it is usually easy to choose the first-order model that will agree fairly well with the observed anomaly, the refinements needed to accomplish a closer match of theoretical and observed values raise both theoretical and practical difficulties. In the end it is not always clear that the well-defined model is significantly more useful than the first-order one.

With the above discussion in mind, we have used what may be called the maximum-depth, minimum-vertical-extent approach. First, it has long been established that there is a relationship between the gradient-amplitude characteristics of an anomaly and the maximum depth to the top of the source mass. In earlier works, this fact is expressed by the "half-width relations" (Nettleton, 1940, p. 123). More recently, Bott and Smith (1958) generalized this concept for two basic forms, the point mass and the line mass, while Bancroft (1960) extended the generalization to the sheet mass. Quite simply, the results state that the ratio of the amplitude of an anomaly to the maximum gradient is proportional to the maximum depth to the top of the mass. The result obtained by Bancroft is used below and is given by:

$$D_{\max} = \frac{A}{3.14 S_{\max}}$$

where $D_{\max}$ = maximum depth
A = anomaly amplitude
$S_{\max}$ = maximum gradient

It should be noted that $D_{\max}$ is independent of density and is a function only of the mapped field; thus a very simple computation yields a result which is not conditioned by any factor other than the measured data. The usefulness of the solution depends on whether it puts in a meaningful restriction on the choice of the anomaly source.

The second part of the interpretation consists of using the "infinite-sheet" approximation where the minimum vertical extent of the source mass is proportional to the anomaly amplitude. The method implies the following: If the mass does not extend to infinity in the horizontal directions, then the source mass must be thicker; thus the "minimum-vertical-extent." The approximation is made useful by the fact that most geologic bodies are wider than they are thick. The constant of proportionality depends on the choice of a density factor, but the minimal characteristic is retained by choosing a maximum density contrast. The relation is given by:

$$T_{\min} = \frac{A}{.013 d}$$

where $T_{\min}$ = minimum thickness in feet
A = anomaly amplitude in milligals
d = density contrast in g/cm³

Lastly, we make use of the fact that the horizontal outline of the source mass is similar to the trace of the anomaly contours, especially for tabular masses. The above considerations, together with a knowledge of rock densities, lead to a fairly clear picture of the more easily resolved features of the gravity field.

To further simplify analyses we have adopted a single density contrast of 0.2 g/cm³ to derive a general relation between the anomaly amplitude and the minimum vertical extent for any source-mass. The derived relation is 1 mgal per 400 feet (120 m). Accordingly, as an example, a gravity anomaly of 25 mgal indicates a mass extending at least 10,000 feet (3,000 m) in depth. The adopted density contrast includes mafic rock versus host rock, felsic versus host rock (with change in sign), and various combinations of source rock of sedimentary origin.

Correlation of Gravity and Geology

A complex group of negative and positive gravity anomalies is superimposed on a northeast-striking, northwest-decreasing regional gravity field (Fig. 31–1). Most of the individual anomalies correlate with mapped plutonic rocks. The anomalies range in amplitude from 5 to more than 40 mgal, and in areal extent from 10 to more than 1,000 square miles (25 to more than 2,500 square km). In intervening areas underlain by rocks of sedimentary origin, the gravity field is much flatter.

The parts of the gravity field that are relatively flat are of considerable importance as, presumably, they define the gravity field undisturbed by the masses giving rise to the anomalies. In the analysis, we assume that the gravity field was relatively flat prior to the intrusion of plutonic rocks; igneous activity gave rise to the group of highs and lows that are now evident over the intrusions.

The field between the igneous bodies is in fact affected by the lateral extent of the associated anomalies; the magnitude of the effect cannot be precisely estimated inasmuch as it depends on the depth to which the anomalous mass extends— an unknown factor. It is certain, however, that the effect becomes less with increasing distance from the anomalies and that the "best" values for the undisturbed field are found at points farthest from the anomalies. In the end, we rely on the reasonableness of the geologic model and on the consistency of the regional values to support the validity of this approach. Once accepted, the model simplifies the discussion of the regional field.

Regional Gravity Field

The regional gravity field can be divided into four northeast-trending zones which are separated by gravity gradients and a fifth zone that occupies much of the southwestern tip of the State. The two central zones are the largest and comprise most of the State. On the northwest, a large gravity low of regional proportions occupies much of the area southeast of the United States–

Canada border. To the southeast, along the coastline, the gravity field rises to values considerably higher than those in the interior zones. In the southwest (southwest of line G–H, Fig. 31–1) a northwest-dipping regional gradient dominates the gravity field so that local anomalies are much less apparent than elsewhere.

The regional gravity low on the north is in the area underlain primarily by cyclically bedded slate and metasandstone of Devonian age. Volcanic rocks apparently are absent. The gravity field in the interior of the low is fairly flat at a level of about −60 mgal. A steep but irregular gravity gradient (marked approximately by A–B, Fig. 31–1) separates the low from a gravity level of about −25 mgal to the southeast. The amplitude-gradient relation (35 mgal/3.14 × 3.6 mgal/mile) indicates that the maximum depth to the top of the source of the low is about 3.4 miles (5.4 km). It is clear from this computation that the source is in the upper part of the crust (at least in part) and is less dense than rock strata at comparable depths in the crust in the region southeast of the gradient. The source could be a thick accumulation of quartzose rocks of sedimentary origin (presumably contrasted to older rocks to the southeast which are less quartzose in composition) or a large tabular felsic mass. In either case the vertical extent of the source mass is greater than 14,000 feet (4,300 m).

To the southeast, the regional gravity level of about −25 mgal correlates with a wide variety of rocks of sedimentary origin. Values of the gravity field between areas of intrusive or volcanic rock are surprisingly consistent considering the wide variety of rock types and the lateral influence of nearby anomalies. This region is bounded to the southeast by another gravity gradient (marked by C–D, Fig. 31–1) which is less well defined because of its lower amplitude and slope. This change in regional gravity is seen most clearly in the contrasting gravity field values on either side of C–D. The gradient is best defined near long. 69° W., lat. 45° N. (see Osberg and others, this volume), where the gravity field, in an area underlain wholly by rocks of sedimentary origin, changes from −18 to −2 mgal. The average gravity level in the area to the southeast is about −5 mgal. The amplitude-gradient relation (20 mgal/3.14 × 2 mgal/mile) indicates a maximum depth to the top of the source of about 3.2 (5.1 km) miles. Although the source is upper crustal, its nature is not clear. The minimum vertical extent of the source mass to the northwest is 8,000 feet (2,600 m).

The gravity level southeast of the gradient is present over different lithologic units of sedimentary origin. Near the coast, this gravity level ends, passing into a zone of steep gradients. The gradients are irregular and appear to be absent in the central part of the coastal area. The approximate location of the zone is marked by E–F in Figure 31–1.

The gravity field over the coastal area culminates in closed highs in a few areas, but more generally increases outward into the Gulf of Maine. Maximum Bouguer values of 40 mgal are reached in eastern-most Maine. In the northeast part of the coastal area, the high gravity values correspond closely to areas underlain by mafic rock; the sharp gravity gradients at the boundaries of the mafic masses in this area show that the mafic rocks are the principal source of the gravity highs. To the southwest, the rise in gravity takes the shape of a broad arcuate high, opening seaward. Here there is no clear-cut association of gravity highs and mafic rocks, although the gradients suggest a crustal source. Unfortunately, the lack of anomaly closure precludes a significantly quantitative conclusion on the source of the arcuate high.

The distinct change in the character of the gravity field which takes place southwest of line G–H (Fig. 31–1) corresponds to the region of high-grade metamorphism and complex structure (see Hussey, this volume). Distinctive anomalies and gradients are subdued or masked in a regional gradient dipping northwest toward the large felsic plutons in New Hampshire. It appears as if the sources giving rise to the distinctive character of the gravity field to the northeast either have not developed or have been absorbed by a profound disturbance of the earth's crust. Such a change in gravity does not yield to a critical interpretation. It could be caused by a gradual lowering of the density of the crustal rocks (presumably towards the density of felsic rock) or by a gradual increase in the thickness of a felsic plutonic unit. Felsic plutonic rocks are abundantly exposed in this area, but sharp gradients at the boundaries of these rocks are conspicuously absent.

The total change in the regional field northeast of line G–H (Fig. 31–1) is about 100 mgal (−60 to 40 mgal). If the observations of the regional field are correct, then its geometry consists of relatively broad areas of low gradients separated by narrow belts of steep gradients. The geometry of source-masses giving rise to such a field is that of adjacent blocks of contrasting density, whose boundaries are indicated by the gravity gradients.

The boundaries may be vertical or sloping and sharp or transitional. The gradient-amplitude relation shows that the contrast in density across the boundaries must take place, at least in part, in the shallow level of the crust. It should be noted that the analysis (1) permits exposed rock masses to be considered as a source, and (2) does not rule out part of the source being as deep as the mantle. The important conclusion to be drawn from the analysis is that a considerable part of the regional field must arise from shallow, intracrustal sources. Usually, regional gravity changes are attributed to mass changes at the crust-mantle interface, which in this area is at a depth of about 25 miles (40 km).

Local Gravity Anomalies

The clearest correlation between gravity and geologic features is that of large negative anomalies associated with intrusive felsic masses; only a few of these masses lack a corresponding negative gravity anomaly. An equally good correlation exists between positive anomalies and mafic intrusions, but fewer mafic bodies are present. Both highs and lows exist in areas underlain by volcanic rocks, but highs predominate; the amplitudes and gradients are usually much lower than those associated with intrusive masses. No evidence was found associating local anomalies with purely sedimentary phenomena (e.g., with local thickening of rocks of anomalous density in an ancient depositional basin).

To illustrate the correlation of igneous masses with gravity anomalies, the generalized geology and gravity of an area near Bingham in western Maine is shown in Figure 31–2. The primary gravity feature is a large low, bounded approximately by the −30 isogal and occupying most of the area. Higher gravity values, corresponding to a gravity level of about −25 mgal, are present around the low except at the southwest. Superimposed on the broad low is a secondary low with steeper gradients, starting at about the −40 isogal and ending in a central closure at the −62 isogal. Just to the northeast of this low is a secondary high peaking at −26 mgal.

In accordance with the above description, the anomalous area will be considered for purposes of analysis to consist of (1) a primary broad low at level −40 mgal within a background level of −25 mgal, and (2) two secondary anomalies, the low culminating at −62 mgal and the high at −25 mgal. Additional justification for the identification of the broad low as a primary continuous feature is a gravity gradient which can be seen in the vicinity of the −30 isogal in the upper

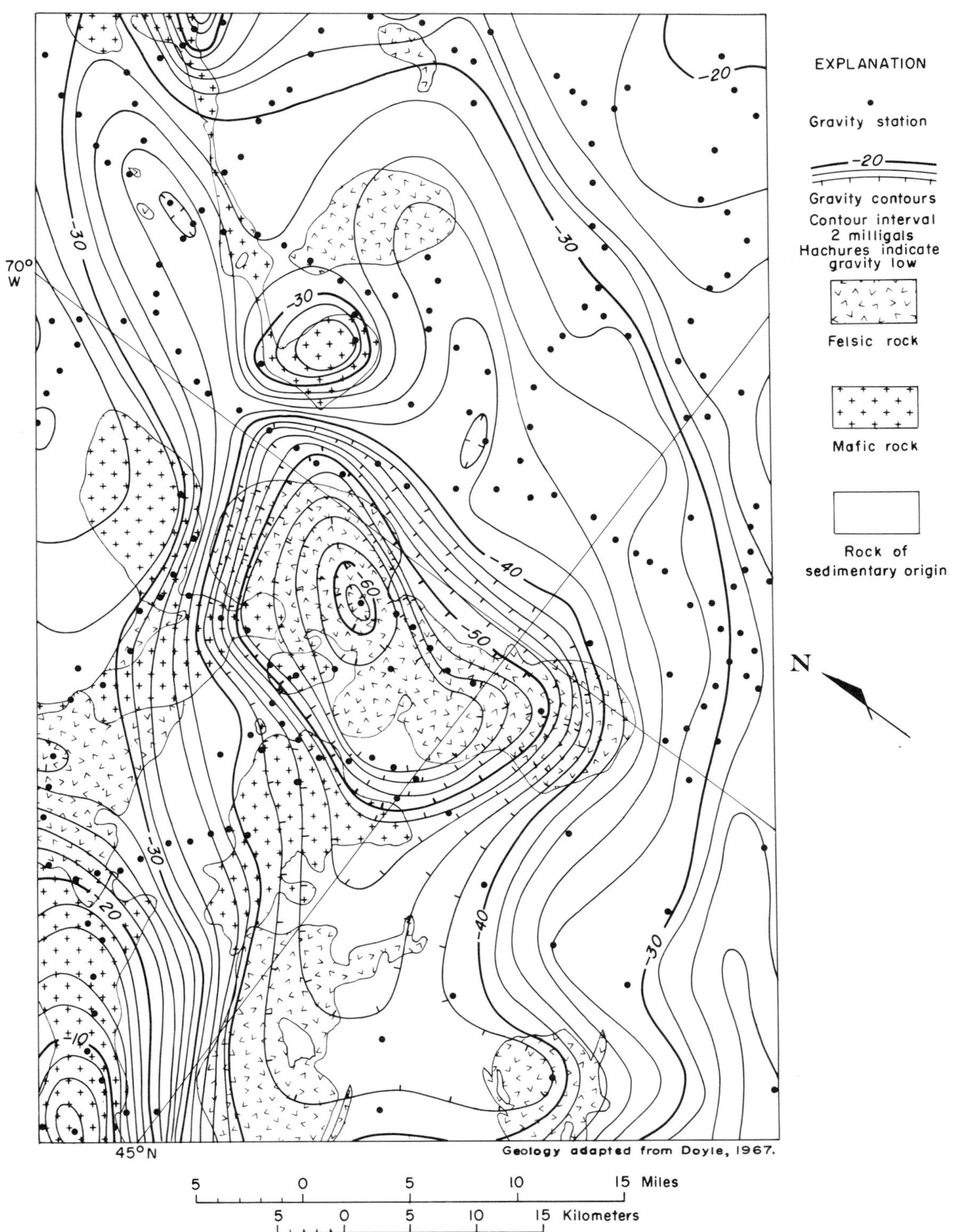

FIGURE 31–2. Gravity and geology of an area in western Maine.

part of the figure. The secondary low has an amplitude (relative to the −40 mgal level of the primary gravity low) of −22 mgal and the secondary high an amplitude of +15 mgal.

The secondary gravity low is caused by the felsic intrusion that crops out near the center of the area (Fig. 31–2). The minimum-vertical-extent relation indicates that the felsic mass extends to a depth of at least 9,000 feet (2,700 m). The mass must extend laterally at least to the half-amplitude points of the anomaly, which are approximately delimited by the −50 isogal. Thus, the felsic mass broadens outward with depth to the west and east but probably not to the north and south. The general dimensions and shape of the felsic intrusion are (1) a vertical extent of about 2 miles (3.2 km); (2) an average diameter of about 12 miles (19 km); (3) a broadening outward from surface outcrop to the east and west; and (4) a horizontal shape somewhat similar to the general trend of the gravity contours. A similar analysis of the secondary gravity high indicates that the mafic intrusion associated with it has (1) a vertical extent of about 1 mile (1.6 km); (2) an average diameter of about 5 miles (8 km); (3) an extension outward from the outcrop; and (4) a shape similar to the gravity contours.

An important conclusion of the analysis is that both masses are much wider than they are thick. This relationship could be changed by lowering the assumed density contrast (−0.2 and +0.2 g/cm³), but lowering one density contrast implies an increase in the other so that at least one of the two will retain a tabular form.

The source of the broad primary gravity low is not as evident as those of the secondary anomalies because it does not obviously correspond to a mapped geologic pattern. The gravity field relations give a maximum depth to the top of the source of 1.6 miles (2.5 km) and a minimum vertical extent of about 1 mile (1.6 km). Two factors appear to indicate the source of this broad low. First, the peripheral position to the secondary gravity low suggests that the two anomalies are related. Second, an igneous felsic source is implied by two small isolated outcrops of felsic rock near the center of the elongate low in the northernmost part of the area, and by the large outcrop area of felsic rocks northeast of the secondary gravity high.

The combined analysis indicates that the entire area of the primary low is underlain at depth by a large felsic intrusion. The felsic mass appears to comprise two parts: (1) a broad sheet-like mass with average diameter greater than 30 miles (48 km), average thickness of about 1 mile (1 km) and an average maximum depth to the top of 1 to 2 miles (1.6 to 3.2 km); (2) a cupola-like central mass with dimensions and shape as given before. Undoubtedly there are departures from this idealized form as witnessed by the local outcrops to the north but any interpretation will yield gross dimensions similar to those given here.

The two-fold nature of the felsic mass and, in particular, the wide lateral extent of the thinner part, has significance beyond the particular area described here. Similar gravity lows peripheral to sharp central lows associated with outcrops of felsic rocks are observed in other parts of Maine. The relationship suggests that in these areas the mode of intrusion includes a widespread lateral migration of felsic material from a central area.

The remaining anomalies on the gravity map of Maine may be interpreted in a relatively straightforward way. Almost every hachured low is the site of a felsic intrusion; every gravity high with two or more closed contours is the site of a mafic pluton. The anomaly amplitudes indicate the relative vertical dimensions of the source-masses, whereas the horizontal extent and shape of the anomaly contours indicate their horizontal dimensions. Thus the map reflects most strongly the igneous plutonic complex within the basement.

GENERAL COMMENTS

The chief characteristics of the regional gravity low northwest of line A–B (Fig. 31–1) are its large extent, the flatness of the gravity field in the interior of the anomaly, the amplitude of the anomaly, and the steep boundary gradient to the southeast. In addition, aeromagnetic data show a flat magnetic field (Boucot et al., 1964). It is clear that the depth to the top of the source is less than 17,000 feet (5,200 m), probably as little as 5,000 feet (1,500 m), and it may even be at the surface. If the anomaly is due to sedimentary rocks, then the applicable density contrast is close to 0.1 g/cm³, and the minimum vertical extent close to 28,000 feet (8,500 m); the flatness of the field suggests a gross lithologic homogeneity. A thick sequence of deltaic deposits (Jacobs et al., 1959, p. 61) would reasonably satisfy all the factors mentioned above. A felsic igneous source would require a mass 3 by 25 by 100 miles (8 by 40 by 160 km) with an amazingly regular thickness; presumably the magnetic field over it should be anomalous. An interpretation based on rocks of sedimentary origin appears more plausible from the evidence at hand.

The region between A–B and C–D is marked by a complex group of anomalies with large amplitudes and steep gradients. Most of the plutons underlying the anomalies widen considerably with depth. A notable exception is the Katahdin batholith (long. 69° W., just south of lat. 46° N.), which is topographically the most impressive felsic mass in the state. The associated anomaly does not have a very large amplitude compared to those of other plutons of this region, suggesting a shallower depth extent. Also, the batholith does not appear to widen with depth. Most of the felsic plutons appear to be concentrated on the southeast side of the region. Five of them aline in fairly regular succession just northwest of the gradient marked by C–D.

Between C–D and E–F, the anomalies are smaller in amplitude, and the gradients, as a rule are less steep (although the most complex group of anomalies in the state is found in the eastern part of this area). Again the felsic plutons, at least the ones with greatest depth extent, are concentrated on the southeast side. A major contrast to the region on the northwest is that most of the felsic masses appear to be largely exposed.

The gradients indicated by A–B and C–D are persistent and well defined but differ considerably. The northern gradient (A–B) is steeper, has a larger amplitude, and is irregular. It is tempting to identify a system of faults with it, but a steep warp or monocline would be sufficient to satisfy the gravity field. It is noteworthy that the flexures in the northern gradient persist far southeast of the gradient and that at least one flexure (near the small closed low) is associated with a felsic pluton (Boucot et al., 1964). The southern gradient (C–D) is less pronounced but straighter. Where a clear change in the direction of this gradient occurs (at long. 70° W.), it is due, at least in part, to the presence of two closely spaced felsic plutons. A very long diabase dike, identified in large part by an aeromagnetic anomaly (Boucot et al., 1964), occurs along the northeastern part of the gradient.

SUMMARY

The gravity field in Maine reflects major lithologic changes in the upper part of the crust, on the one hand distinguishing large regional divisions, and on the other indicating the size, location, and shape of igneous plutons. The underlying cause of the gravity variation that differentiates regional divisions is at present enigmatic, but it seems apparent that it is related to regional geologic features and not necessarily to deep crustal sources. The gravity anomalies associated with plutons are distinctive in appearance and yield considerable evidence on the subsurface nature of the plutons. It should be noted that the present gravity map represents a comprehensive regional survey (i.e., none of the plutons were selected for extensive gravity mapping). Therefore, considerably more information should be produced by detailed surveys of particular plutons (e.g., Kane (1960)).

In the present report we have not attempted to incorporate the time sequence of geological events. At this point it seemed preferable to concentrate on the results that clearly define the major geometric-mass relations. Presumably, a closer synthesis of the time-sequence provided by geology will lead to a clearer understanding of the mode of occurrence of the igneous plutons and of the development of major crustal units.

REFERENCES

Bancroft, A. M., 1960, Gravity anomalies over a buried step: Jour. Geophys. Research, v. 65, p. 1630–1631

Bott, M. H. P., and Smith, R. A., 1958, The estimation of the limiting depth of gravitating bodies: Geophys. Prosp., v. 6, p. 1–10

————, and Smithson, S. B., 1967, Gravity investigations of subsurface shape and mass distributions of granite batholiths: Geol. Soc. America Bull., v. 78, p. 859–878

Boucot, A. J., Griscom, Andrew, and Allingham, J. W., 1964, Geologic and aeromagnetic map of northern Maine: U. S. Geol. Survey Geophys. Inv. Map GP-312

Doyle, R. G., ed., 1967, Preliminary geologic map of Maine: Maine Geol. Survey, scale 1:500,000

Garland, G. D., 1965, The earth's shape and gravity: New York, Pergamon Press, 183 p.

Jacobs, J. A., Russell, R. D., Wilson, J. Tuzo, 1959, Physics and geology: New York, McGraw–Hill, 424 p.

Kane, M. F., 1960, Structure of plutons from gravity measurements: U. S. Geol. Survey Prof. Paper 400-C, p. C258–C259

————, and Bromery, R. W., 1966, Simple Bouguer gravity map of Maine: U. S. Geol. Survey Geophys. Inv. Map GP-580

————, Bromery, R. W., Pavlides, Louis, and Frischknecht, F. C., 1966, Geophysical and geological reconnaissance of the Oakfield Hills area, Maine, in Mining case histories, v. 1: Soc. Explor. Geophysicists, p. 94–104

Nettleton, L. L., 1940, Geophysical prospecting for oil: New York, McGraw–Hill Book Co., 444 p.

Simmons, Gene, 1964, Gravity survey and geological interpretation, northern New York: Geol. Soc. America Bull., v. 75, p. 81–98

Chapter 32

Geologic Interpretation of Aeromagnetic Data for New England*

ANDREW GRISCOM AND
RANDOLPH W. BROMERY

INTRODUCTION

A WEALTH OF AEROMAGNETIC data is available for the New England area. The U.S. Geological Survey has published over 150 aeromagnetic quadrangle maps in this region as well as an aeromagnetic compilation of northern Maine (Boucot et al., 1964). The Maine Geological Survey has published much aeromagnetic data, mostly in the southern half of the state (Doyle et al., 1961; Forsyth, 1955; Trefethen, 1953; Wing, 1958a, 1958b, 1958c, 1959). These data have received relatively little geologic interpretation; however, the relevant references are cited in this report.

The magnetic anomalies of these maps reflect variations in the amount of magnetic minerals, primarily magnetite, contained in the rock masses and can be used to trace various rock units or structural features across areas of poorly known geology. The degree of correspondence with geologic mapping depends on the shapes of the subsurface structures as well as on the distribution and relative amounts of magnetite in the geologic units. Attitudes of contacts and depths to concealed magnetic units may be calculated from a consideration of the magnetic gradients (Vacquier et al., 1951). The aeromagnetic data of this report were flown at a height of 500 feet (150 m) above ground. Because of the large amount of aeromagnetic data, this paper will discuss only selected aeromagnetic features which

may be of general interest to New England geologists. The discussion, in general, proceeds from simple to more complex and speculative interpretations.

SOUTHEASTERN CONNECTICUT

In southeastern Connecticut a distinct east-west lineament (Bromery, 1967a) occurs on aeromagnetic maps (Fig. 32–1) and aeroradioactivity maps. In the vicinity of the lineament the geologic units also strike east-west, but the lineament does not correspond with any mapped geologic unit or structure. The lineament has been traced for a distance of 40 miles (64 km) by means of another aeromagnetic survey in Rhode Island, and consists of a line of parallel elongate magnetic highs and lows. Although the lineament data of Figure 32–1 are restricted to two flight lines, the map is based on a 1-mile-spaced aeromagnetic survey with the intermediate half-mile-spaced traverses filled in later; the two lines of particular interest were flown several years apart. Aeromagnetic, ground magnetometer, and gravity profiles established along several north-south traverses across the aeromagnetic lineament in the area of Figure 32–1 show that an east-trending gravity low and a magnetic anomaly zone correlate with the lineament. Along the western edge of the map (Fig. 32–1), the gravity low is about 5 miles (8 km) wide and more than 10 mgal in amplitude; it decreases in width and amplitude towards the east. High-amplitude ground-magnetic anomalies correlate with indi-

* Publication authorized by the Director, U. S. Geological Survey.

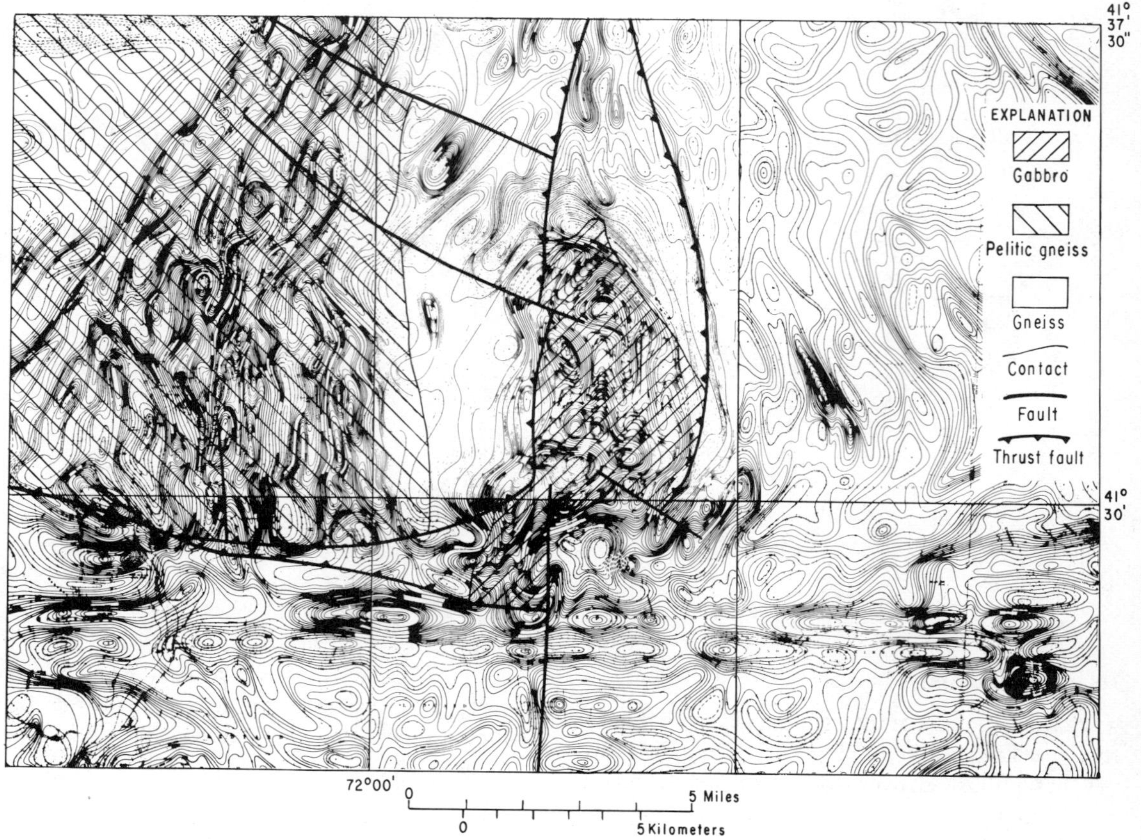

FIGURE 32–1. Aeromagnetic and geologic map of southeastern Connecticut. Contour interval 20 gammas. Geology from Dixon and Lundgren (this volume). Aeromagnetic data by U. S. Geological Survey.

vidual aeromagnetic anomalies mapped along the lineament. Additional airborne and ground geophysical surveys are being conducted to trace the western extension of the lineament. More work is needed to explain this feature.

A major thrust fault in eastern Connecticut changes its strike (Fig. 32–1) where it wraps around the Preston Gabbro (Sclar, 1958; Dixon and Lundgren, this volume). The magnetic data are in good agreement with the position of the fault. The pronounced magnetic minimum which wraps around the east and north sides of the gabbro pluton proves that the dip of this contact is at a low angle to the west. In addition, the gentle magnetic gradient on the west side indicates that this western contact also dips gently west, confirming the geologic interpretations of the gabbro as a west-dipping laccolithic pluton (Sclar, 1958, p. 122).

Interpretation of detailed gravity data along a north-south line of stations across the gabbro pluton indicates that the central part of the mass is about 4,000–6,000 feet (1.2–1.8 km)

thick, and the south contact dips north and the north contact dips south at moderately low angles. Therefore, interpretation of the magnetic and gravity data suggests that the general form of the pluton is that of a nearly circular basinlike body, now tilted down to the west.

GNEISS DOMES IN NEW HAMPSHIRE

The surface distribution and structure of rock units along the Bronson Hill anticlinorium in western New Hampshire are particularly well expressed by a series of aeromagnetic maps extending the length of the state. Between central New England and the Green Mountain anticlinorium, the stratified rocks of Ordovician or older age are predominantly very magnetic, whereas stratified rocks of Silurian and Devonian age are for the most part weakly magnetic. Figure 32–2 shows the typical magnetic expression of this region in New Hampshire, and contains parts of four gneiss domes of the Oliverian Plutonic Series, of presumed Ordovician age (Thompson et al., this volume; Naylor, this volume): the Croydon dome (C), the Lebanon dome (L), the Mascoma dome (M), and the Smarts Mountain dome (S). These domes are composed locally of magnetic rocks, especially the Lebanon dome, and, except for the Lebanon dome, are almost completely rimmed by a mile-wide border of strongly mag- netic Ordovician rocks, the Ammonoosuc Volcanics, which dip outward from the domes. Along the east side of this row of domes, Devonian granitic rocks of the New Hampshire Plutonic Series comprise the Mt. Clough pluton (Billings, 1955). The Mt. Clough pluton has been

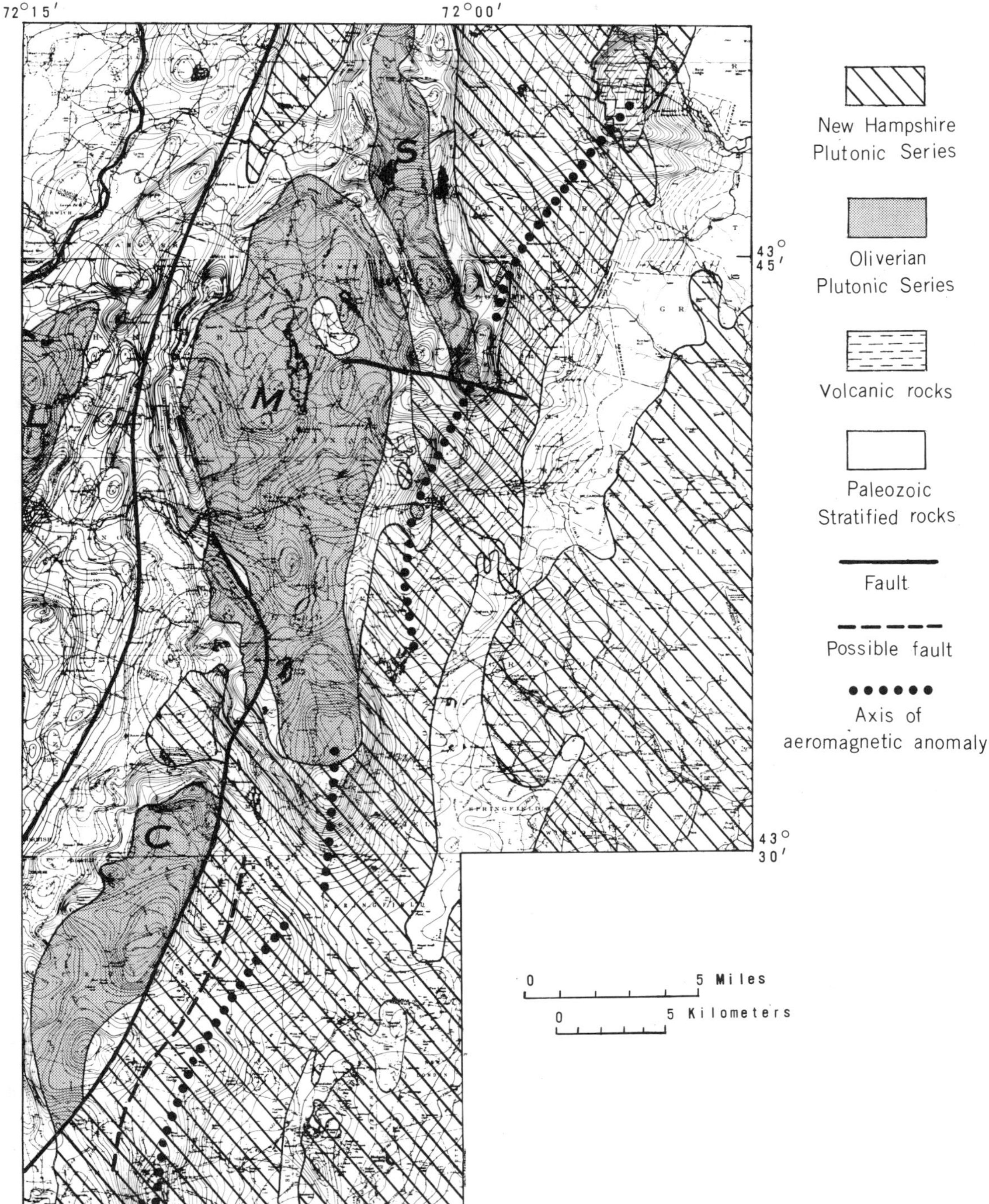

FIGURE 32–2. Aeromagnetic and geologic map of western New Hampshire. Contour interval 10 gammas. Geology from Billings (1955). Aeromagnetic data by U. S. Geological Survey.

shown by Chapman (1939) and Billings (1945) to have a sill-like form which originally extended as a mile-thick sheet over the southern end of the Mascoma dome. By means of the magnetic anomalies generated by the Oliverian gneisses and the overlying volcanic rocks (see also Thompson et al., this volume), the aeromagnetic data of Figure 32–2 very clearly indicate substantial extensions of these gneiss domes beneath the nonmagnetic rocks of the Mt. Clough pluton. The buried eastern half of the Croydon dome thus indicated is larger than the exposed part, and depths to magnetic rocks range from 0.5 to 1.0 mile (0.8–1.6 km). This concealed dome was inferred by Billings (1955, section E–E') from a foliation anticline mapped in this area by Chapman (1952), but the magnetic data indicate the dome is much broader and shallower than shown in the section. A second fault paralleling the eastern border fault of the Croydon dome is inferred from the magnetic data and is shown on Figure 32–2 located approximately 1.5 miles (2.4 km) east of the border fault. The buried southern extension of the Mascoma dome continues at least 3 miles (5 km) south of the mapped contact, and ends only 2 miles (3 km) from the buried northern extension of the Croydon dome. The magnetic anomalies of the two domes are separated by a pronounced northwest-trending magnetic depression, possibly bounded by northwest-trending faults (Thompson et al., this volume). A narrow north-trending magnetic high about 2 miles (3 km) east of the Mascoma dome is the expression of another anticline which contains a small exposure of the Oliverian Plutonic Series. According to the magnetic pattern, this anticline connects with the south end of the Smarts Mountain dome and then continues toward the northeast where it strikes into a prominent magnetic anomaly caused by a patch of volcanic rocks mapped by Page (Fowler-Billings and Page, 1942) as part of the Devonian Littleton Formation. However, the magnetic gradient on the southeast side of these volcanic rocks is continuous with the magnetic gradient along the east flank of the Mascoma dome. As an alternative hypothesis, it is suggested that these volcanic rocks may be Ammonoosuc Volcanics.

GNEISS DOMES IN MASSACHUSETTS

An aeromagnetic map of the Shelburne Falls dome (Balk, 1946; Segerstrom, 1956) is shown in Figure 32–3a. The rim of magnetic anomalies correlates well with magnetite-rich leucocratic volcanic rocks on the southwest side (Hatch and Hartshorn, 1968), but the correlation is poor and difficult to understand on the northeast side. The moderate outward dips of the flanks of the dome are well expressed by the gentle magnetic gradients, and a rather accurate shape for the magnetic mass can be calculated (see also Hatch et al., this volume).

South of the Shelburne Falls dome in the Woronoco quadrangle (Fig. 32–3b) a large circular magnetic anomaly, caused by a magnetic rock mass about 2 miles (3 km) in diameter and buried approximately 2,000 feet (600 m) below the surface (Bromery, 1967b), is interpreted as a concealed gneiss dome mantled by magnetic rock, presumably the Ordovician volcanic rocks. On this figure the usual magnetic contrast between exposed stratified rocks of Cambrian and Ordovician age and stratified rocks of Silurian and Devonian age is well demonstrated.

Two other domes, near Goshen and near Southwick, Massachusetts (Hatch et al., this volume), are well expressed by the Massachusetts aeromagnetic data. The dome at Goshen is in an area of metamorphosed volcanic rocks which give rise to substantial magnetic anomalies. The more subdued central magnetic pattern of the other dome is rimmed by magnetic highs and is very similar to the pattern of the Shelburne Falls dome, so it seems likely that, despite a lack of exposure, the core gneisses are present at the surface and are rimmed by volcanic rocks.

Detailed gravity surveys over the Shelburne Falls and Goshen domes show that pronounced gravity lows correlate with the domes. Measurements of rock samples collected from the Shelburne Falls dome show that the granite and diorite gneisses contained in the central parts of the dome are of relatively low density.

WHITE MOUNTAIN PLUTONIC–VOLCANIC SERIES

Abundant aeromagnetic data are available over rocks of the White Mountain plutonic-volcanic series (Griscom, 1962). In general, the plutonic rocks of the White Mountain batholith and immediately adjacent granite stocks do not display magnetic anomalies, but the numerous other stocks and ring dikes give rise to anomalies as large as 4,000 gammas as measured 500 feet (160 m) above the ground. Analysis of the anomalies has provided considerable information concerning the subsurface structure of the stocks and ring dikes. The marginal contacts of the outermost ring intrusions at Mount Tripyramid, N. H. (lat. 43°58'N.; long. 71°27'W.), Mount Pawtuckaway, N. H. (lat. 43°07'N.; long. 71°11'W.), and Cuttingsville, Vt. (lat. 43°29'N.;

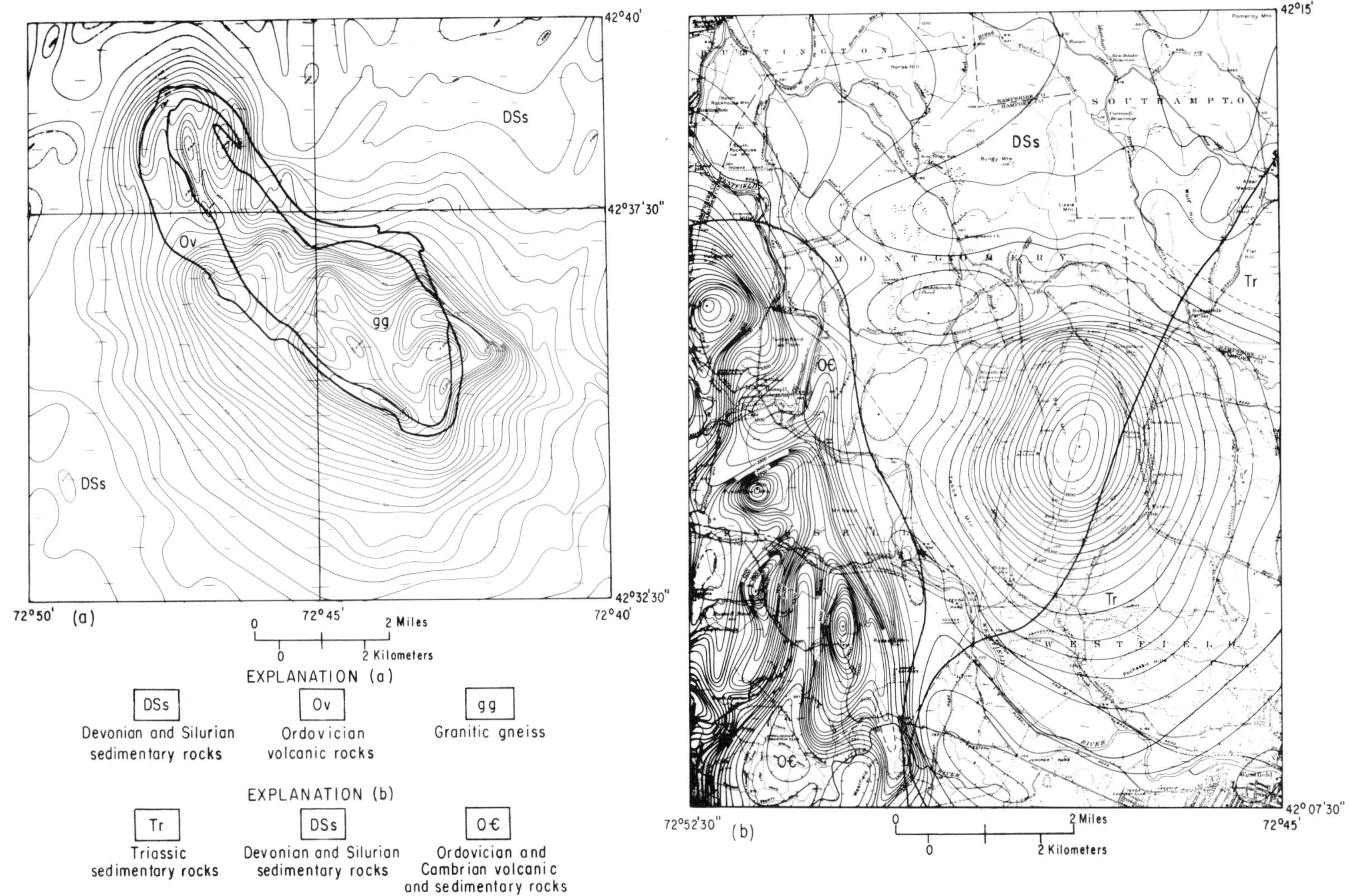

FIGURE 32–3. (a) Aeromagnetic and geologic map of the Shelburne Falls dome, Massachusetts. Contour interval 20 gammas. Geology from Segerstrom (1956) and N. L. Hatch, Jr. (written communication, 1966). Aeromagnetic data by U. S. Geological Survey. (b) Aeromagnetic and geologic map of the Woronoco quadrangle, Massachusetts, showing buried gneiss dome. Contour interval 20 gammas. Geology from N. L. Hatch, Jr. (written communication, 1966). Aeromagnetic data by U. S. Geological Survey.

long. 72°53′W.), appear to be nearly vertical for the upper 5,000 feet (1.6 km). The intrusive complex at Red Hill, N. H. (lat. 43°45′N.; long. 71°27′W.), and the stock at Cherry Mountain, N. H. (lat. 44°21′N.; long. 71°30′W.), both dip outward at about 20°, and the diameters of the buried parts of the bodies are at least two and three times, respectively, the width of the exposed parts. The intrusive complex at Mount Ascutney, Vt. (lat. 43°27′N.; long. 72°27′W.), has steeply dipping sides for the first few thousand feet and then opens out to a stocklike body at least three times as wide as the exposed part (see also Diment, this volume). The central stocks of Conway Granite in the intrusive complexes at Mount Ascutney, the Ossipee Mountains (lat. 43°45′N.; long. 71°15′W.), Green Mountain (lat. 43°46′N.; long. 71°03′W.), and Merrymeeting Lake (Fig. 32–4) coincide with large magnetic highs, although sampling of each granite stock has shown that the granites are essentially nonmagnetic. In each stock, therefore, a large mass of magnetic rock must underlie the granite at depths no greater than a few thousand feet.

The presence of high-density magnetic rock at shallow depth beneath the granite stock at Merrymeeting Lake was first noted by Joyner (1963) who described a 12-mgal gravity high and 1000-gamma magnetic high associated with the low-density nonmagnetic granite. New magnetic data for the north half of the pluton are presented in Figure 32–4a, and a possible shape for the concealed magnetic mass has been calculated from a north-south profile (Fig. 32–4b). The maximum depth of the mass is 20,000 feet (6 km), but a depth of 15,000 feet (4.5 km) may be more reasonable. The presence of a small stock of older gabbroic rocks on the east margin of the granite and the relationships of the associated magnetic anomalies strongly suggest that the same gabbro underlies the granite. Using a density contrast of +0.15 for the gabbro and −0.17 for the granite (Joyner, 1963), and the configuration of Figure 32-4b, the depth to the bottom of the high density mass is at about 13,500 feet (4 km) below sea level. This result is very approximate because the density contrast of the buried gabbroic mass can only be inferred from that of the exposed gabbro stock.

A major problem in matching the computed gravity profile to the observed data is provided by the smooth gravity gradient at the contact between the granite and the surrounding country rock. The cross section (Fig. 32–4b) computed from the magnetic data implies a vertical thickness of 3,000–5,000 feet (1–1.5 km) of granite at the contacts. A thickness of only 1,000 feet

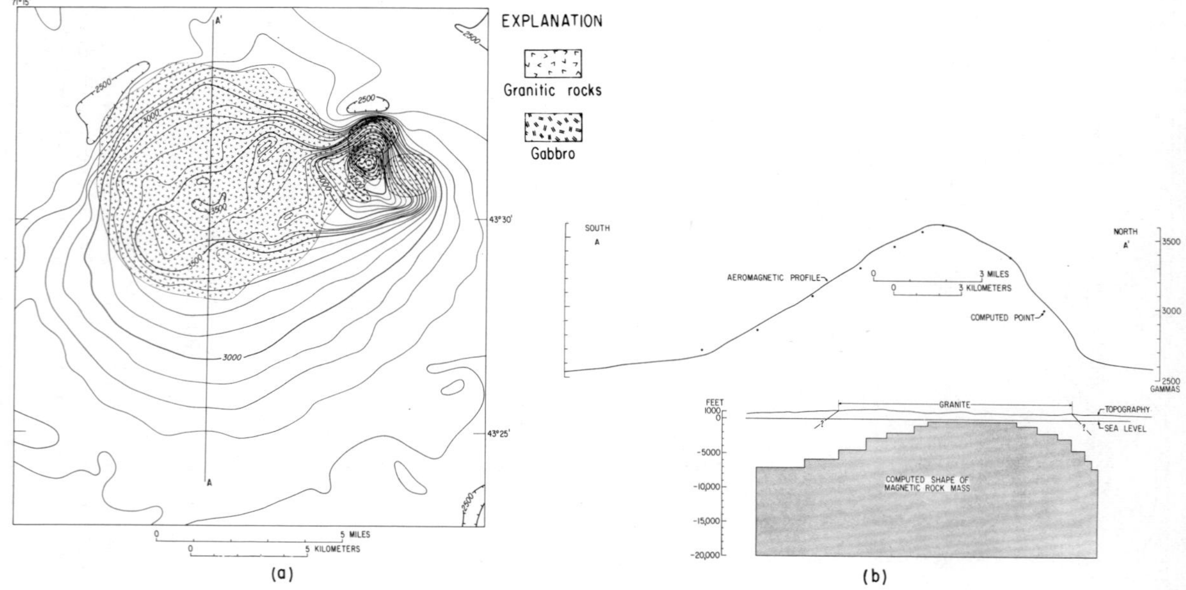

FIGURE 32–4. (a) Aeromagnetic and geologic map of the intrusive rocks of the Merrymeeting stock, New Hampshire. Contour interval 100 gammas. Geology from Billings (1955). Aeromagnetic data by U. S. Geological Survey. (b) Magnetic profile and computed shape of magnetic body along A–A′. Each step in the computed shape represents a cylindrical slab with its axis in the plane of the section. Profile extends beyond ends of A–A′ on Figure 32–4a.

(0.3 km) of granite should provide a local minimum of at least 2 mgal superimposed on the sloping sides of the gravity high. A detailed gravity profile from northwest to southeast across the stock with a 500-foot station spacing (M. F. Kane, written communication, 1962) proves that this predicted minimum is not present at either contact; on the contrary, a small local gravity high of 2 mgal is found near the southeast contact. A more dense, mafic granodiorite border is known along the northwest quadrant of the stock contact, and the local small gravity high at the southeast contact may represent a similar situation although no granodiorite was mapped there by Quinn and Stewart (1941). As a way of reconciling the gravity expression of this pluton with the computed magnetic cross section, it is suggested that at depth, the material near the contacts above the magnetic mass is not granite, but rather a nonmagnetic rock of intermediate composition such as granodiorite or possibly a breccia of gabbro in granite, and that the bulk density of this material is similar to that of the surrounding country rock. The gravity and magnetic data indicate that the total thickness and volume of granite must be relatively small. It is not clear whether the granite was emplaced by cauldron subsidence of a paraboloidal gabbro block into a floored polymagmatic magma chamber (Chapman, 1966), or whether the granite is a light-weight differentiate which was derived from and caps the main gabbro mass.

WEST FLANK OF GREEN MOUNTAIN ANTICLINORIUM IN QUEBEC AND NEW ENGLAND

Aeromagnetic profiles across the west flank of the Green Mountain anticlinorium in Quebec have been constructed from aeromagnetic maps (Canada Geological Survey, 1954a, 1954b, 1954c) and are displayed in Figure 32–5. In this part of Quebec, the magnetic data clearly show the transition from a contact dipping gently northwest at about 15° for many miles (Fig. 32–5 profile A–A′) to a contact which dips 45° (profile C–C′), and is probably cut off by a thrust fault at a depth of 7,000–9,000 feet (2.1–2.7 km). The contact becomes vertical near the international border and is overturned to dip east in northern Vermont (Dennis, 1964).

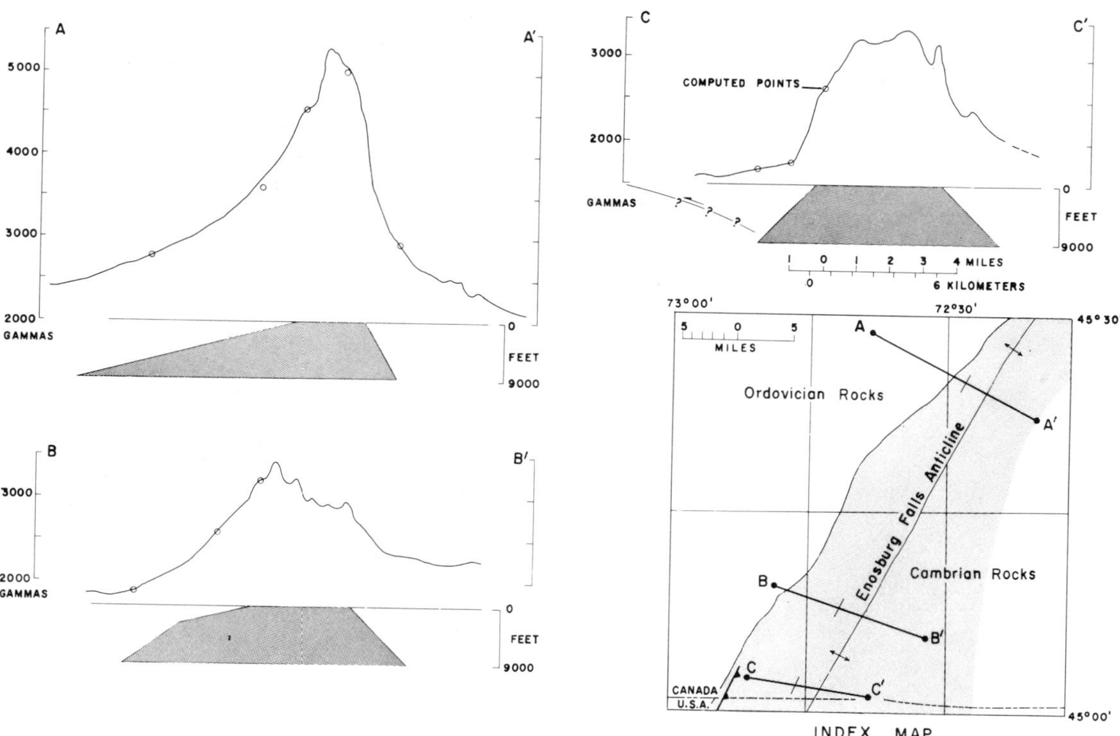

FIGURE 32–5. Aeromagnetic profiles across the west flank of the Green Mountain anticlinorium in Quebec. Index map geology from Cady (1960). Aeromagnetic data from Canada Geological Survey (1954a, b, c).

At the border between the United States and Canada (profile C–C') the magnetic unit is clearly the Cambrian(?) Tibbit Hill Volcanic Member of the Pinnacle Formation. The west contact of the magnetic unit is marked on the index map of Figure 32–5 by a short bar normal to and approximately in the center of each profile.. This contact parallels the axis of the Enosburg Falls anticline (Cady, 1960), and appears to merge with the contact between Cambrian and Ordovician rocks at the north edge of the index map (Fig. 32–5). A major thrust fault is shown by Cady (1960) as extending about 4 miles (6.5 km) into Canada. The magnetic data of profile C–C' suggest that this fault may intersect the magnetic unit at a depth of 7,000–9,000 feet (2.1–2.7 km) and cut it off, thus superimposing older magnetic rocks on younger non-magnetic rocks. The aeromagnetic map of this area (Canada Geological Survey, 1954c) suggests that

the magnetic unit may be offset at depth by a thrust fault as far north at lat. 45°05'N., perhaps a mile (1.6 km) beyond where Cady (1960) extends the fault.

In southwestern Vermont, Precambrian rocks on the west flank of the Green Mountain anticlinorium dip gently westward beneath lower Paleozoic sedimentary rocks (Doll et al., 1961). The magnetic pattern (Fig. 32–6) demonstrates that the upper surface of these Precambrian rocks dips west for a horizontal distance of at least 3 miles (5 km) beneath the sedimentary rocks. At this point a possible fault (dotted line of Fig. 32–6) appears to terminate the buried Precambrian rocks and may represent the subsurface continuation of the thrust fault shown 3 miles (5 km) to the west on Figure 32–6. This structure is in good agreement with section F–F' of Doll et al. (1961).

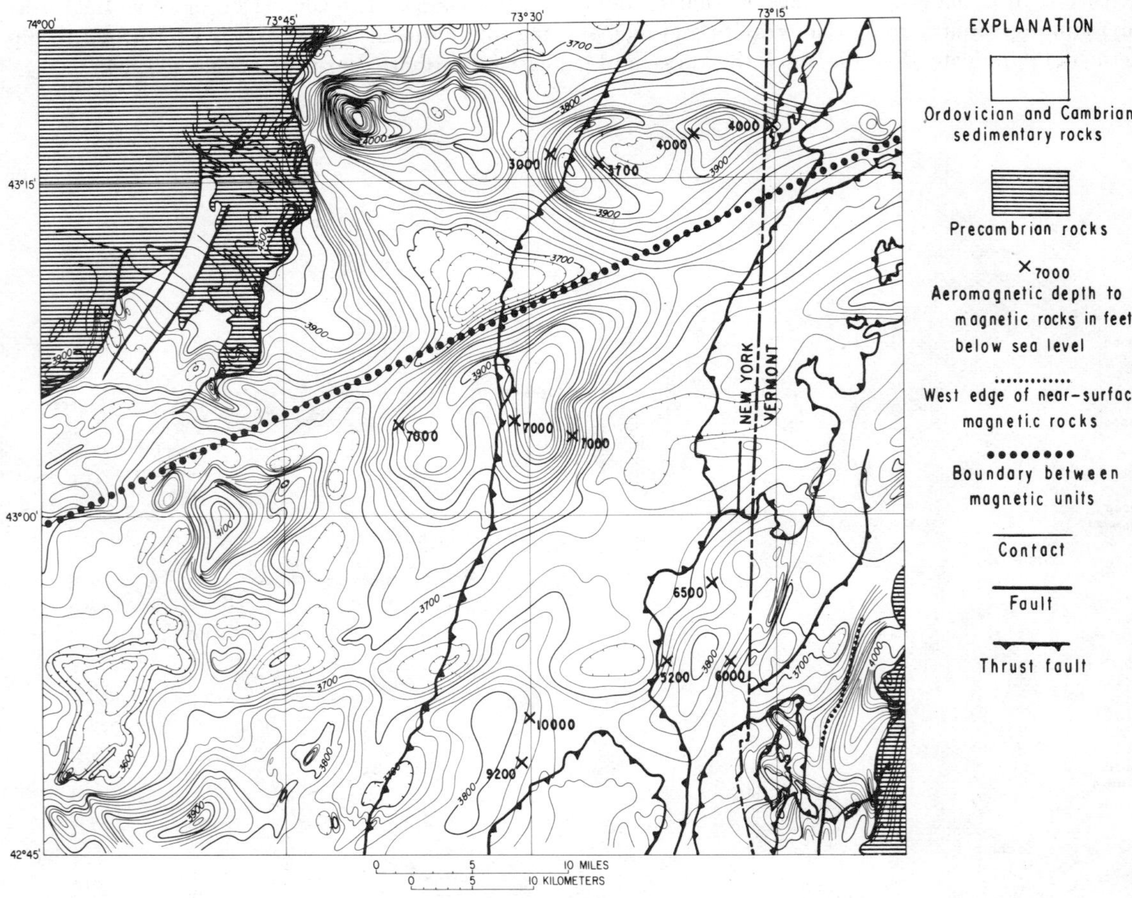

FIGURE 32–6. Aeromagnetic and geologic maps of an area between southern Vermont and the Adirondack Mountains, N. Y. Contour interval 20 gammas. Geology generalized from Fisher et al. (1962), Doll et al. (1961), and Zen (1967). Aeromagnetic data by U. S. Geological Survey.

The Appalachian gravity high and its gradient (Diment, this volume; Griscom, 1963) must also be considered in the interpretation of the structure of the west flank of the Green Mountain anticlinorium in Vermont and in Massachusetts. In general, the exposed Precambrian rocks are associated with the gravity high, and to the west of the high a belt of east-dipping rocks involved in thrust faults and overturned folds are associated with the gravity gradient and gravity low. In southern Vermont, however, the Precambrian rocks extend west into the gravity low (Diment, this volume, Fig. 30–3), and therefore may overlie Cambrian and Ordovician rocks as far east as long 73°02′W.

Farther south, the U.S. Geological Survey aeromagnetic maps for western Massachusetts provide much new information concerning the structure of the western margin of the Precambrian rocks. (The reader should refer to the regional geologic map accompanying this volume in order to follow the discussion of this paragraph.) The weak magnetic expression of the Precambrian rocks north of lat. 42°30′N. suggests that they may be relatively thin. Magnetic data indicate that at lat. 42°39′N. the west contact of the Precambrian rocks is strongly overturned and dips east. The nearly isolated area of Precambrian rocks at lat. 42°15′N., long. 73°15′W. appears to be a relatively thin sheet whose south contact dips north, whereas the north contact dips northeast beneath the adjacent Cambrian and Ordovician rocks and connects at depth with the southwest-dipping contact of the main mass of Precambrian rocks lying 1–3 miles (1.6–5 km) to the northeast. The lack of magnetic expression for the two isolated patches of Precambrian rocks in the southwestern corner of Massachusetts indicates that they also are thin sheets. Taking into account the gravity gradient in western Massachusetts, the data suggest that all the Precambrian rocks west of a line drawn between long. 73°04′W. at the northern border of the state and long. 73°08′W. at the southern border may be thrust over Paleozoic sedimentary rocks.

WESTERN NEW ENGLAND AND EASTERN NEW YORK

An aeromagnetic map of the area between the Adirondack Mountains and southwestern Vermont is shown in Figure 32–6. Other than the Precambrian rocks, only a small area of the Cambrian St. Catherine Formation of Doll et al. (1961) in the southeast corner of the figure appears to be magnetic; the many magnetic anomalies on this map are probably caused by Pre-

cambrian rock units below the Cambrian and Ordovician sedimentary rocks. Depths to these magnetic anomalies, calculated by the method of Vacquier et al. (1951), are shown on Figure 32–6 (see also Griscom and Zietz, 1964). Such depths are generally accurate to within 15 percent. The only magnetic depth determination in serious disagreement with depths estimated from geologic considerations is the 4,000-foot (1.2 km) result at lat. 43°17′N., where section E–E′ of Doll et al. (1961) indicates a basement depth slightly in excess of 10,000 feet (3 km) below sea level. The cause of the discrepancy is not clear. Because of the extension of the anomaly to the Adirondack Mountains, it is unlikely that this anomaly is caused by a thrust slice of Precambrian rock below the Taconic klippe. Of possible signifiance is the steep 10-mgal gravity gradient sloping up between lat. 43°00′N. and lat. 43°15′N. in the same area (Diment, this volume, Fig. 30–4); the gradient may represent an abrupt local shallowing of the basement toward the north although other interpretations of the gravity data are possible.

Structural trends in the Precambrian rocks of the southern Adirondacks tend to be east-west, and the broad magnetic trends of the northwest corner of Figure 32–6 tend to be also approximately east-west. The trends presumably represent structures of Precambrian age. Structural trends are north-south and northwest-southeast in the Precambrian rocks in Massachusetts; they are more diverse in Vermont, although east-west trends are not uncommon. Unfortunately, structural trends in Precambrian rocks of New England are greatly modified, often to an unknown extent, by Paleozoic deformation. The magnetic anomalies in the central and southeastern parts of Figure 32–6, therefore, are of considerable importance because of the pronounced northeasterly trends. A diagonal dotted line has been drawn on Figure 32–6 separating the two areas of different magnetic trend. This line is subparallel to the gravity contours in this area (Diment, this volume, Fig. 30–3), but its significance is uncertain.

CAPE COD, MASSACHUSETTS

There are areas in New England where the bedrock geology is so little known that geophysical data may provide the only clue of structural trend and rock type. For example, the magnetic and gravity data of Figure 32–7 indicate that magnetic rock units concealed under this part of Cape Cod, Mass., strike about N.70°E., similar to the eastward strike inferred

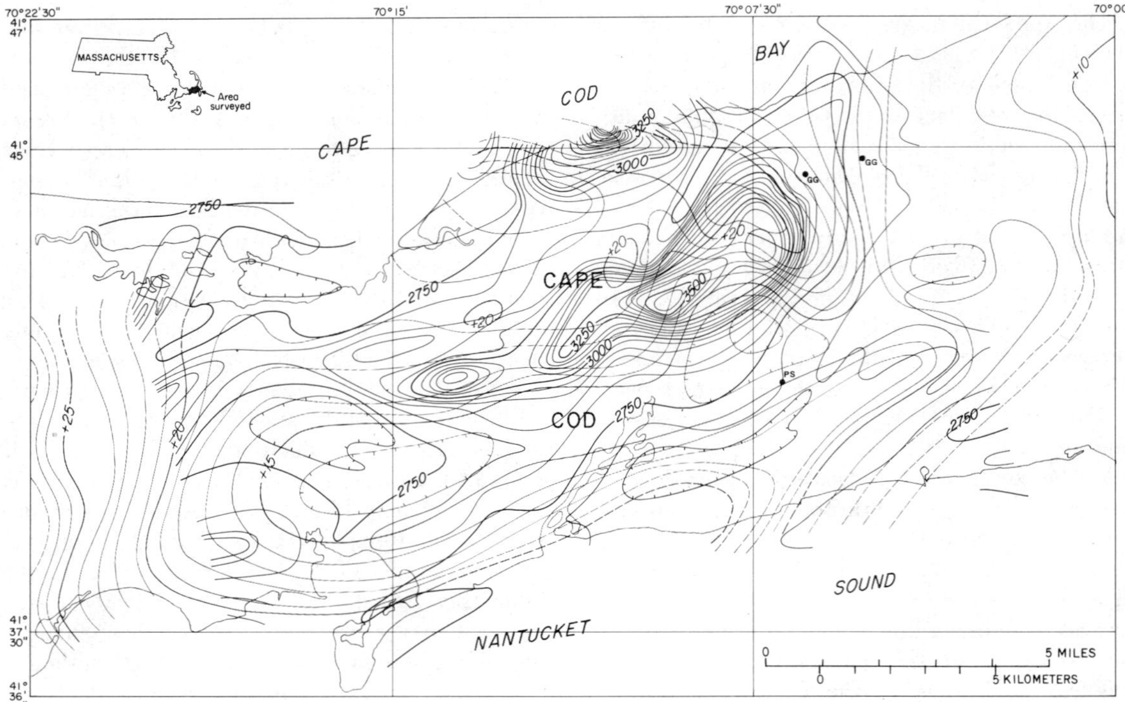

FIGURE 32–7. Aeromagnetic and simple Bouguer gravity map of central Cape Cod, Mass. Dots are wells to basement rocks: GG, granite gneiss; PS, phyllitic schist. Contour intervals are 50 gammas and 1 mgal, respectively. The gravity map is screened gray. Magnetic map flown 500 feet above ground. Geophysical data by U. S. Geological Survey. Geologic data discussed in text.

for the outer Cape by Oldale and Tuttle (1964, p. D120) from variations of seismic velocities with azimuth. The linear magnetic high is associated with a gravity high. Judging by the absence of a well-defined magnetic low on the north side, and by the northerly offset of the gravity high relative to the magnetic anomaly, the anomaly represents a tabular body that dips north. Depth determinations made on the magnetic anomaly indicate that the top of the magnetic mass is buried less than 750 feet (0.2 km) below the ground surface and that the magnetic rocks are presumably exposed at the basement surface below the unconsolidated deposits. The geophysical data along the north shore of Cape Cod further suggest the possibility that the linear anomaly is the south limb of a synclinal mass of magnetic, high-density rocks, presumably mafic volcanics or a folded mafic intrusion. Drill holes (Fig. 32–7) to the basement east of the anomaly penetrated granite gneiss (Koteff, written communication, 1962, *in* Oldale and Tuttle, 1965) while a drill hole south of the anomaly penetrated over 400 feet (0.1 km) of micaceous phyllitic schist (Koteff and Cotton, 1962). These rocks presumably underlie the mafic body.

The gravity high shown on Figure 32–7 appears to be an eastward branch of a major north-trending gravity high whose axis is 5 miles (8 km) farther west. This major gravity feature is approximately similar in size, amplitude, and trend to gravity highs that occur over mafic plutonic rocks in northeastern Massachusetts and south coastal Maine (Joyner, 1963), and that are associated with the mafic plutonic rocks of the Bays-of-Maine Complex (Kane and Bromery, 1966; Chapman, 1962, and this volume).

Chapman (this volume) describes circular, stocklike granitic plutons in the southwestern part of the Bays-of-Maine Complex. Southeast of Mount Desert Island, Maine, about 20 miles (30 km) offshore at lat. 44°05′N., long. 67°55′W., there is a circular magnetic high (Malloy and Harbison, 1966) about 8 miles (13 km) in diameter enclosing a central low. Malloy and Harbison interpret this feature as a ring dike, but an equally possible interpretation is that it represents a granitic stock with a more mafic and magnetic border facies similar in structure and age to those discussed by Chapman. The aeromagnetic patterns and structure of other plutons in Maine, including zoned stocks, have been described by Allingham (1960, 1961).

BROAD MAGNETIC ANOMALIES IN MAINE

In the northern half of Maine, broad synclinoria of relatively nonmagnetic gray slate and sandstone, locally containing small amounts of interbedded volcanic rocks, display broad magnetic highs which, judging by the width and smoothness of the anomaly gradients, must be generated by magnetic rocks buried at depths of at least several miles. A magnetic high of 200 gammas amplitude, 40 miles (65 km) long and over 10 miles (16 km) wide (Hurley and Thompson, 1950; Boucot et al., 1964) is associated with the central part of the Moose River synclinorium (Boucot, 1961) near Moosehead Lake. Another magnetic high, of 400 gammas amplitude (Boucot et al., 1964; Canada Geological Survey, 1953) is associated with the synclinorium of Devonian slate in extreme northwestern Maine and is 55 miles (90 km) long and at least 25 miles (40 km) wide. A possibly similar feature is located in eastern Maine near lat. 45°40′N. and long. 68°15′W. This magnetic high (Griscom and Larrabee, 1963) is at least 20 miles (30 km) long and 12 miles (20 km) wide, and is associated with a monotonous gray slate and siltstone unit. Grantz and Zietz (1960) have described broad magnetic and gravity highs over belts of moderately deformed sedimentary rocks, and consider that the anomalies are generated by large masses of magnetic igneous rocks buried 5 to 10 miles (8–16 km) below the surface; however, gravity highs are not associated with these magnetic anomaly belts in Maine (Kane and Bromery, 1966). An alternative explanation is that a regional depression of the isothermal surfaces beneath these belts has locally increased the amounts of magnetic material because of increased depth to the Curie temperature for the magnetic minerals; but the anomalies are probably too narrow compared with the likely depth to the Curie temperature of magnetite.

REFERENCES

Allingham, J. W., 1960, Use of aeromagnetic data to determine geologic structure in northern Maine: U. S. Geol. Survey Prof. Paper 400–B, p. B117–B119

————, 1961, Aeromagnetic interpretation of zoned intrusions in northern Maine: U. S. Geol. Survey Prof. Paper 424–D, p. D265–D266

Balk, Robert, 1946, Gneiss dome at Shelburne Falls, Massachusetts: Geol Soc. America Bull., v. 57, p. 125–159

Billings, M. P., 1945, Mechanics of igneous intrusion in New Hampshire: Am. Jour. Sci., v. 243–A (Daly Volume), p. 41–68

————, 1955, Geologic map of New Hampshire: U. S. Geol. Survey, scale 1:250,000

Boucot, A. J., 1961, Stratigraphy of the Moose River synclinorium, Maine: U. S. Geol. Survey Bull. 1111–E, p. 153–188

————, Griscom, Andrew, and Allingham, J. W., 1964, Geologic and aeromagnetic map of northern Maine: U. S. Geol. Survey Geophys. Inv. Map GP–312, scale 1:250,000

Bromery, R. W., 1967a, Geophysical evidence of a pronounced east-west lineament in southern New England [abs.]: Geol. Soc. America, Northeastern Sec., Program 1967 Ann. Mtg., Boston, p. 16

————, 1967b, Preliminary results of aeromagnetic and aeroradioactivity surveys in western Massachusetts, *in* Farquhar, O. C., *ed.*, Economic Geology in Massachusetts, Univ. of Massachusetts, Graduate School, Amherst, p. 391–404

Cady, W. M., 1960, Stratigraphic and geotectonic relationships in northern Vermont and southern Quebec: Geol. Soc. America Bull., v. 71, p. 531–576

Canada Geological Survey, 1953, Aeromagnetic map, Ste. Justine–Dorchester, Bellechasse, and Montmagny Counties, Quebec: Canada Geol. Survey Geophysics Paper 117, scale 1:63,360

————, 1954a, Aeromagnetic map, Granby–Shefford, Brome, Rouville, Bagot, St. Hyacinthe, and Missisquoi Counties, Quebec: Canada Geol. Survey Geophysics Paper 171, scale 1:63,360

————, 1954b, Aeromagnetic map, Orford–Shefford, Sherbrooke, Brome, Richmond, and Stanstead Counties, Quebec: Canada Geol. Survey Geophysics Paper 173, scale 1:63,360

————, 1954c, Aeromagnetic map, Sutton–Missisquoi and Brome Counties, Quebec: Canada Geol. Survey Geophysics Paper 183, scale 1:63,360

Chapman, C. A., 1939, Geology of the Mascoma quadrangle, New Hampshire: Geol. Soc. America Bull., v. 50, p. 127–180

————, 1952, Structure and petrology of the Sunapee quadrangle, New Hampshire: Geol. Soc. America Bull., v. 63, p. 381–425

————, 1962, Bays-of-Maine igneous complex: Geol. Soc. America Bull., v. 73, p. 883–888

————, 1966, Paucity of mafic ring-dikes—evidence for floored polymagmatic chambers: Am. Jour. Sci., v. 264, p. 66–77

Dennis, J. G., 1964, The geology of the Enosburg area, Vermont: Vermont Geol. Survey Bull. 23, 56 p.

Doll, C. G., Cady, W. M., Thompson, J. B., Jr., and Billings, M. P., *compilers* and *editors*, 1961, Centennial geologic map of Vermont: Vermont Geol. Survey, scale 1:250,000

Doyle, R. G., Young, R. S., and Wing, L. A., 1961, A detailed economic investigation of aeromagnetic anomalies in eastern Penobscot County, Maine: Maine Geol. Survey Spec. Econ. Studies Ser., no. 1, 69 p.

Fisher, D. W., Isachsen, Y. W., Rickard, L. V., and Offield, T. W., *compilers* and *editors*, 1962, Geologic map of New York—1961: New York State Mus. and Sci. Service, Geol. Survey Map and Chart Series no. 5, scale 1:250,000

Forsyth, W. T., 1955, Airborne magnetometer investigations in eastern Maine: Maine Geol. Survey Rept. State Geologist 1953–1954, p. 31–45

Fowler-Billings, Katharine, and Page, L. R., 1942, The geology of the Cardigan and Rumney quadrangles, New Hampshire: Concord, New Hampshire Planning and Devel. Comm., 31 p.

Grantz, Arthur, and Zietz, Isidore, 1960, Possible significance of broad magnetic highs over belts of moderately deformed sedimentary rocks in Alaska and California: U. S. Geol. Survey Prof. Paper 400–B, p. B342–B347

Griscom, Andrew, 1962, Aeromagnetic evidence for the structure of stocks and ring dikes in New Hampshire and Vermont [abs.]: Geol. Soc. Amer. Spec. Paper 68, p. 186

——————, 1963, Tectonic significance of the Bouguer gravity field of the Appalachian system [abs.]: Geol. Soc. America Spec. Paper 73, p. 163–164

——————, and Larrabee, D. M., 1963, Aeromagnetic interpretation and preliminary geology of the Danforth area, Maine: U. S. Geol. Survey Geophys. Inv. Map GP–423, scale 1:62,500

——————, and Zietz, Isidore, 1964, Differences between aeromagnetic and geologic depths to basement in the Appalachian Basin [abs.]: Geol. Soc. America Spec. Paper 76, p. 68–69

Hatch, N. L., Jr., and Hartshorn, J. H., 1968, Geologic map of the Heath Quadrangle, Massachusetts–Vermont: U. S. Geol. Survey Geol. Quad Map series, GQ 735 (with explanatory text)

Hurley, P. M., and Thompson, J. B., Jr., 1950, Airborne magnetometer and geological reconnaissance survey in northwestern Maine: Geol. Soc. America Bull., v. 61, p. 835–841

Joyner, W. B., 1963, Gravity in north-central New England: Geol. Soc. America Bull., v. 74, p. 831–858

Kane, M. F., and Bromery, R. W., 1966, Simple Bouguer gravity map of Maine: U. S. Survey Geophys. Inv. Map GP–580, scale 1:500,000

Koteff, Carl, and Cotton, J. E., 1962, Preliminary results of recent deep drilling on Cape Cod, Massachusetts: Science, v. 137, no. 3523, p. 34

Malloy, R. J., and Harbison, R. N., 1966, Marine geology of the northeastern Gulf of Maine: U. S. Coast and Geodetic Survey Tech. Bull. 28, 15 p.

Oldale, R. N., and Tuttle, C. R., 1964, Seismic investigations on Cape Cod, Massachusetts: U. S. Geol. Survey Prof. Paper 475–D, p. D118–D122

——————, 1965, Seismic investigations in the Harwich and Dennis quadrangles, Cape Cod, Massachusetts: U. S. Geol. Survey Prof. Paper 525–D, p. D101–D105

Quinn, A. W., and Stewart, G. W., 1941, Igneous rocks of the Merrymeeting Lake area of New Hampshire: Am. Mineralogist, v. 26, p. 633–645

Sclar, C. B., 1958, The Preston gabbro and the associated metamorphic gneisses, New London County, Connecticut: Connecticut Geol. Nat. History Survey Bull. 88, 136 p.

Segerstrom, Kenneth, 1956, Bedrock geology of the Shelburne Falls quadrangle, Massachusetts: U. S. Survey Geol. Quad. Map GQ–87, scale 1:31,680

Trefethen, J. M., 1953, Airborne magnetometer survey, Forest City area, eastern Maine map: Maine Geol. Survey, scale 1:62,500

Vacquier, Victor, Steenland, N. C., Henderson, R. G., and Zietz, Isidore, 1951, Interpretation of aeromagnetic maps: Geol. Soc. America Mem. 47, 151 p.

Wing, L. A., 1958a, Aeromagnetic and geologic reconnaissance survey of portions of Hancock and Penobscot Counties, Maine: Maine Geol. Survey GP. and G. Survey no. 1, 10 sheets, scale 1:62,500

——————, 1958b, Aeromagnetic and geologic reconnaissance survey of Atkinson and vincinity, Piscataquis County, Maine: Maine Geol. Survey GP. and G. Survey no. 2, 6 sheets, scale 1:62,500

——————, 1958c, Aeromagnetic and geologic reconnaissance survey of portions of Penobscot, Hancock, and Washington Counties, Maine: Maine Geol. Survey GP. and G. Survey no. 3, 6 sheets, scale 1:62,500

——————, 1959, An aeromagnetic and geologic reconnaissance survey of portions of Penobscot, Piscataquis, and Aroostook Counties, Maine: Maine Geol. Survey GP. and G. Survey no. 4, 7 p.

Zen, E-an, 1967, Time and space relationships of the Taconic allochthon and autochthon: Geol. Soc. America Special Paper 97, 107 p.

Chapter 33

Heat Flow and Thermal History
in New England and New York

FRANCIS BIRCH, ROBERT F. ROY,
AND EDWARD R. DECKER

INTRODUCTION

PRELIMINARY VALUES for the first determinations of heat flow in New England were reported in 1965 by Roy and Decker (1965). The most conspicuous feature was the high heat flow at three sites in the Conway Granite, of which the unusually high radioactivity had been discovered by Billings and Keevil (1946) and extensively sampled by Adams et al. (1962). A total of 22 determinations of heat flow in New England and New York (Table 33–1) have now been completed: 4 in the Precambrian rocks of the Adirondacks, 3 in the Precambrian core of the Green Mountain anticline, Vermont, 8 in rocks of the New Hampshire Plutonic Series, 3 in the Conway Granite of the White Mountain Plutonic Series, with 4 others; locations and generalized geology are shown on Figure 33–1. A relationship with age is suggested by these measurements; the lowest values are in the Adirondack rocks, of Grenville age (1100 m.y.), the highest in the Conway Granite, of Triassic age (185 m.y.). The correlation is strong between heat flow and local bedrock radioactivity. Before proceeding with the interpretation, we first inquire whether corrections related to age or geological history may seriously affect these correlations.

The techniques of measuring thermal gradients and conductivities, the physical data which are combined to determine heat flow, will be discussed elsewhere (Roy and Birch, preparation). Briefly, temperatures were measured in vertical, diamond-cored holes of about 5-cm diameter; the useful range of depth was usually from 100 to 300 m. The core was sampled for numerous conductivity measurements and for analysis by gamma-ray spectrometry of the uranium, thorium and potassium contents (Adams, 1964); the measurements of radioactivity, by J. A. S. Adams and J. J. Rogers of Rice University, will be described elsewhere. The estimates of radioactive heat generation used in the following discussion are based on their determinations, except for the Chelmsford Granite, for which uranium values are given by Phair and Gottfried (1964).

We also defer discussion of the topographic correction, except to recall that the purpose of this correction is to take account of the effect on heat flow of departures of the real surface of the earth from the local plane of reference, a horizontal surface at or near the level of the well collar. The correction has been applied to the values of Table 33–1, on the assumption that the topography and the surface temperature have always existed as at the present time. The evolution of the topography, understood here as the departures from the reference plane, is not of importance in the terrain under consideration: the topographic correction is dependent upon the relief within a radius of a few tens of kilometers for these shallow holes, and corrections for evolution are substantial only for the topography at greater distances (Birch, 1950). But the temperature of the reference plane has evidently undergone major changes, which we examine under the heading of the "geological" correction to heat flow.

We are thus led to an effort to trace, in highly schematic fashion, the thermal history since Lower Devonian time. The New England–New York region is especially favorable for this purpose as a result of intensive geological study and excellent mapping, circumstances which indeed led to the initial selection of many of the heat flow sites. The discussion will deal principally with New Hampshire, where most of the measurements were made and the corrections are largest; the geological history assumed for purposes of calculation is then essentially a simplified version

TABLE 33-1.   Heat flow in New England and New York

Station	N. Lat.	W. Long.	Collar elevation, meters	Depth,* meters	Heat flow,† 10^{-6} cal/cm^2 sec		
					unc.	topo.	geol.
Maine							
Blue Hill	44°24′	68°37′	30	150–350	1.46 .01	1.44	1.30
Casco	44°03′	70°37′	110	150–300	1.86 .03	1.80	1.63
Massachusetts							
Brewster	41°45′	70°05′	20	200–300	1.14 .02	1.16	1.29
Cambridge	42°23′	71°07′	6	160–260	1.21 .01	1.20	1.20
Chelmsford	42°38′	71°25′	24	130–160	1.61 .02	1.63	1.48
Millers Falls	42°37′	72°27′	310	210–280	1.51 .01	1.67	1.51
New Hampshire							
Bradford	43°16′	71°59′	250	200–260	1.63 .03	1.59	1.44
Concord	43°12′	71°32′	85	200–320	1.76 .03	1.73	1.57
Durham	43°07′	70°55′	6	260–305	1.09 .02	1.08	0.98
Fitzwilliam	42°47′	72°08′	330	100–300	1.65 .01	1.63	1.48
Kancamagus	44°02′	71°29′	730	170–305	2.40 .02	2.27	2.13
North Conway	44°04′	71°10′	195	120–215	2.04 .04	1.89	1.95
North Haverhill	44°06′	72°00′	180	150–240	1.41 .01	1.34	1.21
Waterville	43°56′	71°32′	400	240–320	2.53 .06	2.15	2.21
New York							
Elizabethtown	44°13′	73°32′	150	400–600	0.83 .01	0.81	0.80
Glens Falls	43°18′	73°37′	80	220–265	1.06 .01	1.05	1.04
Riverview	44°35′	73°54′	360	260–330	1.26 .02	1.22	1.21
Saranac Lake	44°20′	74°16′	490	100–360	0.81 .01	0.81	0.80
Wadhams	44°14′	73°28′	100	210–260	0.81 .01	0.79	0.78
Vermont							
Londonderry	43°15′	72°50′	370	160–240	1.32 .01	1.23	1.16
North Springfield	43°20′	72°33′	180	160–240	1.29 .01	1.20	1.13
Weston	43°17′	72°49′	530	350–430	1.20 .01	1.22	1.15

* Depth: this is the interval used for the computation of heat flow.

† Heat flow: "uncorrected" from combination of gradients and conductivities, with statistically determined standard errors. "topo." with steady state correction for topography to 100 km. "geol." with geological corrections as described in this paper. The values at North Conway and Waterville, before correction for contrast of radioactivity, are 1.77 and 2.01, respectively.

of the account given by Billings (1956) with some later modifications by Wilson (1965).

As geological history has no clear-cut beginnings or well-defined initial conditions, we may at best begin at a time of relative quiescence, when no geologically rapid changes were taking place. In New Hampshire, we have chosen to begin with the Lower Devonian, after some 10 km of Ordovician and Silurian sediments had been laid down over a broad strip of north-northeasterly strike, during a period of perhaps 100 million years (m.y.). Beginning in the Lower Devonian, the rate of deposition increased, with 15 km or more deposited in some 50 m.y. Deformation, uplift and erosion followed, accompanied by emplacement of the New Hampshire Plutonic Series approximately 360 m.y. ago (Wilson, 1965, p. 83; Handford, 1965) and regional metamorphism of the surrounding rocks, chiefly represented at present by the Littleton Formation. Erosion and renewed uplift then continued until the Triassic, when a new outbreak of volcanism began, and the White Mountain Plutonic Series

was emplaced with the Conway Granite as the final phase. Erosion then continued up to the present. We defer discussion of the climatic fluctuations of the Pleistocene; they must have affected the various part of the small area we are considering in much the same way, and, except for a small correction for the deposition of sand and gravel at Brewster, Mass., no further attention will be given to Pleistocene events here.

The principal questions pertain to the measurements in the granitic rocks of the New Hampshire and White Mountain series; these rocks must have been originally at the freezing temperature of granite, say 650°C. (Tuttle and Bowen, 1958, p. 117 et seq.). Had they been exposed to the low surface temperature, they would have long ago lost this initial heat, and no significant correction would be necessary. But the present surface may have had as much cover as 15 km in central New Hampshire when the granites and quartz monzonites of the New Hampshire series were emplaced, and as much as 6 km when the Conway Granite was emplaced. The erosion of

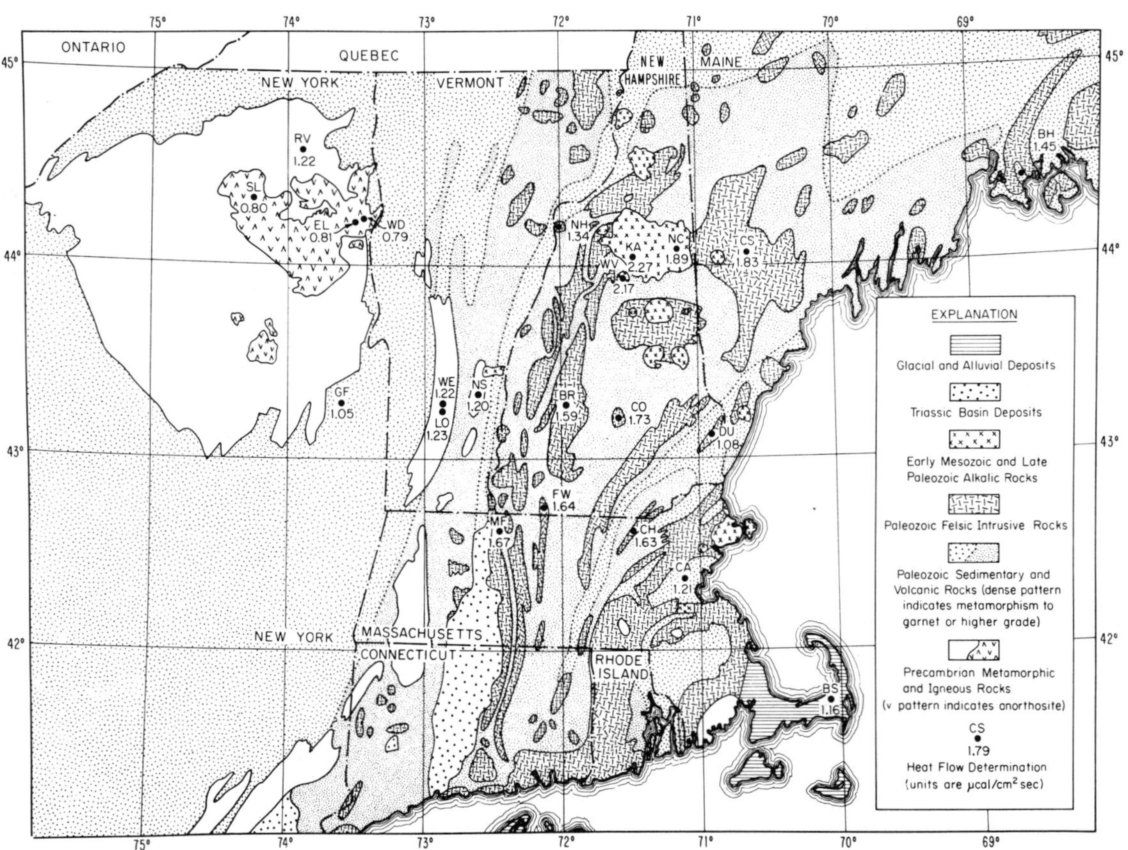

FIGURE 33–1. Generalized geologic map showing locations of heat flow sites, values of heat flow (with topographic correction only) in microcal/cm² sec. Based on Tectonic Map of United States.

this cover required, in some fashion, all the subsequent time up to the present; thus cooling was slower and some residual effects may persist.

In the following discussion, various more or less arbitrary choices of data are introduced, each of which might be the subject of prolonged debate; the numerical results, consequently, are to be judged in the light of the general correctness of these choices, and it might be desirable to examine the effect of alterations in the various numbers at greater length. For brevity, this has been avoided here; it will also turn out that the corrections, which were our first objective, do not greatly change the nature of the correlations based on uncorrected data.

ACKNOWLEDGMENTS

We acknowledge with thanks the assistance of Henry N. Pollack, D. Blackwell, R. Hirst, J. Abbott, J. Albright, R. De Simone, T. Massello, M. MacKenzie in the field operations; of C. Bickel, K. Knight, Joan Abbott, M. Chessman, and N. H. Mao in the measurements of conductivity and the map readings; of Harold Ames, Arthur Ames and J. Plourde in the construction and maintenance of the logging equipment and laboratory apparatus. J. Warner kindly furnished a copy of his unpublished geologic map of southern Maine. The measurements of radioactivity, done at the Rice University through the cooperation of Professor John A. S. Adams and Professor John J. W. Rogers, add an indispensable factor for the interpretation.

We are indebted to Marland P. Billings and J. B. Thompson for valuable guidance in reconstructing the geologic history, and to Professor Billings for his generous permission to reproduce his unpublished map of bedrock radioactivity.

This work has been supported by the Committee on Experimental Geology and Geophysics of Harvard University and by a grant (GP–701) of the National Science Foundation.

THE RISE OF TEMPERATURE DURING DEPOSITION OF A SEDIMENTARY LAYER

Deposition at a uniform rate has been treated by Benfield (1949, 1950; Carslaw and Jaeger, 1959, p. 388) for the semi-infinite medium in which the initial temperature increases linearly with depth; this solution gives reasonably good results for the sedimentary layer, and even for the rise of temperature in the basement, but predicted temperatures become much too high at depths greater than 10 km or so if an initial gradient suitable for the sediments is adopted. Sudden deposition of a sedimentary layer has been discussed by Jeffreys (1931) and Grossling (1959), and the solution for this problem may be taken as the starting point for a more general solution which allows for gradual deposition. The notation of Carslaw and Jaeger is followed.

At time $t = 0$, a layer of thickness H of uniform, non-radioactive sediment is deposited, with zero temperature, upon a semi-infinite basement in which the temperature is assumed to have reached a steady-state temperature distribution with surface temperature zero, $T_0(x)$. The thermal diffusivity k is uniform throughout the half-space, including the layer H. The initial conditions are then: at $t = 0$,

$$T = 0,\ 0 < x < H;\ T = T_0(x)$$
$$\text{for } x > H;\ \text{for } t > 0,$$
$$T = 0 \text{ at } x = 0.$$

The solution is readily found by standard methods to be,

for $0 < x < H$,

$$T(x,t) = \alpha\sqrt{kt}\left[\text{ierfc}\,\frac{H - x}{\sqrt{4kt}} - \text{ierfc}\,\frac{H + x}{\sqrt{4kt}}\right]$$

for $x > H$,

$$T(x,t) = T_0(x) + \alpha\sqrt{kt}\left[\text{ierfc}\,\frac{x - H}{\sqrt{4kt}}\right.$$
$$\left. - \text{ierfc}\,\frac{x + H}{\sqrt{4kt}}\right]$$

Here α is the gradient which would be reached in the sedimentary layer as the limit for infinite time; if Q is the equilibrium heat flow, then $\alpha = Q/K_s$, where K_s is the thermal conductivity of the sediments.

It is now convenient to replace x, measured from the surface of the sedimentary layer, by a new variable y, measured from the surface of the basement; $y = x - H$. For y greater than zero, the solution for fixed H suddenly deposited at $t = 0$ becomes:

$$T(y,t) = T_0(y) + \alpha\sqrt{kt}\left[\text{ierfc}\,\frac{y}{\sqrt{4kt}}\right.$$
$$\left. - \text{ierfc}\,\frac{y + 2H}{\sqrt{4kt}}\right]$$

From this we have

$$\frac{dT}{dH} = \alpha\,\text{erfc}\,\frac{y + 2H}{\sqrt{4kt}};$$

if now H is a function of time, we may write

$$T(y,t) = T_0(y)$$
$$+ \alpha\int_0^t \text{erfc}\left[\frac{y + 2H(\tau)}{\sqrt{4k(t - \tau)}}\right]dH(\tau).$$

For a uniform rate of deposition U, beginning with zero depth at time $t = 0$, we put $H = U\tau$, $dH = U d\tau$. The temperature in the basement is then

$$T(y,t) = T_0(y) + \alpha U t \int_0^t \operatorname{erfc} \frac{y + 2U\tau}{\sqrt{4k(t - \tau)}} \frac{d\tau}{t}$$

Though awkward in appearance, this integral is readily evaluated numerically. Results are given in Table 33–2 and Figure 33–2 for 30 and 50 m.y., respectively, after the beginning of deposition, at the rate $U = 0.3$ km/m.y., or 15 km in 50 m.y. The diffusivity k has been taken as 32 km^2/m.y. With the conditions in central New Hampshire in mind, the equilibrium heat flow from the basement is taken as 1.6 microcal/cm^2 sec, the conductivity of the sedimentary layer as 0.004 cal/cm sec deg and thus $\alpha = 40°$/km.

The initial temperature T_0, assumed to be in equilibrium with the radioactive heat generation in the basement and below, depends upon the details of the distribution of heat sources and conductivities. While the present heat flow furnishes a first approximation to the total heat flow from below the present reference level as of 400 m.y. ago, the distribution of sources and indeed the nature of the rocks at that time must be conjectural, and the following conjectures have been adopted for calculation: the total heat flow was 1.6 microcal/cm^2 sec, of which 0.3 originated below 30 km, and 1.3 from uniform sources through a 30-km layer, of uniform conductivity 0.005 cal/cm sec deg. Below 30 km, the gradient was uniformly 6°/km within the depths of interest, say to 100 km. The corresponding temperatures are given in Table 33–3 as T_0. This 30-km layer is intended to resemble an undifferentiated crust preceding the formation of the Paleozoic granites; the average rate of heat generation was 4.3×10^{-13} cal/cm^3 sec, roughly half that of the granites of the New Hampshire series.

The adoption of a uniform conductivity 0.005 and of diffusivity 0.01 cm^2/sec or 32 km^2/m.y.

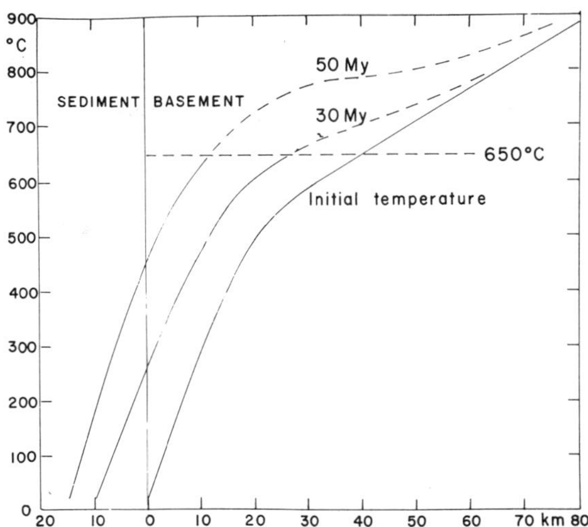

FIGURE 33–2. Temperatures during sedimentation. Zero depth is presently exposed surface. Initial temperature corresponds to zero depth of sediment; sedimentary thicknesses indicated by horizontal scale, increasing toward the left. Curves for 30 m.y. and 50 m.y. after beginning of deposition at rate of 0.3 km/m.y. The continuations above 650°C are meaningful only in the absence of melting and movement of magma.

is of course a drastic simplification. These parameters are functions of temperature; diffusivity in the granitic rocks at low surface temperatures is closer to 50 km^2/m.y. than to 32; on the other hand, the diffusivity of the sediments was probably lower than 32 km^2/m.y. when they were first deposited. We are discussing a process in which the temperatures are everywhere rising; after deposition of a few kilometers of sediments, the diffusivity of the basement rocks falls and that of the sediments near the basement rises. Short of a completely numerical calculation, which would also require details of which we are ignorant, a uniform diffusivity of about 32 km^2/m.y. seems a convenient and reasonable first approximation.

The radioactivity of the basement has been accounted for in the initial distribution T_0, but

TABLE 33–2. Temperatures in basement during deposition of sediment.
$k = 32$ km^2/m.y.; $U = 0.3$ km/m.y., no melting

Depth in basement, km	Initial temperature, T_0, °C	After 30 m.y.		After 50 m.y.		
		ΔT	T, °C	ΔT_1	ΔT_2	T, °C
0	20	243	263	378	62	460
15	403	143	546	250	32	685
30	590	77	667	158	24	772
45	680	38	718	95	16	791
60	770	18	788	56	8	834
75	860	7	867	32		892
90	950			16		966

TABLE 33-3.　Radioactivity, heat generation and heat flow

Locality and rock type	No. of samples	Th, ppm	U, ppm	K,%	A*	Q†
White Mountain series, Conway Granite						
Kancamagus, N. H.	557	59	15.8	4.0	20.5	2.13
North Conway, N. H.	145	52	12.6	4.3	17.5	1.95
Waterville, N. H.	349	61	15.9	4.1	20.9	2.21
Cambridge Formation, argillite						
Cambridge, Mass.	23	9.4	1.2	2.5	2.9	1.20
New Hampshire series						
Casco, Maine, binary granite	32	40	8.2	4.3	12.7	1.63
Chelmsford, Mass. granite-migmatite	††	(30)	9	(4)	(11.5)	1.48
Fitzwilliam, N. H. binary granite	145	23.5	7.6	3.9	9.7	1.48
North Haverhill, N. H. French Pond Granite	24	26.7	4.0	3.2	7.7	1.21
Precambrian						
Riverview, N. Y. granitic gneiss	61	15.3	3.4	5.1	6.1	1.21
Weston, Vt. biotite gneiss	279	5.9	1.0	1.9	2.1	1.15
Saranac Lake, N. Y. anorthosite		(below level of detection)				0.80

* A in 10^{-13} cal/cm³sec.

† Q, with geological correction, in 10^{-6} cal/cm²sec.

†† Uranium for Chelmsford Granite after Phair and Gottfried (1964); Th and K estimated.

as yet, the radioactivity of the sediment has been neglected. The rise of temperature at the base of the sediments may be calculated as follows: the solution for steady heating of a layer, $0<x<H$, with the surface $x = 0$ held at $T = 0$ (Carslaw and Jaeger, 1959, p. 79) gives for the rise of temperature after a time t,

$$\Delta T = (AH^2/2K)(2kt/H^2)\left[1 - 4i^2 \text{ erfc } \frac{x}{\sqrt{4kt}} \right.$$
$$\left. + 2i^2 \text{ erfc } \frac{H + x}{\sqrt{4kt}} - 2i^2 \text{ erfc } \frac{H - x}{\sqrt{4kt}}\right],$$
$$\text{for } x \leq H.$$

At the base of the layer, $x = H$, the rise is

$$\Delta T(H) = (AH^2/2K)[1/2 - 4i^2 \text{ ierfc } w$$
$$+ 2i^2 \text{ erfc } 2w]/(2w^2),$$

with $w = H/\sqrt{4kt}$.

$AH^2/2K$ is the steady state rise at $x = H$, the factor giving the fraction of this reached after a time t. With $H = 15$ km, $t = 50$ m.y., $k = 32$ km²/m.y., this fraction is 0.79.

This solution may be transformed to give the temperature at the base of a radioactive layer which increases in thickness with time by differentiating with respect to H, then integrating with respect to a time variable τ, as before. The result for a uniform rate of thickening, starting from zero thickness at time $t = 0$, is

$$\Delta T(H) = \left(\frac{AU^2t^2}{2K}\right)\int_0^t \frac{2\text{ierfc } z - 2\text{ierfc } 2z}{z} \frac{\tau d\tau}{t^2}$$

where $\quad z = U\tau/\sqrt{4k(t - \tau)}$.

The thickness H at time t is Ut, and thus the first factor is just the equilibrium temperature rise for a permanent layer of thickness H; the integral gives the fraction of this reached at t. For $k = 32$ km²/m.y. and $t = 50$ m.y., this fraction is 0.55 for $U = 0.3$ km/m.y.

Of the Devonian sediments, there remains today only the Littleton Formation, the radioactivity of which may have changed during metamorphism. The surveys of radioactivity discussed by Billings (1964) show little contrast between the Littleton Formation and the intrusive rocks of the New Hampshire Plutonic Series. For the latter, the mean heat production in the granites is about 10×10^{-13} cal/cm³ sec, while the extensive Kinsman Quartz Monzonite and Bethlehem Gneiss are less radioactive, with heat production about 6 to 7×10^{-13} cal/cm³ sec (Lyons, 1964). For the Cambridge Argillite, the mean is 2.9×10^{-13}, for "common shale," about 4.6×10^{-13} (Clark, 1966, p. 538–539). We adopt 4×10^{-13} cal/cm³ sec as the average for the 15-km layer. With $K_s = 0.004$ cal/cm. sec deg, $AH^2/2K_s = 113°$, and the rise at the base of the 15-km layer in 50 m.y., caused by its own radioactivity, is $62°$. The contribution is shown in Table 33–3 as $\triangle T_2$; it has not been calculated for 30 m.y. Evidently, the radioactivity of the sediments is not critical for this discussion.

The sum $T_0 + \triangle T_1 + \triangle T_2$ then gives the temperatures which follow from our assumptions. As shown in Figure 33–2, the temperature was initially 650°C at the depth of about 40 km, where presumably there was no granitic material. After 30 m.y. and the deposition of 9 km of sediments, this temperature moved up to a depth of about 28 km below the original surface, and after 50 m.y., if there were no melting, to about 12 km. It seems probable that the latter distribution could never have been reached; melting of a granitic fraction would have begun sometime between 30 and 50 m.y. after the start of this cycle of sedimentation. Complete melting requires an appreciable time after the melting point is reached; with the assumed 4.3×10^{-13} cal/cm^3 sec, or about 5 cal/g m.y., a heat of fusion of some 60 cal/g would be generated in 12 m.y.

Of the New Hampshire series, Billings (1956, p. 147) remarks: "The origin of the magmas . . . is obscure. Although they may be differentiates from basalt, they may equally well be melted up older rocks or granitized sediments that moved up from greater depths." He also notes the absence from the rocks of this series of olivine and pyroxene and of related volcanic rocks. The most mafic members appear to be diorites, in small quantity. The granites have compositions close to that of the ternary eutectic in the Q–Ab–Or system (Chayes, 1952). Structurally, these rocks are generally concordant, forming wide sheets and lenses. Complete or partial fusion in place appears to be consistent with these circumstances, as pointed out by Quinn (1944) and others. The Littleton Formation was probably too cool for general fusion but the volcanics and older sediments beneath it may have been adequate as the source of magma. The Ordovician and Silurian sediments may add up to as much as 10 km (Billings, 1956, p. 7). The thickness of the granites of the New Hampshire series is about 6 km, if we may judge from evidence of the heat flow (see below).

The 15-km thickness assumed for the sedimentary layer above the present surface is based on Newton's revised pressure (Newton, 1966) of the triple-point in the kyanite-sillimanite-andalusite system, and Wilson's discussion of metamorphism in the Ossipee region (Wilson, 1965). With the higher value of this pressure then current, Wilson concluded that the sedimentary layer must have reached a depth of 30 km. It seems likely that melting would have taken place before this thickness could have been reached. A thickness of 15 to 20 km above the present surface seems plausible.

In regions where the heat flow from the basement was appreciably less than 1.6 microcal/cm^2 sec at the beginning of sedimentation, either a longer time, thicker deposits, or lower conductivity in the sedimentary layer would have been required to generate melting temperatures.

When the thickness of the sedimentary layer reached 15 km, the total heat generation with our assumptions was equivalent to 2.2 microcal/cm^2 sec; this was not in equilibrium with the heat flow at the surface. We may find the heat flow at the surface of the sediment by an application of the method already employed. The flow to the surface resulting from the heat flow from the basement is

$$Q\left(1 - \frac{2}{\sqrt{\pi}} \int_0^t z \exp(-z^2)\, d\tau/\tau\right)$$

here Q is 1.6 microcal/cm^2 sec and the factor is 0.69. The radioactivity of the sediments contributes

$$Q_s \int_0^t \mathrm{erfc}\, z\, d\tau/t$$

where now $Q_s = 0.6$ microcal/cm^2 sec, and the factor equals 0.75. Thus the heat flow at the surface was 1.55 microcal/cm^2 sec, only 70 percent of the "equilibrium" flow. If we had made our measurements 360 m.y. ago, a correction of 40% would have been needed to obtain the equilibrium flow. There are obvious implications here with regard to measurements currently being made in areas of rapid recent sedimentation.

THERMAL CHANGES DURING EROSION

We cannot follow in detail the redistribution of material, radioactivity and temperature through the complex history of the Acadian orogeny though certain trends can be dimly visualized. The rise of granitic melts must have tended to bring much of the crust close to the melting temperature; concentration of radioactive elements nearer the surface tended toward reduction of "equilibrium" temperatures, while erosion began on the uplifted surface. Our principal objective is to follow the history of temperature on the presently exposed surface, now the datum plane for our measurements. Over large areas, this surface truncates the granitic rocks of the New Hampshire series and the Littleton Formation which they intrude. As a rough approximation to the distribution of temperature below this surface 360 m.y. ago, we assume a constant temperature of 650° C to a depth of 53 km below the present surface where this temperature is found on the "steady state" temperature distribution corresponding to the present distribution of heat sources specified below. Below this the initial and "steady state" temperature are assumed to be identical, a linear rise at the rate of 6°/km.

The problem may then be conceived as one of cooling from this initial temperature, with the temperature on the present surface ($y = 0$) gradually diminishing as the Acadian mountains were eroded. Little is known about the rates of erosion, though it is natural to think of erosion proceeding most rapidly in the early stages of uplift when presumably the least metamorphosed rocks were exposed. A number of cycles of uplift and planation may have occurred. We discuss three simple assumptions which lead to easy calculations and provide rough estimates for the correction to the present thermal gradients.

These three assumptions are as follows. First, suppose that all the erosion took place in a few millions of years, or even a few tens of millions of years, in Upper Devonian time (Fig. 33–3, line b). In terms of temperature, this corresponds to a sudden change of temperature on the present surface of some 630°, nearly 360 m.y. ago. This leads to the smallest correction, the temperatures below the present surface having had in this case the longest time for readjustment. Second, suppose that erosion has continued at a uniform rate over 360 m.y., continuing at this same average rate today (Fig. 33–3, line d). This seems certain to be an over-estimate of the present rate and leads to the largest correction. Third, as an intermediate case, suppose that the temperature on the present surface has decreased proportionally to the square root of time, measured from 360

m.y. ago; the present rate is then not zero, but it is much smaller than at earlier times. (Fig. 33–3, line c).

In each case, we have two effects to consider, the initial temperature and the change of surface temperature. In the linear theory of conduction, with k everywhere uniform, these may be treated separately and combined to give the desired solution. The solutions for change of surface temperature are to be found in Carslaw and Jaeger (1959, p. 63); the solution for initial temperature may be found from the Laplace solution (see appendix).

The solution given in the appendix corresponds to the sudden reduction of temperature on $y = 0$ to zero at $t = 0$ (360 m.y. ago for the New Hampshire series). The gradient at $y = 0$ is for later times:

$$(\delta T/\delta y)_0 = \frac{V_0}{\sqrt{\pi kt}} + m \operatorname{erfc} \frac{h}{\sqrt{4kt}}$$
$$+ \frac{2A\sqrt{kt}}{K}\left[\operatorname{ierfc} 0 - \operatorname{ierfc} \frac{H}{\sqrt{4kt}}\right]$$

The steady-state gradient is $m + AH/K$, and thus the difference, or disturbance of gradient at $y = 0$ is

$$(\delta T'/\delta y)_0 = \frac{V_0}{\sqrt{\pi kt}} - m \operatorname{erf} \frac{h}{\sqrt{4kt}}$$
$$- \frac{AH}{K}\left(1 - \frac{\operatorname{ierfc} 0 - \operatorname{ierfc} H/\sqrt{4kt}}{H/\sqrt{4kt}}\right)$$

For t sufficiently great, we have approximately,

$$(\delta T'/\delta y)_0 = \frac{1}{\sqrt{\pi kt}}\left(\frac{mh}{3}\frac{h^2}{4kt} + \frac{AH^2}{12K}\frac{H^2}{4kt}\right)$$

since $V_0 - mh - AH^2/2K = 0$. This is of the order of 0.1°/km for the problems with which we are concerned here; the initial excess of heat would have dissipated almost completely if erosion had reached the present surface as long ago as 300 m.y.

If, however, the fall of temperature on the present surface was more gradual, corresponding to a gradual removal of the original cover, the residual effect on the gradient might be much larger. If the temperature at $y = 0$ varied linearly with time, according to $V_0(1 - t/t_0)$, we must add to the solution of the appendix a term in the temperature, $V_0[\operatorname{erfc} y/\sqrt{4kt} - (t/t_0)\, 4i^2 \operatorname{erfc} y/\sqrt{4kt}]$, and to the gradient at $y = 0$, $V_0/\sqrt{\pi kt}\,(2t/t_0 - 1)$. If, somewhat more plausibly, we take the temperature on $y = 0$ to decrease according to $V_0(1 - \sqrt{t/t_0})$, we add to the solution of the

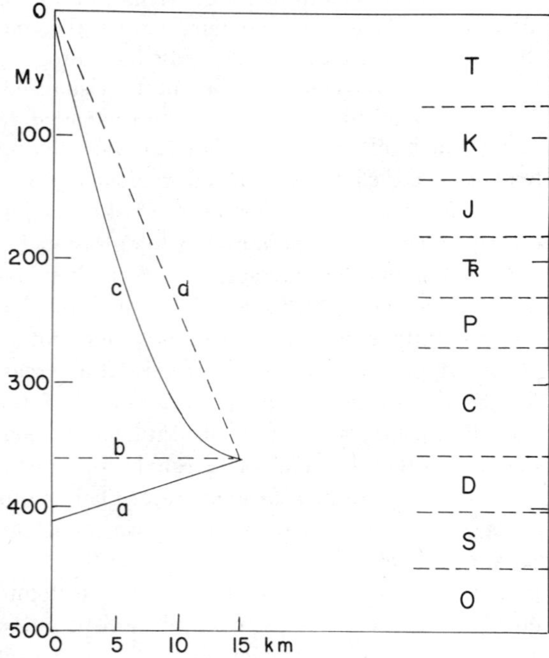

FIGURE 33–3. Schematic representation of assumed sedimentary (a) and erosional (b, c, d) histories. Geologic periods at right, thickness of cover at bottom.

appendix the temperature $V_0[\mathrm{erfc}\ y/\sqrt{4kt} - \sqrt{\pi t/t_0}\ \mathrm{ierfc}\ y/\sqrt{4kt}]$, and to the gradient at $y = 0$, $V_0/\sqrt{\pi kt}\ ([\pi/2]\sqrt{[t/t_0]} - 1)$.

For $t = t_0 = 360$ m.y., $k = 32$ km²/m.y., $V_0 = 630$, the factor $V_0/\sqrt{\pi kt}$ equals 3.3°/km. This would be the disturbance of gradient if the temperature had fallen linearly with time. For the fall according to the square root of the time, this is multiplied by $\pi/2 - 1$, to give 1.9°/km. We take this to be an upper limit for the correction for the measurements in the rocks of the New Hampshire Plutonic Series.

This correction is most suitable for the stations in central New Hampshire: Bradford, Concord, Fitzwilliam, perhaps also Chelmsford, Mass., and Casco, Maine. At all of these, the uncorrected gradients lie between 18 and 22°/km. Application at Blue Hill, Maine, North Haverhill, N. H., Millers Falls, Mass. and Durham, N. H., is more questionable, the latter in particular lying outside the area of deep sedimentation and the inclusion of the Exeter Diorite in the New Hampshire series remaining doubtful. We lack the information needed for individual treatment of these stations. Instead, the gradients have been averaged, to give a mean value of 20.2°/km. The corrected mean is then 18.3°/km, or a reduction of 9.4%. This proportional reduction has been applied to the topographically corrected values of heat flow to give the "geologically corrected" values of Table 33–1.

WHITE MOUNTAIN PLUTONIC SERIES

The White Mountain Plutonic Series differs in many respects from the New Hampshire Plutonic Series. Billings (1956, p. 186) mentions the discordant structural relations, associated volcanic rocks including basalt, the presence of olivine and pyroxene throughout the series and the presence of gabbroic bodies. Wilson (1965, p. 165) estimates the original depth of the present exposed surface in the Ossipee area as about 6 km, which may include several kilometers of the associated Moat Volcanics. The concentrations of uranium and thorium in the youngest member, the Conway Granite, are exceptionally high, roughly three or four times as great as in "average" granite, and about twice as great as in the granites of the New Hampshire series. Higher than average radioactivity is characteristic of all of the members of this series (Billings and Keevil, 1946, p. 821; Rogers and Ragland, 1961; Adams et al., 1962; Rogers, 1964; Butler, 1961). The areal exposure of rocks of this series is much smaller than that of the New Hampshire series,

though small widely scattered bodies have been mapped as "White Mountain" on the basis of lithologic similarities. The main exposure is an elliptical area about 50 by 30 km centered on Mt. Tremont, with other bodies along a nearly north-south line about 200 km long, which cuts across the trend of the Merrimack synclinorium. Chapman and Williams (1935) conclude that basaltic magma, after invading and assimilating the country rocks, then produced the various types of rock by fractional crystallization.

For the Conway Granite, Tilton and Davis (1959, p. 196) show radioactive ages based on five independent decay series for samples from Redstone, N. H., a few km from Conway (see also Toulmin, 1961); the range is from 182 m.y., by K^{40}/A^{40}, to 190 m.y. by Th^{232}/Pb^{208}. With so long an interval between this date and the date of emplacement of the New Hampshire series, it might appear useless to look for relationships between them. On the other hand, the times required for deep penetration of thermal effects originating near the surface are also long, and we must in any case look for basaltic magma at considerable depths. We have consequently sought to trace the thermal history from the time of consolidation of the New Hampshire series to the time of emplacement of the White Mountain series, taking into account the downward diffusion of temperature while erosion diminished the cover above the now-exposed surface.

As a rough guess at the temperature distribution following the emplacement of the New Hampshire series, we have supposed that the temperature was uniformly 650°C between the depths of 0 and 53 km (zero depth corresponding to the present surface) and that below 53 km, the temperature increased by 6°/km. On the present surface ($y = 0$), the temperature is assumed to have fallen to the present temperature, according to $(t/t_0)^{1/2}$, with $t_0 = 360$ m.y. The assumed "equilibrium" temperature distribution, after emplacement of the New Hampshire series with its upward concentration of radioactivity, is plotted in Figure 33–4; though calculated for a two-layer crust, it corresponds closely to a one layer solution with $AH/K = 20.8°/\mathrm{km}$, $H = 30$ km, and for simplicity the latter has been used in calculating the transient effect.

For "White Mountain" time, we put $t = 180$ m.y.; the corresponding temperatures are shown in Figure 33–4. The temperatures within the upper crust are now everywhere below the granitic melting temperature, but at deeper levels there has been a rise of 160° at 60 km, and nearly 100° as deep as 120 km. Whether such

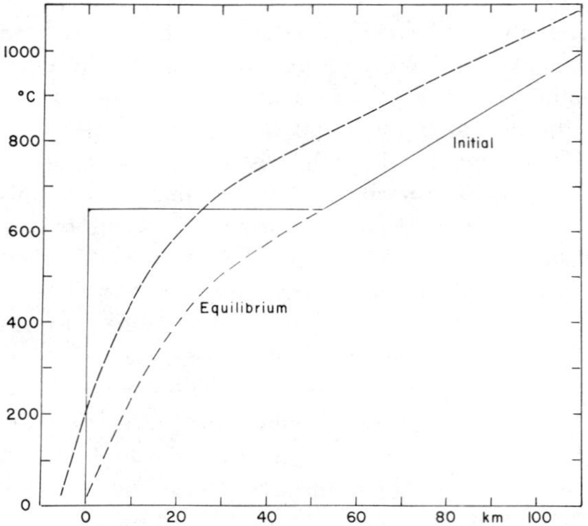

FIGURE 33–4. Temperature distribution before intrusion of White Mountain Plutonic Series (upper broken line). Equilibrium curve corresponds to assumed distribution of radioactivity, with cover removed to present surface; initial curve (solid lines) corresponds to assumed temperature after emplacement of New Hampshire series.

increments could have been significant in the formation of the White Mountain series magmas is uncertain. As these estimates are based on a one-dimensional theory, with no allowance for the finite width of the syncline, they are likely to be too high at depths as great as 100 km. We have to look for basaltic magmas at a temperature of $\sim$ 1200°C, thus to deal with a range of depth and temperature within which the thermal properties are virtually unknown. Nevertheless it is tempting to speculate that this long delayed consequence of the sedimentary cycle may have aided the formation of the White Mountain magma; at the least, we may see in the warming-up of this thick layer a process favorable to the ascent of magma and to the assimilation of crustal rocks.

The high radioactivity of the Conway Granite is not easily accounted for in terms of fractionation of basalt; the quantitative difficulty noted by Chapman and Williams (1935) for the major components is much greater for the radioactive elements, for which enrichments of at least 20-fold and possibly 200-fold are required. Selective assimilation of "pre-heated" crustal rocks may conceivably have provided the needed enrichments.

If the granites now exposed in the White Mountains crystallized at depths of some 6 km the gradients must have been roughly 100°/km at the time of consolidation, and the cooling relatively rapid. The equilibrium gradient in the sediments can not have exceeded 60°/km, and

thus, even with no further erosion, the temperature at a depth of 6 km would have fallen close to 360° or so in no more than 10 m.y. If then erosion proceeded in such a way as to reduce the temperature on the present surface according to $t^{1/2}$, the correction to the gradient at present would be approximately $360°(\pi/2 - 1)/(\pi kt)^{1/2}$, with t 170–180 m.y., or about 1.6°/km. This correction has been subtracted from the measured values for North Conway, Waterville, and Kancamagus; a further correction for these sites is discussed below.

OTHER SITES

The corrections at the other sites may be briefly summarized. In Vermont, two (Weston, Londonderry) are close to the axis of the Green Mountain anticline, the other (North Springfield) in the Chester Dome, all in Precambrian gneisses. The reconstructed sections of the geologic map of Vermont (Doll et al., 1961) show large folds involving lower Paleozoic rocks with as much as 10 km or more of cover above the present surface. Temperatures of some 500°C appear to be implied by the metamorphism, and the time will again be taken as 360 m.y. The correction with temperature falling according to $t^{1/2}$ is 1.6°/km, to be subtracted from the measured gradients.

The Adirondack sites are in rocks of Grenville age, from which perhaps 2 km has been removed since Devonian times, with a total lowering of surface temperature of some 80°, and a correction, for $k = 50$ km²/m.y., of 0.2°/km. We may take account of the lower conductivity and diffusivity in the anorthosite by multiplying this by $2^{1/2}$, since the conductivity is about one-half of that of the granites. A correction of 0.2°/km has been applied at Riverview, 0.3°/km at Saranac Lake, Wadhams and Elizabethtown. At Glens Falls, where the low gradient reflects the high conductivity of quartzites, a nominal correction of 0.1°/km has been applied. The measurements in Cambridge are in the Cambridge Formation, consisting mainly of argillite with lenses of quartzite, of Pennsylvanian (?) age. The grade of metamorphism suggests a temperature no higher than 200°C. The erosional history is obscure. A correction of as much as a few tenths of a degree per kilometer might be appropriate, but none has been applied. An upward correction of about 10% has been applied to the measurement at Brewster, Mass.; this is intended to correct for the deposition of some hundred meters of sand, at a time estimated to be $\sim$5000 years ago.

DISTRIBUTION OF HEAT SOURCES

One of the principal objectives of studies of heat flow is to discover the distribution of heat sources, presumably radioactive, both laterally and vertically. The most striking outcome of the present work is the close correlation between heat flow and radioactive heat generation in the near-surface bedrock. This is shown by the values of Table 33–3, which are plotted in Figure 33–5.

Whether corrected or uncorrected values are used, the nature of the correlation is essentially the same. The simplest interpretation is that the observed contents of radioactive elements are representative of bodies having a thickness of from 5- to 7-km. A layer of Conway Granite 10 km thick with the radioactivity found near the surface would generate all the heat now coming to the surface in the Conway area; this is an upper limit of thickness on the assumption of uniform radioactivity. If the values in the Adirondacks, where the anorthosites furnish no measurable contribution to the flux, are taken as indicating the amount supplied by the lower crust and mantle, 0.8 microcal/cm²sec, then in the Conway area, about 1.2 microcal/cm²sec is supplied by the granite, which is then found to be about 6 km thick if the radioactive content remains uniform. This, in turn, is in reasonable agreement with Joyner's estimate of 15,000 feet (4,500 m) (Joyner, 1963) for the thickness of Conway Granite required to explain the White Mountain gravity low.

Least-squares straight lines of the form, $Q = Q_0 + hA$ have been fitted to the data, with Q and Q_0 in microcal/cm² sec and A, the heat genera-

tion, in 10^{-13} cal/cm³ sec; h is then in units of 10^7 cm, or 100 km. With the geologically corrected values of heat flow, we find $Q = 0.85 + 0.0625$ ($\pm$ 0.0038) A, with the standard deviation of heat flow of 0.08 microcal/cm²sec. With heat flow corrected only for topography, the corresponding line is $Q = 0.93 - 0.0612$ (± 0.0047) A, with a standard deviation of heat flow of 0.10 microcal/cm²sec. The variation of Q with A suggests the presence of a layer approximately 6 km thick having the radioactivity found at the surface, with about 0.9 microcal/cm²sec originating below this layer.

A crustal origin for a substantial fraction of the heat flow is in any case required by the rate of horizontal variation; the extreme example is the change from 1.2 microcal/cm² sec at North Haverhill to 2.1 at Kancamagus, about 40 km away. From Bradford, with $Q = 1.4$, to North Springfield, $Q = 1.1$, the distance is less than 50 km. But these observations would not, by themselves, imply so remarkable a degree of upward segregation of radioactivity as do the regression lines of Q versus A.

We may speculate on the origin of the 0.9 microcal/cm² sec which appears to come from below this 6-km layer. Suppose that 0.6 microcal/cm² sec originates in the remaining 30-odd km of crust; the average A for 30 km is then 2×10^{-13} cal/cm³ sec; for the Weston gneiss, the measured value is 2.1×10^{-13}. If the 0.9 microcal/cm² sec is distributed through the 30-km crust, the average is 3×10^{-13} cal/cm³ sec, still an acceptable value in terms of a conceivable crustal composition. But some contribution from below the crust seems likely, and the average A for the Weston gneiss, based on 279 samples from more than a thousand feet of core, provides a valuable estimate of basement heat production. It is in good agreement with the value from the anorthosite areas, when we allow for the virtually zero heat production in the anorthosite and its probable thickness of a few kilometers (Simmons, 1964). Only in the exposed Precambrian core of the Green Mountain anticline are the surface rocks conceivably representative with respect to radioactivity of a 30–35 km crust. A 30-km crust of anorthosite is as unacceptable as a 30-km crust of Fitzwilliam Granite. The layer of high radioactivity is clearly missing from the Precambrian areas; some 6 km of granitic rocks like those of the New Hampshire series would be needed to bring the heat flow in the anorthosite area up to 1.4 microcal/cm² sec. Values of heat flow between 0.7 and 1.0 microcal/cm² sec are common in the continental shields (Lee and Uyeda, 1965) and

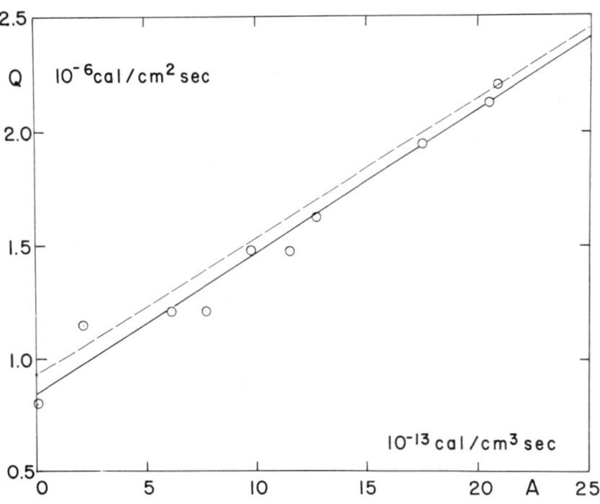

FIGURE 33–5. Heat flow Q, with geological correction (circles) versus local heat generation A. Solid line is least-squares line fitting these points; broken line, least-squares line for topographically corrected data (not plotted).

probably reflect the deep erosion of the more radioactive rocks.

More complex explanations of these correlations are not excluded by the existing evidence. Since all members of the White Mountain Plutonic Series are relatively high in radioactivity, the denser members underlying the granite may contribute more to the heat flow than do the corresponding levels below the New Hampshire series, for example. This possibility might be tested by drilling to 5 km in the Conway Granite; if the rate of heat generation continues through a depth of 5 km as in the shallow samples, the heat flux will fall from about 2 to about 0.9 microcal/cm^2 sec at 5 km.

No measurements of heat flow have as yet been made in such widely-exposed formations as the Littleton Formation or the Bethlehem Gneiss. There are, however, a number of determinations of uranium and thorium in samples of the Bethlehem Gneiss and Kinsman Quartz Monzonite

(Lyons, 1964); the mean values for these two units do not differ significantly, but are slightly lower than those for the granites of the New Hampshire series. It is difficult to deal with the Littleton Formation in this way, but a relationship can be established by way of air-borne and truck-borne gamma-ray counters. In a careful analysis of all available forms of information concerning radioactivity, Billings (1964, unpublished) concluded that the Littleton Formation has nearly the same gamma-ray activity as the granites of the New Hampshire series; the mean thickness of this formation, moreover, is about 5 km (Billings, 1956, p. 7). Thus we may conjecture that heat flow in the Littleton Formation is in the range 1.4 to 1.5 microcal/cm^2 sec in central New Hampshire, with slightly lower values in the Bethlehem Gneiss.

In the work just mentioned, Billings prepared a map (Figure 33–6) showing contours of "equivalent uranium," eU, for the states of

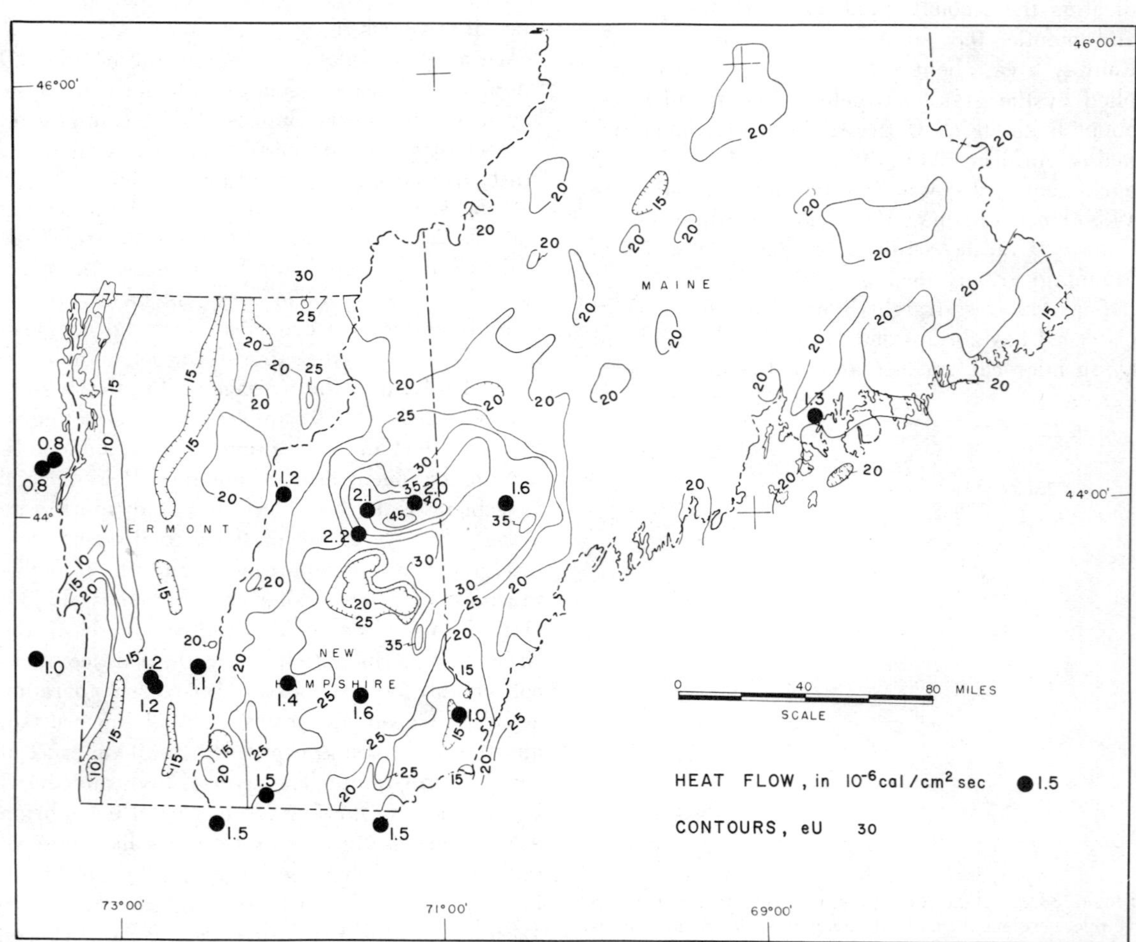

FIGURE 33–6. Contours of bedrock equivalent uranium, eU, after Billings; heat flow determinations with geological corrections.

Maine, New Hampshire and Vermont, all of which had been traversed to some extent with air-borne or truck-borne counters. The unit, eU, was intended to indicate gamma activity, from uranium, thorium and potassium, in terms of the amount of uranium and daughters in equilibrium necessary to produce the same effect. Calibration of field instruments was established by reference to standards containing known amounts of uranium or radium. When analyses are available, eU may be calculated. The formula used by Billings was eU = U + 0.42Th + 3K, where eU, U and Th are in ppm, and K or (K_2O) is in percent. A more recent determination of this expression gives eU = U + 0.48Th + 2.15K (Gustafson and Brar, 1964). Neither of these is equal to the equivalent uranium for heat generation, which may be written in the same form, as eU (thermal) = U + 0.27Th + 0.37K; potassium contributes a much smaller share of the heat production than of the gamma activity. Nevertheless, eU (gamma) is roughly proportional to the quantity A, the rate of heat generation per cm^3, and thus a correlation of heat flow with A means also a correlation with eU.

To show this, equivalent uranium was read from the contoured map for the heat flow stations located within the contoured area, to the nearest unit of eU; the results are shown in Figure 33–7. Except in the Conway Granite, the correlation is remarkably good; the discrepancies here arise from variations of radioactivity within the Conway Granite which were not known when Billings assembled his data. The good agreement elsewhere suggests that the contours of bedrock eU may be converted to contours of heat flow, with errors of the order of 0.2 microcal/cm^2 sec, for this region. This also suggests that simple gamma logs might contribute to the interpretation of heat flow.

The correlation between heat flow and bedrock heat generation is improved by a further correction for the finite size of the body of Conway Granite. The sites in the New Hampshire series are surrounded for considerable distances by rocks having on the average about the same radioactivity as the granites in which the measurements were made, but the Conway Granite is roughly twice as radioactive as the surrounding rocks, and the sites are within less than 5 km of the mapped contacts with less radioactive formations. Thus the heat flow at these points near the margin is less than it would be at stations nearer the center of the outcrop. If we suppose the outcrops as mapped to represent vertical cylinders with these cross-sections, with the radioactive contrast reaching to a depth of 5 km, the values of heat flow at North Conway and at Kancamagus should be raised about 10% and at Waterville, nearly 30%. But this would throw the values at both Kancamagus and Waterville seriously out of alignment with the other readings, and it seems more likely that the detached outcrop mapped around Waterville is separated from the main granitic mass only by a thin remnant of the Kinsman Quartz Monzonite. If the correction is calculated for a single large cylinder whose periphery includes the Waterville outcrop, the correction at Kancamagus becomes negligible and at Waterville and North Conway is reduced to 10%. In making this correction, differences between the Conway Granite, and the Mount Osceola Granite and other bodies within the main White Mountain batholith have been ignored; Richardson (1964, p. 43) and Rogers (1964, p. 57) give some information on the radioactivity of the Mount Osceola Granite, but as yet the sampling seems inadequate to establish a significant difference. These corrected values are employed in Figure 33–5 and Table 33–1, and are evidently in good accord with the linear relation suggested by the other measurements.

APPENDIX

At $t = 0$, let the temperature be $20° + V_0$, for $0 < x < h$, $20° + V_0 + m(x - h)$ for $x > h$. At equilibrium ($t = \infty$), the temperature is $20° + mx + (A/K)(xH - x^2/2)$ for $0 < x < H$, $20° + mx + AH^2/2K$ for $x > H$. The temperature is $20°$ on $x = 0$ for $t > 0$. The depth h is determined by $mh = V_0 - AH^2/2K$. The temperature for $t > 0$ is then:

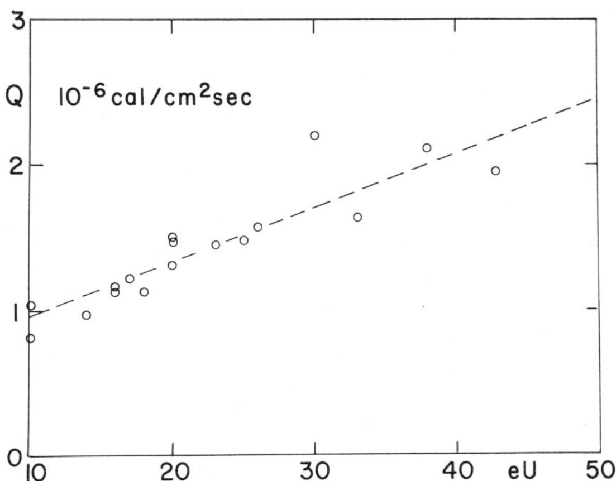

FIGURE 33–7. Heat flow Q, with geological correction, versus local bedrock eU read from Figure 33–6.

for

$$0 < x < H: T(x,t) = 20° + V_0 \text{ erf } x/\sqrt{4kt} + m\sqrt{kt}\left(\text{ierfc } \frac{h-x}{\sqrt{4kt}} - \text{ierfc } \frac{h+x)}{\sqrt{4kt}}\right)$$

$$+ \frac{Akt}{K}\left(1 - 4i^2\text{erfc }\frac{x}{\sqrt{4kt}} + 2i^2\text{erfc }\frac{H+x}{\sqrt{4kt}} - 2i^2\text{erfc }\frac{H-x}{\sqrt{4kt}}\right)$$

for

$$H < x < h: T(x,t) = 20° + V_0 \text{ erf } x/\sqrt{4kt} + m\sqrt{kt}\left(\text{ierfc }\frac{h-x}{\sqrt{4kt}} - \text{ierfc }\frac{h+x}{\sqrt{4kt}}\right)$$

$$+ \frac{Akt}{K}\left(2i^2\text{erfc }\frac{x-H}{\sqrt{4kt}} + 2i^2\text{erfc }\frac{x+H}{\sqrt{4kt}} - 4i^2\text{erfc }\frac{x}{\sqrt{4kt}}\right)$$

for

$$h < x: T(x,t) = 20° + V_0 \text{ erf } x/\sqrt{4kt} + m(x-h)$$

$$+ m\sqrt{kt}\left(\text{ierfc }\frac{x-h}{\sqrt{4kt}} - \text{ierfc }\frac{x+h}{\sqrt{4kt}}\right)$$

$$+ \frac{Akt}{K}\left(2i^2\text{erfc }\frac{x-H}{\sqrt{4kt}} + 2i^2\text{erfc }\frac{x+H}{\sqrt{4kt}} - 4i^2\text{erfc }\frac{x}{\sqrt{4kt}}\right)$$

REFERENCES

Adams, J. A. S., 1964, Laboratory x-ray spectrometer for geochemical studies, *in* J. A. S. Adams and W. M. Lowder, *eds.*, The natural radiation environment: Chicago: Univ. Chicago Press, p. 485–497

————, Kline, M. C., Richardson, K. A., and Rogers, J. J. W., 1962, The Conway granite of New Hampshire as a major low-grade thorium resource: Nat. Acad. Sci. Proc., v. 48, p. 1898–1905

Benfield, A. E., 1949, A problem of the temperature distribution in a moving medium: Q. Appl. Math, v. 6, p. 439–443

————, 1950, The temperature in an accreting medium with heat generation: Q. Appl. Math, v. 7, p. 436–439

Billings, M. P., 1956, The geology of New Hampshire, Part II, Bedrock geology: N. H. Planning and Dev. Commission, Concord, N. H., 203 p.

————, 1964, Areal distribution of natural radioactivity from rocks in northern New England. (unpublished manuscript)

————, and Keevil, N. B., 1946, Petrography and radioactivity of four Paleozoic magma series in New Hampshire: Geol. Soc. America Bull., v. 57, p. 797–828

Birch, Francis, 1950, Flow of heat in the Front Range, Colorado: Geol. Soc. America Bull., v. 61, p. 567–630

Butler, A. P., Jr., 1961, Ratio of thorium to uranium in some plutonic rocks of the White Mountain plutonic-volcanic series, New Hampshire: U. S. Geol. Survey Prof. Paper 424–B, p. 67–69

Carslaw, H. S., and Jaeger, J. C., 1959, Conduction of heat in solids, (2nd edition): Oxford Clarendon Press, 510 p.

Chapman, R. W., and Williams, C. R., 1935, Evolution of the White Mountain Magma Series: Am. Mineralogist, v. 20, p. 502–530

Chayes, F., 1952, The finer-grained calcalkaline granites of New England: Jour. Geology, v. 60, p. 207–254

Clark, S. P., Jr., *ed.*, 1966, Handbook of Physical Constants: Geol. Soc. America Memoir 97, 587 p.

Doll, C. G., Cady, W. M., Thompson, J. B., Jr., and Billings, M. P., *compilers* and *editors*, 1961, Centennial geologic map of Vermont: Montpelier, Vermont Geol. Survey, scale 1:250,000

Grossling, B. F., 1959, Temperature variations due to the formation of a geosyncline: Geol. Soc. America Bull., v. 70, p. 1253–1282

Gustafson, P. F., and Brar, S. S., 1964, Measurement of γ–emitting radionuclides in soil and calculation of the dose arising therefrom: *in* J. A. S. Adams and W. M. Lowder, *eds.*, The natural radiation environment: Chicago: Univ. Chicago Press, p. 499–512

Handford, L. S., 1965, Rb–Sr whole rock age study of the Andover and Chelmsford granites, Mass.: 13th Ann. Prog. Rept. to Atomic Energy Commission, Massachusetts Inst. of Tech. Dept. of Geology and Geophysics (unpublished), p. 11–14

Jeffreys, Harold, 1931, The thermal effects of blanketing by sediments: Monthly Notices, Roy. Astron. Soc., Geophys. Suppl. v. 2, 323–329

Joyner, W. B., 1963, Gravity in North Central New England: Geol. Soc. America Bull., v. 74, p. 831–858

Lee, W. H. K., and Uyeda, S., 1965, Review of heat flow data, *in* W. H. K. Lee *ed.*, Terrestrial heat flow: Geophys. American Geophys. Union Mon. no. 8, p. 87–190

Lyons, J. B., 1964, Distribution of thorium and uranium in three early Paleozoic plutonic series of New Hampshire: U. S. Geol. Survey Bull. 1144–F, 43 p.

Newton, R. C., 1966, Kyanite–andalusite equilibrium from 700° to 800°: Science, 1. p. 170–172

Phair, George, and Gottfried, David, 1964, The Colorado Front Range, Colorado, U. S. A. as a uranium and thorium Province, *in* J. A. S. Adams and W. M. Lowder, *eds.*, The natural radiation environment: Univ. Chicago Press, chapt. 1

Quinn, Alonzo, 1944, Magmatic contrasts in the Winnipesaukee region, New Hampshire: Geol. Soc. America Bull., v. 55, p. 473–496

Richardson, K. A., 1964, Thorium, uranium and potassium in the Conway Granite, New Hampshire,

U. S. A., *in* J. A. S. Adams and W. M. Lowder, *eds.*, The natural radiation environment: Univ. Chicago Press, chapt. 2

Rogers, J. J. W., 1964, Statistical tests of the homogeneity of the radioactive components of granitic rocks, in J. A. S. Adams and W. M. Lowder, *eds.*, The natural radiation environment: Univ. Chicago Press, p. 51–52

————, and Ragland, P. C., 1961, Variation of thorium and uranium in selected granitic rocks: Geochim. and Cosmochim. Acta, v. 25, p. 99–109

Roy, R. F., and Decker, E. R., 1965, Heat flow in the White Mountains, New England [abs.]: Am Geophys. Union Trans., v. 46, p. 174–175

Simmons, Gene, 1964, Gravity survey and geological interpretation, Northern New York: Geol. Soc. America Bull., v. 75, p. 81–98

Tilton, G. R., and Davis, G. L., 1959, Geochronology: *in* P. H. Abelson, *ed.*, Researches in geochemistry: New York, John Wiley & Sons, 511 p.

Toulmin, P., III, 1961, Geological significance of lead-alpha and isotopic age determinations of "alkalic" rocks of New England: Geol. Soc. America Bull., v. 72, p. 775–780

Tuttle, O. F., and Bowen, N. L., 1958, Origin of granite in the light of experimental studies in the system $NaAlSi_3O_8$–$KAlSi_3O_8$–SiO_2–H_2O: Geol. Soc. America Mem. 74, 153 p.

Wilson, J. R., 1965, Bedrock geology of the Ossipee Lake Area, New Hampshire: Ph.D. Thesis, Harvard Univ., 177 p.

Number in parentheses following name is the first page of an article in this volume of which the individual is author or coauthor. Index does not include citations of articles in this volume.

Junior authors included by the expression "et al." are cited, even though their name may not appear on the page listed.

Editors of guidebooks are cited where the title of the guidebook is given under "References," but not where mentioned in the text.

Italicized page numbers distinguish particularly detailed or definitive discussions. Italics are not always used where only a few page numbers follow the name of a topic.

Names of specific geologic features or events are indexed under appropriate major headings, such as: Anticlinoria; Faults, major; Geosynclinal belts; Orogenies; Unconformities.

Names of States and Provinces are subheadings under the following primary headings: Names of geologic periods; Fossil lists, areas; Fossils; Geologic structure; Gravity surveys; Maps; Stratigraphic columns.

Discussions of certain processes, abstract concepts, theories, etc., are indexed even though the author himself may not have used the common name of the process, concept, or theory, etc., on any page cited. More commonly, however, the index follows the author's terminology.

Italics identify detailed or definitive descriptions.

List does not include names from far outside northern Appalachian region. It does not include series, stage, and zone names. Units of less than formation rank are not generally listed where only mentioned on one page. Some units merely listed in tables, charts, or diagrams, and not mentioned in text, are also omitted.

Authors have not been required to follow uniform nomenclature, so names in index may differ slightly from those used on a given page; e.g., Northfield Formation vs. Northfield Slate.